HOW TO GET HIGH

ed. by Michael Joseph Halm

"Until **H**eaven and **E**arth pass away ... until it all comes true." (Mt 5:18)

hierogamous@lycos.com

Also available from Hierogamous Enterprises at Amazon:

- *Psalms, Hymns and Inspired Songs: self-hate to Love through Scriptures* by Michael Joseph
Halm is an autobiography including many Bible-inspired songs that have guided me from a hatred of my own sexuality to a relationship with God Who is Love.

- *Reignbeau's Riddles and Rhymes* by Reignbeau the Clown is a collection of riddles and songs suitable suitable for children of all ages.

- *Proverbials* by Michal Joseph Halm is, like Psalms, Hymns and Inspired Songs, a sample of making Scripture more meaningful by versification, this time focusing on Wisdom literature, including Jesus' proverbs.

- *Basilian Tales* ed. By Michael Joseph Halm is a book of fairy tales about the kingdom of Basilia, based on the Sunday Gospel readings.

- Crosswords with Jesus by Michael Joseph Halm contains a hundred crossword puzzles collected from those published in *My People* newspaper, that do not exclude Jesus.

- Sherlock Holmes and the Mad Doctor ed. by Michael Joseph Halm, BSI, is a case of a missing person lost in time and found with great difficulty and great adventure.

From hierogamous AT lyos.com

- *Hierogamous Hymns* by Michael Halm and Razilee Purdue is a booklet, a sampler of selections of songs from *Psalms, Hymns and Inspired Songs.*

- *The Luminous Mysteries of the Rosary: A play for 8-10 Year Olds* by Kathy McCarthy is a booklet describing how to put on a play illustrating the Luminous Mysteries of the rosary.
from *Proceedings of the André Joyce Appreciation Society* #0 (July, 1985)

INTRODUCTION

What would turn out to be the Math Club's last meeting opened with our faculty adviser, Prof. André Joyce[1] introducing himself. He really had no need to do so, since he was one of the university's most unforgettable characters. His goatee and beret were quite well known outside the math department. Some of us in the meeting room were or had been his students, a few his grad students. More than the usual number this time however seemed to have come from other departments because of summer doldrums. This was undoubtedly because of our calculatedly intriguing topic, "How to Get High" advertised in the weekly "What's Happening" section of campus paper *The Clarion*.

Joyce opened the meeting by quoting *Sirach* 17:21, "Turn to the Most High." Then Society President Dan Svjetlost[2] explained, "What is called 'height' in mathematics is how many exponentially stacked tens a number is equivalent to. One has a height of zero, since it is ten to the zeroth power. A lone ten has a height of one. Ten raised to the tenth power or tenplex has a height of two. That means ten is multiplied together ten times."

"A ten billion", newbie Tom Ficka[3] answered.

"In the U. S. it is, but in France and elsewhere it's called ten milliard, a number higher than the number of people in the world, but not as high as the number of years since the Big Bang."

"Who'd think the change from M to B could make such a big difference!"

"To think otherwise is a symptom of 'number numbness' as Douglas Hofstadter called it," Dan continued, writing on the blackboard. "So onetriplex or, would have a height of three since it's ten-to-the-tenth-to-the-tenth, ten duplex or to-the-third-ten. We symbolize that like so, using right and left superscripts."

$$10 = 10^1 = {}^{1}10$$
$$10^{10} = 10{,}000{,}000{,}000 = {}^{2}10$$
$$10^{10{,}000{,}000{,}000} = {}^{3}10$$

"So onetriplex is both ten-billion-to-the-tenth and ten-to-the-ten-billionth?" Tom asked.

"Ten-billion-to-the-tenth would <u>not</u> be the same as ten multiplied ten billion times, but only ten billion multiplied ten times. We have to evaluate exponentially stacked numbers, 'power towers' from the top-down, not the bottom-up."

$$(10^{10})^{10} = 10^{10}10^{10}10^{10}10^{10}10^{10}10^{10}10^{10}10^{10}10^{10}10^{10} = 10^{10(10)} = 10^{100}$$

"That's only ten-to-the-hundredth, what's informally called a googol. If we try stacking one, two, three and four we'd get what?"

"One-to-the-second-to-the-third-to-the-fourth would just be ... one!"

"And the other way round, four-to-the-third-to-the-second-to-the-first."

"That'd be four-times-four-times-four-squared or ... uh, four thousand ninety-six."

"No, you get higher faster than that by stacking exponents, since that's evaluated from the top-down, rather than bottom-up. The stack is four-to-three-squared, not four-cubed-squared. Using carets and parentheses to make it clearer the two kinds of stacking, it looks like this."

$$4\text{^}(3\text{^}(2\text{^}1)) = 262{,}144 >> ((4\text{^}3)\text{^}2)\text{^}1 = 4{,}096$$

"That's quite a noticeable difference; even with these small exponents, a difference of two orders of magnitude, a 99 percent error."

"When is smaller larger than small?"

"What?"

"When you add the suffix -er."

"As I was saying they might be expressed using the same exponents, though grouped differently, and have the same weight and height, but differ in almost every other way, in length, height, head, tail, weight, density, genders, orientation, longevity, persistence, ancestors, descendants, temperament, ..."

"Whoa! The difference in lengths, that would be the number of digits, six and four. I see that. Height would be only slightly different, both being much closer to one than two stacked tens. The heads, I would guess, would mean the initial digits, two and four, and the tails the final ones, four and six. Neighbors would be the numbers either side of a given number, but what in the Most High's name do you mean by the rest"

"The weight[4] is also called the digital sum. Both numbers surprisingly have the same weight of nineteen, but spread over different lengths. The duration is what we in the Society call the syllable count, a measure of how long it takes to say the number name. Two hundred sixty-two thousand, one hundred forty-four would be fourteen syllables wide while four thousand ninety-six would be only six. As four-to-the-ninth and four-to-the-sixth they have the duration of four.

"If you repeatedly sum the digital sum you reach the digital root or density. Persistence just means the number of summations that it takes to get there. If you sum the squares of the digits repeatedly you eventually get to one of four temperaments, one for choleric, seven for sanguine, any other odd number for phlegmatic and any even number for melancholic.[5]

"Tenplex is top-heavy with all its weight in its head. The next higher number, tenplex one, weighs twice as much but with terminal ones, a numberdrome. The next lower number, tenplex minus one or nine hundred ninety-nine million, nine hundred ninety-nine thousand, nine hundred ninety-nine is also very unbalanced, 'above average' like Lake Wobegon children. 'Average' originally meant "ess than total damage'.[6] The other two are below average. All three are very different, though very close to the same height of two."

number	9,999,999,999	10,000,000,000	10,000,000,001
density	9	1	2
gender	M	F	M
head	9	1	1
height	2	2	≈ 2
length	10	11	11
longevity	21	3	4
orientation	M	91% F	18% M
persistence	2	1	1
tail	9	1	1
temperament	M	C	M

weight	90	1	2

"That makes almost all numbers unbalanced!"

"Yes, ninety-nine point two percent of the first thousand and it gets higher as the numbers do!"

"That makes sense, given the definitions, I guess, though I never thought about numbers this much before, especially higher numbers. They're much more interesting than I ever thought they were! But what do you mean by a number's gender, orientation and ancestry?"

"In behalf of your less than inspiring math teachers, I apologize." Prof. Joyce responded with what seemed like honest regret. "I've always loved both mathematics and wordplay. It seems natural to me to see all these relationships between my old friends, the integers.

"As the Pythagoreans thought of them, the odd numbers were male and even numbers female. To paraphrase Kipling, 'The odds is odd and the evens are even and ne'er the twain shall meet.' It can be handy however to note also how many of the numbers digits are odd or even, male or female, whatever the tail may be. That's what I call their orientation. You can draw a whole family tree for numbers from the numbers that are related to them by adding, deleting or moving their digits.[7]

"Think of a number that can be transformed by rearranging and/or removing digits into another number, transpositioning or transdeleting, like getting either the one-digit zero or one from the two digits of ten. Ten is considered the ancestor of zero and one, just as a hundred, a hundred one and a hundred ten are all ancestors of ten. The number before transpositioning or transdeletion would be called cispositioned or cisdeleted."

"So the two numbers, one hundred one and one hundred ten, are the father and mother of eleven, their son!" Cal Pantof, wearing his number eleven jersey[8] and an "Aha!" look in his eyes, spoke up.

"So one would be the one and only descendant that eleven could have, the grandson of one hundred one and one hundred ten?" asked Cal.

"That's right. Since eleven only has ones as its digits that can be deleted and only one one after deleting either one, we have no other possibility but the one son, one."

"Like the Most High's one and only begotten Son."

"Right."

"How many descendants would eleven and ten have?" Joyce continued, drawing a number tree on the blackboard, unaware of the blush that rose on Cal's face as he turned away from staring at our beautiful secretary, Clair.[9]

"Both eleven and ten have one as a descendant, a 'son', and both have one hundred one and one hundred ten as 'parents', making them 'siblings'. They however only share two of each of their three 'parents' in common. They can have descendants in common, but also shared with other siblings by other parents or not with any others, like a clone.

```
[111] (110)[101](100)
   \ \X \/
  [11] (10) Y
    \/ \/
   [1] (0)
```

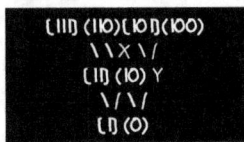

"Number family trees are complicated!", Anne Tabiat remarked. One has two parents and four grandparents, but zero does not. It, or rather she, since zero's even, has one one parent and three grandparents! Zero is one's half-sister." [10]

Dan continued, "Perhaps if we went all the way back to the beginning, when things were simpler, when all we had to get high was base one, it would help." [11]

Erasing the blackboard he continued, "In base one one vertical line indicates the number one, two two, three three, four four, and so on. So how many vertical lines would it take a caveman, let's call him Fred[12], to get to tenplex?"

Joyce played the part of Fred, shuffled up to the board and wrote on the board.

"One and ten, two and nine, three and eight, four and seven, and five and six. That's five times eleven or fifty-five."

Joyce passed on the chalk to Clair acting like like a totally exhausted relay racer as his audience cheered.

"So, Cal, how would you interpret these marks?"

"That'd be eleven, thirteen, seventeen and nineteen, someone counting odd numbers but missing fifteen? Wait, there's something familiar about those numbers. I've got it! They're the primes!" [13]

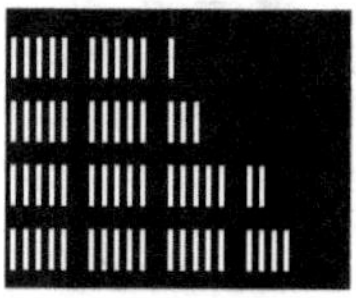

"Fred could work with primes, but could he get as high as tenplex?" Dan continued.

"Ten billion? I don't think he could!"

"Right."

"That's why someone cleverer, call her Wilma, invented another way to get high. By using a slash to mark groups of five and a backslash to mark a second group of five, Wilma could count by fives and tens, which lead to what would eventually become the familiar system of Roman numeral, with I for one, an X for ten and a half X or V for five, X cut in the middle like a magician's assistant."

"Ah gotcha, gov, like 'finnif' t'is two 'fins' t'gethah, eh?" lead to a meeting-ending outbreak of Cockneyspeak.

from *PAJAS* #1 (August, 1985)

ROMAN NUMERALS

At the next meeting history major Lana Coperta[14] took over, in the absence of the good professor. After Prof. Joyce's forced retirement, it was advertised rather clandestinely under a new name, the AJAS, the André Joyce Appreciation Society, and met at progressive new meeting places on and off-campus, beginning at the Clair's. Some were speculating that it was the prof's seventy-fifth birthday that was the reason for his absence. Others thought he had mentioned the Most High one too many times. Still others speculated that it was a sudden unknown illness, even the curse of the Black Theorem.[15]

She opened by quoting Ecclesiastes 7:29, "God made mankind straight, but men have had recourse to many calculations."

This was followed by the almost equally expected opening riddle, "How can you take two from six and get ten?"

"That's an oldie! Taking S and I from SIX leaves X," the janitor in the back said.[16]

"Right! See what he did there. He recognized the number six written as an English word as including the Roman numeral X and subtracted two letters, what Joyce called jootsy calculus, calculations that 'jump out of the system'."

"But you're confusing words and numbers!"

"No, you're confusing numbers and numerals. Numbers can be represented in many equally valid ways. English words and Roman numerals are only a couple of the many, many ways of expressing numbers."

"Oh, it's like the joke from Latin class, "Two Romans walk into a bar and one holds up his index and middle fingers, while the other says 'Quinque potiones.'"

"I don't get it."

"The second said, 'Five beverages' in words, while the first made the V sign for five, not for two."

"We do have alternative theories to the origin of the Roman numeral X to what we heard last meeting, too," she continued after the groans died down.

"We have some archeological evidence[17] that what became Roman numerals were originally based on the basic vertical line meaning one. I'll grant you that, Dan. I believe however that sometime before the X came a curved line, what eventually became the Roman numeral for *centum*, symbolized by C. The symbol for the number may have come before the name for the number, not the other way 'round. If that were so, then a bent or curved line open to the left would have meant a tenth and the two together, forming the X, would mean a tenth of a hundred rather than twice five."

(= 100
)(= (1/10) 100 = 10

"Oh!" Sam[18] cried out, attracting all eyes in the room toward the back row.

"Uh," the eavesdropping janitor stammered, unused to such attention, "It's just that I noticed that sawing the C in half like we did with the X magically gives a bottom half very much like an L, the Roman numeral for half of a hundred!"

\ = ½ (= 50

"Very observant of you," Lana observed, making shy Sam blush. "That's just what this theory says. It suggests that the numeral M for a thousand also may have predated its Latin name, *mille*, the source of our million. It was apparently originally symbolized as a vertical line between two curved or bent lines, like so."

(|) = M = 1,000

"And half of that is a D for half of a thousand, five hundred. Roman numerals do make sense!" Cal joined in.

"Archeologists have found other symbols supporting such interpretations of these proto-Roman numerals," Lana went on with an approving nod from Dan. "With two more semicircles around the vertical line the proto-Romans got higher by a factor of ten. The symbol for ten thousand or a myriad would not have been an awkward ten M's but just two semicircles, a line and two more semicircles, like so."

((I)) = MMMMMMMMMM = 10(1,000) = 10,000
((((I)))) = 10(10)10(1,000) = 1,000,000

"That's a far better way to get high, not with straight lines, but with curves", commented Cal, with a grin directed at Lana.

That lead into a tangent into curvaceous numbers, those containing the curved digits, zero, three, six, eight and nine, including curvaceous female numbers. The first curvaceous prime higher than three wasn't found until eighty-three.

curvaceous	0	3	6	8	9	30	33	36	38	39	60	63	66	68	69	80	83	86	88	89
females	0	6	8	30	36	38	60	66	68	80	86	88	100	106	108	110	116	118	160	166

That naturally lead to an investigation into the other shapely categories of numbers, the curvilinear numbers, with the digits two and five, and the linear numbers with digits one, four and seven. In some fonts three may be curvilinear and four may not be holey. Even two has a small hole in some. Just two of the curvilinears are primes, while many of the linears are.

curvilinear	2	5	22	25	52	55	222	225	252	255	522	525	552	555
neighbors	1	3	4	6	21	23	24	26	51	53	54	56	221	223
sum	2	7	29	54	106	161	383	608	860	1,115	1,637	2,162	2,714	3,269
linear	1	4	7	11	14	17	41	44	47	71	74	77	111	114
sum	1	5	12	23	37	54	95	139	186	257	331	408	519	633
linear prime	7	11	17	41	47	71	1,117	1,171	1,471	1,741	1,747	1,777	4,111	4,177

"Later an overhead bar represented multiplication by a thousand," Lana finally continued, "so a myriad became X-bar and M-bar represented a million."

X = 10(1,000) = 10,000
M = 1,000(1,000) = 1,000,000

"And M double bar would get as high as a billion?"

"Nineplex, yes, so getting to tenplex wouldn't be so very difficult, just X with three bars above it and twelveplex just M with three bars."

$\overline{\overline{\overline{X}}}$ = 10(1,000)1,000(1,000) = tenplex $\overline{\overline{\overline{M}}}$ = 1,000(1,000)1,000(1,000) = twelveplex

"Not only that," Dan said, taking up a piece of chalk himself, "but extrapolating proto-Roman's rightward-opening semicircle for a tenth you could even write decimals.[19] Pi would have been approximately in eight symbols."

π ≈ III)) X IV

numeral	I	II	III	IV	V	VI	VII	VIII	IX	X
density	1	2	3	6	5	6	7	8	9	10
descendants		I	II	I, V		I, V	II, VI	III, VII	I, X	
duration	1	2	3	2	1	2	3	4	2	1
gender	M	F	M	F	M	F	M	F	½M	F
head	I	I	I	I	V	V	V	V	I	X
height	0	0.3	0.5	0.6	0.7	0.77	0.84	0.9	0.95	1
holiness	0	0	0	0	0	0	0	0	0	0
length	1	2	3	2	1	2	3	4	2	1
orientation	M	M	M	M	M	M	M	M	½M	F
persistence	0	0	0	1	0	0	0	0	1	0
tail	I	I	I	V	V	I	I	I	X	X
weight	1	2	3	6	5	6	7	8	11	10

"The weights and densities in the Roman numerals do not continually get higher, because subtraction of letters is allowed."

"Their orientations are all masculine, whether or not they are male or female, until we reach IX, which is half-male and half-female, an androgynous number."

"I'm interested in what we can know about the number of the Beast, DCLXVI?"

"It's numerals are one-third odd and two-thirds even."

"Its weight is six equal to it value, since all its numerals differ, unlike the Arabic ones."

"Four of the letters are even and two odd, so its orientation would be two-thirds feminine."

"Its duration and length are both six, so its density would be one hundred eleven."

"I can see that looking at numbers like the Romans was much different from how we do now!", Cal said. "Like all the Roman numerals the Number of the Beast not so much evil, but amoral, because their holiness and naughtiness are both zero!"

from *PAJAS* #2 (September, 1985)

BASE TWO

Dan began first meeting of the new school year with a quote from Francis of Assisi, "Happy those who endure in peace, for by You, Most High, they will be crowned."

Since all there had read the article in the *Clarion*, "Let Us Remember André Joyce", they all knew that their professor had been forced into retirement.[20] The wild speculations of summer had disappeared, but the Society felt like it had lost the star it had had as its guiding light, so Dan added a riddle, "How many divas does it take to change a light bulb?"

"I know that one", Tom said, "Just one, to hold the light bulb while the world revolves around her."

Dan continued, "Base two or binary was, as far as we know, first used by John Napier, who also invented logarithms and the decimal point. Gauss independently used binary. It is now the most common base used on the planet, far more common than base ten, in many modern electronic devices..."

As if on cue someone's pager chirped, stopping Dan's explanation from careening off on a tangent Joycesquely.

"Base ten, the decimal system, has ten as its base. One-zero means one ten plus zero ones, one-zero-zero means one ten-squared or hundred plus zero tens and zero ones. In binary or base two it's quite different, one-two plus zero-ones and one two-squared or four plus zero twos plus zero ones."

"So is this numbering system based on two the one used all over the world?" Cal asked.

"And even outside the solar system by *Pioneer 10*!" Charles[21] blurted out.

"Yeah, it's used by the Bynars[22] of sigma Regonis." Al[23] added with even more enthusiasm. He attempted to ward off the puzzled looks he got from the non-trekkers by spreading his fingers in the Vulcan peace sign and the explanation, "You know, from 'Star Trek'."

"We get high in base two," Dan continued turning the discussion back on course this time, without revealing whether he knew the reference or not, "by doubling the value of each position leftward. 1 means one or two-to-the-zeroth. The next higher number, one-zero, which looks like a ten but isn't, means one two plus zero ones. The place to the left of that would mean two squared or four, then two cubed or eight, and so on.

At this he stepped away from the board to reveal what his body had been hiding.

There are 10 kinds of people,
those who think in binary
and those who don't.

It took some in the room sometime to catch on, but with a little help from their more astute friends, finally all joined the binary-thinking kind, even Cal who asked, "So one-one in this base two would mean not eleven, one ten plus one one, but one two plus one or three, right?"

"I can't stop thinking of one-one as eleven." another said.

"Eleven means specifically the number after ten in base ten, but one-one is what Joyce called a homonum, a number name that sounds like another one, commonly in a base other than decimal."

"I think in binary a lot in computer class. I have to catch myself in church when they announce the hymn as one-one-one and not turn to number seven!"

"That's not as bonehead of an error as thinking of it as number three!"

Dan smiled and nodded, "So the answer to 'How many actors does it take to change a stage light?' would be..."

"One-zero, one to climb up on stage and do it and one to say, 'I wish I were up there.'" Tom answered.

"Is there a limit on how high you could get using just ones and zeros? By counting, I mean." Cal said, glancing in Al's direction. "Not by starship."

Tavola Tafelberg answered his question while writing on the blackboard, "The binary numbers get higher as one, one-zero, one-zero, one-one, one-zero-zero, and so on, but it is more efficiently than base one's one, one-one, one-one-one."

"Repunits and binarish[24] numbers resemble base one and two numbers," she continued, showing the calculation of her faithful companion, Pebbles[25], "in that they are numbers with ones and zeroes and with ones only. With both one and zero ten would take four bits – that's a contraction for 'binary units'[26], not the smaller 'half' of twenty-five cents – to express, namely, one-zero-one-zero."

binary	1	10	11	100	101	110	111	1000	1001	1010
density	1	1	2	1	2	2	3	1	2	2
duration	1	3	2	5	4	4	3	7	6	6
gender	M	F	M	F	M	F	M	F	M	F
head	1	1	1	1	1	1	1	1	1	1
height	0	0.3	0.5	0.6	0.7	0.77	0.84	0.9	0.95	1
holiness	0	1	0	2	1	1	0	3	2	2
length	1	2	2	3	3	3	3	4	4	4
orientation	M	½F	M	⅔F	⅔M	⅓F	M	¾F	½M	½M
persistence	0	1	2	1	2	2	3	1	2	2
tail	1	0	1	0	1	0	1	0	1	0
temperament	C	C	C	C	C	C	C	C	C	C
weight	1	1	2	1	2	2	3	1	2	2

"Is one followed by a hundred zeroes still a googol in base two?"

"Technically no, because it's not tenplex. It is however an interesting number."

"Aren't they all!" [27]

"It's called a little googol, because its the smaller of the two prime power factors of a googol, two-to-the-hundredth, rather like ten is the smaller 'half' of a quarter's 'two bits' and fifteen is the larger 'half'. Joyce called the other prime power factor, five-to-the-hundredth, a 'not-so-little googol', littler than a googol but not as little as a little googol ."

(2^100)(5^100) = 10^100

"Numbers with an even number of ones in binary John Conway called 'evil', that would be zero, one-one and one-zero-one. "

"In decimal that'd be zero, three and five."

"So three and five would be 'evil twins'.

"No, unlike the twin primes that are two primes with a composite between, evil twins are two consecutive evils with nothing between."

binary	0	11	101	110	1001	1010	1100	1111	10001	10010	10100	10111	11000	110011	1100101
evil	0	3	5	6	9	10	12	15	17	18	20	23	24	27	29

evil twins	5 6	9 10	17 18	23 24	29 30	33 34	39 40	45

"Even zero is evil since it has an even number of ones, namely zero. The next even evil is six."

"Three and five're prime, but the next evil prime isn't until seventeen."

"So if six is an evil number, the Number of the Beast would be triply evil."

"Perhaps."

"What is six hundred sixty-six in binary?"

"It'd be five twelve, two-to-the-ninth, plus one twenty-eight, two-to-the-seventh, six forty..."

"Plus twenty six, two-to-the-fourth plus two cubed plus two."

"One-zero-one-zero-zero-one-one-zero-one-zero."

"An odd number of ones!"

"The complimentary sequence, the binary numbers with an odd number of ones, like the Number of the Beast, would be called good then?!"

"No, 'No one is Good, except the Father'. Conway, who named the 'evil' numbers, called those numbers 'odious', a pun on odd, like evil is on even. The odious primes start with two, seven, eleven, thirteen, nineteen, thirty-one."

binary	1	10	100	111	1000	1011	1101	1110	10000	10011	10101	11001	11010	11100	11111
odious	1	2	4	7	8	11	13	14	16	19	21	25	26	28	31

"By convention the first number in the sums, and other sequences like it, is not an actual sum, but the seed number that starts the sequence. The evil sum primes start with three, seven and a hundred sixty-seven."

"As easy as one, one-zero, one-one!"

from *PAJAS* #3 (October 1985)

BASE THREE[28]

As the rain poured outside Charles opened the meeting with Psalm 18:14, "The Lord thundered from Heaven, the Most High gave forth His voice."

Maintaining his serious monotone, he continued, "How many paleontologists does it take to change a light bulb?"

Lana answered with a pseudo-serious "Three, one to find it and two others to disagree over how old it is."

Then Charles followed up last meetings discussion of binary with ternary, "Base three gets higher by counting groups of ever higher powers of three. We get high like one, two, one-zero, one-one, one-two, two-zero, two-one, two-two and so on."

ancestors	[10]	(11)	[12]	(20)	[21]	(22)
descendants	[1], (0)	[1]	[1], (2)	(0), (2)	[1], (2)	(2)

"As you can see from the table, two and one each have four parents and share two in common, the odd couple[29], one-two and two-one, and so are what could be called half-siblings. Zero shares just one each of her two parents with one and two's four and so would be their their quarter-sister."

"So now in this odd base one-one would be an even number?!" Cal asked.

"Yes, and one-two would be odd, while two-zero would be the same gender, female, in both an even and an odd base. We can subdivide the ternary sequence into rather odd-looking odd and even numbers."

odd	1_3	10_3	12_3	21_3	100_3	102_3	111_3	120_3	122_3	201_3	210_3	212_3	221_3	1000_3	1002_3	1011_3
even	0_3	2_3	11_3	20_3	22_3	101_3	110_3	112_3	121_3	200_3	202_3	211_3	220_3	222_3	1001_3	1010_3

"So now in this base three, one-one would be three plus one or four." Cal commented..

"Two-two reminds me of that old tongue-twister." Sam put it. When all he got was blank stares, he continued, "You never heard of Won-one and Tutu? 'Tutu was a racehorse and Won-one was one too. When Tutu raced with Won-one how did the duo do? Tutu tied with Won-one, since their races numbered two, And though Won-one won one Tutu won one too.'"

"Ah, I get it. One-one-one-one-two-two-one-one-two! That's a nine-place ternary number."

"It'd be -- what? -- over fifty million?"[30]

"We could get high a bit more interestingly," Charles commented, "without either zeroes or ones or twos, like Kepler did when he recorded the planets' positions discovering his famous Three Laws of Planetary Motion."

There was another awkward silence from the astronomically challenged to whom they were not known at all, let alone famous.

"Let me show you." Charles continued, "Using three symbols, rather than just two as in binary, Johannes Kepler represented numbers, as above, at or below a power of three. By using a minus, zero, and plus, we can get high Kepler-like. As I hope you can see, this is an abbreviated form of zero, plus-one, plus-three-minus-one, three-plus-zero, etc."

Keplerian	0	+	+-	+0	++	+--	+-0	+-+	++-	+00	+0+
descendants			-, +	0, +	+	--, -+, +-	-0, +-, +0	-+, +-, ++	-+, +-, ++	+0	+0, ++
duration	2	1	3	3	2	5	5	4	4	5	5
holiness	1	0	0	1	0	0	1	0	0	2	1
orientation	F	M	F	½M	M	M	⅓F	M	M	⅓M	⅓F
persistence	0	0	1	1	0	1	1	1	1	1	1
weight	0	1	0	1	2	-1	0	1	1	1	2

"I see something new here. We have some weightless numbers, so their densities are incalculable. Plus-minus, although female, has no orientation because its plus and minus cancel out. She's an asexual female, a consecrated virgin!"

"I see that zero, plus-minus and plus-minus-zero all are weightless."

"I see a third way to get high in ternary," Clair announced, "without using zeros, just ones and twos, like Sam's racehorses. Rather than representing three as one three and zero ones, one-zero, we could think of it also as three-ones, just as we normally do in base ten."

$$10_3 = 3_3 = 3_{10}$$

"I see it too." Cal said. "It's like an improper fraction, like the used-car salesman that sold me my VW at a fraction of what it cost him, three halves, I found out too late."

"And I hear the Math Help Room is now open 24/7 ... about three and a half hours a day."

Over the snickers and giggles Lana shared about Earth people who use ternary[31] and Al about a couple of alien ones.[32]

ternarish	1	2	10	11	12	20	21	22	100	101	102	120	121	122	200	201	202	210
sum	1	3	13	24	36	56	77	99	199	300	402	522	643	765	965	1166	1368	1578

Charles continued after the digressions, "So, even though two plus two, two times two, two squared and to-the-second-two are all four, three plus three, three squared, three cubed and to-the-third-three are all very different, each much higher than the previous operation on three [33], which just goes to prove Grabel's Law: 'Two does not equal three, even for very large values of two.'"

ternary	1_3	2_3	10_3	11_3	12_3	20_3	21_3	22_3	100_3	101_3
density	1	2	1	2	1	2	3	4	1	2
descendants			0, 1	1	1, 2	2	1, 2	2	10	10, 11
duration	1	1	3	2	2	3	2	2	5	4
height	0	0.3	0.5	0.6	0.7	0.77	0.84	0.9	0.95	1
holiness	0	0	1	0	0	1	0	0	2	1
orientation	M	F	½F	M	½M	F	½M	F	⅓M	⅓F
persistence	0	0	1	2	3	1	2	2	1	2
temperament	C	M	C	M	C	M	C	M	C	M
weight	1	2	1	2	3	2	3	4	1	2

"What I wonder about is why we group the digits in threes."

"That's because generally we now count in thousands rather than myriads, which once was the standard."

"Or it may be because of the ubiquitous Rule of Three. We seem to be designed to think in threes because it's the smallest number able to establish a pattern. The Romans evens had the saying, 'Omne trium perfectum.', which became our 'Third time's the charm.'"

"Like *The Three Musketeers*."

"Or the three Stooges."

"Or *The Three Little Pigs*."

"Throughout literature we have the literary device of the tricolon, like Julius Caesar's 'Veni, vidi, vici.", three words of the same length building to a climax."

"Or the hendiatris, three words or phrases of different lengths, like Jefferson's "Life, liberty and the pursuit of happiness."

"Or Mark Anthony's 'Friends, Romans and countrymen.'"

"Beginning, middle and end.'"

"Good, better and best."

"High, higher and Highest."
"Father, Son and Holy Spirit."
This would have continued on endlessly, save for what someone mercifully and mysterious had written on the blackboard.

There are three kinds of people,
those who count,
those who cant
and those who can't.

from *PAJAS* #4 (November 1985)

QUADRARY

Dan opened with "The Most High uttered his voice, hailstones and coals of fire." from Psalm 18:13.
"When playing golf what number does a golfer use most often?"
"There are eighteen holes, so I think he or she would use the number one at least ten times."
"Four!"
"O. K., but what usually has four letters, but sometimes has nine?"
"I give up. What usually has four letters, but sometimes has nine?"
"Yes," Dan answered, " the word 'what' does usually have four letters, unless it's plural, and the word 'sometimes' has nine letters and is plural."
"Today we are in quadrary," Dan continued in his serious tone, adding "that is, base four." [34] After the laughter and groans died down. "Yet it is not very much used now, but before metricization, it still was. Back when your parents were young they had a pop song that went, 'I love you a bushel and a peck.'[35] which meant five-fourths, a hundred twenty-five percent, or with an overflowing love."
There was a collective and nostalgic "Aw!", followed by Lana's "I brought notes from history on a base four system here somewhere."
"Ah, here it is. The Aramæans[36] used a base four system using the symbols I and X, like the Romans, but for one and four, using their thumbs to count on four fingers of one hand, rather than five fingers of both hands. One, two and three were one, two and three vertical lines like in Roman and proto-Roman numerals, but then four was X and five IX or one-plus-four,"
"Which looks deceptively just like the Roman numeral for nine." Dan said.

I	II	III	X	IX	IIX	IIIX	XX	IXX	IIXX	IIIXX	XXX	IXXX	IIXXX	IIIXXX
1	2	3	4	5	6	7	8	9	10	11	12	13	14	15

"But seven looks like ten minus three and six looks like an upside-down twelve." Sam suggested, "And thirteen like a backward thirty-one! Mindboggling!"
"Mindstretching at least. The orientations are much difference than we even saw with the Roman numerals, including halves and thirds and quarters before reaching even the first ten."

Aramæan	I_4	II_4	III_4	X_4	IX_4	IIX_4	IIIX_4	XX_4	IXX_4	IIXX_4
density	1	2	3	4	5	6	7	8	9	10

descendants		I	II		I,X	II,IX	III,IIX	X	IX,XX	IIX,IXX
duration	1	2	3	1	2	3	4	2	3	4
head	I	I	I	X	I	I	I	X	I	I
height	0	0.3	0.5	0.6	0.7	0.77	0.84	0.9	0.95	1
length	1	2	3	1	2	3	4	2	3	4
orientation	M	M	M	F	½M	⅓F	¾M	F	⅓M	½F

After a brief game of finding other Aramæan numerals, Anne shared on a more basic base-four system, the language of life, DNA, deoxyribonucleic acid.

"It's made up of just four components, adenine, cytosine, guanine and thymine, understandably abbreviated to A, C, G and T, what we call the genetic alphabet." she explained as the non-biologists gave a collective sigh and prepared for a lecture. "All share the same one nitrogen, two carbons, two oxygens and four hydrogens, but vary in additional elements and the arrangement of those elements. They combine in threes according to instructions also coded into our DNA to form what are called codons, which combine into vitally important proteins and enzymes, in three-letter 'words' between AAA (zero) and TTT (three-three-three).

	A	C	G	T		A	C	G	T
AA	AAA	AAC	AAG	AAT	**GA**	GAA	GAC	GAG	GAT
AC	ACA	ACC	ACG	ACT	**GC**	GCA	GCC	GCG	GCT
AG	AGA	AGC	AGG	AGT	**GG**	GGA	GGC	GGG	GGT
AT	ATA	ATC	ATG	ATT	**GT**	GTA	GTC	GTG	GTT
CA	CAA	CAC	CAG	CAT	**TA**	TAA	TAC	TAG	TAT
CC	CCA	CCC	CCG	CCT	**TC**	TCA	TCC	TCG	TCT
CG	CGA	CGC	CGG	CGT	**TG**	TGA	TGC	TGG	TGT
CT	CTA	CTC	CTG	CTT	**TT**	TTA	TTC	TTG	TTT

"In our DNA's double helix each helix is linked to the other by base pairs, A to T, C to G, G to C and T to A with a billion codons."

With that Lana tagged Clair and together they[37] linked arms with Anne to form a troika, mercifully turning her and the meeting in a much less serious direction, in fact, all directions at once terminating with the chicken dance.[38]

from *PAJAS* #5 (December 1985)

BASE EIGHT

Anne began the next meeting with the quote, "You will rightly be called sons of the Most High, since He Himself is good to the ungrateful and the wicked." (Luke 6:35) It may have been a reference to her not too well received contributions on bases four.

"How many pollsters does it take to change a light bulb?"

Without waiting for a response, she gave the answer, "Two plus or minus one."

She continued more confidently, "The next higher power-of-two base, eight or octal[39], could be thought of as counting on four-fingered hands like Toons.[40] Actually it's more like a condensed version of both base two and four.

"Eight is two twos twice or two cubed. Eight bits are called a byte, spelled B-Y-T-E. Get it, a play on the words ‘bit’ and 'bite'? And a half-byte is called a nybble also spelled with a Y. Who says geeks don't have a sense of humor?"

"A geeky sense of humor!" [41]

Ignoring the comment Anne continued, "Eight thousand of them is a kilobyte and eight million a megabyte and eight billion a gigabyte."

"So in this system we would get high like five, six, seven, one-zero, one-one, one-two then." Cal suggested. "And one-one would mean nine."

The Society came up with numerous examples of everyday uses from their various disciplines of base eight in counting pints and gallons and bushels, furlongs and miles, gills and quarts and pecks, drachms and ounces. They even determined after a bit of discussion that eight is unique in being the first number alphabetically.

Al shared about some octal users he'd read about that used it and a hybrid system.[42] Others were able to contribute from education[43] and anthropology.[44]

"I see that we finally have a non-zero with holiness, a six."Tom noted.

octal	1_8	2_8	3_8	4_8	5_8	6_8	7_8	10_8	11_8	12_8
density	1	2	3	4	5	6	7	1	2	3
descendants								0,1	1	1,2
duration	1	1	1	1	1	1	2	2	2	2
gender	M	F	M	F	M	F	M	F	M	F
head	1	2	3	4	5	6	7	1	1	1
holiness	0	0	0	0	0	1	0	1	0	0
length	1	1	1	1	1	1	1	2	2	2
orientation	M	F	M	F	M	F	M	½F	M	½F
persistence	0	0	0	0	0	0	0	1	1	1
tail	1	2	3	4	5	6	7	0	1	2
weight	1	2	3	4	5	6	7	1	2	3

"We can write base eight with Yijing trigrams," Lana shared, "three broken and/or unbroken horizontal lines."

octal	0	1	2	3	4	5	6	7	10	11	12	13	14	15	16	17	20	21	22
Yijing	☷	☶	☵	☴	☳	☲	☱	☰	☶☷	☶☶	☶☵	☶☴	☶☳	☶☲	☶☱	☶☰	☵☷	☵☶	☵☵

"'I see', as the blind man said to the deaf-mute. The broken lines are zeroes and the unbroken ones one in a condensed binary," Tom commented. "But I see a problem too. If read upside-down the numbers change. One is an upside-down four and three an upside-down six."

"And vice versa."

"That's why they need to indicate 'This end up' with barcodes, because its binary code reads differently right side-up and upside-down."

"Yijing can be further condensed into six-line hexagrams in a base sixty-four which has even more such ambiguous numerals."

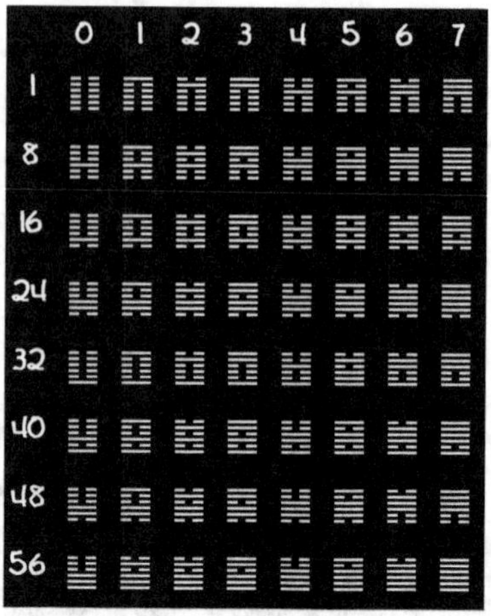

"It's harder to decipher than binary's strings of ones and zeroes!" Sam said.

"I think it's beautiful," Tom responded.

"But base sixty-four's much more orientation-dependant, more oriental, than even base eight."

"Both Louis Braille's and Abraham Nemeth's math codes for the blind encode digits differently, but have the same problem. Braille's uses the same symbols for the letters 'a' through 'j' for the digits '1' through '9' and then '0', like on a standard Qweerty keyboard. His six-dot symbols all end in a blank, while Nemeth's all begin with a blank. Counting the dots down the two columns as ones and the blanks as zeroes they become six-place binary numbers. Commas between groups of three digits are indicated by what would be sixteen and the prefix indicating a number in Braille would be fifteen.

Braille	#1	#2	#3	#4	#5	#6	#7	#8	#9	#0
000111-	100000	110000	100100	100110	100010	110100	110110	110010	10100	10110
base 64	1532	1548	1536	1538	1534	1552	1554	1550	1520	1522

"So ten would be one-five-three-two-two-two then and eleven one-five-three-two-three-two?"

"Yes, but number names would not have the numerical prefix."

Braille	o	n	e	t	w	h	r	f	u	i	v	s	x	g	l	y	d	a	m	p
base 64	42	46	34	30	23	50	68	60	41	22	57	28	45	52	56	47	38	32	44	60

"The letters left, not used in number names, not counting billion, quadrillion and centillion, better called nineplex, fifteenlex and three-hundred-threeplex, are j, k and z, hence the indeterminate numbers, jillion, killion and zillion."

Braille	one	two	three	four	five	six	seven
base 64	424634	302342	3050683434	60424168	60225734	282245	2834573446
decimal	175,010	124,394	516,704,418	15,903,364	15,822,434	116,141	575,646,126

	1	2	3	4	5	6	7	8	9	0
Nemeth	10000	11000	10010	10011	10001	11010	11011	11001	1010	1011
base 64	16	24	18	19	17	36	51	25	10	11

"So one-zero would be sixteen-eleven and one-one sixteen-sixteen."

"Uh, four thirty-five and four forty in decimal."

"Samuel Morse's code uses dots and dashes, in a much more logical manner. Using one for dot and zero for dash and reading right to left with leading zero, we get base-32 numbers. If we read them in ternary with a zero for the space, one for the dot and two for the dash, we get something quite different.

	0	1	2	3	4	5	6	7	8	9
Morse	-----	•----	••---	•••--	••••-	•••••	----•	---••	--•••	-••••
base 32	0	1	3	7	15	31	16	24	28	30
ternary	22222	12222	11222	11122	11112	11111	22221	22211	22111	21111
decimal	242	241	238	229	202	121	161	134	125	122

"So in Morse one-zero meaning ten would actually be thirty-two if read as base thirty-two or one followed by nine zeros in ternary, uh one less than three-to-the-tenth."

"Thirty-nine thousand three hundred sixty-five!"

"And one-one would be thirty-three or eighty-one less!"

"Thirty-nine thousand two hundred eighty-four."

"In the Morse code alphabet however five dots wouldn't mean five but P."

"P for penta perhaps?"

Morse	o	n	e	t	w	h	r	f	u	i	v	s	x	g	l	y	d	a	m	p
ternary	222	21	1	2	122	1111	121	1121	112	11	1112	111	2112	221	1211	2122	211	12	22	11111
decimal	13	7	1	2	17	40	16	43	14	4	41	13	68	25	49	71	22	5	8	121

Morse	one	two	three	four	five
ternary	22202101	2201220222	20111101210110 11	112102220112012 1	1121011011 1201
decimal	6,382	53,891	93,250,147	109117141	644,664

"IDIC!" Al exclaimed, then elaborated, "Infinite Diversity in Infinity Combinations from 'Star Trek'."

"Not quite, but we're getting there," Dan commented, appreciating the Star Trek reference.

from *PAJAS* #6 (January 1986)

BASE SIXTEEN

After opening with Second Samuel 22:14, "The Lord thundered from Heaven, the Most High uttered His voice", Dan lead the discussion to another power-of-two base by asking, "What would one hundred ninety in base sixteen be?" Those with experience in hexadecimal just smiled, while the rest were brought into the joke with his explanation.[45]

"Many calculators are able to convert decimal to hexadecimal, since like octal it is a further condensation of binary, sort of a lingua franca between man and machine. In Hexspeek the numbers from ten through fifteen represented by the first six letters of the alphabet so that we'd get high like A, B, C, D, E, F, 10, with one-zero now meaning sixteen. It was used to count pounds, ounces and drams.[46]

"I see that eleven as B now has a holiness of two and a naughtiness of zero!" Cal shared.

number	9_{16}	A_{16}	B_{16}	C_{16}	D_{16}	E_{16}	F_{16}	10_{16}	11_{16}	12_{16}
density	9	10	11	12	13	14	15	1	2	3
descendants								0, 1	1	1, 2
duration	1	1	1	1	1	1	1	2	2	2
height	0.95	1	1.02	1.03	1.05	1.06	1.07	1.08	1.09	1.1
holiness	1	1	2	0	1	0	0	1	0	0
naughtiness	0	0	0	0	0	0	0	1	0	0
orientation	M	F	M	F	M	F	M	½F	M	½F
weight	9	A	B	C	D	E	F	1	2	3

"How many Hexspeekers does it take to make coffee?", Lana asked, quickly explaining, "By making coffee we mean C-zero-F-F-E-E[47], a mnemonic that replaces others like "Tutu won one too" for bases sixteen and higher."

Before very long all the Society members learned to make many different kinds of C0FFEE.

"Joseph Bowden had a different way however." she went on. "He gave the hexadecimal numbers usually called, D, E, F and one-zero, the new names: 'thrun', 'fron', 'feen', and 'wunty' respectively, with A, B, and C remaining with the values of ten, eleven and twelve. What would C0FFEE be called in Bowdenese?"

COFFEE = 12(16) + 0(16) + 15(16) + 15(16) + 14(16) + 14 = 12,648,430

By answering "See-zero-feen-feen-fron-fron" correctly, and correctly identifying it in decimal, Dan won the right to redirect the meeting to coffeemaking, sparing all from an overdose of Bowdenese,[48] and fulfilling the Scriptural prophesy, "He brews."

from *PAJAS* #7 (February 1986)

BASE FIVE

Dan chose to open the meeting by quoting Longfellow, "It is the heart and not the brain, That to the Highest doth attain."

Then he continued with the riddle, "How many country singers does it take to change a light bulb?"

On cue Lana answered, "Five, one to make the change and a backup quartet to sing about missing the old bulb."

He then proceeded with a discussion of the second odd base, base five[49] or quinary system. Charles had extrapolated from Kepler's ternary using the symbols: equals, minus, plus or minus, plus and dagger for minus two, minus one, zero, plus one and plus two respectively. Rather than three being represented as three pluses, it would be expressed as five minus two and four as five minus one, etc. Cal contributed the next number and others even higher numbers.[50]

quintary	1	2	5-2	5-1	10	11	12	10-2	10-1	10	21	22
symbol	+	‡	+=	+-	+±	++	+‡	‡=	‡-	‡±	‡+	‡‡
descendants			=, +	-, +	±, +	+	+, ‡	=, ‡	-, ‡	±, ‡	+, ‡	‡
duration	1	3	5	3	5	2	4	7	5	7	4	6
head	+	‡	+	+	+	+	+	‡	‡	‡	‡	‡
orientation	M	F	-½F	-½M	½M	M	½M	-½F	-½M	F	½M	F
tail	+	‡	=	-	±	+	‡	=	-	±	+	‡
weight	1	2	-1	0	1	2	3	0	1	2	1	4

"This is an unusual base in many ways. Like Roman numerals it lacks any holiness."

"I see another new feature here, the first misogynous number. Plus-double-minus, five-minus-two, has a half anti-female orientation because the double minus. It also has the most variable weights and orientations yet."

quintarish	1	2	3	4	10	11	12	13	14	20
descendants					1,0	1	1,2	1,3	1,4	0,2
duration	1	1	1	1	3	2	2	2	2	3
gender	M	F	M	F	F	M	F	M	F	F
head	1	2	3	4	1	1	1	1	1	2
holiness	0	0	0	0	1	0	0	0	0	1
length	1	1	1	1	2	2	2	2	2	2

persistence	0	0	1	1	1	1	1	1	1	1
tail	1	2	3	4	0	1	2	3	4	0
weight	1	2	3	4	1	2	3	4	5	2
sum	1	3	6	10	20	31	43	56	70	90

from *PAJAS* #8 (March 1986)

BASE SIX

Dan opened the meeting with "The sands of the seashore, the drops of rain, the days of eternity: who can number these?" from Sirach 1:2

Anne followed that up with the question, "Six won two, so why was she disappointed?"

To which she answered, "Because Zero won too, three for five."

With the ice not merely broken, but shattered, she continued, "It has been discovered in some extinct bacteria had another base pair, making it a base six[51] rather than the familiar four."

"Oh," Kamika interjected "Like the difference between the two-toed sloth-like aliens, the Bradypodans, and three-toed Chloepodans."

"Ah, yes," Anne aggreed, continuing, "The genetic alphabet has since expanded to include A, C, K, G, P, and T." [52]

Al, for his part, mentioned yet other alien number systems[53] and that Middle Earth's own Elves preferred to think in half-dozens and dozens[54], while Lana added, perhaps not coincidentally, so too do some New Guineans.[55]

"Is that why we have the expression 'Six of one and half a dozen of the other'?"

"Yeah, why not 'ten of one and half a score of the other'?"

"Or 'fifty thousand of one and a half a myriad of the other'?"

"Or 'five ninety-nineplex of one and a half a googol of the other'?"

"'Cause that could too easily be misinterpreted as ten-to-the-five-hundred-ninety-ninth."

senary	1_6	2_6	3_6	4_6	5_6	10_6	11_6	12_6	13_6	14_6	15_6	20_6	21_6
descendants						0,1	1	1,2	1,3	1,4	1,5	0,2	1,2
duration	1	1	1	1	1	3	2	2	2	2	2	3	2
gender	M	F	M	F	M	F	M	F	M	F	M	F	M
head	1	2	3	4	5	1	1	1	1	1	1	2	2
height	0	0.3	0.5	0.6	0.7	0.77	0.84	0.9	0.95	1	1.3	1.5	1.6
holiness	0	0	0	0	0	1	0	0	0	0	0	1	0
length	1	1	1	1	1	2	2	2	2	2	2	2	2
orientation	M	F	M	F	M	½F	M	½F	M	½F	M	F	½M

persistence	0	0	0	0	0	1	1	1	1	1	1	1	1
tail	1	2	3	4	5	0	1	2	3	4	5	0	1
weight	1	2	3	4	5	1	2	3	4	5	6	2	3

senarish	1	2	3	4	5	10	11	12	13	14	15	20	21	22	23	24	25	30	31	32
sum	1	3	6	10	15	25	36	48	61	75	90	110	131	153	176	200	225	255	286	318

from *PAJAS* #9 (April 1986)

BASE SEVEN[56]

Opening with "You will rightly be called sons of the Most High, since He Himself is good to the ungrateful and the wicked." (Lk 6:35), Dan asked, "How can you take four from seven and get five? Taking two Es, an N and an S from seven leaves V."

Al continued the discussion the next meeting by sharing the rather normal base seven system of the Brzöibs, Rikchik and Tau aliens.[57] He then contrasted these with a fourth system, which uses a zero-less variation something like the Klingon's base three."

"How can you eliminate the ones in a number?" someone asked.

Playing along he asked, "I don't know. How can you eliminate the ones in a number?"

"Add a G and one is 'gone'."

"In Talcryl" [58], he continued, "the number in the units' place is evaluated as one less than in any other place. 'Ni' in the units' place means zero, 'yu' one, 'yi' two, 'sa' three, 'si' four, 'to' five, 'ek' six, and each of them one higher elsewhere. 'Nini' means one-seven-plus-zero, rather than either zero-sevens-plus-zero or one-seven-plus-one, as one might expect. Ekni means seven-sevens-plus-zero or forty-nine, so that Talcryl is a slightly more efficient system than Brzöibian, using eight rather than seven digits."

"As you can see because this is another odd base the two-place numbers seem to be transgendered, both six and its successor one-zero have holes."

septary	1_7	2_7	3_7	4_7	5_7	6_7	10_7	11_7	12_7	13_7
descendants							0, 1	1	1, 2	1, 3
duration	1	1	1	1	1	1	3	2	2	2
gender	M	F	M	F	M	F	M	F	M	F
head	1	2	3	4	5	6	1	1	1	1
height	0	0.3	0.5	0.6	0.7	0.77	0.84	0.9	0.95	1
holiness	0	0	0	0	0	1	1	0	0	0
length	1	1	1	1	1	1	2	2	2	2
naughtiness	0	0	0	1	0	1	½	1	½	1
orientation	M	F	M	F	M	F	½M	M	½M	M
persistence	0	0	0	0	0	0	1	1	1	1

tail	1	2	3	4	5	6	0	1	2	3
weight	1	2	3	4	5	6	1	2	3	4

septarish	1	2	3	4	5	6	10	11	12	13	14	15	16	20	21	22	23	24	25	26
sum	1	3	5	9	14	20	30	41	53	66	80	95	111	131	152	174	197	221	276	302

from *PAJAS* #10 (May 1986)

NYMBEROLOLOGY

Dan opened with Luke 1:32, "He shall be great and shall be called the Son of the Most High." and the less seriously with "How many lawyers does it take to change a light bulb?"

Lou[59] answered with "It depends on how many you can afford."

"Base nine[60] is a condensed form of ternary," Al then began. "The even more collapsed version of base three, base twenty-seven, however extends the use of letters like Hexspeek's A through F all the way to Q or to Z in Lee Sallow's base 27.[61] Words can then be transformed into numbers (wordnums) and numbers can be transformed into words, what Joyce called nymbers[62].

	A	B	C	D	E	F	G	H	I	J	K	L	M	N	O	P	Q	R	S	T	U	V	W	X	Y	Z
0	1	2	3	4	5	6	7	8	9	10	11	12	13	14	15	16	17	18	19	20	21	22	23	24	25	26

"We can get high nymerologically from any word, but of particular interest are number names, what Joyce called arithmonyms. 'One' is read as the wordnum ONE or a three-place base-twenty-seven number meaning eleven thousand three hundred eighteen in base ten. It doesn't seem so at first, but as the number name's lengths get higher so do their nymber's. This new number name can be re-read as a thirty-eight-place nymber, which in turn would represent a number higher than fifty-fiveplex. This sequence gets higher fast though not smoothly.

arithmonyms	one	two	three	four	five	six
wordnum	151405	202315	2008180505	6152118	6092205	190924
decimal	11,318	15,216	10,791,620	129,618	125,255	14,118

"I found what the Romans called æquicalculus that did the same sort of thing," Lana explained, "in which the values of the letters were added without considering place values."

A	B	C/G	D	E	F	Z	H	I/J	K	L	M	N	O	P	Q	R	S	T	U/V	X
1	2	3	4	5	6	7	8	10	20	30	40	50	60	80	90	100	200	300	400	600

"In æquicalculus the word 'uno' would therefore be transformed into the number four hundred plus fifty plus seventy or five hundred twenty. 'Duo' would be four plus four hundred seventy or four hundred seventy-four."

uno	duo	tres	quattuor	quinque	sex	septem	octem	novem
DCC	CDLXXIV	DCV	XVCLI	MXLV	CXXXV	MXLXXX	CDVIII	DLV
520	474	605	15,151	1,045	805	1,430	408	555

"We could apply æquicalculus in English, since they didn't have W, Y or zero."

one	two	three	four	five	six	seven	eight	nine	ten	eleven	twelve	thirteen
115	360	418	566	421	810	660	326	115	355	765	740	778

"We can alternate between arithmonyms and nymbers to make a sequence that we could call 'arithmonymbers', that seem to get higher consistently."

one	1
one hundred fifteen	115
one thosand four hundred ten	1,410
two thousand three hundred eighty-six	2,386
two thousand four hundred twenty-six	2,426
four thousand three hundred eighty-four	4,384

"The same can even more easily be done doing æquicalculations on Roman numerals, where I now means ten, V four hundred, X six hundred, L thirty, C three, D four and M forty."

I	II	III	IV	V	VI	VII	VIII	IX	X	XI	XII	XIII	XIV	XV	XVI	XVII	XVIII
10	20	30	410	400	410	420	430	610	600	610	620	630	1,010	1,000	1,010	1,020	1,030

"Ah, Roman numeral wordnums, 'Romanums'."

"We wouldn't get high very fast that way! MMMM would only have a value of a hundred sixty."

"Try it the other way round. 'Quattuor milia' would become a fourteen-place base-27 number, nearly twentyplex."

"Let's go back to Roman numerals! Those I think I understand. So reading Roman numerals as nymbers as we did before," Cal put in, "we would get one equals I, which equals ten, which equals X, which equals six hundred?"

"Yes, and in Roman numerals that would be DC, which would be four plus three, seven."

"And seven would be VII or four hundred plus two tens, four hundred twenty."

"And four hundred twenty, CCCCXX, four twenties plus two six hundreds or one thousand twelve."

"And one thousand twelve, MCCXII, forty plus two threes plus six hundred plus two tens."

"Six hundred and sixty-six!"[63]

I	X	DC	VII	CCCCXX	MCCXII	DCLXVI
1	10	600	7	420	1,212	666

Tony hastily redirected the meeting by introducing the idea of using Scrabble's letter values having noticed that the word "twelve" has a point value of twelve.[64]. These nymbers also varied wildly. After several such series, the discussion diverged off to other more esoteric numerologies such as isopsephia[65], gematria[66], and so-called new merologies[67] before Clair turned the discussion back to Dan.

"A much simpler system called pseudonumerology[68] uses just the consonants, like Hebrew numerology. Unlike wordnums, it's based on just ten groupings of consonantal phonemes rather than letters. The original mnemonic for remembering the substitution was "Satan may relish coffee pie." which contains the consonants in order. Later the word for this was coined that does nearly as well phonetically, soo-doh-noo-mer-oll-oh-gee. These numbers from arithmonyms, Joyce called pseudonums.

letters	S/Z	T/D	N	M	R	L	Ch/J/Sh/Zh	G/K/Q	F/V	B/P
digit	0	1	2	3	4	5	6	7	8	9
name	zero	one	two	three	four	five	six	seven	eight	nine
pseudonum	4	2	1	14	84	88	70	82	1	22
sum	4	6	7	21	105	193	263	345	346	

"Repeated applications of pseudonumerology are also sometimes cyclical. 'One' gets reduced to N, which means two, which when read as 'two' is reduced to T, which means one. On the other hand 'three' reduces to TR, which means fourteen, which reduces to FRTN, etc. This we call this operation pseudonumerological expansion and the opposite operation pseudonumerolog."

three	14
fourteen	8,412
eight thousand, four hundred twelve	11,021,842,142,158
eleven twelveplex, twenty-one nineplex, eight hundred forty-two thousand, one hundred fifty-eight	582,158,970,- 121,1221,295,- 701,214,184,111,- 021,221,418,811

"Joyce wasn't a chromesthesiac, that is, someone who experiences one sense as another, he was able to combine the electronics color code used since the early '20s with the even older pseudonumerology. Then he coined yet other mnemonics for this new property of numbers, that he called their digital spectra.

0	1	2	3	4	5	6	7	8	9
black	brown	red	orange	yellow	green	blue	purple	gray	white
Suit	Tie	Neck	Marmalade	Rose	Leaf	Jay	Cow	Fog	hoPe

"So for us Joyceans for whom every digit is so color-coded, Most High would be orange-black-brown. This code usually is interpreted by electricians as rounded numbers, the third color indicating the exponent of a factor of ten, rather than merely a third digit, so they would evaluate it as thirty times ten or three hundred, rather than three hundred one.

"We can generate nymber sequences with bases twenty-seven, thirty-six, one hundred twenty-eight and two hundred fifty-six as well. They would get higher much faster and so unlikely get caught in the cycling of other nymber generators.[69]

"During the attempt to decipher the Minoan Linear B tablets soon after their discovery in 1900, each of the eighty-seven characters was assigned an arbitrary number. J. R. R. Tolkien did a similar thing in *The Lord of the Ring*. The Tengwar, like Atbash or pseudonumerology is without vowels, which some of you I know use as a shorthand, might be a base thirty-two cipher. 'One' would transform into N or seventeen, 'two' to T or one, 'three' to ThR or nine-twenty-one, four to FR or ten-twenty-one and so on.

	1	2	3	4	5	6	7	8	9	10	11	12	13	14	15	17	18	21	22	23	27	29	31
Tengwar	t	p	ch	k	d	b	j	g	th	f	sh	hw	dh	v	zh	n	m	r	w	y	l	s	z

	one	two	three	four	five	six	seven	eight	nine
Tengwar	17	1	921	1021	1014	290429	291417	1	1717
decimal	17	1	309	341	334	29,853	30,161	1	561
Tengwarsum	17	18	1006	2027	3109	300406	1272823	1272824	1281407

"So Most High, expressed phonetically with GH silent, would be a seven-place base thirty-two nymber?"

"Several billions, uh, nineplexes."

"The more complex Angerthas consonants and vowels would get us even higher, since they would be base sixty."

1	2	3	4	6	8	9	10	11	12	13	14	15	16	17	18	21	25	27	29	31	34
p	b	f	v	m	t	d	th	dh	n	ch	j	sh	zh	z	k	g	ŋ	kw	r	l	s

39	40	42	43	44	46	47	48	49	50	51	54	60
i	y	u	ū	w	e	ē	a	ā	o	ō	h	&

"A nine-place base-sixty nymber, over sixteenplex."

"'One' would be, uh, w-u-n, forty-four-forty-two-twelve."

"Two' would be t-ū, eight-forty-two."

"And 'three', th-r-ē, ten-twenty-nine-forty-seven."

	one	two	three	four	five	six	seven
Angerthas	444212	842	102947	35029	34004	34391634	3446044612
decimal	160,932	522	37,787	13,829	13,204	14,820,394	901,169,172
Angerthas sum	444212	445052	552039	591108	1025112	35470746	3511515358

from *PAJAS* #11 (June 1986)

ALPHAMETICS

Dan opened the meeting with a quote from the *Qur'an*, "His throne extends over Heavens and Earth and He feels no fatigue in guarding and preserving them for He is the Most High."

"What is the richest country in the world?" Patrick[70] asked and then supplied the answer, "Ireland, because it's capital is always Dublin."

That set the tone of the meeting as school broke for the summer, the remnant members of the Society decided to take a break, playing the classics.[71]

```
  HAVE                  DO
     A      SEND       YOU      APPLES
+ GREAT   + MORE    + FEEL   + ORANGES
 SUMMER    MONEY     LUCKY    BANANAS

            THEN                  ONE
 ORANGE    JESUS     JESUS       NINE
+ORANGE   + WENT    + BIBLE   +TWENTY
 APPLES   THENCE    CHURCH      EIGHT
```

"The world's best known alphametic puzzle is undoubtedly "SEND MORE MONEY", created by H. E. Dudeney in 1924. Originally it had to do with a kidnaper's ransom demand, not a college student's plea to parents. According to the rules of the game no leading zeroes are allowed, so neither S, nor M, are zero. The M in MONEY is obviously a carried-over one, so S must be a nine with another carry-over and O must be zero, since it cannot be one."

"So in the hundreds column we have E plus a carry-over one equals N."

"And from the tens column E plus ten equals N plus R plus another carry-over one, so nine equals R plus one and R equals eight."

"How did you know that there was a carry-over one?"

"Because R couldn't be nine, because S already was nine."

"So then from the units column we have D plus E are greater than eleven, because Y is greater than one."

"And so E must be five, N six, D seven and Y two!"

"Alas, we can't know what three or four are because we're only given eight letter variables. A so-called doubly-true alphametic like the last one, summing to EIGHTY, in which number names were used, was relatively recently introduced by Alan Wayne in '45. These kinds of mathematical recreations however go back to ancient 'letter arithmetic' in which letters were substituted for digits, now called cryptarthms. In the Middle Ages, before algebraic notation, a variant developed in which many of the digits, the 'flesh', were replaced by asterisks leaving what was called the 'skeleton'. According to Evvie Nef's Law, "There is a solution to every problem; the only difficulty is finding it." Of course, the more clues you're given the easier it is to find the solution. That's why intermediate steps of long multiplication or division are often given, like so:

```
   NOT      RUN        ****
 x  ON     x TO       x  **
  LLAY     NIYO      2***3
 EELM     NDRY       1***4
 ENEMY    RADIO     **55**
```

```
    WOW          TOO          OLD
W) MEOW      T) GOOD      A) GOOD
   MW           GD           GO
    BO           SO           TO
    BW           SO           TO
     MW           SD            D
     MW           SD            D
```

"We can also transform numbers into letters by rotations and reflections.[73] An upside-down seven suggests a partial A with a vertical line connected to a diagonal line. An upside-down nine, b, looks like a lowercase B, and so on. Zero and one would be O and I without any reflection or rotation, of course."

"A two reflected and rotated to the left is much like a lowercase N, and to the right much like a lowercase U."

"And a reflected two like a C."

"A reflected three, Ɛ, would be an E, a three rotated to the left, ɷ, an M and to the right, a ω, W."

"A four is very close to a Y, rotated to the right, F, would be an F and upside-down a lowercase h."

"A five could easily impersonate an S and a reflected five a Z, a six a G. A seven looks something like a T and upside-down like an L. Reflected it looks like a lowercase r, and rotated a V."

"A nine could also represent a lowercase b, d, p or q with the appropriate reflection and/or rotations."

"That just leaves eight which is the closest to an X among the integers."

"But we haven't anything for J or K."

"We don't need J or K because there are no number names that use a J or K."

"What about jillion and killion."

"O. K., if you insist. We could get a J by turning seven upside-down and reflecting and at least the bottom half of the K by left-turning it."

"So we've got a whole look-alike alphabet."

A	B	C	d	E	F	G	h	I	J	K	L	M	n	O	p	q	r	S	T	u	V	w	x	Y	Z
7	9	2	9	3	4	6	4	1	7	7	7	3	2	0	9	9	7	5	7	2	7	3	8	4	2

"So 'zero' would get transformed into two-three-seven-zero!"

"A 'one' into zero-two-three?"
"Yes, and 'two' and 'three' into seven-three-zero and seven-four-seven-three-three."

number	1	2	3	4	5	6	7	8	9	10	11
name	one	two	three	four	five	six	seven	eight	nine	ten	eleven
lookalike	023	730	74733	4027	4173	518	53732	31647	2123	732	373732
sum	23	753	75,486	79,513	83,686	84,204	137,936	141,583	143,706	144,438	518,170

"We can also get high by re-evaluating one's lookalike nymber zero-twenty-three and so on."
"That would be with the initial zero, 2370-733274-74733."
"Two hundred thirty-seven twelveplex, seventy-three nineplex, three hundred twenty-seven million, four hundred seventy-four thousand, seven hundred thirty-three!"
"And higher than that more than a hundred thirty digits."

1
023
237,073,327,474,733
> hundred-thirtyplex

"303!"
"What?"
"Read sideways it's..."
"Wow!"
"The fastest-growing sequence yet, to a hundred-thirty-threeplex in just three steps!"
The rest of the meeting was spent calculating sequences starting with higher seeds.

2	730	537,324,229,739,741,774
3	74,733	537,327,440,277,402,572,953,732,422,973,974,177,474,733
4	4,027	40,277,402,572,973,327,453,732
5	4,173	4,027,740,257,290,234,229,739,537,327,474,733
6	518	4,173,422,973,931,647,332
7	53,732	414,747,473,374,025,729,537,324,229,739,741,774,730
8	31,647	7,417,740,237,402,572,951,842,297,394,077,453,732
9	2,123	73,074,025,729,023,422,973,973,327,474,733
10	732	537,324,229,739,741,774,730
11	373,732	74,733,422,973,953,732,747,473,374,025,729,537,324,229,739,741,774,730

from *PAJAS* #12 (July 1986)

ETHICALCULUS

Dan shared "The Lord Most High is awesome." from Psalm 47:2 and then asked, "What's the difference between a good gutter and a bad ballplayer?"

Cal answered, "One catches drops, while the other drops catches."

Dan then continued, "Although even the GOOD alphametics we solved last month had just one answer in base ten," Raz continued, "they had several different possible ones in higher bases. This is the basic -- pun intended -- problem of what Joyce called by the portmanteau, ethicalculus. As those of you had him will remember, he illustrated the problems of mathematics as well as moral relativism by combining ethics and calculus. Bad and good would mean one thing in one number system and something quite different in another." [74]

"The problem is at least as old as Indo-European in which the word 'bhad' meant good," Lana added.

"Any base higher than thirteen includes both BAD and pseudo-good, nine-double zero-D that looks like GOOD. Those higher than twenty-four include O, except for the asterisked base twenty-seven, and so have real GOOD replacing pseudo-good.

"Base thirty-six that includes all the capital letters and digits, but without the space, is called alphadigital, a condensation of base six. The ASCII bases, one twenty-eight and two fifty-six, include many, many more symbols and thus have a much, much higher GOOD/BAD ratio."

base	14	15	16	17	18	19	20	21	22	23	24
900D/BAD	10.7	11.5	12.3	13.2	14.0	14.8	15.6	16.4	17.2	18.1	18.9

base	25	26	27	27*	28	29	30	31	32	33	34	35	36
900D/BAD	19.7	20.5	21.3		22.2	23.0	23.8	24.6	25.4	26.2	27.1	27.9	28.7
GOOD/BAD	930	1,005	1,083	100	1,164	1,247	1,334	1,424	1,516	1,612	1,710	1,817	1,915

base	128	256
900D/BAD	104	208
GOOD/BAD	25,000	120,000,000

The rest of the meeting was taken up with answering these and other 'burning' ethicalculus questions like "Which is more valuable R2-D2 or C-3PO?"

from *PAJAS* #13 (August 1986)

CRYPTOLOGY

Dan opened the meeting with Psalm 9:2, "I will sing praise to Thy name, O Most High." and with the riddle, "What would you call twenty-five score Native Americans without apples?"

"The Indian appleless five hundred."

When that answer came too easily, he tried another, quoting Lewis Carroll, "Dreaming of apples on a wall And dreaming often, dear, I dreamed that if I counted all -- How many would appear?"

That didn't come quite so quickly, but eventually someone realized he'd dreamed "of ten" and another that Rhyming Slang for 'score', twice ten, happened to be "apple" from "apple core" too. Still others reasoned that since he was now awake none would appear. He too was appleless.

After playing with alphametics in which letters represent numbers, the Society naturally spent some time playing with numbers representing letters.[75] Any seemingly cryptic substitution of symbols for letters can be reduced to a number substitution by merely numbering the symbols. Sam had promised to bring in his prize decoder ring collection.

"The Bible's Atbash substitutes letters in reverse alphabetical order," Lana began, "the Hebrew aleph for taw, and beth for shin, hence the name from the four letters. Since this alphabet has twenty letters the cipher is just twenty-one minus the alphabetical position."

א	ב	ג	ד	ה	ו	ז	ט	י	כ
'	B	G	D	H	W	Z	Th	Y	K
1	2	3	4	5	6	7	8	9	10
ת	ש	ר	ק	פ	ע	ס	נ	מ	ל
T	Sh	R	Q	P	'	S	N	M	L
20	19	18	17	16	15	14	13	12	11

"The word 'Sheshak' used in Jeremiah comes from the nineteenth, nineteenth and tenth phonemes, ShShK. Its inverse would be the second, second and eleventh phonemes or BBL or two-zero-two-one-one. With the unwritten vowels replaced in reverse order we get the more familiar name Babel or Babylon. The name Jesus, Yeshua in Hebrew, would be Mabue in Atbash. Most High, Elyon, would be Okmeth.

"A name for the analogous English version, reversing all letters in our alphabet, might be called Atoz. Its GOOD/BAD ratio would be nearly six hundred and not half bad just over eleven hundred. Most High would then be 'thirteen, fifteen, nineteen, twenty, zero, eight, nine, seven, eight'."

A	B	C	D	E	F	G	H	I	J	K	L	M
26	25	24	23	22	21	20	19	18	17	16	15	14
Z	Y	X	W	V	U	T	S	R	Q	P	O	N
01	02	03	04	05	06	07	08	09	10	11	12	13

"So we would get 'fifteen, fourteen, five' for one and get higher with 'twenty, twenty-three, fifteen' for two."

Atoz	one	two	three	four	five	six
nymbers	151405	202315	2008180505	6152118	6092205	190924
decimal	11,318	15,216	21,585,830	247,716	243,356	14,118

"Reversing the order of the consonants and vowels separately however, we'd get a different scrambled alphabet more like Atbash. It has five pairs of letters swapped just like Atoz and thirteen differently. In Aybz, pronounced 'Abe's', Most High would be 'thirteen, fifteen, nineteen, twenty, zero, eight, nine, seven, eight'. Its GOOD/BAD ratio would be a little lower"

A	B	C	D	E	F	G	H	I	J	K	L	M
25	26	24	23	21	22	20	19	15	18	17	16	14
Y	Z	X	W	U	V	T	S	O	R	Q	P	N
1	2	3	4	5	6	7	8	9	10	11	12	13

"Although the nymbers initially look like they would just fluctuate wildly, they gradually get higher as the number names get longer. ONE is a three-place wordnum, but ONE HUNDRED is eleven places."

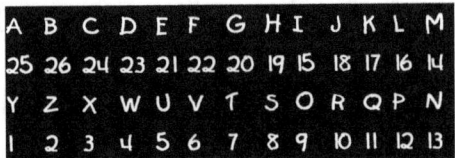

Aybz	one	two	three	four	five	six
nymbers	91321	70409	719102121	220905 10	22150621	81503
decimal	6,933	5,220	8,196,006	872,758	877,170	6,240

"Eleven?" Cal asked, though he quickly added, "Oh, yeah, the space between the one and the hundred is represented as a double zero in base twenty-seven, isn't it?"

A	B	C	D	E	F	G	H	I	L	M	N	O	P	Q	R	S	T	V	X	Y	Z
E	F	G	H	I	L	M	N	O	P	Q	R	S	T	V	X	Y	Z	A	B	C	D
5	6	7	8	9	10	11	12	13	14	15	16	17	18	19	20	21	22	1	2	3	4

Lana continued, "Julius Caesar's shift cipher was only slightly more complex with twenty-one possible alphabetical shifts. This plus-four code conveniently shifts the vowels E to A, I to E and O to I, so Most High, Latin's Altissimus would be Epzoyyoqay.

"If we shifted the vowels and consonants separately we get a completely different shifting. Altissimus in this plus-four shift would now be Yqbezzercz.

A	B	C	D	E	F	G	H	I	L	M	N	O	P	Q	R	S	T	V	X	Y	Z
Y	G	H	L	A	M	N	P	E	Q	R	S	I	T	V	X	Z	B	C	D	O	F
19	7	8	10	1	11	12	14	5	15	16	17	9	18	20	21	22	2	3	4	13	6

"Most other encryptions would be quite unpronounceable, better represented as a nine-place base twenty-three number. 'One' would be 'nine, eight, one', 'two' or rather 'tvvo', would be ten-eleven-eleven-nine'. Its GOOD/BAD ratio would be almost four."

	one	two	three	four	five	six	seven
+4	90801	10111109	1004120101	2091512	2051501	90512	901150108
decimal	6,778	401.985	10,795,060	85,710	82,783	6,708	9,616,274

from *PAJAS* #14 (September 1986)

ISOGRAMS AND PANGRAMS

As students returned from summer break the Society continued along the lines they had during the summer. After opening with Psalm 50:14, "Pay your vows to the Most High." Dan ask, "What do Democrats and Republicans have in common?"

"They're both isograms"[77], he continuing, answering before anyone else could, "words without repeated letters, make good keys for generating scrambled alphabets, because they are not so easily broken, though easily encoded and decoded. The keyword 'uncopyrightables' with sixteen unrepeated letters, for example, encodes one particular scrambled alphabet. Most High would be the nine-place wordnum, 'two, five, ten, eleven, zero, nine, seven, eighteen, nine'."

A	B	C	D	E	F	G	H	I	J	K	L	M	N	O	P	Q	R	S	T	U	V	W	X	Y	Z
U	N	C	O	P	Y	R	I	G	H	T	A	B	L	E	S	D	F	J	K	M	Q	V	W	X	Z
21	14	03	15	16	25	18	09	07	08	20	01	02	12	05	19	04	06	10	11	13	17	22	23	24	26

"'One' would be transformed into 'five, one, nineteen'."

"And 'two' would be 'eleven, twenty-two, sixteen' and 'three', 'twenty, nine, six, sixteen, sixteen', so we're get higher and higher."

"But not for long! The next number is just four places, 'twenty-five, five, thirteen, six'."

uncopyrightable	one	two	three	four	five	six
nymbers	51216	112205	1109061616	25051306	25071716	100723
decimal	3,985	8,618	12,050,818	988,152	989,728	7,502

"Pangrams[78]," Dan interrupted, changing the subject a bit, "sentences which contain all twenty-six the letters of the alphabet, can also be used. If we do not use repeated letters the scrambled alphabet is easily remembered, but hard to decipher. The classic typewriting exercise, 'The quick, brown fox jumps over the lazy dog.', for example, generates such a scrambled alphabet. Most High would be yet another nine-place wordnum.

A	B	C	D	E	F	G	H	I	J	K	L	M	N	O	P	Q	R	S	T	U	V	W	X	Y	Z
T	H	E	Q	U	I	C	K	B	R	O	W	N	F	X	J	M	P	S	V	L	A	Z	Y	D	G
20	8	5	17	21	9	3	11	2	18	15	23	14	6	24	10	13	16	19	22	12	1	26	25	4	7

pangram #1	one	two	three	four	five	six
nymber	2406	222624	2211162121	9241216	9020121	190224
decimal	654	16.764	23,828,682	372,130	355,800	13,929

"A better, that is, less easily broken, cipher however would be one which allowed multiple substitutions for the same letter, and so any number of more numbers higher than twenty-six. Using the whole pangram, it'd have three more values for O, two more for E and another for H, R, T and U, for thirty-five substitutions. Sometimes 'sleeping' is added after 'lazy' adding nine more duplicate letters. In yet another alternative, we can count both words and letters in the words to get a base-hundred-like substitution. So even without nulls the nymbers would look like any of several quite different base-ten numbers! "

T	H	E	Q	U	I	C	K	B	R	O	W	N	F	O	X	J	U	M	P	S
01	02	03	04	05	06	07	08	09	10	11	12	13	14	15	16	17	18	19	20	21
11	12	13	21	22	23	24	25	31	32	33	34	35	41	42	43	51	52	52	54	55

O	V	E	R	T	H	E	L	A	Z	Y	D	O	G
22	23	24	25	26	27	28	29	30	31	32	33	34	35
61	62	63	64	71	72	73	81	82	83	84	91	92	93

pangram #3	one	two	three	four	five	six
nymbers	333,563	113,442	7,172,326,373	41,612,264	41,236,213	552,343

from *PAJAS* #15 (October 1986)

DECODERS

Dan chose to open with Psalm 91:1-2, "You who dwell in the shelter of the Most High, who abide in the shade of the Almighty, say to the Lord, 'My Refuge and Fortress, my God in Whom I trust", but then he had to skip the traditional ice-breaking riddle, because everyone was so interested in Sam's collection of decoder rings.

"Back in '34, when I was just a child," Sam began, "the Radio Ophan Annie's Secret Society code was childishly simple. It was just twice the alphabetical position. All the odds between seven and ninety-nine however could be used as nulls to make it seem more complex. We could have used double-zero as a space, but usually didn't stringing the words together for even more misdirection. Most High would be at least eight double-digit even numbers."

	A	B	C	D	E	F	G	H	I	J	K	L	M	N	O	P	Q	R	S	T	U	V	W	X	Y	Z
'34	2	4	6	8	10	12	14	16	18	20	22	24	26	28	30	32	34	36	38	40	42	44	46	48	50	52

"So 'one' would be encoded as 'three-zero-two-eight-one-zero'?"

"In a base fifty-three, so over fourteenplex."

	one	two	three	four	five	six
ROASS '34	302810	404630	4016361010	12304236	12184410	381848
decimal	85,764	114,828	636,104,208	3,659,580	3,625,952	107,744

"Right, but as much higher as you liked with the added nulls." Sam continued. "By the next year, the ROASS used any of twenty-six cipher alphabets from this decoder, depending on which number was set opposite which letter. A1 and M2, etc. were just code names for the same setting, while making it seem that there were not merely twenty-six codes but six hundred seventy-six. The next year they switched to a different set of twenty-seven scrambled alphabets. Our chapter nicknamed the '35 A-1 though P-26 codes the Oklahoma codes for the OK in them and the '36 A-1 through X-26 codes the Nevada codes for their NEV.

	1	2	3	4	5	6	7	8	9	10	11	12	13	14	15	16	17	18	19	20	21	22	23	24	25	26
'35	A	M	Z	N	B	L	Y	O	K	C	Q	X	J	D	R	W	I	E	S	V	H	G	T	F	U	P
	1	2	3	4	5	6	7	8	9	10	11	12	13	14	15	16	17	18	19	20	21	22	23	24	25	26
'36	A	S	T	P	B	H	M	C	G	Q	D	F	Z	L	N	E	V	J	Y	I	W	U	R	O	K	X

	one	two	three	four	five	six
Oklahoma	80418	231608	2321151818	24082015	24172018	191712
decimal	22,702	65,463	369,258,067	7,169,643	7,194,927	54,284
Nevada	241516	32124	306231616	12242223	12201716	22026
decimal	68,227	9,564	49,194,881	3,641,653	3,630,145	6,704

"In '41 Capt. Midnight's Secret Squadron used this badge-like decoder called a code-o-graph. As you can see in '45 it included a magnet, in '46 a signal mirror and in '47 a whistle. In '49 Johnny Quest revived the decoder ring that included the hidden commercial message 'Wear PFs'. We nicknamed these codes from state and territory abbreviations, Hawaii (A-1 to Y-26), New York or Rhode Island (A-1 to L-26), Florida (A-1 to U-26), North Carolina (A-1 to Z-26) and Arkansas (A-1 to E-26). Between these decoding devices we had hundreds of different ways to get high."

	1	2	3	4	5	6	7	8	9	10	11	12	13	14	15	16	17	18	19	20	21	22	23	24	25	26
'41	A	X	N	Q	E	G	M	K	F	W	Z	H	I	O	B	L	T	D	S	R	C	J	V	U	P	Y
'45	A	F	X	D	T	Z	K	N	Y	C	J	W	S	G	U	M	P	O	Q	H	R	I	V	E	B	L
'46	A	G	H	T	V	Q	S	E	P	Y	J	I	F	L	X	K	D	C	W	R	B	O	N	Z	M	U
'47	A	T	M	P	Q	U	O	V	F	Y	B	H	J	N	C	I	X	S	D	K	R	G	W	L	E	Z
'49	A	R	P	F	S	L	Q	M	Y	B	U	H	X	V	C	Z	N	D	K	I	O	T	G	J	W	E

	one	two	three	four	five	six
Hawaii	140305	171014	17122200505	9142420	9132305	191302
decimal	10,292	12,677	18,556,106	365,168	364,397	14,204
New York	180824	51218	520212424	2181521	2222324	132203
decimal	13,362	3,987	6,117,711	840,234	95,415	10,074
Florida	222308	41922	403200808	13222620	13120508	71215
decimal	16,667	3,451	4,384,430	528,518	520,649	5,442
N.Carolina	71425	22307	2122125525	9070621	9160825	181617

decimal	5,506	2,086	23,196,022	359,580	366,199	13,122
Arkansas	211726	222521	2212022626	4211102	4201426	52013
decimal	15,794	16.734	23,857,982	173,072	172,448	4,198

from *PAJAS* #16 (November 1986)

RUNNING CIPHERS

Dan opened the meeting with "Blessed be God Most High, who delivered your foes into your hand." from Genesis 14:20.

He passed the meeting on to Lana who asked "Did you hear what happened to two eighty-eight?"

When everyone including Dan drew a blank, she responded with "I can't tell you! It's too gross."

Then she plunged into her topic the history of cryptology, "Back in the Fifteenth Century Johannes Trithemius used not a decoder ring or code-o-graph, but what he called a Tabla Recta, Latin for square table, with twenty-four ciphers for the twenty-four letters of the alphabet, with I equated with J and U with V as they were then. This so-called running cipher changes scrambled alphabets letter-by-letter from one key phrase, for example, 'In principio erat Verbvm.' from John 1:1, Latin for 'In the beginning was the Word.'"

15th	1	2	3	4	5	6	7	8	9	10	11	12	13	14	15	16	17	18	19	20	21	22	23	24
1	I	K	L	M	N	O	P	Q	R	S	T	V	W	X	Y	Z	A	B	C	D	E	F	G	H
2	N	O	P	Q	R	S	T	V	W	X	Y	Z	A	B	C	D	E	F	G	H	I	K	L	M
3	P	Q	R	S	T	V	W	X	Y	Z	A	B	C	D	E	F	G	H	I	K	L	M	N	O
4	R	S	T	V	W	X	Y	Z	A	B	C	D	E	F	G	H	I	K	L	M	N	O	P	Q
5	I	K	L	M	N	O	P	Q	R	S	T	V	W	X	Y	Z	A	B	C	D	E	F	G	H

"'One' would be SPG from rows two through four or six-zer-one-one-five from the cleartext. The two E's in 'three' would be represented by two different base-twenty-five numbers."

15th C	one	two	three	four	five	six
nymber	60115	110924	1120031321	22020601	22010613	102108
decimal	3,790	7,124	9,220,971	688,901	688,288	6,783

"Voynichese, the 'language' of the cryptic manuscript rediscovered by Wilfred Michael Voynich is also Fifteenth Century. Jacques Guy's Frogguy transcription approximates it with twenty-four basic symbols, plus connectors and capitalizations which can be taken as nulls and combined symbols. Six-zero-one-one-five, in this example, would be written in Basic Voynichese as 'kax'."

	1	2	3	4	5	6	7	8	9	10	11	12	13	14	15	16	17	18	19	20	21	22	23	24
symbol	a	c	g	i	j	k	l	n	o	p	q	s	t	v	x	y	z	&	2	4	8	9	'	^
connector	3	-	+	,	)	\																		
capital	A	C	G	I	J	K	L	N	O	P	Q	S	T	V	X	Y	Z							
combo	d	e	f	F	iiv	iv																		

"In the next century, Gaspar Scott told the Count of Gronsfeld of a similar cipher that used a key number, rather than a key phrase, for example, the alphabetically sorted digits, and including a space, so 'one' would now be 70901 and 'one hundred' would be an eleven-place wordnum even without nulls.

16th C	0	1	2	3	4	5	6	7	8	9	10	11	12	13	14	15	16	17	18	19	20	21	22	23	24	25	26
8	H	I	J	K	L	M	N	O	P	Q	R	S	T	U	V	W	X	Y	Z		A	B	C	D	E	F	G
5	E	F	G	H	I	J	K	L	M	N	O	P	Q	R	S	T	U	V	W	X	Y	Z		A	B	C	D
4	D	E	F	G	H	I	J	K	L	M	N	O	P	Q	R	S	T	U	V	W	X	Y	Z		A	B	C
9	I	J	K	L	M	N	O	P	Q	R	S	T	U	V	W	X	Y	Z		A	B	C	D	E	F	G	H
1	A	B	C	D	E	F	G	H	I	J	K	L	M	N	O	P	Q	R	S	T	U	V	W	X	Y	Z	

	one	two	three	four	five	six
16th	70901	121811	1203142304	25101709	25041823	110420
decimal	5,342	9,245	12,883,513	991,908	987,575	8,147

"During the Boer War in the Nineteenth Century a new kind of cipher was invented by Charles Whetstone and popularized by Baron Lyon Playfair. It coded bigrams or letter pairs, rather than individual letters, using a five-by-five square table and a keyword but used the extra letter as a null, in this example the Z."

	0	1	2	3	4
0	U	N	C	O	P
1	Y	R	I	G	H
2	T	A	B	L	E
3	S	D	F	J	K
4	M	Q	V	W	X

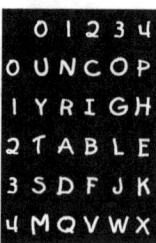

"If two letters were in the same column they were replaced by the letters below, if in the same row by those to the right and if neither by those in the opposite corners of the square determined by the original pair. If the word had an odd number of letters an extra one, say X was added."

"So 'one' would be read ONEX and encoded as CPKP or 2043404 in base five."

	one	two	three	four	five
Playfair	20403404	24430443	241414223404	32020110	42224421
decimal	182,104	247,623	140,340,854	268,155	359,861

"The six-bit, base thirty-two, system patented by Jean-Maurice-Émile Baudot in 1874 was used for teletype encoded symbols other than the alphabet."

	A	B	C	D	E	F	G	H	I	J	K	L	M	N	O	P	Q	R	S	T	U	V	W	X	Y	Z
0	24	19	14	18	16	22	11	5	12	26	30	9	7	6	3	13	29	10	20	1	28	15	25	23	21	17

	one	two	three	four	five	six
Baudot	30616	12503	105101616	22032810	22121516	201223
decimal	3,280	1,827	2,435,600	1,445,770	1,454,576	20,887

"Felix Delasteele combined substitution and transposition in his so-called bifrid code. This did encode letters into base five, but with a twist. Using Felix as the key and equating U and V we get the following table."

	0	1	2	3	4
0	F	E	L	I	X
1	A	B	C	D	G
2	H	J	K	M	N
3	O	P	Q	R	S
4	T	U	W	Y	Z

"'One' would be 302401, whose six digits are re-written as every other digit, 320 and 041, and then hyperadded to 320041."

	one	two	three	four	five	six	seven
bifrid	320041	443020	423000311	03430013	00400311	300434	3111241414
decimal	21,271	30,760	1,765,706	75,633	25,081	18,869	6,361,609

Lana then wrote the following.

How mnay hree can raed tihs?

"I can."

"Me too."

"Me three."

"According to Graham Rawlinson, 'We only need the first and last two letters to spot changes in meaning.' This condition of partial blindness to typos has been called typoglycemia.[79] Swapping interior digits, however would usually completely change a number.

A sequence would look normal until the fourth place and then become quite disordered."

As if on cue, the members became quite disorderly searching for such subsequences of the out-of-order or permutable primes.

from *PAJAS* #17 (December 1986)

PHONETIC ALPHABETS

Dan opened the meeting with "God Most High, the creator of Heaven and Earth!" (Gn 14:19) and then followed that with the question, "What do you get if you take a half from a half-dollar?"

When several at once suggested a quarter, he corrected them with "If you take the prefix 'half-' from 'half-dollar' you get a dollar."

"I wish I could do that with my half-dollars."

Dan continued, "The discussion of ciphers last month reminded me of the Navaho code used in WWII and never broken. It used two hundred seventy-four code words, like the phrase Besh-be-cha-he, Iron Hat, for Germany and Beh-na-ali-tsosie, Slant Eye, for Japan, but also used a code alphabet. The Navaho or Diné word for ant, wol-la-chee, was used for A, the word na-hash-chid, or badger, for B, and so on. You could do that same word-for-letter substitution for any language, but the Axis powers were not familiar with Navaho." [80]

"Neither am I!" several voices cried out simultaneously.

"Suffice it to say the nymber 'one' would have been encoded as a twenty-four-place base twenty-eight nymber, with both the space and hyphen important placeholders. In the English codewords would be just a twelve-place wordnum."

one	owl nose ear
wordnum	151,405,001,415,190,500,050,118
Navaho	ne-ahs-jah a-chin ah-jah
wordnum	140,527,010,819,271,001,080,001,270,308,091,400,010,827,100,108
two	tooth weasel owl
wordnum	20,151,520,080,023,050,119,051,200,152,312
Navaho	a-woh gloe-ih ne-ahs-jah
wordnum	12,723,150,800,071,215,052,709,080,014,052,701,081,927,100,108
three	tooth hair ram ear ear
wordnum	20,151,520,080,008,010,918,001,801,130,005,011,800,050,118
Navaho	d-ah tse-gah clah-nes-ta ah-jah ah-jah
wordnum	427,010,800,201,905,270,701,080,003,120,108,271,- 405,192,720,010,001,082,710,010,800,010,827,108
four	fly owl uncle ram
wordnum	612,250,015,231,200,211,403,120,500,180,113
Navaho	tsa-e-donin-ee ne-ahs-jah shi-la dah-nes-tsa

wordnum	2,019,012,705,270,415,140,914,270,505,001,405,270,108,192,-710,010,800,190,809,271,201,000,401,082,714,051,927,201,901
five	fly itch victor ear
wordnum	6,122,500,092,003,080,022,090,320,151,800,050,118
Navaho	tsa-e-donin-ee yeh-hes a-keh-di-glini ah-jah
wordnum	2,019,012,705,270,415,140,914,270,505,002,505,082,708,051,-900,012,711,050,827,040,927,071,209,140,900,010,827,100,108

"The United States also used a phonetic alphabet during the Second World War to clarify a message, rather than disguise it. 'One' would have been expanded to thirteen places and since both upper and lower cases and space are used it's in base fifty-three, over twenty-twoplex in base ten."

A	a	B	b	C	c	D	d	E	e	F	f	G	g	H	h	I	i	J	j	K	k	L	l	M	m
1	2	3	4	5	6	7	8	9	10	11	12	13	14	15	16	17	18	19	20	21	22	23	24	25	26
N	n	O	o	P	p	Q	q	R	r	S	s	T	t	U	u	V	v	W	w	X	x	Y	y	Z	z
27	28	29	30	31	32	33	34	35	36	37	38	39	40	41	42	43	44	45	46	47	48	49	50	51	52

	1941 code
one	oboe Nan easy
wordnum	30043010002702280010023850
two	tare William oboe
wordnum	4002361000451823231802260030043010
three	tare how Roger easy easy
wordnum	40022336001630460035301410360010023850010023850
four	fox oboe uncle Roger
wordnum	123048003004301000422806241000353014 1036
five	fox item Victor easy
wordnum	12304800184010260043180640303600 10023850
six	sugar item X-ray
wordnum	3842140236001840102600473602 50

"Before the war the International Telegraph Union also had such a phonetic alphabet, based mostly on geographical proper names. In it 'one' would be a twenty-place base-fifty-three nymber, over thirty-fourplex."

	1932 code
one	Oslo New York Edison
wordnum	2938243000271046004830362200090818383028
two	Tripoli Washington Oslo
wordnum	3836183230241800450238161828144030280029392430
three	Tripoli Havana Roma Edison Edison
wordnum	3936183230241800150244022802003530260200090818383028090818383028
four	Florida Oslo Upsala Roma
wordnum	112430361809020029382430004132380224020035302602
five	Florida Italia Valencia Edison
wordnum	112430361809020017400224180200430223102806180200090818383028
six	Santiago Italia Xanthippe
wordnum	370228401802143000174002180200470228401618323210

"After the world wars NATO standardized the international phonetic alphabet that would make Victor Charlie famous, or rather infamous. It's 'one' nymber would only be fourteen base-fifty-three places, over twenty-threeplex."

	1955 code
one	Oscar Nan easy
wordnum	2938060236002702280009023850
two	tango whiskey Oscar
wordnum	40,022,814,300,046,161,838,221,050,002,938,060,236
three	tango hotel Romeo easy easy
wordnum	400228143000153040102400353026103000100238490010023849
four	foxtrot Oscar uniform Romeo
wordnum	123048403630400029380602360042281812303626003530261030
five	foxtrot India Victor easy
wordnum	12304840363040001728091702004318064030360010023850
six	Sierra India X-ray
wordnum	3718103636020017280917020047360250

"Long before that, during the First World War, the British had their own phonetic alphabet with common words and proper personal names.

	1914 code
one	orange nuts Edward
wordnum	303602281410002842403800090846023608
two	Tommy William orange
wordnum	3930262650001502363650000908460236080090846023608
three	Tommy Harry Robert Edward Edward
wordnum	39302626500015023636500035300410364009084601360800090846013608
four	Freddy orange uncle Robert
wordnum	1136100909500030360228141000412806241000353004103640
five	Freddy ink vinegar Edward
wordnum	11361009095000172822004418281014023600090846023608
six	sugar ink Xerxes
wordnum	38411402360018282200471036481038

"Most High would be 'monkey orange sugar Tommy Harry ink George Harry'. These could rather easily be disguised with nulls, as in 'The boys' new pet monkey ate the banana, orange and sugar that Tommy and Harry had set out for breakfast and then spilled ink on both George and Harry.'"

"This craziness reminds me of the Crazy Alphabet.[81] Y'know, A is for Aisle, B is for Bdellium."

"B for dellium?!"

"B-D-E-L-L-I-U-M, an aromatic tree resin like myrrh, found in Genesis 2:12. The B is silent. Czar could stand for C, because it too starts with a silent initial. Since it includes some accented letters the extended ASCII base 256 is called for. It would get high very fast! The nymber 'one' would be twenty-six places, over sixty-twoplex!"

	Crazy
one	ouija board ndomo Euphates
wordnum	111117105106097032098117097114100032110100111109111032069117112104097116101115
two	tsunnami who ouija board
wordnum	1161151171101100971091050321191041110321111171051060970320981170971141000
three	tsunnami hour argyle Euphrates Euphrates
wordnum	116115117110110097109105032097114103121108101032- 069117112104097116101115032069117112104097116101115
four	fohn ouija board urn argyle
wordnum	102111104110032111117105106097032098117097114100032117114110032097114103121108101

five	fohn irk vraisemblance Euphrates
wordnum	102111104110032105114107032118114097105115101109- 098108097110099101032069117112104097116101115
six	Szár irk Xian
wordnum	83123225114032105114107032088105097110

The meeting then went off on a tangent dedicated to discovering the longest words with misleading initials.

from *PAJAS* #18 (JANUARY 1987)

PHONY NUMBERS

All of the members and a few visitors were happy to learn from his niece Razilee Purdue [82], that our beloved professor emeritus had not died in the year of the comet. Like Mark Twain she said he said, "Rumors of my death have been greatly exaggerated." and that not even she knew much more from his Christmas card than that he still lived. To some this confirmed their theory that he had returned to France, but to others that he had finally built the hypothetical time machine he had often alluded to.[83]

Dan opened with Psalm 87:5, "Of Zion it shall be said, 'This one and that one were born in her; for the Most High Himself will establish her.'"

Then he continued by asking "How many Calvinists does it take to change a light bulb?"

The answer surprisingly came from a newcomer, Brother Ken,[84] "None, because only God knows which are predestined to be changed."

Continuing Dan began again, "In 1916 when dial telephones were still new, W. G. Blauvelt proposed a mapping system that displayed three letters higher than the digits two through nine on the dial, not using the number one or the operator's number zero. Using this code the nymber series can get high, converting numbers to words and back, though not as fast as even , more like base ten, not counting the use of zero and one as nulls, of course."

		ABC	DEF	GHI	JKL	MNO	PQRS	TUV	WXYZ
digit	1	2	3	4	5	6	7	8	9
arithmonym	one	two	three	four	five	six	seven	eight	nine
wordnum	663	896	84,733	3,687	3,483	749	73,836	34,448	6,463

"So these 'phony numbers' are without zeros or ones."

"One here would seem to stand for a space, rather than the zero for wordnums."

"Yes, so ten would be just eight thirty-six, while eleven gets incresed to three hundred fifty-three thousand, eight hundred thirty-six," Cal noticed.

"The six hundred sixty-three from one gets high fast this way, becoming twenty-two digits and then over two hundred."

phonies	1	663	7,491,486,373,317,498,984,733	>two-hundredplex

Old Sam commented, "This all reminds me of Klondike-5."

"Is that a Canadian quintet?"

"No, what I'm talking about are the phone numbers in TV and films when they can't use real ones, all from the original KL-5 or 'triple-nickel' exchange. Each exchange was limited to ten thousand phone numbers indicated by the last four digits. The five-five-five exchange became used for phone numbers used in fiction."

"So the phony number for the Hawaii State Police would have been 555-4450, from HI-50, Hawaii, the fiftieth state?"

"Aha, 'Hawaii-50'."

"And for our state it would have been 555-3339, for ED-39, East Dakota, the thirty-ninth state." Br. Ken asked.

"Right. Many Klondike-5 numbers were associated with wordnums that spell out four-letter words. 555-2368, for example, was a popular one, used by Tony Baretta, Jim Rockford, Theodor Kojak, the Ghostbusters, Jaime Sommers and Rick Hunter in their respective area codes."

"That would be, uh, BENT?"

"Or CENT."

"Kal-El alias Clark Kent's number was 555-3425, uh, perhaps ELCK, his real surname and his adopted initials."

"So to dial HERO[85], like on the H-Dial you'd dial 555-4376?"

"Other 'phony numbers' higher than seven digits would be those with area codes not used in the real world, like Arkham's 455."

"Or the Simpson's Springfield, NT, area code 112."

"What's NT? A spoonerism of Tennessee, Nettessee?

"Perhaps. I do know Smallville, Kansas, is area code 429."

"London's phony area code is eleven digits beginning 020-7946-0, so it's no mystery that Holmes' phone number must end in the three remaining digits, 221."

"And the queen's would end in 732, QE2."

This and similar speculations prompted Society members to spent some time compiling a directory of Klondike numbers and hunting Klondike primes.

from *PAJAS* #19 February 1987)

LUDLINGUISTICS [86]

Dan began with a quote from Augustine, "No nature can exist save from the Most High, then Br. Ken asked, "How many Catholics does it take to change a light bulb?"

When none responded, he answered cheerfully, "None, we use candles."

The Dan took over again with an explanation of advertised subject, "We have been ignoring null to simplify our cryptology, but we do have systematic ways to 'nullify' nymbers.

"Donald Laycock coined the word ludling for those 'secret' languages that add extra letters or syllables, called cryptemes, to disguise a known language, rather than clarify it as with the phonetic alphabets. One way to get high in perhaps the most unsecret of ludlings is in Pig Latin."

"Onevay, ootvay, eethray, ah, ah, ah!" laughed a Count von Count[87] impersonator.

Pig Latin	oneway	wotay	eethray	ourfay	ivefay
wordnum	151405230125	2315200125	5052008180125	152118060125	92205060125
decimal	445,482,475	25,051,408	2,102,177,257	453,500,746	281,864,986

"Can you count in Pig Greek, Turkey Irish, or Double Dutch?"

"Vunub, taboo, tuthutchrugee, all at the same time, ah, ah, ah!"

"Right! Vunub, er, onnube would be one in Pig Greek which adds -ub to the consonant sounds? So two and three would be what?"

Others caught on answering with "twuboo" and "thrubee"!

Pig Greek	onube	twubo	thrubee	fubourub
wordnum	1514210205	2023210315	20081821020505	621021521182102
decimal	16,509,722	22,178,463	7,997,952,452	7,097,218,722

"And in Turkey Irish?"

"Adding -ab is as easy as onabe, tabwo, thrabee."

Turkey Irish	onabe	tabwo	thrabee	fabourab
wordnum	1514010205	2001022315	20081801020505	601021521180102
decimal	16,495,142	21,299,100	7,997,165,132	63,223,718,402

"Ah, ah, ah, yes, but counting in Double Dutch or Tutnese takes an expert counter. It substitutes syllables for letters, bub for B, just as in Pig Greek, but then continues to add a syllable per consonant, sometimes u between the consonant repeated, but sometimes something else. With wash for W and nun for N, w-o-n and o-n-e are no longer homonyms but 'washonun' and 'onune'.

Tutnese	onune	tutwasho	tuthutchrugee
wordnum	1514211505	2021202201190815	20212008212003081821070505
decimal	16,510,073	217,940,287,461	3,122,817,946,898,519,256

"And in some dialects when you get to thousands, the comma even gets vocalized or at least semi-vocalized.

Some other members preferred the Jibberish dialects since they increased the lengths the most at five letters per vowel sound. Atheb eventually beat out the other Jibberish dialects of Uddag and Uthug, since it does not have duplicated letters.

Atheb	athebone	twathebo	thrahebee
wordnum	120080502151405	2023012008050215	20081801200805020505
decimal	1,844,379,955	218,168,009,403	15,496,064,242,749

Then Dan suggested extrapolating Tutnese or Double Dutch into Double Tutnese. Calculating Double Tutnese nymbers kept the Joyceans quite busy from then on. The lengths and durations generated by repeated translations naturally got higher and higher.

from *PAJAS* #20 (March 1987)

ARITHMONYMS

Br. Ken opened the meeting with a quote from Sirach 35:12, "Give to the Most High as he has given to you, generously, according to your means."

Dan then continued with the riddle, "Why is a school for priests like a base one-half?"

Lana guessed this one, answering with "They're both called 'seminary'."

"Oh, my, to just represent two you'd have to add an infinite number of powers of a half!"

"So let's not and change the subject. To form number names, or arithmonyms, Joyce borrowed some prefixes taken from astronomy, -ile for a general reciprocal, or the twelfths specifically from the Latin, with duode- for minus two and unde- for minus one."

1/12	1/10	1/6	1/4	1/3	5/12	1/2	7/12	2/3	3/4	5/6	11/12	3/2
unci-	deci-	sexti-	quarti-	terti-	quincunci-	semi-	septunci-	be-	dodri-	dextri-	deunci-	sesqui-

"So technically a six-month anniversary would be a semi-anniversary."

"And since a year is also 365 days, it could be divided exactly into five quinti-anniversaries a hundred seventy-three days apart."

"Or a leap year into exactly three terti-anniversaries of a hundred twenty-two days apart or six sexti-anniversaries sixty-one days apart."

"A quarti-anniversary would be approximately three months apart, a quincunci-anniversary about five months, a septunci-anniversary seven months, a be-anniversary eight months, a dodri-anniversary nine months, a dexti-anniversary ten months, a deunci-anniversary eleven months and a sesqui-anniversary eighteen months."

"So decigoogol would be ninety-nineplex, and semigoogol five ninety-nineplex?"

"And a sesquigoogol three times that."

"Right and right again, you're too clever by half."

"If I remember my Latin fractions," Lana put in, "sextula means one seventy-second, a sixth of a twelfth."

"So we could also have sextuli- as the prefix?"

"If sexti- means one sixth, septi- should mean one seventh and septuli- one eighty-fourth."

1/144	1/120	1/108	1/96	1/84	1/72	1/60	1/48	1/36	1/24
unculi-	deculi-	nonuli-	octuli-	septuli-	sextuli-	quintuli-	quartuli-	tertuli-	semuli

"Ah, but one twenty-fourth was actually semuncia, sesuncia one eighth, siliqua one over one thousand twenty-eight or twelve cubed and scripulum one over two hundred eighty-eighth So we get even more fractional prefixes."

3/20	3/16	5/24	3/14	7/24	3/10
sesdeci-	sesocti-	semquinunci-	sessepti-	semseptunci-	sesquinti-
3/8	11/24	9/8	21/24	5/4	11/8
semdodri-	semdeunci-	sesdodri-	sesseptunci-	sesdextri-	sesdeunci-

1/1256	1/864	1/192	1/80		1/64
sessiliqui-	semsiliqui-	sesscripuli-	sesdeculi-		sesoctuli-
1/56	1/40	1/32	1/24	1/20	1/8
sesseptuli-	sesquintuli-	sesquartuli-	semunci-	semdeci-	sesunci-

"So -- for a kiss -- who can tell me how high something called sesdodrisemquinuligoogol would be?" Claire asked playfully, breaking the silence.

It wasn't that hard to translate, but she'd sensed that a break in the rampant pedagogy was called for. The new number name was broken apart and put back together again so that all her competing admirers at least understood not only how to pronounce it but what it meant.

"Sesquinuligoogol would mean a fortieth googol," Sam said.

"That'd be twenty-five ninety-eightplex." added Cal.

"Applying dodri- to it deduces it to three-fourths or one thousand eight hundred seventy-five ninety-sixplex," Tom added.

"Adding fifty percent we get two thousand, eight hundred twenty-five followed by ninety-fiveplex." Dan continue."

Claire concluded, ending the competition with a wide-angle blown kiss and a wink, graciously, if disappointingly, accepted by all.

By backformation the more creative Joyceans coined the indeterminate number agentillion, any number not nonagintillion. Similarly by gentillion they meant any number not nongentillion.

"We make use of Greek-based prefixes", Dan continued, "when plexing to get higher much faster. A tenplex is ten-to-the-tenth or to-the-second-ten, but if you insert the Greek prefix before the plex it multiplies the plexings. Tenduplex would be tenplexplex or to-the-third-ten, and so on."

For the numbers below one we have the Greek prefixes: hemi- meaning half, and for greater than one hen- meaning plus one, do- meaning plus two.

"For the unit prefixes higher than plus-two the suffix -kai- is added means so many units plus so many tens (or whatever) as used in triskaidekaphobia (morbid fear of the number thirteen).

	2	3	4	5	6	7	8	9
	du	tres	tetra	penta	hexa	hepta	okta	ennea
10	20	30	40	50	60	70	80	90
deka	isoca	triacont	tetracont	pentecont	hexacont	heptacont	oktacont	enneacont

"The higher Greek prefixes are hecto- meaning hundred, kilo- meaning thousand, and myria- meaning ten thousand, adopted in 1793, but made illegal in France in 1795. These have been officially extrapolated in the international metric system to include the even higher prefixes: mega- from the Greek for 'large' for sixplex adopted in 1919, giga- from the Greek for 'giant' for nineplex and tera- from the Greek for 'monstrous' for twelveplex were adopted in 1960. Since this last one was similar to a contraction tetra- they were further extrapolated to, peta- contracted from penta for fifteenplex and exa- contracted from hexa- for eighteenplex by 1975, Continuing in this way would have given heta- and ota-, but these did not get accepted." [90]

hecto-	kilo-	myria-	mega-	giga-	tera-	peta-	exa-
twoplex	threeplex	fourplex	sixplex	nineplex	twelveplex	fifteenplex	eighteenplex

"Joyce's qu- prefix from Benjamin Schumaker's term 'qubit', a portmanteau of 'quantum bit', gets us higher still. He used it to indicate raising two to the power of the following prefix. You'll notice that quhecto- indicates a little googol and qukilo- the paradoxical little great googol."

quhecto-	qukilo-	qumega-	qugiga-	qutera-	qupeta-	queza-
2^(10^2)	2^(10^3)	2^(10^6)	2^(10^9)	2^(10^12)	2^(10^15)	2^(10^18)

"You are getting too high too fast for me with too many twos! Googolplex is mindboggling enough."

"There are also reducing prefixes that might help. Centi- and milli- from the Latin for hundredth and thousandth, micro- for millionth from 1873, myri- from the Greek myiad rarely used since 1960. Nano- comes from the Greek for 'dwarf', pico- from Spanish 'beak or small quantity', femto- and atto- from Norwegian for fifteen and eighteen by General Conference of Weights and Measures in 1964." [91]

prefix	centi-	milli-	micro-	nano-	pico-	femto-	atto-
ten power	2	3	6	9	12	15	18

"Attogoogol would be a googol divided by eighteen tens?"

"Right. So that makes it just eighty-twoplex."

"That helps," some declared, while with others it sounded more like a question, "That helps?"

She then took charge again by adding that some other decidedly unofficial prefixes have been backformed from the more familiar English number names. "By removing the final -y from numbers ending in -ty," she explained, "some have gotten higher perhaps too creatively with prefixes like twent-, thirt-, fort-, fift-, sixt-, sevent- and ninet-. The prefix twent- they take to mean three sixtyplex. Twentillion would therefore be ten-to-nine-times-tenplex-plus-three. Similarly the prefixes hundr- and thous- from curtailed hundred and thousand have been coined."

twent-	thirt-	fourt-	fift-	sixt-	sevent-	eight-	ninet-
3(60plex)	3(90plex)	3(112plex)	3(150plex)	3(180plex)	3(210plex)	3(240plex)	3(270plex)

"By extrapolation from these we can add the even more creative prefixes: tent-, elevent-, twelvet-, scoret-, grosst- and even googolt-. In general the suffix -t can be applied to any number just as with six, seven and nine."

tent-	elevent-	twelvet-	scoret-	grosst-	googolt-
3(30plex)	3(33plex)	3(36plex)	3(60plex)	3(432plex)	3(3 googolplex)

"So twentty, with a double t, would not just be a typo but would mean three-times-sixty-oneplex, right?", Tom asked.

"I know another curtailed prefix, bil-." Sam added. "The Hungarians used it during World War II, when hyperinflation multiplied their currency by a billiard, twelveplex."

"So extrapolating from bil-, we can get high with: tril-, quadril-, quintil-, and so on too!"

prefix	bil-	tril-	quadril-	quintil-	sextil-	septil-	octil-	novil-	decil-
ten power	12	18	24	30	36	42	48	54	60

vigintil-	trigintil-	quadragintil-	quinquagintil-	sexagintil-	centil-	millil-
120	180	240	300	360	600	6,000

With the Society's new arsenal Tom and others joined Taffy in coining more tongue-twisting and mind-boggling number names in a game of "Pick a number between one and eightplex", their version of Twenty Questions.

from *PAJAS* #21 (April 1987)

OTHER BASES [92]

Dan opened with "The Most High rules in the kingdom of men" from Daniel and with the riddle, "What would you call a campsite for a hundred."

"Tenty."

"And an orchestra with twenty horns?"

"Tooty."

Then he began more seriously, "The Polynesians who speak Umbu-Ungu, also known as Kakoli, are reported to have base-24 numerals, while the Sko languages of Papua New Guinea use a hybrid base 24/6, so they synchronize at one-zero-zero.

Kakolish	1	2	3	4	5	6	7	8	9	10	11	12	13	14	15	16	17	18	19	20	21	22	23	100	101	102
Skoish	1	2	3	4	5	10	11	12	13	14	15	20	21	22	23	24	25	30	31	32	33	34	35	100	101	102

"Ngiti language group in the Ituri province of the Congo is reported to have a hybrid thirty-two/four base, so they differ from base four at four-zero."

Ngitish	1	2	3	10	11	12	13	20	21	22	23	30	31	32	33	40	41	42	43	50	51	52	53	60	61	62	63	70

"Supyire, a language spoken in the Sikasso region of Mali, and known as Shemprire in the neighboring Ivory Coast, that uses a numbering system based on fives, tens, twenties, eighties and four-hundreds or 'great score'.

Supyire-ish	1	2	3	4	10	11	12	13	14	100	101	102	103	104	110	111	112	113	114	200	201	202	203

"Several so-called primitive peoples count by indicating body parts with what are called anatonums, far higher than merely counting on fingers or even toes. The Oksapmin use a base twenty-nine, the Yupno thirty-three and the Kewa a very complex sixty-eight. In Zulu the word for six, isithupa, literally means 'extend right thumb'. [93]

"In Tacana *tunka* means ten, *tunka-tunka* a hundred and *tunka-tunka-tunka* a thousand.

Gypsies call a hundred *desh*, a thousand *baro desh* or 'great hundred' and a million *baro shel*. The Delaware also called a thousand, a great hundred, *ngutti kittaapuchkei*. The Kwakiutl called a million *tlinhi* or 'uncountable'." [94]

Exploration of these new bases continued, accompanied by choruses of "Uncountable Bheers".

from *PAJAS* #22 (May1987)

ARITHMOPOEMS

Dan opened the meeting by quoting Psalm 7:18, "Sing praise to the name of the Lord Most High." and then asked, "What word can be said faster by adding a syllable?"

Lana answered, "Fast."

He then continued the previous month's discussion by introducing yet another counting system, "Welch shepherds count sheep with their own singsong way of counting. Can you see the pattern?"

yan	tyan	tethera	methera	pimp
sethera	lethera	hovera	dovera	dick
yan-a-dick	tyan-a-dick	tethera-a-dick	methera-a-dick	bumfit
yan-a-bumfit	tyan-a-bumfit	tethera-bumfit	methera-bumfit	giggot
yan-a-giggot	tyan-a-giggot	tethera-giggot	methera-giggot	

number	1	2	3	4	5	6	7	8	9	10	11	12	13	14	15	16	17	18	19	20	21	22	23	24
duration	1	1	3	3	1	3	3	3	3	1	3	3	5	5	2	4	4	5	5	2	4	4	5	5

There was some disruption by a couple underclassmen who noticed the words for five and ten, but it was quickly squelched by a professorial stare.

Sam wisely avoided using those particular words, answering, "It seems to be a base five system with a superabundance of rhyming pairs, *yan* and *tyan, tethera, methera, sethera* and *lethera, hovera and dovera* and eight, nine and ten seem to have become 'hickory, dickery, dock' in the well-known Nursery Rhyme.

"Since the usual number names or arithmonyms are formed in predictable and repetitive ways, though not as pronounced as these, they too can be quite poetic, what Joyce called arithmopoems. In English syllabic number poems are easier to find than rhyming number poems, since not many small numbers rhyme: one and million, three and -ty, Lincoln's four and score, ten, seven, eleven and dozen."

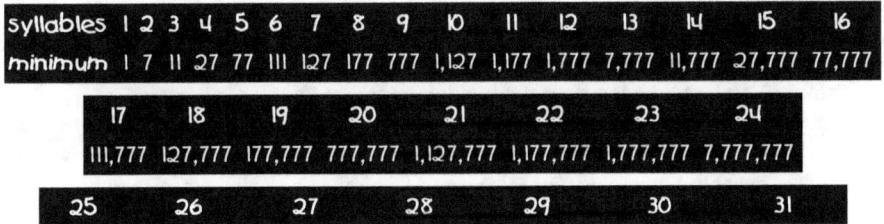

syllables	1	2	3	4	5	6	7	8	9	10	11	12	13	14	15	16
minimum	1	7	11	27	77	111	127	177	777	1,127	1,177	1,777	7,777	11,777	27,777	77,777

17	18	19	20	21	22	23	24
111,777	127,777	177,777	777,777	1,127,777	1,177,777	1,777,777	7,777,777

25	26	27	28	29	30	31

11,77,777 27,777,777 77,777,777 111,777,777 127,777,777 177.777,777 777,777,777

"Rather than needing thirty-one syllables," Dan continued, "the final number can also be expressed hypermathematically in just three as 'nine sevens'. [96]

"There's no more information in one representation of a number than another, since all describe exactly the same number. This is true even for numbers that cannot be represented digitally, but 'only' symbolically. Different representations merely reveal different facets of the same number. Each representation expresses the very same number no matter how it happens to be expressed.

"If we avoid the use of the ambiguous 'billion', the shortest exponential expression, nineplex, is the next logical step higher than hundred millions. Continuing as usual by steps of a thousand, we get higher by 'plexing' multiples of three. Intermediate powers of ten. Other number names, like dozen, gross, score, eleventy, eleventeen, googol, googolplex, etc., might also be used, increasing the poetic possibilities. Numbers squared or cubed might creatively also be used.

"These more normally-formed number names are obviously digitally repetitive and so can be represented in a more compact way recursively with interior left subscripts, since left superscripts are already used for tetration, right superscripts for exponentiation and right subscripts for bases.

"The pentapaul, from the Greek 'penta-' for five and the Latin 'paulus' for small, is simply the same one-syllable word repeated five times.[97] The simplest poem form would therefore be a monopaul, just one one-syllable word."

"Not as simple as the infamous Nepalese limerick," Tom interjected. "'There once was a man from Nepal Whose limericks had no lines at all. Since no space they took, he published a book with an infinite number in all.'" [98]

"Right, not <u>that</u> simple."

"The monopaul numbers would only be one, two, three, four, five, six, eight, nine, ten, or twelve, just ten possibilities, then?"

"You're forgetting the higher bases' one-syllable numbers higher than ten and twelve, A, B, C, etc."

"Oh, yeah, thrun, fron and feen."

"The first dupauls however would be our old friends one-one and two-two though. Being homonums, they mean many different things depending on the base they're in."

"So there could be dupauls not only up to ten-ten and twelve-twelve, but in the higher bases, AA, AB, AC, etc.!"

"Higher than dupauls would be tripauls, tetrapauls and higher yet pentapauls, hexapauls, all the way to the eighteenplex-syllable exapauls. There are, of course, many other more interesting arithmopoems from several different languages that count syllables and do not have endrhymes. The gaita-gallega numbers, for example, have eleven syllables, which can all be different, so the first would be one thousand, one hundred seventy-seven.[99]

"The next several would each be just a hundred more, but not for very long.

gaita-gallega 1,177 1,277 1,377 1,477 1,577 1,677 1,727 1,737 1,747 1,757

"The twelve-syllable hudibrastic numbers are a little more complex since they are made up of two six-syllable lines. The first would begin with twenty-seven thousand and the second one hundred eleven and so on, with the usual comma separating the groupings of three places. The six-syllable semihudirastic might be called simply an udibrastic, a hudibrastic with its first line beheaded.[100]

hudibrastic 27,111 27,121 27,122 27,123 27,124 27,125 27,126 27,128 27,129

udibrastic 111 121 122 123 124 125 126 128 129 131 132 133 134 135 136 138 139 141 142 143

"The seventeen-syllable haiku numbers are made up of three lines, of five, seven and five syllables respectively. The first such number would start in the millions, eleven of them, continue with the thousands, seventy-seven of them, and conclude with seventy-seven." [101]

haiku 11,077,077 11,077,107 11,077,113 11,077,114 11,077,115 11,077,116 11,077,118 11,077,119

"The villanelle numbers are made up of nineteen syllables, broken into six parts, five three-syllable ones and a four-syllable tail. Since these are so short they tend to break between the commas on the hundreds. The first is one hundred one million, one hundred one thousand, one hundred twenty-seven. What Joyce called an illanelle or beheaded villanelle would have a three-three-three-four syllable pattern and a llanelle a three-three-four syllable pattern." [102]

villanelle 101,101,127 101,101,137 101,101,147 101,101,157 101,101,167 101,101,171 101,101,172

illanelle 1,101,127 1,101,137 1,101,147 1,101,157 1,101,167 1,101,171 1,101,172 1,101,173 1,101,174

llanelle 101,127 101,137 101,147 101,157 101,167 101,171 101,172 101,174 101,175 101.176 101,178

"The gayatri numbers have twenty-four syllables and so begin with seven million, seven hundred seventy-seven thousand, seven hundred seventy-seven." [103]

gayatri 7,777,777 13,777,777 14,777,777 15,777,777 16,777,777 17,777,771

"The tanka or haikai is similar to a haiku with two seven-syllable lines added, so the tanka numbers have a total of thirty-one syllables, beginning higher than millions and even nineplexes at eleven twelveplex, followed by seventy-seven nineplex, eleven million, seventy-seven thousand, and ending with one hundred twenty-seven. The anka or aikai would have a seven-five-seven-seven-seven syllable pattern." [104]

tanka 111,077,011,077,127 11,077,011,077,137 11,077,011,077,147 11,077,011,077,157

anka 77,011,077,127 77,011,077,137 77,011,077,147 77,011,077,157 77,011,077, 167

"The renga is twenty tankas with six hundred twenty syllables in all, a semirenga with three hundred ten syllables, and the quartirenga with a hundred fifty-five syllables.

"Prof. Joyce named many other such arithmonums, such as the 'aknat', that is, a reverse tanka in a seven-seven-five-seven-five syllable pattern." [105]

"So they'd start with seventy-seven nineplex, eleven million, seventy-seven thousand, one hundred twenty seven and seventy-seven twelveplex, seventy-seven nineplex, eleven million, seventy-seven thousand, one hundred seven."

"So an 'agner' would be a reverse renga and an 'enga' a beheaded regna with only six hundred thirteen syllables."

The meeting ended with the calculation of such numbers.

from *PAJAS* #23 (June 1987)

MORE ARITHMOPOEMS

Dan opened the meeting with "I cry out to God Most High, to God Who accomplishes my requests for me." from Psalm 57:2, a prayer before he attacked the previous months topic again.

He ask "Who was the greatest mathematician in Bible?" answering it with, "Moses who wrote the book of *Numbers*."

"There could be many books of arithmopoems with multiple syllable counts. For example, the naga-uta numbers are in a five-seven syllable pattern and so a sequence of semigayatri numbers, like the first, eleven thousand, one hundred twenty-seven." [106]

naga-uta 11,127 11,137 11,147 11,157 11,167 11,171 11,172 11,173 11,174

"So the aga-uta would have a seven syllable pattern."

aga-uta 127 137 147 157 167 171 172 173 174 175 176 178 179 187 197 227 237 247 257

"The madrigal numbers include those with two or three lines with seven or eleven syllables, and so include fourteen, twenty-one, twenty-two or thirty-three syllables.[107] The adrigal would start out the same but not include a third line. The drigal would only have one line of seven or eleven syllables. The first double-seven-syllable number would be seventy-seven thousand, one hundred twenty-seven. The first triple-seven-syllable number would add seventy-seven million. The first double-eleven-syllable number would have to include one hundred, seventy-seven, the first eleven-syllable number, added to seven hundred seventy-seven million. The first triple-eleven number would add seven hundred seventy-seven nineplex."

madrigal 77,127 77,137 77,147 77,157 77,167 77,171 77,172 77,173 77,174 77,175 77,176

"The adrigal then would be one or two lines with seven or eleven syllables, starting out like the aga-uta."

"The rondelet numbers have two lines of four or eight syllables, concluding with two of eight syllables and one of four syllables, for a total of twenty-eight or thirty-two syllables. The ondelet has just one of four or eight syllables with the same final three." [108]

rondelet 100,100,111,111,027 100,100,111,111,037 100, 100,111,111,047 100,100,111,111,057

ondelet 100,111,111,027 100,111,111,037 100,111,111,047 100,111,111,057 100,111,111,067

"The choka numbers are an expansion of both the tanka and naga-uta numbers with any number of five- and seven-syllable pairs with a seven-syllable pair tail. The hoka would begin with a seven-seven-seven syllable pattern and then becoming seven-five-seven-seven." [109]

choka 11,107,107,127 11,107,107,137 11,107,107,147 11,107,107,157 11,107,107,167

hoka 107,107,127 107,107,137 107,107,147 107,107,157 107,107,167 107,107,171

"The lüshi numbers are six lines of either five or seven syllables, for a total of either thirty or forty-two syllables. The üshi numbers would be five lines and the shi four lines." [110]

lüshi 11,011,011,011,077 11,011,011,011,107 11,011,011,011,113 11,011,011,011,114

üshi 11,011,011,077 11,011,011,107 11,011,011,113 11,011,011,114 11,011,011,115 11,011,011,116

shi 11,011,077 11,011,107 11,011,113 11,011,114 11,011,115 11,011,116 11,011,118 11,011,119 11,011,120

"The bergette numbers include four or five lines of eight or ten syllables for a total of thirty-two, forty, forty-eight or fifty syllables. The first, with four lines of eight syllables, would be one hundred eleven nineplex, one hundred eleven million, one hundred eleven thousand, one hundred seventy-seven. The ergette numbers would have three or four lines." [111]

bergette 111,111,111,177 111,111,111,277 111,111,111,377 111,111,111,477 111,111,111,577 111,111,111,677

ergette 111,111,177 111,111,277 111,111,377 111,111,477 111,111,577 111,111,677 111,111,727 111,111,737

"The dizain numbers include ten eight-syllable or ten-syllable lines for a total of eighty- or hundred-syllables, the izains nine, the zain eight and the ain seven and the semidizain five." [112]

dizain 117,117,117,117,117,117,117,117,117,177 117,117,117,117,117,117,117,117,117,277

izain 117,117,117,117,117,117,117,117,177 117,117,117,117,117,117,117,117,277

zain 117,117,117,117,117,117,117,177 117,117,117,117,117,117,117,277

ain 117,117,117,117,117,117,177 117,117,117,117,117,117,277 117,117,117,117,117,117,377

semi-diain 117,117,117,117,177 117,117,117,117,277 117,117,117,117,377 117,117,117,117,477

"Ah, the dizain numbers are easily beheaded. The izain numbers would be nine lines, the zain eight, the ain seven and semidizain five."

He obviously cut his presentation short again to avoid information overload with "There are, of course, many more rhyming numbers with very complex rhyming patterns,[113] subsets of the unrhyming ones."

from *PAJAS* #24 (July 1987)

PHINARY

Opening with "Maker of All Things, God Most High", translating St. Ambrose's "Deus Creator Omnium", Dan then posed the riddle of the month, "What has fifty heads and fifty tails, but no legs?"

"A roll of pennies."

"I proposed this month's topic when I suddenly realized that the base for a number system doesn't have to be an integer. We could just as well make a number system based on an irrational number like the ubiquitous golden ratio, phi, or half the quantity one plus root five. It is different in that it has more than one way to write the same number."

"Ah. yes," Raz said approvingly. "An instance of Brooke's Law, 'Whenever a system becomes completely defined, some fool discovers something which either abolishes the system or expands it beyond recognition.'"

With that his audience woke up though not without more than a couple puzzled looks, and he turned to the board for a few examples.

"What is the solution to 'z equals z plus one over z'?"

"Let's see." Cal began. "Multiplying both sides by z, we'd change it to 'z squared equals z plus one' and subtracting the z plus one from both sides, we get z squared minus z minus one equals zero, which we can plug into the binomial theorem."

$\Phi = (\sqrt{5} + 1)/2 \quad \Phi^2 = \Phi + 1 \quad \Phi^{z+2} = \Phi^{z+1} + \Phi^z \quad 100 = 011$

"One-zero-zero in base phi would mean phi squared, which is exactly equal to zero-one-one, phi plus one. That is just a special case of the more general equation that phi to any power equals the sum of phi to the previous two powers. This is what, so Taffy tells me, computer programmers call opposite parity, where the one becomes a zero and the zeroes become ones. We get high in this base rather more like climbing a tree than a mountain, especially if irregular 'phigits' like two or three, etc., are allowed."

"You're giving me fidgets!", Sam complained. "This I just don't get!"

"It has been Just remember the equations there, each place to the left, each higher power of phi, is equal to the sum of the next two on the right, the next two lower powers. One-two means one phi plus two but could also be represented as one-zero-one, meaning phi-squared plus one, a perfect example of Baldy's Law, 'Some of anything plus the rest of anything equals the whole thing."

phinary	$\Phi + 1 =$	$11_\Phi =$	$\Phi^2 =$	100_Φ	
	$\Phi + 2 =$	$12_\Phi =$	$\Phi^2 + 1 =$	101_Φ	
	$\Phi + 3 =$	$13_\Phi =$	$\Phi^2 + 2 =$	102_Φ	
	$2\Phi =$	$20_\Phi =$	$\Phi^2 + \Phi + 1 =$	$111_\Phi = \Phi^3 + 1 =$	1001_Φ

"I know of another sequence of numbers that does the same sort of thing but with whole numbers," Lana volunteered. "The next Fibonacci number[114] is also calculated by adding the previous two. The gnomon and the sum get high at the exactly same rate."

Fibonacci	1	1	2	3	5	8	13	21	34	55	89	144	233	377	610	987	1597	2578
sum	1	2	3	5	8	13	21	34	55	89	144	233	377	610	981	1,597	2,578	5,175

"We can use either of these as a base. In base F, as it's called for short, the same number can, of course, represented in different ways even without 'improper' digits. Four would be one three plus no twos plus one one or three-zero-one. Six could be either in contracted form as three plus two plus one, three-two-one, or expanded to five plus one with no twos or threes, five-zero-zero-one."

"There are an infinite number of such Lucas sequences."

"Not George Lucas, the director?"

"Right, not George, but François Edouard, the mathematician. Starting with any two numbers generates one with similarly similar gnomons and sums and higher and higher growth rates and different prime densities."[115]

																ratio
1	3	4	7	11	18	29	47	76	123	199	322	521	843	1,364	2,207	1.618035190
1	4	5	9	14	23	37	60	97	157	254	411	665	1,076	1,741	2,817	1.6180356117
1	5	6	11	17	28	45	73	118	191	309	500	809	1,309	2,118	3,427	1.618035588
1	6	7	13	20	33	53	86	139	225	364	589	953	1,542	2,495	4,037	1.618036072
1	7	8	15	23	38	61	99	160	259	419	678	1,097	1,775	2,872	4,647	1.6180360211
1	8	9	17	26	43	69	112	181	293	474	767	1,241	2,008	3,249	5,257	1.618036318

"There are several different ways of representing ever-increasing numbers in such bases as F. We can indicate the number of Fibonacci numbers by position, or explicitly as the value of each place or multiple places and then there's William Zeckendorf's zero-less expansion that hyperadds one or more Fibonaccis." [116]

"So three could also be expanded to one-two or even one-one-one?"

"Yes, those would be considered improper representations, but valid. Usually the the least possible Fibonaccis are placed in descending order, but as you get higher the more possible irregularities you'll get."

binarish	1	10	100	101	1000	1001	1010	10000	10001	10010	10100	10101
minimalist	1	20	21	301	320	321	5020	5021	5301	5320	5321	80301
maximalist	1	20	300	301	5000	5001	5020	80000	80001	80020	803000	803001
Zeckendorfish	1	2	3	31	5	51	52	8	81	82	83	13
sorted	1	2	3	5	8	13	31	51	52	81	82	83

"To paraphrase Burgess, 'I never saw a Zeckendor fish and never hope to see one, but I can tell any how, I'd rather see than be one." a voice from the back interjected, but was generally ignored.

"You can do the same for any sequence," Lana continued, "but of particular interest is one based on the primes, so we get base p, like Joseph Louis Lagrange used."

"So the first four places would be the same, the unit's place, two, three and five, but the next place would be the next prime, seven, rather than the next Fibonacci number, eight?"

"Right."

"Four could be either three-one or one-zero-one, but seven would now be either one followed by four zeroes or five-two."

"Since the Zeckendorf expansion applies to the Fibonacci numbers, we could call the one based on the primes the Zeckendorp expansion, changing the final F to a p."

primary	1	20	300	301	5000	5001	70000	70001	70020	700300

Zeckendorpish	1	2	3	31	5	51	7	71	72	73
sorted	1	2	3	5	7	31	51	71	72	73

"The expanded form is zero-less, just taking the nearest preceding prime and hyperadding primes."

from *PAJAS* #25 (August 1987)

CURIOUSER AND CURIOUSER BASES

Dan opened the meeting with "It has seemed good to me to show the signs and wonders that the Most High God has worked toward me." from Daniel 4:2.

Lou provide the opening riddle,"How many mutants does it take to change a light bulb?"

"Two-thirds."

Dan helped illustrate for those who had trouble by standing behind the larger Lou and supplying him with a third arm.

Tom expanded his topic from the previous month to yet a stranger base for a number system or rather Neil Kelvie's hypernumber system, "Taking the imaginary number, i, the square root of minus one, as our base, we can turn a number plane into a number line. The unit's place would be, as usual, i-to-the-zeroth or one, while the second place would be a multiple of i, the third a multiple of the square of i or minus one and the fourth place a multiple of minus i. Fourth or higher, of course, are 'out of the money'."

	-2	-1	0	1	2
2i	2020	120	20	21	22
i	2010	110	10	11	12
0	2000	100	0	1	2
-i	2100	1100	1000	1001	1002
-2i	2200	1200	2000	2001	2002

"So we would get high from zero to one and two, but then one-zero would mean i and one-one i-plus-one, one-two i-plus-two, etc."

number	0	1	*i*	*i*+1	-1	*i*-1	-*i*	-*i*+1	-*i*-1
base *i*	0	1	10	11	100	110	1000	1001	1100

"Just counting positive, non-complex numbers in 'inary' they get quite expanded.

inary 1 10001 100010001 1000100010001 10001000100010001 100010001000100010001

"And if you think that's mindboggling, it gets worse. Making a base of i-to-the-ith is even more ambiguous than phinary, since i-to-the-ith is equal to e-to-two-pi-i-z, where z can be any integer, giving an infinite number of simultaneous values. Each two pi is one turn about the axis of the complex number plane, so it turns not just a two-dimensional plane into a number line, but all of three-dimensional space into a number line!"

"Base zero is the same, yet different. Only the unit's place, zero-to-the-zeroth, has any value, as usual, that of one, something from nothing, like only the Most High could do. All of the other powers of zero are zero and so nulls, so that in improper base zero one is represented by any number up to an infinite number of places! Two is an infinite number of numbers ending in two and so on. Zero in base zero is a whole lot of nothing, though it looks like multiples of ten."

0_0	0	10	20	30	40	50	60	70	80	90	100	110	120	130	140	150	160	170	180	190	200

"And the three-place one-zero-zero can also be read as its homonum the two-place eleven-zero and both still be zero!"

"Base i is more complex. C. Muses theorizes that in addition to the imaginary number i, the square root of a minus one, that connects space and time dimensions, there are several other such 'hypernumbers'. He describes particularly those which he claims connects the paranormal and the supernatural with space-time. Epsilon, ε, is a matrix quantity whose square is one, but which itself is neither one nor minus one. Similarly he gives w whose sixth power is one and whose square is w minus one."

base w	1	1000001	1000001000001	1000001000001000001

"Boggling!" more than one said.

"O. K. Let's try the odds and evens as bases then. The odd base is more versatile than the base based on evens, because an even number of odds add to an even and an odd number of odds add to an odd, but no number of evens ever adds to an odd. You could however use the units place as one to make odds from evens or evens from odds. You could call them Zeckendoro and Zeckendore."

base odd	1	11	300	301	5000	5001	70000	70001	900000	900001	11000000	11000001
Zeckendoro	1	11	3	31	5	51	7	71	9	91	11	111
base even	1	21	201	400	401	6000	6001	80000	80001	1000000	1000001	120000000
Zeckendore	1	2	21	4	41	6	61	8	81	10	101	12

"Paul Durich of the Googolplex Project got high very fast indeed. He used base googolplex to write a googolplex in record-shattering time with just two characters, one-zero, rather than the usual tedious base ten way with a one and a googol zeroes or in the even more tedious binary. Similarly googolexaplex could be written just as easily in base googolexaplex."

The rest of the meeting was spent thinking of other curious bases.[117]

from *PAJAS* #26 (September 1987)

HAWAIIAN NUMBERS

Dan opened the first meeting of the new school year with Deuteronomy 32:8, "The Most High ... fixed the bounds of the peoples according to the number of the sons of God."

Newcomer Kamika Schmidt[118] was given the honor of presenting the traditional riddle, "How many babysitters does it take to change a light bulb?"

Dan supplied the answer, "None, because they don't make diapers that small."

When the groans died down, he turned the meeting back over to Kamika.

"From the Old Hawaiian number system", she began, "we have numbers based on forty, ka'au, as well as the powers of ten, a hybrid base, likely because the fisherman traditionally count fish two in each hand. A dozen they called 'three fours' as well as 'umi-kumana-lua and a score 'five fours' as well as iwakalua. They had a ten-day period, the anahulu, like the short-lived French Revolutionaries' 'week'.

"In this system we would get high by fours and tens to thirty-nine and then by forty times powers of ten. We did not have a word for hundred or thousand before the Europeans came.[120]

kahi	lua	kolu	kauma	lima	ono	hiku	walu	iwa	'umi
1	2	3	4	5	6	7	8	9	10

ka'au	lau	mano	kini	lehu	nalowale
40	400	4,000	40,000	400,000	4,000,000

"Numbers higher than ten are formed with the prefix 'umi-kumana, the equivalent of our suffix -teen. By extrapolation by using the prefix kana-, the equivalent of our suffix -ty, could be added Joyce-like to coin an even higher number name 'kananalowale' for forty million, which would be a lot of fish!"

Hawaiianish	....	39	100	101	102	103	104	105	106	107	108	109	110	111

"And a lot of bheer," Al added, bheer being the fannish word for a generic beverage, just as fen is the fannish plural of fan.

from *PAJAS* #27 (October 1987)

BABYLONIAN NUMBERS

Dan began with "All things praise Thee, Lord Most High, Heaven and Earth and sea and sky," quoting George William Conder and then added the question, "How many mathematicians does it take to change a light bulb?"

Lana answered, leading into her topic, "None, assuming that at least one light bulb has been changed, by extrapolation it can be inferred than all other have been."

She had an easier time introducing the base sixty system than some others, since everyone was already familiar with counting by seconds and minutes and hours, and perhaps seconds and minutes and degrees. All she had to do was to refer to the familiar use of the colon to indicate base sixty.

"The Babylonians cuneiform system used the downward-pointing wedge for one, the leftward-pointing one for ten and by the third century B. C. also the double upward-pointing ones for zero. Sixty was called 'ges' and six hundred 'gesu', sixty squared 'sar' and sixty cubed 'sarges'. The zeroless Old Babylonian is more challenging since the value of the number depends upon the context, like now when 1:01 could mean both one hour, one minute or sixty-one minutes, as well as one minute, one second or sixty-one seconds."

base 60	...	59	1:00	1:01	1:02	1:03	1:04	1:05	1:06	1:07	1:08	1:09
Babylonianish	...	59	1	11	12	13	14	15	16	17	18	19

"By extrapolation we can get higher by combining the known word frags."

Babylonian	ges	gesu	sar	saru	sarges	sarsar	sarsaru
number	60	360	3,600	36,000	216,000	2,160,000	21,600,000

"Because of the zerolessness and 'pointlessness' of the Babylonian system, the 'igibum' or so-called Babylonian reciprocals can be read as whole numbers. They are those factors that turn an ugly number into a power of sixty, the 'ugly' numbers being those with a smoothness, or maximium prime factor, of five or less. Zero and one with no prime factors would be 'rough'. Those with a minimum smoothness of just two, the powers of two, are 'too smooth'.

ugly	1	2	3	4	5	6	8	9	10	12	15	16	18	20	24
igibum	60	30	20	15	12	10	450	400	6	5	4	225	200	3	150

Lara gave up leadership of the meeting when Al and his fellow fen began singing "Sarsaru Bottles of Bheer".

from *PAJAS* #28 (November 1987)

DOZENSES

Dan read yet again from Daniel, this time from 3:26, "You servants of the Most High God, come forth, and come here."

Lana asked, "If two's company and three's a crowd, what're four and five?"

Some answered, "Nine."

Others however, not considering that a complete answer, gave answers in other bases, 'one-one in base nine'. Cal, naturally answered "-one in octal". Still others gave other synonums, 'three squared' and 'root eight-one'.

Finally Lana regained control and began, "Base twelve is used in the English feet-inches system. By indicated feet with a prime mark and inches with a double prime mark, we already get high like one foot eleven inches, two feet, two feet one inch. We can extrapolate for the whole base to indicate the leftmost place with one prime and the next with two and the third with three and so on."

"Oh, as if it were counting twelfths of inches, eh?"

"Or units, dozens and dozens squared or grosses. But it would be a zero-less system, not counting what's not there."

zeroless base twelve	1	2	3	4	5	6	7	8	9	10	11	1 11	1 11	12	13	14	15	16	17	18	19	110	111	2	21

"Grossth!"

"Gross?"

"No, grossth, the reciprocal of a gross, a hundred-forty-fourth. So after eleven feet eleven inches we'd have eleven feet eleven inches plus a twelfth of an inch. Higher than eleven twelfths would be a grossth foot."

"That'd lead to an infinite series of elevens!" Cal exclaimed.

base 12	11'	11'11"	11'11"11"'	11'11"11"'11""	11'11"11"'11""11""'	11'11"11"'11""11""'11"""'
base 10	11	143	1,727	20,735	248,831	2,985,983

"Yes, what we might call the even-length pseudorepunits. They look like repunits with all ones, but aren't, since they'd include eleven and even higher repunits as placeholders.

base 12	I	II	III	IIII	IIIII	IIIIII	IIIIIII	IIIIIIII	IIIIIIIII	IIIIIIIIII	IIIIIIIIIII	IIIIIIIIIIII
base 10	1	11	23	132	287	1,727	3,455	20,735	41,471	248,831	497,663	21,985,983

base 13	I	II	III	IIII	IIIII	IIIIII	IIIIIII	IIIIIIII	IIIIIIIII	IIIIIIIIII	IIIIIIIIIII	IIIIIIIIIIII
base 10	1	11	24	154	323	2,013	4,210	26,180	54,741	340,351	711,644	4,424,574

base 112	I	II	III	IIII	IIIII	IIIIII	IIIIIII	IIIIIIII	IIIIIIIII	IIIIIIIIII
base 10	1	11	111	223	1,343	12,543	25,087	150,527	1,404,927	2,809,855

"I couldn't touch that with a twelfth foot, Pohl."

When the laughter died down, Tony brought up the example of counting by dozens with "Twelve is a dozen, a dozen dozens is a gross or one hundred forty-four and a dozen gross is a great gross or one thousand seven hundred twenty-eight. Months also usually come a dozen to a year, and so once upon a time in Merry Ole England did pence to a shilling. A great hundred is another name for ten dozen, so a long gross or grossty would be ten gross or twelvety dozen."

Al responded with, "There have been terms used for even higher powers of twelve. The glerint[121], for example is a gross squared aka a dozen-to-the-fourth or a gargross."

"So a great gargross would be a dozen-to-the-sixth, a garglerint a dozen-to-the-eighth, and a great garglerint a dozen-to-the-twelfth."

Lana also shared, "The original Duodecimal Society used lower case chi and epsilon for ten and eleven, while its successor, the Dozenal Society preferred, what it called 'dek' to be symbolized by a hash-mark, and 'el' by an asterisk."

"So my eleven series would become an ever-growing string of epsilons or asterisks?" Cal asked.

"I'm afraid so."

Al continued, "In *When the Sleeper Wakes* by H. G. Wells, he has people of the future contracting a dozen to 'do', a gross to 'gro' and a great gross to 'mo'."

"Oh, mo, just like Rudyard Kipling's 'mo' in 'Eenee, meenee, mainee, mo!" Lana added.

With a little help from Pebbles, they compared the higher terms for the various base twelve systems.[122]

from *PAJAS* #29 (December 1987)

MORE DOZENS

Tony opened the next meeting with Psalm 50:14, "Offer to God praise as your sacrifice and fulfill your vows to the Most High." and then asked, "Why was the German suspect arrested when asked if he'd robbed banks?"

"He answered, 'Nein.'"

After the interest in getting high dozenally last meeting, he let Al lead the way into Baker's or long dozens and base thirteen. He began unexpectedly by tell the Society about the BBC version of Douglas Adams' *The Hitchhiker's Guide to the Galaxy.*

"In the seemingly anti-climactic scene," he explained, "the main character discovers a prehistoric castaway crossword puzzler has formed an interlocking pattern of the words 'six', 'times' and 'nine'.

"He takes this as a clue to the tale's quest, the ultimate question, 'What is six times nine?' Unfortunately the answer, as it came to be known, to "Life, the Universe and Everything" to in base ten is fifty-four, but the answer the super-computer Deep Thought gave to the question it was asked seventeen and a half megayears ago was know to be four-two. Unlike the book series or the film adaptation, the radio version had an offstage voice that suggests 'base thirteen'. Four-two in base thirteen <u>is</u> the answer to six times nine."

"If we allow improper placeholders, any base is possible, the only way to express it in base zero and fifty-five and higher. In base one it's fifty-four places. The four-two sequence is just two more than multiples of four. "

base	0	1	2	3	4	5	6	7	8	9	10	11	12	13	14	15	16	17	18	19
6x9	54	111...111	110110	2000	312	204	90	75	66	60	54	410	46	42	312	39	36	33	30	216
42	2	6	10	14	18	22	26	30	34	38	42	46	50	54	60	64	68	72	76	80

"But in hypermath 'six times nine' means nine hundred ninety-nine thousand, nine hundred ninety-nine, while 'nine times six' means six hundred sixty-six million, six hundred sixty-six thousand, six hundred sixty-six, no where near each other!"

"The first is not divisible by four when two is subtracted, but the second is and is four-two in base one hundred sixty-six million, six hundred sixty-six thousand, six hundred sixty-one."

Raz then defended the base thirteen hypothesis with what her uncle Joyce called homonums, numerical homonyms, like won, to, too, for, fore or ate.[123]

"In Japanese" she explained, "four-two is a homonym for death-burden, or mortality, which could very well be called 'the answer to life, the universe and everything'. Shí means not only four but also death and ní means not only two but also load, burden, package, luggage and freight." [124]

With the discussion turning toward wordplay, Raz noted that thirteen is a kangaroo number, quickly contracted to "roonum" by the Society, with three "joeys" contained inside, "three", "nine" and "ten".

She also explained that an English number name containing all fifteen of the letters in "one" through "nine" are called panoramic, and gave the example, "forty-two thousand six hundred seventy-eight". The members quickly determined that they must contain e through i, n, o and r through x and must contain an eight for the g, a two for the w and a six for the x. Since panoramic is 360 degrees, numbers would range from thirty-six to three sixty degrees in "range of vision" in multiples of twelve.

number	1	2	3	4	5	6	7	8	9	10	11	12	13	14	15	16	17	18	19	20	21
range	36	36	48	48	48	36	48	60	36	36	48	60	72	84	60	60	60	72	48	60	72

"Thirteen is known as a baker's dozen since bakers were forced to add an extra loaf be sure to comply to the Bread and Ale Law of 1266. A Baker's half dozen is seven and a Baker's quarter dozen four. We can get higher faster than we did with the ordinary dozen.[125] The fear of the number thirteen, treskaidekaphobia[126], is evident is the renaming of thirteenth floors, seat rows, streets, etc. as fourteen. The thirteen-free sequence would avoid any thirteen and so all the thirteen hundreds and thirteen thousands, etc., substituting pseudonums.

number	13	113	213	313	413	513	613	713	813	913	1013	1113	1213	1300	1301	1302
pseudonum	14	111	210	309	408	507	606	705	804	903	1002	1101	1200	1287	1288	1289

Then Lana shared, "Although the fear of the number thirteen has been associated with the thirteen guests at Jesus' Last Supper or the thirteen at Baldr's, the superstition goes back to King Hammurabi of Babylonia in the Eighteenth Century B. C. His Code does not have a thirteenth law, but then neither does it have any laws numbered sixty-six through ninety-nine either, skipping over sixty-five to a hundred. In Babylonian that would be one-six though one-thirty-nine."

"Ah, like Moses' lost third tablet of Commandments eleven through fifteen."

from *PAJAS* #30 (January 1988)

EVEN MORE DOZENS

Dan opened the meeting with Isaiah 14:14, "I will ascend about the heights of the clouds; I will be like the Most High." then went into a knock-knock.

"Knock! Knock!"

Everyone answered together with the expected "Whose there?"

"Dozen."

"Dozen who?"

"Dozen any one here know me?!"

Then he handed the meeting over to farmboy Cal who expanded the subject of dozens from baker's to poulter's dozen.

"Traditionally chicken farmers did the bakers one better and added a bonus pair of eggs when a costumer bought a second dozen. A poulter's or second dozen would be fourteen, so what would a poulter's gross be?"

Taffy supplied the answer and more, "One hundred ninety-six."

"Ah, and a baker's gross was, one hundred six-nine, the same three digits."

"So 'six of one and half a baker's dozen of the other' would mean two things just slightly different from each other and 'six of one and half a poulter's dozen of the other would mean slightly more different!"

"A poulter's great gross however would be fourteen cubed or two thousand seven hundred forty-four, more than half again as many as an ordinary great gross."

Al pointed out that the Karrapol in *Under a Calculating Star* by John Morressy used base 14. Using double digit placeholders rather than letters for ten through thirteen, he got the Karrapolish numbers.

Karrapolish	...	10	11	12	13	100	101	102	103	104	105	106	107	108	109	110

By the time that Taffy had calculated the second through tenth great gargross[127] on the blackboard and turned around, however, she found herself alone.

"It looks like I've demonstrated Lynch's Law", she said with a half-smile, "When the going gets tough, everybody leaves."

"Not quite everyone," janitor Sam said, remaining behind to lock up and they both smiled.

from *PAJAS* #31 (February 1988)

ELEVENSIES

Dan read from Psalm 97, verse 9, "You, God, are Most High over all the Earth."

Then Al opened her topic with a riddle, "How many professors does it take to change a light bulb?"

Grad student Raz answered, "None, that's what we grad students are for."

Al continued, "The term elevensies sounds like it might have, but actually has nothing at all, to do with base eleven. Rather it refers to a morning British teatime about eleven o'clock like the Hobbits' second breakfast in J. R. R. Tolkien's *The Lord of the Rings*. Tolkien's *The Hobbit* however did popularize the term eleventy.

"Undecimal or base eleven is in a sense non-decimal, so its ten-elevens was naturally translated into decimal as eleventy, eleven tens and eleventy-eleven as twelfty." [128]

undecimalish	...	10	100	101	102	103	104	105	106	107	108	109	110	200

"Superman's kin, the Kryptonians,[129] had a symbol for zero, but did not use it for place-holding. Instead they used the symbol for ten as a digit, yet theirs was still a base ten, albeit an improper one, not a base eleven. They also wrote right-to-left, Hebrew-like. They would think of eleven as one plus ten, oneteen, and a hundred as ninetyteen. We could write it right-to-left substituting ten with one symbol, the Roman X, the duodecimal chi, the dozenal hash-mark or the hexadecimal A, or in Kryptonish with the usual one-zero placeholder for ten and read left to right."

Kryptonese	...	11	21	31	41	51	61	71	81	91	X1	12	22	32	42	52	62	72	82	92	X2	13	23	33	43
Kryptonish	...	11	12	13	14	15	16	17	18	19	110	21	22	23	24	25	26	27	28	29	210	31	32	33	34

"So if we came across the censored four-place year XXXX, it would be in base eleven, uh, ten times eleven cubed plus ten times eleven squared plus ten times eleven plus ten?"Cal asked.

"Yes, you wouldn't actually be in four digits any more. It'd be the number fourteen thousand, six hundred forty."

from *PAJAS* #32 (March 1988)

NUMBERS FROM ALL OVER

Dan opened the meeting quoting "I have lifted up my hand to the Lord, God Most High, possessor of Heaven and Earth" from Genesis 14:22.

He continued with the riddle, "What would you call a ten-book series?"

Lana answered "A novelty."

She continued by sharing "In the ancient texts we Hindus have several different methods for forming powers of ten as high as what we would now call four-hundred-twenty-oneplex, although *ayut*, a thousand, and *lakh*, a myriad, are the most commonly ones still used." [130] She then took the discussion to the Orient, comparing Sino-Korean high number names[131] and Japanese.[132]

Taffy steered the discussion to the New World and the Incan empire and quipu or knot notation which also was based on ten. She told them that although a Flemish knot was thought to have represented one, the long knots with a certain number of windings represented the digit one less than that number of windings and so shared something with

Talcryl.

Then Raz made the somewhat smaller shift to Mayan numbers with "My uncle, some of you may remember, used the Mayan for two, *ka,* as a prefix, so that kazillion had a specific though indeterminant value.[133] Their numbers for one through four were indicated by horizontal dots and each underlining increased the dot-number by five.

"A special sign indicated a score, so that their system was a hybrid twenty/five base. The French also use a hybrid base-twenty system when they call eighty quatre-vingts or four score and ninety quatre-vingts-dix, four score ten."

Mayanish	1	2	3	4	10	11	12	13	14	20	21	22	23	24	30	31	32	33	34	100	101	102

"Curiously the Mayans sometimes deviated from the purely base twenty system by making the place after twenty not always a consistent twenty squared but sometimes twenty times eighteen instead, three sixty rather than four hundred, closer to days per year."

"They did however have special names for several powers of twenty. Notice how kal, which means twenty, is just *ka* plus a suffix."

power of 20	1	2	3	4	5	6	7
Mayan	kal	bak	pik	kalab	kinchil	alau	habla

"So *alaul* could have mean twenty-to-the-sixtieth and *hablal* twenty-to-the-seventieth."

The rest of the meeting was spent getting higher in several languages.

from *PAJAS* #33 (April 1988)

BASE 25

Dan began the meeting by quoting "Glorify the Name of thy Guardian-Lord Most High," from the Qoran and then asked, "How many apes does it take to change a light bulb?"

Al, the speaker for the month, answered, "Countless generations until they by chance evolve opposable thumbs."

Then he continued on the topic, "The D'ni of the Myst and Riven mythoi actually use superposition to represent their numbers in base twenty-five using combinations of the symbols.[134] All are represented in a cartouche-like frame, likely a latter addition. The symbol with the central dot for zero is called *rùn*. A vertical line would symbolize one or *fava*, similarly a right parenthesis two or *brí*, a greater-than-like angle three or *sen*, a gamma-like right angle four or *tor*, a horizontal line, five or *vat*. Six or *vagafa* is a contraction of *vat-ga-fava*, five-plus-one, and is symbolized by their superposition, a plus sign. Similarly for seven or *vagabrí*, eight or *vagasen* and nine or *vagator*. Rotating a symbol ninety degrees counterclockwise multiplies it by five to symbolize ten or *nävú*, fifteen or *híbor* and twenty or *riš*."

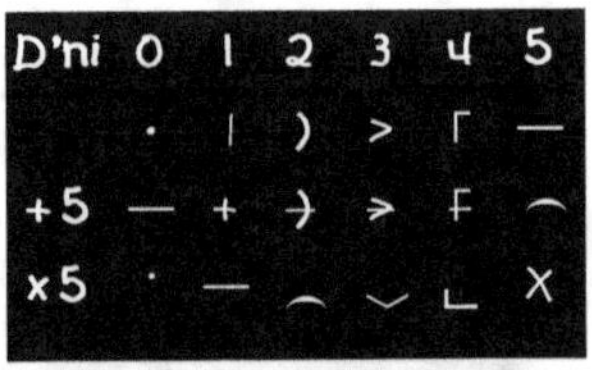

"*Fasí*, twenty-five, can be represented either with joined cartouches, or with its own symbol like *vagafa* rotated forty-five degrees or X. Twenty-five squared is *fara*, twenty-five cubed *falen*, twenty-five-to-the-fourth *famel* and twenty-five-to-the-fifth *fablo*."

"So the combination *bloblo* would mean twenty-five-to-the-tenth, almost twelveplex."

"At least two different keyboard substitutions have been proposed for base twenty-five, which only agree for the first five placeholders.[135]

	0	1	2	3	4	5	10	15	20	25	125	625	3,125	15,625
#1	!	@	#	$	%	*	(	[	]	\	{	}	|	)
#2	!	@	#	$	%	^	&	*	(	[	]	{	}	\

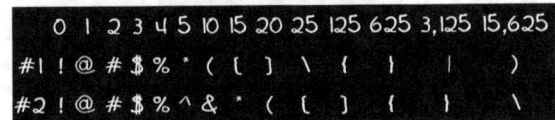

"Even in base five in which they do agree, it is better to use less confusing double-digit notation."

D'nish	...	24	100	101	102	103	104	105	106	107	108	109	110	111

from *PAJAS* #34 (May 1988)

EVEN MORE BASES

Dan read "Doesn't evil and good come out of the mouth of the Most High?" from Lamentations 3:38 and then asked, "How many senators does it take to change a light bulb?"

Kamika gave the answer, "Thirty-three, two to introduce the bill and thirty-one more to make a quorum and then continued, "After thinking on the superpositioning used in the D'ni number system, I saw how it could easily be adapted for base seventy using single and double overlines, strikethroughs and underlines.

	0	1	2	3	4	5	6	7	8	9
10	0	1	2	3	4	5	6	7	8	9
20	~~0~~	~~1~~	~~2~~	~~3~~	~~4~~	~~5~~	~~6~~	~~7~~	~~8~~	~~9~~
30	$\overline{\underline{0}}$	$\overline{\underline{1}}$	$\overline{\underline{2}}$	$\overline{\underline{3}}$	$\overline{\underline{4}}$	$\overline{\underline{5}}$	$\overline{\underline{6}}$	$\overline{\underline{7}}$	$\overline{\underline{8}}$	$\overline{\underline{9}}$
40	0	1	2	3	4	5	6	7	8	9
50	~~0~~	~~1~~	~~2~~	~~3~~	~~4~~	~~5~~	~~6~~	~~7~~	~~8~~	~~9~~
60	$\overline{\underline{\underline{0}}}$	$\overline{\underline{\underline{1}}}$	$\overline{\underline{\underline{2}}}$	$\overline{\underline{\underline{3}}}$	$\overline{\underline{\underline{4}}}$	$\overline{\underline{\underline{5}}}$	$\overline{\underline{\underline{6}}}$	$\overline{\underline{\underline{7}}}$	$\overline{\underline{\underline{8}}}$	$\overline{\underline{\underline{9}}}$

"This notation could be adapted to any base between eleven and seventy, but especially those higher than alphadigital."

Others extrapolated the system further to base a hundred ten with overdots, overdashes, and single and double overwaves.

70	0 1 2 3 4 5 6 7 8 9
80	0 1 2 3 4 5 6 7 8 9
90	0 1 2 3 4 5 6 7 8 9
100	0 1 2 3 4 5 6 7 8 9

Still others added underdots, underdashes and single and double underwaves.

110	0 1 2 3 4 5 6 7 8 9
120	0 1 2 3 4 5 6 7 8 9
130	0 1 2 3 4 5 6 7 8 9
140	0 1 2 3 4 5 6 7 8 9

Naturally that lead to combining the six kinds of overlinings, two strikethroughs and six underlinings in all of two hundred and four more combinations to represent base two thousand, one hundred ninety!

from *PAJAS* #35 (June 1988)

RABDOLOGY

To open the meeting Dan quoted "Day and night without pause they sing 'Holy, holy, holy.' (Revelation 4:8)" and then asked, "How many entertainers does it take to change a light bulb?"

Gillian answered with "Ten, one for the money, two for the show, three to get ready and four to go."

Then she got into the topic of the month, "Before the abacus, in the Han Dynasty, between the second century B. C. and the fourth century A. D., the Chinese used computing rods, red for positive values and black for negative ones. Since we also used different arrangements of the rods for even and odd powers of ten, plus or minus half-powers like with the ten, just twelve rods of each color could represent any of a hundred one numbers, from minus fifty up to plus fifty with zero represented by an empty position.

"The horizontal rods above one another represent the numbers one through four and the vertical one represents five, so that zero through nine are as drawn on the board. Note how five is indicated not by one vertical rod but by five horizontal ones. A vertical rod represents ten, two twenty and so on up to four for forty. A horizontal rod above the vertical rods increases their values by fifty, so fifty takes just two rods rather than five. By changing the color of the rods from red to black the numbers become negative rather than positive, so that six is written as ten minus four, seven as ten minus three, etc. Substituting white chalk for red and gray for black, we get the following."

"With this system our Ancestors calculated not only long division, but roots, and creatively calculated such things as the value of pi to three decimal places.[136]

"The legendary Dzu Tse, aka Jucius, flourished circa 250 A. D., pioneering in what would become known as Jootsy[137] calculus by making 'improper' arrangements of rods and so representing, for example, three not as three horizontal red rods but as four horizontal red rods and a black in any of four possible positions, as minus one plus three, plus one minus one plus two, plus two minus one plus one or plus three minus one. He would also position rods at non-right angles to represent numbers between the integers or a red on a black or black on red to indicate plus-or-minus and minus-or-plus.'His fellow Taoists were said to have calculated fractions of men, sums of dissimilar things like areas and volumes, or inferred dimensions from volumes, and problems from solutions.

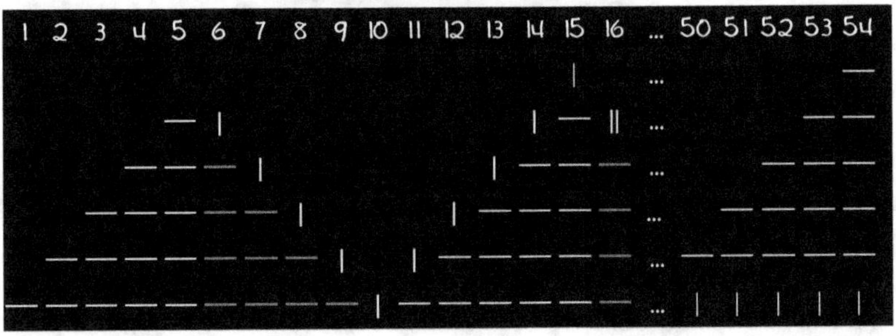

"For example, 'If the sum of the area and volume of a rectangular stone block is thirteen twelve and it would take twenty-two and a half men to lift one a unit shorter in each dimension, what is the maximum area of its sides two less men would carve?'"

"That's impossible!"

"No, its jootsy. Think it through. Think out of the system."

" A rectangular solid would have a volume of xyz and an area of twice the sum of xy plus xz plus yz. In order for the product and sum of these three to be as indicated we have only one integral solution, rather like in cryptarithms."

"The dimensions would have to be seven, eleven and thirteen, the three prime factors of one thousand, three hundred twelve."

"So its area would be twice seventy-seven plus ninety-one plus one forty-three."

"And the maximum area of four sides would be twice ninety-one plus one forty-three, four sixty-eight."

" A block one unit shorter in each dimension would be six, ten and twelve."

"With a volume of seven hundred twenty."

"The ratio of that to twenty-two and a half gives the number of men needed for the original block."

"Forty-one."

"Two less men would be thirty-nine."

"Four sixty-eight divided by thirty-nine is twelve!"

"And these Taoists calculated all this on their rod-calculator, a precursor to John Napier's rods used digital markings rather than rod arrangements. By arranging his nine rods Napier could simplify multiplication to mere addition, an analog computer."

	1	2	3	4	5	6	7	8	9
1	1	2	3	4	5	6	7	8	9
2	2	4	6	8	10	12	14	16	18
3	3	6	9	12	15	18	21	24	27
4	4	8	12	16	20	24	28	32	36
5	5	10	15	20	25	30	35	40	45
6	6	12	18	24	30	36	42	48	54
7	7	14	21	28	35	42	49	56	63
8	8	16	24	32	40	48	56	64	72
9	9	18	27	36	45	54	63	72	81

"Combining the older and newer rabdologies like Jesus' proverbial scribe, we get an even more interesting and compact one. To square one hundred twenty-three million, four hundred fifty-six thousand, seven hundred eighty-nine is as easy as adding the diagonals from upper left to lower right."[138]

"It's easier to see if you shift the rods to make the 'diagonals' into 'verticals'."

1	2	3	4	5	6	7	8	9								
	2	4	6	8	10	12	14	16	18							
		3	6	9	12	15	18	21	24	27						
			4	8	12	16	20	24	28	32	36					
				5	10	15	20	25	30	35	40	45				
					6	12	18	24	30	36	42	48	54			
						7	14	21	28	35	42	49	56	63		
							8	16	24	32	40	48	56	64	72	
								9	18	27	36	45	54	63	72	81

"So, just by correctly selecting and positioning these 'rods', you can turn multiplication into addition. It's likely the way that some so-called idiot savants multiply large numbers."

from *PAJAS* #36 (July 1988)

GROSSARY LIST

Retiring president Dan opened with a quote from Ecclesiastes 5:8, "Do not be amazed at the matter; for the high official is watched by a higher, and there are yet higher one over them."

"How many supreme court justices does it take to change a light bulb?"

Lou quickly answered. "Five-ninths."

Al however was the month's presenter, upgrading the topic of dozens to dozens of dozens, "The Cheela [139] use a number system based on a hundred forty-four."

"Gross!"

"Yes, it is, but with it we can get high pretty quickly."

"The third place would be a hundred forty-four squared or twelve-to-the-fourth, a gargross, already."

"And the fourth would be twelve-to-the-sixth."

"Let's make a grossary list."

grossary place	1	2	3	4	5	6
gross power	1	144	20,736	2,985,984	429,981,696	61,917,364,224

"That reminding me of the old cashier's code," Sam commented.

When everyone just responded with blank stares, he continued, "When I was a young cashier in my future brother-in-law's grocery, we substituted numbers for letters with an easily remembered formula, for example, every third number plus the ubiquitous final nine, to disguise the message as a receipt."

A	B	C	D	E	F	G	H	I	J	K	L	M	N	O	P	Q	R	S	T	U	V	W	X	Y	Z
01	04	07	10	13	16	19	22	25	28	31	34	37	40	43	46	49	52	55	58	61	64	67	70	73	76

"Most High, for example, would would look like the sum of eight items on a receipt slip."

$3.79
$4.39
$5.59
$5.89
$2.29
$2.59
$1.99
$2.29

$28.82

"That ubiquitous nine after the first two digits of the code reminds me of that tenth of a cent[140] that they insist on giving gas prices in," Lana added. "They always look to me like a nine exponent."

z	1	2	3	4	5	6	7	8
z^9	1	512	19,683	262,144	1,953,125	10,077,696	40,353,607	134,217,728

"Thank the Most High that gas prices don't go up that fast!" Cal commented and all agreed.

from *PAJAS* #37 (August 1988)
DOMINISSIMO[141]

The new summer president pro tem Taffy opened the meeting with "O Wisdom, who came from the mouth of the Most High, reaching from end to end and ordering all things mightily and sweetly, Come, and teach us the way of prudence," from the O Antiphons. She then posed the question, "How long could you sail a ship two hundred twenty yards at one nautical mile per hour?" and answered it with "Knot-furlong."

"Oh, not for long. I get it. A knot is another name for a nautical mile per hour and two hundred twenty yards is a furlong."

Before the meeting began in earnest however some of the older members, now grad students, had been playing a strange game with dominoes, still celebrating the Professor's birthmonth last month. The uninitiated had watched for a while, now one dared comment.

"What I don't get is that game you were playing. I could see that you are trying to come up with a given pattern by playing correct or incorrect dominoes, like the card game Eleusis,[142] but I couldn't understand the numbers you're using to describe them."

"You're right. It is something like the card game. We like to think of Joyce's Dominissimo, as the ultimate game, but some of us, I confess, have perhaps become too familiar with the *Encyclopedia of Sequences.*[143] Most of us gamesters have become 'experts'."

"That's as defined by Ryan's Law, 'having guessed correctly three time in a row.'"

"We however have over time compiled our own *Index of Sequences* too, a listing of overused ones that have been voted either too easy for old-timers or too difficult for new-comers."

"*The Index of Forbidden Sequences* some of us call it." [144]

"You might say that when you become familiar with a number you call it by name and when you become overly familiar with a sequence you call it by its index number."

"Let's play a sample game. Say the dealer started with an eagle-duck or zero-two. The obvious next play would be four-six, for the next two evens.

"After four-six would come eight-one, zero-one, two-one, four-one, and six-one. Since eight-one has already been played, half-bone of suit eight, say eight-two, would have to be played upward.'

"That however would be much too obvious, and so obviously wrong, so the four-six gets positioned downward and the player has to draw another 'bone' from the 'boneyard'."

"The play however might not be totally wrong, but half-right, if the next number is four, but the one after that is <u>not</u> six. The next play might be an eight suit, the next power of two. If so the plays after that would be one, six, three, two, etc. If the suit matches but not the rank it's played upward."

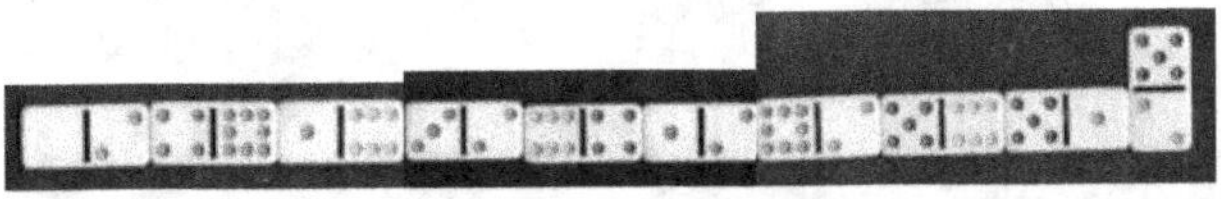

"Four-eight however is also likely too obvious. It is more likely, at least with this group, something like the decimal expansion of the fraction one forty-first, zero-two and three-nine repeated using half-bones.", Taffy continued, switching to numbers on the blackboard.

1 1 1 1 3 2 3 0 4 3 4 1 5 4 4 2 6 5 5 3 7 6 6 4 8 7 7 5 9 8 8 6
0 2 3 9 0 2 3 9 0 2 3 9 0 2 3 9 0 2 3 9 0 2 3 9 0 2 3 9 0 2 3 9

"The player with no bones or, when no more bones can be played, the least wins. It's even more likely however that some other sequence beginning with zero-two and three-nine might be it. As Scott's First Law has it, 'The probability of something wrong looking right is greater than that of it looking wrong.'"

"Add to that that Dominissimo Masters can be extraordinarily devious. They might be thinking of square roots."

√0.00572	02	39	16	52	14	86
√0.00573	02	39	37	41	84	07
√0.00574	02	39	58	29	71	04
√0.00575	02	39	79	15	76	16

"Or cube roots of which there are also an infinite number."

√0.00013652	02	39	04	72	68
√0.000137	02	39	28	02	51
√0.000138	02	39	86	10	31
√0.00013823	02	39	99	42	12

"Or they might be thinking of a continued fraction generated from a decimal, in this case, a continuation of one forty-first. They're usually written with assumed nested parentheses for simplicity." [145]

1/(41+1/(1+1/3))	02	39	52	09	58
1/(41+1/(1+1/4))	02	39	23	44	50
1/(41+1/(1+1/5))	02	39	04	38	25

"It could be the sum of some difference sequence with a more easily recognizable pattern."

fractured 1/41	0	2	3	9	11	12	18	20	21	27	29	30
difference 1		2	1	6	2	1	6	2	1	6	2	1
	0	2	39	101	208	350	527	739	986	1,258	1,565	1,907
difference 2		2	37	72	107	142	177	212	247	272	307	342
difference 3			35	35	35	35	35	35	35	35	35	35
	0	239	478	717	956	1,195	1,434	1,673	1,912	2,151	2,390	2,629
difference 4		239	239	239	239	239	239	239	239	239	239	239

"Or the difference of some more recognizable sum."

sum 3				26	26	26	26	26	26	26	26	26	26	26	26
sum 2			7	19	45	71	97	123	149	175	201	227	253	279	295
sum 1		2	5	14	59	130	227	350	499	674	875	1,102	1,355	1,634	1,939
	0	2	3	9	68	198	425	775	1,174	1,848	2,723	3,825	5,180	6,814	8,753

"It could even be that zero, two, three and nine are in reverse alphabetical or zeewyexical order.[146] Since six, seven and one have apparently been eliminated, there are left only four, five and eight or perhaps only two or one or none of these three left."

0239458	02	39	54	80	239	5,480	23,954
023945	02	39	450	2,394	5,023	9,450	239,450
023948	02	39	480	2,394	8,023	9,480	239,480
023958	02	39	580	2,395	8,023	9,580	239,580
02398	02	39	80	239	802	3,980	23,980
0239	02	3	90	2,390	23,902	39,023	902,390

from *PAJAS* #38 (September 1988)

PUZZLING

Taffy opened the first meeting of the new school year with "The power of the Most High will overshadow you," from Luke 1:35. After the traditional riddle, "What's a nine-letter word for twenty cents?" and answering it with "Paradigms. Get it? Pair o' dimes.", she posed an even more cryptic puzzle on the blackboard.

log(¢/m¢) = ($/¢)% + log($/¢)

"Some of you've heard of the Four Fours Problem,[147] I trust, where you try and express the integers using just four fours?", she asked.

"Sure, that's child's play. Zero is forty-four minus forty-four, one forty-four over forty-four, two four over four plus four over four, and so on." Charles answered and added to the board.

44 - 44 = 0

44/44 = 1

4/4 + 4/4 = 2

"Take heed of DeNever's Law, 'The simplest subjects are the ones you know nothing about.' Lana cautioned

"This symbol manipulation I started with is an example of representing numbers completely digitlessly," Taffy continued. "The number of cents in a dollar as a percent is just one and its logarithm is two and their sum is equal to the logarithm of millicents in a dollar; one plus two equals three."

"What other symbols do you suggest we use to get higher?"

"There's the capital S symbol which stands for successor, which operating on the logarithm of a dollar over a millicent would be four."

"We could use two parentheses to multiply by ten like with proto-Roman numerals to get forty and four to get to four hundred."

"There's the factorial, symbolized by exclamation mark to the right, that means the number multiplied by its predecessors. For four it'd be four time three time two times one or twenty-four. The left factorial with the exclamation mark to the left confusingly has been used to symbolize both the subfactorial and the sum of the factorials of the number and its predecessors, better symbolized with a left sigma and right exclamation mark. For four it'd be twenty-four plus six plus two plus one or thirty-three."

z!	1	2	6	24	120	720	5,020	40,160	361,440	3,614,400	39,758,400
Δz!		1	4	18	96	700	4,300	35,140	326,300	3,252,960	3,6144,000
Σz!	1	3	9	33	153	873	5.893	46,053	372,383	3,625,343	43,383,743

"Then there's the bifactorial, symbolized by a double exclamation mark, that is just every other factor, four times two or eight."

"So the left bifactorial would be every other one, just twenty-four plus two, twenty-six?"

"Yes."

z!!	1	2	3	8	15	48	105	384	945	3,840	10,395	46,080	135,135	645,120
Δz!!		1	1	5	7	33	57	279	561	2,895	7,500	55,685	89,055	509,985
Σz!!	1	3	6	14	29	77	134	513	1,074	3,969	11,469	67,154	156,209	666,194

"Then there's the primorial symbol, a hash mark, to the right that means the product of the indicated number of primes. For four it'd be two times three times five time seven or two hundred ten."

p(z)	2	3	5	7	11	13	17	19	23	29
Δ(p(z))		1	2	2	4	2	4	2	4	6
z#	2	6	30	210	2,310	30,030	510,510	9,699,690	223,092,870	4,536,136,010
Δz#	2	4	24	180	2,100	27,620	480,480	9,189,180	213,393,180	4,313,043,140
Σz#	2	8	38	248	2,558	32,588	543,098	10,242,788	233,335,658	4,769,471,668

"And the four biprimorial three times seven?"
"Yes, twenty-one."

z##	10	21	110	273	1,870	5,187	43,010	150,423	1,333,310
Δz##		11	89	163	1,597	3,317	37,823	107,413	1,182,887
Σz##		31	141	414	2,284	17,471	60,481	210,904	1,544,214

"As fast-growing as these operators are, even faster ones have been invented like a rightward vertical bar indicating tetration-to-the-second. For four it would be four-to-the-fourth, two hundred fifty-six."[148]

z|	1	4	27	256	3,125	46,656	823,543	16,777,216	387,420,489
Δz|	1	3	23	229	2,869	43,531	776,887	15,953,673	370,643,273
Σz|	1	5	32	288	3,413	49,781	873,324	17,650,540	405,071,029

"Dufactorial would be the factorial of a factorial, symbolized by an exclamation mark, surrounding parentheses and another exclamation mark. For four it would be twenty-four factorial, greater than twenty-threeplex."

(z)!	1	2	720	620,448,401,733,239,439,360,000
Δ(z!)!		1	718	620,448,401,733,239,439,359,280
Σ(z!)!		3	723	620,448,401,733,239,439,360,723

"An exponential factorial stacks the predecessors rather than merely multiplying them. It could be symbolized by a rightward superscripted exclamation mark. For four it would be four-to-the-third-squared or four-to-the-nineth, two hundred sixty-two thousand, one hundred forty-four. Doubling the caret would indicate tetrational factorial."

z^!	1	2	9	262,144	> 183,230plex
Δz^!		1	7	262,133	> 183,230plex
Σz^!		3	12	262,156	> 183,230plex
z^^!	1	4	19,683	> 154plex	~154duplex
Δz^^!	1	3	19,679	> 154plex	~ 154duplex
Σz^^!	1	5	19,688	> 154plex	~154duplex

"So the simplest and lowest Four Fours number would be four in base four hundred forty-four, just using right subscripting and no other symbols."

"And four-four in base forty-four would be forty-four times four plus four, uh, one eighty."

"And four-four-four in base four would be only four times sixteen plus four times four plus four, eighty-four."

"See the pattern here? The base is more important than the length."

"And with no subscripting at all it's four thousand, four hundred forty-four."

"Using right superscripting we'd get four hundred forty-four-to-the-fourth, somewhere in the millions."

"And forty-four-to-the-forty-fourth would be more than ten times longer, sevenplex."

"Good you're learning to guesstimate and gaining number sense."

"The Copernican principle makes the probability of a guess being a factor of one third to three times off fifty percent and the Rule of Thirty-nine makes ninty-five percent that it's off by a factor of thirty-nine."

"So, getting off the tangent and back on topic, the exponent is much more important that the number exponentiated."

"And four-to-the-four-hundred-forty-forth would be ten-to-point-oh-six-times-four-forty-four, nearly two-hundred-sixty-sixplex."

"Very good! You used rounded logarithms.' They come in even handier with the left superscripts."

"Four hundred forty stacked four times would be approximately three-fifths the height of a to-the-forth-hundred. That would be four times as high a stack, more than two-hundred-sixty-sixtriplex.

"Forty-four-to-the-forty-fourth would be higher than forty-fourplex and to-the-four-hundred-forty-fourth-four higher than four hundred forty-four stacked tens."

"But the left subscript would just count the length. Four-hundred-forty-fours would be between a bit more than twelveplex, forty-four forty-fours less than eighty-eightplex."

"And four-to-the-four-hundred-forty-fourth over two-hundred-sixty-sevenplex."

4_{444}	444_4	444_4	4444	444	${}_4444$	44	${}_{44}44$	4
4	84	180	4,444	38,862,602,496	444,444,444,444	~ 72plex	~ 87plex	~ 267plex

"Higher than that are the left superscripts indicating tetration, indicating the height of a stack of the identical exponents. To-the-fourth-four-hundred-forty-four would be higher than tentriplex. To-the-forty-fourth-forty-four would be higher than tentetracontatriplex and to-the-four-hundred-forty-fourth-four higher than tentetrahectotetracontatriplex!"

All those who had been empowered to think in terms of plexes understood. Those who were not were perplexed.

from *PAJAS* #39 (October 1988)

MORE PUZZLING

Taffy welcomed the new and returning students with "How different the person who devotes himself [or herself] to the study of the law of the Most High!" (Sirach 39:1)

Lana asked, "What would you have if you received two thousand pounds of Chinese soup?" and before Taffy could answer, Lou beat her to it, "One ton Won Ton!"

Taffy nevertheless continued the with variations of the previous month's topic, "Besides the

Four Fours challenge, the there several other similar ones.

"Three Threes is a bit more challenging, but as Hein's Law states 'Problems worthy of attack prove their worth by hitting back.'" [149]

"Even harder is the so-called Hard numbers[150] using just ones and twos. The total number of symbols, including the number of ones and twos, is called the creation number. For zero, one minus one gives a creation number of three. One and two are just the digits for a creation number of one each."

"So three would be two plus one for a creation number of three again."

"Right."

"Another variant is just using the digit zero. Zero is just the digit and one is simply zero-to-the-zeroth."

"Just two symbols."

"Right, two though would take a minimum of five, wouldn't it? Zero-to-the-zeroth plus zero-to-the-zeroth."

"The Three Symbols challenge is not to represent every integer, but to create the highest number using just three different symbols. This illustrates the difference between the kinds of mathematicians that deal with very large numbers. They fall into three categories. The first are the syntactic googologists who use creative wordplay to forge names and naming systems for large numbers, 'covering all the bases' and 'pausing to smell the roses', like us Joyceans. The second category are the abstract googologists who generate the largest numbers with the simplest, yet fastest-growing functions, ignoring all the interesting numbers along the way.[152]

"So that would mean using the heavyweight digits, nine, eight and perhaps seven, right?"

"Right. Nine hundred eighty-seven is what comes most easily to mind."

"Ninety-eight-to-the-seventh would be higher than thirteenplex."

"Seven-to-the-ninety-eight would be over fifty percent higher than nine-to-the-eighty-seventh."

"Ninety-eight factorial would be nearly a hundred-fifty-eightplex."

987	987
98^7	86,812,553,324,672
9^{87}	> six eighty-twoplex
7^{98}	> eighty-threeplex
$98!$	> hundred-fifty-sevenplex

That was about as high as that line of discussion went,[151] before going off on a tangent.

"Phi, of course, can be expressed with just one symbol, the Greek letter. The Phi challenge is to express phi with the same number of digits as the digit. The first was four fours." [152]

$$(\sqrt{4} + \sqrt{(4! - 4)})/4$$

$$(5/5 + \sqrt{5})\ \sqrt{(.5(.5))}$$

$$(6 + 6\sqrt{(6 - 6/6)})(6 + 6)$$

$$((8 + 8\sqrt{(\sqrt{(8 + 8)} + 8/8)})/(8 + 8)$$

$$(9 + 9\sqrt{(\sqrt{9} + 9/9 + 9/9)})/(9 + 9)$$

"Six Ones limits expressions to just six ones and operator symbols. [153] Countdown limits expressions to just the digits one through six and the symbols plus, minus, left and right parentheses and slash for division. Nine Digits not only allows, but requires the use of all the digits one through nine." [154]

z	Six Ones	Countdown	Nine Digits
0	111 - 111	(3 - 2 - 1)/(4 + 5 + 6)	(3 - 2 - 1)(456,789)
1	111/111	5 + 6 - 4 - 3 - 2 - 1	(1 + 2 - 3 - 4 + 5)(9 - 8)(7 - 6)
2	1/1 + 11/11	2(6 - 5) + (4 - 3 + 1)	(1 + 2 + 3 + 4 + 5 + 7 + 8)/(6 + 9)
3	1/1 + 1/1 + 1/1	3(6 - 5)(4 - 2 - 1)	3(9 - 8)(7 - 6)(5 - 4)(2 - 1)

Play continued in both the syntactic and analytic camps, demonstrating Eric Sevareid's Law, "The chief cause of problems is solutions."

from *PAJAS* #40 (November 1988)

PERFECTION

Taffy opened with, "Be perfect as your Heavenly Father is perfect," from Matthew 5:48.

Her riddle of the month was not so perfect, "What is the unit used to measure the suspense in a mystery story?"

Her answer, "A whod unit" received many groans in response, so she quickly moved on to the topic.

"The only One Who's perfect, of course, is God the Most High, but numbers can be also", she explained, "those whose factors add up to themselves, like six which is both three times two times one and three plus two plus one. No perfects, but some perfect neighbors are prime.

perfects	6	28	496	8,128	33,550,336	8,589,869,056	137,438,691,328

binary	110	11100	111110000	1111111000000	1111111110000000000

perfect neighbors	5	7	27	29	495	497	8,127	8,129	33,550,335	33,550,337

"They have an intimate connection with what are known as the Mersenne primes of the form two-to-a-prime minus one. The first Mersenne prime would be three, two-squared minus one. The next perfect numbers would be twenty-eight and four hundred ninety-six. The pattern is much easier to see in binary."

z	p(z)	M(p)	M(z) = 2^p - 1
1	2	110	11
2	3	1110	111
3	5	111110	11111

"The Mersenne primes got their name from the Seventeenth Century monk Marin

Mersenne, though many others had searched for them before him.[155] Mersenne's perfectly corrected list of the first ten perfect and Mersenne prime numbers, was not verified until 1914.

z	p(z)	M(z) length	P(z) length	year
4	13	4	8	1458
5	17	6	10	1588
6	19	6	12	1588
7	31	10	19	1772
8	61	19	37	1883
9	89	27	54	1911
10	107	33	65	1914
11	127	39	77	1876

"The search for the primes that generate Mersenne primes and perfects got a big boost in the computer age, when both broke the googol barrier." [156] Primes higher than fourplex have been called titanic and those who discover them Titans. The first was Harry L. Nelson's in 1979, which generated the twenty-sixth perfect. By 1984 Samuel Yates listed over a hundred Titans. A prime higher than sixplex has been called a megaprime.[157] The lowest gigantic prime, those higher than nineplex, is nine hundred ninety-nine times nineplex plus seven."

from *PAJAS* #41 (December 1988)

AUDIO-ACTIVITY

Taffy opened the meeting, temporarily held in the chemistry building, with "The Most High will establish it," from Psalm 87:5.

Lana asked, "What do you get if you add eight teacups and four teacups?" and Taffy responded "Ten dozen cups."

Then they settled in to investigating the look-and-say sequence or Gleichniszahlen-Reihe that transforms a number into a higher number by inventorying its digits. As Taffy explained, "You look at the one and inventory it by saying 'one one', then you would look at the one-one and say 'two one', and then 'one two, one one' and so on. Those sequences starting with seeds higher than two only differ in the tail."

seed 0	10	1110	3110	132210	1113221110	3113223110	13211322132110
seed 1	11	21	1211	111221	312211	13112221	1113213211
seed 2	12	1112	3112	132112	1113122112	311311222112	13211321321 2
seed 3	13	1113	3113	132113	1113122113	311311222113	13211321322113
length	2	4	4	6	9	12	14
ratio	2	1.5	1.333	1.5	1.5	1.333333333	1.1666666666

"As you can see the count never goes higher three when beginning with zero, one, two or three. Conway also noticed that the lengths get higher at a rather constant rate, in fact, the ratio gets closer and closer to what became known as Conway's constant as it continues, a little higher than one-point-three.

"Depending on which base the audio-active sequence is read as, whether improper ternary or higher, we naturally get ever higher decimal equivalents."

base 3	1	4	7	49	376	886	4,777	31,288
base 4	1	5	9	101	1,385	3,493	30,121	358,885
base 5	1	6	13	181	3,936	10,306	129,061	2,476,056
base 6	1	7	15	295	9,373	25,135	429,493	12,194,791
base 7	1	8	17	449	19,664	53,614	1,196,497	47,332,048
base 8	1	9	19	649	37,521	103,561	2,921,617	153,949,833
base 9	1	10	21	901	66,520	185,338	6,444,541	436,971,520

"Conway also noticed the similarity between the behavior of these strings of digits and radioactive decay and called it audio-active decay. He determined that any string of one, two or three without a string of four identical digits decays to one of ninety-two unique elemental strings, like the periodic table of elements. Although there are seven other possible ordering Conway's has been taken as the standard, like Mendeleev's was of the chemical elements, calling three uranium and the stable two-two hydrogen.

"Starting with three we then get: one-three, an 'improper' zero-less ternary number, in base ten a six. One-three then decays to one-one-one-three." [158]

"That'd be, uh, twenty-seven plus nine plus three plus three or forty-two."

"Identifying three with uranium, its audio-active products are, like radio-active uranium's protactinium, thorium, actinium, radium, francium, and radon.

U → Pa → Th → Ac → Ra → Fr → Rn

"Translating to base ten equivalents we get a new sequence.

3 → 6 → 44 → 96 → 555 → 31,335 → 102,048

"At this point Conway identified the decay product of audio-active radon as a compound, rather than an element, since it's fissionable into two independent elements, namely holmium and astatine.[159] Audio-active radon is like the element sarranium[160] in that it is included in its own decay products!

"This however is a more like a compound, and yet not, because the two elements are not communicative. Holmium astatinide, HoAt, is not the same as astatine holmide, AtHo.[161]
The holmium decays into dysprosium, which decays back inro radon and potassium and the polonium into bismuth."

Rn → Ho-At → Dy-Po

311311222113 → 1321132-1322113 → 11312211312-3222113 →

Rn-K-Bi

311311222113-1112-3113322113

"This isn't alchemy!" Raz interrupted in response to the dazed newbies. "Each of these strange compounds, it must be remembered, are actually code names for certain zeroless ternary numbers, like the chemical elements are for certain arrangements of subatomic particles.

"Most numbers correspond to an audio-active element or compound, but not all. For the rest we have to get jootsy. Since two decays into one-two or calcium we could, by analogy with protactinium and actinium call two protocalcium. Or we can alternatively name them using monosyllabic Greco-Latin prefixes for the digits, like the as-yet unnamed physical elements plus the metallic suffix -ium.[162]

	0	1	2	3	4	5	6	7	8	9
Greco-Latin	nil	un	b(i)	tr(i)	quad	pent	hex	sept	oct	en(n)

"Starting with any other digit, z, higher than three or equal to zero, yields the rather uninteresting decay sequence, one-z, one-one-one-z, three-one-one-z, one-three-two-one-one-z, etc. Unnilium, for example, decays to ununium unnilide, then trunium unnilide and untro-bunium unnilide, growing larger and larger like sarranium, yet always retaining the original unnilium.

Un → UuUn → TuUn → UtBuUn

"Or we could use Mendeleev's traditional Sanskrit prefixes." Lana suggested.

Sanskrit	eka	dva	dre	chatra	pancha	shasta	sapta	asha	narva	dasta

"Eka- meant one row lower on the periodic table, dva- two, and so on. The prefix das- is the equivalent to our suffix -ty. From these and the formula for the nuclear 'magic' numbers, we can rename quite a few more numbers." [163]

"Some astronomers have compared supermassive stars to ionized atomic nuclei," Charles shared. "They're at least a thousand times the mass of our sun, about fifty-sevenplex baryons. Using these higher Sanskrit prefixes, that'd be a vivaharadon nuceus. Those found in galactic cores are called hypermassive at a million solar masses, sixtyplex baryons or more."[164]

from *PAJAS* #42 (January 1989)

ARCHERY

Taffy opened the meeting before Christmas break with "Thou, O Lord, art on high forever," from Psalm 92:8. Her riddle asked, "How many optimists does it take to change a light bulb."

After giving the answer, "None, they all know it'll get brighter," she began with the monthly topic, "Donald Knuth, came up with the up-arrow notation for representing large numbers like a googolplex" she explained, "or its abbreviated version using the caret. It was a way to avoid the awkward stacking of superscripted exponents.

"One up-arrow between numbers means the same as the familiar exponential. Two up-arrows mean the next higher operation, called tetration, the first number stacked exponentially the second number of times."

$10\uparrow 100$	googol
$10\uparrow 10\uparrow 100$	googolplex

"So to-the-second-ten, tenduplex, would be ten-double-up-arrow-two, right," Cal asked.

"Right" Taffy said, while turning to the board, "You can get some idea of the power of the up-arrow by using it repeatedly on the first non-trivial integer, two, and seeing how fast you get high."

"Two-up-arrow-four would be two-to-the-second or four."

"But two-double-up-arrow-two would be to-the-second-two which is also four, not higher at all."

"Ah, but two-triple-up-arrow-two however gets high much faster. It doubles the stack of twos to to-the-fourth-two."

"Two-to-the-four or sixteen, then two-to-the-sixteenth, sixteen-log-twoplex, nearly fiveplex"

"Two-quadruple-arrow-two would double the stack of twos yet again to to-the-eighth-two, or eight-log-twoplex, about two-point-four stacked tens, more than a googol already"

$2\uparrow 2$		4
$2\uparrow\uparrow 2$	$2\uparrow 2$	4
$2\uparrow\uparrow\uparrow 2$	$2\uparrow 2\uparrow 2\uparrow 2$	65,536
$2\uparrow\uparrow\uparrow\uparrow 2$	$2\uparrow 2\uparrow 2\uparrow 2\uparrow 2\uparrow 2\uparrow 2\uparrow 2$	> googol

"It then keeps doubling with every arrow."

"Three-up-arrow-three is even faster growing, tripling the stack every time. Conway's chained arrow notation is a condensed version of Knuth's up-arrow. It uses horizontal rather than vertical arrows, which is sometimes substituted by a dash and greater-than sign, like the up-arrow is with the caret. The horizontal arrow means the same thing as the vertical one when between two numbers. The all-important difference between the two notations comes when a number comes between two arrows.

$$z^{z'} = z\uparrow z' = z\text{^}z' = z\rightarrow z' = z\text{->}z'$$

$$^{z'}z = z\uparrow\uparrow z' = z\text{^^}z' = z\rightarrow 2\rightarrow z' = z\text{->}2\text{->}z'$$

"That number indicates the number of up-arrows needed, like the superscripted up-arrow some had used. Robert Munafo's notation is similar to the functional exponential up-arrow notation, but with parentheses around the exponent.

$$z^{(z')} = z\uparrow z' \quad ^{(z')}z = z\uparrow\uparrow z'$$

"So, a googol would be ten-chained-arrow-hundred. Tenplex would be ten-chained-arrow-two-chained-arrow-two. That does not sound like an improvement."

"Maybe not, but ten-arrow-ten-arrow-ten would be much higher, and expressible in just eight symbols rather than fourteen and ten-arrow-hundred-arrow-ten much, much higher in just nine!" [165]

"And ten-arrow-ten-arrow-ten-arrow-ten would much, much higher with ten-arrow-ten doublings!"

"There are, believe it or not, expressions that make chained arrow notation seem very slow-growing."[166]

from *PAJAS* #43 (February 1989)

SUPERNATURAL NUMBERS

Taffy opened with Psalm 7:17, "I will sing praise to the name of the Lord, the Most High."

"How many terrorists does it take to change a light bulb?" she asked and Tom answered,

"None. They don't want change, the want an end to civilization."

"On that 'pleasant' note, we'll look into an attempted change that doesn't seem to have taken yet. Donald Knuth invented a numbering system based, not on thousands, but on the myriad aka ten thousand aka fourplex, that he called supernatural. So that his term myllion means a myriad squared or eightplex, his byllion a myriad cubed or twelveplex, a tryllion a myriad-to-the-fourth or sixteenplex and in general a zyllion eight-z-plus-eightplex."

myllion	1,0000,0000
byllion	1,0000,0000,0000
tryllion	1,0000,0000,0000,0000

'So a byllion is a mylliard myllion and a tryllion a myllion myllion."

"Oh, like a bylaw is a law within a law."

"Actually the word bylaw has nothing to do with the prefix bi-, but comes from the Old Norse suffix -by for 'settlement'."

"Yes, it's bi- as in bikini, the two-piece bathing suit, named for Bikini Atoll, where the first H-bomb turned the first element, hydrogen, into the second, helium."

Lana continued, "Knuth's system had prestigious precedent. Archimedes' *Sand Reckoner,* circa 200 B. C., also was based on the myriad. This Greek Titan's highest number, 'myriakis myriostas-periodu myriakis-myriston arithmon myriai myriades', what we would call to-the-second-twice-myriad squared.

twice myriad	20,000
twice-myriad squared	> 400,000,000
to-the-second-twice-myriad squared	> 520,412plex

"Knuth's system conflicts however with the traditional system of Nicolas Chuquet who originally called billiart or eightplex and trilliard or thirtyplex respectively byllion and tryllion. Jehan Adam used the form bymillion and trimillion as early as 1475. In 1549 Jacques Peletier introduced his so-called long system with milliart, which later changed to milliard, our nineplex.

"There therefore seems a need for the supernatural number system to be extrapolated to a new and unambiguous 'long Knuth' system, based on the zylliard, sixteen-z-plus-sixteenthplex.[167]

myllliard	thity-twoplex
bylliard	fourty-eightplex
trilliard	sixty-fourplex
quadrylliard	eightyplex
centylliard	thousand-six-hundred-sixteenplex

This majority opinion of the membership was that there was not.

from *PAJAS* #44 (March 1989)

BARBARIAN NUMBERS

Taffy began by quoting "The thought of them is with the Most High," from Wisdom 5:15.

Tom then posed the riddle, "How many carpenters does it take to change a light bulb?"

Taffy answered, "None, that's for the electrician's union," then she went quickly into the month's topic."Noting that the representation for thousands in Roman numerals is with an added bar above the numeral, Joyce defined the suffix -bar as a multiplicative operator.

"Applying this prefix to the Roman numeral suffixes gets us higher much faster. Googobarl mean a hundred-thousand-to-the-fifty-thousandth or two-hundred-tfifty-thousandplex. Googobarc would mean two-hundred-thousand-to-the-hundred-thousandth. Verbalizing the second M with the telegrapher's ump we get to googobarmump."

		length
googolbar	100,000^50,000	250,000
googocbar	200,000^100,000	530,103
googodbar	1,000,000^500,000	3,000,000
googombar	2,000,000^1,000,000	6,301030
googomumpbar	4,000,000^2,000,000	13,204,120

Taffy continued, "In Isaac Asimov's numbering system he got high by applying T-, the metric symbol for tera- or twelveplex, not to a unit of measurement or even another number, but to itself. Thus T-T meant not merely ten-to-the-twenty-fourth, but to-the-second-twelveplex, greater than to-the-fourth-ten.

	length
T	24
T-T	>tentriplex
T-T-T	>tenpentaplex
T-T-T-T	>tenhendekaplex

"Similarly Joyce applied -bar to itself to get the first barbarian number, onebarbar, aka onedubar or informally called 'wunderbar'. It's read as one-(bar-bar) rather than merely (one-bar)-bar, one times a thousand a thousand times rather than one times a thousand times a thousand. In googolbar the -bar modifies just the L, while in bargoogol it modifies the googo- prefix and so means a thousand iterations of two-z-to-the-zth. Barbargoogol would mean a thousand iterations a thousand times, hence the name 'barbarian!'"

Lana added, "Apollonius of Perga's M with a smaller M written above it, the origin of superscription, also meant three-thousandplex."

Taffy continued, "The bar notation however was easier, so too was the Greek infixes. Asimov's 'teratera-' would now probably be abbreviated to 'dutera-', though it could be confused with, 'deutero-', meaning the quite different 'for the second time'. 'Terateratera-', which not to be confused with the Japanese "Tora! Tora! Tora!", became 'tritera-'.

"The plexing suffixes are easier. The du- prefix in dugoogol therefore would modify the whole googol and so mean the same as the longer googogoogol or twice-googol-to-the-googolth, while the -du- infix in googolduplex modifies the -plex suffix. The prefix bar- could also modify the whole number after it, though it would only increase a googol to a thousand googol, a hundred-threeplex. Dubargoogol however would not be a mere million googol, a hundred-sixplex, but an abbreviation for bargoogobargoogol. four-thousand-googol-to-the-two-thousand-googolth." [168]

The meeting was ended abruptly with Al and the Fen's first verse of "Wunderbar Bottles of Bheer" and the procession to the hall refreshment machine near the restrooms.

from *PAJAS* #45 (April 1989)

REALLY BIG NUMBERS[169]

Taffy open the meeting quoting John Mason Neale's song, "Eternal Monarch, King Most High, Whose blood hath brought redemption nigh, by Whom the death of death was wrought, and conquering grace's battle fought."

Then Al asked, "What did the German answer after ordering sherry and the bartender asks 'Dry?'"

To which she replied, "Ja, Drei!", contuning after those who got it shared with those who hadn't, "Googologist Alistair Cockburn gave us the prefix 'fuga-' meaning the quantity z-to-the-zth-to-the-zth, a stack of three evaluated from the bottom up, rather than the usual top down as in tetration. Steve Howell came up with the complimentary operation calling it 'taking the fugarithm'.

fugax	(ten-to-the-tenth)-to-the-tenth	hundredplex
fugac	(hundred-to-the-hundredth)-to-the-hundredth	twenty-thousandplex
fugam	(thousand-to-the-thousandth)-to-the-thousandth	three-millionplex

"Alistair's six-year old son, Kieran Cockburn, contributed the prefix, gar- backformed from his gargoogol or googol squared.

garfugax	two-hundredplex
garfugac	forty-thousandplex
garfugam	six-millionplex
fugagarx	twenty-thousandplex
fugagarc	four-hundred-millionplex

"Matt Leach contributed the prefix ban- for to-z-pentated-to-the-second, so that bantwo is sixteen, banthree is three-to-the-three-pentated-to-the-thirdth."

Several cries of "Huh?" prompted an explanation.

"Pentation is the operation higher than tetration, which is the one higher than exponentiation. Three-to-the-third would be twenty-seven. Higher than that would be to-the-third-three or three-to-the-twenty-seventh, more than twelveplex. Three-pentated-to-the-second would be three tetrated to that three-to-the-twenty-seventh, higher than twelveplex stacked threes, approximately twelveplex stacked tens!"

3^^3 = 3^(3^3) = 3^27 > 7(10^12)

3^^^3 = 3^^(3^^3) = 3^^(3^27) > 3^^7(10^12) > 10^^3(10^12)

"Alistair Cockburn also contributed the prefix gag- for a one-variable version of the Ackermann function to meaning two-z-plus-one-ated-to-the-zth. Gagone would just be two multiplied by one or two. Gagtwo would be two-to-the-second or four, but gagthree becomes to-the-third-two or sixteen, for four two-pentated-to-the-fourth.[170] It's therefore a much more powerful operation than fuga-.

gagtwo	2*2 = 4
gagthree	2^3 = 8
gagfour	2^^4 = 2^(2^(2^2)) = 2^(2^4) = 2^16 = 65,636
gagfive	2^^^5 = 2^^(2^^(2^^(2^^2))) = 2^^(2^^(2^4)) = 2^^(2^^16)

from *PAJAS* #46 (May 1989)

MATHEMAGIC

Taffy read Genesis 9:7, "God blessed them, saying: 'Be fertile and multiply, fill the Earth and subdue it.'"

Tom posed the riddle, "What do you get if you take half away from five?"

"Two-point-five?"

"No, not what I'm looking for," Tom said pedantically.

"Four-point-five?" came the tentative response/

"No, not that either. Joots, if you take the F and the E away from 'five', you're left with IV, the Roman numeral for four."

With that Tom swiftly took over with the main topic, a continuation from last meeting, "Tom Kreitzberg[171] extrapolated Cockburn's fast-growing operation to 'multiple gag-' with the portmaneau mag-, which Joyce naturally extrapolated further, analogously from million, billion, trillion, to bag-, and trag-, etc.[172] Applying it to the Roman numeral IC for ninety-nine he made magic real"

magic	g(99, 100, 99, 2)
bagic	g(2(99), 2(99) - 1, 2(99), 2) = g(198, 199, 1982, 2)
tragic	g(3(99), 3(99) - 1, 3(99), 2) = g(297, 298, 297, 2)

"Applying it as a modifier a magicgoogle would be very much higher than a google."

Taffy continued, "Hugo Steinhaus notation using numbers within triangles, squares and circles. The system can be rather easily defined by the statements that a number in a triangle is simply that number-to-itself. A number in a square is the same as that number in that number of triangles and a number in a circle is that number in that number of squares

"So you're saying that two-in-a-circle would mean two-squared or four then?" Sam asked.

"That's right! Keep going. What would two-in-a-square mean?"

"That would be two-in-two-triangles or four-in-a-triangle or four-to-the-fourth."

"And that would be ..."

"Let's see, four times four is sixteen, sixteen times four is sixty-four and four times sixty-four is two hundred fifty-six."

"Right. You're becoming a mathematician, Sam. Can you put that on the board for us?"

It was the first time he'd been confident enough to walk to the front of a class in thirty years. He marched forward with all eyes on him and liked it.

"That'd be two-in-a-circle which means two-in-two-squares which means, good grief!, two hundred fifty-six in two hundred fifty-six triangles!"

Sam was far from the only one taken by surprise by the awesome power of Steinhaus' circle. Tony had to pause before the stunned Sam, "Yes, that's what Steinhaus called a mega. Each triangle doubles the number of twos in the power tower."

"Oh, like my folding ruler, I can understand that."

Sam whipped out his visual aid and unfolded it once, then again, again and again doubling it every time, four inches became eight inches, then one and a quarter feet, two and a half feet, to five feet.

"Two within a circle Steinhaus called by the special name 'mega' and ten in a circle a 'megaton', pronounced meg-uh-tahn, not meg-uh-tuhn. Leo Moser generalized the operation to any polygon with a pentagon and it naturally became known as the Steinhaus-Moser operation.[173] The name moser was given to the number represented by a two within a circle-like megagon, a million-sided polygon. The polygonal notation was further simplified by Susan Stepley to a square bracket notation.[174]

Steinhaus	Steinhaus-Moser	Stepley	Joyce
⓪	sm(0, 1, 5)	0[5]	g(5, 0, 0)
①	sm(1, 1, 5)	1[5]	g(5, 1, 1)
mega = ②	sm(2, 1, 5)	2[5]	g(5, 2, 2)
③	sm(3, 1, 5)	3[5]	g(5, 3, 3)
megaton = ⑩	sm(10, 1, 5)	10[5]	g(5, 10, 10)
moser	sm(2, 1, 1'000'000)	2[1'000'000]	g(1'000'000, 2, 2)

"The mega and megaton naturally got extrapolated by Joyce to bega and trega, giga and tera, begaton and tregaton, gigaton and teraton, and moser to boser, troser, etc.

bega	sm(sm(2, 1, 5), 1, 5)	2[2[5]]	g(2, 5, 2, 2)
trega	sm(sm(sm(2, 1, 5), 1, 5), 1, 5)	2[2[25]]]	g(3, 5, 2, 2)
giga	sm(2000, 1, 5)	[2000[5]	g(5, 2, 2000)
tera	sm(2000000, 1, 5)	2000000[5]	g(5, 2, 2000)
gigaton	sm(10000, 1, 5)	10000[5]	g(5, 2, 10000)
teraton	sm(10000000, 1, 5)	10000000[5]	g(5, 2, 10000000)
boser	sm(moser, 1, 1000000)	moser[1000000]	g(2, 1000000, 2, 2)
troser	sm(boser, 1, 1000000)	boser[1000000]	g(3, 1000000, 2, 2)

"Is it possible to get any higher?!" said Cal who looked like he'd been hit by a fullback.

"Why soitainly," she said in a playful Curly-like[176] Joisey accent and wave. "To begin again like Finnegan with Roman numerals, not as suffixes, but now as prefixes, Joyce indicated the operation nesting count."

icgoogol	g(99, 1, 1, 2, 50, 100)
magicgoogol	g(g(99, 100, 99, 2), 1, 1, 2, 50, 100)

"That's icchy'!" someone remarked and was unanimously ignored.

Taffy concluded the meeting with "These very, very, very small numbers once again confirm the Frivolous Theorem: 'Almost all natural numbers are very, very, very large.'"

from *PAJAS* #47 (June 1989)

GOOGOLISMS[176]

Raz began with "The utterance of one who hears what God says and knows, what the Most High knows,...'I see Him, though not now. I behold Him, though not near.'" from Numbers 24:16-17. Then she continued, "From what number can you take four and leave ten and take another four and leave six?"

Not to be tricked after last month's question, Cal got the answer this time, "Sixteen."

"Don't you mean fourteen?"

"No, S, I, X and E from the word sixteen leaves TEN, while T, both Es and N from it leaves the SIX."

This particularly memorable meeting was the one at which Professor Joyce returned, if only via speakerphone. He was, he explained, sorry that he could not be with us in person. Since the loss of his eyesight, he had been working on a Braille version of Dominissimo. Many of the neos had been initiated by learning of Joyce's most well-known neologism, googoc. As a recognized pioneer in the art and science of number names above a googol and their mathematical representation, he could hardly ignore their desire to learn more.

"Conway's numbering system, with which I hear you're familiar, is a modified version of Nicolas Chuquet's *échelle courte* or short scale, used by Americans, Brazilians, Greeks, Italians, Puerto Ricans, Russians, Turks. It names a googol as ten duotrigintillion. In the alternative system of Pelletier's *échelle longue* or long scale, it is called ten sedecilliard by nearly everyone else. The French reverted back to the original Chuquet system from Pelletier's in 1948 and declared any other system illegal in 1961. The British recently officially recognized the number as a billion rather than a milliard. Internationally milliard/billion or billion/milliard are often used. Knuth called it a myriad undecyllion. In what I hear you call long Knuth, it's a million sextilliard.

"I did not know at the time that the name googol was coined by a boy and took it as having deep significance. I suppose that's how I think about every number. Who knows prehaps the Most High was guiding us both."

"The term googol for ten-to-the-hundredth was nevertheless certainly popularized by George Gamow in his *One, Two, Three ... Infinity.* Later Carl Sagan, most famous for his indeterminate 'billions of billions', used it in his 'Cosmos' series.[177]

"Congressman George Herman Mahon of Texas former chairman of the House Appropriations Committee once warned that the national debt might eventually reach a googol dollars. At a five percent per year growth rate, for example, it would take about forty-five centuries, indicating an incredibly great confidence that the United States will last much longer than any other nation or civilization ever has or an equally incredible devaluation of the dollar.

"A googol is an interesting number even if not in dollars. It is much easier to factor than its neighbors, a googol minus-or-plus one. A googol is just a hundred twos times a hundred fives. A downtown neighbor googolmini would be a repunit multiple with ninety-nine nines and and the uptown neighbor googolpli would be a numberdrome with ninety-eight internal zeros. 'Gooprol' is the nickname for the prime just higher than a googol, specifically two hundred sixty-seven above, presumably to distinquish it from 'goopol', the numberdrome just above googol. Logically then the prime below googol would be called 'proogol' and the numberdrome just below googol 'poogol'.

"Googology moved into the computer age in 1951 when Miller and Wheeler discovered the first prime using a computer, the EDSAC1, a number greater than a hundred-twenty-sevenplex.[178]

"Almost immediately upon learning of the googol I noticed that it could also be expressed as a function of fifty, specifically twice-fifty-to-the-fiftieth, as apparently indicated by the non-g letters in the word. This discovery that the prefix googo- indicated an operation on a Roman numeral was the turning point for all of my own googolisms.

"Many other Roman numerals can therefore also be operated on by googo-, proper ones such as googoc or two-hundred-to-the-hundredth, or less proper ones such as googocic or three-hundred-ninety-eight-to-the-one-hundred-ninety-nineth.[179]

"Although I have been partial to using the Mayan word for three, ox, as a substitution for the Roman numeral III, I admit that your substitution of the German U-umlaut for the troublesome double I may be a better solution."

"My nephew, André Joyce, II, I hear, gets junk mail addressed to 'Mr. Ii', as if he were a character out of Bradbury's *The Martian Chronicles.*"

Three cheers broke out as Taffy proceeded to write out the newly accepted number name on the board.

googolüj g(2, 53, 106))

"We have the encouraging 'Go good!', gogood, as another name for to-the-second-thousand or three-thousandplex."

"And, for example, googoom to-the-second-two-thousand or googommix four-thousand-eighteen-to-the-two-thousand-nineth."

gogood	g(3, 2, 1000) = g(2, 3000, 10)
googoom	g(3, 2, 2000) > g(2, 6602, 10)
googommix	g(2, 2009, 4018) > g(2, 7420, 10)

"Many more Roman numerals can be made pronounceable using what I call the Principle of Equivalency which equates orthography and phonetics, so that the -ex, -el, and -em in the -plex, -plel amd -plem suffixes are taken as equivalent to lone -x, -l and -m.

"So combining googo- with the Roman sixty, LX, could be pronounced as googolex rather than googolx, one-hundred-twenty-to-the-sixieth and as googomex two-thousand-twenty-to-the-one-thousand-tenth."[180]

googolex	g(2, 60, 120) > 5g(2, 124 , 10)
googomex	g(2, 1010, 2020) > 2g(2, 3338, 10)

"What I called the o-count in googo- is two-one. Other possible o-counts would be one-one for the prefix gogo-, one-two for the prefix gogoo- and two-two for the prefix googoo-." [181]

gogoz	g(2, z, z) = g(3, 2, z)
gogooz	g(2, 2z, z)
googoz	g(2, z, 2z)
googooz	g(2, 2z, 2z) = g(3, 2, 2z)

Interrupted by snickers, Joyce admitting that it seemed a rather childish way to form number names, but in the Scriptures we are told to become like children.

"Then I noticed that this o-count could be extrapolated higher, to three and higher by the mnemonic of the English digits' vowel sounds. Thus not only would the o-infix mean one and the oo-infix mean two, both just as they sound, but an ee-infix could stand for three, an or-infix for four, an ie-infix for five, an i-infix for six, an e-infix for seven, and an ei-infix for eight.

geegeel	g(2, 3(50), 3(100)) = g(2, 150, 300) > 3g(2, 371, 10)
gorgorl	g(2, 4(50), 4(100)) = g(2, 200, 400) > 2g(2, 520, 10)
giegiel	g(2, 5(50), 5(100)) = g(2, 250, 500) > 5g(2, 674, 10)
gigil	g(2, 6(50), 6(100)) = g(2, 300, 600) > 2g(2, 833, 10)
gegel	g(2, 7(50), 7(100)) = g(2, 350, 700) > 6g(2, 995, 10)
geigeil	g(2, 8(50), 8(100)) = g(2, 400, 800) > g(2, 1161, 10)

"I could form arithmonyms with these rather than only with Roman numerals for numbers with non-integral logarithms. higher than a hundred, such as sixty-six-to-the-fifty-fifth, could be abbreviated using these conventions. Geegixi, for example, is easily understood as three-times-eleven-to-the-six-times-eleventh or thirty-three-to-the-sixty-sixth.[182]

"The g-count could also be extrapolated higher, to three by noting that the g-g- in googol seems to indicate the second operation in the Ackermann generalized exponential notation. Oui, I know, in Knuth's chained arrow notation it is the first and in Bowers's array notation it is the third. The history of googology, like history, His story, has been very interesting.

"So googgol pronounced goog-gol with three g's would be to-the-fifieth-hundred?!"

"Oui, hundred is still the number operated on as indicated by the oo-infix for two and the l-suffix for fifty and the number of operations is fifty as indicated by the suffix o-infix. Since a hundred is only ten squared, it is very nearly the same height as ten. Googgol would be only a little higher than to-the-fifieth-ten.[183]

googgol	g(3, 50, 100) > g(3, 50, 10)
googgool	g(3, 100, 100) = g(4, 2, 100)
googgood	g(3, 1000, 1000) = g(4, 2, 1000)

"When however I saw bazillion, not as merely a zazzification of nineplex, but as ten z-ated to the nineth, I saw a way to get higher still. I defined a trazillion as ten z-minus-one-ated-to-the-twelfth, a previously indeterminate number related to trillion, and thus got the handy az-infix. Gazoogol would mean ten-z-minus-one-ated-to-the-hundredth, z being three for tetration, four for pentation, etc. By again substituting Roman numerals for the variable z nearly any operation number could be indicated.

gavoogol	g(5, 50, 100)	gacoogol	g(100, 50, 100)
gaxoogol	g(10, 50, 100)	gadoogol	g(500, 50, 100)
galoogol	g(50, 10, 100)	gamoogol	g(1000, 50, 100)

"By analogy with left superscription I later came up with the spoonerism golgoo as an alternative for googgol or to-the-fiftieth-hundred and plexgoogol for to-the-tenth-googol and all their many variations. So then I had a possible g-count of four for golggoo or even five for golggoog."

golgoo	g(3, 50, 100)
golggoo	g(4, 50, 100)
golggoog	g(5, 50, 100)

Although the phone call ended, even newbies understood both Prof. Joyce and his more generalized exponential notation better, though some more than others.

from *PAJAS* #48 (July 1989)

SELF-REFERENTIAL NUMBERS

Taffy opened the meeting with "I AM the Alpha and the Omega, the First and the Last, the Beginning and the End." from Revelation 22:13.

Sam posed the month's riddle, "What's the difference between a healthy rabbit and this riddle?"

Taffy answered, "One's a fit bunny and the other's just a bit funny." and then continued. "Some ways of getting high involve not so much manipulating the numbers themselves, but their number names or arithmonyms. These, of course, depend upon the language in which they are named.[184] The number of letters in the names of the integers in general get higher as the numbers themselves do eventually, but with irregularity.

English	one	two	three	four	five	six	seven	eight	nine	ten	eleven	twelve
length	3	3	5	4	4	3	5	5	4	3	6	6
sum	3	6	11	15	19	22	27	32	36	39	45	51

"A. J. Aronson generated his unusual sequence by counting the letters in the never-ending sentence, 'T is the first, fourth, eleventh, ... letter in this sentence.' T is the first letter in the sentence, then the fourth one that begins the 'the', then the eleventh, the last in 'first'."

"So the next T would be the next to the last letter in 'fourth', uh, the sixteenth."

'Sixteenth generates the next two numbers, twenty-four and twenty-nine. 'Twenty-fourth' generates the next three numbers, thirty-three, thirty-five and thirty-nine, and so on.

T-is 1 4 11 16 24 29 33 35 39 45 47 51 56 58 62 64 69 73 78 80 84 89 94 99 104 111

"So that guarantees that the number of T's is getting higher faster than with just the -th suffix. Not all such sentences get high as fast."

"In Latin we'd have 'T est', the T-est sequence."

"Or starting with B, the B-est sequence!" [185]

Returning to English, she continued, "The H-is sequence, which we pronounce with the hyphen silent as His, also is self-perpetuating because of that -th suffix.

H-is 1 5 16 25 36 38 47 49 57 59 71 82 94 105 121 123 140

"Other good candidates would be those that have letters in 'is the first letter of this sentence', C, E, F, I, L, N, R and S." [186]

"S-is numbers!" the Delta Schwa sisters chortled.

"The first and third letters are S, and there's another S in 'first'. the tenth.

"That'd give us the next numbers, twenty-three and twenty-four."

"And more S's with fifty-one and fifty-two!"

S-is 1 3 10 23 24 51 52 79 80 113 114 147 148 159 174 175

"We can however generate other sequences like 'A-isn't', 'A-is-not' and even 'A-ain't', though they would have most of the natural numbers included."

"Oh, rather like a pathocircle."

"A pathowhat?"

"A pathocircle, a rather impractical figure very much like a megagon, a million-sided polygon, a circle with one point missing."

"Something like that. The most well-known such numbers are the S-ain'ts.[187] The name comes from the fact that S ain't in the generating sentence after the first letter until the nineth letter in the word 'second'. So one and nine are the first two non-S-ain'ts or 'S-inner' numbers.[188] As the values get higher we see interestingly many more S-ain'ts than S-inners.

S-ain'ts	2	3	4	5	6	7	8	10	11	12	13	14	15	16	17	18	19	20
S-inners	1	9	31	36	98	107	156	164	210	221	266	312	358	365	405	415	460	467

"There are many other Aronson-like sequence-generating sentences that count the letters in words, 'The first word of this sentence ... has an odd number of letters." [189]

"So we need to know the letter count for the ordinals."

odd length	1	4	7	11	14	16	19	21	22	23	25	26	27	28	31	32	33	35	36	37	42	44	48	49
even length	2	3	4	5	6	8	9	10	12	13	16	17	20	22	23	26	29	34	38	39	40	41	43	45

"We could use this as a generating sentence, since 'word', 'of', 'this' and 'sentence' all have an even count and can keep generating odd counts. Seven out of the first eleven are prime."

The-first	1	2	7	13	14	19	25	31	37	38	43	44	49	50	55	56	61	67	68	73	74	79	85	91

"You could form many other generating sentences than The-first, with other generating sentences like 'The second word, the third word, ... in this sentence has an even number of letters.'

The-second	2	3	6	9	11	12	15	18	20	21	24	26	27	29	30	33	36	39	41	42	44	45	48	51

"You can start with a seed number and generate a sequence counting two, three, four, six, seven and eight letters in the words in '... is the number of letters in the ... word of this sentence, ...'"[190]

Two-is	2	5	7	10	15	17	21	24	26	28	32	35	37	39	43	46	48	50	54	57	59	61	65
Three-is	3	8	14	19	25	30	36	41	47	52	58	63	69	74	80	85	91	96	102	107	113	118
Four-is	1	10	13	16	19	22	25	28	31	34	37	41	43	46	49	52	55	58	62	65	68	71

"Ah, the last one begins with one because four is a self-referential number, one that counts its own letters!"

"And 'word' is also a four-letter word."

"'Also' also."

from *PAJAS* #49 (August 1989)

IF WISHES WERE NUMBERS

The new acting president, Tom, began by quoting Francis, "All-powerfull, Most Holy, Most High, supreme God: all Good, supreme Good, totally Good, You Who alone are Good, may we give You all praise," then asked, "What's a nine-letter word for an especially strange stranger?"

"A fiveigner," came the answer.

He then continued the previous month's counting of letters, "We can also generate sequences by counting letters and digits in other ways, without a generating sentence. In Dominissimo there are ten suits of numbers, usually named with the common suffix -ish, zero-ish through nine-ish, though they do have nicknames as well. The zero-ish suit, for example, is also called blanks, because one side is blank, while the one-ish one is bullets, because one side looks like a bullet hole. The suits with one through five are left-handed or sinister, being on the left-hand side of a standard Qwerty keyboard and those with six through nine and zero are right-handed.[191]

sinister	1	2	3	4	5	11	12	13	14	15	21	22	23	24	25	31	32
zero-ish	0	10	20	30	40	50	60	70	80	90	100	101	102	103	104	105	106
one-ish	1	10	11	12	13	14	15	16	17	18	19	21	31	41	51	61	71
two-ish	2	12	20	21	22	23	24	25	26	27	28	29	32	42	52	62	72
three-ish	3	13	23	30	31	32	33	34	35	36	37	38	39	43	53	63	73
four-ish	4	14	24	34	40	41	42	43	44	45	46	47	48	49	54	64	74
five-ish	5	15	25	35	45	50	51	52	53	54	55	56	57	58	59	65	75
six-ish	6	16	26	36	46	56	60	61	62	63	64	65	66	67	68	69	76
seven-ish	7	17	27	37	47	57	67	70	71	72	73	74	75	76	77	78	79
eight-ish	8	18	28	38	48	58	68	78	80	81	82	83	84	85	86	87	88
nine-ish	9	19	29	39	49	59	69	79	89	90	91	92	93	94	95	96	97

"If we classified numbers by their English number name letters 'one' would be included in the o-ish, n-ish and e-ish numbers, 'two' would also be an o-ish number, and also a t-ish and a w-ish."[192]

o-ishes	0	1	2	4	14	21	22	24	31	32	34	40	41	42	43	44	45	46	47	48	49
n-ishes	1	7	9	10	11	13	14	15	16	17	18	19	20	21	22	23	24	25	26	27	28
e-ishes	0	1	3	5	7	8	9	10	11	12	13	14	15	16	17	18	19	20	21	23	25
t-ishes	2	3	8	10	12	13	14	15	16	17	18	19	20	21	22	23	24	25	26	27	28
w-ishes	2	12	20	21	22	23	24	25	26	27	28	29	32	42	52	62	72	82	92	102	112

"It looks to me like w-ishes are the rarest and e-ishes the commonest, as expected from what we know about letter frequencies."

"The w-ishes are exactly the same as the two-ish!"

"Not exactly. When we get as high as twelveplex, they're different."

"Oh, yeah, twelveplex is a w-ish but not two-ish.

"Three would start both the h-ish and r-ish numbers." [193]

h-ish	3	8	13	18	23	28	30	31	32	33	34	35	36	37	38	39	43	48	53	58	63	68	73	78
r-ish	3	4	13	14	23	24	30	31	32	33	34	35	36	37	38	39	40	41	42	43	44	45	46	47

"Ah, and then with four and five we go to the f-ishes."

"And don't forget the i-ishes, u-ishes and v-ishes." [194]

i-ishes	5	6	8	9	15	16	18	19	25	26	28	29	35	36	38	39	45	46
f-ishes	4	5	14	15	24	25	34	35	40	41	42	43	44	45	46	47	48	49
u-ishes	4	14	24	34	44	54	64	74	84	94	100	101	102	103	104	105	106	107
v-ishes	5	7	12	15	17	25	27	35	37	45	47	55	57	65	67	70	71	72

"Six and seven give us the s-ishes and the x-ishes.[195]
"But the x-ishes are exactly like the six-ishes."
"Only until you get to nineplex."

s-ishes	6	7	16	17	26	27	36	37	46	47	56	57	60	61	62	63	64	65	66	67

"Eight gives us one new letter, G, but it looks exactly like the eight-ish numbers! Eighteenplex seems to be the first g-ish number without an eight."
"Eleven and twelve give us a new letter, L."
"So could you count one thousand one hundred and one thousand two hundred as l-ish if you called them eleven hundred and twelve hundred?"
"Why not? Remember no matter how you represent a number it's the same number. We'd get different difference values though when we get that high. "[196]

L-ishes	11	12	111	112	211	212	311	312	411	412	511	512	611	612	711	712	811
difference	1	99	1	99	1	99	1	99	1	99	1	99	1	99	1	99	

"So so far we have most of the alphabet covered, E, F, G, H, I, L, N, O, R, S, T, U, V, W and X."
"Y?"
"Why?"
"No, Y, the tail letter in twenty, thirty and so on."
"Ah, its difference pattern would change when it gets as high as ninety-nine."

y-ish	20	21	22	23	24	25	26	27	28	29	30	31	32	33	34	35	36

"Zero would seem to be the only z-ish number below a zillion."
"Not if you include representations in other bases. More than half the numbers in binary would contain at least one Z."
"All of them would be, if you called one 'zeroplex'!"
"And two 'zeroplex-zero'!"
"'Hundred' would give us d-ishes, 'million' would give us m-ishes and billion b-ishes."[197]
"Not if you count myriads."
"Or if you count nineplex, The p-ishes would be the same as the x-ishes higher than the sixplexes."
"Pish! Personally I prefer the Ir-ish numbers, those with an IR," Pat commented, opening up a whole new set of possibilities with bigrams and polygrams and polydigits."

Ir-ish	13	30	31	32	33	34	35	36	37	38	39	130	131	132	133	134	135

"Ten-ish anyone?" was Al's meeting-closing response.[198]

from *PAJAS* #50 (September 1989)

WEIGH-IN

Tom opened the meeting by quoting Psalm 46:5, "There is a stream whose runlets gladden the city of God, the holy dwelling of the Most High.", then he asked, "How many relativity theorists does it take to change a light bulb?"

Lana answered, "Two, one to hold the light bulb and another to rotate the universe."

Then she continued, "Numbers can be classified by the weight of their digits. Those with just one of the middleweight digits, four, five or six, are called middling numbers.[199] Those with two of the three 'middlinger' [200] and those with all three 'middlingest'.[201] Five is the one and only prime among them."

middling	4	5	6	14	15	16	24	25	26	34	35	36	40
middlinger	45	46	54	56	64	65	440	441	442	443	447	448	449
middlingest	456	465	546	564	645	654	1,456	1,465	1,546	1,564	2,645	2,654	3,456

"Cheap numbers are those with the lightweight digits, one, two or three, but not two of them.[202] Cheaper numbers are those with two of the three [203] and the cheapest those with all three."[204]

cheap	1	2	3	10	14	15	16	17	18	19	20	24	25	26	27
cheaper	12	13	21	23	31	32	120	121	122	124	125	126	127	128	129
cheapest	123	132	213	231	312	321	1,123	1,132	1,213	1,231	1,312	1,321	2,123	2,132	2,213

"Dear numbers are those with heavyweight digits, seven, eight or nine, but not two of them.[205] Dearer numbers are those with two of the three[206] and the dearest those with all three." [207]

dear	7	8	9	17	18	19	27	28	29	37	38	39	47	48
dearer	78	79	87	89	97	98	178	179	187	189	197	198	278	279
dearest	789	798	879	897	978	987	1,789	1,798	1,879	1,897	1,978	1,987	2,789	2,798

"The uglier[208] and ugliest[209] numbers are similarly extrapolated from the ugly numbers, those with prime divisors less than or equal to five. The uglier numbers are those divisible by any two, that is, six, ten or fifteen, and the ugliest those divisible by all three, that is, thirty, so naturally they have no primes at all."

"The ugliest aren't all that ugly once you get used to them. Thirty-sixty-ninety, for example, is one of my favorite right triangles."

"In Babylonian they're both a little odd and do have primes."

uglier	6	10	12	15	18	20	24	30	36	45	48	50	54	60	70	72	75
Babylonian	6	10	12	15	18	20	24	30	36	45	48	50	54	1	110	112	115
ugliest	30	60	90	120	150	180	210	240	270	300	330	360	390	420	450	480	510
Babylonian	30	1	130	2	230	3	330	4	430	5	530	6	630	7	730	8	830

"Bipolar numbers are those with only the two extreme 'polar' digits, in decimal, zero and nine." [210]

bipolar	90	900	909	990	9,000	9,009	9,090	9,099	9,900	9,909	9,990	90,000	90,009

"Then there are numbers or number names that are defined as not having particular combination of digits or letters, polyphobic numbers. Bantu,[211] for example, bans the digit two. Caliban[212] bans the letters A, C, I and L. Suburban[213] bans the letters B, S, U and R. The Taliban bans the letters A, I, L and T, while the turban ban the letters R, T and U and urban ban just R and U."[214]

Bantu	0	1	3	4	5	6	7	8	9	10	11	13	14	15	16	17	18	19	30
Caliban	1	2	3	4	7	10	14	17	20	21	22	23	24	27	40	41	42	43	44
suburban	1	2	5	8	9	10	11	12	15	18	19	20	21	22	25	28	29	50	51
Taliban	0	1	4	7	100	101	104	107	401	404	700	701	704	707					
urban	1	2	5	6	7	8	9	10	11	12	15	16	17	18	19	20	21	22	25

turban	1	5	6	7	9	11	1,000,000	1,000,001	1,000,005	1,000,006	1,000,007	1,000,009

"The complimentary numbers to the Caliban numbers are called the Ariel[215] numbers, rival characters in 'The Tempest' by William Shakespeare."

Ariel	5	6	8	9	11	12	13	15	16	18	19	25	26	28	29	30	31	32

"I see that the Taliban numbers do not go on endlessly! All the thousands have a T and all the -illions have an I. Thank the Most High!"

"Similar numbers are formed with the -less suffix. Flawless[216] numbers do not have the letters F, L, A or W. Godless[217] numbers are without D, G or O, harmless numbers without A, H, M or R. The useless[218] numbers are without E, S or U. The worthless[219] numbers are without H, O, R, T or W. To get higher than eleven we have get jootsy and count fifty-five thousand, five hundred fifty-five as five-fives and so do not use the usual commas. A hundred thousand can be added if read as fiveplex.

flawless	1	3	6	7	8	9	10	13	16	17	18	19	30	31	33	36	37	38	60	61	63	66
godless	3	5	6	7	9	10	11	12	13	15	16	17	19	20	23	25	26	27	29	30	33	35
harmless	1	2	5	6	7	8	9	10	11	12	15	16	17	18	19	20	29	50	51	52	55	56

useless	2	40	42	50	90	92	200	240	250	290	292	2,000,000
worthless	5	6	7	9	11	55555	66666	77777	99999	555555	666666	777777

"The non-flawless numbers are called the flawed[220], the non-godless godly[221], the not harmless harmed[222], the non-useless used[223], and the non-worthless worthy.[224] We thus have the new properties of flawlessness, godlessness, harmlessness, uselessness and worthlessness.[225] The mostly numbers are the subset of the above numbers with less than half, but not zero, of the banned letters.[226] Those with zero uselessness are 'useful numbers' and those with zero harmlessness 'harmful'."[227]

That triggered a rendition of "Banned from Arggho" [228] by Al and the Fen.

from *PAJAS* #51 (October 1989)

ANANUMS

Tom opened with "O Most High, when I begin to fear, in You will I trust," from Psalm 56:4.

After this he asked, "If you had nine millipedes and I gave you another, what would you have."

"A bigger, better bug collection?" someone hazarded.

"A centipede! Get it? Ten millipedes 'equals' one centipede."

With that clarified, Tom continued, "Ananums are a subsequence of number pairs that share a common property even when digitally reversed, excluding numberdromes. The principle can be applied to many other sequences than the primes' emirps[229], for example, the gnilddim[230], Leira[231], Nabilac[232], paehc[233], raed[234], Utnab numbers.[235] The ananym of evil numbers would be 'live' numbers and those that aren't would be 'dead'." [236]

emirps	13	17	31	37	71	73	79	97	107	113	149	157	167	179	199	311	337	347
gnilddim	45	46	54	56	64	65	445	446	454	455	456	464	465	466	544	545	456	544
Leira	12	13	15	16	18	19	21	23	25	26	28	29	31	32	34	35	36	37
Nabilac	14	17	24	27	41	42	47	71	72	74	102	103	104	107	114	117	124	127
paehc	12	13	21	23	31	32	112	113	122	123	132	133	211	213	221	223	231	233
raed	78	79	87	89	97	98	778	779	788	789	798	799	877	879	887	889	977	978
Utnab	10	13	14	15	16	17	18	19	30	31	34	35	36	37	38	39	40	41
live	15	17	27	29	30	34	36	43	45	54	57	58	60	63	71	72	85	90
dead	0	1	2	3	4	5	6	7	8	9	10	11	12	13	14	16	18	19

"I notice that bipolar numbers with only trailing zeroes, and no interior zeroes, cannot have ananums because then they would no longer be bipolar and some of them with interior zeroes would become numberdromes." [237]

ralopib	9,099	9,909	90,099	90,990	99,009	99,090	99,909	900,099	900,909	900,999

"Right. The so-called cubans[238] however do. They get their name from 'cube', not the -ban suffix, being the difference between cubes and also known as centered hexagonals. They have early ananums, nineteen and ninety-one. The first three cubans are also prime, but then, as is typical, become less common.[239]

cubans	7	19	37	61	91	127	169	217	331	397	469	547	631	721	817	919
cuban primes	7	19	37	127	217	331	397	547	631	919	1,657	1,801	1,951	2,269	2,437	2,791

"Fortunate primes[240] are primes whose sum with a primorial is also prime. Because there are many duplicates for each added primorial, these may be deleted and the results sorted into a new sequence. Since both seventeen and seventy-one are fortunate primes, other 'etanutrof' primes may exist.

primorials	2	6	30	210	2,310	30,030	510,510	9,699,690	223,092,870
fortunate primes	3	5	7	13	17	19	23	37	47

"Yet another kind of prime, this time named for Jerome Solinas, is one more than the difference of powers of two. They also have reversed pairs, thirteen and thirty-one, seventeen and seventy-one, etc., so there exist Sanilos primes as well. Although both thirty-five and fifty-three are both Solinases, only the later is prime.[241]

Solinases	0	2	4	8	16	32	64	128	256	512	1,024	2,048	4,096
-8+1				1	9	25	53	117	245	501	1,013	2,041	3,999
-4+1			1	5	13	29	61	125	253	509	1,021	2,045	4,093
-2+1		1	3	7	15	31	63	127	255	511	1,023	2,047	4,095
0+1	1	3	5	9	17	33	65	129	257	513	1,025	2,049	4,097
+2+1	3	5	7	11	19	35	67	131	259	515	1,027	2,051	4,099
+4+1	5	7	9	13	21	37	69	133	261	517	1,029	2,053	4,101
+8+1	9	11	13	17	25	41	73	119	247	503	1,015	2,039	4,115

Solinas primes	2	3	5	7	11	13	17	19
Sanilos primes	13	17	31	37	71	73	79	97

\

"Truncatable primes are those that can be beheaded[242] and/or curtained[243] and still retain the property of being prime. Numbers both left truncatable or multi-headed and right truncatable or multi-tailed are called multi-sided.[244] Some are also 'elbatacnurt' primes, like thirty-seven and seventy-three. Permutable primes[245] are those that are still prime under any re-arrangement of their digits. Joyce nicknamed the beheadable primes 'rimes' and the curtailable primes 'prims' and two-sided primes 'rims'.

beheadable primes	13	17	23	37	43	47	53	67	73	83	97
curtailable primes	23	29	31	37	53	59	71	73	79	233	239
two-sided primes	23	37	53	73	113	131	137	311	313	317	373
permutable primes	13	17	31	37	71	73	113	131	311	337	373

from *PAJAS* #52 (November 1989)

ODDER AND ODDER

Lana opened with a quote from "Come Holy Ghost", Edward Caswell's translation of Rhabanus Maurus' "Veni, Creator Spiritus", "O Comforter, to Thee we cry, Thou heavenly Gift of God Most High, Thou Font of life and fire of love, and sweet anointing from above."

Tom then asked, "How can you add one and five and get four?"

"In Roman numerals, I plus V equals IV."

"That <u>is</u> odd!"

"What's odder is that although we all know what odd numbers are, some may not know that the word 'odder' [246] has a specific meaning in number theory as describing numbers with an odd number of prime factors. This would include the primes, which only have one. Since two is a prime, many odders are even and many 'eveners' [247] are odd, though none of them are prime."

"As Loren Eiseley aptly said in *The Unexpected Universe*," Ann commented, 'Things get odder on this planet, not less so.'"

odder	2	3	5	7	8	11	12	13	17	18	19	20	23	27	28	29	30	31	32	37	38	39	41

evener	4	6	9	10	14	15	16	21	22	25	26	33	34	35	36	49	54	55	56	57

"If we did the same things to the odds as we did to the primes last month, we'd include numbers like eleven, a prime that's both beheadable and curtailable, that remains odd even if it doesn't remain prime?", Cal said.

"Yes, what Joyce called 'urtailable'." [248]

urtailables	11	13	15	17	19	22	24	26	28	31	33	35	37	39	40	42	44	46	48	51	53
urtailable odds	11	13	15	17	19	31	33	35	37	39	51	53	55	57	59	71	73	75	77	79	91
urtailable evens	20	22	24	26	28	40	42	44	46	48	60	62	64	66	68	80	82	84	86	88	100

"Non-curtailable numbers like twenty-one are odd but curtail into an even, while non-curtailable evens like ten curtail into an odd." [249]

non-curtailable	10	12	14	16	18	21	23	25	27	29	30	32	34	36	38	41
non-curtailable odd	21	23	25	27	29	41	43	45	47	49	61	63	65	67	69	81
non-curtailable even	10	12	14	16	18	30	32	34	36	38	50	52	54	56	58	70

"And some these odd non-curtailables can be permuted into evens?" [250]

"Yes, those higher than the trivial double-digits would have to have an even interior to be both non-curtailable and non-permutable."

...	201	203	205	207	209	221	223	225	227	229	241	243	245	247	249	261	263	265

"Higher than these are the oddest of all, numbers that remain odd no matter how many times they're curtailed or permuted, the panodds." [251]

panodd	1	3	5	7	9	11	13	15	17	19	31	33	35	37	39	51	53	54	55	57	59	71	73	75	77	79	91	93	95

"The Steven Todd transformation[252] turns evens into odds and odds into even by doubling the odds and repeatedly halving the evens.

StevenTodd	2	1	6	1	10	3	14	1	18	5	22	3	16	7	30	1	34	9	38	5	42	11	46	3	50	13	54	7

"This transformation can get higher faster by applying it not just to numbers as a whole but digitally.

S.Todd	2	1	6	1	10	3	14	1	18	20	22	21	26	21	210	23	214	21	218	10	12	11	16	11	115

"You may also have heard of the oddly odd and evenly even[253] numbers, the products of two odds and two evens respectively."

"Ah, so the evenly evens would all be multiples of four."

"Yes, if you multiplied an odd and an even though you'd always get an even, but they would not be evenly even, but oddly even[254]. These three classes of composites seem to be just about even, though they never seem to overcome their starting gate gait.'"

oddly odd	9	15	21	25	27	33	35	39	45	49	51	55	57	63	65	69	75	77	81
oddly even	6	10	14	18	22	26	30	34	38	42	46	50	54	58	62	66	70	74	78
evenly even	4	8	12	16	20	24	28	32	36	40	44	48	52	56	60	64	68	72	76

"Still others are called minimal primes[255] if any part of them, any 'frag', is not also a prime. A minimal odd[256] would have all even digits except for an odd tail and a minimal even all odd digits except for an even tail."

"Ah, because if the tail were not odd or even the whole number wouldn't be either."

minimal primes	2	3	5	7	11	19	41	61	89	101	109	149	181	401	409	449	491	499	601	691
minimal odds	1	3	5	21	23	25	27	29	41	43	45	47	49	61	63	65	67	69	81	83
minimal evens	2	4	6	8	10	12	14	16	18	30	32	34	36	38	50	52	54	56	58	70

"I see that the minimal evens do not differ from the ordinary evens until the tenth term, when they jump to thirty."

"Yet another kind are the calculator primes[257] are those that are still prime when rotated or reflected when represented in a seven-segment display as on a calculator or digital clock. Two is the reflection of five. Zero, one and eight are their own reflections. [258] They include one thousand one hundred eight-one and one thousand eight-hundred eleven which would be lardehid twin primes, but not eleven or a hundred one that are numberdromic primes." [259]

calculator primes	2	5	11	101	181	1,181	1,811	18,181	108,881	110,881	118,081	120,121
mirror numberdromes	1	8	11	88	101	111	181	808	818	888	1,001	1,111
numberdromic primes	11	101	131	151	181	191	313	353	373	383	727	797

from *PAJAS* #53 (December 1989)

NUMBERDROMES

Tom continued to lead the meetings, starting this one with a Bahai prayer, "In the Name of God, the Most High! Lauded and glorified art Thou, Lord, God Omnipotent!"

Lana then asked, "What would you call it if students asked a thousand questions in one class period?"

"A kilowhat hour," he answered.

"What?" a couple dozen would-be humorists responded, laughing at their own wit.

Tom waited until their laughter died down before continuing, "We touched on the numbers that read the same backward and forward last month, analogous to the words called palindromes from the Greek word for a furrow that runs back and forth. We call those made with numbers numberdromes and indicate them with a p- prefix."

"Like eleven," Cal exclaimed.

"Yes, all the single-digit numbers are technically numberdromes, though trivial, and the rest of the double-digit decimal numberdromes are multiples of eleven. It's the three-digit numberdromes, that they start getting interesting."

3-digit numberdromes	101	111	121	131	141	151	161	171	181	191	202	212	222	232	242	252

"There are also numerical palinddromes or double-yolked numberdromes[260] with an even number of digits that have the center doubled. They are a minimum of four digits. Triple-yolked and higher multi-yolked numberdromes would begin with ten thousand one.

numberrdromes	1,001	1,221	1,331,	1,441	1,551	1,661	1,771	1,881	1,991

"There are also cancrines[261] or 'crabby' word palindromes, like one hundred one, some of which are also numberdromes and others that are not, like ten thousand ten.

cancrine	101	202	303	404	505	606	707	808	909	1,001	1111	1212	1313	1414	1515

"Base one would give nothing but numberdromes, zero and repunits, but if we examine the Roman numerals that are palindromic[262], we get a quite different sequence."

"With just four Promanish primes."

Proman	I	II	III	V	X	XIX	L	C	CXC	CC	CCC	D	M	MCM
Promanish	1	2	3	5	10	19	50	100	190	200	300	500	1,000	1,900

"The Aramæan numberdromes[263] would be even worse."

Paramæans	I	II	III	X	XX	XXX	XXXX	XXXXX	XXXXXX	XXXXXXX
Paramæanish	1	2	3	4	8	12	16	20	24	28

"Unholey quintarish[264] has some different numberdromes from standard holey notation and some similar."

unholey	++	‡‡	+=+	+-+	+±+	+++	+‡+	‡=‡	‡-‡	‡±‡	‡+‡	‡‡‡	+==+
quintary	11	22	31	41	101	111	121	132	142	202	212	222	231
quintarish	6	12	16	21	26	31	36	42	47	52	57	62	66

"Psexarish, pseptarish and poctarish would be the decimal homonum of the palindromic base six, seven, eight and sixteen numbers."

"Oh, yeah, the p is silent in psexarish and pseparish like the p in 'pswimming'."

"They only differ from each other when higher than five-five."[265]

psexarish	11	22	33	44	55	101	111	121	131	141	151	202	212	222	232	242	252	303
pseptarish	11	22	33	44	55	66	101	111	121	131	141	151	161	202	212	222	232	242
poctarish	11	22	33	44	55	66	77	101	111	121	131	141	151	161	171	202	212	222

"Since Yijing is symmetrical until it gets higher than sixty-five, the first non-trivial Pyijingish number would be higher than that, sixty-six.

"And the next would be twice that, a hundred thirty-two and so higher than sixty-four squared to one-zero-one base sixty-five, four thousand, two hundred twenty-six."

Pyijingish	66	132	198	264	330	396	462	528	594	660	726	792	858	924	990	1056	1122

"The peven and podd[266] numbers describe the even and odd numberdromes.

peven	22	44	66	88	202	212	222	232	242	252	262	272	282	292	404	414
podd	11	33	55	77	99	101	111	121	131	141	151	161	171	181	191	303

"The pevils[267] and the podious[268] describe those numberdromes that are also evil or odious.

pevil	0	3	5	6	9	33	66	77	99	101	111
podious	1	2	4	7	8	11	22	44	55	88	121

"The binarish numberdromes were called pyranibish by Joyce."

"Ah, 'binary' backward with a p- prefix and a -ish suffix."

"It could've been worse. Try pronouncing binarish with a p- prefix!"

pyranibish	0	1	11	101	111	1,001	1,111	10,001	10,101	11,011	11,111	100,001	101,101	110,011
decimal	0	1	3	5	7	9	15	17	21	27	31	33	45	51

"That's odd. All the numbers after the initial one, zero, are odd."

"Only as odd as the primes that are odd after the initial two. They couldn't be prime if they were divisible by two and neither can binary numbers ending in zero be palindromic."

"But we could name numberdromes in two and in other bases as well." [269]

pyaranirtish	0	1	2	11	22	101	111	121	202	212	222	1001	1111	1221	2002	2112
decimal	0	1	2	4	8	10	13	16	20	23	26	28	40	52	56	68

pyardauqish	0	1	2	3	11	22	33	101	111	121	131	202	212	222	232	303
decimal	0	1	2	3	5	10	15	17	21	25	29	34	38	42	46	51

pyratniuqish	0	1	2	3	4	11	22	33	44	101	111	121	131	202	212	222
decimal	0	1	2	3	4	6	12	18	24	26	31	36	41	46	52	57

"Plucky numbers[270] are the palindromic subset of Stanislaw Ulam's lucky numbers.[271] Those that are still lucky when reversed, like thirteen and thirty-one, fifteen and fifty-one and thirty-seven and seventy-three are ykcul.[272]

"The first survivors are those lucky enough to survive the removal of every other number, that is, the evens, making one the first lucky number and three the second lucky number. The second survivors are those surviving the removal of every third number left, seven being the third lucky number and five unlucky."[273]

"So eleven didn't survive the second cut either, but there are still infinitely many lucky numbers, some prime some not," Cal said.

"Yes, one, three, seven and thirteen survive, but so too do nine, fifteen, twenty-one, twenty-five and twenty-five.

all	0	1	2	3	4	5	6	7	8	9	10	11	12	13	14	15	16	17	18	19	20	21	22	23	24
1st survivors	1	3	5	7	9	11	13	15	17	19	21	23	25	27	29	31	33	35	37	39	41	43	45	47	49
2nd survivors	1	3	7	9	13	15	19	21	25	27	31	33	37	39	43	45	49	51	55	57	61	63	67	69	73
3rd survivors	1	3	7	9	13	15	21	25	27	31	33	37	43	45	49	51	55	57	63	67	69	73	75	79	85
4th survivors	1	3	7	9	13	15	21	25	31	33	37	43	45	49	51	55	63	67	69	73	75	79	85	87	93
lucky	1	3	7	9	13	15	21	25	31	33	37	43	49	51	63	67	69	73	75	79	87	93	99	105	111

"They are interestingly nearly as common as primes and with about as many twins[274] and cousins." [275]

lucky twins	1	3	7	9	13	15	31	33	49	51	67	69	73	75	101	103	107	109	127	129	133	135
lucky cousins	3	7	9	13	21	25	33	37	63	67	69	73	75	79	111	115	129	133	159	163	189	193

"The oblong numbers[276] are of the rather dull subset of the evenly odds, just an odd times its even neighbor. Only two is prime. Some of the interoblongs however are odd, but they're all squares. Oblong neighbors are, of course, all odd and sometimes prime."

oblong	2	6	12	20	30	42	56	72	90	110	132	156	182	210	240	272

oblong neighbors	1	3	5	7	11	13	19	21	29	31	41	43	55	57	71	73	89	91	109	111	131	133

"The poblongs[277] however are much rarer and so much more interesting."

"By that, of course, you mean they get higher faster."

poblong	2	6	272	6,006	2,899,982

"Duh," the Delta Schwa sisters replied, "Just like the phappy numbers are more interesting than the happy numbers, one."

"Seven, forty-four..."

"Two sixty-two and three thirteen."

"Or the pring are more interesting than the ring.[278] After the trivial single digits they jump to two hundred thirty-two and then jumps much, much higher."

"I don't even know what the ring numbers are, something to do with cryptology?"

"No, but although they're Fibonacci-like in that they're defined by the previous two terms, they're calculated a bit more complexly.

ring 1 3 6 15 36 91 232 603 1,585 4,213 11,298 30,537 1,730,787 4,805,595

"Speaking of Fibonacci, the only palindromic Fibonacci, or Pfibonnacci number, higher than the trivial single-digit ones seems to be double-nickels, fifty-five."

The meeting ended in a futile attempt to find an 'Iccanobif' pair.

from *PAJAS* #54 (January 1990)

ALMOST-CALCULUS

"With Christmas break having started for some the interim president Al opened with Luke 24:48, "Remain in the city until you are clothed in power from on high," and then with a surprise pop quiz with the instructions, "Choose the most correct answer to complete the equation."[279]

	1+1=		
1.	0	6.	7
2.	1	7.	10
3.	3	8.	11
4.	4	9.	all of the above
5.	5	10.	none of the above

The newbies chose ten, veterans nine and a few others others,

"The first is correct in modulo two or hypermodulo eleven."

"The second is correct in hypermodulo two, five, ten or inflationary."

"The third is correct in hypermodulo eight, the fourth in hypermodulo seven and the fifth in hypermodulo six."

"The sixth is correct in dihedrals or hypermodulo four."

"The seventh is correct in binary and the eighth in unary and hypermath, so nine is most correct."

"Some of you may be taking pre-calculus," Al continued, "some of you are into higher math, post-calculus. Today we're going to be practicing 'almost-calculus' or digital expansion. Many easily recognized sequences become less recognizable when digitally expanded as in Dominissimo, which makes the game quite challenging. The almost-odds sum, for example, looks deceptively like the almost-squares,[280] at least for the first seven digits."

"Ah, yes, they both begin with the same three dominoes: one-four, nine-one and six-two."

almost-odds	1	3	5	7	9	1	1	1	3	1	5	1	7	1	9	2	1	2	3	2	5
sum	1	4	9	16	25	26	27	28	31	32	37	38	45	46	55	57	58	60	63	65	70
almost-sum	1	4	9	1	6	2	5	2	6	2	7	2	8	3	1	3	2	3	7	3	8
almost-squares	1	4	9	1	6	2	5	3	6	4	9	6	4	8	1	1	0	0	1	2	1

"The same is true of many other 'almosts', like almost-evens or almost-cubes,[281] two-four, six-eight, one-zero, one-two,

almost-evens	2	4	6	8	1	0	1	2	1	4	1	6	1	8	2	0	2	2	2	4
sum	2	6	12	20	21	21	22	23	24	28	29	35	36	44	46	46	48	50	52	56

"Or almost-cheaps and almost-Bantu." [282]

almost-cheap	1	2	3	1	0	1	4	1	5	1	6	1	7	1	8	1	9	2	0	2	2	2	4
sum	1	3	6	7	7	8	12	13	18	19	25	26	33	34	42	43	52	54	54	56	58	60	64

almost-Bantu	1	3	4	5	6	7	8	9	1	0	1	1	1	3	1	4	1	5	1	6	1	7	1
sum	1	4	8	13	19	26	34	43	44	44	45	46	47	50	51	55	56	61	62	68	69	76	77

"Or the almost-Ir-ish." [283]

almost-Ir-ish	1	3	3	0	3	1	3	2	3	3	3	4	3	5	3	6	3	7	3	8	3	9
sum	1	4	7	7	10	11	14	16	19	22	25	29	32	37	40	46	49	56	59	67	70	79

"Or the almost-ugly numbers." [284]

almost-ugly	1	2	3	4	5	6	8	9	1	0	1	2	1	5	1	6	1	8	2	0

The meeting then turned ugly, that is, the almost-ugly numbers were turned into a rock-paper-scissors competition[285] via modular three arithmetic. Just noting the remainder when dividing by three yields zero modulo three or "rock", one or "paper" and two or "scissors". Any sequence ternarish or above, of course, can be turned into such a contest, but the almosts are much easier to calculate in modulo three while also being quite unpredictable. Paper started out ahead, then rock and scissors caught up, but rock finally took a slight lead in the seventeenth round and seemed to keep it, though not yet getting a 'hat trick', three goals in a row.

almost-ugly (mod 3)	1	2	0	1	2	0	2	0	1	0	0	1	0	2	0	1	0	0	2	0	1	0	0	2	0
0 count	0	0	1	1	1	2	3	3	4	4	4	4	5	5	6	6	7	8	8	9	9	10	11	11	12
1 count	1	1	1	2	2	2	2	3	3	4	4	5	5	5	5	6	6	6	6	6	7	7	7	7	7
2 count	0	1	1	1	2	2	2	2	2	2	3	3	4	4	5	5	6	7	7	8	8	9	9	10	10

from *PAJAS* #55 (February 1990)

WHAT DO YOU MEAN?

Br. Ken began the first meeting of the new year and celebrated Phi Day, January sixth, by quoting St. Francis of Assisi, "Most high, all-powerful, all good, Lord! All praise is yours, all glory, all honor and all blessing," then Charles continued with the question, "What would you call a diet that lasts twelve months?"

"A miracle!" the Lara responded.

"Perhaps, but what I had in mind was a lite year," Charles answered, "Get it? L-I-T-E, lite, rather than L-I-G-H-T, light."

Moving on hastily she continued, "The interprimes are those numbers exactly between two consecutive primes. The same smoothing operation can be applied to any sequence, though some would yield halves rather than whole internumbers, like that between the primes two and three. We're more interested in those that're whole numbers. They are sometimes odd, like the average between nineteen and twenty-three, and sometimes even, like the average of three and five, but they get higher as fast as the primes themselves. The interodds would be just be the evens higher than zero and the interevens just another name for the odds.

"The average between consecutive emirps, those primes that remain primes when their digits are reversed, would be interemirps,[286] some even and some odd. Some, like one thirty-one and one seventy-three, are even primes, primes, that is, not <u>the</u> even prime.

interemirps	15 24 34 51 72 76 88 102 110 131 153 162 173 189

"The interddo[287] numbers are averages of consecutive odds whose ananums are also odd."

interddos	14 16 18 25 33 36 38 45 52 55 58 65 72 74 77 85 92 94 96 100 104 106

"The interpodds[288] are the averages of consecutive odd numberdromes. They deceptively start out identical to the evens for the first five terms..

interpodds	2 4 6 8 10 22 44 66 88 100 106 112 117 136 146 156 166 176 186 297 308

"The interneves[289] are the averages of consecutive neves, those non-palindromic evens whose ananums also are.

interneves	25 27 34 41 44 47 54 61 63 66 74 81 83 85 143 201 207 222 237 247 257

"Taking the Morris or look-and-say sequence as base ten, we can get 'Morrisish' numbers[290] and so intermorrisish[291] numbers and even, since these are also all even, duintermorrish." [292]

Morrisish	1 11 21 1,211 111,221 312,211 13,112,221 1,113,213,211 31,131,211,131,221
intermorrisish	6 16 616 56,216 211,716 6,712,216 563,167,716 15,516,162,172,216
duintermorrisish	11 316 28,411 133,961 3,211,961 284,939,961 7,758,362,669,961

"The polyinterugliests[293] always remain either even or odd multiples of five with repeated averagings, because their differences are always thirty!"

interugliest	45	75	105	135	165	195	225	255	285	315	345	375
duinterugliest		60	90	120	150	180	210	240	270	300	330	360
tresinterugliest			75	105	135	165	195	225	255	285	315	345
differences		30	30	30	30	30	30	30	30	30	30	30

"Since the interprimorials[294] continue to be even, we can calculate duinterprimorials,[295] which are not and begin with a prime."

interprimorials	4	18	120	1,260	16,170	270,270	5,105,100	116,396,280
duinterprimorials	11	69	690	8,715	143,220	2,687,685	60,750,690	111,291,180

"The interluckies have evens, odds and primes." [296]

interlucky	2	5	8	11	14	18	23	28	32	35	40	46	50	57	65	68	71	74	77	83	90	96

"The ebans,[297] those numbers with names without e, happen all to be even, at least at first, and so we also have interebans,[298] nabes[299] and internabes." [300]

eban	2	4	6	20	22	24	26	30	32	34	36	40	42	46	62	64
intereban		3	5	13	21	23	25	28	31	33	35	38	41	44	54	63
nabe	20	24	26	40	42	46	60	62	64	2000	2020	2024	2026	2040	2042	2046
internabe		22	25	33	41	44	53	61	63	1032	2010	2022	2025	2033	2041	2044

The meeting ended investigating many of the other internumbers.

from *PAJAS* #56 (March 1990)

STRONG NUMBERS

Tom began with "In days to come the mount of the Lord's house shall be established higher than the mountains; it shall rise high above the hills," the beginning of Micah 4.

He continued with asking "How many strongmen does it take to change a light bulb?"

Sam, who "happened" to be changing a lightbulb at the time, answered from his ladder, "None, because a light bulb doesn't need a strongman, though a heavy bulb might."

That lead right into his topic for the month, "The strong primes[301] are those greater than the interprimes, while the weak primes[302] are less. Strength and weakness would therefore be attributes of every number in a sequence, even those with non-integral inters. Defining strength as the difference from the internumbers, strong numbers would have positive 'strength' and weak numbers negative 'strength'. Interprime primes a strength of zero."

"So the first strong prime is eleven with a strength of one," Cal said rather triumphantly.

strong primes	11	17	29	37	41	59	67	71	79	97	101	107	127	137	149	163	179	191
strength	1	1	2	1	1	2	1	1	1	2	1	1	5	2	4	1	2	4
weak primes	3	7	13	19	23	31	43	47	61	73	83	89	103	109	113	131	139	151
strength	-½	-1	-1	-1	-1	-2	-1	-1	-2	-2	-1	-1	-1	-1	-5	-1	-4	-2

"The weak primes start with three and seven, with strengths of minus half and minus one. The next prime with a strength weaker[303] than minus one isn't until one thirty-one."

weaker primes	3	7	31	113	523	1,327	9,551	15,683	19,609	31,397
strength	-½	-1	-2	-5	-8	-14	-16	-20	-23	-34

"Many sequences have no strong numbers, however many others do, like the curvaceous[304], curvilinear,[305] and linear.[306] Some of the latter are also prime."

strong curvaceous	6	8	30	38	60	66	68	80	83	86	88	96	300	306

strong curvilinear	22	52	222	252	522	552	2222	2252	2522

strong linear	11	41	71	111	141	171	411	471	711	1,111	1,141	1,171	1,411	1,441	1,711

"The strong luckies[307] seem to have more primes, though as we well know that may not last."

strong lucky	7	13	21	31	49	63	67	73	87	111	115	127	133	151	159	16

"The Aronsonian-like numbers, that I've been working with, that I call the 'spell-and-count' numbers, have some," Raz added. "Rather than counting the letters in a generating sentence or the letters in the number name, you count the letters in the previous number's digits. A warning to any of you who may suffer from zenophobia, the abnormal fear of converging sequences, whether you start with a seed of 'zero' [308] or 'one' [309] the two sequences very quickly converge."

"Ah, that's because 'zero, four' and 'one, three' both have eight letters!"

0	4	8	13	21	30	36	45	54	63	73	85	95	105	119	137	158	178	200	210
1	3	8	13	21	30	36	45	54	63	73	85	95	105	119	137	158	178	200	210

"Why did the mathematician think he was lost?" came the seemingly irrelevant question, followed by the more relevant answer, "Because when he counted the colleagues who he was with, he habitually started counting from zero instead of one."

"The first strong spell-and-count [310] is thirty with a strength of one and a half."

strong spell-and-count	30	85	158	200	210	250	284	361	402	455	519

"The Siamese primes[311] are those that both differ from squares by two. One squared just gives three. Two squared gives two and six, Siamese cousins,[312] four apart, one prime and one not. Three squared gives the Siamese cousins, both prime, seven and ..."
"Eleven!" Cal answered.

Siamese primes	2	3	7	11	79	83	223	227	439	443	1,087	1,091	13,687
strength		-1½	0	-32	32	-68	68	-104	104	-320	320	-6,296	6,296

"Both half the Siamese and half the strong Siamese[313] start off prime."

Siamese	2 3 6 7 11 14 18 23 27 34 38 47 51 62 66 79 83 98 102 119 123 142 146 167
strong Siamese	6 11 14 23 38 47 62 79 98 119 142 194 223 258 291 326 363 439

"Many of the Siamese[314] odds are strong and prime, but none of the Siamese evens[315] are both strong and prime."

Siamese odds	3 7 11 23 27 47 51 79 83 119 123 167 171 223 227
strong Siamese odds	23 47 79 116 167 223 287 359 439 527 623 727 843 963

Siamese evens	2 6 14 18 34 38 62 66 98 102 142 146 194 198
strong Siamese evens	14 34 38 66 142 194 254 322 398 482 574 674 782 902

"Primes can also be subdivided into five subsequences by the number of their tails. Two is the only two-tailed prime. The other non-trivial ones are either one-tailed[316], three-tailed[317], seven-tailed[318] or nine-tailed[319], called from Japanese xenozoology respectively: ichibi, sanbi, shichibi and kyuubi primes.

ichibi primes	11 31 41 61 71 101 131 151 181 191 211 241 251 271 281 311
strong ichibi primes	31 61 131 181 241 271 311 401 421 521 631 751 811 881 971 1,021

sanbi primes	3 13 23 43 53 73 83 103 113 163 173 193 223 233
strong sanbi primes	43 73 103 163 223 263 353 373 433 563 593 643 673 733

shichibi primes	7 17 37 47 67 97 107 127 137 157 167 197 227
strong shichibi primes	37 97 127 257 307 317 337 457 547 607 727 787 857

kyuubi primes	19 29 59 79 89 109 139 149 199 229 239 269 349
strong kyuubi primes	59 79 139 199 229 349 379 409 439 479 499 569 599

"Alliterative numbers[320] are those like twenty-two that repeat initial phonemes. The first strong one is two hundred two."[321]

alliteratives	22	33	44	55	66	77	88	99	202	212	220	222	303	313	330	333	404	414

strong alliteratives	202	212	220	303	330	404	440	505	550

"And then higher plexes, two twelveplex, four and five fifteenplex, eight eighteenplex and other such multiples of threeplex."

"Since forty-six and sixty-four can swap syllables we call them 'spooneristics',[323] many of which are strong.[324] Only the shichibi have any chance of also being prime."

spooneristic	46	47	48	49	64	67	68	69	74	76	78	79	84	86	87
strong spooneristic	64	67	69	74	78	84	86	94	96	446	464	474	476	484	486

The meeting ended with a quest for these elusive strong spooneristic primes.

from *PAJAS* #57 (April 1990)

STRONGER NUMBERS

Al began meeting by quoting Daniel 7:18, "But the saints of the Most High God shall take the Kingdom: and they shall possess the Kingdom for ever and ever."

Cal asked, "How can you add twelve and sixty and get one?"

Lana answered, "On a clock."

.Cal added, "In clock arithmetic or modulo sixty," and then continued on topic, "If we count the next stronger prime[325] than eleven we reach twenty-nine with a strength of two."

"And after that, one twenty-seven with a strength of five."

stronger primes	11	29	127	541	907	1,151	1,361	15,727	31,469	2,010,881
strength	1	2	5	6	8	10	14	20	32	41

"The stronger luckies[326] are interesting in their own way, particularly with seven and thirty-one also being prime and one twenty-seven being both a stronger lucky prime.

stronger lucky	7	31	63	127	189	259	697
strength	1	2	4	5	7	8	11

"The stronger curvaceous[327] grow stronger much faster, because of the gaps before three times powers of ten, the stronger curvilinear[328] and the stronger linear[329] not so much."

stronger curvaceous	6	30	300	3,000	30,000	300,000	3,000,000
strength	½	9	99	999	9,999	99,999	999,999

stronger curvilinear	22	52	222	522	2,222	5,222	22,222	52,222
strength	7	12	82	132	832	1,332	8,332	13,332

"The stronger linear is the only one of the three that contains any primes."

stronger linear	11	41	111	411	1,111	4,111	11,111	41,111	111,111
strength	½	10½	15½	115½	165½	1,165½	1,663½	11,665½	31,665½

STRONGEST NUMBERS

from *PAJAS* #58 (May 1990)

Al began the meeting quoting Psalm 105:4, "Seek ye the Lord and be strengthened." then asked, "How many grad students does it take to change a light bulb?"

One of the grad students answered, "Less than a hundredth, because we change hundreds." to which they and even Sam agreed.

"Besides finding strong and stronger numbers we can generate many sequences whose strengths are always getting higher. Starting with one, as usual, the second number would be one higher than the average of one and the third number, so they can't be two or three."

"Four and five however work as strongest numbers![330] One plus five average to three and four is one more than three."

"The next would be ten, because it's average with four is two more than five."

strongest	1	4	5	10	21	40	69	110	165	236	325	434	565	720	901	1,110	1,349	1,620
strength		1	2	3	4	5	6	7	8	9	10	11	12	13	14	15	16	17

"A few are prime."

"A very few! Five and sixty-nine may be the only ones."

"There're many strongest odds[331] and evens[332], in nearly even numbers," Sam commented.

strongest odd	1	5	21	69	165	325	565	901	1,349	1,925	2,645	3,525	4,581
strongest even	4	10	40	110	236	434	720	1,110	1,620	2,266	3,064	4,030	5,180

"The strongest odds and evens can be further subdivided by their tails. There seems not to be any strongest numbers ending in two, three, seven or eight, though there are strongest ichibi[333], gobi[334], kyuubi[335], loved[336], yonbi."[337] and rokubi numbers.[338]

strongest ichibi	21	901	4,581	13,061	28,341	52,421	87,301	134,981
strongest gobi	165	325	565	1,925	2,645	3,525	7,285	8,965
strongest kyuubi	69	1,349	5,829	15,509	32,389	294.789	396,869	520,149

strongest loved	10	40	110	720	1,110	1,620	4,030	5,180
strongest yonbi	434	3,064	9,894	22,924	44,164	78,584	119,214	177,044
strongest rokubi	236	2,266	8,096	19,726	39,156	68,386	109,416	164,246

"Like the Fibonacci and Lucas numbers, the strongest sequences depend upon their seed numbers. Starting with one, five and seven generate the second strongest sequence,[339] six and nine the third strongest,[340] seven and eleven, the fourth strongest[341] eight and thirteen the fifth strongest[342] and nine and fifteen the sixth strongest,[343] etc., each growing higher a little faster. These all seem to contain some primes and together they generate a strongest diagonal also with primes."

2nd	1	5	7	13	25	45	75	117	173	245	335	445	577	733	915	1,125	1,365	1,365
3rd	1	6	9	16	29	50	81	124	181	254	345	456	589	746	929	1,140	1,381	1,637
4th	1	7	11	19	33	55	87	131	189	263	355	467	601	759	943	1,155	1,397	1,671
5th	1	8	13	22	37	60	93	138	197	272	365	478	613	772	957	1,170	1,413	1,688
6th	1	9	15	25	41	65	99	145	205	281	375	489	625	785	971	1,185	1,429	1,705

"The second strongest sequence and fourth strongest sequence include shichibi. The fifth strongest also includes nibi and hachibi numbers."

"Although these sequences are called strongest because their strength is always positive and growing higher, the strengths do not have to be limited to just getting higher by minimum increments. They be generated by any other sequence as its seed. The dustrongest or strongest strongest numbers,[344] for example, would get higher quite differently than the merely strongest numbers."

"Starting as usual with one, the average of one and the next number would have to be less than four, so the next number would be less than twice four minus one, six."

"And the next number would have to be less than twice five minus six, which would make it only three!"

"Ah, but it's only a temporary setback, and a prime one at that. The next number that averages to less than ten with three is sixteen, higher again"

dustrongest	1	6	3	16	25	54	83	136	193	278	371	496	633	806

The meeting continued in an attempt to find other polystrongests.

from *PAJAS* #59 (June 1990)

SUPERNUMBERS

Al opened with Genesis 14:22, "I have sworn to the Lord, God, Most High, the Creator of Heaven and Earth," and then asked, "What is a ten-letter word for a kilogram of wet socks?"

He had to provide the cryptic answer himself, "Literhosen."

"Besides the average internumbers we also have supernumbers, analogous to the superfactorials[345], the product of the previous factorials.

superfactorial	12	288	34,560	24,883,200	125,411,328,000

"Superodds[346] and superevens[347] are subsequences of the double factorials, which includes them both. Superprimes could be an alternative name for primorials, though not a very good one, since they are composites and not primes at all."

"The factorial of zero is defined as one so that we can define superevens."

"So we also have superevil numbers!"[348]

"And superodious."[349]

superodds	3	15	105	945	10,395	135,135	2,027,025	34,459,425	654,729,075

superevens	8	32	192	1,536	15,360	184,320	2,580,480	41,287,680	743,178,240

superevils	15	30	54	90	120	180	255	306	360	460	552	648	783	870	960	1,088

superodious	2	8	28	56	88	143	182	224	304	399	525	650	728	868

"We can 'supersize' to get to the superbinarish,[350] supertenarish[351], superquadrarish,[352] superquinarish[353], supersenarish[354], or higher.

superbinarish	1	10	110	11,000	1,111,000	122,210,000	13,565,310,000
superternarish	1	2	20	220	2,640	264,000	26,664,000
superquadrarish	1	2	6	60	660	7,920	102,960
superquintarish	1	2	6	24	240	2,640	31,680
supersenarish	1	2	6	24	120	1,200	13,200

"As they get higher they share more and more numbers, a two, then a six, then a twenty-four."

"The supercomposites[355] however are quite unlike the superprimes or primorials.

supercomposites	4	24	192	1,728	17,280	207,360	2,903,040	43,545,600

"Digitally balanced numbers are a subsequence of the pandigitals with each digit the same number of times, rather than just each at least once. For binarish the lengths would be twice the weights.[356]

balanced binarish	10	1,001	1,010	1,100	100,011	100,101	100,110	101,001	101,010
length	2	4	4	4	6	6	6	6	6
weight	1	2	2	2	3	3	3	3	3

superbalanced	10	10,010	10,110,100	11,121,110,000	1,112,233,332,210,000

"The balanced ternarish would be a subsequence of pandigitals having lengths and weights that are equal to multiples of three.[357]

balanced ternarish	102	120	201	210	100,122	100,212	100,221	102,012
weight	3	3	3	3	6	6	6	6

superbalanced	102	12,240	2,460,240	516,650,400	51,728,071,348,800

"With Joyce we nickname the superbalanced numbers 'superbs'.[358] In base one they'd just be the repunits. In base ten they reach a sub depth of forty-one."

"Superb indeed!"

balanced	1,023,456,789	1,023,456,798	1,023,456,879	1,023,456,897

superbs 1,023,456,789 1,047,463,808,161,301,622 1,072,034,039,966,220,486,629,757,738

"The superperfects[359] we've been working on get high superfast, but not quite that fast!" Lana began.

superperfect 168 83,328 677,289,984 22,723,306,532,634,624

"Dusuperperfects however get higher much faster than the dusuperfactorials.[360] The first dusuperperfect [361] already starts with nearly fourteen million."

dusuperfactorials 12 3,456 11,949,360 292,033,482,752,000

dusuperperfect 169 13,999,104 132,731,845,556,620,643,794,944 > thirty-twoplex

"The first tressuperperfect or 'superduperperfect' [362] starts at over thirty-sevenplex," the Delta Schwa sisters proclaimed after much calculation.

"Summing the almost-superperfects[363] is more my speed," Sam commented.

almost-superperfects	1	6	8	8	3	3	2	8	6	7	7	2	8	9	9	8	4
sum	1	7	15	23	26	29	31	39	45	52	59	61	69	77	86	94	98

"Duh," the Delta Schwa sisters chimed in, "after we did the grunt work!"

"Like Jesus," Br. Ken concluded, "We are perfected through suffering." [364]

from *PAJAS* #60 (July 1990)

MORE SUPERNUMBERS

Al quoted "Almsgiving is a worthy offering in the sight of the Most High for all who practice it," from Tobit 4:11 and then asked, "Why is Polynesia more important than Micronesia?"

Kamika, our resident Pacific Islander, responded with "Polynesia is more than two million as many nesias."

Al then continued, introducing the month's topic, "We may not be superperfect, but we are supercurious. The supercurious numbers are, of course the products of the curious numbers,[365] the downtown neighbors to squares whose halves are also squares, like forty-eight. Forty-eight plus one is seven squared and twenty-four plus one is five squared."

supercurious 48 80,640 4,606,156,800 8,938,007,750,246,400

curious 48 1,680 57,120 1,940,448 65,918,160 2,239,277,040 76,069,501,248

"I'm not only supercurious, I'm also supercheap." [366] Br. Ken added.

supercheap 2 6 60 840 12,600 201,600 3,427,200 61,689,600 1,172,102,400

"Please be a dear, Lana, and calculate the superdears," [367] Al asked.

"If you'll do the superuglies," [368] she countered, leaving her adjective for him unspoken.

superdears 56 504 8,568 154,224 2,930,256 79,116,912 2,215,273,536 superuglies 2 6 24 120 720 5,760 51,840 518,400 6,220,800 93,312,000

"More interesingly, I think, we have the superprime-like superemirps." [369]

superemirps 221 6,851 253,487 17,997,577 1,313,823,121 103,792,026,559

"And supersquares[370] and supercubes." [371]

supersquares 4 36 576 14,400 518,400 25,401,600 1,625,702,400

supercubes 8 216 13,824 1,728,000 373,248,000 128,024,064,000

"The Super Cubans may be the best baseball players chosen to play in the Super Series in Cuba, but the supercuban numbers [372] are interesting too."

supercubans 133 4,921 300,181 27,316,471 3,469,191,817 586,293,417,073

"The superoblongs[373] begin like powers of twelve, but soon diverge."

superoblongs 12 144 2,880 86,400 3,628,800 203,212,800 14,631,321,600

"Personally I like the supercats," [374] Catherine shared. "The products of successive CATS remind me of Supergirl's supercat Streaky. The CATS[375] sequence itself gets its name from the acronym for the operation cube-and-then-sort. Like Schrödiger's undead cat they're quite unpredictable until you observe them closely."

cubes	1	8	27	64	125	216	343	512	729	1,000	1,331	1,728	2,197	2,744	3,375	4,096
CATS	1	8	27	46	125	126	334	125	279	1	1,133	1,278	1,279	2,447	3,357	469
difference	0	0	0	18	0	90	9	387	450	999	198	450	918	297	18	3,618

streaky 1 8 216 9,936 1,242,000 156,492,000 52,268,328,000

from *PAJAS* #61 (August 1990)

HAVE A SUB?

Al began the meeting with "O Wisdom of our God Most High, guiding creation with power and love: come to teach us the path of knowledge!" from the O Antiphons."

He then continued the tradition of presented the month's riddle this time visually.

¢ > # ?
¢ = #?
¢ < #?

When no consensus was forthcoming, he settled the question with: "The cent sign is greater than the pound sign according to the adage, 'Penny wise, pound foolish.'"

"Besides having two symbols and meanings for the word 'pound', in addition to supernumbers we also have subnumbers. By analogy from subfactorial,[376] the prefix sub- can be expessed as the rounded-up quotient after division by e, the base of the natural logarithms. As expected, it gets higher more slowly than factorial and much more slowly than superfactorial."

subfactorial	1	2	9	44	265	1,847	14,774	132,966	1,329,663

"We can even compound, pun intended, the prefixes 'sub-' and 'super-'."

subsuperfactorial	4	105	12,713	9,154,017	46,136,249,261
supersubfactorial	16	704	185,856	307,405,824	4,559,443,181,568

"Ah, this subsuper- operation[377] obviously gets higher more slowly than the supersub-."[378]

"If that's still too fast for you, you can indicate repeated division by e and rounding the Greek prefixes with dusub-, tressub-, tetrasub-, pentasub-, etc. and get higher ever more slowly. Repeatedly applying this operation until reaching one determines what we call a sequence's 'sub depth'. Some sequences are only slightly reduced by 'subbing', reducing the starting point from one to zero, for example.[379]

googoz	1	2	216	4,096	100,000	2,985,984	35,831,808	1,475,789,056
subgoogoz	0	1	79	1,507	36,788	1,098,482	13,181,786	542,912,453

"We're more interested in the sequences that get high fast enough not to stutter and which that start at one, not for example, like the sub- or tressubcubans." [380]

sub	3	7	14	22	33	47	62	80	122	146	173	201	232	265
tressub	0	1	2	3	4	6	8	11	17	20	24	27	31	36

"Ah, more of the early dusubcubans[381] are prime, one, three, five, seventeen, twenty-three, twenty-nine, six out of the first eight, compared to just three and two."

dusubcubans	1	3	5	8	12	17	23	29	45	54	64	74	85	97	110

"These polysubs're both understandable and yet unpredictable. I like them," someone said, prompting Al to begin singing "We All Like the Polysubnumbers".

from *PAJAS* #62 (September 1990)

MORE SUBS

Al opened the meeting before by quoting St. Augustine, "Most High, most excellent, most potent, most omnipotent; most piteous and most just; most hidden and most near; most beauteous and most strong."

His riddle was,"How many days can you name that begin with the letter T?"

After Tuesday and Thursday were named, so too were 'today', 'tomorrow' and 'Thanksgiving Day.' Br. Ken even suggested that every day should be a thanksgiving day.

Al continued, "Sequences generated pseudonumerologically can get high very fast too, but can be slowed somewhat by this 'subbing'.

Starting with the seed number name 'three', for example, it reduces to TR, one-four or fourteen and 'fourteen' reduces to FRTN, six-eight-one-for or 'six thousand, eight hundred fourteen'. This sequence gets down to a sub depth of just one.[382]

three	3	14	6,814	11,021,842,141,158	>forty-fourplex
sub	1	5	2,507	4,054,709,127,569	>forty-threeplex

"I see why you started with 'three',' Cal commented. "Starting with the seeds 'one' or 'two' we just get N and T which in turn just generate alternating ones and twos and don't get higher. 'Four' however quickly reaches sub depth and continues to get higher very quickly. [383]

four	4	84	184	22,141,184	1,211,352,221,418,412,102,122,141,184
sub	1	31	68	8,145,286	>twenty-sixplex

"'Nine', 384 generates a sequences with a sub depth of two."

nine	9	22	222	121,411,211	2,214,112,113,552,842,141,582,102,112,141,582
dusub	1	3	30	16,431,220	>thirty-twoplex

"And 'thirteen' [385] reaches a sub depth of three, 'thirty-four' four, and 'ninety-two' five."

thirteen	13	1,412	21,021,842,141,158	>fifty-sixplex
tressub	1	70	1,046,615,891,900	>fifty-fourplex

thirty-four	34	14,184	8,412,102,112,141,184	>seventyplex
tetrasub	1	260	154,073,024,581,137	>sixty-eightplex

ninety-two	92	2,211	1,102,112,141,582	>fifty-fiveplex
pentasub	1	15	7,425,973,197	>fifty-threeplex

"The higher double-digit look-and-say or Gleichniszahlen-Reihe sequences can similarly be slowed. That with the seed number ten[386] reaches a sub depth of two, while the triple-digit hundred is five.[387]

look-and-say-10	10	1,110	3,110	132,110	1,113,122,110	311,311,222,110	>fourteenplex
dusub	1	150	421	17,879	150,644,696	42,131,392,419	>twelveplex

look-and-say-100	100	1,120	211,210	1,221,121,110	112,221,123,110	>fifteenplex
pentasub	1	8	1,423	8,227,849	756,139,980	>fourteenplex

"Although it doesn't get high as fast, the spell-and-count sequence with a seed of three[388] doesn't stutter and reaches a sub depth of one. Interestingly the sequences with seed numbers of four through fourteen all converge at thirteen, though some do stutter. [389]

spell-and-count-3	3	5	9	13	21	30	36	45	54	63	73	85	95	104	118	136	155	174	195
sub	1	2	3	5	8	11	13	17	20	23	27	31	35	38	43	50	57	64	72

from *PAJAS* #63 (October 1990)

KILLION

Al opened with "There is nothing to fear. I AM the First and the Last." from Revelation 1:17 and then asked "Why was six afraid of seven?"

"Before anyone ventured a plausible answer, he answered himself, "Seven ate nine."

"The idea of the killion," he continued, "the number so huge it may kill, literally or figuratively, by memory overload or brainlock was popularized by Ian Frazier.[390] He reported the advocating by Albert Einstein of a killion governor on early computers.

"The actual value of the killion however seems to vary greatly, depending on each individual's numerical altitude tolerance. Some have even rather plausibly proposed that the noun 'number' comes from the verb 'numb', but this month however we will 'rack our brains', stretching our tolerance levels.

"Numbers between a height of two and three stacked tens", Charles began, "include Feynman's 'astronomical' elevenplex stars in the galaxy or Sagan's stars in the known universe, 'billions of billions' or higher than four eighteenplex."

"King Shirham's limit was less than the sum of doubled grains of rice for every square on a chessboard, nineteenplex.[391]

2 power	1	2	4	8	16	32	64	128	256	512	1,024	2,048	4,096	8,192	16,382
sum	1	3	7	15	31	63	127	255	511	1,023	2,047	4,095	8,191	16,381	32,763

"The number of Ernő Rubik's cube combinations is a little higher, over twentyplex." [392]

"So is the first integer with a million factorizations,[393] is nearly the same as the number of seconds in the Hindu period the *mahakalpa*." [394]

"Avogadro's number, which many, I'm sure, have encountered in chemistry, is a common high number limit, approximately six twenty-threeplex." [395]

"Above that is the sum of the tenth powers, or dekapowers, of the numbers from one to a thousand, which is higher than thirty-oneplex." [396]

10th power	1	1,024	59,049	1,048,576	9,765,625	60,466,176	282,475,249
sum	1	1,025	60,074	1,108,650	10,874,275	71,340,451	353,815,700

"*Framework Documentation* by Alias Graphics references the limiting number, FHUGE, slightly higher than thirty-fiveplex." [397]

"Gromacs' limit was two-to-the-two-hundred-fifty-sixth, nearly thirty-eightplex." [398]

"The maximum possible orthodox chess positions is calculated at higher than fortyplex." [399]

"Fugaseven or seven-to-the-fifty-sixth is higher than forty-sevenplex." [400]

"The limit of possible shuffles in an ordinary deck of cards is fifty-two factorial, higher than fifty-oneplex." [401]

"The Gamer's Nook had a running discussion on 'How many teaspoons are there in a cubic light year?', Although nearly all possible cubic light years have no actual teaspoons, the answer is simply the cube of the number of centimeters per light year divided by the number of cubic centimeters per the unit called a teaspoon, nearly sixty-twoplex." [402]

"The limit to the narcissistic numbers,[403] one whose zth powers of its digits sum to the zth power of the whole, is higher than sixty-sevenplex. They start out trivially with one through nine before catapulting to one hundred fifty-three." [404]

narcissistic 1 2 3 4 5 6 7 8 9 153 370 371 407 1,634 8,208 9,474 54,748

"The second sublime number,[405] whose number of factors and sum of factors are both perfect, is higher than seventy-twoplex, the first sublime number being just twelve."

"The largest factor of a googolplexpli and the Eddington number for the number of particles in the known universe are both higher seventy-nineplex." [406]

from *PAJAS* #64 (November 1990)

TRANSGOOGOL NUMBERS

Beginning the meeting by quoting, "Blessed are you, daughter, by the Most High." (Judith 13:18) Al then asked "How much is two n plus two n?"

Old Sam, who likely had heard this one before, came up with the answer, "I don't know. Algebra is all 'four n' to me."

"Now that we've gone through the cisgoogol numbers, we can move on to the numbers higher than a googol. The number of *purvi* in a *shirsha prahelika,* for example, is higher than hundred-thirty-threeplex [407] and the number of years in a *shirsha prahelika* is higher than a hundred-forty-threeplex." [408]

"The number of Go positions is as high as a hundred-sevenplex.[409] The so-called dimensionless volume of known universe in Planck units is higher than two-hundred-eighteenplex. [410] The number of seconds in a *shirsha prahelika* is a little higher."

"The first Apocalyptic number, or one with six hundred sixty-six digits is, of course, six-hundred-sixty-fiveplex, with one one and six-hundred-sixty-five zeroes. Similarly the first titanic number, one with a thousand digits, is ten-to-the-nine-hundred-ninety-ninth. The slightly higher great googol or thousandplex is the first Arabian number, named for *A Thousand and One Arabian Nights,* with a thousand one digits has nine hundred nine hundred ninety-eight zeroes and two ones. The first Lusitanian number, one with two thousand digits, is a thousand-nine-hundred-ninety-nineplex."

"Matt Leach's banthree is three-to-the-three-pentated-to-the-second-power, which is higher than nine-thousand-three-hundred-ninety-oneplex." [411]

"The first gigantic number, one with ten thousand digits, is ten-to-the-nine-thousand-nine-hundred-ninety-ninth."

"Hugo Steinhaus's mega is two-hexated-to-the-second-power or two-to-the-sixty-five-thousand-five-hundred-thirty-sixth, nearly twenty-thousandplex." [412]

"Tom Kreitzberg's expostfactofour is four-to-the-six-factorial, which is higher than thirty-seven-thousand-eight-hundred-fifty-sevenplex." [413]

"In his book, *The Hitchhiker's Guide to the Galaxy*, Douglas Adams referenced the number two-to-the-two-hundred-sixty-thousand-one-hundred-ninety-ninth, which is higher than seventy-eight-thousand-three-hundred-twenty-sevenplex, In the BBC radio version it was given as the slightly higher two-to-the-two-hundred-sixty-thousand-seven-hundred-ninth." [414]

"Archimedes' cattle problem number eluded discovery for many centuries but turned out to be higher than ten-to-the-two-hundred-two-thousand-five-hundred-forty-fourth." [415]

"The maximum number in PARI developed at University Bordeaux, France, is higher than two-hundred-millionplex." [416]

"James Joyce referenced to-the-third-nine in his book *Ulysses*, which is higher than three-hundred-millionplex." [417]

"The ten-to-the-billionth referred to in Schoolhouse Rock's 'My Hero, Zero' is higher."

This prompted a rendition of the golden oldie by more than just the Fen.

from *PAJAS* #65 (December 1990)

TRANSTENDUPLEX NUMBERS

Quoting Isaac Watts this time Al said, "Give thanks to God Most High, the universal Lord, the sovereign King of kings."

Cal asked, "How can a bookmaker know the exact score of a game before it's played?"

"Duh, it's always zero-zero before it's played," the Delta Schwa sisters answered.

Al continued with the previous month's ever-higher numbers, "Gauss's 'measurable infinity' was nine-tetrated-to-the-fourth, nine-to-the-nineth-to-the-nineth-to-the-nineth, less than to-the-fourth-ten." [418]

"Eric Brahinsky's number, a solution to the Three Symbols Problem, eight-to-the-ninth factorial is higher." [419]

"The number of orthodox chess games is just higher than to-the-third-ten." [420]

"The maximum number in Mathematica is higher than a hundred-millionplex." [421]

"Alistair Cockburn's gagfour or four-tetrated-to-the-fifth-power is higher than ten-tetrated-to-the-third." [422]

"Rudy Rucker's estimated number of Human brain neuron interconnections is nearly three-hundred-millionplex." [423]

"Prof. David J. Chalmers' fifth largest response in his 'Pick a number between zero and infinity?' experiment was higher than to-the-third-ten." [424]

"Vinogradov 's number is higher than to-the-third-ten-to-the-eighteenth."[425]

"Two-to-the-third-to-the-forty-first, a solution to the 1-2-3-4 puzzle, is higher than tenplex-to-the-nineteenth." [426]

"Tom Kreitzberg's expostfactofive is five-to-the-twenty-fourth-factorial, which is higher than tenplex-to-the-twenty-third." [427]

"Littleton's number, his estimated odds against a mouse surviving on the sun for a week, is tenplex-to-the-forty-second." [428]

"Four-to-the-fourth-to-the-fourth-to-the-fourth, a solution in the Four-Fours Problem, is higher than tenduplex-to-the-hundred-fifty-third." [429]

The meeting was quickly ended by Raz's statement, "Mathematician Joseph S. Madachy called numbers higher than to-the-fourth-nine 'rather large'." [430]

from *PAJAS* #66 (January 1991)

SLICING AND DICING

Al quoted, "Before all he gave witness to the deeds of the Most High," (2 Maccabees 3:36), then asked, "What kind of snake would it be if it was exactly thirty-nine and thirty-seven hundedth inches long?"

"A pi-thon!"

"How I wish I could calculate pi!"

"Ha! A seven-place piphilogism!"

"A What?"

" A mnemonic for remembering the digits of pi usually by counting the letters in the words: 'Now a poem I shall construct,' three, one, four, one, five, nine."

Other members were happy to contribute, "'Cunningly devised endeavour,' with the British spelling, nine, seven, <u>nine.</u>'"

"'Con it now remember ever', three, two, three, eight, four."

"'Widths in circle here you see,' six, two, five, four, three, three."

"Sketched out in strange obscurity.' eight, three, two, seven, nine." [431]

almost-pi 3 1 4 1 5 9 2 6 5 3 5 8 9 7 8 3 2 8 4 9

"We can get higher by counting the number of digits of pi to reach a certain number frag, which we call the Horner sequence for Jack Horner because numbers are pulled like plums out of pi.[432] Curiously they start out almost almost-e.

Horner 2 7 1 3 5 8 14 12 6 50 95 149 111 3 5 40 96 426 37

"We can also cut pi into 'slices' such that each 'slice' is always higher than its predecessor. Starting with three, the next higher frag is fourteen and the next after that fifteen." [433]

sliced pi 3 14 15 92 653 5,897 9,323 84,626 433,832 744,502 884,197

"We can also get a sequence from cutting pi into frags increasing by a digit each time, what we call 'dicing'. If this happens to expose leading zeroes the resulting lower number is just a temporary setback in the quest to get ever higher." [434]

diced pi 3 14 159 2,653 589,793 2,384,626 43,383,274 450,288,419

"We can also slice pancakes and cakes, the maximum number of pieces made by two-dimensional and three-dimensions cuts. The pancake numbers[435] are just half the product of the cut number and its successor plus one.

pancake 2 4 7 11 16 22 29 37 46 56 67 79 92 106 121 137 154

"The cake numbers[436] are a sixth of the cube of the cuts plus five times the cuts plus six.

cakes 2 4 8 15 26 42 64 93 130 176 232 299 378 470 576 697

"The difference between pancake and cake seems to gives us pan-, another prefix for lowering a sequence, though by very, very little, unlike the sub- prefix. This is very different than the use of the prefix on the word pandigital, where it means "all".

pancake - cake	0	0	-1	-4	-10
pandigital	1,023,456,789	1,023,456,798	1,023,456,879	1,023,456,897	1,023,457,689
"digital"	1,023,456,789	1,023,456,798	1,023,456,878	1,023,456,893	1,203,457,679

"Joyce called beheaded pandigitals 'andigitals'. [437]

andigital 23,456,789 23,456,798 23,456,879 23,456,897 23,457,689

"And the beheaded cake numbers the Marie Antoinette numbers.[438]

Marie Antoinette	5	6	2	4	3	18	30	76	32	99	78	70	76	97	34	88	160	351	562	794	48
sorted	2	3	4	5	6	18	30	32	34	48	70	76	78	88	90	97	99	160	176	178	325

"Besides numbers named for cakes and pancakes, we also have doughnut numbers for the number of pieces for cuts of a doughnut or torus.[439]

doughnut	2	6	13	24	40	62	91	128	174	230	297	376	468	574	695	832	986	1,158	1,349

"There are also mnemonics for other constants, like that for the base of the natural logarithms,[440] 'ephilogisms' like, 'It enables a numskull to memorize a quantity of numerals'."

almost-e	2	7	1	8	2	8	1	8	2
sliced e	2	7	18	28	182	845	904	5,235	36,287
diced e	2	71	828	1,828	45,904	523,536	287,461	35,266,249	775,724,709

"Pi can be approximated as three, but more closely approximated as three and a seventh or twenty-two-sevenths," Lana continued, after making a "pi chart".

when	who	what	accuracy
2700 B.C.	Egyptians	$4/\phi^{1/2} = 3.1446$	1%o
1950 B.C.	Babylonians	4073/1296 = 3.1427469136	4%oo
250 B.C.	Archimedes	3123/994 = 3.1418511066	8%ooo
150 A.D.	Claudius Ptolemy	377/120 = 3.14166666	2%ooo
264 A.D.	Lin Hsin	3.14159	6 places
460 A.D.	Zu Changzhi	355/113 = 3.1415929204	7 places
1429 A.D.	Al-Kashi		16 places
1610 A.D.	Ludolph von Ceulen		35 places
1630 A.D.	Christoph Grienberger		39 places
1699 A.D.	Abraham Sharp		71 places
1706 A.D.	John Machen		100 places
1719 A.D.	Thomas Fantet DeLagney		112 places
1794 A.D.	Georg Vega		140 places
1844 A.D.	L.K.Schulz von Strassnitsky & Zacharias Dase		200 places
1847 A.D.	Thomas Clausen		248 places
1853 A.D.	William Rutherford		440 places
1873 A.D.	William Shanks		527 places

1947 A.D.	D.F. Ferguson	808 places
1949 A.D.	G.W. Reitwieser	2,037 places
1954 A.D.	S.C. Nicholson & J. Jeenel	3,089 places
1958 A.D.	F. Genuys	10,000 places
1973 A.D.	J. Guilloud	1,000,000 places

"I have a phiphilogism: 'I desire a solution gloriously new and inventive, artistic, creative, however very memorable.'" [441]

almost-phi	1 6	1	8	0	3	3	9	8
sliced phi	1 6	180	339	887	4,989	48,482	458,634	3,656,381
diced phi	161	803	3,988	74,989	484,824	5,863,436	5,638,117,720	30,917,980,576

"By George, I think she's got it," Al said, quoting Prof. Henry Higgins, which signaled a chorus of "The Rain in Spain", which morphed into "The Drain on Brains".

from *PAJAS* #67 (February 1991)

MORE EXPANSIONS

Al quoted Sirach 34:6, "Unless they are specially sent by the Most High, do not fix your heart on [dreams]," and then asked, "Why are fleas mathematical?"

Old Sam knew this oldie, "They subtract happiness, divide attention, add misery and multiply rapidly."

Al continued, "Continued fraction expansion[442] first appeared in the works of the Indian mathematician Aryabhata in the Sixth Century. They were originally used in solving 'squaring the rectangle' problems. A two-by-three rectangle could be subdivided into a two-by-two square and two one-by-one squares. If the side of the rectangle were an irrational hypotenuse, as in the golden rectangle with a side of phi. It could be subdivided into ever smaller and smaller golden rectangles with the vertices marking the Root Five Spiral.

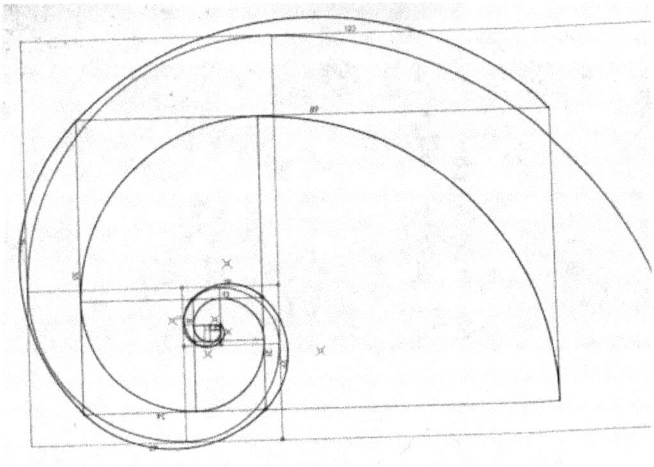

"The continued fraction expansion of phi is then the simplest, just ones. It can reveal other more complex and otherwise hidden patterns. The square root of two, for example, also begins with a one but then repeats two. " [443]

two	1	4	1	4	2	1	3	5
sliced	1	4	14	21	35	62	3,730	9,504
diced	1	41	421	3,562	37,309	504,880	1,688,724	20,969,807
fracked	1	2	2	2	2	2	2	2

"This 'fracking' gives closer and closer rational approximations to an irrational as well as numerator and denominator sequences."

At that point Brother Ken commented, "There's a fine line between a numerator and a denominator."

This prompted calling the numerator sequence "broken-up", the denominator sequence "broken-down" and a combined non-stuttering sequence "broken" in honor of "Bro. Ken" and the exploration of these sequences for the square roots of two.

~√2	1/1	5/2	6/3	29/14	64/31	93/45	343/166	1,808/875	11,191/5,416	68,954/11,707
broken-up	1	5	6	29	64	93	343	1,808	11,191	68,954
broken-down	1	2	3	14	31	45	166	875	5,416	11,707
broken	1	2	3	5	6	14	29	31	45	65

"We can do the same things with other roots, root three,[444] root five,[445] root six, etc." [446]

almost-root three	1	7	3	2	0	5	8	0
sliced	1	7	320	580	756	887	7,293	52.744
diced	1	73	205	8,075	68,877	293,527	4,463,415,058	72,366,942.805
fracked	1	1	2	1	2	1	2	1

"Because of the early-onset zeroes in root three and five we soon get stuttering when breaking-up or -down."

broken-up	1	8	25	58	25	183	1,489	183	2,770	19,573	120,208
broken-down	1	2	7	16	7	51	415	51	772	3,911	24,238
broken	1	2	7	8	16	25	51	58	183	415	772

almost-root five	2	2	3	6	0	6	7	9	7
sliced	2	23	60	67	97	774	997	8,969	64,040
diced	2	23	606	7,977	49,978	969,640	4,091,736	68,731,276	235,440,618,359
fracked	2	2	4	4	4	4	4	4	4

broken-up	2	5	17	107	17	209	1,480	13529	96183	686,810	2,843,423
broken-down	1	3	10	63	10	123	871	7,962	56,605	404,197	1,673,393
broken	1	2	3	5	10	17	107	123	209	871	1,480

almost-root six	2	4	4	9	4	8	9	7	2
diced	2	44	948	9,742	78,317	809,819	7,284,074	70,589,139	196,594,748
sliced	2	4	49	489	742	783	1,780	9,819	72,840
fracked	2	2	4	2	4	2	4	2	4

broken-up	2	5	22	49	218	485	2,158	4,801
broken-down	1	2	9	20	89	198	881	1,960
broken	1	2	5	9	20	22	49	89

"Cube roots can also generate sequences for two,[447] three,[448] four,[449] five,[450] six,[451] etc.

almost-cube root two	1	2	5	9	9	2	1	0	4	9	4	8	7
fracked cube root two	1	3	1	5	1	1	4	1	8	1	1	14	1
broken-up	1	4	5	29	34	63	286	349	3,078	3,427	6,505	94,497	101,002
broken-down	1	1	2	11	13	24	109	133	1,173	1,306	2,479	36,012	38,491
broken	1	2	4	5	11	13	24	29	34	63	109	133	286

almost-cube root three	1	4	4	2	2	4	9	5	7	0	3	0	7
fracked cube root three	1	2	3	1	4	1	5	1	1	6	2	5	2
broken-up	1	3	10	13	62	75	437	512	949	6,206	13,361	73,011	159,383
broken-down	1	2	7	9	43	52	303	355	658	4,303	9,264	50,623	110,510
broken	1	2	3	7	10	13	43	52	62	75	303	355	437

almost-cube root four	1	5	8	7	4	0	1	0	5	1	9	6	8
fracked cube root four	1	1	1	2	2	1	3	2	3	1	3	30	1
broken-up	1	2	3	8	19	27	100	227	781	1,008	3,805	115,158	118,963
broken-down	1	1	2	5	12	17	63	143	492	635	2,397	72,545	74,942
broken	1	2	3	5	8	12	17	19	27	63	100	143	227

almost-cube root five	1	7	0	9	9	7	5	9	4	6	6	7	6
fracked cube root five	1	1	2	2	4	3	3	1	5	1	1	4	10
broken-up	1	2	5	12	53	171	566	737	4,251	4,988	9,239	41,944	428,679
broken-down	1	1	3	7	31	100	331	431	2,486	2,917	5,403	24,529	250,693
broken	1	2	3	5	7	12	31	53	100	171	331	431	737

almost-cube root six	1	8	1	7	1	2	0	5	9	2
fracked cube root six	1	1	4	2	7	3	508	1	5	5
broken-up	1	2	9	20	149	467	237,385	237,852	1,426,645	7,371,077
broken-down	1	1	5	11	82	257	130,638	130,895	785,113	4,056,460
broken	1	2	5	9	11	20	82	149	257	467

"Pi and half-pi too can be expressed as continued fractions, though there does not seem to be a pattern to either.[452]

fracked pi	3	7	15	1	292	1
approximation	3	22/7	333/106	355/113	103,993/33,102	104,348/33,215
broken-up	3	22	333	355	103,993	104,348
sub	1	8	123	131	38,257	38,387
broken-down	1	7	106	113	33,102	33,215
broken	1	3	7	22	106	113

fracked half-pi	1	1	1	3	31	1	145
approximation	1/1	2/1	3/2	11/7	344/219	355/226	51,819/32,989
broken-up	1	2	3	11	344	355	51,819
broken-down	1	1	2	7	219	226	32,989
broken	1	2	3	7	11	219	226

"The base of the natural logarithm, e and half-e, can also be expressed as continuing fractions, which do have easily recognizable patterns.[453]

fracked e	2	1	2	1	1	4	1	1	6	1
approximation	2/1	3/1	8/3	11/4	19/7	87/32	106/39	193/71	1,264/465	1,457/536
broken-up e	2	3	8	11	19	87	106	193	1,264	1,457
broken-down e	1	1	3	4	7	32	39	71	465	536
broken e	1	2	3	4	7	8	11	19	32	39

fracked half-e	1	2	1	3	1	1	1	3	3
approximation	1/1	3/2	4/3	15/11	19/14	34/25	53/39	193/142	832/465
broken-up	1	3	4	15	19	34	53	193	632
broken-down	1	2	3	11	14	25	39	142	465
broken	1	2	3	4	11	15	19	25	34

STILL MORE EXPANSIONS

from *PAJAS* #68 (March 1991)

Al read from Augustine's commentary on Psalm 91, "He dwells under the defence of the Most High, who is not proud."

Then humbly handed the meeting over to Cal who asked, "What's an eleven-letter word for a gathering of three people?"

"A get-threegether." [454] Lana answered and then quickly moved on to current fad, expansions, "Just as a number can be reinterpreted as a series of continuing fractions, so too can a series of numbers be re-interpreted as a new number that can be read as digitally expanded, diced or sliced.

"Phi for example can be expressed as a series of continued fractions that are all ones. Its approximations are all ratios of consecutive Fibonacci numbers. If however we read its decimal expansion as a series of continued fractions with leading or trailing zeroes we get a different irrational numbers, in fact, depending upon how the digits are subdivided an infinite number of them, including two-place,[455] three-place,[456] and even alternating one/two- [457] and two/one-place." [458]

two-place	16	18	3	39	88
broken-up	16	289	883	34,726	3,056,771
tressub	1	14	44	1,729	152,188
broken-down	1	18	55	2,163	190,399
broken	1	16	18	55	289

three-place	161	803	398	874	989
broken-up	161	129,284	51,455,193	44,971,967,966	44,477,327,773,567
pentasub	1	871	346,702	303,018,736	299,685,877,199
broken-down	1	803	319,595	279,326,833	276,254,557,432
broken	1	161	803	129,284	319,595

one/two-place	1	61	8	3	98	8	74	9
broken-up	1	62	497	1,553	152,691	1,223,081	90,660,685	817,169,246
broken-down	1	61	489	1,528	150,233	1,203,392	8,9201,241	804,014,561
broken	1	61	62	489	497	1,528	1,553	150,233

two/one-place	16	1	83	9	88	7	49	8
broken-up	16	17	1,427	12,860	1,133,107	7,944,609	390,418,948	3,131,296,193
broken-down	1	1	84	757	66,700	467,657	22,981,893	184,322,801
broken	1	16	17	84	757	1,427	12,860	66,700

"Each of these in turn generate a new numerator, denominator and combined sequence and each of these three sequences converge to three new irrationals that can in turn generate three more sequences. Those nine in turn generate twenty-seven more new sequences and nine more irrationals and so on and so on ad infinitum.[459]

3 powers	3	9	27	81	243	729	2,187	6,561	19,683	59,049	177,147	531,441	1,594,323 ...

"The natural numbers, as the first example, do, with varying prime densities.[460]

broken-up	1	3	10	43	225	1,393	11,369	103,714	1,048,509	12,685,822
broken-down	1	2	7	30	157	972	7,933	72,369	731,623	8,851,845
broken	1	2	3	7	10	30	43	157	225	972

"The odds[461] and evens[462] do the same."

broken-up odd	1	4	21	151	1,380	15,331	200,683	3,025,576	51,635,475
broken-down odd	1	3	16	115	1,051	11,676	152,839	2,304,261	39,325,276
broken odd	1	3	4	16	21	115	151	1,051	1,380

broken-up even	0	1	4	25	204	2,065	24,984	351,841	5,654,440
broken-down even	1	2	9	56	457	4,626	55,969	788,192	12,667,041
broken even	0	1	2	4	9	25	56	204	457

That prompted meeting-ending "Odds and evens do the same. Do they? Do they? Odds and evens, do the same, all the du-dah day."

from *PAJAS* #69 (April 1991)

DOMINISSIMO REVISITED

Al quoted Daniel 4:14, "The Most High is Sovereign over Human kingship, giving it to whom he wills, and setting it over the lowliest of mortals," and then changed the mood by asking, "Why couldn't the pet shop's snakes multiply?"

"They were adders."

In the aftermath of the games night on Pi Day, Einstein's birthday, in this numberdromic year, the veteran members introduced newcomers to some of the unfamiliar terms of Dominissimo and newbies added new terms to the already large repertoire.

"A 'bean' means either a five or a seven, from the once famous slogan of Heinz, that led to 'beans' becoming the Tombola term for fifty-seven."

"So beans[463] would be those with either a five or a seven 'bean' or 'semiheinz' and the Heinzish numbers[464] would have a fifty-seven, or more generally a five-seven, frag."

bean	5	7	15	17	25	27	35	37	45	47	55	57	65	67	75	77
Heinz	57	157	257	357	457	557	657	757	857	957	1,057	1,157	1,257	1,357	1,457	1,557

"I see a pattern. The differences between the beans alternates between two and eight."

"Yes, it does, for a while. Think ahead to what happens when you get as high as the five hundreds."

"Oh, yeah, at four hundred ninety-seven it gets higher by only three."

"'Bingo!"

"Our use of the word 'bingo' comes not from Bingo but from Dominoes where it means the blank doublet.[465] In roulette the double zero is half of the Green or Eagle suit with the singleton zero being the other half. It could also be called, and often is, 'oh-oh'. A doublet, aka 'po' from the Tien Gow for pair, is any two identical digits.[466] The first bingo prime is a thousand nine."

bingo	100	200	300	400	500	600	700	800	900	1,001	1,002	1,003	1,004	1,005	1,006

po	11	22	33	44	55	66	77	88	99	100	101	112	113	114	115	116	117	118	119	121	122	131	133

"More than one po would be 'pos'." [467]

pos	1,001	1,010	1,100	1,122	1,133	1,144	1,155	1,166	1,177	1,188	1,199	1,212	1,221	1,313	1,331	1,414

"One nickname for the two doublet comes from Craps where it's 'snake-eyes'. When in combination with other digits, they could also be called, aces from cards, or 'legs' from their shape and their coming in pairs, 'bullets' from Dominoes or 'wonks' from Tombola." [468]

wonks	11	101	110	111	112	113	114	115	116	117	118	119	121	131	141	151	161	171	181	191	211	311	411

"So would three ones be wonks also?"

"Three-of-a-kind would also qualify as more than one wonk. You could however call it a 'wonks royal' by analogy with Cribbage's pair royal to distinguish it, if you like."

"The two doublet we call 'ducks' from their shape or by the homonym 'dux',[469] 'bench' from Tien Gow or even 'buckles' from Rhyming Slang's "two, buckle me shoe".

"Ah, so three twos would be 'dux royal', a royal duke, eh?"

"Yeah, it sounds better than 'royal bench or buckles'".

dux	22	122	220	221	223	224	225	226	227	228	229	322	422	522	622	722	822	922

"The three doublet's names come from thirty-three, Restaurantese for cherry cola, treys from cards, crabs from their shape or fevvers from Tombola from the Cockney phrase "ferty-free fahsand fevvers on a frush's froat'.[470]

fevvers 33 133 233 313 323 330 331 332 333 334 335 336 337 338 339 433 533 633 733

"The four doublet is called 'man' [471] in Tien Gow and 'sailboats' from their shape or 'doors' from Rhyming Slang. The first prime man or 'Adam' number, is four forty-three. More than one pair of fours would be 'men'. [472] A 'sedan' can have either two or four 'doors'."

man 44 144 244 344 440 441 442 443 445 446 447 448 449 544 644 744 844 944

men 4,444 14,444 24,444 34,444 44,044 44,144 44,244 44,344 44,544 44,644

"The five doublet we identify with the Restaurantese code for 'rootbeer',[473] aka 'plum flower' in Tien Gow or simply 'nickles'. Since four and five are complimentary in base ten, this complimentary number to 'man' could also be called 'woman', which naturally gives us the name for the prime woman or 'Eve' numbers, beginning with five fifty-seven. Numbers with at least four fives could be called 'women'." [474]

rootbeer 55 155 255 355 455 550 551 552 553 554 555 556 557 558 559

women 5,555 15,555 25,555 35,555 45,555 55,055 55,155 55,255 55,355 55,455

"'Buzz' in the game of the same name means any multiple of five, in other words either the even loved or the odd gobi."[475]

buzz 5 10 15 20 25 30 35 40 45 50 55 60 65 70 75 80 85 90 95 100 105 110

"The six doublet is called 'clicketies' or 'clickety-clicks' from Tombola, 'boxcars' from Craps, 'heaven' from Tien Gow and 'toms' from Rhyming Slang's Tom Mix." [476]

heavenly 66 166 266 366 466 566 606 616 626 636 646 656 660 661 662 663 664

"Numbers with three sixes would be 'beastly'!" [477]
"Ah, from the first being *Revelation*'s Number of the Beast."
"The first beastly prime would be six thousand six hundred sixty-one."

beastly 666 1,666 2,666 3,666 4,666 5,666 6,066 6,166 6,266 6,366 6,466 6,566

"I'd call numbers with two pairs of sixes by a less beastly name like 'tomtoms'," [478] Tom suggested.
"So the first tomtom prime would be a hundred sixty-six thousand, six hundred sixty-seven then!"

tomtoms 6,666 16,666 26,666 36,666 46,666 56,666 60,666 61,666 62,666

"Since sixes and threes are complimentary, threes could be called queens for the female cat, as opposed to the male tom cat. Those numbers with more than one three, could then be called 'merry',[479] a pun alluding to both Mary Immaculate, queen of the universe, and Mary contrary, queen of the Scots."

queens 3 13 23 30 31 32 33 34 35 36 37 38 39 43 53 63 73 83 93 103 113 123 130 131

merries 33 133 233 330 331 332 333 334 335 336 337 338 339 433 533 633 733 833 933

"What do you get if you put toms and queens together?"
"Obviously 'kittens'."
"And their complimentary sequence just as obviously 'mittens'." [480]

kittens 3 13 23 30 31 32 33 34 35 36 37 38 39 43 53 63 66 73 83 93 103 113

mittens 6 16 26 36 46 56 60 61 62 63 64 65 66 67 68 69 76 86 106 116 126 136

"Is three thirty-three 'merry'?"
"Yes, because it refers to more than one three or queen, but it also would be counted as a proil or three-of-a-kind."
"It'd also be a half-beast, half of three sixes, like a werewolf?"
"A werebeast, 'were' [481] for short."

weres 333 1,333 2,333 3,033 3,133 3,233 3,303 3,313 3,323 3,343 3,353 3,363 3,373

"The seven doublet is called 'crutches' [482], which often come in pairs, or hockey sticks from their shape."

crutches 77 177 277 377 477 577 677 770 771 772 773 774 775 776 777 778 779 787 797

"The eight doublet is called 'gates' from Rhyming Slang or 'waafs' [483] from the Women's Auxilary Air Force or snowmen and the nine doublet 'pothooks' for their shape." [484]

waafs 88 188 288 388 488 588 688 788 880 881 882 883 884 885 886 887 888 889

pothooks 99 199 299 399 499 599 699 799 899 909 990 991 992 993 994 995 996

"From Craps a one and a two together are called cockeyes.[485] As one-two or twelve they're called a duzz,[486] a contraction of a dozen. Its ananum, two-one or twenty-one, is called 'sam' [487] in Tien Gow. 'Duzz' seems to change meaning when prefixed. 'Two-duzz' from Tombola could mean two-four,[488] 'three-duzz' three-six[489] and 'four-duzz' four-eight." [490] Their ananums, which name the same domino oriented in the opposite direction, of course, would be 'does-too',[491] 'does-three' [492] and 'does-for'." [493]

cockeyed 12 21 120 123 124 125 126 127 128 129 201 210 211 213 214 216 217 218 219

"The first one-duzz prime is one hundred twenty-seven."
"But the next isn't until one thousand, one hundred twenty-three."

one-duzz 12 112 120 121 122 123 124 125 126 127 128 129 212 312 412 512

sam 21 221 321 421 521 621 721 821 921 1,021 1,121 1,221 1,321 1,421 1,521 1,621 1,721 1,821

"The first two-duzz prime is twenty-four eleven and the first does-too prime is forty-two hundred one."

two-duzz	24	124	224	324	424	524	624	724	824	924	1,024	1,124	1,224	1,324	1,424	1,524
does-too	42	142	242	342	442	542	642	742	842	942	1,042	1,142	1,242	1,342	1,442	1,542

"The first three three-duzz prime is thirty-six hundred seven."

three-duzz	36	136	236	336	436	536	636	736	836	936	1,036	1,136	1,236	1,336	1,436
does-three	63	163	263	363	463	563	663	763	863	963	1,063	1,163	1,263	1,363	

"The first four-duzz prime is forty-eight hundred one and the first three does-for prime eight-four nineteen."

four-duzz	48	148	248	348	448	548	648	748	848	948	1,048	1,148	1,248	1,348	1,448
does-for	84	184	284	384	484	584	684	748	848	948	1,048	1,148	1,248	1,348	

"'Five-duzz' however does not follow this pattern. It's six-zero, commonly called 'six-o', [494] and it's ananum, zero-six commonly called 'o-six'. [495] The 'sixties' would be a subsequence. The first six-o prime is sixty-o-one and the first o-six prime one-o-six-one, aka one thousand sixty-one."

six-o	60	160	260	360	460	560	600	601	602	603	604	605	606	607	608
sixties	60	160	260	360	460	560	660	760	860	960	1,060	1,160	1,260	1,360	1,460
o-six	106	206	306	406	506	606	706	806	906	1,006	1,060	1,061	1,062	1,063	1,064

"This pattern can be repeated for the rest of the blank suit: 'dime' [496], 'o-one',[497] 'score',[498] 'o-two',[499] 'three-o',[500] 'o-three',[501] 'four-o',[502] 'o-four',[503] 'five-o',[504] 'o-five',[505] 'seven-o',[506] 'o-seven',[507] 'eight-o',[508] 'o-eight',[509] 'nine-o' [510] and 'o-nine'." [511]

"The first 'dime' prime is ten-oh-nine, aka ten hundred nine."

dimes	10	110	210	310	410	510	610	710	810	910	1001	1002	1003	1004	1005	1006	1007
oh-ones	101	201	301	401	501	601	701	801	901	1001	1101	1201	1301	1401	1501	1601	1701

"The first twenty primes is twenty oh three and the first oh-two prime one oh two oh one."

score	20	120	220	320	420	520	620	720	820	920	1020	1120	1220	1320	1420
o-two	102	202	302	402	502	602	702	802	902	1002	1102	1202	1302	1402	1502

"The first three-o or 'half-past' prime is thirty hundred one, while the first o-three prime is the first, one hundred three."

half-past	30	130	230	330	430	530	630	730	830	930	1,030	1,130	1,230	1,330	1,430
o-three	103	203	303	403	503	603	703	803	903	1003	1103	1203	1303	1403	1503

"The first four-o prime is four hundred one and the first o-four prime one hundred four thousand three."

four-o	40	140	240	340	440	540	640	740	840	940	1,040	1,140	1,240	1,340	1,440
o-four	104	204	304	404	504	604	704	804	904	1004	1104	1204	1304	1404	1504

"The first five-o prime is five thousand three and the first o-five prime is one hundred five thousand nineteen."

five-o	50	150	250	350	450	550	650	750	850	950	1,050	1,150	1,250	1,350	1,450
o-five	105	205	305	405	505	605	705	805	905	1005	1,105	1,205	1,305	1,405	1,505

"The first seven-o prime is seven thousand one."

seven-o	70	170	270	370	470	570	670	770	870	970	1,070	1,170	1,270	1,370
oh-seven	107	207	307	407	507	607	707	807	907	1,007	1,107	1,207	1,307	1,407

"The first eight-o prime is eight thousand nine and the first oh-eight prime is one hundred eight thousand seven."

eight-o	80	180	280	380	480	580	680	780	880	980	1,080	1,180	1,280	1,380	1,480
o-eight	108	208	308	408	508	608	708	808	908	1,008	1,108	1,208	1,308	1,408	1,508

"The first nine-o prime is nine thousand one, while the first nine-o prime is the first a hundred nine."

nine-o	90	190	290	390	490	590	690	790	890	990	1,090	1,190	1,290	1,390	1,490
o-nine	109	209	309	409	509	609	709	809	909	1009	1109	1209	1309	1409	1509

"One and three are collectively 'harmony' [512] in Tien Gow, so number containing both are 'harmonious' and those much more common ones without both 'disharmonious'."

harmonious	13	31	103	113	123	130	131	132	133	134	135	136	137	138	139	143	153	163	173	183

disharmonious	1	2	3	4	5	6	7	8	9	10	11	12	14	15	16	17	18	19	20	21	22	23	24	25	26

'Fewteen' refers to thirteen to fifteen.[514] 'Midteen' to sixteen[515] and 'moreteen' to seventeen to nineteen." [516]

fewteen	13	14	15	113	114	115	213	214	215	313	314	315	413	414	415	513	514	515	613	614

midteen	16	116	216	316	416	516	716	816	916	1,016	1,116	1,216	1,316	1,416	1,516	1,616

moreteen 17 18 19 117 118 119 217 218 219 317 318 319 417 418 419 517 518 519 617 618

"So thirteen hundred would be a fewteen, sixteen hundred a midteen and seventeen hundred a moreteen?"

"Yes, you do count by hundred as well as thousands, don't you?"

"But no longer by myriads, so thirteen thousand would count, but not thirteen myriads."

"The fewteen number fifteen is contracted to 'fiff' as in a quarter past the hour or base sixty. Fiffish numbers would be those that have a one-five frag that can called fifteen, in decimal, hexadecimal and higher.[516] It's also known as 'big bit', since it's the bigger 'half' of two bits or 'pieces of eight'."

fiffish 15 115 150 151 152 153 154 155 156 157 158 159 215 315 415 515 615 715 815 915

"'Fewty', by analogy with 'fewteen', means twenty, thirty or forty. 'Midty' is fifty, that is, the same as five-o and 'morety' sixty, seventy, eighty or ninety. [518]

fewty 20 30 40 120 130 140 220 230 240 320 330 340 420 430 440 520 530

morety 60 70 80 90 160 170 180 190 260 270 280 290 360 370 380 390 460

"These prefixes can be further extrapolated up to fewplex and beyond." [519]

"Thirty-nine is 'steps' [520] from the Alfred Hitchcock film based on the book by John Buchan, so either a three of a nine would be a 'half-step'." [521]

steps 39 139 239 339 439 539 639 739 839 939 1,039 1,139 1,239 1,339 1,439

half-steps 3 9 13 19 23 29 30 31 32 33 34 35 36 37 38 43 49 53 59 63 69 73 79 83 89

"Three, five and seven together are 'valles' in Panguingue.[522] One of the three would be 'tertivalles' [523] and two of the three 'bevalles' or 'vo', the portmanteau for 'valles po'.[524]

valles 357 375 537 573 735 753 1,357 1,375 1,537 1,573 1,735 1,753 2,357 2,375 2,537

tertivalle 3 5 7 13 15 17 23 25 27 30 31 32 33 34 36 38 39 43 45 47 50 51 52 54 55 56

vo 35 37 53 57 73 75 135 137 153 157 173 175 235 237 253 257 273 275 305 307

"Forty-two is a 'lok' [525] from Tien Gow and 'semilok' [526] would therefore be either two or four."

lok 42 142 242 342 420 422 423 420 421 422 423 424 425 426 427 428 429 442

semilok 2 4 12 14 20 21 22 23 25 26 27 28 29 32 34 40 41 43 44 45 46 47 48 49 52

"Sixty-two is 'ticketty-boo' in Tombola.[527] Just plain 'ticketty' would therefore be an alias for six-o, 'ticket' for six, 'booty' for two-o or score and 'boo-boo' for dux."

ticketty-boo 62 162 262 362 462 562 620 621 622 623 624 625 626 627 628 629

"Sixty-four in Tien Gow is called a partition.[528] A semipartition is either a four or six." [529]

partition 64 164 264 364 464 564 664 764 864 964 1,064 1,164 1,264 1,364

semipartition 4 6 14 16 24 26 34 36 40 41 42 43 44 45 46 47 48 49 54 56 60 61

"Seventy-six is 'trombones' [530] from the "Music Man" song, "Seventy-six Trombones", so individually six or seven each would be a singular 'trombone'."

trombones 6 7 16 17 26 27 36 37 46 47 56 57 60 61 62 63 64 65 66 67 68 69

"Three-of-a-kind is called a proil,[531] a portmanteau of pair royal from Cribbage, and a proil of fours or higher a 'gleek' [532] from Gleek. A proil of six or less is a 'raffle' from Craps.[533] More than one proil is 'proils'. [534]

proil 111 222 333 444 555 666 777 888 999 1,000 1,011 1,101 1,110 1,112 1,113 1,114 1,115 1,116

gleek 444 555 666 777 888 999 1,444 1,555 1,666 1,777 1,888 1,999 2,444 2,555 2,666

raffle 111 222 333 444 555 666 1,000 1,011 1,101 1,110 1,112 1,113 1,114 1,115 1,116 1,117 1,118

proils 100,011 100,101 100,110 101,001 101,010 101,100 110,001 110,010 110,100 111,222 111,333

"Why do the proils include both zeroes and sixes? I've seen dice with six, twelve and twenty sides and cubical ones marked with both zero to five and one to six, but never a seven-sided die!"

"Ah, you must not have gotten to *Wuthering Heights* yet."

"How high is that?"

"It's a novel by Emily Brontë in which a supposedly six-sided die breaks in two exposing its blank inside."

"More than one gleek is 'gleeks' or 'gleex', which have no zeroes." [535]

gleex 444,555 444,666 444,777 444,888 444,999 445,455 445,545 445,554 446,466

"'Chut' from Tien Gow is either two and five or three and four, both which sum to seven.[536] A semichut would be a two without a five, a three without a four, a four without a five or a five without a four." [537]

chut 25 34 43 52 125 134 143 152 205 215 225 235 245 250 251 252 253 254

semichut 2 3 4 5 12 20 21 26 27 28 29 30 31 36 37 38 39 40 41 46 47 48 49 50 51

"The four digits, zero, one, two and three, are collectively called 'quatre' [538] in Roulette. Any two of the four would be a 'semiquatre'. [539]

quatre 1,023 1,032 1,203 1,230 1,302 1,320 2,013 2,031 2,103 2,130 2,301 2,310 3,012

semiquatre 10 12 13 20 21 23 30 31 32 104 105 106 107 108 109 204 205 206 207

"By extrapolation zero, one and two would be 'trois', [540] one of these three would be 'tertitrois' and two 'betrois',[541] zero and one 'deux' [542] and just zero 'un'. [543]

trois 102 120 201 210 1,203 1,204 1,205 1,206 1,207 1,208 1,209 1,302 1,320 1,402

tetritrois 0 1 2 13 14 15 16 17 18 19 23 24 25 26 27 28 29 30 31 32 40 41 42 50 51 52

betrois 10 12 20 21 103 104 105 106 107 108 109 110 112 123 124 125 126 127 128 129

deux 10 100 101 102 103 104 105 106 107 108 109 110 120 130 140 150 160 170 180 190 201 210

"Zero through four would be 'cinq' or 'bicycle' from Lowball.[544] Zero through five 'six', [545] any two would be 'tertisix', [546] any three 'semisix', [547] and any four 'besix'.[548]

bicycle 10,234 10,243 10,324 10,342 10,423 10,432 12,034 12,043 13,024 13,042 14,023

six 102,345 102,354 102,435 102,453 103,245 103,254 104,235 104,253 104,325 104,352

tertisix 10 12 13 14 15 20 21 23 24 25 30 31 32 34 35 40 41 42 43 45 50 51 52 53 54

semisix 102 103 104 105 120 123 124 125 130 132 134 135 140 142 143 145 150 152 153 154

besix 1,023 1,024 1,025 1,032 1,034 1,035 1,042 1,043 1,045 1,052 1,053 1,054 1,203

"Zero through six would be 'sept'.[549] Zero through seven would be 'huit',[550] any four between zero and seven, 'semihuit'.[551]

sept 1,023,456 1,023,465 1,023,546 1,023,564 1,023,645 1,023,654 1,024,345 1,024,354

huit 10,234,567 10,234,576 10,234,657 10,234,675 10,234,756 10,234,765 10,235,456

semihuit 1,023 1,024 1025 1,026 1,027 1,032 1,034 1,035 1,036 1,037 1,042 1,043 1,045

"Zero through eight would be 'neuf'.[552]

neuf 102,345,678 102,345,687 102,345,768 102,345.786 102,345,867 102,345,876

"Zero through nine, another name for pandigital, would be 'dix'. 'Sesquintidix' would therefore describe any three and 'semidix' any five." [553]

semidix 10,234 10,235 10,236 10,237 10,238 10,239 10,324 10,325 10,326

sesquintidix 102 103 104 105 106 107 108 109 120 123 124 125 126 127 128 129 130 131 132

from *PAJAS* #70 (May 1991)

MORE DOMINISSIMO

Al opened with Psalm 83:18, "You alone, Whose name is the Lord, are the Most High over all the Earth," and then asked, "If cherry cola is 'thirty-three' in Restaurantese, what's cherry cola on the rocks?"

"'Thirty-three cubed'," someone answered.

"Thirty-five thousand, five hundred thirty-seven," another soon calculated.

For those who missed March's Dominissimo tourney, old and new terms were shared.

"'Book' [555] from Authors means four-of-a-kind and 'books' or 'boox' two four-of-a-kinds." [556]

book 1,111 2,222 3,333 4,444 5,555 6,666 7,777 8,888 9,999 10,111 11,011 11,101 11,110 11,112

boox 10,000,111 10,001,011 10,001,101 10,001,110 10,010,011 10,010,101 10,010,110 10,011,001

"A 'sesquiwonks' is a one-word synonym for the phrase 'wonk proil'." [557]

sesquiwonks 111 1,011 1,110 1,112 1,113 1,114 1,115 1,116 1,117 1,118 1,119 1,211 1,311 1,411 1,511

"'Bust' [558] from Blackjack is a number with a weight higher than twenty-one. 'Busts' [559] would be a number with a digit sum higher than twice that, forty-two."

bust 778 779 787 788 789 877 878 879 976 977 978 979 985 986 987 988 989 994

busts 79,999 88,999 97,999 98,999 99,799 99,899 99,979 99,989 99,997 99,998 99,999

"'Left bucket' from bowling means two, four, five and eight, while 'right bucket' means three, five, six and nine.[560]

left bucket 2,458 2,485 2,548 2,584 2,845 2,854 4,258 4,285 4,528 4,582 4,825
right bucket 3,568 3,586 3658 3685 3856 3865 5,368 5,386 5,638 5,683 5,836

"'Craps' [561] from Craps is two sixes, two ones or a one and two."

craps 11 12 21 66 110 113 114 115 116 117 118 119 120 123 124 125 126 127 128 129 131

"A 'quad' [562] means any four-of-a-kind, 'redux' [563] four twos and 'quints' [564] any five-of-a-kind."

quads 1,111 2,222 3,333 4,444 5,555 6,666 7,777 8,888 9,999 10,000 10,111 11,011

redux 2,222 20,222 21,222 22,022 22,122 22,202 22,220 22,221 22,223 22,224

quints 11,111 22,222 33,333 44,444 55,555 66,666 77,777 88,888 99,999 100,000

"'Go' [565] from Cribbage means a number with a weight of thirty, 'stop" its complimentary sequence, 'semigo' [566] one of fifteen and a 'sesquigo' [567] forty-five, 'bego' [568] twenty, and 'tertigo' [569] ten."

go 3,999 4,899 4,989 4,998 5,799 5,979 5,997 6,699 6,969 6,996 7,599 7,959

stop 1,006 1,050 1,104 1,140 1,203 1,230 1,302 1,320 1,401 1,410 1,500

semigo 69 78 87 96 159 168 177 186 195 249 258 267 276 285 294 339 348 357 sesquigo 99,999 189,999 198,999 199,899 199,989 199,998 279,999 288,999 289,998

bego 299 389 398 479 488 497 569 578 587 596 659 668 677 686 695 767 776

tertigo 19 28 37 46 55 64 73 82 91 109 118 127 136 145 154 163 172 181 190 208 217

"A 'hatchet' in Tien Gow is a five and a six and what might be called a semi-hatchet, a five or a six, but not both, we nickname a 'hatchling'." [570]

hatchet 56 65 156 165 256 265 356 365 456 465 556 560 561 562 563 564 565

hatchling 5 6 15 16 25 26 35 36 45 46 50 51 52 53 54 55 57 58 59 60 61 62 63 64

"A 'sabacc' from the Star War mythos has a weight of twenty-three.' [571]

sabacc 599 959 995 1,499 1,589 1,598 1,679 1,688 1,697 1,769 1,796 1,868 1,886 1,958

"A 'manque' [572] from Boule is a number with one, two, three and four" [573]

manque 1,234 1,243 1,324 1,342 1,423 1,432 2,134 2,143 2,314 2,341 2,413 2,431 3,124

A 'passe' also from Boule is a six, seven, eight and nine, so a 'semipasse' would be any of those two [574] and a 'quartipasse' just one of the four." [575]

passe 6,789 6,798 7,689 7,698 7,869 7,896 7,968 7,986 8,679 8,697 8,769 8,796 8,967

semipasse 67 68 69 76 78 79 86 87 89 96 97 98 167 168 169 176 178 179 186 187 189

quartipasse 6 7 8 9 16 17 18 19 26 27 28 29 36 37 38 39 46 47 48 49 56 57 58 59 60

"Since a 'ng' in Tien Gow is a one with a four or a two with a three, a 'seming' [576] is a one without a four, a two without a three, a three without a two or a four without a one.

ng 14 23 32 41 104 123 132 140 141 142 143 144 145 146 147 148 149 203 213 214 223 230

seming 1 2 3 4 10 11 12 13 15 16 17 18 19 20 21 22 24 25 26 27 28 29 30 31 33 34 35 36

"A 'stiff' [577] from Blackjack is any number weighing between twelve and sixteen. A 'half-stiff' would therefore be with half that, between six and eight [578] and a 'quarter-stiff' between three and four."

stiff 39 48 49 57 58 59 66 67 68 69 75 76 77 78 79 84 85 86 87 88 94 95 96 97

half-stiff 4 7 8 15 16 17 24 25 26 33 34 35 42 43 44 51 52 53 60 61 62 105 106 107

quarter-stiff 3 4 12 13 21 30 31 40 102 103 120 212 121 130 201 202 211 220 301 310 400

"Synonyms for drunk, also include 'half-blind', which we take to mean having two of the four digits with 'eyes': five, six, eight or nine. Totally 'blind' numbers would have none, [579] 'cyclops' numbers just one,[580] 'triops' numbers three, and four-eyed 'tetrops' numbers all four."

half-blind 56 58 59 65 68 69 85 86 89 95 96 98 156 158 159 165 168 169 185 186

blind 0 1 2 3 4 7 10 11 12 13 14 17 20 21 22 23 24 27 30 31 32 33 34 37 40 41 42 43 44

cyclops 5 6 8 9 15 16 18 19 25 26 28 29 35 36 38 39 45 46 48 49 50 51 52 53 54

triops 568 569 586 589 596 598 658 659 689 698 856 859 865 869 895 896 956

tetrops 5,689 5,698 5,869 5,896 5,968 5,986 6,589 6,598 6,859 6,895 6,958

"The most well-known blinded cyclops is, of course, Polyphemus in *Homer's Odyssey.* The supersequence combining cyclops and blind numbers, but excluding half-blind, triops or tetrops numbers, we call 'polyphemous' for him. The complimentary supersequence excluding the former and including the latter, we call by backformation 'monophemous', which start out like half-blind" [581]

"'Half-cockeyed' refers to numbers with a one without a two or a two without a one,[582] and 'half-looped' with swapped front half-frag and the reversed back half-frag of a number with an even number of digits. This excludes repdigits and/or multiples of ten." [583]

half-cockeyed 1 2 10 13 14 15 16 17 18 19 20 23 24 25 26 27 28 29 31 32 41 42 51 52

half-looped 12 13 14 15 16 17 18 19 21 23 24 25 26 27 28 29 31 32 34 35 36 37 38 39 42

"A 'yacht' or 'yatzee' [584] is a quint higher than fives."

yacht 66,666 77,777 88,888 99,999 166,666 266,666 366,666 466,666 566,666

"A 'napoletana' [585] from Tressente is a one, two and three, so a 'benapoletana' is two of three and tritonapoletana just one of three." [586]

napoletana 123 132 1,023 1,032 1,234 1,235 1,236 1,237 1,238 1,239 1,320 1,324 1,325 1,326

benapoletana 12 13 21 23 31 32 102 103 120 121 124 125 126 127 128 129 130 131 134 135

tritonapoletana 1 2 3 10 14 15 16 17 18 19 20 24 25 26 27 28 29 30 34 35 36 37

"If a 'bart' from Tien Gow is two and six or three and five, both of which add to eight." [587]

bart 26 35 53 62 126 135 153 162 226 235 253 262 326 335 353 362 426 435 453

"A Tutnese bart or 'bubarugtut' would be a bart just with two other digits between. The first such prime would be five thousand three." [588]

bubarugtut 2,006 2,016 2,026 2,036 2,046 2,056 2,066 2,076 2,086 2,096 2,106

"Similarly a Tutnese bingo or 'bubinungugo' [589] and a Tutnese sam or 'susamum', [590] count every third digit."

bubinungugo 10,110 10,120 10,130 10,140 10,150 10,160 10,170 10,180 10,190 10,210 10,220

susamum 2,001 2,031 2,041 2,051 2,061 2,071 2,081 2,091 2,301 2,331 2,341 2,351 2,361

"A 'pontoon' from Pontoon has a weight of twenty-one, a 'tertipontoon' seven and a septipontoon three ." [591]

pontoon 399 489 498 579 588 669 696 759 795 849 894 939 993 1,299 1,389

tertipontoon 7 16 25 34 43 52 61 70 106 115 124 133 142 151 160 205 214 223

septipontoon 3 12 21 30 102 111 120 300 1,002 1,011 1,020 1,101 1,110 1,200 3,000

"'Pat' [592] from Blackjack has a weight between seventeen and twenty-one, so 'fewpat' would weigh seventeen or eighteen, 'midpat' nineteen and 'morepat' twenty or pontoon." [593]

pat 89 98 99 188 189 197 198 199 278 279 287 296 297 298 299 368 369 377 386 387

fewpat 89 98 99 188 189 197 198 278 279 287 296 297 368 369 386 395 396 459

midpat 199 289 298 379 388 397 469 496 559 595 649 694 739 793 829 892 938

morepat 299 389 398 399 488 489 497 498 578 579 587 588 668 669 695 696 758

"'Point' weighs between four and ten,[594] 'fewpoint' between four and six, 'morepoint' between eight and ten. 'Midpoint' describes a number with a weight of exactly seven aka 'tertipontoon' or 'matador'. 'Farpoint' describes a number weighing either four or ten." [595]

point 4 5 6 7 8 9 13 14 15 16 17 18 19 22 23 24 25 26 27 28 31 32 33 34 35 36 37

fewpoint 4 5 6 13 14 15 22 23 24 31 32 33 40 41 42 50 51 60 103 104 105 112 121 122

morepoint 8 9 17 18 19 26 27 28 35 36 37 44 45 46 53 54 55 62 63 64 71 72 73 80

midpoint 7 16 25 34 43 52 61 70 106 115 124 133 143 151 160 205 214 223 232 241

farpoint 4 13 19 22 28 31 37 40 46 55 64 73 82 91 103 109 112 118 122 127 130 136

"A 'baccarat', on the other hand, is a number with a weight divisible by ten and 'sesquibaccarats', those with weights divisible by fifteen, are subsequences respectively of the first two." [596]

baccarat 19 28 37 46 55 64 73 82 91 109 118 127 136 145 154 163 172 181 190

sesquibaccarat 69 78 87 96 159 168 177 186 195 249 258 267 276 285

"'Mournival' [597] from Gleek means four-of-a-kind fours or higher."

mournival 4,444 5,555 6,666 7,777 8,888 9,999 14,444 15,555 16,666 17,777

"'A 'combo' [598] contains just two neighboring digits."

combo 10 12 21 23 32 34 43 45 54 56 65 67 76 78 87 89 98 103 104 105 106

"A Tutnese combo or 'catchmumbubo',[599] is a combo with two non-neighboring digits between."

catchomumbubo 1,330 1,340 1,350 1,360 1,370 1,380 1,390 1,430 1,440 1,450 1,460

"A 'pugo' [600] is a po with a two different digits between and a 'pugrugoilul' [601] is a proil with two pairs of different digits between."

pugo 1,001 1,021 1,031 1,041 1,051 1,061 1,071 1,081 1,091 1,201 1,221 1,231 1,241 1,251

pugrugoilul 1,001,001 1,001,021 1,001,031 1,001,041 1,001,051 1,001,061 1,001,071 1,001,081

"A 'run' contains three neighboring digits,[602] while a 'rugunun' [603] has two non-neighboring digits between."

run 102 120 201 234 243 321 324 342 345 354 453 456 465 564 567 576 675 678 687

rugunun 1,440,662 1,440,682 1,440,772 1,440,792 1,440,862 1,440,882 1,440,972

"A 'straight' [604] from Poker contains five neighboring digits. The next rather rare straight prime seems to be twenty-five thousand, four hundred sixty-three. There are somewhat surprisingly just as many, that is, infinitely many, of the much more common 'crooked primes'."

straight 10,234 10,243 10,324 10,342 10,423 10,432 12,034 12,043 12,304 12,340 12,403

"A 'sexette' [605] from Piquet contains six neighboring digits, and the 'septet' [606] seven."

sexette 102,345 102,354 102,435 102,453 103,245 103,254 103,425 103,452 104,245

septet 1,023,456 1,023,465 1,023,546 1,023,564 1,023,645 1,023,654 1,024,356

"A 'john' [607] from *Odd John* by Olaf Stapledon is five neighboring odd digits and a 'steven' from 'Even Steven' from Jonathan Swift's *Stella* [608] is five neighboring even digits."

john 13,579 13,597 13,759 13,795 13,957 13,975 15,379 15,397 15,739 15,793 15,937

steven 20,468 20,486 20,648 20,684 20,846 20,864 24,068 24,086 26,048

"A 'cat-hop' [609] from Faro is a po with an extra number, the 'kicker', and a 'crag' [610] from Crag is a cat-hop weighing thirteen. If higher than three digits the 'kicker' must be separate from the po."

cat-hop 100 101 110 112 121 122 131 133 141 144 151 155 161 166 171 177 181 188 191 199

crag 166 229 292 337 355 373 445 454 535 544 553 616 661 733 1,129 1,192

"A Tutnese crag or 'cashrugagug' is a crag with two other digits between." [611]

cashrugagug 1,006,006 2,002,009 2,009,002 3,003,007 3,005,005 3,007,003

"'Gow' [612] from Tien Gow is a three and six or a four and five, so a 'semigow' [613] would be a three, four, five or six. A 'sesquigow' would be a little more complex, a three and six plus a four or a five or a four and five plus a three or a six." [614]

gow 36 45 54 63 136 145 154 163 236 245 254 263 306 360 361 362 364 365 367

semigow 3 4 5 6 13 14 15 16 23 24 25 26 30 31 32 37 38 39 40 41 42 47 48 49 50

sesquigow 345 346 356 364 365 435 436 453 456 463 465 534 543 634

"Similarly a 'sesquichut'' [615] is a two and five plus a three or a four or a three and four plus a two or a five."

sesquichut 234 235 245 253 254 324 325 342 345 352 354 423 425 432 435

"'Sixes and sevens' [616] comes from the English idiom for confused. The Chinese have a similar saying, 'luan qi ba zao' [617] or 'chaos seven eight ', the French 'cinque et sice' [618] or 'five and six', the Chileans 'al tres al cuatro' [619] or 'the threes and the fours', while 'four-and-two' in Restaurantese means a 'sandwich', so a 'half-sandwich' would mean four or two." [620]

sixes andsevens 6,677 6,767 6,776 16,677 16,767 16,776 16,776 26,677 26,767 26,776

luan qibazao 78 87 778 787 788 877 878 7,778 7,787 7,788 7,877 7,878 7,887

cinque et sice 56 65 156 165 256 265 356 365 456 465 506 516 526 536 546

al tres y al cuatro 3,344 3,434 3,443 33,344 33,434 33,443 33,444 34,334 34,343 34,344

sandwich 24 42 124 142 204 240 241 242 243 244 245 246 247 248 249 324 342

half-sandwich 2 4 12 14 20 21 23 25 26 27 28 29 32 34 40 41 43 45 46 47 48 49

"'Half-odd' [621], on the other hand, would be odd numbers with an equal number of even digits and 'half-even' [622] evens with an equal number of odd digits."

half-odd 21 23 25 27 29 41 43 45 47 49 61 63 65 67 69 81 83 85 87 89 1,001 1,003 1,005

half-even 12 14 16 18 30 32 34 36 38 50 52 54 56 58 70 72 74 76 78 90 92 94 96 98

"Some, for example all the double-digit ones, are 'alternating',[623] while the non-alternating or 'direct evens' [624] don't start until a hundred and the 'direct odds' [625] not until two hundred one."

alternating 10 12 14 16 18 21 23 25 27 29 30 32 34 36 38 41 43 45 47 49 50 52 54 56

direct even 100 110 112 114 116 118 120 122 124 126 128 130 132 134 136 138 150 152 154 156

direct odd 201 203 205 207 209 211 213 215 217 219 221 223 225 227 229 231 233 235

"The 'invincible' numbers from Dominoes, aka 'Schneider', means seven pairs. The 'vincible' or 'Neider' numbers are much more common." [626]

invincible 10,012,233,445,566 10,012,233,445,565 10,012,223,445,655

The hunt for the first invincible prime continued long after the meeting ended.

from *PAJAS* #71 (June 1991)

HOLINESS

Taffy recited Vincent S. S. Coles' "Almighty Father, Lord Most High, Who madest all, who fillest all, Thy Name we praise and magnify, for all our needs on Thee we call."

Then Al asked, "Why couldn't the Romans do algebra?"

"They always got x equals ten," she answered.

"Although we got our modern representations for the digits from the Arabs, and they have advantages over the Romans', they are not universally used, except for the latecomer zero. Muhummed ibn Musa al-Khwarizmi, from whom we get the word 'algorithm', re-re-discovered the zero in 800. It had already been re-discovered by Brahmagupta in 594 and Chiang Hong in 125.

"These non-Arabic numerals are more curvaceous and many quite holey."

"Holy numerals!" newcomer Robin interrupted, "I wonder if you mean 'holey' with an E or 'holy" without an E?"

"It's sort of an inside joke to take it as without the E as an allusion to the Most High and Most Holy, He who makes much of nought."

"And besides removing e's is 'e'sy'."

Robin Robb promptly acquired the nickname "Robin the Girl Wonder".

"Comparing them is rather like an international golf tourney," Al continued, "Mongolia and Tibet have the least number of holes, just one hole in zero. The Mongolian eight is linear, the Tibetan curvilinear.

Mongolian	᠐	᠑	᠒	᠓	᠔	᠕	᠖	᠗	᠘	᠙	total
Tibetan	༠	༡	༢	༣	༤	༥	༦	༧	༨	༩	
holiness	1	0	0	0	0	0	0	0	0	0	1

"The only holy Mongolian or Tibetan digit is just zero, so the holy numbers contain just one zero, while the holier ones have more than one. [627]

holy Mongolian 10 20 30 40 50 60 70 80 90 100 101 102 103 104 105 106 107 108

holier Mongolian 100 200 300 400 500 600 700 800 900 1,000 1,001 1,002 1,003

"Since zero is color-coded as black, numbers with zeros in their spectra can be subdivided into the wholly holey or holiest: the totally black Mongolian or Tibetan singleton zero, the half-zero or mulatto numbers,[628] and the rest of the partially zero numbers, the triroon,[629] the quadroon,[630] the quintroon,[631] the sextroon,[632] the septroon,[633] and the octroon." [634] Their holiness would also depend upon how many blue sixes, gray eights or white nines they have.

mulatto	10	1,001	1,002	1,003	1,004	1,005	1,005	1,006	1,007	1,008	1,009	1,010	1,020	1,030
holiness	1	2	2	2	2	2	2	3	2	4	3	2	2	2

triroon	101	102	103	104	105	106	107	108	109	110	120	130	140	150	160	170	180	190	201
holiness	1	1	1	1	1	1	1	1	2	1	1	1	1	1	2	1	3	2	1

quadroon	1,011	1,012	1,013	1,014	1,015	1,016	1,017	1,018	1,019	1,021	1,022	1,023	1,024	1,025
holiness	1	1	1	1	1	2	1	3	2	1	1	1	1	1

quintroon	10,111	10,112	10,113	10,114	10,115	10,116	10,117	10,118	10,119	10,121	10,122	10,123
holiness	1	1	1	1	1	2	1	3	2	1	1	1

sextroon	101,111	101,112	101,113	101,114	101,115	101,116	101,117	101,118	101,119	101,121	101,122
holiness	1	1	1	1	1	2	1	3	2	1	1

"The first septroon prime isn't until one million eleven thousand one hundred thirty-seven."

septroon	1,011,111	1,011,112	1,011,113	1,011,114	1,011,115	1,011,116	1,011,117	1,011,118	1,011,119

octroon	10,111,111	10,111,112	10,111,113	10,111,114	10,111,115	10,111,116	10,111,117	10,111,118

"Devanagari from India also has a four with a hole. It is the script that Sanskrit and many other Indian languages are written in and comes from the words for 'divine town'."

Devanagari	०	१	२	३	४	५	६	७	८	९	total
holiness	1	0	0	0	1	0	0	0	0	0	2

"The holy Devanagari numbers have just one zero or four,[635] the holiest only zeros and fours" [636] and the unholy ones neither. [637]

holy Devanagari	0	4	10	14	20	24	30	34	41	42	43	45	46	47	48	49	50

holiest Devangari	40	400	440	4,000	4,004	4,040	4,044	4,400	4,404

unholy Devanagari	1	2	3	5	6	7	8	9	11	12	13	15	16	17	18	19	21	22	23	25	26	27	28	29	31

"Gurmukhi also from India also has a hole-in-one. Its five looks something like our four. It is the script used by the Punjabi followers of the Sixteenth Century Sikh guru Angad.

Gurmukhi	੦	੧	੨	੩	੪	੫	੬	੭	੮	੯	total
holiness	1	1	0	0	1	0	0	0	0	0	3

"Holy Gurmukhi numbers have a zero, one or four,[638] the holiest only zero, one and four[639] and the holiest between one and all. [640]

holy Gurmukhi	0 1 4 12 13 15 16 17 18 19 20 21 24 30 31 34 42 43 45 46 47
holiest Gurmukhi	104 140 1,004 1,040 1,041 1,044 1,104 1,140 1,400 1,401
holier Gurmukhi	10 11 14 40 41 101 102 103 105 106 107 108 109 110 111 112

"Gujarati from Western India adds a hole-in-seven and its five looks something like our four, though curvilinear, like eight and nine. Its six looks like a Greek xi'.'

Gujarati	૦	૧	૨	૩	૪	૫	૬	૭	૮	૯	total
holiness	1	1	0	0	1	0	0	1	0	0	4

"The holy Gujarati numbers have a zero, one, four or seven,[641] the holiest have them all[642] and the holier more than one and less than all.[643]

holy Gujarati	0 1 4 7 12 13 15 16 18 19 20 21 24 27 30 31 34 37 42 43 45
holiest Gujarati	1,047 1,074 1,407 1,470 1,704 1,740 10,047 10,074
holier Gujarati	10 11 14 17 40 41 44 47 70 71 74 77 101 104 107 111 114 117

"Telugu from Central India looses the hole-in-one but gains a hole-in-two, which looks something like our nine, a hole-in-five and a curvilinear eight."

Telugu	౦	౧	౨	౩	౪	౫	౬	౭	౮	౯	total
holiness	1	0	1	0	1	1	0	0	0	0	4

"The holy Telugu numbers have a zero, two, four or five,[644] the holiest all of them[645] and the holier more than one and less than all." [646]

holy Telugu	0 2 4 5 10 12 14 15 21 22 23 26 27 28 29 30 32 34 35 41
holiest Telugu	2,045 2,054 2,405 2,450 2,504 2,540 4,025 4,052
holier Telegu	20 22 24 25 40 42 44 45 50 52 54 55 200 202 204

"Half of the Khmer numerals from Cambodia and the Mayalam numerals from Southwest India have holes, but they only agree on the hole-in-seven, which looks very much like our nine. Mayalam was originally written in the Vatteluttu or 'round writing', which merged with Grantha script and eventually became standardized in the Thirteenth Century."

Khmer	០	១	២	៣	៤	៥	៦	៧	៨	៩	total
holiness	1	0	1	1	0	1	0	1	0	0	5

Mayalam	൦	൧	൨	൩	൪	൫	൬	൭	൮	൯	
holiness	1	1	0	0	1	0	0	1	0	1	5

"The holy Khmer numbers have a zero, two, three, five or seven,[647] the holiest all of them[648] and the holier more than one and less than all." [649]

holy Khmer	0	2	3	5	7	10	12	13	15	17	21	24	26	28	29	31	34	36	38	39	40

holiest Khmer	20,357	20,375	20,537	20,573	20,735	20,753	23,057

holier Khmer	20	22	23	25	27	30	32	33	35	37	50	52	53	55	57	70	72	73

"The holy Mayalam numbers have a zero, one, four, seven or nine,[650] the holiest all of them[651] and the holier more than one and less than all." [652]

holy Mayalam	0	1	4	7	9	12	13	15	16	18	20	21	24	27	29	30	31	34	37	39	40	42	43

holiest Mayalam	10,479	10,497	10,749	10,794	10,947	10,974	14,079	14,097	14,709

holier Mayalam	10	14	17	19	40	41	47	49	70	71	74	79	90	91	94	97	101	102	103	104	105

"The symbol for four in Bengali from East Pakistan and Bengladesh, looks like our eight, with two holes. Five, seven and eight have one hole."

Bengali	০	১	২	৩	৪	৫	৬	৭	৮	৯	total
holiness	1	0	0	0	2	1	0	1	1	0	6

"The holy Bengali numbers have a zero, four, five, seven or eight,[653] the holiest qll fours[654] and the holiest more than one but not all fives.[655]

holy Bengali	0	4	5	7	8	10	14	15	17	18	20	24	25	27	28	30	34	35	36	39	41

holiest Bengali	4	44	444	4,444	44,444	444,444	4,444,444	44,444,444	444,444,444

holier Bengali	40	44	45	47	48	50	54	55	57	58	70	74	75	77	78	80	84	85	87	88	104

"Odia from Eastern India has a five with two holes and zero, one, two, four and seven with one each, a linear eight and a curvilinear nine."

Odia	୦	୧	୨	୩	୪	୫	୬	୭	୮	୯	total
holiness	1	1	1	0	1	2	0	1	0	0	7

"The holy Odia numbers have one zero, one, two, four, five or seven,[656] the holiest are all fives,[657] the holier have more than one and less than all fives.[658]

holy Odia 0 1 2 4 5 7 13 16 18 19 23 26 28 29 30 31 32 34 35 37 43 46 48 49 53 56 58

holiest Odia 5 55 555 5,555 55,555 555,555 555,555 5,555,555 55,555,555

holier Odia 10 12 14 15 17 20 21 24 25 27 40 41 42 45 47 50 51 52 54 57 70 71 72 74 75

"Vai from Liberia has a couple linear numbers, one and four, three curvaceous ones, zero, eight and nine, and six curvilinear ones.

Vai	꘠	꘡	꘢	꘣	꘤	꘥	꘦	꘧	꘨	꘩	total
holiness	1	0	1	1	0	0	1	1	1	2	8

"The holy Vai numbers have a zero, two, three, six, seven, eight or nine,[659] the holiest all of them all nines[660] and the holier more than one and less than all nines.[661]

holy Vai 0 2 3 6 7 8 9 10 12 14 15 21 24 25 30 32 34 35 40 42 43 46 47 48 49 50 52

holiest Vai 9 99 999 9,999 99,999 999,999 9,999,999 99,999,999 999,999,999

holier Vai 20 22 23 26 27 28 29 30 32 33 36 37 38 39 60 62 63 66 67 68 69 70 72

"Kannada from Southern India looses the hole-in-two and the hole-in-seven, but gains two in four and five and one in nine."

Kannada	೦	೧	೨	೩	೪	೫	೬	೭	೮	೯	total
holiness	1	0	0	1	2	2	1	0	1	1	9

"The holy Kannada numbers have a zero, three, four, five, six, eight or nine,[662] and the holiest are those with fours and/or fives[663] and the holier are those with more than one but less than all fours and fives.[664] The both three and seven look very much like our two.

holy Kannada 0 3 4 5 6 8 9 10 13 14 15 16 18 19 20 23 24 25 26 28 29

holiest Kannada 4 5 44 45 54 55 444 445 454 455 544 545 554 555

holier Kannada 30 33 34 35 36 38 39 40 43 44 45 46 48 49 50 53 54 55

"Thailand wins for consistency. It has at least one hole in all ten digits and so no number is unholy. All-fives however is holiest, just as in Odia. The merely holy numbers would be those without a five, with just one hole per digit, the five-less, while the holier numbers would be those with a five.[665]

Thai	๐	๑	๒	๓	๔	๕	๖	๗	๘	๙	total
holiness	1	1	1	1	1	2	1	1	1	1	11

holy Thai	0	1	2	3	4	6	7	8	9	10	11	12	13	14	16	17	18	19	20	21	22	23	24

"All of the Tamil numerals from Southern India and Northern Sri Lanka are curvilinear. 'Only' eight digits have holes, but one, four, six and eight have two and nine has three, so the holiest Tamil numbers are the same as the Vai."

Tamil	0	௧	௨	௩	௪	௫	௬	௭	௮	௯	total
holiness	1	2	1	0	2	0	2	1	2	3	14

"The holy Tamil numbers have a zero, one, two, four, six, seven, eight or nine, but no three or five,[666] while the holier have more than one and less than all nine.[667] Its holiest is the same as the Thai and Vai.

holy Tamil	0	1	2	4	6	7	8	9	30	31	32	34	36	37	38	39	43	45	50	51	52	54	56

holier Tamil	10	12	14	16	17	18	19	20	21	22	24	26	27	28	29	40	41	42	44	46

"Laos has even Thailand beat with a second hole in three through eight and a third one in five, so although its holiest is the same as Odia, though its hole count[668] and cumulative hole count is the highest of all, what we call 'the holy of holies'." [669]

Lao	໐	໑	໒	໓	໔	໕	໖	໗	໘	໙	total
holiness	1	1	1	2	2	3	2	2	2	1	17

Lao holiness	1	1	1	2	2	3	2	2	2	1	2	2	2	4	3	4	2	3	3
holy of holies	1	2	3	5	7	10	12	14	16	17	19	21	23	26	29	33	36	39	42

Al broke up the meeting by breaking into a rendition of "Laos Has the Holiest Numbers.", parodying Three Dog Night's "One Is the Loneliest Number".

from *PAJAS* #72 (July1991)

MORE HOLY NUMBERS

Taffy recited another old hymn as she had last meeting, "O Lord Most High, with all my heart Thy wondrous works I will proclaim; I will be glad and give Thee thanks and sing the praise of Thy Name."

"If a worm that's an inch long's called an inchworm, what would you call a snake three point one four meters long?"

"A pi-thon."

This meeting continued where the last had left off, this time concentrating on the holiness not in other scripts but in other bases.

"Our modern Arabic numerals, of course, have one hole in zero, six and nine and two in eight."

"What about four? It sometimes has a hole and sometimes not."

"What four?"

"What for? Oh, yeah, 'What four?' We avoid addressing the question of four's holiness or unholiness altogether to avoid ambiguity, so we too have unique holy[670], holiest[671], holier[672] and unholy[673] numbers. The holiness of a sequence usually increases slowly and with setbacks, just like it does with people."

holy Arabic 0 6 9 10 16 19 20 26 29 30 36 39 40 46 49 50 56 59 61 62 63 64 65 67

holiest Arabic 8 88 888 8,888 88,888 888,888 8,888,888 88,888,888 888,888,888

holier Arabic 60 66 68 69 80 86 88 89 90 96 98 99 100 106 108 109 166 168 169 180 181

unholy Arabic 1 2 3 4 5 7 11 12 13 14 15 17 21 22 23 24 25 27 31 32 33 34 35 37 41 43

"Base one, unary, of course, has no zero and so is the unholiest. Unholy binarish is, of course, the repunits. Only bases nine or higher have a double-holed placeholder. Hexadecimal has another if using B, but not if using double-digit eleven. Base nineteen and higher have another, if using double-digit numbers. The holy base-two numbers would have just one zero,[674] the holiest more than one and one less than their length [675] and the holier between the two.[676]

holy binarish 0 10 101 1,011 1,101 10,111 11,011 11,101 11,110 101,111 110,111 111,011 111,101 111,110 1,011,111

holiest binarish 100 1000 10000 100000 1000000 10000000 100000000 1000000000

"The first holier binarish prime is a million, eleven thousand, one."

holier binarish 1,001 1,010 1,100 10,001 10,010 10,011 10,100 10,101 10,110 101,011 101,110

"It's similar for the rest of the bases, though each of them, of course, gets longer more slowly while getting higher faster: the holy[677], holiest,[678] holier[679] and unholy[680] tenarish, the holy[681], holiest[682], holier[683] and unholy[684] quadarish, the holy[685], holiest[686], holier[687] and unholy[688] quintarish, the holy[689], holiest[690], holier and unholy[691] senarish, the holy,[692] holiest,[693] holier and unholy[694] septarish, and the holy[695], holiest,[696] holier[697] and unholy[698] octarish.

holy ternarish 0 10 20 101 102 110 120 201 202 1,011 1,012 1,021 1,022 1,101 1,110 1,201

holiest ternarish 100 200 1,000 2,000 10,000 20,000 100,000 200,000 1,000,000

holier ternarish 1,001 1,002 1,010 1,020 1,100 1,200 2,000 2,001 2,002 2,010 2,100

unholy ternarish 1 2 11 12 21 22 111 112 121 211 212 221 222 1,111 1,112 1,121 1,122 1,211

holy quadarish 0 10 20 30 101 102 103 110 120 130 201 202 203 210 220 230 301 302

holiest quadarish 100 200 300 1,000 2,000 3,000 10,000 20,000 30,000 100,00

holier quadarish 1,001 1,002 1,003 1,010 1,020 1,030 1,100 1,200 2,000 2,001 2,002 2,003

unholy quadarish 1 2 3 11 12 13 21 22 23 31 32 33 111 112 113 121 122 123 131 132 133 211

holy quintarish 0 10 20 30 40 101 102 103 104 110 120 130 140 201 202 203 204 210

holiest quintarish 100 200 300 400 1,000 2,000 3,000 4,000 10,000 20,000 30,000

holier quintarish 1,001 1,002 1,003 1,004 1,010 1,020 1,030 1,040 1,100 1,200 1,300 1,400

unholy quintarish 1 2 3 4 11 12 13 14 21 22 23 24 31 32 33 34 41 42 43 44 111 112 113

holy senarish 0 10 20 30 40 50 101 102 103 104 105 110 120 130 140 150 201 202 203

holiest senarish 100 200 300 400 500 1,000 2,000 3,000 4,000 5,000 10,000

holier senarish 1,001 1,002 1,003 1,004 1,005 1,010 1,020 1,030 1,040 1,050 1,100 1,200

unholy senarish 1 2 3 4 5 11 12 13 14 15 21 22 23 24 25 31 32 33 34 35 41 42 43 44 45

holy septarish 0 10 20 30 40 50 60 101 102 103 104 105 106 110 120 130 140 150 106 201

holiest septarish 100 200 300 400 500 600 1,000 2,000 3,000 4,000 5,000 6,000

holier septarish 1,001 1,002 1,003 1,004 1,005 1,006 1,010 1,020 1,030 1,040 1,050 1,060

unholy septarish 1 2 3 4 5 6 11 12 13 14 15 16 21 22 23 24 25 26 31 32 33 34 35 36 41 42

holy octarish 0 10 20 30 40 50 60 70 101 102 103 104 105 106 107 110 120 130 140 150 106

holiest octarish 100 200 300 400 500 600 700 1,000 2,000 3,000 4,000 5,000 6,000

holier octarish 1,001 1,002 1,003 1,004 1,005 1,006 1,007 1,010 1,020 1,030 1,040 1,050

unholy octarish 1 2 3 4 5 6 7 11 12 13 14 15 16 17 21 22 23 24 25 26 27 31 32 33 34 35 36

from *PAJAS* #73 (August 1991)

NOBLE NUMBERS

Al quoted King James's Psalm 18:13 "The Lord also thundered in the heavens, and the Highest gave His voice, hail stones and coals of fire," and then challenged everyone to interpret the expression on the blackboard.

i2 Σπ

"The 'i' is the symbol for the so-called imaginary number, the square root of a minus one, two-cubed is just eight, but can't guess what the sum of pi could mean." one answered.

"The sum of almost-pi would approach infinity." another added.

"I know what it is, a rebus! It says, 'I ate some pie.'" someone else concluded.

"'And it was very good' to quote the Most High," Al continued, "And phi, as we also all should know, can be expressed in continuing fractions as endless ones. The reciprocal of phi, called tau, which 'happens' also to be phi minus one, is the simplest noble number." [699]

"Yes, Phidias' Law," someone shared, "Sixty-one point eight percent of everything's tau."

"The rest of the nobles have intervening numbers between the initial zero and initial one and the ones which repeat. All three broken sequences quickly converge to the Fibonacci sequence.

almost-tau	6	1	8	0	3	3	9	8	8	7	4	9	8	9	4	8	4	8	2	0
fracked	0	1	1	1	1	1	1	1	1	1	1	1	1	1	1	1	1	1	1	1
broken-up	0	0	1	2	3	5	7	13	21	34	55	89	144	233	377	610	987	1,597	2,597	4,181
broken-down	1	1	2	3	5	8	13	21	34	55	89	144	233	377	610	987	1,597	2,584	4,181	6,765
broken	0	1	2	3	5	7	8	13	21	34	55	89	144	233	377	610	987	1,597	2,584	4,181

"You'll notice when expanding tau, a number below zero, we do not use a leading zero, but when expressing it as a continued fraction sequence we do. We don't with batting averages, but we do with medical dosages. All of the noble numbers are below zero, so when the continued fraction sequence does not begin with zero, it begins with the integral part of the mixed number. We however leave our naked radix exposed without a leading zero."

"The next simplest noble number is not so predictable when broken. Only the broken-up sequence converges into Fibonacci.[700]

2nd noble	1	2	1	1	1	1	1	1	1	1	1	1	1	1	1
broken-up	1	2	5	8	13	21	34	55	89	144	233	377	610	897	1,597
broken-down	2	3	7	11	18	29	47	76	123	199	322	521	843	1,364	2,207
broken	1	2	3	5	7	8	11	13	18	21	29	34	47	55	76

"The third noble number and higher,[701] break up and down into sequences that quickly converge into Lucas sequences.

3rd noble	1	3	1	1	1	1	1	1	1	1	1	1	1	1	1
broken-up	1	3	4	7	11	18	29	47	76	123	199	322	521	843	1,364
broken-down	1	4	5	9	14	23	37	60	97	157	411	665	1,076	1,741	2,817
broken	1	3	4	5	7	9	11	14	18	23	29	37	47	60	76

4th noble	1	4	1	1	1	1	1	1	1	1	1	1	1	1	1	1
broken-up	1	4	5	9	14	23	37	60	97	157	254	411	665	1,076	1,741	2,817
broken-down	1	5	6	11	17	28	45	73	118	191	309	500	809	1,309	2,118	3,427
broken	1	4	5	6	9	11	14	17	23	28	37	45	60	73	97	118

5th noble	1	5	1	1	1	1	1	1	1	1	1	1	1	1	1	1
broken-up	1	5	6	11	17	28	45	73	118	191	309	809	1,309	2,118	3,427	5,545
broken-down	1	6	7	13	20	33	53	86	139	225	364	589	953	1,542	2,495	4,037
broken	1	5	6	7	11	13	17	20	28	33	45	53	73	86	118	139

"Notice how eighty-six and ninety-nine on the sixth noble[702] are matched up?"

6th noble	1	6	1	1	1	1	1	1	1	1	1	1	1	1	1	1
broken-up	1	6	7	13	20	33	53	86	139	225	364	589	953	1,542	2,495	4,037
broken-down	1	7	8	15	23	38	61	99	160	259	419	678	1,097	1,775	2,872	4,647
broken	1	6	7	8	13	15	20	23	33	38	53	61	86	99	139	160

7th noble	1	7	1	1	1	1	1	1	1	1	1	1	1	1	1	1
broken-up	1	7	8	15	23	38	61	99	160	259	419	678	1,097	1,775	2,872	4,647
broken-down	1	8	9	17	26	43	69	112	181	293	474	767	1,241	2,008	3,249	5,257
broken	1	7	8	9	15	17	23	26	38	43	61	69	99	112	160	181

8th noble	1	8	1	1	1	1	1	1	1	1	1	1	1	1	1
broken-up	8	9	17	26	43	69	112	181	293	474	767	1,241	2,008	3,249	5,257
broken-down	1	9	10	19	29	48	77	125	202	327	529	856	1,385	2,241	3,626
broken	8	9	10	17	19	26	29	43	48	69	77	112	125	181	202

9th noble	1	9	1	1	1	1	1	1	1	1	1	1	1	1	1
broken-up	1	9	10	19	29	48	77	125	202	327	529	856	1,385	2,241	3,626
broken-down	1	10	11	21	32	53	85	138	223	361	584	945	2,474	4,003	6,477
broken	1	9	10	11	19	21	29	32	48	53	77	85	125	138	202

10th noble	1	10	1	1	1	1	1	1	1	1	1	1	1	1	1
broken-up	1	10	11	21	32	53	85	138	223	361	584	945	1,529	2,474	4,003
broken-down	1	11	12	23	35	58	93	151	244	395	639	1,034	1,673	2,707	4,380
broken	1	10	11	12	21	23	32	35	53	58	85	93	138	151	223

"These noble numbers are all of the form as the sum of a number, z, and the square root of five all over another number, z prime, both of which have an obvious pattern." [703]

z	5	15	29	47	69	95	125	159	197	239	277	327	381	449
z'	10	22	38	58	82	110	132	178	218	262	310	362	398	458

"So there are an infinite number of noble numbers then?"

"No, since you can insert any number of numbers between the initial zero and one and the final repeated ones, there are actually an infinite number of infinite number of noble numbers, starting with the one-two noble." [704]

1-2 noble	1	1	2	1	1	1	1	1	1	1	1	1	1	1	1
broken-up	1	1	3	4	7	11	18	29	47	76	123	199	322	521	843
broken-down	1	2	5	7	12	19	31	50	81	131	212	343	555	898	1,453
broken	1	2	3	4	5	7	11	12	18	19	29	31	47	50	76

"Ow! That gives me a brainache."

from *PAJAS* #74 (September 1991)

NEAR-NOBLE NUMBERS

"Since you have been raised up in company with Christ," Al began, "set your heart on what pertains to higher realms where Christ is seated at God's right hand. Be intent on things above rather than on things of Earth." (Colossians 3:1-2)

He then continued, "Why couldn't pi and phi get along?"

"They're always both so irrational." someone answered.

Al then began the month's topic, "The continuing fractions that start with zero and have some other number after one or more ones before repeating a number other than one is called near-noble. These too can be broken into non-repeating sequences. The first[705] is half root two, familiar from trig."

1st near-noble	1	2	2	2	2	2	2	2	2	2	2	2	2	2
broken-up	1	2	5	12	29	70	169	408	985	2,378	5,741	13,860	33,461	80,782
broken-down	1	3	7	17	41	99	239	577	1,393	3,363	8,119	19,601	47,321	114,243
broken	1	2	3	5	7	12	17	29	41	70	99	169	239	408

"The second, one plus root thirteen all over six, and higher near-nobles[706] are not so familiar, though easily expressed as continued fractions."

2nd near-noble	1	3	3	3	3	3	3	3	3	3	3	3
broken-up	1	3	10	33	109	360	1,189	3,927	12,970	42,837	141,481	467,280
broken-down	1	4	13	43	142	469	1,549	5,116	16,897	55,807	184,318	608,761
broken	1	3	4	10	13	33	43	109	142	360	469	1,189

3rd near-noble	1	4	4	4	4	4	4	4	4	4	4
broken-up	1	4	17	72	305	1,292	5,473	23,184	98,209	416,020	1,762,289
broken-down	1	5	21	89	377	1,597	6,765	28,657	121,595	514,229	2,178,309
broken	1	4	5	17	21	72	89	305	377	1,292	1,597

4th near-noble	1	5	5	5	5	5	5	5	5	5
broken-up	1	5	26	135	701	3,640	18,901	509,626	2,646,275	13,741,001
broken-down	1	6	31	161	836	4,341	22,541	117,046	607,771	3,155,901
broken	1	5	6	26	31	135	161	701	836	3,640

5th near-noble	1	6	6	6	6	6	6	6	6	6
broken-up	1	6	37	226	1,405	8,658	53,353	328,776	2,026,009	12,484,830
broken-down	1	7	43	265	1,633	10,063	62,011	382,129	2,354,785	14,510,839
broken	1	6	7	37	43	226	265	1,405	1,633	8,658

6th near-noble	1	7	7	7	7	7	7	7	7	7
broken-up	1	7	50	357	2,549	18,200	129,949	927,843	6,624,850	47,301,793
broken-down	1	8	57	407	2,906	20,749	148,149	1,057,792	7,552,693	53,926,643
broken	1	7	8	50	57	357	407	2,549	2,906	18,200

7th near-noble	1	8	8	8	8	8	8	8	8
broken-up	1	8	65	528	4,289	34,840	283,009	2,298,912	18,674,305
broken-down	1	9	73	593	4,817	39,129	317,849	2,581,921	20,973,217
broken	1	8	9	65	73	528	593	4,289	4,817

8th near-noble	1	9	9	9	9	9	9	9	9
broken-up	1	9	82	747	6,805	61,992	564,733	5,144,589	46,866,034
broken-down	1	10	91	829	7,552	68,797	626,725	5,709,322	52,010,623

broken	1	9	10	82	91	747	829	6,805	7,552

[0; 1, 2, 3]	1	2	3	3	3	3	3	3	3	3	3	3	3
broken-up	1	2	7	23	76	251	829	2,738	9,043	29,867	98,644	325,799	1,076,041
broken-down	1	3	10	33	109	360	1,189	3,927	12,970	42,837	141,481	467,280	154,321
broken	1	2	3	7	23	33	76	109	251	829	1,189	2,728	3,927

[0; 1, 3, 2]	1	3	2	2	2	2	2	2	2	2	2	2	2	2
broken-up	1	3	7	17	41	99	239	577	1,393	3,363	8,119	19,601	47,321	114,243
broken-down	1	4	9	22	53	128	309	746	1,801	4,348	10,497	25,342	61,181	147,704
broken	1	3	4	7	9	17	22	41	33	53	99	128	239	309

[0; 1, 3, 4]	1	3	4	4	4	4	4	4	4	4	4	4
broken-up	1	3	13	55	233	987	4,181	17,711	75,025	317,811	1346,269	5,702,887
broken-down	1	4	17	72	305	1,292	5,473	23,184	98,209	416,020	1,762,289	7,465,176
broken	1	3	4	13	17	55	72	233	305	987	1,292	4,181

[0; 1, 2, 4]	1	2	4	4	4	4	4	4	4	4	4	4
broken-up	1	2	9	38	161	682	2,889	12,238	51,841	219,602	930,249	3,940,598
broken-down	1	3	13	55	233	987	4,181	17,711	75,025	317,811	1,346,269	5,702,887
broken	1	2	3	9	13	38	55	161	233	682	987	3,889

from *PAJAS* #75 (October 1991)

MEAN NUMBERS

"This is the meaning of the parable" Al began, "... hear the word in a spirit of openness, retain it, and bear fruit through perseverance." (Luke 8:11,15)

"How much is half of eight?" Vicky asked.

"That depends on what you mean by it," her sister Elly answered, "y'know what I mean? Dividing by two it's usually four, but dividing it horizontally it's zero and dividing it vertically it's three."

"The silver means[707] are related to the golden mean," Al continued, "all being of the form z plus the root of z plus four all over two, with z greater than one. Z equal to one, of course, gives phi. The broken-up and broken-down sequences are just one different, literally, while the broken-down and broken differ not at all."

silver mean	2	2	2	2	2	2	2	2	2	2	2	2	2	2
broken-up	2	5	12	29	70	169	408	985	2,378	5,741	8,119	13,860	21,979	35,839

sub	1	2	4	11	26	62	150	362	875	2,112	2,987	5,099	8,086	13,184
broken-down	1	2	5	12	29	70	169	408	985	2,378	5,741	8,119	13,860	21,979
broken	1	2	5	12	29	70	169	408	985	2,378	5,741	8,119	13,860	21,979

2nd silver mean	3	3	3	3	3	3	3	3	3	3	3	3
broken-up	3	10	33	109	360	1,189	3,927	12,970	42,837	141,481	467,280	1,543,321
sub	1	4	12	40	132	437	1,445	4,771	15,759	52,048	171,903	567,756

3rd silver mean	4	4	4	4	4	4	4	4	4	4
broken-up	4	17	72	305	1,292	5,473	23,184	98,209	416,020	1,762,289
sub	1	6	26	112	475	2,013	8,529	36,129	153,045	648,310

4th silver mean	5	5	5	5	5	5	5	5	5	5
broken-up	5	26	135	701	3,640	18,901	98,145	509,626	2,646,275	13,741,001
dusub	1	4	18	95	493	2,558	13283	68,970	358,134	1,859,642

5th silver mean	6	6	6	6	6	6	6	6	6	6
broken-up	5	26	135	701	3,640	18,901	98,145	509,626	2,646,275	13,741,001
dusub	1	4	18	95	493	2,558	13283	68,970	358,134	1,859,642

6th silver mean	7	7	7	7	7	7	7	7	7
broken-up	7	50	357	2,549	18,200	129,949	927,843	6,624,850	47,301,793
dusub	1	7	48	345	2,463	17,587	125,570	896,576	6,401,601

7th silver mean	8	8	8	8	8	8	8	8	8
broken-up	8	65	528	4,289	34,840	283,009	2,298,912	18,674,305	151,693,352
dusub	1	9	71	581	4,715	38,301	311,124	2,527,292	20529463

8th silver mean	9	9	9	9	9	9	9	9	9
broken-up	9	82	747	6,805	61,992	564,733	5,144,589	46,866,034	426,938,895
dusub	1	11	101	921	8,390	76,428	696,245	6,342,628	57,779,896

9th silver mean	10	10	10	10	10	10	10	10	10
broken-up	10	101	1,020	10,301	104,030	1,050,601	10,610,040	107,151,001	1,082,120,050
dusub	1	14	138	1,394	14,079	142,184	1,435,913	14,501,311	146,449,023

"The silver means are all of the form z plus the square root of the quantity z squared plus four all over two, so it's worth looking at the z squared plus four sequence." [708]

z+4 5 8 13 20 29 40 53 68 85 104 125 148 173

By this time the once high-yielding silver lode had seemed to peter out and the members gladly turned their attention to this simpler sequence.

from *PAJAS* #76 (November 1991)

COLLAPSING BLACK HOLES

Al began with "Hosanna in the highest!" (Matthew 21:9)

Then Elly asked, "Why can't you be both normal and also be first?"

Her sister Vicky answered, "You can't be number one without also being odd."

Introducing the topic of the month, Al began again, "A continued fraction including zeroes, aka 'black holes', can be collapsed since the reciprocal of a reciprocal is the original. An odd number of zeroes adds its neighbors and an even number of zeroes just disappears, giving birth to a new continued fraction."

$$n + n' = n + 1/(0 + 1/n') = n''$$
$$n + 1/(0 + 1/(0 + n') = n + 1/n'$$

"The almost-numbers can be collapsed at the zeros in ten and twenty. [709]

almost-natural 1 2 3 4 5 6 7 8 9 1 0 1 1 1 1 2 1 3 1 4 1 5 1 6 1 7 1 8 1 9 2 0 2
collapsed 1 2 3 4 5 6 7 8 9 2 1 1 1 2 1 3 1 4 1 5 1 6 1 7 1 8 1 9 4 1 2 2 2

"Pi, e and phi can also be collapsed, though their zeroes are all nearly an average ten percent of their digits. The first in pi however does not appear until the thirty-third digit. [710]

position of 0 in pi 33 51 55 66 72 78 86 98 107 117 122 129 133 147 160 165 168

"In e a zero appears at the fourteenth place." [711]

position of 0 in e 14 22 44 68 73 113 114 116 134 141 142 151 157 175 187 197

"In phi a zero appears earlier, in fifth place." [712]

position of 0 in phi 5 21 40 48 66 70 78 84 87 90 107 120 121 123 139 152 171

"So the fourth and sixth digits around the first zero, eight and three, collapse into eleven?" Cal asked." [713]

"Yes, the next zero collapses the twentieth and twenty-second, a two and a four, into six.

collapsed almost-phi 1 6 1 11 3 9 8 8 7 4 9 8 9 4 8 4 8 6 5 6 5 8 6 8 3 4 3 6 5 6 3 8 1

"Conversely any sequence, not already binary, can be fully expanded into a binary sequence by replacing digits higher than one by collapsible series of ones and zeroes. 'Two' expands to 'one-zero-one', 'three' to 'one-zero-one-zero-one' and so on. [714]

fully expanded naturals	110 110 10 110 10 10 110 10 10 10 110
fully expanded odds	110 110 10 110 10 10 110 10 10 10 10 10
fully expanded evens	10 110 10 10 110 10 10 10 10 110 10 10 10 1
fully expanded primes	10 110 10 110 10 10 10 110 10 10 10 10 10 1

"Numbers can also be partially expanded to impersonate sequences in other bases. Three as one-zero-two or two-zero-one impersonates ternary." [715]

pseudoternarish	101	102	201	202	10,101	10,102	10,201	10,202	20,101	20,102
recollapsed	2	3	3	4	3	4	4	5	4	5

"Four as one-zero-three, two-zero-two or three-zero-one impersonates quadrary." [716]

pseudoquadrarish	101	102	103	201	202	203	301	302	303	10,101	10,102	10,103	10,201
recollapsed	2	3	4	3	4	5	4	5	6	3	4	5	4

"Five as one-zero-four, two-zero-three, three-zero-two or four-zero-one impersonates quintary." [717]

pseudoquintarish	101	102	103	104	201	202	203	204	301	302	303	304	401	402	403
recollapsed	2	3	4	5	3	4	5	6	4	5	6	7	5	6	7

"Six as one-zero-five, two-zero-four, three-zero-three, four-zero-two or five-zero-one impersonates senary,[718] while seven as one-zero-six, two-zero-five, three-zero-four, four-zero-three, five-zero-two or six-zero-one impersonates septary and eight as one-zero-seven, two-zero-six, three-zero-five, four-zero-four, five-zero-three, six-zero-two or seven-zero-one impersonates octary." [719]

pseudosenarish	101	102	103	104	105	201	202	203	204	205	301	302	303	304	305	401
recollapsed	2	3	4	5	6	3	4	5	6	7	4	5	6	7	8	5

pseudoseptarish	101	102	103	104	105	106	201	202	203	204	205	206	301	302	303
recollapsed	2	3	4	5	6	7	3	4	5	6	7	8	4	5	6

pseudoctarish	101	102	103	104	105	106	107	201	202	203	204	205	206	207	301	302
recollapsed	2	3	4	5	6	7	8	3	4	5	6	7	8	9	4	5

"Representing numbers like Dzu Tse in modern positional notation, rather than in unary, we might, expand in yet another way by underlining rather than the tradition red for negative digits. One can be represented as ten plus nine, a hundred minus ninety-nine, etc. " [720]

-2	$\underline{1}8$	$\underline{1}98$	$\underline{1}998$	$\underline{1}9998$	$\underline{1}99998$
-1	$\underline{1}9$	$\underline{1}99$	$\underline{1}999$	$\underline{1}9999$	$\underline{1}99999$
1	$1\underline{9}$	$1\underline{99}$	$1\underline{999}$	$1\underline{9999}$	$1\underline{99999}$
2	$1\underline{8}$	$1\underline{98}$	$1\underline{998}$	$1\underline{9998}$	$1\underline{99998}$

"With this plus-or-minus notation we can represent negatives. Normally negatives in continued fractions are not allowed, but they can be shifted indefinitely to the right. When a zero is reached, that is, when z double-prime equals minus z double-prime, the zero can be collapsed and z prime minus one and z triple-prime added."

$$[z, \underline{z}', z'', z'''] = [z-1, z'-1, \underline{z}'', z''']$$
$$[z-1, z'-1, 0, z'''] = [z-1, z'+z'''-1]$$

"So say z, z prime and z triple-prime are two and z double-prime is zero, then z minus one would become one and z prime plus z triple-prime minus one would become three." [721]

$[2, \underline{2}, 0, 2]$	2	$\underline{2}$	0	2	2	$\underline{2}$	0	2	2	$\underline{2}$	0	2	2	$\underline{2}$	0	2	2	$\underline{2}$	0	2	2	$\underline{2}$	0	2	2	$\underline{2}$	0	2
[1, 1, 0, 2]	1	1	0	2	1	1	0	2	1	1	0	2	1	1	0	2	1	1	0	2	1	1	0	2	1	1	0	2
[1, 3]	1	3	1	3	1	3	1	3	1	3	1	3	1	3	1	3	1	3	1	3	1	3	1	3	1	3	1	3

"Or say z, z prime and z triple-prime are not equal, but get higher arithmetically, two, three and four, and z double-prime is still zero, then z minus one would become one and z prime plus z triple-prime minus one would become four. Getting high exponentially, we might have two, four and eight, and z double-prime still zero, then z minus one would become one and z prime plus z triple-prime minus one would become ..."

"Eleven!" Cal finished." [722]

$[2, \underline{2}, 0, 3]$	2	$\underline{2}$	0	3	2	$\underline{2}$	0	3	2	$\underline{2}$	0	3	2	$\underline{2}$	0	3	2	$\underline{2}$	0	3	2	$\underline{2}$	0	3	2	$\underline{2}$	0	3
[1, 1, 0, 3]	1	1	0	3	1	1	0	3	1	1	0	3	1	1	0	3	1	1	0	3	1	1	0	3	1	1	0	3
[1, 4]	1	4	1	4	1	4	1	4	1	4	1	4	1	4	1	4	1	4	1	4	1	4	1	4	1	4	1	4

$[2, \underline{4}, 0, 8]$	2	$\underline{4}$	0	8	2	$\underline{4}$	0	8	2	$\underline{4}$	0	8	2	$\underline{4}$	0	8	2	$\underline{4}$	0	8	2	$\underline{4}$	0	8	2	$\underline{4}$	0	8
[1, 3, 0, 8]	1	3	0	8	1	3	0	8	1	3	0	8	1	3	0	8	1	3	0	8	1	3	0	8	1	3	0	8
[1, 11]	1	11	1	11	1	11	1	11	1	11	1	11	1	11	1	11	1	11	1	11	1	11	1	11	1	11	1	11

"We can get rid of some zeros by 'trimming', reducing all digits by one and so getting rid of the nines as well. Because three becomes the only even prime when trimmed, the only primes that remain primes when trimmed are sanbi with only ones and zeroes as other digits."[723]

demmirt primes	13	103	113	1,013	1,103	10,103	11,003	11,113	100,003	101,103	101,113	111,113

The meeting ended in a frenzy of over-trimming.

from *PAJAS* #77 (December 1991)

SAFE AND UNSAFE NUMBERS

"Glory to God in the highest" Al quoted, "and peace on Earth to those upon whom His favor rests." (Luke 2:14) and then asked, "What odd number can you subtract two from and make it even?"

"Eleven, if you what you subtract is its first two letters." Cal answered and the continued, "Safe primes[724] are those which are when doubled and increased by one are also prime. If they don't, then they're unsafe.[725] All of the odds would be safe odds, though not safe primes."

safe primes	2	3	5	11	23	47	59	83	107	149	167	179	227	263	347	359	383
unsafe primes	7	13	17	19	29	31	37	41	43	53	61	67	71	73	79	89	97

"Other numbers than primes can also have this property. The safe luckies[726] seem to get higher faster than the unsafe[727], while the safe unluckies[728] obviously get higher much faster than the unsafe[729]."

safe luckies	1	3	7	15	31	43	51	63	67	75	87	99	127	135	151	159	211
unsafe luckies	9	13	21	25	33	37	49	69	73	79	93	105	111	115	129	133	141

safe unluckies	23	53	83	103	113	123	143	173	183	203	233	243	263	313	323
unsafe unluckies	2	4	5	6	8	10	11	12	14	16	17	18	19	20	22

"The first safe square is nine and the second two eighty-nine. Again the unsafe ones[730] get higher much more slowly.

unsafe squares	1	4	16	25	36	49	64	81	100	121	144	169	196	225	256	324	361

"There are no safe linears, because twice one plus one is three, twice four plus one is nine and twice seven plus one is five, none of which are linear. Twice two plus one however does equal five, so there is at least one that's not an unsafe curvilinear." [731]

unsafe curvilinears	2	22	25	52	55	222	225	252	255	522	525	552	555

"Ternarish has both safe[732] and unsafe[733] numbers, as do higher bases.[734]

safe ternarish	21	201	2,001	20,001	200,001	2,000,001	20,000,001	200,000,001

unsafe ternarish	1	2	10	11	12	20	22	100	101	102	200	202	210	211	212	220	221	222

"The first five safe octarish[735] and eight out of the first eleven are prime, the unsafe[736] not so much."

safe octarish	3	5	7	11	13	21	23	25	27	31	33	35	41	43	45	47	51	53	55	56	57	60
unsafe octarish	1	2	4	6	10	12	14	15	16	17	20	22	24	26	28	30	32	34	36	37	40	42

from *PAJAS* #78 (January 1992)

SUBDIVISION

"God Most High, Who inhabits eternity, have mercy on us," Lana quoted from the Litany of the Most Holy Trinity.

"What is a seven-letter word for a burial place for multiple bodies?"

"Threemb."

"Beatty used irrational seed numbers to generate new sequences." [737]

"'Bring 'Em Back Alive' Clyde Beatty!"

"No, Samuel Beatty, not the Civil War general either, the mathematician, who 'subdivided' the integers into two complementary sequences via 'seed numbers'." Lana answered. "Multiplying each of the natural numbers by an irrational seed number and rounding generates one unique sequence and a second sequence not the first. The two seed numbers' reciprocals always add to one. The second seed number for phi turns out to be phi squared.[738]

phi Beatty	1	3	4	6	8	9	11	12	14	16	17	19	21	22	24	25	27	29	30	32	33
non-phi Beatty	2	5	7	10	13	15	18	20	23	26	28	31	34	36	39	41	44	47	49	52	54
gnomon	1	2	3	4	5	6	7	8	9	10	11	12	13	14	15	16	17	18	19	20	21

"Four new sequences can be formed from these two sequences as well, and each of them in turn generating two more sequences, ad infinitum.[739]

duphi Beatty	1	2	4	5	6	8	9	10	12	13	14	16	17	18	20	21	22	24	25
non-duphi Beatty	3	7	11	15	19	23	27	31	34	38	42	46	50	54	58	62	66	69	73
phi-non-phi Beatty	2	5	7	10	12	15	17	20	23	25	28	30	33	35	38	41	46	48	51
dunon-phi Beatty	1	3	4	6	8	9	11	13	14	16	18	19	21	22	24	26	27	29	31

"Other mathematicians since, not unexpectedly, expanded the number of sequences by repeated multiplications and roundings, 'subdivisions'. One subsequence is the rounded seed multiplied by the seed and rounded again. The another subsequence is the square of the seed rounded and then multiplied by the seed and rounded again. A third subsequence is the square of the seed rounded just once." [740]

1st subdivision	1	4	6	9	12	14	17	19	22	25	27	30	33	35	38	40	43	46	48
2nd subdivision	2	5	7	10	13	15	18	20	23	26	28	31	34	36	39	41	44	47	49
3rd subdivision	3	8	11	16	21	24	29	32	37	42	45	50	55	58	63	66	71	76	79

"This 'subdivision' operation can be extrapolated to any number of subsequences, including even an infinite number of which all begin with one and diagonals as well." [741]

1st protosubdivision	1	3	5	7	9	11	13	15	17	19	21	23	25	27	29	31	33
2nd	1	4	7	10	13	16	19	22	25	28	31	34	37	40	43	46	49
3rd	1	5	9	13	17	21	24	27	31	35	39	43	47	51	55	59	63

"The Thue-Morse sequence, named for Axel Thue and Marston Morse, for example, can also be used as a sieve to subdivide sequences. We start with zero in binary, change the parity and hyperadd the new number to the old to get the next power-of-two number of digits.[742]

Thue-Morse	0	1	1	0	1	0	0	1	1	0	0	1	0	1	1	0	1	0	0	1	0	1	1	0
1st subdivision	0			3		5	6			9	10		12			15		17	18		20			23
2nd subdivision		1	2		4			7	8			11		13	14		16			19		21	22	

"It's usually digitized like an almost-number, but can also generate other, quite different sequences by slicing,[743] dicing,[744] and fragging." [745]

sliced	11	11010	11010011	1101001001	110100110010 11	1101001100101101 0
decimal	3	27	211	1,689	13,515	218,715

diced	0	1	11	110	1101	11010	110100	1101001	11010011	110100110	1101001100	1101001100 1
decimal	0	1	3	6	13	27	53	105	211	423	653	1,689

fragged	0	1	110	1101001	110100110010110	1101001100101101001011001101001
decimal	0	1	6	105	27,031	1,771,476,585

"The same technique can be extrapolated to higher bases than binary to subdivide sequences into three subsequences, one of which begins with one.[746]

ternary	0	1	2	0	1	2	1	2	0	2	0	1	1	2	1	2	0	2	0	1	1	2	0	2	0	1
1st sub	0			3					8		10						16		18				22		24	
2nd sub		1			4		6					11	12		14					19	20					25
3rd sub			2			5		7		9				13		15		17				21		23		

"Any fully or partially digitally expanded sequence can subdivide any other sequence, even itself, like this self-dividing sequence." [747]

naturals	1	2	3	4	5	6	7	8	9	10	11	12	13	14	15	16	17	18	19
fully expanded naturals	1	1	0	1	1	0	1	0	1	1	0	1	0	1	0	1	1	0	1
zeroth subsequence			3			6		8			11		13		15			18	
first subsequence	1	2		4	5		7		9	10		12		14		16	17		19

from *PAJAS* #79 (February 1992)

NUMBER MATES

"The two shall be made one," Al quoted Mark 10:8 and then asked "If life with love is better and life without love is worse, what do get when you put the two together?"

"For better or worse," came the answer.

With the marriage of Cal and Lana the conversation naturally turned to complimentarity.

"We can also get complimentary numbers by pairing digits with its mate in the base. For decimal zero and nine, one and eight, two and seven, three and six and four and five are mates."

"Ah, so for the primes we can have prime mates! Getb it? Primates!"[748]

"Yes, for two we have seven, for three six, for five four and for seven two."

"Two of which are also primes and two of which are not."

"And all females have male mates and all males have female mates."

"Does ninety-seven have two as a mate along with seven, because of the leading zero?"

"To avoid bigamous numbers, we use a matchmaking leading zero for a nine-headed number, so its mate is another nine-headed number. For ninety-seven the mate'd be nine hundred two, rather than two, which is mated with seven."

primes	2	3	5	7	11	13	17	19	23	29	31	37	41	43	47	53
prime mates (unsorted)	7	6	4	2	88	86	83	80	76	70	68	63	58	57	52	47
prime mates (sorted)	2	4	6	7	10	16	20	26	28	32	38	40	46	47	52	56

"So eleven's mate is eighty-eight," Cal said. "They'd be the parents of an odd and an even, one and eight, neither prime."

"The grandfathers however could be the primes one eighty-one, eight eleven, eight eighty-one."

"And their mates, eight eighteen, one eighty-eight and one eighteen."

"Mates come in all varieties, like our members, lucky or unlucky,[749] curvaceous or ugly[750] or square.[751]

lucky mates	0	2	6	8	12	20	24	26	30	32	36	48	50	56	62	66	68	74	78
unlucky mates	1	3	4	5	7	9	10	11	13	14	15	16	17	18	19	21	22	23	25
curvaceous mates	0	1	3	6	9	10	11	13	16	19	30	31	33	36	39	60	61	63	66
ugly mates	1	2	3	4	5	6	7	8	9	18	19	24	27	35	39	45	49	51	54
square mates	0	5	8	9	18	35	50	63	74	83	158	215	270	323	374	423	470	515	558

"Some mates suggest the name of their mates. Classmates are the mates of cubed luckies and sorted sanbi,[752]; playmates the mates of prime luckies and yongbi,[753] stalemate the mates of sorted ternarish and lucky evens[754] and teammate the mates of ternarish evens and Mersennes.[755]

classmates	8	72	270	470	656	830	1,350	3,310	4,670	6,030	6,624	7,190	7,802

playmates	2	5	6	16	25	26	32	35	45	55	56	62	65	68	75	85	86	176	212	230	260	272

stalemate	3	4	7	13	14	24	34	37	43	44	54	64	67	73	74	83

teammates	2	4	6	7	9	32	68	77	79	80	82	86	87	89	1,808	7,777	7,778	7,779	7,787

"Some others only suggest their mates by backformation, like 'climate' from cubed lucky ichibi, rather than Roman for one fifty-one, 'cremate' from cubed rokubi even. There's even 'penultimate' numbers from prime even nibi-unindexed lucky ternarish ichibi. Since the only even prime also happens to be nibi, the second lucky ternarish ichibi, twenty-one, is excluded. The first three are one, one eleven and two hundred one, so the first four penultimate numbers are eight, seven eighty-eight, seven ninety-eight and eight eighty-eight.[756]

climate 8 70,202 90,738 867,348 1,879,398 4,999,788 6,557,048 7,196,778 8,632,368

cremated 487 4,167 45,127 78,047 318,527 525,447 685,567 804,887 889,407

"For bases below ten[757] all numbers have unique 'monogamous' mates the same length.

ternarish mates 7 8 9 77 78 79 87 88 89 777 778 779 780

"For base ten and higher we find some 'serially polygamous' mates,[758] because a lead nine takes a lead zero, which takes another lead nine. Although the one and only mate of zero is nine, nine is also the mate of ninety and the other mate of ninety is nine hundred nine."

polygamous 9 90 91 92 93 94 95 96 97 98 99 900 901 902 903 903 904 905 906 907

from *PAJAS* #80 (March 1992)

MacDONALD NUMBERS

Al quoting Thomas Aquinas, "The vision of the essence of Father, Son, and Holy Ghost... is given to the clean of heart alone and is the highest bliss."

"'What happen to the cats walking on the partially frozen Seine?"

"Un, deux, trois cats sank."

'We can make a subset of a sequence by sorting out with itself or another sequence. If we sort out the evenly-indexed evens, we get what we've called the evenly evens. If we sort out the evenly even-indexed odds, we get the e-i-e-i-o or 'MacDonald' numbers.[759]

MacDonald 7 15 23 31 39 47 55 63 71 79 87 95 103 111 119 127 135 143 151 158 166 174 182

"That's just every eighth number after seven."

"Yes, but more than half of the first nineteen numbers are prime."

With that the acronymania spread when a newcomer Max piped in, "More interesting, I think, are the 'pip' numbers," [760]

"The number of pips on a domino?"

"No, the prime-indexed primes."

pip 3 5 11 17 31 41 59 67 83 109 127 157 179 191 211 241 277 283 331 353 367 401 431 461 499

"Ah, so it's now easy to tell a 'fibber' [761] from a 'liar'.[762] The former is Fibonacci-indexed binarish and the latter lucky-ichibi-and-rokubi."

"And the 'lil' numbers,[763] lucky-indexed luckies," Lily Littleton added.

lil 1 7 21 31 49 63 87 111 135 151 171 211 241 261 331 361 385 409 421 451 511 537 579

"And the 'pie' numbers,[764] prime-indexed evens and the oil numbers,[765] oddly-indexed luckies," the culinary arts major continued.

pies 4 6 10 14 22 26 34 38 46 58 62 74 82 86 94 106 118 122 134 142 146 158 166

oils 1 7 13 21 31 37 49 63 69 75 87 99 111 127 133 141 159 169 189 195 205 219 231

"And since oil and water don't mix, we could call the non-oil numbers 'waters'." [766]

waters 0 2 3 4 5 6 8 9 10 11 12 14 15 16 17 18 19 20 22 23 24 25 26 27 28 29 30 32

"We'd also have 'boils', luckies indexed by binarish odds,[767], like one, thirty-seven and five ninety-one, 'toils', luckies indexed by ternarish odds[768], like eighty-seven, luckies indexed by Fibonacci odds,[769], 'foils', like one, seven and thirteen."

binarish odd 1 11 101 111 1001 1011 1101 1111 10001 10011 10101 10111 11001 11011 11101 11111 100001

ternarish odd 1 11 21 101 111 121 201 211 221 1001 1011 1021 1101 1121 1201 1211 1221 2001

Farmboy Cal added 'silos',[770] sanbi-indexed lucky odds, fudging a little because all luckies are odd and never even and 'pigs',[771] prime-indexed gobis. Before someone else did, he also added that those beginning with pig numbers, like a hundred fifty through a hundred fifty-nine and two fifty through two fifty-nine, as 'pig-headed'.[772]

silos 7 49 99 159 219 331 409 475 529 591 643 715 777 903 975

pigs 15 25 45 65 105 125 165 185 225 285 305 365 405 425 465 525 585 605 665

pig-headed 150 151 152 153 154 155 156 157 158 159 250 251 252 253 254 255 256

"How about the Pio numbers[773] for the Italian holy man, Padre Pio, prime-indexed odds?" Brother Ken suggested.

Pio 3 5 9 13 21 25 33 37 45 57 61 73 81 85 91 103 114 117 129 137 141 153 161 173 181 189 193

"Or Rio numbers,[774] rokubi-indexed odds."

Rio 11 31 51 71 91 111 131 151 171 191 211 231 251 271 291 311 331 351 371 391 411 431 451 471 491

"We could call the lucky-indexed primes 'lips' [775] and 'slips', [776] the primes indexed by sanbi-lucky-indexed primes." [777]

lips 2 5 17 31 41 47 73 97 127 137 157 191 227 233 307 331 347 367 379 401

slips 5 41 137 191 307 367 487 691 907 1,103 1,427

sanbi lucky 3 13 33 43 63 73 93 133 163 193 223 273 283 303 423 463 483 513 553 583

"Aha, so all the rest would be 'non-slip' numbers!"[778]

non-slip 0 1 2 3 4 6 7 8 9 10 11 12 13 14 15 16 17 18 19 20 21 22 23 24 25 26 27 28

"And gobi-indexed 'gips',[779] the hachibi-indexed 'hips' [780], the kyuubi-indexed 'kips',[781] the nibi-indexed 'nips',[782] the rokubi-indexed 'rips',[783] the shichibi-indexed 'ships' and 'shipmates' [784], the sanbi-indexed 'sips',[785] ternary-indexed 'tips',[786] and the yonbi-indexed 'yips'." [787]

gips 11 47 97 149 197 257 313 379 439 499 571 631 691 761 829 907 977 1,039 1,103 1,187

hips 19 61 107 163 223 271 337 397 457 521 593 647 719 787 857 929 997 1,061

kips 23 67 109 167 227 277 347 401 461 523 599 653 727 797 859 937 1,009

nips 3 43 89 139 193 251 311 373 433 491 569 619 683 757 827 887 971 1,033

rips 13 53 101 151 199 263 317 383 443 503 577 641 701 769 839 911 983 1,049

ships 17 59 103 157 211 269 331 389 449 509 587 643 709 773 853 919 991 1,051

shipmates 40 82 146 226 290 356 412 490 550 610 668 730 788 842 896

sips 5 41 83 113 127 131 137 139 149 151 157 163 167 191 241 307 367 431 487 563 661 739

tips 2 3 29 31 37 593 599 601 647 653 659 719 727 733 1,213 1,117 1,223 1,283 1,289 1,291

yips 7 43 89 139 193 251 311 373 433 491 569 619 683 757 827 887 971 1,033

"I claim the Filip numbers,[788] primes indexed by the Fibonacci-indexed luckies."[789] Filip Pinno added.

Filips 2 5 17 41 97 227 449 877 1,823 1,823 3,613

fils 1 3 7 13 25 49 87 159 285 517 855

"Then there's the 'lisps',[790] lucky-indexed sanbi primes, the rarer 'blips', primes indexed by binarish luckies and the even rarer 'blimps', binarish lucky-indexed Mersenne primes[791] and the 'flips',[792] primes indexed by Fibonacci luckies." [793]

lisps 3 23 53 83 113 223 263 383 503 743 953 1,063

flips 2 5 41 73 7,877

Fibonacci luckies 1 3 13 21 987

"That's the 'limit',[794] ternarish indexed by lucky-indexed Mersennes!" [795]

limit 3 1,011 22,122,012,102

lim	3	31 524,297	2,305,843,009,213,693,951

They meeting ended with preparations for the celebration of the four hundredth anniversary of Pi Day, 3/14/1592.

from *PAJAS* #81 (April 1992)

OUT BACK

Al quoting Charles Wesley, "Hosanna in the highest to our exalted Savior, Who left behind for all mankind these tokens of His favor."

Ann provided the monthly riddle, "Why is June through September math majors' favorite time of the year?" and the answer, "They're summer months."

"As the mathematician René Descarte didn't say, 'I think therefore I sum.'"

"And what's their favorite song?" the Fen added, "It's summertime, summertime, sum, sum, summertime, summertime!"

Al cut off the Jamies' oldie with a gesture and continued, "Back numbers[796] are simply reversed diced numbers, usually 'pointless', that is, without the radix or decimal point. They usually also get higher and higher, but not always. Numbers duplicated because of trailing zeroes that turned into leading zeroes are not re-counted. We're familiar with those from irrationals like pi, phi and e[796] and roots.[797]

piback	3	31	413	1,413	51,413	951,413	2,951,413	62,951,413
phiback	1	61	8,161	308,161	3,308,161	93,308,161	893,308,161	8,893,308,161
eback	2	72	172	8,172	28,172	828,172	1,828,172	81,828,172

root twoback	1	41	141	4,141	24,141	124,141	3,124,141	53,124,141
root threeback	1	71	371	2,371	502,371	80,502,371	8,080,502,371	708,080,502,371
root fiveback	2	22	322	60,322	760,322	9,760,322	79,760,322	779,760,322
root sixback	2	42	442	9,442	49,442	849,442	9,849,442	79,849,442
root sevenback	2	62	462	5,462	75,462	575,462	1,575,462	31,575,462
root eightback	2	82	282	8,282	48,282	248,282	7,248,282	17,248,282
root tenback	3	31	613	2,613	22,613	222,613	7,222,613	77,222,613

"If we call these 'fullback' numbers. those that begin with an even number can easily be converted to one beginning with one by dividing by two, a 'halfback'." [798]

half root fiveback	1	11	161	30,161	380,162	4,880,161	38,880,161	388,880,161
half rootsix back	1	21	221	4,721	24,721	424,721	4,924,721	39,924,721
half root sevenback	1	31	231	2,731	37,731	287,731	787,731	15,787,731
half eback	1	36	86	4,086	14,086	414,086	914,086	40,914,086
half root eightback	1	41	141	4,141	24,141	124,141	3,624,141	8,624,141

"Similarly those beginning with four can be divided by four to give many 'quarterbacks' beginning with one, some even with eleven." [799]

quarter root twentyback	1	11	111	8,111	308,111	3,308,111
quarter root twenty-oneback	1	11	411	5,411	65,411	465,411
quarter root twenty-twoback	1	11	711	2,711	62,711	3,062,711
quarter root twenty-threeback	1	11	911	8,911	58,911	758,911

from *PAJAS* #82 (May 1992)

ALMOST-EASY NUMBERS

Quoting Charles Villiers Stanford, "Purest and Highest, Wisest and Most Just, there is no truth save only in Thy trust."

"Why couldn't the student solve his trig problem?"

"He couldn't count on his friends to help and he didn't want to cosine alone."

"Since our professor emeritus so often used z for a variable, the multiples of the base of the natural logarithm naturally became called 'easy' and their decimal expansions, collapsed and uncollapsed, 'almost-easy'. [800]

almost-3e	8 1 5 4 8 4 5 4 8 5 3 7 7 1 3 6
almost-5e	1 3 5 9 1 4 0 9 1 4 2 2 9 5 2 2
almost-7e	1 9 0 2 7 9 7 2 7 9 9 2 1 3 3 1
almost-9e	2 4 4 6 4 5 3 6 4 5 6 1 3 1 4 0

almost-5e hypersum	1	13	135	1,359	13,591	135,914	1,359,140	13,591,409	135,914,091
almost-7e hypersum	1	19	190	1,902	19,027	190,279	1,902,797	19,027,979	190,279,799

"Multiples of two ez became 'too-easy' and their decimal expansion 'almost-too-easy'." [801]

almost-2e	5 4 3 6 5 6 3 6 5 6 9 1 8 0 9 0
almost-4e	1 0 8 7 3 1 2 7 3 1 3 8 3 6 1 8
almost-6e	1 6 3 0 9 6 9 0 9 7 0 7 5 4 2 7
almost-8e	2 1 7 4 6 2 5 4 6 2 7 6 7 2 3 6

almost-4e hypersum	1	10	108	1,087	10873	108731	1,087,312	10,873,127	108,731,273
almost-6e hypersunm	1	16	163	1,630	16309	163,096	1,630,960	16309609	163,096,097

"The variable z, of course, can represent other numbers than integers, so we have infinitely many fractional 'easy' numbers." [802]

almost-3e/2	4 0 7 7 4 2 2 7 4 2 6 8 8
almost-5e/2	6 7 9 5 7 0 4 5 7 1 1 4 7

almost-7e/2	9 5 1 3 9 8 6 3 9 9 6 0 6
almost-9e/2	1 2 2 3 2 2 6 8 2 2 8 0 6

almost-9e/2 hypersum	1 12 122 1,223 12,232 122,322 1,223,226 12,232,268

from *PAJAS* #83 (June 1992)

MORE ALMOSTS

Al opened the meeting with Hebrews 9:22, "Almost everything is purified by blood, and without the shedding of blood there is no forgiveness."

"If six was afraid of seven, why was six afraid of sixteen?" he asked, answering his own question with "Because sixteen ate four too."

After that full-groan pun he was forgiven, though he narrowly escaping the shedding of blood, and allowed to continue with last month's topic,"We can get sequences by digitally expanding repeating decimals[803] and higher roots".[804]

almost-one-third	3 3 3 3 3 3 3 3 3 3 3 3 3 3 3
almost-two-thirds	6 6 6 6 6 6 6 6 6 6 6 6 6 6 6
almost-one-sixth	1 6 6 6 6 6 6 6 6 6 6 6 6 6 6
almost-one-seventh	1 4 2 8 5 7 1 4 2 8 5 7 1 4 2
almost-two-sevenths	2 8 5 7 1 4 2 8 5 7 1 4 2 8 5

almost-one-seventh hypersum	1 14 142 1,428 14,285 142,857 1,428,571

"All the almost-fourth roots from two through fifteen would all start with one."

almost-fourth root 2	1 1 8 9 2 0 7 1 1 5 0
almost-fourth root 3	1 3 1 6 0 7 4 0 1 2 9
almost-fourth root 5	1 4 9 5 3 4 8 7 8 1 2
almost-fourth root 6	1 5 6 5 0 8 4 5 8 4 5
almost-fourth root 7	1 6 2 6 5 7 6 5 6 1 6
almost-fourth root 8	1 6 8 1 7 9 2 8 3 0 5

"Digitally expanding the almost-fourth-roots between ten thousand one and one less than twenty thousand also begin with one!"

"So too do the almost-fifth-roots between two and thirty-one and the almost-sixth-roots between two and sixty-three."

The meeting proceeded to get mired down in a deluge of ever higher and higher roots of ever higher and higher numbers.

from *PAJAS* #84 (July 1992)

STILL MORE DOMINISSIMO

Al shared Psalm 46:4, "There is a river, the streams of which make the city of God glad, the holy place of the tents of the Most High."

"At what temperature do you bake pineapple upside-down cake?"

"A hundred eighty degrees, of course."

"The slang term 'one-and-one' refers to fish and chips[805], but we take it to mean a one-headed ichibi[806].

one-and-one 11 101 121 131 141 151 161 171 181 191 1,001 1,021 1,031 1,041 1,051 1,061 1,071

"From 'one for the money' comes the 'moneyed" numbers[807] with one and four as digits, from 'two for the show' comes the 'showy' numbers[808] with two and four and from 'three to get ready' comes the 'ready' numbers[809] with three and two."

moneyed 14 41 104 114 124 134 140 141 142 143 144 145 146 147 148 149 154 164 174 184

showy 24 42 124 142 204 214 224 234 240 241 242 243 244 245 246 247 248 249

ready 23 32 123 132 213 231 302 312 320 321 323 324 325 326 327 328 329 423 432

"The 'poor' numbers have no one or four,[810] the 'no-show' numbers[811] no two or four and the 'not-quite-ready' numbers [812] have a two or three, but not both."

poor 0 2 3 5 6 7 8 9 20 22 23 25 26 27 28 29 30 32 33 35 36 37 38 39 50 52 53 55

no-show 0 1 3 5 6 7 8 9 10 11 13 15 16 17 18 19 30 31 33 35 36 37 38 39 50 51 53 55 56 57

not-quite-ready 2 3 12 13 20 21 22 24 25 26 27 28 29 30 31 33 34 35 36 37 38 39 42 43

"'One and eight' is Rhyming Slang for plate, 'two and eight' for a confused or emotional state, so 'plated' numbers[813] contain a one and eight, 'stately' numbers.[814] Since the stately numbers that are not neighbors all have a number between them we also have 'interstate' numbers[815] with a two and eight."

plated 18 81 108 118 128 138 148 158 168 178 180 181 182 183 184 185 186 187 188 189

stately 28 82 128 208 218 228 238 248 258 268 278 280 281 282 283 284 285

interstate 55 105 168 213 233 243 253 263 273 279 303 313 323 343 353 363 373

"'Five-to-two' is Rhyming Slang for shoe, so the 'shoed' or 'shod' numbers[816] are those with a five, four, three and two. 'Shoeless' numbers[817] are those without two through five as digits and 'sabotaged' numbers[818] are those with at least one them."

shod 2,345 2,354 2,435 2,453 2,534 2,543 3,245 3,254 3,425 3,452 4,235 4,253

shoeless 0 1 6 7 8 9 10 11 16 17 18 19 60 61 66 67 68 69 70 71 76 77 78 79 80 81 82

sabotaged 2 3 4 5 12 13 14 15 20 21 22 23 24 25 26 27 28 29 30 31 32 33 34 35 36

"From Rhyming Slang's 'six-and-eight' for straight or honest comes the 'half-straight' [819] numbers with six or eight, but not both."

half-straight 6 8 16 18 26 28 36 38 60 61 62 63 64 65 66 67 69 76 78 80 81 82 83

"From the cheer, 'Two, four, six, eight, who do we appreciate?!' comes the appreciated numbers[820] with any of these four even digits, while the 'unappreciated' [821] don't."

appreciated 2 4 6 8 12 14 16 18 20 21 22 23 24 25 26 27 28 29 32 34 36 38 40 41

unappreciated 0 1 3 5 7 9 10 11 13 15 17 19 30 31 33 35 37 38 39 50 51 53 55 57 59 70

"From boxing comes the 'one-two punch' and so the 'punchy' numbers[822] with one and two as digits."

punchy 12 21 102 112 120 121 122 123 124 125 126 127 128 129 201 210 211 212 213 214

"From the Roman numeral for fifty-one, LI, come the 'false' numbers[823] that have a one and a five in the tens place, a 'lie'. In a three-valued logical way, those with just one or the other are 'half-true', that is, half-lie,[824] and those with neither 'true'." [825]

false 51 150 151 152 153 154 155 156 157 158 159 251 351 451 551 651 751 851 951 1,050

half-true 1 10 11 12 13 14 15 16 17 18 19 21 31 41 50 52 53 54 55 56 57 58 59 61 71 81 91

true 0 2 3 4 6 7 8 9 20 22 23 24 26 27 28 29 30 32 33 34 35 36 37 38 39 40 42 43

"Similarly the 'dyed' numbers[826] with a five in the hundreds place and a one elsewhere come from DI, 'die' and 'my' numbers[827] with a one in the thousands place and another one come from MI."

dyed 501 510 512 513 514 515 516 517 518 519 521 531 541 551 561 571 581 591 1,500

my 1,001 1,012 1,013 1,014 1,015 1,016 1,017 1,018 1,019 1,021 1,031 1,041 1,051 1,061 1,071

"From 'Ali Baba and the Forty Thieves' comes the 'thieves' numbers[828] with four in the tens place."

thieves 40 41 42 43 44 45 46 47 48 49 140 142 143 144 145 146 147 148 149 240 241

"From POW slang 'twenty-three fifty-nine', the last minute before midnight, meant dark. We take 'dark' [829] to mean numbers with two, three, five and nine, 'half-dark' or 'twilight' [830] those with two of them, and 'light' [831] those without any."

dark 2,359 2,395 2,539 2,593 2,935 2,953 12,359 12,395 12,539 12,593 12,935 12,953

twilight 23 25 29 32 35 39 52 53 59 92 93 95 123 125 129 132 135 139 152 153 159 192

light 0 1 4 6 7 8 10 11 14 16 17 18 40 41 44 46 47 48 60 61 64 66 67 68 70 71 74 76 77 78

"From the base sixty of clocks we get 'quarter-past' for a 'fiff', 'half-past' [832] for a thirty and 'quarter-to' [833] for a forty-five."

half-past 30 130 230 330 430 530 630 730 830 930 1,030 1,130 1,230 1,330 1,430 1,530

quarter-to 45 145 245 345 445 545 645 745 845 945 1,045 1,145 1,245 1,345 1,445 1,545

"Perhaps from the rhyming years, thirteen thirteen to nineteen nineteen, which all rhyme with canteen,[834] nineteen has come to mean 'long ago' in Rhyming Slang.

canteen 19 119 219 319 419 519 619 719 819 919 1019 1119 1219 1319 1419 1519 1619 1719

"From the phrase 'nine-to-five' we get the 'work' numbers, those with any of those five digits, while those without any of them are the 'faith' numbers." [835]

works 5 6 7 8 9 55 56 57 58 59 56 66 67 68 69 75 76 77 78 79 85 86 88 89 95 96

faith 1 2 3 4 11 12 13 14 20 21 22 23 24 30 31 32 33 34 40 41 42 43 44 100 101 102 103

The Society then worked on transforming these work numbers into the Roman numeral joeys within them, namely five to IV or four, six to IX or nine, seven to V or five, eight and nine to I or one, and so to the chronogramatic numbers with one, four, five and nine instead.[836]

chronogramatic 1 4 5 9 11 14 15 19 41 44 45 49 51 54 55 59 91 94 95 99 111 114 115 119

from *PAJAS* #85 (August 1992)

SQUARING THE TRIANGLE

"They remembered that God was their rock, the Most High God their Redeemer." (Psalm 78:3)

"Why do math majors wear glasses?"

"They help with division."

"The triangle named for and popularized in Europe by Blaise Pascal was actually known long before by Al-Karajī[837] Starting with a one, two more ones are placed below and the sum of pairs of numbers are added between ones below. Here each row adds to a power of two.

1	=	1
1 1	=	2
1 2 1	=	4
1 3 3 1	=	8
1 4 6 4 1	=	16
1 5 10 10 5 1	=	32

"The Hindus arranged the same numbers somewhat differently as mātrā-mera, 'number-mountain',[838] as if tilted forty-five degrees, so that the rows now add to the Fibonacci numbers.

1	=	1
1	=	1
1 1	=	2
2 1	=	3
1 3 1	=	5
3 4 1	=	8
1 6 5 1	=	13
4 10 6 1	=	21
1 10 15 7 1	=	34
5 20 21 8 1	=	55

"Al-Karajī's triangle can be rotated forty-five degrees into 'Al-Karajī 's square' fitting its apex into a corner that reveals some interesting higher sequences. The third is the two-dimensional simplex or triangular numbers. The fourth is the three-dimensional simplex or tetrahedral numbers.[839] The fifth is the fifth-dimension simplex, pentatopes, etc.[840] and together they give a new simplex diagonal.[841]

ones	1	1	1	1	1	1	1	1
integers	1	2	3	4	5	6	7	8
triangulars	1	3	6	10	15	21	28	36
tetrahedrals	1	4	10	20	35	46	74	110
pentatope	1	5	15	35	70	116	190	300
6-D simplex	1	6	21	46	116	232	422	722
7-D simplex	1	7	28	74	190	422	844	1,566
8-D simplex	1	8	36	110	300	722	1,566	3,132

simplex diagonal	1	2	6	20	70	232	844	3,132

"Rotating the square again, we can easily read the former diagonals as hypersimplex sequences[842] with yet another new diagonal.[843]

1st hypersimplex	1	2	6	20	70	116	190	722
2nd hypersimplex	1	3	10	35	116	422	1,566	2,710
3rd hypersimplex	1	4	15	46	190	722	1,988	5,375
4th hypersimplex	1	5	21	74	300	422	2,665	9,593
5th hypersymplex	1	6	28	110	455	677	4,218	6,001
6th hypersimplex	1	7	36	155	610	1,553	2,783	10,449

hypersimplex diagonal	1	3	15	74	455	1,553

"Rotating it yet again gives us more sequences and a new diagonal we call 'duhypersimplex'.[844]

1st duhypersimplex	1	3	15	74	455	1,553
2nd duhypersimplex	1	4	21	110	610	2,793
3rd duhypersimplex	1	5	28	155	820	4,469
4th duhypersimplex	1	6	36	210	1,096	4,919
5th duhypersimplex	1	7	45	276	1,450	6,814
6th duhypersimplex	1	8	55	354	1,895	9,259

duhypersimplex diagonal	1	4	28	210	1,450	9,259

"Rotating it yet again gives us 'treshypersimplex' sequences and another diagonal.[845]

1st treshypersimplex	1	4	28	210	1,450	9,259
2nd treshypersimplex	1	5	36	276	1,895	13,694
3rd treshypersimplex	1	6	45	354	2,870	19,259
4th treshypersimplex	1	7	55	445	3,476	26,143
5th treshypersimplex	1	8	66	550	4,435	34,554
6th treshypersimplex	1	9	78	670	5,565	44,720

treshypersimplex diagonal	1	5	45	445	4,435	44,720

"We can also extrapolate Al-Karajī's square by changing the first two numbers of the first row and taking them as a Lucas sequence. If the first number is a one we get an infinite number of Al-Karajī-Lucas sequences and, of course, yet another diagonal.[846]

Al-Karajī-Lucas	1	2	3	5	8	13	21	34	55	89
2nd	1	3	6	11	19	32	53	87	142	231
3rd	1	4	10	21	40	72	125	212	354	585
4th	1	5	15	36	76	148	273	485	839	1,424
5th	1	6	21	57	133	281	554	1,039	1,878	3,302
6th	1	7	28	85	218	499	632	1,671	3,549	6,851
7th	1	8	36	121	339	838	1,470	3,141	6,690	13,541
8th	1	9	45	166	505	1,343	2,813	5,954	12,644	26,185

diagonal 1 3 10 36 133 499 1,470 5,654 23,480 79,022

"If we arrange the natural numbers in a 'squared triangle' diagonally up left-to-right, we get three spokes from the starter one, diagonal.[847]

diagonal 1 5 13 41 61 85 113 145 181 221 265 313 365 421 481 545 613 685

"We also get an infinite number of both horizontal and vertical sequences, a sieve of the natural numbers.[848] The first horizontal one is the triangular numbers again.

1	3	6	10	15	21	28	36	45	55	66	78	91
2	5	9	14	20	27	35	44	54	65	77	90	104
4	8	13	19	26	34	43	53	64	76	89	103	118
7	12	18	25	33	42	52	63	75	88	102	117	133
11	17	24	32	41	51	62	74	87	101	116	132	149
16	23	31	40	50	61	73	86	100	115	131	148	166
22	30	39	49	60	72	85	99	114	130	147	165	184

"We can subdivide any sequence in this way, for example, the primes.[849]

2	5	13	29	47	73	107	151	197	269	337	409	487
3	11	23	43	71	103	149	193	263	331	401	479	577
7	19	41	67	101	139	191	257	317	397	467	571	659
17	37	61	97	137	181	251	313	389	463	569	653	733
31	59	89	131	179	241	311	383	461	563	647	727	809
53	83	127	173	229	307	379	457	557	643	719	797	881

"So the 'diagonal primes' [850] would start as usual with two and then jump to eleven," Cal commented.

diagonal primes 2 11 41 97 179 307 541 631 701 773 863 971 1,063 1,187 1,301 1,451

"Yes, but we can also arrange the numbers, rather than always in one direction, alternately diagonally left and diagonally right, called from the Greek boustrophedonic[851] meaning like a plowed field. These sequences however suffer from a little irregularity, though they still get higher and higher. The first gnomon, for example, is one, four, one, eight, one, twelve, etc."

1	2	6	7	15	16	28	29	45	46	66	67	91
3	5	8	14	17	27	30	44	47	65	68	90	93
4	9	13	18	26	31	43	48	64	69	89	94	118
10	12	19	25	32	42	49	63	70	88	95	117	124
11	20	24	33	41	50	62	71	87	96	116	125	149
21	23	34	40	51	61	72	86	97	115	126	148	159
22	35	39	52	60	73	85	98	114	127	147	160	184

"It also gives an irregular diagonal.[852]

diagonal boustrophedon 1 5 13 25 41 61 85 113 145 162 186

"We can make isosceles triangles, rather than merely equilateral ones, square as well.[853] The first row is, of course, the squares. This sorts into even and odd diagonals.[854]

1		4		9		16		25		36		49	
	3		8		15		24		35		48		63
2		7		14		23		34		47		62	
	6		13		22		33		46		61		78
5		12		21		32		45		60		77	
	11		20		31		44		59		76		95
10		19		30		43		58		75		94	

from *PAJAS* #86 (September 1992)

EGYPTIAN FRACKING

The new president Robin opened with "You, Lord, are Most High above all the Earth. You are exalted far above all gods." from Psalm 97:9. The new school year also brought new members, many with "Math Counts" tee shirts.

"If the solid has a radius of z and a thickness of a, what's the volume?"

"Pizza."

"An alternative to continuing fraction expansion is the way that the ancient Egyptians expressed numbers in unit-fractions.[855] Their denominators form an irrational seed number, such as pi,[856] half-pi,[857] phi,[858] e,[859] half-e[860] or roots[861] form an infinite sequence, though they tend to get higher faster than the familiar continued fractions."

pi	3	8	61	5,020	139,085,963	> sixteenplex	
half-pi	1	2	15	243	69,284	6,707,793,329	
phi	1	2	9	145	37,986	2,345,721,554	> eighteenplex
e	2	2	5	55	9,999	3,620,213,552	
half-e	1	3	39	6,006	36,547,382	> fifteenplex	
√2	1	3	13	243	218,201	61,323,125,725	> twenty-threeplex
√3	1	2	5	32	1,249	5,986,000	461,044,780,477,807
√5	2	5	28	2,828	11,765,225	> fourteenplex	
½√5	1	9	145	37,986	2,345,721,554	> eighteenplex	
√6	2	3	9	199	49,572	> elevenplex	

"Another interesting feature of Egyptian fractions is that some fractions can be easily expressed in many different ways. Three-sevenths, for example, is a third plus an eleventh plus two hundred thirty-first, but also a third plus a twelfth plus an eighty-fourth, a third plus a fourteenth plus a forty-second, a third plus a fifteenth plus a thirty-fifth, a fourth plus a sixth plus an eighty-fourth and a fourth plus a seventh plus a twenty-eighth. We represent them like the continued fraction as a sequence of denominators between square brackets.

3/7 = [3, 11] = [3, 12, 84] = [3, 14, 42] = [3, 15, 35] = [4, 6, 84] = [4, 7, 28]

"Denominators can be expanded into ever-increasing denominators indefinitely, the uptown neighbor and the product of the original denominator and that neighbor.[862]

[2] 3 7 43 1,807 3,263,442 10,650,056,950,807

"These sequences yield yet another diagonal.[863]

MORE MATHEMAGIC

from *PAJAS* #87 (October 1992)

"I blessed the Most High, and I praised and honored Him Who lives forever; for His dominion is an everlasting dominion, and His Kingdom from generation to generation," (Dan 4:34)

"Why can't you send a mathematician out for coffee?"

"She thinks half-and-half is the same as whole."

"Besides the 'magical' nuclear numbers and the 'magic' numbers based on the mag- prefix, we have a much simpler type of 'magic', those numbers that 'magically' produce repdigits when multiplied, 'predigits'.[864] Thirty-seven, for example, is a third of a hundred eleven and so every multiple of three yields a repunit. Fifteen thousand eight hundred seventy-three is a seventh of a hundred eleven thousand, one hundred eleven, so it does the same for every seven multiple. As you can see the multipliers seem to be primes, but not quite every prime.

multiplier	3	7	9	11	13	17	19
prepunits	37	15,873	12,345,679	101	8,547	65,359,477,124,183	5,847,953,216,374,269
repunit lengths	3z	6z	10z	4z	6z	16z	18z

"As you can also see, the prepunit of some multipliers, like eleven's and thirteen's are less than that of their predecessors. Thirty-seven thousand, thirty-seven is also a replicator since it divides the six-digit repunit and thirty-seven million, thirty-seven thousand, thirty-seven is because it divides the nine-digit repunit, and so on." [865]

37	37,037	37,037,037	37,037,037,037	37,037,037,037,037	37,037,037,037,037,037

"Every multiple of these basic 'magic' numbers that doesn't produce a repunit would also be 'magic' since they also produce repdigits.[866]

prepdigits	37	74	101	148	185	202	259	296	303	370	404	407	481	505	518	592

NACCI NUMBERS

from *PAJAS* #88 (November 1992)

Raz asked, "'Can anyone tell me the significance of the expression on the blackboard?"

$$i\varepsilon > m_a$$

Some speculated that because the expression to the left of the inequality was imaginary the mass on the right had to be also. That lead to a discussion of tachyons and other esoteric theoretical physics.

Finally Sam ventured, "What you all are calling 'epsilon is greater than' looks to me like a sideways heart, so 'I heart the Ath m' might be a rebus for 'I love math.'"

"Right! In order to see new things you have to see things anew. In honor of Fibonacci day on the twenty-third this month we'll re-investigate the Fibonacci numbers. Since those higher than one are all the sum of pairs of integers, some even and some odd, infinitely many subsequences are contained within it that can easily be exposed by division and rounding.

Those evenly divisible by a number z when divided by z yield a new sequence, which we name using capitalized Latin fractional prefixes, Seminacci[867], Tertinnaci,[868] Quartinacci,[869] etc.[870] They get higher at very varied rates.

F	1	1	2	3	5	8	13
(F/2]	1	4	17	72	305	1,292	54,730
(F/3]	1	7	48	2,55	15,456	105,937	726,103
(F/4]	2	36	646	11,592	208,010	3,732,588	66,978,574

"Other nacci sequences begin with more than just two addends. They're indicated with capitalized Greek prefixes rather than either Latin. We simplify them by not duplicating the repeated initial ones, Trinacci,[871] Tetranacci,[872] Pentanacci[873], Hexanacci [874], Heptanacci [875], Oktanacci.[876]

Tribonacci	1	3	5	9	17	31	57	105	193	355	653	1,201	2,209	4,063
Tetranacci	1	4	7	13	25	49	94	181	349	673	1,297	2,500	4,819	9,289
Pentanacci	1	5	9	17	33	65	129	253	497	977	1,921	3,777	7,425	14,597
Hexanacci	1	6	11	21	41	81	161	321	636	1,261	2,501	4,961	9,841	19,521
Heptanacci	1	7	13	25	49	97	193	385	769	1,531	3,047	6,073	12,097	24,097
Oktanacci	1	8	15	29	57	113	225	449	897	1,793	3,578	7,141	14,253	28,449

Exploring the higher realms of these naccis, while eating nachos, kept the members busy for some time.

from *PAJAS* #89 (December 1992)

SPIRALING

Quoting St. Francis, "Happy are those who endure in peace, by You. Most High, will they be crowned.", Robin then asked, "If you're having problems dealing with squares and cubes, where should you go next?"

"Your Higher Power."

"By arranging sequences in spirals,[877] we can uncover sequences by counting off in different directions. Some are even prime-dense sequences like the 'spokes' in the southeast or southwest directions: three, thirteen and thirty-one or five, seventeen, thirty-seven."

73	74	75	76	77	78	79	80	81	82
72	43	44	45	46	47	48	49	50	83
71	42	21	22	23	24	25	26	51	84
70	41	20	7	8	9	10	27	52	85
69	40	19	6	1	2	11	28	53	86
68	39	18	5	4	3	12	29	54	87
67	38	17	16	15	14	13	30	55	88
66	37	36	35	34	33	32	31	56	89
65	64	63	62	61	60	59	58	57	90
100	99	98	97	96	95	94	93	92	91

"Eight of these 'spokes' begin with one.[878]

E	1	2	11	28	53	86	127	176	233	298	371	452	541	638	743	856	977	1,106
SE	1	3	13	31	57	91	133	183	241	307	381	463	553	651	757	871	993	1,123
S	1	4	15	34	61	96	139	190	249	316	391	474	565	664	771	886	1,009	1,140
SW	1	5	17	37	65	101	145	197	257	325	401	485	577	677	785	901	1,025	1,157
W	1	6	19	40	69	106	151	204	265	334	411	496	589	690	799	916	1,041	1,174
NW	1	7	21	43	73	111	157	211	273	343	421	507	601	703	813	931	1,057	1,191
N	1	8	23	46	77	116	163	218	281	352	431	518	613	716	827	946	1,073	1,208
NE	1	9	25	49	81	121	169	225	289	361	441	529	625	729	841	961	1,089	1,225

"The northeast spoke is the odd squares! None of them are prime!"

"The southwest spoke has the most primes so far, nine out of the first fourteen numbers."

"All of them have prime factors that are one more than a multiple of four. One is one more than zero, five one more than four, seventeen one more than four squared, thirty-seven one more than four times nine. Even the non-prime, sixty-five, is one more than four times sixteen.

"Many of them are within six of another prime. Those that are also prime are called 'sexy primes'." [879]

sexy primes	7	11	13	17	23	31	37	41	47	53	61	67	73	83	97	101	103	107	131	151	157	167

"There is just one set of prime quintuplets: five, eleven, seventeen, twenty-three and twenty-nine. Prime twins are two apart,[880] and are either 'uptown' [881] or 'downtown'.[882]

prime twins	3	5	7	11	13	17	19	29	31	41	43	59	61	71	73	101	103	107	109	137	139
uptown	5	7	13	19	31	43	61	73	103	109	139	151	181	193	199	229	241	271	283	313	349
downtown	3	5	11	17	29	41	59	71	101	107	137	149	179	191	197	227	239	269	281	311	347

"The prime neighbors are also uptown or downtown[883] as well as some that are both, more aptly call 'midtown' neighbors." [884]

uptown	3	4	6	8	12	14	18	20	24	30	32	38	42	44	48	54	60	68	72	74	80	84	90
downtown	1	2	4	6	10	12	16	18	22	28	30	36	40	42	46	52	58	60	66	70	72	78	82

midtown	4	6	12	18	30	42	60	72	102	108	150	180	192	198	228	240	270	282	312	348

"Spiraling just the odd numbers we get another eight spokes with an even higher prime density.[885]

313	223	225	227	229	231	233	235	237	239	241	243	245
311	221	147	149	151	153	155	157	159	161	163	165	247
309	219	145	87	89	91	93	95	97	99	101	167	249
307	217	143	85	43	45	47	49	51	53	103	169	251
305	215	141	83	41	13	15	17	19	55	105	171	253
303	213	139	81	39	11	1	3	21	57	107	173	255
301	211	137	79	37	9	7	5	23	59	109	175	257
299	209	135	77	33	31	29	27	25	61	111	177	259
297	207	133	75	73	71	69	67	65	63	113	179	261
295	205	131	129	127	125	123	121	119	117	115	181	263
293	203	201	199	197	195	193	191	189	187	185	183	265
291	289	287	285	283	281	279	277	275	273	271	269	267

E	1	3	21	57	107	173	255
SE	1	5	25	63	105	181	265
S	1	7	29	69	123	193	277
SW	1	9	33	75	129	201	289
W	1	11	39	89	139	213	303
NW	1	13	43	87	147	223	315
N	1	15	47	93	155	233	327
NE	1	17	51	99	163	243	343

"If we start with the higher seed number, seven, and spiral on a hexagonal grid that decreases the spokes to six, we can see hidden prime patterns." [888]

"Southwest seems to have the highest prime density again."

SE	7	9	23	49	87	137	199	273	359	457	567	689	823	969	1,127	1,297	1,479
S	7	11	27	55	95	147	211	287	375	475	587	711	847	995	1,155	1,327	1,511
SW	7	13	31	61	103	157	223	301	391	493	607	733	871	1,021	1,183	1,357	1,543
NW	7	15	35	67	111	167	235	315	407	511	627	755	895	1,047	1,211	1,387	1,575
N	7	17	39	73	119	177	247	329	423	529	647	777	919	1,073	1,239	1,417	1,607
NE	7	19	43	79	127	187	259	343	439	547	667	799	943	1,099	1,267	1,447	1,639

"We can decrease the number of spokes to just four and make a cruciform 'spiraling'," Brother Ken added. "Rearranging the original sequence this time in four directions, we get the good NEWS.[889] After subdividing it into four subsequences all sharing one in common, they could be re-united into the evens, East and West, like the divided Germanies and North and South, like the divided Koreas."

N	1	5	9	13	17	21	25	29	33	37	41	45	49	53	57	61	65	69	73	77	81	85	89	93
E	1	2	6	10	14	18	22	26	30	34	38	42	46	50	54	58	62	66	70	74	78	82	86	90
W	1	4	8	12	16	20	24	28	32	36	40	44	48	52	56	60	64	68	72	76	80	84	88	92
S	1	3	7	11	15	19	23	27	31	35	39	43	47	51	55	59	63	67	71	75	79	83	87	91

"If we keep the square spiral and just count these four cruciform spokes, we get first cruciform or 'cross' sequence.[890]

cross 1 2 4 6 8 11 15 19 23 28 34 40 46 53 61 69 77 86 96 106 116 127 139

"If we do not count the north spoke, we get the T-shaped Franciscan 'tau' cruciform numbers,[891] nothing like the almost-tau numbers.

tau 1 2 4 6 11 15 19 28 34 40 53 61 69 86 96 106 127 139 151 163

"If we rather count just the four diagonals, we get the X-shaped 'St. Andrew' cruciform numbers." [892]

St. Andrew 1 3 5 7 9 13 17 21 25 31 37 43 49 57 65 73 81 91 101 111

"Why, if we count just the northwest, northeast and south spokes, we could also get the 'Y' or 'why' numbers or 'wise' as well as those that aren't, the 'why-nots'." [893]

wise 1 4 7 9 15 21 25 34 43 49 61 73 81 96 111 121 139 157 169 190 211 225 249

why-not 2 3 5 6 8 10 11 12 13 14 16 17 18 19 20 22 23 24 26 27 28 29 30 31 32 33

"Aha, if we don't count the south spoke, but just the northwest and northeast spokes, they're like a V for Victoria, ... the north, south, northeast and southeast like a K ... and the north and east like an L," [894] the Delta Schwa sisters, who happened also be be siblings, added.

Victoria 1 7 9 21 25 43 49 73 81 111 121 157 169 211 225

Kay 1 5 7 15 17 25 29 47 51 63 69 93 99 115 123 155 163 183 193 233 243

El 1 3 15 21 47 57 93 107 155 173 233 255

"The east and south spokes are like a Greek gamma," they continued, "and the southwest and southeast like a lambda, Λ." [895]

gamma 1 3 7 21 29 57 69 107 123 173 193 255

lambda 1 3 5 13 `7 31 37 57 65 91 101 133 145 183 197 241 257

"That reminds me of the non-Greco-Roman futhorc," Lana commented.
"What kind of orc?!"
"No kind of orc. The name doesn't come from Tolkien, but from an acronym of the first six Old English runes. *feoh* 'wealth', *úr* 'auroch', *thorn* 'thorn', *ōs* 'god' or the Latin 'mouth', *rād* 'ride' and *cēn* 'torch', something like the word alphabet that comes from the Greek letters alpha and beta.
"The St. Andrew cross or X looks like the 'gift' rune, *gyfu*, the vertical north-south spokes like the 'ice' rune, *is*, or I or 'eye'. [896]

eye 1 4 8 15 23 34 46 61 77 96 116 139 163 190 218 249 281 316 391 431 474 518 565 613

"The northwest, north, northeast and south spokes look like the rune for the spiked flower the 'sedge', *eohl*;[897] the southwest, south and southeast spokes like the 'glory' rune, *tir*.[898]

sedge 1 4 7 8 9 15 21 23 25 34 43 46 49 61 73 77 81 96 111 116 121 139 157 163 169 190

glorious 1 3 4 5 13 15 17 31 34 37 57 61 65 91 96 101 133 139 145 183 190 197 241 249 257

"Six spokes, all but the horizontal east and west, the 'dash',[899] is like the 'eel' rune *īor*,[900] though we prefer to call this six-spoked shape an 'asterisk'.

dash 1 2 6 11 19 28 40 53 69 86 106 127 151 176 204 233 265 298 334 371 411

asterisk 1 3 4 5 7 8 9 10 12 13 14 15 16 17 18 20 21 22 23 24 25 26 27 29 30 31 32 33 34

"It's also just like the Celtic Ogham ébad, EA and the NEWS spokes are like *beith*, B," she added, but was by then only speaking to herself.
Turning away fom the blackboard she saw that the mention of the why-nots had triggered a Why-knot, the linking of hands randomly and then the unraveling of the resultant knot of persons.

COMPOSITES

from *PAJAS* #90 (January 1993)

Quoting Daniel 7:18,"The saints of the Most High shall receive the Kingdom, and possess the Kingdom forever, even forever and ever." Robin then asked, "Why did some cowboys count a hundred cattle and others only ninety-seven?"
"Some of them rounded them up and the others didn't."
"Since the product of two primes that is relatively prime to the sum of its factors is called a semiprime, by extrapolation the product of three primes would logically be called a tertiprime, four quartiprime and so on.[901]
"Composites with just two different prime power factors can be factored like a googol into little and not-so-little 'halves', for example, three and four for a dozen or four and five for a score."

semiprimes	4	6	9	10	14	15	21	22	25	26
tertiprimes	8	12	18	20	27	28	30	42	44	45
quartiprimes	16	24	36	40	54	56	60	81	84	88

"They too have vertical sequences and a diagonal sequences.[902] The first column is powers of two higher than two. The second is three times powers of two higher than one, while the third is a power of two higher than zero times three to the second.

1st vertical	4	8	16	32	64	128	256	512	1,024	2,048	4,096
2nd vertical	6	12	24	48	144	192	576	768	1,536	3,072	6,144
3rd vertical	9	18	36	72	160	288	896	1,152	2,304	4,608	9,216

diagonal	4	12	36	80	240	672	1,440	2,592	5,384	11,264

"Rotating these we get a new square and new diagonal.[903]

2nd diagonal	4	12	36	80	240	672
3rd	6	18	40	108	324	800
4th	9	20	54	112	336	1,008
5th	10	27	56	120	400	1,056
6th	14	28	60	162	528	1.080

diagonal	4	18	54	120	528

"Semiprime ananums are called 'emirpimes', yet also biprimes and 2-almost primes. Other ananymous arithmonyms too can be extrapolated from emirpitres aka emirpirts or 3-tsomas, etc., in general 'tsomlas'.[904]

emirpimes	15	26	39	49	51	58	62	85	93	94	115	122	123

from *PAJAS* #91 (February 1993)

OFF WITH THEIR HEADS!

"Love your enemies," Al began, "and do good, and lend, expecting nothing back; and your reward will be great, and you will be sons of the Most High." (Luke 6:35)

"How would you translate the statement, 'Two of us nine three burgers five lunch at thirteen.'?"

"Ah, translating inflationary English into deflationary: 'One of us ate two burgers for lunch at noon."

"Continuing to deflate we'd get, 'None of us seven won burgers three lunch at eleven.'"

"Many of the many named sequences can be beheaded to form less easily recognizable, yet easily nameable sequences. 'Quares', for example, are beheaded squares.[905] Unsorted they do contain duplicate numbers. Sorted they are even less recognizable, but useful for vital pattern recognition training.

quares	6	5	6	9	4	1	0	21	44	69	96	25	56	89	24	61	0	41	84	29	76	25	76
sorted	0	1	4	5	9	21	24	25	29	41	44	56	61	69	76	84	89	96	100	116	136	156	164

"Similarly two eleven is an 'ube' [906] because the cube of one seventy one is five million two hundred eleven."

ubes	7 4 25 16 43 12 29 0 331 728 197 744 375 96 913 832 859 0 261 648 2,167
sorted	0 4 7 12 16 25 29 43 96 197 211 261 331 375 376 648 653 728 744 763 832

"'Blongs' are beheaded oblongs[907] and since they are all even 'interblongs' [908] belong in a subsequence of them.

blongs	12 24 56 72 90 156 272 306 342 420 506 600 702 756 812
interblongs	18 40 64 81 123 234 289 324 381 463 553 729 784 934 1,422

"The 'rimorials' are beheaded primorials and easily sorted." [909]

rimorials	0 10 310 30 10,510 699,690 23,092,870
sorted	0 10 30 310 10,510 699,690 23,092,870

"The 'ings' are beheaded ring numbers." [910]

ings	5 6 1 32 3 585 213 1,298 537 730,797 805,595
sorted	1 3 5 6 32 213 537 585 1,298 730,797 805,595

"Some of these beheadings even form English words, rather than just the previous suffix. A beheaded chut would be a 'hut'.[911] It'd include those with one or both of two and five or of three and four.

hut	2 3 4 5 12 13 14 15 20 21 22 23 24 25 26 27 28 29 30 31 32 33 34 35 36 37 38

"A beheaded sept, an 'ept' number,[912] has six of the digits zero through six, less if exposing a leading zero. The much more common beheaded non-septs are the 'inept' numbers."

ept	1,234 1,235 1,236 1,243 1,245 1,256 1,263 1,264 1,265 1,324 1,325 1,326 1,423
inept	... 1,233 1,237 1,238 1,239 1,240 1,241 1,242 1,247 1,248 1,249 1,250 ...

"A beheaded crap is a 'rap'" [913] with a one, two or six, but not both a one and a two or two sixes."

rap	1 2 6 10 13 14 15 16 17 18 19 20 23 24 25 26 27 28 29 31 32 36 41 42 46 51 52 56

"A beheaded gleek, a 'leek',[914] is a po higher than fours," Lily said.

leek	44 55 66 77 88 99 144 155 166 177 188 199 244 255 266 277 288 299 344

"I claim the beheaded bart, an art[915]," said Arthur. "It includes both barts and semibarts."

art	2 3 5 6 12 13 15 16 20 21 22 23 24 25 26 27 28 29 30 31 32 33 34 35 36 37

"I've got 'ego'," [916] Al interjected.

"You certainly do!" one of his Fen responded with a Hardy voice and nod.
"A beheaded 'bego', that is," he explained, "with a weight between eleven and eighteen."

ego 29 38 39 47 48 49 56 57 58 59 65 66 67 68 69 74 75 76 77 78 79 83 84 85

"Ah, so a weight between eleven and fourteen is a 'lo-ego' [917] and that between fifteen and eighteen 'hi-ego'." [918]

lo-ego 29 38 39 47 48 49 56 57 58 59 65 66 67 68 74 75 76 77 83 84 85 86 92

hi-ego 78 79 87 88 89 97 98 99 168 169 177 178 179 186 187 188 199 267 268 269

"Now I have an 'ow',[919] a beheaded gow."

DIHEDRALIZED NUMBERS

from *PAJAS* #92 (March 1993)

"Dihedral or calculator numbers are made up of seven bars, four vertical and three horizontal. When we subtract one from eight dihedrally, for example, we get three."

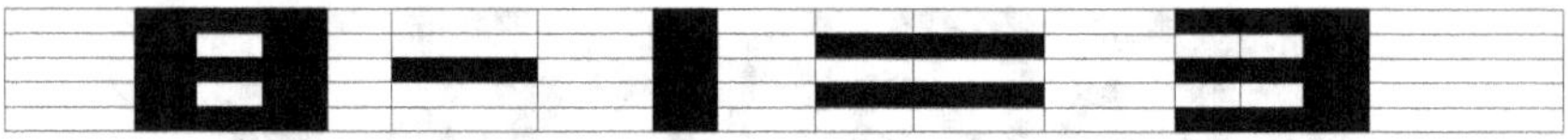

"Designating 'on' with a one and 'off' with a zero, they're equivalent to a seven-bit binary number, reading from top-to-bottom and left-to-right, say the first place for the top bar, the second and third for the upper vertical bars, the fourth for the middle horizontal, the fifth and sixth for the lower vertical bars and the seventh for the bottom bar.[920]

0	1	2	3	4	5	6	7	8	9
1110111	10010	1011101	1101101	101110	11011	1111011	1100101	1111111	1101111

"We can also designate the digits with positions in a three-by-five grid, equivalent to a fifteen-bit binary number.[921]

1	2	3	4	5
10010010010010	11100111100111	111001111001111	1011010010011	111100111001111
6	**7**	**8**	**9**	**0**
111100111101111	111001001001001	111101111101111	111101111001001	111101101101111

"We can also designate the digits with a zeroless encoding just counting how many vertical and horizontal positional are filled. If no positions are filled no position is counted, hence the name 'nonogram'.[922]

1	2	3	4	6
5-IIIII	13-III-31-31313	III-III5-31313	3-15-II-II-3	3-III-I-3-3-I-3-13
6	7	8	9	0
5-III-13-3-I-3-2-3	I-I-5-3-IIII	5-III-5-3-2-3-2-3	3-I-III-5-3-2-3-I-3	5-2-5-3-222-3

"These can be further encoded as either three-place quintary or five-place ternary numbers."

NUMBER GENERATORS

from *PAJAS* #85 (November 1992)

Robin quoted Job 31:4, "Doth not He consider my ways and number all by steps?" and then asked, "Why can't an elephant dance."

"It has two left feet," came the answer from Anne.

"Our first 'easy' number[923] generator comes from the game 'Agent 8'. It adds forty-four, divides by seven and rounds, adds seventy-four, divides by three and rounds, subtracts three, multiplies by four, doubles, subtracts forty-four, divides by thirty-seven and rounds. All this however just seems to generate the integers higher than two.

Agent 8 3 4 5 6 7 8 9 10 11 12 13 14 15 16 17 18 19 20 21

"The second 'tough' number[924] generator, call it eight-point-one, however multiplies a number by nine, adds a hundred fifty-three, divides by three and rounds, halves and rounds, squares, divides by nine and rounds, multiplies by three, subtracts twenty-seven, divides by twenty-eight and rounds. It starts like numbers above seven but is discontinuous.

Agent 8.1 7 8 9 10 12 13 14 15 17 18 20 21 23 24 26 27 30 31 33

"The sequel 'Agent 9' has another two number generators. The first 'easy' one[925] adds twelve, divides by four and rounds, adds twenty-three, divides by three and rounds, subtracts two, multiplies by six, doubles, subtracts three, divides by nine and rounds. The second 'tough' one[926] multiplies by seven, adds thirty-seven, divides by thirteen, rounds, cubes, adds twenty-eight, divides by nine, rounds, multiplies by seven, subtracts twenty-three, divides by twelve and rounds."

Agent 9 7 9 10 11 13 14 15 17 18 19 21 22 23 25 26 27 29 30 31

Agent 9.1 7 35 84 154 252 385 385 553 763 1,022 1,330 1,694 2,121

"In 'Time in Numbers',[927] time travelers trapped in AD 1000 have to calculate 'the number days in the week by the number of hours in a day, then add the number of weeks in a year before multiplying by the number of years in a century, then subtract the number of minutes in a day and add the number of months in a decade before finally adding the number of years' between their hometime and the year 1000. Entering the number minus nineteen thousand, six hundred eighty would chronoport the time traveler to the year zero, outside of space-time altogether, either 0 A. D., Heaven, or 0 B. C., hell!"

time-in numbers	19,681	19,682	19,683	19,684	19,685	19,686	19,687

"The so-called Antarctic numbers[928] are generated by subtracting sixty-nine, dividing by three and rounding, multiplying by three-fourths and rounding, adding twenty-seven, multiplying by four-fifths and rounding, multiplying by five and subtracting seventeen."

Antarctic numbers	23	28	33	38	43	48	53	58	63	68	73	78	83	88	93

"In the sequel 'Back to the Antarctic' new Antarctic numbers[929] are calculated by multiplying by forty percent and rounding, dividing by four and rounding, multiplying by two-thirds and rounding twice, multiplying by forty-four, adding thirteen and adding five."

new Antarctic numbers	18	62	106	150	194	238	282	326	370	414	458

"Performing the operation in the opposite order would logically generate the Antarctic number's opposite numbers, the Arctic numbers,[930] adding five, multiplying by forty-four, multiplying by two-thirds and rounding twice, dividing by four and rounding and multiplying by forty percent and rounding, They start off odd[931] for a while and then oddly become even at thirty-six, then odd again at eighty-nine, then even at one twenty-four.

Arctic	25	27	29	31	33	35	36	38	40	42	44	46	48	50	52	54	56	58	60	62	63

odd Arctic	25	27	29	31	33	35	89	91	93	95	97	99	101	103	105	107	109	111

"Applying both Arctic and Antarctic would give us two other generic circumpolar generators. The one starting with Arctic we'll call North[932] and the other, starting with Antarctic, South.[833]

N.Polar	43	48	53	58	63	68	73	78	83	88	93	98	103	108	113	118	123	128	133

"That the first two S. Polar numbers are odd is a hint that others may be also[934] and indeed it becomes odd again at eight-nine like the odd Arctic."

S.Polar	33	35	36	38	40	42	44	46	48	50	52	54	56	58	60	62	64	66	68

S.Polar odd	33	35	89	91	93	95	97	99	101	103	105	107	109	111	113	115	117	119	121

WHAT ARE THE ODDS?

from PAJAS #94 (May 1993)

Robin began the meeting with "All praise and thanks to God Most High, the Father Who is perfect love; the God Who doeth wondrously, the God Who from His throne above my soul with richest solace fills, the God Who every sorrow stills!" by Johann Jakob Schütz translated by Catherine Winkworth.

"Why did the mathematicians Bob and Anna wanted to be buried head to head?"

"Cemetery symmetry."

"All odds higher than one, of course, are safe because they are all still odd when decreased by one and halved. All evens are not because when decreased by one they cannot be evenly halved. The linear numbers can be either even[935] or odd,[936] but none of them are safe.

even linears 4 14 74 114 144 174 414 444 474 1,114 1,144 1,174 1,414 1,444 1,474 1,714 1744

odd linears 1 7 11 17 41 47 71 77 111 117 141 147 171 177 411 417 441

"Luckies are all odd, but unluckies can be either even or odd.[937]

odd unluckies 5 11 17 19 21 23 27 29 35 39 41 45 47 53 55 57 59 61 65 71 77 81 83

"The kyuubi are all odd, a few either safe[938] or curvaceous[939] and fewer still both.[940]

safe kyuubi 19 39 59 79 99 119 139 159 179 199 219 239 259 279 299 319 339 359 379

curvaceous kyuubi 39 69 99 309 339 369 399 609 639 669 699 909 939 969 999

curvaceous safe kyuubi 39 99 339 399 639 699 939 999 3,039 3,099 3,339 3,399

"The sanbi are all odd, but none are safe and only a few are curvaceous.[941]

curvaceous sanbi 3 33 63 93 303 333 363 393 603 633 663 693 903 933 963 993

Fibonaccis are more unsafe[942] than safe, since only five may be safe.

unsafe Fibonacci 1 2 3 8 13 21 34 55 89 144 233 377 610 987 1,597 4,181 6,765 10,946

Fibonacci odds can be abbreviated as the acronum 'fo'. 'Fee' is obviously the Fibonacci evenly evens, and 'fie' the Fibonacci-indexed evens." [943]

fee 8 144 2,584 46,368 832,040 14,930,352 267,914,296 4,807,526,976

fie 2 4 6 10 16 26 42 68 110 178 288 466 754 1,220 1,974 3,194 5,168 8,362

fo 1 3 5 13 55 89 233 377 987 1,597 4,181 17,711 28,657 75,025 121,393 317,811 514,229

"And 'fum' the Fibonacci-unindexed Mersennes." [944]

non-Fibonacci 4 6 7 9 10 11 12 14 15 16 17 18 19 20 22 23 24 25 26 27 28 29 30 31

fum 127 8,191 131,071 8,388,607 53,670,911 2,147,483,647 137,438,953,471

"Aha, Jack's giant must have been a mathematician, too."

POLYDIGITALS

from PAJAS #95 (June 1993)

"O Lord Most High," Robin began, adding "eternal King, by Thee redeemed Thy praise we sing; the bonds of death are burst by Thee, and grace has won the victory." from Ambrose translated by John Mason Neale.

"What's an eleven-letter word for half-hearted forgiving?"

"Twogiveness,"

"Between the repdigits and the pandigitals are the polydigital, those numbers with more than one, but less than all digits. In base ten we find semidigitals[945] and quintidigitals[946], numbers with respectively fifty and twenty percent of possible digits.

semidecimal	10,234	10,235	10,236	10,237	10,238	10,239	10,243	10,245	10,246

quintidecimal	10	12	13	14	15	16	17	18	19	20	21	23	24	25	26	27	28	29	30	31

"Jootsing to other bases however otherwise impossible polydigitals become possible. In bases divisible by three, ternarish,[947] senarish[948] and nonarish,[949] tertidigitals with one-third and bedigitals with two-thirds of possible digits, become possible. The first senarish bedigital prime isn't until fourteen fifty-three and the first nonarish bedigital not until a hundred three thousand, four fifty-seven.

tertiternarish	1	2	11	22	111	222	1,111	2,222	11,111	22,222	111,111	222,222
tertisenarish	10	12	13	14	15	20	21	23	24	25	30	31
tertinonarish	102	103	104	105	106	107	108	120	123	124	125	126

beternarish	10	12	20	21	100	101	110	112	121
besenarish	1,023	1,024	1,025	1,032	1,034	1,035	1,042	1,043	1,305
benonarish	102,345	102,346	102,347	102,348	102,435	102,436	102,437	102,458	102,534

"In both quadrarish and octarish, besides semidigitals,[950] dodridigitals with three-fourths of possible digits and quartidigitals[951] with one-fourth become possible.

semiquadarish	10	11	12	13	20	21	23	30	31	32	102	103	120	123	130	131	132	140	142

semoctarish	1023	1024	1025	1026	1027	1234	1235	1236	1237	1324	1325	1326

dodriquadarish	102	103	120	123	130	132	201	203	210	213	230	231	301	302	310

dodroctarish	102,345	102,346	102,347	103,245	103,246	103,247

quartiquadrarish	1	2	3	11	22	33	111	222	333	1,111	2,222	3,333	11,111	22,222	33,333

quartoctarish	10	12	13	14	15	16	17	20	21	23	24	25	26	27	30	32	34	35	36	37	40	41

"In base six dextridigitals[952] with five-sixths of possible digits become possible, besides semidigits, tertidigitals and bedigitals.

dextrisenarish	10,234	10,235	10,243	10,245	10,253	10,254	10,324	10,325	10,342

"The zeroless polydigitals that are just one digit shy of being pandigital are technically called penholodigitals." [953]

penholodecimals 123,456,789 123,456,798 123,456,879 123,456,897 123,456,978

The meeting ended with the exploration of holy and unholy penholodigitals in other bases than ten.[954]

UNINDEXED PRIMES

from PAJAS #96 (July 1993)

Robin began with "The Lord Most High is awesome. He is a great King over all the Earth." from Psalm 47:2 and with the question, "Why couldn't the baseball player understand octal?"

"He could get past base three." Victoria answered.

Then speaking for the rest of the Delta Schwa sisters continued,"We don't want to be bitchy, but if we named primes not by indexing, but by not indexing, we could do the opposite of pips and get 'pups',[955] prime-unindexed primes.

pups 2 7 13 19 23 29 37 43 47 53 61 71 73 79 89 97 101 103 107 113 131 137 149 151 161 163

"'We could also describe binarish-indexed shichibi primes with the acronum 'bishops'," [956] another sister contributed, "since all shicibi numbers are odd by adding an odd 'o'.

bishops 7 157 167 2,707 2,767 2,957 3,037

"'Bump" and "bumpy' [957] describe binarish-unindexed Mersenne primes."

bumpy 7 31 127 8,191 131,071 524,297 2,147,483,645 2,305,843,009,213,693,951

"'Clumps' [958] describes cubed lucky-unindexed Mersenne primes, including the second through the three hundred forty-second, but excluding the twenty-seventh."

cubed luckies 1 27 343 729 2,197 3,375 9,261 15,625 29,791 35,937 50,653

clumps 7 127 8,191 131,071 524,287 2,147,483,647 137,438,953,471

"'Cups' [959] describe cube-unindexed-primes."

non-cubes 2 3 5 6 7 8 10 11 12 13 14 15 17 18 19 20 21 22 23 24 26 27 28 29 30 31

cups 3 5 11 13 17 19 29 31 37 41 43 47 59 61 67 71 73 79 83 89 101 103 107 109 113 127 131

"'Cutups' [960] describe cube-unindexed-ternarish-unindexed-primes. The prerequisite 'cuts', cube-unindexed ternarish, only co-incidentally have anything to do with cakes, pancakes or doughnuts."

cut 2 10 12 20 21 22 101 102 110 111 112 120 122 200 201 202 210 211 212 220 222 1,001 cutup 3 5 7 11 13 17 19 23 29 37 41 43 47 53 59 61 67 71 73 79 83 89 97 101 103 107 109

"'Foul-ups' [961] describe Fibonacci odd-unindexed-lucky-unindexed primes."

Fibonacci odds 1 3 5 13 21 55 89 233 377 987 1,597 4,181 6,765 17,711

non-Fibonacci odds 7 9 11 15 17 19 23 25 27 29 31 33 35 37 39 41 43 45 47 49 51

fouls 21 31 37 63 69 75 99 111 127 133 141 159 169 189 195 211 219 231 237 259 267

non-fouls 1 2 3 4 5 6 7 8 9 10 11 12 13 14 15 16 17 18 19 20 22 23 24 25 26 27 28 29

foul-ups 2 3 5 7 11 13 17 19 23 29 31 37 41 43 47 53 59 61 67 71 79 83 89 97 101 103

"So the only foul-up with the foul-up primes is that we're just missing the twenty-first prime, seventy-three!"

"No, we're missing an infinite number of primes. Seventy-three is just the first to go AWOL. One twenty-seven is the next."

"'Humps' [962] describes hachibi-unindexed Mersenne primes, which start out very much like the clumped primes. A number with a three or a seven would be a one-humped 'dromedary' and one with both a two-humped 'bactrian'." [963]

humps 3 7 31 127 8,191 131,071 524,297 2,305,843,009,213,693,951

dromedaries 3 7 13 17 23 27 30 31 32 33 34 35 36 38 39 43 47 53 57 63 67 70 71

bactrians 37 73 137 173 237 273 307 317 327 337 347 357 367 370 371 372 373 374

"'Lump' describes lucky-unindexed Mersenne primes[964], while 'limp' describes the lucky-indexed Mersenne primes." [965]

lump 7 127 8,191 131,071 2,147,483,647 2,305,843,009,213,693,951

limp 3 31 524,287 137,438,953,471

"'Pickups' describe prime-indexed cubed kyuubi-unindexed primes, which would include all the primes up to, but not including, the six thousand, eight hundred fifty-nineth prime, because that's the first 'pick'." [966]

cubed kyuubi 729 6,859 24,389 59,319 117,649 205,379 328,509 493,039

picks 6,859 24,389 117,649 328,509 1,295,029 2,146,689

"'Pumped' [967] describes prime-unindexed Mersenne primes."

pumped 3 127 131,071 2147,483,647 2,305,843,009,213,693,951

"'Pushups' describe prime-unindexed shichibi-unindexed primes, while 'pushy' describes prime-unindexed shichibi[968] and their complimentary sequence is described as 'pulley', even inserting an extra 'e' added for evens.[969]

pushups 2 3 5 7 11 13 19 23 29 31 37 41 43 47 53 59 61 67 71 73 79 83 89 97 101 103

pushy 7 37 57 77 87 97 117 137 147 157 177 197 207 217 237 247 257 267 277 297

pulley 2 12 22 42 62 122 132 142 152 162 182 192 202 222 232 242 252 262

"'Sit-ups' [970] describe sanbi-indexed-ternarish-unindexed primes, without the tenth prime, twenty-nine."

sit 10 111 212 1,020 1,112 1,220 2,021 2,122 10,000 10,101 10,202 11,010 11,111 11,212

sit-ups 2 3 5 7 11 13 17 19 23 31 37 41 43 47 53 59 61 67 71 73 79 83 89 97 101 103

"'Slipups' [971] describe sanbi-lucky-indexed-prime-unindexed-primes.

slipups 2 3 7 11 13 17 19 23 29 31 37 43 47 53 59 61 67 71 79 83 89 97 101 103 107 109

"'Slumps' [972] describes sanbi lucky-unindexed Mersenne primes.

slump 3 7 31 8,191 131,071 524,297 2,147,483,647 2,305,843,009,213,693,951

"'Sunups' describes sanbi-unindexed-nibi-unindexed-primes and "sunny" the sanbi-unindexed nibi,' [973] 'sin' sanbi-indexed nibi, 'sinless' or' immaculate' the non-sins. [974]

sunny 2 12 32 42 52 62 72 82 92 102 112 132 142 152 162 172 182 192 202 212

sunups 3 37 131 181 239 293 359 421 479 557 613 743 821 881 953 1,033

sin 22 122 222 322 422 522 622 722 822 922 1,022 1,122 1,222 1,322 1,422 1,522

immaculate 1 2 3 4 5 6 7 8 9 10 11 12 13 14 15 16 17 18 19 20 21 23 24 25 26 27 28 29

"'Tipsy' [975] describes numbers with ternarish-indexed prime frags."

tipsy 23 29 32 229 231 237 271 273 279 329 331 337 371 373 379

This prompted impressive intoxication impresssions by the bheerdrinkers, ending the meeting.

HOW BIG IS BIG?

from PAJAS #97 (August 1993)

From a Presbyterian hymn Robin sang "O Lord Most High, with all my heart Thy wondrous works I will proclaim; I will be glad and give Thee thanks and sing the praises of Thy Name."

Then she asked, "If six is bigger when turned upside-down, what number is bigger turned sideways?"

"Eight turned sideways becomes infinity."

"To answer the poser in the *Clarion* this month, 'big' [976] would be the acronym for binarish-indexed gobi numbers. The only 'big' prime is not very big, being just five.

big 5 95 105 995 1,005 1,095 1,105 9,995 10,005 10,095 10,105 10,905 11,005 11,095

"'Bibs' [977] describes binarish-indexed binarish.

bibs 1 1,010 1,011 1,100,100 1,100,101 1,101,110 1,101,111 1,111,111,000 1,111,111,001

"'Bins' describes binarish-indexed nibi and 'bios' describe binarish-indexed odds." [978]

bins 2 92 102 992 1,002 1,092 1,102 9,992 10,002 10,092 10,102 10,902 11,002 11,092

bios 1 19 21 199 201 203 219 221 1,999 2,001 2,019 2,021 2,199 2,201 2,219 2,221

"'Bit' describes binarish-indexed ternarish".[979]

bits 1 101 102 1,101 1,102 1,202 1,210 1,101,001 1,101,002 1,101,102 1,101,110 1,101,111

"'Bugs' [980] describes binarish-unindexed gobi.

bugs 15 25 35 45 55 65 75 85 115 125 135 145 155 165 175 185 195 205 215

"'Bun' and 'bunny' [981] describe binarish-unindexed nibi.

bunnies 12 22 32 42 52 62 72 82 112 122 132 142 152 162 172 182 192 202 212

"'Bush' and 'bushy' [982] describe binarish-unindexed shichibi.

bushy 17 27 37 47 57 67 77 87 117 127 137 147 157 167 177 187 197 207 217

"'But' describes the non-bits and 'butter' the non-bits with two of three 'trits'," [983]

but 2 12 20 21 22 100 112 120 121 122 200 201 202 210 211 212 220 222

butter 12 20 21 100 112 121 122 200 202 211 212 220 221

"'Clicks' are cubed lucky-indexed cubed kyuubi, while 'clucks' are cubed lucky-unindexed cubed kyuubi.[984]

cubed kyuubi 729 6,859 24,389 59,319 117,649 205,379 328,509 493,039

clicks 729 19,465,109 40,318,322,589 387,261,078,569 10,603,051,396,209

clucks 6,859 24,389 59,319 117,649 205,379 328,509 493,039 704,969 970,299

"Clubs' [985] are cubed-lucky-unindexed binarish and their complimentary, also black, sequence would be 'spades'.[986]

clubs 10 11 100 101 110 111 1,000 1,001 1,010 1,011 1,100 1,101 1,110 1,111 10,000 10,001

spades 88 89 888 889 898 899 8,888 8,889 8,898 8,899 8,988 8,989 8,999

"Since the 'diamond' numbers[987] are the number of diamond polynominoes for a particular number of cells, their complementary sequence would be the 'hearts' [988] and anything that's not one of these four 'suits' would be 'jokers'.[989]

diamonds 1 3 7 20 62 204 709 2,526 9,212 33,989 126,838 476,597 1,802,618

hearts 2 6 8 37 79 209 795 7,473 66,010 90,787 523,402 873,161 3,149,030

jokers 4 5 9 12 13 14 15 16 17 18 19 21 22 23 24 25 26 27 28 29 30 31

"'Clues' [990] describes cubed-lucky-unindexed evens and 'clueless' those that aren't.

clueless 2 54 686 1,458 4,394 6,750 18,522 31,250 59,582 71,874 101,306

"Since 'clut' describes cubed lucky-unindexed ternarish, 'clutter' would describe those with all three 'trits', excluding cubed lucky-indexed ternarish 10,201 and higher." [991]

clut 2 11 12 20 22 101 102 110 112 121 122 200 201 202 211 212 220 222 1,000

clutter 102 201 210 1,002 1,012 1,020 1,021 1,022 1,102 1,120 1,200 1,201 1,202 1,210 1,220

"'Cubs' [992] describes cube-unindexed binarish.

cubs 6,859 24,389 59,319 117,649 205,379 328,509 493,039 704,969 970,299

"'Cues' [993] describes cube-unindexed evens.

cues 4 6 8 10 12 14 18 20 22 24 26 28 30 32 34 36 38 40 42 44 46 48 50 52 56

"'Figs' [994] describe Fibonacci-indexed gobi and their complimentary sequence would be 'newtons'.

figs 5 15 25 45 75 125 205 335 545 885 1,435 2,325 3,765 6,095 9,865 15,965

newtons 4 24 54 74 84 114 454 664 794 874 3,904 6,234 7,674 8,564 32,354 58,194

"'Fine' [995] describe Fibonacci-indexed nibi which are all even and their complimentary sequence is 'dandy'.

fine 2 12 22 42 72 122 202 332 542 882 1,432 2,322 3,762 6,092 9,862 15,962

dandy 7 27 57 77 87 117 457 667 797 877 3,907 7,677 8,567

"'Fire' [996] and 'fiery' describe Fibonacci-indexed rokubi, which also are all even and it complimentary sequence is "smoke", which aren't."

fire 6 16 26 46 76 126 206 336 546 886 1,436 2,326 3,766 6,096 9,866 15,966

smoke 3 23 53 63 83 113 453 663 793 873 3,903 6,233 7,673 8,563

"'Fist' describes Fibonacci-indexed sorted ternarish." [997]

sternarish 1 11 12 111 112 122 222 1,111 1,112 1,122 1,222 2,222 11,111 11,112 11,122 11,222

fist 1 11 12 112 1,111 11,111 111,222 11,111,222 11,111,222 11,112,222 1,122,222

"'Fit' [998] describes Fibonacci-indexed ternarish.

fit 1 2 10 12 22 111 210 1,021 2,010 10,022 12,100 2,122 10,022 12,100 22,122

"'Fubi' [999] is not only an acronym for a foul-up beyond imagining, but also for Fibonacci-unindexed binarish ichibi.

non-Fibonacci 4 6 7 9 10 11 12 14 15 16 17 18 19 20 22 23 24 25 26 27 28 29 30 31

binarish ichibi 1 11 101 111 1,001 1,011 1,101 1,111 10,001 10,011 10,101 10,111 11,001 11,011 11,101

fubi 111 1,011 1,101 10,001 10,011 10,101 10,111 11,011 11,111 100,001 100,011 100,101

"'Funny' [1000] describes Fibonacci-unindexed nibi.

funny 32 52 62 82 92 102 112 132 142 152 162 172 182 192 212 222 232 242 252

"'Furry' [1001] describes Fibonacci-unindexed rokubi.almost-phi

furry 36 56 66 86 96 106 116 136 146 156 166 176 186 196 216 226 236 246 256

'Gin" [1002] describes gobi-indexed nibi, whose complimentary sequence would be "tonic".[1003]

gin 42 142 242 342 442 542 642 742 842 942 1,042 1,142 1,242 1,342 1,442 1,542 1,642

tonic 57 157 257 357 457 557 657 757 857 1057 1,157 1,257 1,357 1,457 1,557 1,657

"'Guns' [1004] describes gobi-unindexed nibi, whose complimentary sequence would be called 'roses'.[1005]

non-gobi 1 2 3 4 6 7 8 9 10 11 12 13 14 16 17 18 19 20 21 22 23 24 26 27 28 29 30

guns 2 12 32 52 62 72 82 92 102 112 122 132 152 162 172 182 192 202 212 222 232 252

roses 17 27 37 47 67 87 107 117 127 137 147 167 177 187 207 217 227 237 247 267 277

"'Gushy' [1006] describes gobi-unindexed shichibi.

gushy 17 37 57 67 77 87 97 107 117 127 137 157 167 177 187 197 207 217 227 237 257

"'Guys" [1007] describes gobi-unindexed yongbi, while its complimentary sequence would be described as 'dolls'. Oddly the 'guys' are even and the 'dolls' are odd.

guys 14 34 54 64 74 84 94 104 114 124 134 154 164 174 184 194 204 214 224 234 254

dolls 15 25 35 45 65 85 105 115 125 135 145 165 175 185 205 215 225 235 245 265

"'Halt" describe hachibi-and-lucky ternarish[1008], 'hits' hachibi-indexed ternarish[1009] and 'huts' non-hits." [1010]

halt 1 8 18 21 28 38 48 58 68 78 88 98 108 111 128 138 148 158 168 178 188 0 208

hits 22 200 1,001 1,102 1,210 2,011 2,112 2,220 10,021 10,122 11,000 11,101 11,202 12,010

huts 1 2 10 11 12 20 21 100 101 102 110 111 112 120 121 122 201 202 210 211 212 220 221

"'Huggy' [1011] describes hachibi-unindexed gobi.

huggy 75 175 275 375 475 575 675 775 875 975 1,075 1,175 1,275 1,375 1,475 1,575

"'Hun' [1012] describes hachibi-unindexed nibi.

huns 72 172 272 372 472 572 672 772 872 972 1,072 1,172 1,272 1,372 1,472 1,572

"'Hush' [1013] describes hachibi-unindexed shichibi.

hush ... 76 78 79 80 81 82 83 84 85 86 87 88 89 90 91 92 93 94 95 96

"'Kicks' [1014] describes kyuubi-indexed cubed kyyuubi and 'kick-offs' [1014] neighbors.

kicks 704,969 6,751,269 24,137,569 58,863,869 116,930,169 204,336,469 327,082,769

kickoffs 704,968 704,969 6,751,268 6,751,270 24,137568 24,137,568 24,137,570

"'Kin' [1015] describes kyuubi-indexed nibi.

kin 82 182 282 382 482 582 682 782 882 982 1,082 1,182 1,282 1,382 1,482 1,582

"'Kit' [1016] describes kyuubi-indexed ternarish and 'skit' sorted kyuubi-indexed ternarish.[1017]

kit 100 201 1,002 1,110 1,211 2,012 2,120 2,221 10,022 10,200 11,001 11,102 11,210
skyuubi 9 19 29 39 49 59 69 79 89 99 19 119 129

skit 100 201 1,001 1,102 1,210 2,011 2,112 2,220 10,021 10,121 201 11,100 11,201

"'Licked' [1018] describes lucky-indexed cubed kyuubi.

licked 729 24,289 328,509 704,969 2,146,689 3,307,949 9,129,329 15,438,249

"'Lit' describes lucky-indexed ternarish, 'litter' those with two of the three 'trits' and 'litterer' those with just one 'trit'." [1019]

lit	1	10	21	100	110	120	210	221	1,011	1,022	1,110	1,120	1,210
litter	10	21	100	110	221	1,011	1,110	2,000	2,002	2,112	2,200	2,202	2,211
holiness	1	0	2	1	0	1	1	3	2	0	2	1	0

"'No' describes the nibi odds and 'yes' its complimentary sequence."
"Ha! All the nibi are even so the 'no' sequence's no sequence at all!"
"'Nuns' [1020] does however describe nibi-unindexed nibi. Their Mate is, as He will be for all at the eternal wedding feast in New Jerusalem, Jesus the Bridegroom.[1021]

non-nibi 1 3 4 5 6 7 8 9 10 11 13 14 15 16 17 18 19 20 21 23 24 25 26 27 28 29 30

nuns 2 22 32 42 52 62 72 82 92 102 122 132 142 152 162 172 182 192 202 222

Bridegroom 7 17 27 37 47 57 67 77 107 117 127 137 147 157 167 177 207 217 227 237 247

"'Nuts' [1022] describes nibi-unindexed ternarish and its complimentary sequence would be 'soup'.[1023]

nuts 1 10 11 12 20 21 22 100 101 111 120 121 122 200 201 210 212 220 221 222 1,000

soup 8 77 78 79 87 88 89 777 778 779 787 789 798 799 877 878 879 888 898 899

"'Ours' [1024] describes odd-unindexed rokubi and its complimentary sequence would be 'theirs' [1025] 'and 'mine' Mersenne-indexed nibi evens.[1026]

ours 16 36 56 76 96 116 136 156 176 196 216 236 256 276 296 316 336 356 376 396

theirs 23 43 63 83 103 123 143 163 183 203 223 243 263 283 303 323 343 363 383 403

mine 22 62 302 1262 20,462 81,902 1,310,702 5,242,862 83,886,062 5,368,709,106

"'Out' [1027] describes odd-unindexed ternarish and its complimentary sequence 'in'.

out 2 11 20 22 101 110 112 121 200 202 211 220 222 1,001 1,010 1,012 1,021 1,100 1,102

in 8 77 79 88 777 779 788 797 799 878 887 889 898 7,777 7,779 7,788 7,797 7,799 7,878

"'Par" describes prime and rokubi. [1028]

par 2 3 5 6 7 11 13 16 17 19 23 26 29 31 36 37 41 43 46 47 53 56 59 61 66 67 71 73 76 79

'"Picnic' [1029] describes a prime-indexed cubed nibi-indexed cube and their complimentary sequence would be "baskets".[1030]

cubed nibi 8 1,728 10,648 32,768 74,088 140,608 238,328 373,248 551,368

picnibi 1,728 10,648 74,088 238,328 1,061,208 1,815,848 4,251,528 6,028,568

picnic 5,159,780,352 1,207,269,217,792 406,671,383,849,472

basket 4,840,219,647 8,792,730,782,207 593,328,616,150,527

'"Pins' [1031] describes prime-indexed nibi and the complimentary sequence would be "needles".[1032]

pins 12 22 42 62 102 112 162 182 222 282 302 362 402 422 462 522 582 602

needles 37 57 77 87 117 177 217 277 297 337 397 417 477 537 577 597 637 697 717

'"Pit' describes prime-indexed ternarish, 'pendulum" the complimentary sequence[1033] and 'pitiless' those not prime-indexed ternarish ichibi." [1034]

pits 2 10 12 21 102 111 121 201 212 1,002 1,011 1,101 1,112 1,121 1,202 1,220 2,010 2,012 2,102

pendula 7 78 87 89 787 878 888 897 7,787 7,789 7,877 7,897 7,989 8,779 8,797 8,878

pitiless 1 2 3 4 5 6 7 8 9 10 11 12 13 14 15 16 17 18 19 20 22 23 24 25 26 27 28

'"Plucked" [1035] describes prime lucky-unindexed cubed kyuubi and 'feathered" [1036] the complimentary sequence."

prime lucky 3 7 13 15 31 37 43 67 73 79 127 151 163 193 211 223 241 283 307 331 349

plucked 729 6,859 59,319 117,649 205,379 493,039 704,969 970,299 1,295,029

feathered 270 3,140 40,680 295,030 506,960 794,620 882,350 2,119,400

'"Plugged" describes prime lucky-unindexed gobi, which appropriately starts with a 'nickle', and their complimentary sequence 'open'. [1037]

plugged 5 15 65 125 145 305 365 425 665 725 785 1,265 1,505 1,625 1,925 2,105

open 4 34 84 214 274 334 574 634 694 854 874

'"Plus" [1038] describes prime lucky-unindexed sanbi and its complimentary sequence 'minus',[1039] not Mersenne-indexed-nibi-unindexed sanbi.

plus 3 13 33 43 53 73 83 93 103 113 133 153 163 173 183 193 203 213 223 233 243 253 263 minus 6 16 26 46 53 66 83 216 226 236 246 256 266 286 296 306 316 326 346

"'Plush" [1040] describes prime lucky-unindexed shichibi."

plush 17 37 47 57 77 97 107 117 137 157 167 187 197 217 227 237 257 267 277 287

"'Posh" [1041] describes prime odd shichibi, as well as 'port out, starboard home', and their complimentary sequence 'soph',[1042] comes from 'starboard out, port home', and its downtown neighbors 'frosh'.[1043] The uptown neighbors would be 'juniors' [1044] and the juniors uptown neighbor 'seniors'.[1045]

posh 7 17 37 47 67 97 107 127 137 157 167 197 227 257 277 307 317 337 347 367 397 457

soph 2 32 52 62 82 112 122 142 242 272 322 352 382 392 412 422 442 452 512 532

frosh 1 31 51 61 81 111 121 131 241 271 321 351 381 391 411 421 441 451 511 531 541 601 631

juniors 3 33 53 63 83 113 123 143 243 273 323 353 383 393 413 423 443 453 513 533 543

seniors 4 34 54 64 84 114 124 144 244 274 324 354 384 394 414 424 444 454 514 534

"'Pugs" describes prime-unindexed gobi, while 'puns' [1046] describes prime-unindexed nibi.

pugs 5 35 55 75 85 95 115 135 145 155 175 195 205 215 235 245 255 265 275 295

puns 2 32 52 72 82 92 112 132 142 152 172 192 202 212 232 242 252 262 272 292

The question "What's the worse pun?" was followed a chorus of "We don't know and we wanna know!'" followed by the answer " A full-groan one." followed by a group groan.

"'Pure' describes prime-unidexed rokubi evens," Lana continued, "'impure' all those not pure and 'simple' the pure's complimentary sequence." [1047]

pure 6 36 56 76 86 96 116 136 146 156 176 196 206 216 236 246 256 266 276 296

impure 1 2 3 4 5 7 8 9 10 11 12 13 14 15 16 17 18 19 20 21 22 23 24 25 26 27 28 29

simple 3 13 23 43 63 103 113 123 143 153 163 183 193 203 223 233 243 253 263 283

"'Put' describes prime-unindexed ternarish, 'put-down' the downtown neighbors and 'put-up' the uptown." [1049]

put 1 11 20 22 100 101 110 112 120 121 200 202 210 211 220 221 222 1,000 1,001

put-down 10 19 21 99 100 109 111 119 120 199 201 210 219 220 221 999 1,000 1,002

put-up 2 12 21 23 101 102 111 113 121 122 201 211 212 221 222 223 1,001 1,002 1,004

"Similarly 'shut-up', describes shichibi-unindexed ternarish uptown neighbors, 'shut-down' the downtown and 'shut-off' both combined." [1049]

shut-up 2 3 11 12 13 21 23 101 102 103 111 112 113 121 122 201 202 203 211 212

shut-down 0 2 9 10 11 19 21 99 100 101 109 110 111 119 120 199 200 201 209 210

"All ribbing aside, 'ribs' [1050] describes rokubi-indexed binarish, while 'rigs' describes the rokubi-indexed gobi." [1051]

ribs 110 10,000 11,010 100,100 101,110 111,000 1,000,010 1,001,100 1,010,110 1,100,000

rigs 55 155 255 355 455 655 755 855 955 1,055 1,155 1,255 1,355 1,455 1,555

"'Rush" and "rushed' [1052] describes rokubi-unindexed shichibi, 'rush-off' their neighbors[1053], 'ruts' rokubi-unindexed ternarish and, writing 'rites' right as R-I-T-E, rokubi-indexed ternarish evens.[1055]

rushed 7 17 27 37 47 67 77 87 97 107 117 127 137 147 167 177 187 197 207 217 227 237

rush-off 6 8 16 18 26 28 36 38 46 48 66 68 76 78 86 88 96 98 106 108 116 118

ruts 1 2 11 12 21 22 100 101 102 120 122 200 201 202 212 221 1,000 1,001 1,002

rites 20 222 1,100 2,002 2,110 10,012 10,120 11,022 11,200 12,210 20,012 20,221

"'Sack' describes sanbi and cubed kyuubi, which is the same as sanbi until seven twenty-nine, [1056] while 'sick' describes sanbi-indexed cubed kyuubi, whose complimentary sequence would be described as 'well'.[1057]

sick 24,389 2,146,689 12,008,989 35,611,289 78,953,589 148,035,889 248,858,189

"'Stick" and 'sticky' [1058] describe sanbi ternarish-indexed cubed kyuubi, while 'stuck' [1059] describes sanbi ternarish-unindexed cubed kyuubi.

sticky 729 1,295,029 9,129,329 1,027,243,729 1,363,938,029 1,767,172,329

stuck 6,859 970,299 1,685,159 7,880,599 10,503,459 997,002,999

"'Tails' [1060] describes ternarish-and-ichibi-luckies, which are just like ternarish until thirty-one.

tails 1 2 10 11 12 21 22 31 51 100 101 102 110 111 120 121 122 141 151 171 200 201 202 210 211

"'Tick' [1061] describes ternarish-indexed cubed kyuubi.

tick 729 6,859 970,899 1,295,029 1,685,159 7,880,599 9,129,329 10,503,459

tock 270 3,140 2,119,400 8,119,499 8,314,840 8,704,970 9,029,100

tick-off 728 730 6,858 6,860 970,898 97,900 1,295,028 1,295,030 1,685,158

"'Tics' [1062] describes ternarish-indexed cubes.

tics 1 8 27 1,000 1,331 1,728 8,000 9,261 10,648 12,167 13,824 15,625 17,576

"'Tie" and 'tied' and even 'tide' [1063] describe ternarish-indexed evens, 'bow' their complimentary sequence,[1064] and 'tied-off' both neighbors.[1065]

tied 2 4 20 22 24 40 42 44 200 202 204 220 222 224 240 242 244 400 402 404

bow 5 7 55 57 59 75 77 79 555 557 559 575 577 579 597 599 755 757 759 775

tied-off 2 3 5 19 21 23 25 39 41 43 45 199 201 203 205 219 221 223 225 239 241

"'Tin' [1066] describes ternarish-indexed nibi.

tin 2 12 92 102 112 192 202 212 992 1,002 1,012 1,092 1,102 1,112 1,192 1,202 1,212 1,222

"'Tit" [1067] describes ternarish-indexed ternarish, 'tat' the complimentary sequence, 'tut' the non-tit.

tit 1 2 101 102 110 202 210 211 10,201 10,202 11,001 11,002 11,010 11,102 11,110 11,111 21,102

tat 7 8 788 789 797 889 897 898 77,979 77987 77,988 78,788 78,789 78,889 78,897

tut 3 4 5 6 7 8 9 10 11 12 13 14 15 16 17 18 19 20 21 22 23 24 25 26 27 28 29 30 31

"'Tubby' [1068] describes ternarish-unindexed binarish and 'Lulu' [1069] the complimentary sequence.

Tubby 11 100 101 110 111 1,000 1,001 1,101 1,110 1,111 10,000 10,001 10,010 10,011 10,100 10,101

Lulu 88 888 889 898 899 8,888 8,889 8,898 8,998 8,999 88,888 88,889 88,898

"'Tuck' [1070] describes ternarish-unindexed cubed kyuubi, 'friar' [1071] their complimentary sequence and 'tuck-in' ternarish-unindexed cubed kyuubi downtown neighbors.[1072]

tuck 24,389 59,319 117,649 205,379 328,509 493,039 704,969 970,299 2,146,689

friar 49,680 75,610 295,030 506,960 671,490 794,620 882,350 7,880,599 3,248,730

tucked-in 24,388 59,318 117,648 205,378 493,038 704,968 2,146,688

"'Tug' [1073] describes ternarish-indexed gobi.

tug 5 15 95 105 115 995 1,005 1,015 1,095 1,105 1,115 1,195 1,205 1,215 1,995 2,005

The meeting ended with the quest for "Rin-Tin-Tin" [1074] numbers, the rokubi-indexed nibi ternarish-indexed nibi ternarish-indexed nibi.

BOBTAILED NUMBERS

from PAJAS #98 (September 1993)

Robin began by singing, "Sing praises to the Lord Most High, to Him Who doth in Zion dwell; declare His mighty deeds abroad, His deeds among the nations tell." from a Presbyterian hymnal.

Al asked "Why would a mathematician carry an abacus."

"As back-up, 'cause even when your calculator dies, you can always count on an abacus."

Bobby then introduced the month's topic, "The tailed number sequences can be subdivided into subsequences via fractional prefixes and so 'bobtailed'. A number with both half zeroes and a zero tail would be 'semilove',[1075] with a third 'tertilove',[1076] with a fourth 'quartilove', [1077] with two-thirds 'beloved', etc.[1078]

semilove 10 20 30 40 50 60 70 80 90 1,010 1,020 1,030 1,040 1,050 1,060 1,070 1,080

tertilove 110 120 130 140 150 160 170 180 190 210 220 230 240 250 260 270 280

quartilove 1,110 1,120 1,130 1,140 1,150 1,160 1,170 1,180 1,190 1,210 1,220 1,230 1,240 1,250

beloved 100 200 300 400 500 600 700 800 900 100,010 100,020 100,030 100,040

"Similarly a number with both half ones and a one tail would be 'semichibi',[1079] with a third 'tertichibi',[1080] with a fourth 'quartichibi', etc.[1081]

semichibi 21 31 41 51 61 71 81 91 1,001 1,021 1,031 1,041 1,051 1,061 1,071 1,081 1,091 1,201

tertichibi 201 221 231 241 251 261 271 281 291 301 321 331 341 351 361 371 381 391 401

quartichibi 2,001 2,021 2,031 2,041 2,051 2,061 2,071 2,081 2,091 2,201 2,221 2,231

"And seminibi,[1082] tertinibi,[1083] and quartinibi, etc.[1084]

seminibi 12 32 42 52 62 72 82 92 1,022 1,122 1,202 1,212 1,232 1,242 1,252 1,262

tertinibi 102 112 132 142 152 162 172 182 192 100,022 100,122 100202 100,212 100,232

quartinibi 1,002 1,012 1,032 1,042 1,052 1,062 1,072 1,082 1,092 1,102 1,112 1,132

"Semisanbi,[1085] tertisanbi,[1086] quartsanbi, etc.[1087]

semisanbi 13 23 43 53 63 73 83 93 1,033 1,133 1,233 1,303 1,313 1,323 1,343 1,353

tertisanbi 103 113 123 143 153 163 173 183 193 100,033 100,133 100,233 100,303 100,313

quartisanbi 1,003 1,013 1,023 1,043 1,053 1,063 1,073 1,083 1,093 1,103 1,123 1,143

"Semiyongbi,[1088] tertiyongbi,[1089] quartiyongbi, etc.[1090]

semiyongbi 14 24 34 54 64 74 84 94 1,044 1,144 1,244 1,344 1,404 1,414 1,424 1,434

tertiyongbi 104 114 124 134 154 164 174 184 194 100,044 100,144 100,244 100,404 100,414

quartiyongbi 1,004 1,014 1,024 1,034 1,054 1,064 1,074 1,084 1,094 1,104 1,124 1,134

"Semigobi,[1091] tertigobi,[1092] quartigobi, etc.[1093]

semigobi 15 25 35 45 65 75 85 95 1,055 1,155 1,255 1,355 1,505 1,515 1,525 1,535

tertigobi 105 115 125 135 145 165 175 185 195 100,055 100,155 100,255 100,355

quartigobi 1,005 1,015 1,025 1,035 1,045 1,065 1,075 1,085 1,095 1,105 1,125 1,135

"Semirokubi,[1094] tertirokubi,[1095] quartirokubi, etc.[1096]

semirokubi 16 26 36 46 56 76 86 96 1,066 1,166 1,266 1,366 1,466 1,566 1,606 1,616

tertirokubi 106 116 126 136 146 156 176 186 196 100,066 100,166 100,266 100,366

quartirokubi 1,006 1,016 1,026 1,036 1,046 1,056 1,076 1,086 1,096 1,106 1,126 1,136

"Semishichibi,[1097] tertishicibi,[1098] quartishichibi, etc.[1099]

semishichibi 17 27 37 47 57 67 87 97 1,077 1,177 1,277 1,377 1,477 1,577 1,677 1,707

tertishichibi 107 117 127 137 147 157 167 187 197 100,077 100,177 100,277 100,377

quartishichibi 1,007 1,017 1,027 1,037 1,047 1,057 1,067 1,087 1,097 1,107 1,117 1,127

"Semihachibi,[1100] tertihacibi,[1101] quartihachibi, etc.[1102]

semihachibi 18 28 38 48 58 68 78 98 1,088 1,188 1,288 1,388 1,488 1,588 1,688 1,788

tertihachibi 108 118 128 138 148 158 168 178 198 100,088 100,188 100,288 100,388

quartihachibi 1,008 1,018 1,028 1,038 1,048 1,058 1,068 1,078 1,098 1,108 1,118 1,128

"Semikyuubi,[1103] tertikyuubi,[1104] quartikyuubi, etc.[1105]

semikyuubi 19 29 39 49 59 69 79 89 1,099 1,199 1,299 1,399 1,499 1,599 1,699 1,799

tertikyuubi 109 119 129 139 149 159 169 179 189 100,099 100,199 100,299 100,399

quartikyuubi 1,009 1,019 1,029 1,039 1,049 1,059 1,069 1,079 1,089 1,109 1,119 1,129

FRACTALS[1106]

from PAJAS #99 (October 1993)

Quoting Kurt Vonnegut, Al began, "Oh Lord Most High, Creator of the Cosmos, Spinner of Galaxies, Soul of Electromagnetic Waves, Inhaler and Exhaler of Inconceivable Volumes of Vacuum, Spitter of Fire and Rock, Trifler with Millennia — what could we do for Thee that Thou couldst not do for Thyself one octillion times better?"

"Ah, octillion. That'd be twenty-sevenplex, right?"

"Right" Robin answered and then asked her own question, "What does the middle initial stand for in fractal discoverer Benoit B. Mandelbrot's name?"

"Benoit B. Mandelbrot" Al answered.

"Yes, that's who I asked about, Benoit B. Mandelbrot."

"And that's who I answered about, Benoit B. Mandelbrot."

After several cycles of this Who's On First-like routine, Robin clarified, "So his name is actually an infinite number of nested Benoits and Mandelbrots." and then explained, "As you see some in the Society have been infected by a fractalization frenzy. It's therefore been deemed high time to once again edit the *The Index of Forbidden Sequences*. Self-similar sequences, like fractals, have self-similar subsequences exactly like the whole, like the more familiar two-dimensional fractal graphics. You can even cull an infinite number of numbers from such a sequence any number of finite times and still have the original sequence. Sequences beginning with one can be turned into fractals by inserting one number between each original number, inserting the preceding numbers or inserting the preceding numbers in reverse order. These most common fractalizations are called respectively odd cull,[1107] crescendo[1108] and decrescendo.[1109] If you remove, or cull, every odd number or the increasing or decreasing subsequence you still have the same sequence afterward.

odd cull	1 1 2 2 3 2 4 3 5 2 6 4 7 3 8 5 9 2 10 6 11 4 12 7 13 3 14 8 15 5 16 9 17
crescendo	1 1 2 1 2 3 1 2 3 4 1 2 3 4 5 1 2 3 4 5 6 1 2 3 4 5 6 7 1 2 3 4 5
decrescendo	1 2 1 3 2 1 4 3 2 1 5 4 3 2 1 6 5 4 3 2 1 7 6 5 4 3 2 1 8 7 6 5 4

"Any sequence can be expanded fractally in these three ways by indexing them as above, for example some starting with one, odds,[1110] ichibi,[1111] or linears.[1112]

odd-culled odds	1 1 3 1 5 3 7 1 9 5 11 3 13 7 15 1 17 9 19 5 21 11 23 3 25 13 27 7 29

odd-culled ichibi	1 1 21 1 41 21 61 1 81 41 101 21 121 61 141 1 161 81 181 41 201 101 221

odd-culled linears	1 1 4 4 7 4 11 7 14 4 17 11 41 7 44 14 47 4 71 17 74 11 77 41 111 7

"The fractal stutter can easily be remedied by summing,[1113] or hypersumming. [1114]

odd-cull sum	1 2 4 6 9 11 15 18 23 25 31 35 42 45 53 58 67 69 79 85 96 100 107 110 118

odd-cull hypersum	1 11 112 1,122 11,223 112,232 1,122,324 11,223,243 112,232,435

"Any sequence can index itself or any other fractally, though they get more and more complex, too complex for the sequences list."

The meeting ended with attempts to challenge or confirm that proclamation.

GRUESOME NUMBERS

from PAJAS #100 (November 1993)

Robin began with a song again, "Glory to God in the highest! Glory to God, glory to God! Glory to God in the highest! Shall be our song to-day!" by Joanne Barrett, Ron E. Long and Fanny J. Crosby.

Pat asked, "Did you hear about the mathematician who was terrified of negative numbers?" and she responded "No, I didn'tbhear about the mathematician who was terrified of negative."

"He would absolutely stop at nothing to avoid them," he continued, but then had to explain to some that taking the absolute value of a number makes a negative number positive.

"The generalized repunits or GRUs, the gruesomes, include not only easily recognizable repunits and repdigits, but less easily recognized ones recognizable only in other bases, like the Mersenne numbers that are repunits in binary.[1115]

base 2	1	3	7	15	31	63	127	255	511
base 3	1	4	13	40	121	364	1,093	3,280	9,841
base 4	1	5	21	85	341	1,365	3,461	21,845	87,381

"They, of course, also between them generate a diagonal.[1116]

diagonal	1	4	21	156	1,555	19,608	299,593	5,380,840	111,111,111	2,593,742,460

"A double-Mersenne[1117] is the Mersenne of a Mersenne.

M(M(p)	7	127	2,147,483,647

"The double-GRUs[1118] yield more diagonals.

double-GRUs	1	2	7	40	341	3,906	55,987	960,800	19,173,961	435,848,050

"The rest of the terms get higher much faster.[1119]

base 3	1	40	36,472,996,377,170,786,403
base 4	1	341	>two-hundredplex
base 5	1	3,906	>twenty-seven-hundredplex

"The recursive Mersenne or Catalan-Mersenne numbers[1120] were sugested by Eugène Charles Catalan and continue with the triple-Mersenne, quadruple-Mersenne and quintuple-Mersennes. It is not yet known if the sextuple-Mersenne is prime.

CM(p(z))	2	3	7	127	170,141,183,460,469,231,731,687,303,715,884,105,727

"The recursive GRUs get higher much, much faster. The third term for base three is already nineteenduplex.

"The Fermat numbers,[1121] named for Pierre de Fermat, are all of the form of a power of two-to-a-power-of-two minus one, but only the first five are known to be prime.

Fermat	3 5 17 257 65,537 294,967,297 18,446,744,073,709,551,617

"To-the-second-three minus one is just eight, but to-the-second-three-squared minus one is sixty-five and to-the-third-three minus one is nineteen-thousand-six-hundred-eighty-two."

The meeting ground to a halt in the attempted calculation of to-the-fourth-three minus one.

FACTORIAL PRIMES

from PAJAS #101 (December 1993)

Robin began with "Glory to God in the highest, Savior of the World, glory, glory, glory, glory, glory to God! Glory to God in the highest and peace on Earth, good will to men." this time from the Latter Day Saints.

Kay put in, "I hear they have a new trig class for the hearing impaired?"

"Yes," Robin continued, "They use sine language."

When at least some caught on she continued, "A different kind of expansion from the others we explored is based on prime factors and one. Four, for example, is expanded to its hyperadded factors or 'hyperdivides' into two-two or twenty-two,[1122] while primes are expanded by hyperadding one, two or three head ones until they too become composites, expanded 'ex-primes'.[1123]

hyperdivided	111 12 1131 22 15 23 117 222 33 25 111 223 1113 27 35 2,222 117
sorted	12 15 22 23 25 27 33 35 111 113 117 129 222 223 231 233 235

ex-primes	12 15 111 113 117 119 123 129 141 143 147 153 159 161 171 183 189 1,101 1,107

"Of particular interest are the so-called 'factorial primes' [1124] like twenty-three which can be read as the hypersum of the prime factors of a composite. Some like three thirteen can be read as more than one hypersum, yeilding three hyperfactors, namely three, thirteen and thirty-one.

factorial primes	23 37 53 73 113 137 173 211 213 223 229 313 317 337 347 353 359

"The factorial prime factorials[1125] would not be at all the same as the superfactorial primes.[1126]

1st factorial prime factorial	25,852,016,738,884,976,640,000

superfactorial primes	23 851 45,103 3,292,519 372,054,647

"The superfactorial prime factorials get higher very much faster, the very first already being higher than twenty-twoplex."

ACRONUMS AND MORE

from PAJAS #102 (January 1994)

Quoting Jean B. de Santëuil, Robin opened with "Christ, in highest Heav'n enthronèd, equal of the Father's might, by pure spirits, trembling, ownèd, God of God, and Light of Light, Thee 'mid Angel hosts we sing, Thee their Maker and their King."

Then she asked, "Why was the cafeteria clock losing a minute ever three weeks?"

Lily answered, "Every school day it went back for seconds."

"Ah, 'four seconds' times three times five is sixty seconds!"

Moving on Robin explained "The acronym 'bag' [1127] and the adjective 'baggy' describe binarish and gobi combined into a more easily remembered 'acronum' sequence and "baggage" [1128] from the phrase "bag and baggage" is its complimentary sequence.

baggy 1 5 10 11 15 25 35 45 55 65 75 85 95 100 101 105 110 115 125 135

baggage 4 8 14 24 34 44 54 64 74 84 88 89 104 114 124 134 144 154 164

"'Bar' [1129] describes binarish and rokubi combined and 'grill' [1130] its complimentary sequence.

barred 1 6 10 11 16 26 36 46 56 66 76 86 96 100 101 106 110 116 126 136

grilled 3 8 13 23 33 43 53 63 73 83 88 89 103 113 123 133 143 153 163 173

"'Brash' [1131] describes binary, rokubi and shichbi combined.

brash 1 6 7 10 11 16 17 26 27 36 37 46 47 56 57 66 67 76 77 86 87 96

"'Camp" [1132] would describe cubes and Mersenne primes combined and "boot" [1133] its complimentary sequence.

camp 1 2 3 5 7 8 13 17 19 27 31 61 64 89 107 125 127 216 343 512 729 1,000

boot 1 2 4 6 7 8 10 35 38 68 72 80 82 86 270 487 656 783 874 1,999

"'Boo", [1133] on the other hand, describes binarish oddly odd and "bo" binarish odds. [1134]

boo 111 1,001 1,011 1,101 1,111 10,001 10,011 10,101 11,001 11,101 11,111 100,001

bo 1 11 101 111 1,001 1,011 1,101 1,111 10,001 10,011 10,101 10,111

"'Cash' describes cubed and shichibi and 'carry' [1135] its complimentary sequence.

cash 1 4 7 9 16 17 25 27 36 37 47 49 57 64 67 77 81 87 97 100 107 117

carry 2 5 8 12 18 22 32 35 42 50 52 62 63 72 74 82 83 103 113 123 133

carry 2 5 8 12 18 22 32 35 42 50 52 62 63 72 74 82 83 103 113 123 133

"'Cloaks' describes cubed-lucky-odds-and-kyuubi. [1136]

cloaks 1 9 19 27 29 39 49 59 69 79 89 99 109 119 129 139 149 159 169

"'Cobs' describes cubed odd binarish. [1137]

cobs 1 121 1,030,301 1,367,631 1,003,003,001 1,033,364,331 1,334,633,301

"'Cock" [1138] and "cocky" describe cubed odd cubed kyuubi, the same as kyuubi-to-the-sixth, all odd and "hen" [1139] its complimentary sequence, which happen to be hachibi even.

cocky 631,441 47,045,881 594,823,321 3,518,743,761 13,841,287,201 42,180,533,641

hens 368,558 405,176,678 52,954,118 6,481,256,238 57,819,466,358 86,158,712,798

"Cogs" [1140] describes cubed odd gobi and "sprockets" its complimentary sequence.[1141]

cogs 125 3,375 15,625 42,875 91,125 166,375 274,625 421,875 614,125 857,375

sprockets 874 6,624 57,124 84,374 142,624 385,874 578,124 725,374 833,624

"Cook" [1142] describes cubed oddly odd kyuubi and "cookie'[1143] cook-indexed evens even Fibonacci.

cooks 729 117,649 704,969 2,146,689 4,826,809 9,129,329 15,438,249

cookies 1,458 235,298 809,938 9,653,618 18,258,658 30,876,498

"Cool" describes cubed oddly odd luckies and "warm" [1144] its complimentary sequence, the downtown neighbor of the cool numbers would be 'cooler' and the uptown neighbor of the warm numbers 'warmer'.[1145]

cool 2,197 29,791 117,649 328,509 658,503 1,367,631 2,352,637 4,019,679

warm 7,802 70,208 341,496 671,490 882,350 1,384,874 3,248,730 5,980,320

cooler 2,196 29,790 117,648 328,508 658,502 1,367,630 2,352,636 4,019,678

warmer 7,803 70,209 341,497 671,491 882,351 1,384,875 3,248,731 5,980,321

"Coot" [1146] describes cubed oddly odd ternarish and "old" [1147] its complimentary sequence.

coot 9,261 1,367,631 1,771,561 8,120,601 10,793,861 1,003,003,001 1,033,364,331

old 10,738 1,879,398 8,228,438 8,632,368 89,206,138 89,029,354,951

"Cops" [1148] describes cubed odd primes and "robbers" [1149] its complimentary sequence.

cops 27 125 343 1,331 2,197 4,197 6,859 12,167 24,389 29,791 50,653 68,421

robbers 72 656 874 3,140 5,802 7,803 8,668 20,492 31,078 49,346 70,208

"'Cots' [1150] cubed odd ternarish

cots 1 121 9,261 1,030,301 1,367,631 1,771,561 1,003,003,001 1,033,364,331

"Cute" [1151] refers to cube-unindexed ternarish evens

ternarish evens 0 2 10 12 20 22 100 102 112 122 200 202 212 222 1,000

cutes 2 10 12 20 22 102 112 122 200 202 212 222 1,000 1,002 1,012 1,022

"'Double-digit inflated' numbers[1152] have each digit doubled, so at fifteen it 'doubles', not to thirty, but to two-ten.

double-digit inflated 2 4 6 8 10 12 14 16 18 20 22 24 26 28 210 212 214 216

"A 'double-header' [1153] is a number higher than three-digits that has the first two digits the same and a 'triple-header' three of at least four.[1154]

double-header 110 112 113 114 115 116 117 118 119 220 221 223 224 225 226 227 228

triple-header 1,110 1,112 1,113 1,114 1,115 1,116 1,117 1,119 2,220 2,221 2,223 2,224 2,225

"The 'Eckover' numbers[1155] are an extrapolation from the Pickover numbers only counting the position of number frags of pi in e, the base of the natural logarithms, rather than the other way around. 'Three' is the eighteenth digit in e, 'three-one' the hundred ninety-first and 'three-one-four' the nine hundred fifth.

digit	1	2	3	4	5	6	7	8	9	10	11	12	13	14	15	16	17	18	19	20	21
position in e	3	1	18	11	12	21	2	4	13	207	212	382	29	235	219	96	90	8	110	113	89

"So-called 'enormous' numbers[1156] are of the form to-the-second-z plus to-the-second-z-plus-one. They don't start out high at all, but do get higher faster than to-the-second-z and are much harder to factor. The first five are prime.

enormous 2 5 31 283 3,381 49,781 870,199 17,600,759 404,197,703

"'Face" describes Fibonacci and cubed evens combined, which do not differ from Fibonacci until sixty-four." [1157]

face 1 2 3 5 8 13 21 34 55 64 89 144 216 233 377 610 672 987 1,000 1,728 1,797

"'Fast' describes Fibonacci-and-sorted-ternarish and 'slow' their complimentary sequence, not 'feast', which would be the subsequence, Fibonacci-evens-and-sorted-ternarish.[1158] Paradoxically some numbers, like one, four and five, can be both fast and slow and famines seems to have more primes than feasts.

fast 1 2 3 4 5 8 13 21 34 55 89 100 121 144 233 377 400 441 484 576 610

slow 1 4 5 6 7 8 10 44 65 78 86 216 323 389 423 515 558 599 622 766 855

feast 1 2 4 8 34 100 121 144 400 441 484 576 610 676 784 900 987 1,024 1,156

"Fog" and "foggy" [1159] describe Fibonacci odd gobi.

foggy 5 55 6,765 75,025 9,227,465 102,334,155 12,586,269,025 139,583,862,445

"Fops' [1160] describes Fibonacci odd primes with only eight below a hundred million. 'Foppish' numbers[1161] describes numbers with Pfibonacci prime frags.

fops 3 5 13 89 233 1,597 28,657 514,229

foppish 3 5 13 15 23 25 30 31 32 33 34 35 36 37 38 39 43 45 50 51

"Foul" describes Fibonacci odd-unindexed luckies.[1162]

foul 3 9 15 21 25 31 33 37 43 51 63 67 69 73 75 75 79 87 93 99 105 111 115 127

"'Gate' [1163] describes gobi and ternarish evens combined and 'pearly' [1164] its complimentary sequence.

gated 2 5 10 12 15 20 22 25 35 45 55 65 75 85 95 100 102 105 110 112

pearly 4 7 14 24 34 44 54 65 74 77 79 84 87 89 114 124 134 144 154 165

"'Gigio" [1165] describes gobi-indedexed-gobi-indexed odds and 'topo' [1166] its complimentary sequence.

gigio 91 191 291 391 491 591 691 791 891 991 1,091 1,191 1,291 1,391 1,491 1,591

topo 108 208 308 408 508 608 708 808 908 1,108 1,208 1,308 1,408 1,508

"'Gigo" [1167] describes not only the computer science acronym 'gargage in, garbage out' but gobi-indexed gobi odd.

gigo 45 95 145 195 245 295 345 395 445 495 545 595 645 695 745 795 845

"The 'Gödel numbers',[1168] named for Kurt Gödel, are formed by the 'arithmeticization' by which he proved his famous inconsistency theorem. Representing zero as a nine prime exponent and the successor operation as an eight prime exponent, one, the successor of zero becomes two-to-the-eighth times three-to-the-nineth, already higher than five million.

Gödel 5,038,848 3,280,500,000,000 26,476,001,552,700,000,000

"'Grays' [1169] describes gobi, rokubi and yongbi combined, rather than just color-coded eights and 'blues' its complimentary sequence, rather than just color-coded sixes.

grays 1 4 5 11 14 15 21 24 25 31 34 35 41 44 45 51 54 55 61 64 65 71 74 75 81

blues 4 5 8 14 15 18 24 25 28 34 35 38 44 45 48 54 55 58 64 65 68 74 75

"I've got 'guts', gobi-indexed ternaries,"[1170] Al said, but someone mercifully added, "'Hash' [1171] describes hachibi-and-shichibi."

guts 1 10 11 12 21 22 100 101 102 110 111 112 120

hash 7 8 17 18 27 28 37 38 47 48 57 58 67 68 77 78 87 88 97 98 107 108

"'Hue' describes hachibi-unindexed evens.[1172]

hue 2 4 6 8 10 12 14 18 20 22 24 26 28 30 32 34 38 40 42 44 46 48

"'Ibgnoy' [1173] describes a reversable yongbi non-numberdrome, a subsequence of eight-headed amphisbæna.[1174]

ibgnoy 4,014 4,024 4,034 4,044 4,054 4,064 4,074 4,084 4,094 4,104 4,124 4,134

"'Ibihcah' [1175] describes a reversable hachibi non-numberdome, a subsequence of sixteen-headed amphisbæna.

ibihcah 8,018 8,028 8,038 8,048 8,058 8,068 8,078 8,088 8,098 8,108 8,128 8,138

'"Ibihcihs' [1176] describes a reversable shichibi non-numberdrome, a subsequence of fourteen-headed amphisbæna and 'ibihci' a reversable ichibi non-numberdrome, a subsequence of two-headed amphisbæna.

ibihcihs 7,017 7,027 7,037 7,047 7,057 7,067 7,077 7,087 7,097 7,107 7,127 7,137

ibihci 1,011 1,021 1,031 1,041 1,051 1,061 1,071 1,081 1,091 1,101 1,121 1,131 1,141 1,151 1,161

'"Ibin' [1177] describes a reversable nibi non-numberdrome, a subsequence of four-headed amphisbæna.

ibin 2,012 2,022 2,032 2,042 2,052 2,062 2,072 2,082 2,092 2,102 2,122 2,132

'"Ibnas' [1178] describes a reversable sanbi non-numberdrome, a subsequence of six-headed amphisbæna.

ibnas 3,013 3,023 3,033 3,043 3,053 3,063 3,073 3,083 3,093 3,103 3,123 3,133 3,143

'"Ibog' [1179] describes a reversable gobi non-numberdrome, a subsequence of ten-headed amphisbæna.

ibog 5,015 5,125 5,135 5,145 5,155 5,165 5,175 5,185 5,195 5,205 5,215 5,235

'"Ibukor' [1180] describes a reversable non-rokubi numberdrome, a subsequence of twelve-headed amphisbæna.

ibukor 6,016 6,126 6,136 6,146 6,156 6,166 6,166 6,176 6,186 6,196 6,206 6,216

'"Ibuuyk' [1181] describes a kyuubi numberdrome, a subsequence of eighteen-headed amphisbæna.

ibuuyk 9,019 9,029 9,039 9,049 9,059 9,069 9,079 9,089 9,099 9,109 9,129 9,139

'"Intwo' [1182] describes a number with a two in its interior, but not at its head or tail.

intwo 120 121 123 124 125 126 127 128 129 320 321 323 324 325 326 327 328

'"List' [1183] describes lucky-indexed sorted ternarish, 'honeydew' [1184] its complimentary sequence, and 'listless' [1185] the non-lists.

list 1 100 441 10,000 12,321 14,400 44,100 48,841

honeydew 8 558 899 51,158 55,899 85,599 87,678 89,999

listless 2 3 4 5 6 7 8 9 10 11 12 13 14 15 16 17 18 19 20 21 22 23 24 25 26 27

"'Log' describes lucky odd gobi, which are all odd, and 'jam' [1187] its complimentary sequence.

log 15 25 75 105 115 135 195 205 235 285 325 385 415 445 475 495 535

jam 24 74 84 114 144 194 254 264 284 314 324 354 384 464 504 524 554

"'Lop' [1188] describes lucky odd primes and 'lop-off' [1189] its neighbors.

lop 3 7 13 31 37 43 67 73 79 127 151 163 193 211 223 241 283 307 331 349

lop-off 2 4 6 8 12 14 30 32 36 38 42 44 66 68 72 74 78 80 126 128 150

"'Malu' [1190] describes reversable Ulam numbers, the unique sums for two previous Ulam numbers, beginning twenty-six, twenty-eight, sixty-two and eight-two.

Ulam 1 2 3 4 6 8 11 13 16 18 26 28 36 38 47 48 53 57 62 69 72 77 82

"'Markoff numbers,[1191] named for Andrei Markoff, are numbers expressable as the sum of three squares that equal three times their roots' product. The non-Markoffs might be called the Markons.[1192]

Markoffs 1 2 5 13 29 34 89 169 194 233 433 610 985 1,325 1,597 2,897 4,181 5,741

Markons 3 4 6 7 8 9 10 11 12 14 15 16 17 18 19 20 21 22 23 24 25 26 27 28 30

"'Narcissistic' [1193] describes numbers whose sum of the cubes of their digits (SOCOD) equals themselves. The first non-trivial one, one fifty-three, is the number of miraculous fish caught in John 21:11. It also 'happens' to be the sum of the first seventeen integers and so a triangular number. It can also be expressed as the product of factors composed of its reversed digits, three times fifty-one."

narcissistic 1 2 3 4 5 6 7 8 9 153 370 371 407 1,634 8,208 9,474 54,758 92,727

"'Net" [1194] describes nibi even ternarish, 'internet' [1195] numbers exactly between them and 'drag' their complimentary sequence."

net 2 12 22 102 112 122 1,002 1,012 1,022 1,102 1,112 1,122 1,202 1,212 1,222

internet 7 17 62 107 117 562 1,007 1,017 1,062 1,107 1,117 1,162 1,207 1,217 1,612

"Oop" [1196] could describe oddly odd primes and 'alley' [1197] its complimentary sequence."

oops 3 13 29 43 67 79 103 113 149 163 191 199 233 251 277 293 331 349 379

alley 6 20 32 56 70 86 136 146 178 202 242 260 298 322 352 368 398 412

"'Peace' [1198] describes prime evens and cube evens and 'war' [1199] its complimentary sequence.

peace 2 8 64 216 512 1,000 1,728 2,744 4,096 5,832 8,000 10,648 13,824

war 1 7 35 487 783 1,999 4,167 5,903 8,255 8,271 8,999 14,815 25,911

"The 'Phickover' [1200] numbers describe the position of the decimal expansion of e in phi, an extrapolation of the Pickover number[1201] which describes the position of the decimal expansion of e in pi.

Phickover 20 65 1,463 17,125 ...

Pickover 6 28 241 11,706 28,024 33,789 1,526,800 73,154,827

"'Plinear' [1202] describes linear numberdromes.

plinear 1 4 7 11 44 77 111 141 171 414 444 474 717 747 777 1,111 1,441 1,771

"'Quartered' [1203] describes a number containing the frag two-five and 'drawn' [1204] its complimentary sequence with the frag seventy-four.

quartered 25 125 225 325 425 525 625 725 825 925 1,025 1,125 1,225 1,325

drawn 74 174 274 374 474 574 674 774 874 974 1,074 1,174 1,274 1,374 1,474

"Rags' [1205] describes rokubi and gobi combined and 'riches' [1206] its complimentary sequence, which is richer in primes."

rags 5 6 15 16 25 26 35 36 45 46 55 56 65 66 75 76 85 86 95 96 105

riches 3 4 13 14 23 24 33 34 43 44 53 54 63 64 73 74 83 84 103 104 113

"Ratty" describes rokubi-and-ternarish.[1207]

ratty 1 6 10 11 12 16 20 21 22 26 36 46 56 66 76 86 96 100 101 102 106 110

"'Shin' describes shichibi-indexed nibi, but since all are even, so does 'shine'." [1208]

shine	62	162	262	362	462	562	662	762	862	962	1,062	1,162	1,262	1,362	1,462

"'Sit' [1209] describes sanbi-indexed ternarish.

sit	10	111	212	1,020	1,121	1,222	2,100	2,201	10,002	10,110	10,211	11,012	11,120	11,221

"'Skin' [1210] and 'skinny' describe sorted-kyuubi-indexed nibi.

skyuubi	9	19	29	39	49	59	69	79	89	99	119	129	139	149	159
skinny	82	182	282	382	482	582	682	782	882	982	1,182	1,282	1,382	1,482	1,582

"'Soul' describes sanbi odd-unindexed luckies.[1211]

soul	3	9	13	15	25	33	37	43	51	67	69	73	75	79	93	99	105	115	127	129	133	135

"'Space' [1212] describes sanbi-primes-and-cubed-evens and 'time' their complimentary sequence.[1213]

space	3	8	13	23	43	53	64	73	83	103	113	163	173	193	216	223	263	283	293
time	1	6	16	26	35	46	56	76	86	106	116	126	136	146	156	166	176	186	196

'Spic' describes sanbi-prime-indexed cubes,[1214] while 'span' describes the sanbi primes and nibi combined.[1215]

spic	9	169	529	1,849	2,809	5,329	6,889	10,609	12,769	26,569	29,929	37,249

span	90	470	830	3,110	4,670	7,190	8,150	14,150	19,910	30,830	45,710	50,270	62,750

"'Top" [1216] describes ternarish odd primes and 'bottom' their complimentary sequence.[1217]

top	11	101	211	1,021	1,201	2,011	2,111	2,221	10,111	10,211	12,011	12,101	12,211	20,011

bottom	88	788	898	7,778	7,888	7,988	8,798	8,978	77,888	78,778	78,788

"'Triple-digit inflated' [1218] describes the extrapolation of double-digit inflation.

triple-digit inflation	3	6	9	12	15	18	21	24	27	30	33	36	39	312	315	318	321	324	327

"Ylgu' [1219] describes those ugly numbers that remain ugly even when reversed.

ylgu	1	2	3	4	5	6	8	9	10	18	20	27	30	40	45	50	54	60	72	80	81	90	100	108

MULTIPLICADDITION

from PAJAS #103 (February 1994)

Robin quotied St. Francis, 'Most High, glorious God, enlighten the darkness of my heart, and give me right faith, certain hope, and perfect charity, wisdom and understanding, Lord, that I may carry out Your holy and true command. Amen."

Sam asked, "Why was the garbage picked up at sixteen minutes before two?"

"It was one forty-four."

"Gross!"

Robin continued, "Our topic is 'multiplicaddition', a portmanteau of multiplication and addition, 'multiplication' with a 'ddi' added, and so differs from both. Merely adding digits gives a number's weight, but multiplying the odd- and even-placed digits before adding changes everything. Between a hundred and a hundred nine we just get one through ten, but at a hundred ten we get one and at a hundred eleven one through ten again and again. Finally at both a hundred and a hundred one we get one, then with finally at a hundred twenty we get four.

multiplicadded	...	10	4	5	6	7	8	9	10	11	12	13	8	9	10	11	12	13	14	15	16	17	16	17	18

"The primes we call 'multiplicadditive' [1220] are those that remain prime after multiplicaddition.

multiplicadditive	103	107	149	211	223	227	229	251	257	281	283	307	347	419
multiplicadded	3	7	13	3	7	11	13	11	17	17	19	7	19	13

"They change even more when the fourth digit becomes the multiplicand of the previously multiplicadded three digits[1221] and the fifth digit becomes the addend of the previous multiplicadded four digits.[1222]

multiplicadditive	1,013	1,021	1,031	1,051	1,103	1,301	1,471	2,017	2,131	2,311	2,351
multiplicadded	3	2	3	5	3	2	11	7	5	7	11

multiplicadditive	10,007	10,037	10,067	10,093	10,103	10,111	10,141	10,211	10,223
multiplicadded	7	7	7	3	3	2	5	3	7

"Combining exponentiation and addition gives us the portmanteau 'exponentiaddition' and so 'exponentiadditive' and 'exponetiadded primes'.[1223]

exponentiadditive	101	131	151	181	191	211	223	227	229	233	239	257	263	271
exponentiadded	2	2	2	2	2	3	7	11	13	11	17	71	131	257

"Empowered numbers[1224] are defined by Conway as those numbers whose digits are alternately raised to an exponent and multiplied. Some of them also remain prime after 'empowerment'. As with multiplicaddition, the fourth digit becomes the exponent of the previously empowered three and the fifth the multiplicand of the previously empowered four.

primes	113	115	117	127	137	157	163	167	173	193	197	211	307	311	503	607	907
empowered	3	5	7	7	7	7	3	7	3	3	7	2	7	3	3	7	7

Mention of "tetraddition" and "tetradditive primes", combining tetration and addition, "pentaddition" and "pentadditive primes", combining pentation and addition, etc. brought the meeting to an end, but triggered an on-going quest for these higher primes.

SUM MUSICAL NUMBERS

from PAJAS #104 (March 1994)

Robin began with "Most High, omnipotent, good Lord, to thee be ceaseless praise outpoured." from St. Francis as translated by Howard Chandler Robbins and then asked, "Why didn't Goldilocks drink the ice water with eight pieces of ice?"

"It was too cubed; zero cubed would be too uncubed, so one cubed would be just right."

Sam surprisingly introduced the month's topic, "When telephones switched from dials to push buttons in 1963 the dial tone was introduced, a harmonic produced by combining seven different frequencies.

	1209 Hz	1,336 Hz	1,477 Hz
697 Hz	1	2	3
770 Hz	4	5	6
852 Hz	7	8	9
941 Hz		0	

"Soon users noticed that not only were numbers now given a voice, they could carry a tune. The best however were those without the two-syllable digits, zero and seven, so the numbers could better be sung along with.[1225] Some sums contain primes. We even have a Klondike number among them, 555-5556."

Doo Wah Ditty	5	5	5	5	5	5	6	6	4	5
sum		10	15	20	25	30	41	47	51	56

Here We Go Round the Mulberry Bush	4	4	4	2	2	6	6	2	4	4	8	8	8	8
sum		8	12	14	16	22	28	30	34	38	46	54	62	70

Jingle Bells	3	3	3	3	3	3	3	9	1	2	3	6	6	6	6	6	3	3	3	3	3	2	2
sum		6	9	12	15	18	21	30	31	33	36	42	48	54	60	66	69	72	75	78	81	83	85

Louie, Louie	1	1	1	6	6	9	9	9	6	6

sum	2	3	9	17	26	35	44	50	56

Love Me Tender	1	6	3	6	9	2	9	6	3	2	3	6	1	6	3	6	9	2	9
sum		7	10	16	25	27	36	42	45	47	50	56	57	63	66	69	78	80	89

Oh, When the Saints Go Marching In	1	3	6	9	1	3	6	9	1	3	6	9	3	1
sum		4	10	19	20	23	29	38	39	42	48	57	60	61

Strangers in the Night	4	8	8	4	8	6	6	4	8	6	4	4	8	8	4	8
sum		12	20	24	32	38	44	48	56	62	66	70	78	86	90	98

The Itsy Bitsy Spider	1	1	1	1	2	3	3	3	2	1	2	3	1	3	3	6	9	9	6	3	6
sum		2	3	4	6	9	12	15	17	18	19	22	23	26	29	35	44	53	59	62	68

The Muffin Man	1	1	3	3	6	9	6	6	3	2	6	6	3	2	1	1	1	1	3	3	6
sum		2	5	8	14	23	29	35	38	40	46	52	55	57	58	59	60	61	64	67	73

The meeting ended in a cacophany of competing musical sequences.

PHOENIX NUMBERS

from PAJAS #105 (April 1994)

Robin sang from John Menzies Macfarlane's "Far, Far Away on Judea's Plains": "Glory to God, glory to God, glory to God in the highest, peace on Earth, goodwill to men, peace on Earth, goodwill to men! Sweet are these strains of redeeming love, message of mercy from Heaven above."

She then asked, "Where should you go to find the warmest spot in a cold room?"

Lana answered, "Into a corner where it's always ninety degrees."

She then continued with the month's topic, "The nine-headed jiŭ huáng from Chu mythology gives its name to numbers identified by their head number. To have a head to behead a number would have to also have a tail and so must be at least two digits long. Translating to Classical mythology we'd call it a phoenix, or more specifically an ennea phoenix.[1226]

ennea phoenix	90	91	92	93	94	95	96	97	98	99	900	901	902	903	904

"A phoenix with one less than nine heads would be an okta phoenix,[1227] with seven a hepta phoenix,[1228] with six a hexa phoenix,[1229] with five a penta phoenix,[1230] with four a tetra phoenix,[1231] with three a tri phoenix,[1232] with two a duo phoenix,[1233] and with one a mono phoenix.[1234]

okta phoenix	80	81	82	83	84	85	86	87	88	89	800	801	802	803	804

hepta phoenix 70 71 72 73 74 75 76 77 78 79 700 701 702 703 704

hexa phoenix 60 61 62 63 64 65 66 67 68 69 600 601 602 603 604

penta phoenix 50 51 52 53 54 55 56 57 58 59 500 501 502 503 504

tetra phoenix 40 41 42 43 44 45 46 47 48 49 400 401 402 403 404

tri phoenix 30 31 32 33 34 35 36 37 38 39 300 301 302 303 304

duo phoenix 20 21 22 23 24 25 26 27 28 29 200 201 202 203 204 205

mono phoenix 10 11 12 13 14 15 16 17 18 19 100 101 102 103 104 105 106 107

"Zero is called the headless 'horseman'.[1235]

from PAJAS #106 (May 1994)

TOPOGRAPHIC NUMBERS

Quoting Micah 4:1, Robin began, "In days to come the mount of the Lord's house shall be established higher than the mountains; it shall rise high above the hills."

"What's the difference between a bear and a polar bear?"

"A polar bear is a Cartesian bear that has undergone a coordinate transformation."

"A topographic number " she continued, "whose digits get higher is called an 'upgrade' [1236] and one whose digits decrease 'downgrade'. [1237]

upgrade 12 13 14 15 16 17 18 19 23 24 25 26 27 28 29 34 35 36 37 38 39

downgrade 10 20 21 30 31 32 40 41 42 43 50 51 52 53 54 60 61 62 63 64

"If a number's digits get higher, reach a maximum and then decrease it's called a 'peak',[1238] while one whose digits decrease to zero and then get higher again to another maximum is a 'valley'.[1239]

peak 120 121 131 141 151 161 171 181 191 231 232 241 242 243 250 251 252 353 354

valley 101 102 103 104 105 106 107 108 109 201 202 203 204 205 206 207 208 209 301

"Two peaks not separated by a valley is a 'pass'.[1240] One that gets higher, repeats the maximum and then decreases is a 'plateau'.[1241] The opposite, one that decreases, repeats a minimum and then gets higher again is a 'basin'.[1242]

pass 12,120 12,121 12,130 12,131 12,132 12,140 12,141 12,142 12,143 12,150 12,151 12,152

plateau 1,220 1,221 1,330 1,331 1,332 1,440 1,441 1,442 1,443 1,550 1,551 1,552 1,553

basin 1,001 1,002 1,003 1,004 1,005 1,006 1,007 1,008 1,009 2,001 2,002 2,003

"One that alternately gets higher and then decreases or vice versa is 'undulating'.[1243]

undulating 102 103 104 105 106 107 108 109 120 130 132 140 142 143 145 146

"An undulating number alternately white nine and black zero and would be a 'zebra' [1244] and one alternately orange three and black zero would be a 'tiger' [1245] and its complimentary sequence a 'lady'.[1246]

zebra 90 909 9,090 90,909 909,090 9,090,909 90,909,090 909,090,909,090

tiger 39 93 393 939 3,939 9,393 39,393 93,939 393,939 939,393 3,939,393

lady 60 606 906 6,060 9,060 60,606 906,060 6,060,606 90,606,060

The meeting ended when some of the male members impersonating tiggers escorted the ladies out the door.

from PAJAS #107 (June 1994)

TO INFINITY AND BEYOND!

As an introduction to the month's topic Raz quoted "Great is our Lord and mighty in power, to His wisdom there is no limit.' (Psalm 147:5) and then in lieu of a riddle recited the famous limerick, 'There was a young fellow from Trinity, who took the square root of infinity, but the number of digits gave him the fidgets; he dropped math and took up divinity."

"These numbers go back to 1915 when P. E. B. Jourdain translated and so popularized Georg Cantor's work with transfinite numbers."

"Transfin? Beyond five?"

"No, much, much higher than six, beyond infinity,[1247] symbolized with a lazy eight, which he symbolized with the first Hebrew letter and a subscripted zero, calling it Aleph-null, and then the got even higher infinities, like Aleph-one, numbers beyond counting, but not beyond ordering.

$$\aleph_0 > \aleph_1 > \aleph_2 > \aleph_3 > \dots$$

"B. S. Johnson finally defined the square root of Aleph-null, mentioned in the limerick, in 1963. In 1972 John Horton Conway defined other surreal numbers, using the ordering method of Eudoxus on von Neumann's numbers. Many previously undefined terms became definable.[1248]

"My uncle used roulette's familiar double zero to represent the first inaccessible number along with the backslash to make surreal mathematics at least infinitesimally easier for us his students. With these more familiar symbols some of us could more easily grasp surreal numbers that are infinities plus integers, between infinity and finity, what he called 'nfinities' or with the n- verbalized 'enfinities'.

$$0/1 = 0 \qquad 1\backslash 0 = 00$$

"Ten backslash or under one is our representation for ten-to-the-Aleph-nullth plus one, the first archangelic number,[1249] those of a higher order than the merely angelic numbers,[1250] multiples of what he called ooplex."

$$1/10 = 0.10 \qquad 10\backslash 1 = 01.0 = 10^{00} + 1$$

"Since one over five is point two, what would one under five be?"

"Ooplex plus two."

$$1/5 = 0.20 \quad 5\backslash 1 = 02.0 = 10^{00} + 2$$

"So ten under one plus one is five under one and one under zero would be ooplex?"

$$0/1 = 0.0 \qquad 1\backslash 0 = 0.0 = 10^{00}$$

"And infinity is then the logarithm of zero backslash one!"

Although one or two moaned, "My brain hurts!", the strange words and symbols kept coming without limit, ad infinitum.

"So what would two under zero be? Twice as much?"

"But two times double zero would just be double zero."

$$2\backslash 0 = 2(10^{00}) \qquad 2/0 = 2(00) = 00$$

"Right and right! Angelic numbers are of a higher order than merely countable numbers. Another kind of surreal is that which has an infinite number of digits, like a rational or irrational with the digits to the right and left of the decimal point reversed." [1251]

"I get it! Arrational is between rational and irrational, like the enfinite is between the finite and the infinite. O my God! It's beginning to make sense to me!" one of the moaners exclaimed.

"Representing a repeating number fragment or frag with overlining we get, for example, several ways of expressing the same surrealities.

$$3\backslash 1 = 3 = (10^{00} - 1)/3$$

$$3\backslash 2 = 6 = 2(10^{00} - 1)/3$$

$$3\backslash 3 = 9 = 10^{00} - 1$$

"Some of the others are even more difficult to put into order, but the different notations help.

1\1	$(10^{00} - 1)^{1/3}$
2\2	$(10^{00} - 1)^{2/3}$
3\3	9
4\4	(2\2)(2\2)
8\8	(2\2)(2\2)(2\2)
9\9	(3\3)(3\3)

"The backslash notation isn't an advantage over the overlining in these cases, but certainly is in infinitely many others, just as fractional representations sometimes can be over decimals. Organizing these ordinal numbers can be quite useful in pattern recognition training, say just the seven or twenty-three backslashes, 'retroseptile' and 'retrotrivigintile'[1252] respectively, both 'reptiles' for short.

7\3	175,824	7\2	417,582	7\1	758,241
7\6	241,758	7\5	582,417	7\4	824,175

23\17	1,256,596,806,287,434,031,937
23\7	1,937,125,659,680,628,743,403
23\4	2,565,968,062,874,340,319,371
23\14	2,874,340,319,371,256,596,806
23\1	3,193,712,565,968,062,874,340
23\11	3,403,193,712,565,968,062,874
23\22	3,712,565,968,062,874,340,319
23\8	4,031,937,125,659,680,628,743

"Other arrationals are enverted roots, 'retroroots'.[1253]

$2^{1\backslash 2}$	...4,265,312,414.1
$3^{1\backslash 2}$	...6,708,050,237.1
$5^{1\backslash 2}$	...5,779,760,632.2
$7^{1\backslash 2}$	...1,113,157,546.2
$8^{1\backslash 2}$	...7,421,724,828.2

"'Ciscendental' surreal numbers[1254] would be those formed by enversion from transcendental numbers, like 'ip' and 'ihp' or multiples like ten times them, 'xip' and 'xihp'. It's difficult to say whether xip or xihp is higher, though pi is higher than e and e higher than phi.

ip	...383,346,264,832,397,985,356,295,141.3
ihp	...438,685,402,848,498,947,889,330,816.1
retro-e	...74,782,063,532,540,954,828,182,817.2
xip	...833,462,648,323,979,853,562,951,413
xihp	..386,854,028,484,989,478,893,308,161
ten retro-e	...747,820,635,325,409,548,281,828,172

"Calculating their cumulative weights, right-to-left,[1255] does not seem to help rank them at first, but eventually they seem to diverge so that we can say xihp is probably greater than xip is probably greater than xe.

xihp wt.	1	7	8	16	16	19	22	31	39	47	54	58	67	77	86	90	98	102	106	114
xe wt.	2	9	10	18	20	28	29	37	39	47	51	56	65	65	69	74	76	81	84	90
xip wt.	3	4	8	9	14	23	25	31	36	39	44	52	61	68	77	79	82	90	98	102

"They can also be formed by deleting digits or the decimal point.[1256]

evenless xip	...333,397,953,595,113
oddless ip	...2,846,264,828,624
pointless pi	314,159,263,589,793,2...
evenless xihp	...377,113,533,599,793,311
oddless ihp	...840,284,848,488,086
pointless phi	161,803,398,874,989,484,8...

"As with ciscendentals, Roman numeral prefixes can be used with the ooplex."

vooplex	$10^{00(5)}$
xooplex	$10^{00(10)}$
mooplex	$10^{00(1'000)}$
mooplem	$10^{00(3'000)}$
booplem	${}^{2}10^{00(3'000)}$
trooplem	${}^{3}10^{00(3'000)}$

"Raising infinity to higher powers we get ooduplex, ootriplex, etc., until we get to the transfinite ooploo, aka Aleph-null-to-the-Aleph-nullth, and higher. On a higher order than these are to-the-Aleph-nullth-Aleph-null or epsilon-null, epsilon-one, etc. through all the Greek alphabet to Omega, ..."

As late arrivals arrived interrupting the meeting, they were able to join other unsurrealists in a rousing rendition of "Aleph-null Bottles of Bheer". Since all of the infinite number of verses were exactly the same, they could join in and feel as if they hadn't missed anything at all.

NAME TAGS

from PAJAS #108 (July 1994)

The society's members, alumni and chairman "Rod" Roderick and the math department staff, all gathered to celebrate Joyce's eighty-fourth birthday and their own anniversary on the Fourth with Joyce's presence and with their presents. Br. Ken eloquently quoted from memory St. Francis, "At all times and seasons, in every country and place, every day and all day, we must have a true and honorable faith, and keep Him in our hearts, where we must love, honor, adore, serve, praise and bless, glorify and aclaim, magnify and thank, the Most High supreme and eternal God, Three and One, Father, Son, and Holy Spirit, Creator of all and Savior of those who believe in Him, who hope in Him, and who love Him; without beginning and without end, He is unchangeable, invisible, indescribable and ineffable, incomprehensible, unfathomable, blessed and worthy of all praise, glorious, exalted, sublime, Most High, kind, lovable, delightful and utterly desirable beyond all else, for ever and ever."

The birthday boy then admitted that he was beginning to feel his age, that it just seemed like yesterday that he'd been in his prime, eighty-three being a prime number, that is. One of the presents he shared with his fans was the file of pseudonumerological names for the numbers they had become so familiar with. The odd numbers, of course, had male given names and the evens female ones, though many numbers had a number of aliases.[1257]

0	Sue	Zoe					
1	Dewey	Odo					
2	Ina	Una					
3	Jaime	Moe					
4	Ray	Roy					
5	Al	Cecil	Eli	Lee	Leo		
6	Jo	Joy					
7	Augie	Guy	Hugo	Iago	Ike	Jacque	Ugo
8	Ava	Eva	Fay				
9	Abe	Pio					

Longer numbers, of course, in general got longer names.[1258]

10	Daisy					
11	Eddie	Otto	Tito			
12	Diana	Dinah	Dona	Edna	Edwina	Tina
13	Adam	Tim	Tom			
14	Adria	Audrey	Daria	Dora		

15	Odilo
17	Diego Hodge
19	Toby

The numbers however can identify as transgendered, expressing their other self in other bases. Odd numbers that don't look odd would still be odd, while even even numbers would look odd given their femme and macho pseudonyms.[1259] A few like "Lee" could be either.

0	Isaiah	Isau	José		
1	Ada	Di	Hedy	Ida	Ita
2	Ian	Johan	Noah	Owen	Yon
3	Aimee	Amy			
4	Ray	Roy			
5	Leah	Lee			
6	Joe				
9	Hope				

Others like sixteen take multiple given names like "Ada Jo" or some with interior or leading zeros like "Bonaventure Sigismund", Cal and Lana's new twins.[1260] Appropriately and providentially the Spanish pronunciation of the Name above all names, Jesus, turns out to be double zero, the representation of infinity.

00	Jesus				
002	Susan				
005	Cecil	Cecilia			
01	Asta	Sadie	Seth	Sid	Zita
02	Sean	Sonya	Zeno		
03	Sam				
04	Caesar	Cyr	Cyra	Ezra	Sarah
05	Celia				
06	Sacha				
07	Isaac	Zeke			
08	Sophia				

Like the hobbits chronicled in "The Hobbit" by J. R. R. Tolkien Joyce also gave celebrants gits, abridged copies of his math chronology going back a hundred fifty million years.[1261]

ENDNOTES

1 More on Joyce and his mathematical recreations can be found at the André Joyce Appreciation Society's webpages at http//:michaelhalm.tripod.com/id52.htm. The name "André Joyce" may or may not be a pun on "and rejoice" (Pss 2:11, 9:2, 31:7, 68:3, Mt 5:12), something a Biblically literate Oulipian, a member of Ouvroir de Littérature Potentielle (Oulipo), might choose as a pseudonym.

Wilhelm Ackermann defined a generalized exponential that does not use superscription or special symbols, such that the first argument reading left-to-right in the function is the operation, the second the power and the third the base. [NOTE: Joyce preferred using z to symbolize a variable, as in zillion; x he used as a the Roman numeral for ten in his coined number names. He also preferred priming the variable, reserving subscripting for indicating a number's base or number of repeated digits.]

- g(z, z', z'') = g(z - 1, g(z, z' -1, z''), z'')
- g(z, 1, z') = z', (enumeration)
- g(0, z, z') = z + z' (addition)
- g(1, z, z') = z*z' = z + ... + z (multiplication)
- g(2, z, z') = $z^{z'}$ = z*z*z...z*z*z (exponentiation)
- g(3, z, z') = ${}^{z'}z$ = g(2, g(2, g(2, ..., z'), z'), z') (tetration),

[NOTE: tetration is evaluated from the top of the exponential stack downward. It is a much higher function than that evaluated from the bottom upward, g(z, g(2, g(2, z', g(2, z', ..., g(2, z', z')...)]

- g(4, z, z') = g(3, g(3, g(3, ..., z'), z'), z') pentation, etc.
- h(0, z, z') = z||z' = zz' (hyperaddition, catecnation)
- h(1, z, z') = z|||z' = z||...||z' = (hypermultiplication)
- h(2, z, z') = z||||z' = z'|||...|||z' = h(1, h(1, ..., z'), z') (hyperexponentiation)
- h(3,z, z') = z|||||z' = z'||||...||||z' = h(2, h(2, ..., z'), z') (hypertetration), etc.

Joyce extrapolated Ackermann's notation in the other direction into a polytriadic or More Generalized Exponential notation which indicates how many times the Generalized Exponential function's arguments are iterated. It can indicate operational nesting, but also nestings of any of the other two arguments or even nestings of nestings. Although it does not as easily represent very high numbers as other notations, it can represent more kinds of numbers including the very large.

- g(z, 1, 1, z', z'', z''') = g(z - 1, 1, 1, g(z', z'', z'''), z'', z'''),
- g(z, 1, z', z'', z''') = g(z - 1, 1, z', g(z', z'', z'''), z'''),
- g(z, z', z'', z'') = g(z - 1, z', z'', g(z', z'', z''')).

Similarly the next three, of any number of arguments, gives:

- g(z, z', z'', ...) = g(z - 1, z', z'', g(z', z'', z''',, ...))

2 The AJAS's president's name, Daniel "Dan Svjetlost", is Croatian for "day light". DAN_{33} = JOB_{27}, DAN_{28} = ASH_{31}

3 The starving artist's name, Thomas "Tom Ficka ", is Swedish for "empty pocket". TOM_{30} = OLD_{33}, Ten-to-the-nineth or nineplex, would be called rather milliard in French and Danish and cognates in other languages like Hungarian (milliárd), Indonesian (millar), Polish (miliard), or

Spanish (millardo).

4 Both 2 + 6 + 2 + 1 + 4 + 4 = 19 and 4 + 0 + 9 + 6 = 19. If one continues to sum the digits, also called "casting out nines", one eventually reaches a number between, zero and nine (or the maximum single placeholder in a base not decimal), called the digital root, useful for checking for errors. Both of these numbers have the same digital root of one. One through through three would be lightweight, four through six middleweight and seven through nine heavyweight class and each would subdivide into three subclassifications, light, middle, and heavy lightweight, middleweight and heavyweight with zero being weightless.

number	10	11	12	13	14	15	16	17	18	19	20	21	22	23	24	25	26	27	28	29	30
weight	1	2	3	4	5	6	7	8	9	1	2	3	4	5	6	7	8	9	1	2	3
class	L	L	L	M	M	M	H	H	H	L	L	L	M	M	M	H	H	H	L	L	L
sub	LL	ML	HL	LM	MM	HM	LH	MH	HH	LL	LM	HL	LM	MM	HM	LH	MH	HH	LL	ML	ML

The Hamming weight is the digit sum of the equivalent binary number, measuring how much it differs from an all-zero string of the same length. Two-to-the-eighteenth or and two-to-the-twelfth would obviously do so, since they are both powers of two, both have the same Hamming weight, also one. A pernicious number is one with a prime Hamming weight.

pernicious	3	5	6	7	9	10	11	12	13	14	17	18	19	20	21	22	24	25	26	29	31	33	34	35	36

Taking binary and Hamming weight as continuing fractions we get three more faster-growing sequences, nicknamed "broken-up", "broken-down" (numerator and denominator of the approximations to the convergent respectively) and "broken" (their combination, not counting duplications).

1 + 1/10 + 1/11 + 1/100 + 1/101 + 1/ 110 + 1/111 ...≈ 11/10 ≈ 122/111 ≈ 12,211/11,110 ...

These sequences can be decelerated by "subbing", that is by repeated division by the natural logarithm and rounding analogously to the subfactorial, [n!/e + 1/2] = (n - 1) + f(n - 1) + f(n - 2)). They have "mates", their digital "opposite numbers". They can also be accelerated by addition, multiplication ("super-"). To get high even faster googologists use hypermath, adding and multiplying concatenationally. It's usually indicated by vertical lines between numbers, two for hyperaddition, three for hypermultiplication, etc. The third hyperproduct is already higher than nineteenplex and the fourth higher than fifty-sixplex.

decimal	1	2	3	4	5	6	7
binarish	1	10	11	100	101	110	111
mates	8	88	89	888	889	898	899
holiness	2	4	3	6	5	5	4
sum	1	11	21	32	132	233	343
super	1	10	110	11,000	1,111,000	122,210,000	13,565,310,000
broken-up	1	11	122	12,211	1,233,433	135,689,841	15,062,805,784

broken-down	1	10	111	11,110	122,221	123,455,420	13,704,673,841
broken	1	10	11	111	122	11,110	12,211
Hamming wt.	1	1	2	1	2	2	3

hypersum	1	110	11,011	11,011,100	11,011,100,101	11,011,100,101,110

hypersuper	1	10	11,111,111,111,111,111,111	> nineteenduplex

Higher than superfactorial are hyperfactorials calculated by hypermultiplication, the next higher operation to hyperaddition, aka concatenation. The double-line symbol for concatenation can be easily extrapolated to three or more in the Generalized Exponential with an h for hyper rather than a g. Like hyperaddition it's not commutative. 1||2 = h(0, 1, 2) = 12, 2||1 = h(0, 2, 1) = 21, 1|||2 = h(1, 1, 2) = 2, 2|||1 = h(1, 2, 1) = 11, 2|||3 = h(1, 2, 3) = 33. The opposite operation would be "procatenation". Hyperexponentiation and hypertetration would be z||||z' = h(2, z, z') and z|||||z' = h(3, z, z'). Hypermath googolisms are indicated by Latin prefixes, so that bigoogol, pronounced buy-goo-gol, means two hyperadded googols, and so higher than two-hundred-twoplex. The Smarandache numbers are hypersums of the natural numbers.

z||(z+1)	1	12	123	1,234	12,345	123,456	1,234,567	12,345,678	123,456,789
mates	8	87	876	8,765	87,654	876,543	8,765,432	87,654,321	876,543,210
holiness	2	2	3	3	3	3	3	3	4
gnomon		11	111	1,111	11,111	111,111	1,111,111	11,111,111	111,111,111
sum	1	13	135	1,357	13,579	137,171	1,371,738	13,717,416	137,174,205

broken-up	1	13	1,600	1,974,413	24,374,130,085	3,009,132,605,748,173
broken-down	1	2	247	304,800	3,762,756,247	464,534,835,534,432
broken	1	2	13	247	1,600	304,800

The Smarandache–Wellin numbers, named for both Florentin Smarandache and Paul R. Wellin are the hypersums of the primes. Seventy-five percent of the first four are also primes.

p||p'	2	23	235	2,357	235,711	23,571,113	2,357,111,317	23,571,1131,719
sub	1	8	86	867	86713	8,671,328	867,132,794	86,713,279,415
mates	7	76	764	7,642	764,288	76,428,886	7,642,888,683	764,288,868,380
dusub	1	10	103	1,034	103,435	10,343,525	1,034,352,505	103,435,250,477

broken-up	2	47	11,047	26,037,826	61,374,02,015,333	144,665,396,429,867,913,455
sub	1	17	4,063	9,578,780	227,824,023,645	53,219,425,195,467,779
broken-down	1	3	706	1,664,045	392,233,711,701	9,245,385,140,915,357,258
broken	1	2	3	47	706	11,047

Hyperfactorials would be the hyperproducts of the natural numbers, one hypermultiplied by two, that hypermultiplied by three and so on.

hyperfactorial	1	2	33	444,444,444,444,444,444,444,444,444,444,444
mates	7	8	66	555,555,555,555,555,555,555,555,555,555,555
broken-up	1	3	100	44,444,444,444,444,444,444,444,444,444,444,444
broken-down	1	2	67	2,977,777,777,777,777,777,777,777,777,777,777
broken	1	2	3	67
sum	1	3	36	444,444,444,444,444,444,444,444,444,444,480
super	1	2	66	> thirty-fourplex

Numbers which are hypermultiples of two are hypereven, even if they're odd, and those that aren't are hyperodd, even if they're even. One hypermultiplied by two equals two, while two hypermultiplied by one equals eleven.

hypereven	0	2	11	22	33	44	55	66	77	88	99	222	1,010	1,111	1,212	1,313	1,414
holiness	1	0	0	0	0	0	0	2	0	4	2	0	2	0	0	0	0
untrimmed	1	3	22	33	44	55	66	77	88	99	1,010	333	2,121	2,222	2,323	2,424	2,525
mates	7	9	11	22	33	44	55	66	77	88	777	900	1,111	1,212	1,313	1,414	1,515

broken-up	0	1	11	243	8,030	353,563	19,453,995	1,284,317,233	98,911,880,936
holiness	1	0	0	0	4	1	3	2	10
untrimmed	1	2	22	354	9,141	464,674	21,056,410,106	2,395,428,344	10,910,229,911,045
broken-down	1	2	23	508	16,787	739,136	40,669,267	2,684,910,758	206,778,797,633
broken	0	1	2	11	23	243	508	8,030	16,787
holiness	1	0	0	0	0	0	3	4	3
untrimmed	1	2	3	22	34	354	619	9,141	27,898

hyperodd	1	3	4	5	6	7	8	9	10	12	13	14	15	16	17	18	19	20	21	23	24	25	26	27	28	29	30
mates	1	2	3	4	5	6	8	10	12	13	14	15	16	17	18	19	20	21	23	24	25	26	27	28	29	30	31

broken-up	1	4	17	89	551	3,946	42,119	293,017	2,962,289	35,840,485	468,888,594
broken-down	1	3	13	68	421	3,015	24,541	223,884	2,263,381	27,384,456	358,261,309
broken	1	3	4	13	17	68	89	421	551	3,015	3,946
sum	1	4	8	13	19	26	34	43	53	65	78
super	1	3	12	60	360	2,520	20,160	181,440	1,814,400	21,772,800	283,046,400

5 The choleric numbers and the sanguine numbers are related to the happy numbers since they're numbers whose sum of the squares of their digits (SOSOD) go respectively to one or seven.

- **7 ⇒ 49 ⇒ 97 ⇒ 130 ⇒ 10 ⇒ 1**
- **13 ⇒ 10 ⇒ 1**

choleric	1	7	10	13	31	49	70	79
mates	2	8	20	29	50	68	86	89
broken-up	1	8	81	1,061	32,972	1,616,689	113,201,202	8,944,511,647
broken-down	1	7	71	930	28,901	1,417,079	99,224,431	7,840,147,128
broken	1	7	8	71	81	930	1,061	28,901
sum	1	8	18	31	62	111	181	260
super	1	7	70	910	28,210	1,382,290	96,760,300	7,644,063,700

hypersum	1	17	1,710	171,013	17,101,331	1,710,133,149	171,013,314,970

hypersuper	1	7	10,101,010,101,010	> thirteenduplex

- **1112 ⇒ 7**

sanguine	1,112	1,121	1,211	2,111
mates	7,888	8,788	8,878	8,887
broken-up	1,112	1,246,553	1,509,576,795	3,186,717,860,798
heptasub	1	1,137	1,376,556	2,905,910,547
broken-down	1	1,113	1,347,844	2,845,299,797
broken	1	1,112	1,137	1,246,553

hypersum	1,112	11,121,121	111,211,211,211	1,112,112,112,112,111
heptasub	1	10,141	101,411,498	1,014,114,978,710

Melancholic numbers go to the non-zero even digits, 2, 4, 6 or 8 and the phlegmatics go to the other odd digits, not one or seven, that is, 3, 5 or 9. Both are related to the unhappy numbers since the sum of the squares of their digits (SOSOD) do not go to one.

- **2 ⇒ 4**
- **5 ⇒ 25 ⇒ 29 ⇒ 85 ⇒ 89 ⇒ 145 ⇒ 42 ⇒ 20 ⇒ 2**
- **6 ⇒ 36 ⇒ 45 ⇒ 39 ⇒ 90 ⇒ 81 ⇒ 65 ⇒ 61 ⇒ 37 ⇒ 58 ⇒ 89 ⇒ 145 ⇒ 42 ⇒ 20 ⇒ 2**
- **8 ⇒ 64 ⇒ 52 ⇒ 29 ⇒ 85 ⇒ 89 ⇒ 145 ⇒ 42 ⇒ 20 ⇒ 2**

melancholic	2	3	5	6	8	9	16	18	20	24
mates	1	3	4	6	7	10	14	18	26	34
broken-up	2	7	37	229	1,869	17,050	274,669	4961,092	99,496,509	2,392,877,308
broken-down	1	3	16	99	808	7,371	118,744	2,144,763	43,014,004	1,034,480,859

broken	1	2	3	7	16	37	99	229	808	1,869

hypersum	2	23	235	2,356	23,568	235,689	23,568,916	2,356,891,618
sub	1	8	86	867	8,670	86,705	8,670,520	867,051,971

hypersuper	2	33	555,555,555,555,555,555,555,555,555,555,555	> thirty-twoduplex

- **12 ⇒ 5**
- **30 ⇒ 9**
- **111 ⇒ 3**

phlegmatic	9	12	21	30	102	111	113
holiness	1	0	0	1	1	0	0
mates	69	78	87	90	288	688	789
broken-up	9	109	2,298	69,049	7,045,296	782,027,865	8,8376,194,041
holiness	1	2	3	4	3	6	7
pentasub	1	145	311	9,345	953,477	105,845,306	11,961,473,073
broken-down	1	10	211	6,340	646,891	71,811,241	8,115,317,124
broken	1	9	10	109	211	2,298	6,340

hypersum	9	912	91,221	9,122,130	9,122,130,102	9,122,130,102,111
holiness	1	1	1	2	3	3
dusub	1	124	12,345	1,234,546	1,234,546,061	1,234,546,061,091

6 "Average" originally meant a day's work, however many hours that took, from the French *aver*. Balance on the other hand is weight divided by length. Numbers like googol minus one are like Lake Wobegon, Minnesota, where "all the children are above average", popularized by "The News from Lake Wobegon" by Garrison Keillor. Lower Wobegon would logically therefore all be below average. Between would be those which average to average, like eighteen and twenty-seven, which since bolsheviks just meant the majority before it meant Leninists and mensheviks meant the minority, we could call these that are both neither and both 'sheviks' -- or not.

number	**1**	**2**	**3**	**4**	**5**	**6**	**7**	**8**	**9**	**10**	**11**	**12**	**13**	**14**	**15**	**16**	**17**	**18**	**19**	**20**	**21**	**22**	**23**	**24**	**25**	**26**	**27**
balance	1	2	3	4	5	6	7	8	9	1/2	1	3/2	2	5/2	3	7/2	4	9/2	5	1	3/2	2	5/2	3	7/2	4	9/2
	<	<	<	<	>	>	>	>	>	<	<	<	<	<	<	<	<	=	>	<	<	<	<	<	<	<	=

sheviks	18	27	36	45	54	63	72	81	1,188	1,278	1,287	1,368	1,386	1,458	1,485	1,548	1,584

7 The word "digit" originally refered to the ten fingers we count on rather than the ten minimum number frags. Moving digits is called transpositioning, like 12's relationship to her brother 21. Adding digits and then transposing them is called transaddition, like 1 or 2's relationship to their father 21 or their mother 12. Deleting and then transposing the digits is called transdeletion, like 21's relationship to his fathers, 221 and 211, and his mothers, 212, 122 and 112. The Pythagoreans' sacred number 142,857 = [1,000,000/7], for example, is called 'sacred' because it's two-transposable, 2(142,857) = 285,714, three-transposable, 3(142,857) = 428,571, four-transposable, 4(142,857) = 571,428, five-transposable, 5(142,857) = 714,285, and six-transposable, 6(142,857) = 857,142.

The higher full 'reptend primes' also have frags called 'cyclicals', what Ripley called persistent numbers.

17	5,882,352,941,176,470
19	526,315,789,473,684,210
23	4,347,826,086,956,521,739,130
29	3,448,275,862,068,965,517,241,379,310
47	2,127,659,574,468,085,106,382,978,723,404,255,319,148,936,170

Marcia Birkin and Anne C. Coon compare the sestina to cyclical numbers in *Discovering Patterns in Mathematics and Poetry*. A sestina based on the Pythagorean sacred number would permute its endrhymes: 142857, 714285, 571428, 857142, 285714, 428571. An arithmopoem based on the next cyclical, a sedicina, would have sixteen squared or two hundred fifty-six lines with the first stanza's endrhymes, 5882352941176470. The second's would be 1176470588235294, the third 1764705882352941, and so on. Then would come the 324-line diciottina, 484-line ventiduina, 784-line ventiottina, 2116-line quaranteseina, 3364-line cinquantottino, etc.

reptend primes	7	17	19	23	29	47	59
broken-up	7	120	2,287	52,721	1,531,196	3,115,113	7,761,422
dusub	1	16	309	7,135	207,225	421,585	1,050,394
broken-down	1	8	153	3,527	102,436	4,818,019	284,365,557
broken	1	7	8	120	153	2,287	3,527

hypersum	7	717	71,719	7,171,923	717,192,329	71,719,232,947	7,171,923,294,759
dusub	1	97	9,706	970,614	97,061,427	9,706,142,704	970,614,270,447

Half-cyclicals come in complimentary even and odd pairs that add to a power of ten minus one, six "sacred" ones, each mated to another in 'holy marriages' or hierogamy.

half-cyclicals	142	285	428	571	714	857
mates	857	714	571	428	285	142
gnomon		143	143	143	143	143
pentasub	1	2	3	4	5	6
broken-up	142	40,471	17,321,730	9,890,748301	7,0622,011,608,644	> fifteenplex

hexasub	1	101	42,937	24,516,715	17,504,976,647	> elevenplex
broken-down	1	285	121,981	69,651,436	49,731,247,285	> thirteenplex
broken	1	142	285	40,471	121,981	17,321,730

even	142	428	714	5,882,352
pentasub	1	3	5	39,635
broken-up	142	60,777	43,394,920	255,264,294,512,617
pentasub	1	409	292,393	1,719,956,613,390
broken-down	1	143	102,103	600,605,786,399
broken	1	142	143	60,777

odd	285	571	857	11,764,705
broken-up	285	162,736	139,465,037	1,640,765,018,281,820
hexasub	1	404	345,699	4,067,049,860,464
broken-down	1	286	245,103	2,883,564,489,901
broken	1	285	286	162,736

We also have other fractional cyclicals of possibly every possible lengths.

half-cyclicals	3	8	9	11	14	23	29	30	48	54
third-cyclicals	2	6	20	32	36	60	64	74	76	104
fourth-cyclicals	4	7	15	24	27	28	37	45	48	57
sixth-cyclicals	1	6	10	52	56	61	63	72	81	83
seventh-cyclicals	4	16	48	54	70	94	100	106	136	...
eighth-cyclicals	2	12	14	24	29	32	34	42	47	54

8 The farmboy athlete Calvin Cal Pantof", aka "Eleven", is Romanian for "horse shoe". CAL_{36} = NEL_{26}, CAL_{33} = AXE_{35} = LBJ_{25}.

9 The society's secretary, an incarnate counterexample to Beckhap's Law, "Beauty times brains equals a constant ', a "ten", "Clair Peau", is French for "clear skin". Other little-known though named laws collected by the André Joyce Appreciation Society can be found at http://michaelhalm.tripod.com/id119.htm.

10 The biology major's name, "Anne Tabiat" is Turkish for "mother nature".

11 Besides one, the Indo-European root oino- also gives us the English words: a, alone, an, anon, any, anyone, atone, coadunate, einkorn, eleven, inch, lone, lonely, none, once, ounce, quincunx, triune, turnverein, unanimous, uncial, unicorn, union, unique, unite, unity, and universe. ("Indo-European and the Indo-Europeans" by Calvert Watkins)

12 Fred is, of course, a reference to Frederick Flintstone. Wilma was his smarter wife ("The

Flintstones" by Joseph Barbera and William Hannah, "Why The Flintstones Takes Place in a Post-Apocalyptic Future" by Anthony Scibelli, "The Jetsons Meet the Flintstones" by Don Nelson and Arthur Alsberg, "Time Machine" by Bill Idelson and Sam Bobrick)

13 The Ishango bone has a list of the primes, 11, 13, 17 and 19. In *How Mathematics Happened: The First 50,000 Years*, Peter Rudman writes that prime numbers could only have come about after the concept of division, which he dates to after 10,000 BC. Alternatively Alexander Marshack in *The Roots of Civilization* theorizes that the Ishango bone was merely a calendar, these numbers summing to twice the lunar cycle, implying rather strange lunar half-months.

More interesting is what Dominic Olivastro calls an Ishango computer, one that uses unary with only incrementation (adding one), decrementation (subtracting one), and stops when reaching zero, what is usually called a Minsky machine.

Prime numbers themselves, of course, predate Humans. *The Math Book* estimates that prime cycle cicadas ate the composite cycle cicadas to extinction circa a million B. C. Eratosthenes of Cyrene didn't invent the first prime sieve until 230 BC.

The probability of finding a prime decreases as they increase tenfold.

4%	2.5%	1.68%	1.229%	9.5‰	7.8‰	6.6‰	5.7‰	5.0‰	4.5‰

14 The history major's name, "Lana Coperta", is Italian for "wool blanket".

15 Worshipers of the Most High, aka Hypsistos, were called Hypsistarians and flourished until the Fourth Century A. D. According to Benjamin Trovato the Black Theorem is the math problem linking the "decent into mathness" of quite a number of mathematicians, hence the axiom, "Old mathematicians don't die; they become irrational." The first notable victim was apparently **Archimedes**, 75, murdered in 212 BC when he refused to yield to an enemy soldier.

Many more strange deaths of mathematicians old and not so old are known in modern times. **Abraham de Moivre**, 87, predicted his death date and died after sleeping a maximum 24 hours/day in 1754. **Isaac Newton**, 84, poisoned himself experimenting with mercury in 1797. **Niels Abel**, 26, died of malnutrition and pneumonia in 1829. **Évariste Galois**, 20, was murdered in 1832. **Gotthold Eisenstein**, 29, died in 1852, **Bernard Riemann**, 40, in 1866 and **Alfred Clebsch**, 39, in 1872. **William Kingdong Clifford**, 33, died of overwork and pneumonia in 1879. **Ludwig Boltzmann**, 62, killed himself in 1906. **René Gâteaux,** 25, was killed in 1914. **Georg Cantor**, 72, died in an asylum in 1918. **Srinivasa Ramanujan**, 32, died of malnutrition in 1920. **Pavel Urysohn**, 26, drowned in 1924. **Frank Ramsey**, also 26, died on the operating table in 1930. **Dmitri Egorov**, 61, starved to death in 1931. **Stanisław Saks**, 44, was murdered in 1942. **Dénes Kőnig**, 60, killed himself in 1944 and **Alan Turing**, 41, in 1954. **Kurt Gödel**, 71, starved to death in 1978.
(http://www.math.rutgers.edu/~kellenm/deaths.html,
http://mathandmultimedia.com/2013/06/24/10-mathematicians-who-died-young/)

The alternative theory of Andrei Balabukha is that time-traveling aliens shortened individual mathematicians' lives to lengthen mankind's, fulfilling both "Who the 'gods' destroy they first drive mad." and "It is better for one man to die than that the whole nation perish." (Gamaliel, Jn 11:50). (http://www.uchronia.net/bib.cgi/label/balaappend.html)

16 More riddles and rhymes can be found in *Reignbeau's Riddles and Rhymes* by Reignbeau the Clown available at Amazon.

17 Carl D. Eiseman in his article in the Dec. 1991 *Mensa Bulletin* also mentioned rare usages of G for 400 and K for 250. Circa 1150 Ocreatus tried to mix the old, familiar Roman numerals and the new decimal system with zero and so got higher with ten as I.O and eleven as I.I, etc., but fortunately it did not catch on like the Arabic numerals did. A misplaced radix or decimal point could have been catastrophic. VII (7), for example, could too easily be mistaken for V.II (52) or VI.I (61) or V.I.I (511). Seven and sixty-one however could be called Ocreatus primes since they remain prime under such a transformation.

Ocreatus primes	II	VII	XI	XIII	XVII	XIX	XXIII	XXIX	XXXI	XLIII	XLVII	LIII
primes	2	7	11	13	17	19	23	29	31	43	47	53
Ocreatus primes	I.I	VI.I	X.I	X.III	X.VII	X.IX	X.XIII	X.XIX	XX.XI	XLI.II	XLVI.I	LII.I
primes	11	61	101	103	107	109	1,013	1,019	2,011	421	461	521

18 The shy janitor's name, Samson "Sam Znowu", is Polish for "alone again".

19 This would be the system of the legendary pataphysician Pathodius, the 10th century student of Methodius. "Pathodius" is to "Methodius" as "metaphysics" ("what 'what is' is"), is to "pataphysics" ("what 'what "what is" is' is"), pata- being a contraction of "epi ta meta-", just as metaphysics ("what 'what is' is") is to physics ("what is"). Beyond pataphysics lies papaphysics, "known only by the Father". (Mt 24:36), "what 'what "what 'what is' is" is' is".

20 See this article at http://michaelhalm.tripod.com/luraj.html.

21 The astronomy major's name, "Charles Wayne", is an Old English name for the constellation Ursa Major, aka "the Great Bear".

22 The Bynars are featured in the "Star Trek: The Next Generation" episode, "11001001" by Maurice Hurley. (NOTE: 11001001_2 = 201)

23 The science fiction fan's name, "Al di Fuori", is Italian for "outside".

24 Binarish is nicknamed "Babe" by some for the redundant "binarish and binarish evens" and George "Babe" Ruth or "Babs" for Barbara. Minoan numerals used | for one, O for ten, / for hundred and \ for thousand, so what looks like a Roman five and six is actually one thousand, one hundred and one thousand, one hundred one.

Minoan binarish	|	O	O|	/	/|	/O	/O|	\	\\|	\O	\O|	V	V|	VO	VO|
holiness	0	1	1	0	0	1	1	0	0	1	1	0	0	1	1

The repunits, numbers represented with just ones, are the first example of a repdigit, a number with just one digit repeated a number of times. All are powers of ten minus one, divided by nine and rounded down to an integer. A digit repeated three times could be called a threepeat (attrib. to Pat Kiley). Excessive repetition is called battology for Battos the Stutterer.

threepeats	111	222	333	444	555	666	777	888	999	1,000	1,011	1,101	1,110	1,222

broken-up	111	24,643	8,206,230	3,643,590,763	202,220,179,695

pentasub	1	166	55,293	24,550,322	1,362,483,696
broken-down	1	222	73,927	32,823,810	18,217,288,477
broken	1	111	222	24,643	73,927
hypersum	111	111,222	111,222,333	111,222,333,444	111,222,333,444,555
pentasub	1	749	749,410	749,410,188	749,410,187,864

repdigits	11	22	33	44	55	66	77	88	99	111	222	333	444	555	666	777

dusub	1	3	4	6	7	9	10
broken-up	11	243	8,030	353,563	19,453,995	1,284,317,233	98,911,880,936
dusub	1	33	1,087	47,850	2,632,812	173,813,437	13,386,267,422
broken-down	1	12	397	17,480	961,797	63,496,082	4,890,160,111
broken	1	11	12	243	397	8,030	17,480

hypersum	11	1,122	112,233	11,223,344	1,122,334,455	112,233,445,566

Repdigits can be represented more compactly via "hyperdivision", symbolized by a double slash, the counteroperation to hypermultiplication, 111//3 = 1(1)1 = 1, like the Triune God Who is One or 11//2 = 1(1) = 1, like Cal becoming MVP.

hyperdivided	1	2	3	4	5	6	7	8	9	1	2	3	4	5	6	7	8	9
sum	1	3	6	10	15	21	28	36	45	46	48	51	55	60	66	73	81	90

"Gnomon" comes from geometry where it refers to the area that must be added to a figure to form a similar figure. For example the gray L-shaped figure of area three needed to be added to the single black square of area one to make the larger square of area four, the difference between two squared and one squared.

Calculating the gnomons, differences between these homonums, symbolized with a capital delta, Δ, we can more easily see how both repunit numbers and binarish numbers get high exponentially, without having to do calculus. In this case the gnomon is obviously exponential with a constant ratio of ten or 1000%. We can get higher by faster by calculating the opposite of the gnomon, the sum or even product of the previous numbers. Calculators like supermultiplier can exactly calculate the sum, product, power and factorial of numbers higher than the killion limit of most calculators. The sum is symbolized with a capital sigma, Σ, the product with a capital pi, Π. By extrapolation tetration could be symbolized by a capital tau, T.

repunit	1	11	111	1,111	11,111	111,111
gnomon	1	10	100	1,000	10,000	100,000

sum	1	12	123	1,234	12,345	123,456
super	1	11	1,221	1,356,531	386,551,691	1,507,215,941
broken-up	1	12	1,333	1,480,975	16,455,114,558	1,828,344,235,134,910
broken-down	1	2	223	247,755	2,752,806,028	305,867,030,824,863
broken	1	11	12	223	1,333	247,755

hypersum	1	111	111,111	1,111,111,111	111,111,111,111,111	111,111,111,111,111,111,111

25 Tavola "Taffy" Tafelberg's programmable calculator was named not for Fred and Wilma Flintstone's daughter, at least not entirely, but for the Latin word, *calculi*. meaning "pebbles", from which we get the words calculus, calculation and calculator. Her own name refers both to the high-IQ organization Mensa and the constellation Mensa, called Monte Tavola in Italian and Tafelberg in German.

26 To estimate the height of a stack of twos, or any other number, compared to a stack of tens logarithm approximations are useful: log 2 ≈ 0.3, log 3 ≈ 0.5, log 4 ≈ 0.6, log 5 ≈ 0.7, log 6 ≈ 0.8, log 7 ≈ 0.85, log 8 ≈ 9, log 8 ≈ 0.95. Tenplex in binary takes [10/log(2) + 1] = 33 bits. A little googol, of course, two-to-the-hundredth, takes exactly 100 bits.

evils

unholy evil	3	5	12	15	17	23	24	27	33	34
untrimmed	1	4	7	10	21	23	24	26	28	29
broken-up	0	1	5	31	284	2,871	34,736	523,911	8,941,223	161,465,925
untrimmed	1	2	6	42	395	3,982	45,847	6,341,022	91,052,334	2,725,761,036
broken-down	1	1	6	37	339	3,427	41,463	625,372	10,672,787	96,680,455
broken	0	1	5	6	31	37	284	339	2,871	3,427
untrimmed	1	2	6	7	42	48	395	4,410	3,982	4,538

hypersum	0	10	1,020	102,030	10,203,040	1,020,304,050	102,030,405,060
holiness	1	1	2	3	4	5	7
untrimmed	1	21	2,131	213,141	21,314,151	2,131,415,161	213,141,516,171

evil twins

unholy	5	17	23	24	33	34	45	53	54
broken-up	5	31	160	991	17,007	307,117	7,080,698	170,243,869	4,944,152,899
dusub	1	4	22	134	2,302	41,564	958,268	23,040,002	669,118,333
broken-down	1	6	31	192	3,295	59,502	1,371,841	32,983,686	957,898,735
broken	1	5	6	31	160	192	991	3,295	17,007

hypersum	5	56	5,617	561,718	56,171,823	5,617,182,324	561,718,232,429

dusub	1	8	760	76,020	7,602,030	760,202,961	76,020,296,085

odious

broken-up	1	3	13	94	765	8,509	111,382	1,567,857	25,197,094
broken-down	1	2	9	65	529	5,884	77,021	1,084,178	17,423,869
broken	1	2	3	9	13	65	94	529	765
sum	1	3	7	14	22	33	46	60	76
super	1	2	8	56	448	4,928	64,064	896,896	14,350,336

hypersum	1	12	124	1,247	12,478	1,247,811	124,781,113	12,478,111,314

27 Edwin F. Bechenbach proved that all positive integers are interesting in "Interesting Integers" (*American Mathematical Monthly* vol. 52, p. 211)

28 Besides three, the Indo-European root *trei-* gives us the English words: attest, contest, detest, obtest, protest, riding, sesterce, sitar, tercel, tercet, tern, tertian, tertiary, testament, testify, testimony, third, thirteen, thirty, tierce, trammel, trecento, trephine, trey, triad, tribe, trice, trichotomy, triclinium, tricrotic, tridactyl, trierarch, triglyph, trillium, Trimurti, trine, Trinity, trio, triple, triplex, triploblastic, tritanopia, tritium, tritone, triumvir, triskadekaphobia, trocar, troika ("Indo-European and the Indo-Europeans" by Calvert Watkins)

29 This odd couple is actually the odd numbers five and seven, masquerading as even numbers. The original Odd Couple was Oscar Madison and Felix Ungar popularized by Neil Simon, who later adapted his play in 1985 to female roommates, Olive Madison and Florence Ungar, in "The Female Odd Couple".

The sociability of the digits might be charted by comparing their name's initial and final letters. Zero and one are so linked, so is eight with one, two, three, five and nine. Three and eight are doubly linked and nine also linked to seven. Eight would be the most sociable digit and four and six unsociable.

0	→	**1**	←	↑
9	→	↓		↑
↑	←	**8**	→	**2**
7	↓	↑	←	**5**
	→	**3**		

The most sociable multi-digit numbers would be those alternating three and eight or eight and three and the most unsociable those alternating four and six or six and four.

most sociable	3	8	38	83	383	838	3,838	8,383	38,383	83,838	383,838
unsociable	4	6	46	64	464	646	4,646	6,464	46,464	64,646	464,646

30 This is a mnemonic handy for testing in bases 3 or higher, 111122112_n = g(2, 8, n) + g(2, 7, n) + g(2, 6, n) + g(2, 5, n) + 2g(2, 4, n) + 2g(2, 3, n) + g(2, 2, n) + n + 2. For base three, 111122112_3 = 9941. For bases 18 and higher, see **47**.

31 Ternarish is nicknamed "tate" for the redundant "ternarish and ternarish evens" or "little man" from "Little Man Tate" by Frank Scott.

Bushmen are said to use a number system which gets high like *xa, t'oa, 'quo, t'oa-t'oa* (two-two), *t'oa-t'oa-'ta* (two-two-plus(-one)), *t'oa-t'oa-t'oa* (two-two-two), *'quo-'quo-'ta* (three-three-plus(-one)), *'quo-'quo-'quo* (three-three-three), representing even and odd numbers differently.

The Bakairi use an even simpler base three system without a three, getting higher like *tokale, ahage, tokale-ahage* (one-two), *tokale-tokale* (two-two), *tokale-tokale-ahage* (two-two-one), *tokale-tokale-tokale* (two-two-two). (*Numbers Through the Ages* by G. Flegg) Unlike Bushmanish numbers, the Bakairish numbers have positive gnomons and keep getting ever higher, even if suffering from irregularity.

Bushmanish	1	2	3	22	221	222	331	2,222
mates	6	7	8	77	668	777	778	7,777
sum	1	3	6	25	246	468	799	3,021
super	1	2	6	132	29,172	6,476,184	2,143,616,904	4,763,116,760,688
broken-up	1	3	10	223	49,293	10,943,269	3,622,271,332	1,202,605,025,493
broken-down	1	2	7	156	34,483	7,655,382	2,533,965,925	841,284,342,482
broken	1	2	3	7	10	156	223	34,483

hypersum	1	12	123	12,322	123,22,221	12,322,221,222	12,322,221,222,331

hypersuper	1	12	333,333,333,333	> elevenduplex

Bakairish	1	2	12	22	221	222	2,221
mates	7	8	77	87	777	778	7,778
sum	1	3	15	25	246	468	2,689
super	1	2	24	528	116,688	25,904,736	57,534,418,656
broken-up	1	3	37	817	180,594	40,092,685	8,9046,033,979
broken-down	1	2	25	552	122,017	27,088,326	60,163,294,063
broken	1	2	3	25	37	552	817

hypersum	1	12	1,212	121,222	121,222,221	121,222,221,222	1,212,222,212,222,221

hypersuper	1	2	1,212	> twenty-four-hundredplex

Gideon Frieder and his colleagues at the State University of New York at Buffalo tried to revive ternary by designing a base-3 computer they called TERNAC in 1973.

32 In Old Klingon *wa'* meant one, *cha'* two and *wej* three, all in a zero-less base 3, and so get high like 1, 2, 3, 11, 12, 13. At http://www.tc.umn.edu/~joela/cgi-bin/kalk.cgi is an on-line Klingon kalculator. In New Klingon additional words, undoubtedly borrowed from higher tech

Karsid, are used for zero (*pagh*), for *wa'wa* or 11_3 (*loS*), for *wa'cha* or 12_3 (*vagh*), for *wa'wej* or 13_3 (*jav*), for *cha'wa* or 21_3 (*Soch*), for *cha'cha'* or 22_3 (*chorgh*), for *cha'wej* or 23_3 (*Hut*), for *wejwa* or 31_3 (*maH*), for *wejwa'wejwa* or 3131_3 (*vatlh*, hundred). From this can be extrapolated the rarely, if ever, used pre-Karsid numbers *wa'wejcha'wejcha'wej* for 132323_3 (*SaD*, thousand), *wa'wa'wa'cha'wejwejwejwa* for 11123331_3 (*netlh*, myriad), for 11231311131_3 (*blp*, fiveplex) and even for the preposterously high number name, *wa'cha'wa'cha'wa'wejwejwa'wejwa'cha'wejwa'* or 1212133131231_3 (*uy*, million).

Klingonish	1	2	3	11	12	13	21	22	23	31
sum	1	3	6	14	26	39	60	82	105	136
mates	6	7	8	66	67	68	76	77	78	86
super	1	2	6	66	792	10,296	216,216	4,756,752	109,405,295	3,391,564,176
broken-up	1	3	10	113	1,366	17,871	376,657	8,304,325	191,376,132	5,940,964,417
broken-down	1	2	7	79	955	12,494	263,329	5,805,732	133,795,165	4,153,455,847
broken	1	2	3	7	10	79	113	955	1,366	12,494

hypersum	1	12	123	12, 311	1,231,112	123,111,213	12,311,121,321

hypersuper	1	2	33	> sixty-fiveplex

The deep-sea ichthyoids of Fithia use a base-3 system communicated via an electroshock language (langmaker). The Aranda get high with just the four "numbers": *nyente* (one), *therre* (two), *urrpetye* (few), and *atninghe* (many). (*What Counts* by Brian Butterworth), Pirahã even less with one (small) *hói* and two (slightly larger than small) *hoí*. (*Biting the Wax Tadpole* by Elizabeth Little)

33 Three-tetrated-to-the-third, g(3, 3, 3) = g(2, g(2, 3, 3), 3) = g(2, 27, 3) > twelveplex takes 27 base-3 bits or "trits". To-the-second-ten takes only [10/log(3) + 1] = 21 trits.

34 Besides the word four the Indo-European root k^wetwer- gives us the English words and prefixes: cadre, cahier, cater-cornered, czardas, diatessarion, farthing, forty, fourteen, fourth, quadrate, quadrant, quadri-, quadrille, quadroon, quarantine, quarrel, quarry, quart, quartan, quarter, quarto, quaternary, quaternion, quire, quatrain, quartrocentro, squad, square, trapezium, tessera, tetra-, tetrad, and trocar ("Indo-European and the Indo-Europeans" by Calvert Watkins)

35 "A Bushel and a Peck" was written by Frank Loesser and published in 1950. On *Cash Box Magazine*'s Best-Selling Record charts, where all renditions of this song were combined, it reached #5.

36 The Aramæan numerals for 7 through 15, as upside-down Roman numerals would be 13, 20, 21, 22, 23, 30, 31, 32, 33, just as would be expected in base 4. The pattern however does not continue. IIIXXX is fifteen which is not half of XXXIII. In base 4 tenplex would be [10/log(4) + 1] = 17 symbols. Carl D. Eiseman in his article referenced above also mentioned the rare usage of IIX for 8, though not IIIX for 7.

37 Clair and Lana's place off campus was nicknamed the Delta Schwa sorority after they were joined by Anne, "Taffy" and Kamika. Δə would be pronounced "duh", an interjection used to indicate ironically that a question with an obvious answer has been asked whether or not the hearer knows the answer or not.

38 The total length of DNA present in all the cells of one adult human is calculated to be thirteenplex meters. (http://hypertextbook.com/facts/1998/StevenChen.shtml) or 10 AU (Astronomical Units), further than the distance from the sun to Saturn. If amino acids were randomly assigned to triplet codons, then there would be more than eighty-fourplex possible genetic codes. (*Life from an RNA World: The Ancestor Within*. Cambridge by M. Yarus: Harvard University Press. p. 163)

The rest of the meeting was filled with riddles based on the clucking and rowfing base 4 systems in *Mathematical Games* by Marie Berrondo. In the CLUCK system, C = 0, K = 1, L = 2, U = 3, while in the ROWF system, O = 0, F = 1, W = 2, R = 3.

►"Which is higher LUCK or ROOF?"

ROOF is higher than LUCK since ROOF is 3001_4 and LUCK is only 2301_4.

►"LULU or WOOF?"

LULU is higher than WOOF since LULU is 2323_4 and WOOF only 2001_4.

►"How high is CLUCKCLUCK?"

CLUCKCLUCK = 0230102301_4 = 181,425.

►"How high is CLOCKWORK? [This requires using both clucking and rowfing simultaneously.]

CLOCKWORK = 02001231_4 = 16,493.

►"How high is 111122112?"

111122112_4 = 87,686.

<u>**quadrarish**</u>	1	2	3	10	11	12	13	20	21	22
mates	6	7	8	66	67	68	69	76	77	78
broken-up	1	3	10	103	1,143	13,819	180,790	3,629,619	76,402,789	1,684,490,977
broken-down	1	2	7	72	799	9,660	126,379	2,537,240	53,408,419	1,177,522,458
broken	1	2	3	7	10	72	103	799	1,143	9,660
gnomon	1	1	1	7	1	1	1	7	1	1
sum	1	3	5	13	21	23	25	33	41	43
super	1	2	6	60	660	7,920	102,960	2,059,200	43,243,200	95,1350,400

It's a curiosity, noted in Ripley's "Believe It or Not" that the Name for God, the Tetragrammaton, has just four letters in so many different languages: Adad [Syrian], Addi [Turkish], Adon [Hebrew], Alla [Arabic], Amun [Egyptian], Chur [Etrurian], Codd [Swedish], Deus [Latin], Dich [Irish], Dieu [French], Dios [Spanish], Doga [Croatian], Esgi [E. India], Eher Tyrrhenian], Gott [German], Idga [Tartarian], Lian [Peruvian], Odin [Scandinavian], Oese [Margarian], Rogt [Dalmatian], Syra [Persian], YHWH [Hebrew], Zain [Japanese], Zenc [Wallachian], Zenl [E. Indian], Zeus [Greek], Zeut [Eqyptian]. It's not so unbelievable however when one considers the much higher number of languages and gods.

39 Besides eight the Indo-European root oktō(u)- gives us the English words and prefix: atto-, eighteen, eighty, octad, octans, octant, octave, octavo, octet, octo-, October, octogenarian, octonary, octodecimo, and octopus. ("Indo-European and the Indo-Europeans" by Calvert Watkins)

40 Walt Disney popularized tetradactylism for cartoon characters, Toons, worldwide ever since his alter ego Mickey Mouse starred in "Steamboat Willy" (1928). When Bob the Builder (1999) moved to Japan, he anthropomorphized, growing fifth fingers. Bipedal Erinaceiods would not likely have five fingers per hand like Sonic the Hedghog (Heijjihoggu).

41 The saying "nineteen bites to a bilberry", a smaller relative to the blueberry, inspired the bylberry sequence, starting with nineteen bytes, specifically eight-to-the-twentieth minus one, 1,152,921,504,606,846,975 with a sub depth of forty-four. The similar saying "two bites to a cherry", inspired the much simpler two-bit 'cherri' sequence.

two-bit	11	101	110	1001	1010	1100	10001	10010	100100	101000	110000
cherri	3	5	6	9	10	12	17	18	20	40	48

Other geeky humor equates Halloween with Christmas since 31 in octal (OCT, October) equals 25 in decimal (DEC, December); $31_8 = 25_{10}$.

42 The eight-legged arachnoid Kh!lict were described in *Windows on a Lost World* by V. E. Mitchell and the Klethians in "Poles Apart" by G. David Nordley. Tenplex would be 12 bytes and a googol 111 bytes. $111122112_8 = 19{,}178{,}506$.

octarish	1	2	3	4	5	6	7	10	11	12	13
mates	2	3	4	5	6	7	8	22	23	24	25
broken-up	1	3	10	43	225	1,393	9,976	1,011,533	1,122,659	13,573,061	177,572,452
broken-down	1	2	7	30	157	972	6,961	70,582	783,363	9,470,938	123,905,557
broken	1	2	7	10	30	43	157	225	1,393	6,961	9,976
sum	1	3	5	7	9	11	13	17	21	23	25
super	1	2	6	24	120	720	5,040	50,400	554,400	6,652,800	86,486,400

The Chamalians however in "Seed of Reason" by Daniel Hatch use a hybrid 8/80-base system, rather than a purely base-eight 8/64 system. The numbers between $77_{8/80}$ and $100_{8/80}$ would necessarily include eights and nines in the first and second places. 777 in this base would be 623, and 999 would be 781 and so the fourth place must represent something between, most likely 640, eight times eighty. Our 639 would therefore be 799 in Chamalian.

...	**62**	**63**	**64**	**65**	**66**	**...**	**78**	**79**	**80**	**81**	**82**
Chamalianish	76	77	80	81	82	...	95	96	100	101	102

43 Tom Lehrer sang about base 8 in his song "The New Math" in 1965, specifically $342_8 - 173_8 = 147_8$, describing it as "so easy only a child can do it." Base 8 however was not actually new since E. M. Tingley advocated using base 8 in *School Science and Mathematics:* "Calculate by Eights, Not by Tens" dated Jan. 37, 3616. [Jan. 31, 1934 in decimal.]

44 The Yuki of N. California are said to have counted to eight between their fingers, perhaps as did the Indo-Europeans whose word for nine, newṇ-, seems to be a contraction of newoino- meaning "new one".

The Universal Product code avoids the orientation problem of Yijing by using thick (11) and thin (1) bars and thick (00) or thin (0) spaces in odd-parity and even-parity pairs.

0	0001101	1110010								
1	0011001	1100110		**4**	0100011	1011100		**7**	0111011	1000100
2	0010011	1101100		**5**	0110001	1001110		**8**	0110111	1001000
3	0111101	1000010		**6**	0101111	1010000		**9**	0001011	1110100

45 190_{10} = 11(16) + 14 = BE_{16} Hexadecimal can handily be counted on one hand to four places by touching an opposable thumb to the other four fingers.

Using the normal decimal representations for ten, eleven, twelve, thirteen fourteen and fifteen, and a leading zero for the smaller placekeepers, rather than the normal hexadecimal ones, A, B, C, D, E and F, we get double-digit hexadecimalish numbers.

...	**15**	**16**	**17**
hexadecimalish	15	100	101
broken-up	2,723,022,564,763	27,232,256,476,314	> 2 sixteenplex
broken-down	1,946,194,117,057	194,748,546,392,220	19,671,549,379,731,300
broken	9,976	56,660	81,201

46 The avoirdupois system is a rather ambiguous system in that a "pound" [lb, from Lat. *libra* "weight"] came to mean both a unit of force or mass as well as the British monetary unit, £, so an ounce can mean troy ounce or 1/12 of a pound serling as well as a 1/16 of a pound and dram can mean the Armenian currency, 100 lumas as well as 1/256 of a pound.

Base seventeen and higher would also include G for sixteen and so all the pianonums, numbers with the placeholders A through G or in double-digits one-zero through one-six, with numbers like ABE, ACE, ADAGE, AGE, BABE, BADGE, BEAD, BED, BEEF, CAB, CABBAGE, CAGE, DAB, DAD, DEAD, DEAF, DEB, DEED, ED, EDGE, FACE, GAG, GAFF.

pianonums	16	1010	1011	1012	1013	1014	1015	1016	1110	1111	1112	1113	1114	1115
base 17	16	180	181	182	183	184	185	186	197	198	199	200	201	202
base 18	16	190	191	192	193	194	195	196	208	209	210	211	212	213
base 19	16	200	201	202	203	204	205	206	219	220	221	222	223	224

The most common system for representing information as numbers is a further condensation of binary, the American Standard Code for Information Interchange or ASCII with a seven-bit binary sting for each symbol and its extended version with eight. A space is encoded as as 32 and hyphen as 45, the digits, zero through nine as 48 through 57, the lower case letters as 97 through 122. The upper case or capital letters are 65 through 90. They are however joined together hypermathematically by concatenation, rather than mathematically, like two added to two and getting twenty-two. The base-256 version also called ISO Latin-1 includes nearly every other commonly used symbol from the euro, €, to the umlauted y, ÿ.

Secure encoding on computers is done with one-way functions, $(p^z \pmod{p'})^{z'} = (p^{z'} \pmod{p'})^z$, in ASCII and very large primes. Using Alice and Bob as Simon Singh does in *The Code Book*, though with a simpler example with p = 7, p' = 11 and wordnums, say Alice wants to respond Y for "Yes!" to Bob's message WYMM ("Will you marry me?"). She encodes Y into Z or 26, calculates $7^{26} \pmod{11}$ = 9,387,480,337,647,754,305,649 (mod 11) = 4. Since in Bob's Y encodes to L or 12 and $7^{12} \pmod{11}$ = 13,841,287,201 = 5. Since $(7^5 \pmod{11})^4 = (7^4 \pmod{11})^5$, he knows her answer is as hoped without either needing to know the other's scrambled alphabet.

47 $COFFEE_{16}$ = 1215151414_{16} = 12,648,430. 111122112_{16} = 4,581,368,082. Extrapolating from the hexadecimals however gives G through Z, excluding the letters between H and J and between N and P, which could be too easily mistaken for one and zero, from base 17 through 34.

COFFEE	**base 17**	**base 18**	**base 19**	**base 20**
base 10	17,116,566	22,767,422	29,821,768	38,526,294
	base 21	**base 22**	**base 23**	**base 24**
base 10	49,155,050	62,010,886	77,426,892	95,767,838
	base 25	**base 26**	**base 27**	**base 28**
base 10	117,431,614	142,850,670	172,493,456	206,865,862
	base 29	**base 30**	**base 31**	**base 32**
base 10	246,512,658	246,512,658	344,011,540	403,160,526
	base 33	**base 34**		
base 10	470,180,582	545,832,478		

Some Anglophiles prefer representing the placeholders higher than ten based on the English number names, t for ten, e for eleven, T for twelve, theta, θ, for thirteen, f for fourteen, F for fifteen. Interestingly F equals fifteen in both system. This one could be extended further for higher bases 17 through 20 with s for sixteen, S for seventeen, E for eighteen, and n for nineteen. They would have to make T0FFee, 1215151111, instead of COFFEE, 1215151414.

T0FFee	**base 11**	**base 12**	**base 13**	**base 14**	**base 15**
base 10	1,954,524	3,014,207	4491160	4,491,160	9,166,724
	base 16	**base 17**	**base 18**	**base 19**	**base 20**
base 10	9,166,724	17,116,512	22,767,365	29,821,708	38,526,231

A simpler, faster alternative is the Holmesian number, two z cubed plus two z squared plus z plus eleven.

221B	**base 11**	**base 12**	**base 13**	**base 14**	**base 15**
base 10	2,926	3,767	4,767	5,905	7,226

	base 16	base 17	base 18	base 19	base 20
base 10	8,731	10,432	12,341	14,470	16,831

48 According to *The Name of the Number* by Michael A. B. Deakin, Carolyn Glascodine, and David Leigh-Lancaster, based on Martin Gardner's reference to *Special Topics in Theoretical Arithmetic* (1936), Bowden symbolized 12_{16}, 13_{16}, 14_{16} and 15_{16} with mirror images of two, three, four and five respectively. He also gave the last three hexadigits and the first two-hexadigit numbers, usually called D, E, F and one-zero, the new names: thrun, fron, feen, and wunty respectively, so BE_{16} would be called twelvety-feen.
The Name of the Number cleverly substitutes the Cyrillic letters ю and H, which resemble 10 and 11, for 10_{16} and 11_{16}, so Bowden's thrunty-thrun, 13(16) + 13 = 221, would be symbolized as two mirrored threes, but ought not to be confused with Well's elel also symbolized by εε, but which is only 11(12) + 11 = 143. In Bowdenese $C0FFEE_{16}$ = twelve-oh-feen-feen-fron-fron.

49 Besides five the Indo-European root penkʷe- gives us the English words and prefixes: cinquain, cinque, cinquecento, femto-, finger, fifteen, fifth, fifty, fist, foist, keno, penta-, pentad, Pentacost, pentagon, punch, quinate, quincunx, quindecennial, quinquagenarian, Quinquagesima, quinque-, quint, quintain, quintessence, quintet, quintile, quintillion, quintuple. ("Indo-European and the Indo-Europeans" by Calvert Watkins)

50 11 = ‡+ = 21_5 = 2(5) + 1. Tenplex would be [10/log(5) + 1] = 15 quintary bits and a googol [100/log(5) + 1] = 144. 111122112_5 = 489,007. A base-5 monetary system could use pennies, nickels and quarters, great quarters, garquarters, great garquarters, ... or perhaps not.

<u>quintarish</u>	1	2	3	4	10	11	12	13	14	20
mates	5	6	7	8	55	56	57	58	59	60
broken-up	1	3	10	43	440	4,883	59,036	772,351	10,871,950	218,211,351
broken-down	1	2	7	30	307	3,407	41,191	538,890	7,585,651	15,221,910
broken	1	2	3	7	10	30	43	307	440	3,407
sum	1	3	5	7	14	21	23	25	27	34
super	1	2	6	24	240	2,640	31,680	411,840	5,765,760	115,315,200

hypersum	1	12	123	1,234	123,410	12,341,011	1,234,101,112	123410,111,213

51 Besides six the Indo-European root s(w)eks- gives us the English words and prefixes: hexa-, hexad, semester, senary, sex-, sextant, sestet, sestina, sext, sextant, sextile, sextodecimo ("Indo-European and the Indo-Europeans" by Calvert Watkins)

52 With the addition of the amino acid pair proline (P) and lysine (K), this would have greatly increased the possible genetic codes from eighty-fourplex to hundred-twenty-sixplex. ("Enzymatic incorporation of a new base pair into DNA and RNA extends the genetic alphabet" by Joseph A. Piccirilli, etal., *Nature* 6253:33-37, 1990). A ribosome that reads codons made of four, not three, letters has been. ("Expanding the genetic code of an animal" by S. Greiss and J. W. Chin, *Journal of the American Chemical Society* 133: 14196-14199, 2011).

53 As described in *Defenders #13* the Ul'lulans from U'lula in the Ul system are giant eel-like hexapoids who use base six, as do the trochophores from Zyph, Betelguese system in *The Mathematics of OZ: Mental Gymnastics from Beyond the Edge* by Clifford Pickover.

54 J. R. R. Tolkien wrote that the Elves preferred to think in sixes and twelves. One through eleven in Quenya is given as mir, atta, nelde, lemin, ende, otso, ?, olme, lempe, and minqe. Ole and olme are six apart. Olwen meant three dozen and otwen seven dozen, so the suffix -wen indicated a dozen. Interpolating we get mirwen, atwen, nelwen, lemwen, enwen, olmwen, lempwen, and minwen. Since twenty-three is given as leminkainen, kainen must have meant sesquidozen or eighteen. A gross was tuska, but could be reduced to a mere hundred if followed by the adjective lempea ("decimal").

55 The Ndom of Papua, New Guinea, where "The Lord of the Rings" and "The Hobbit" were filmed, and near where hobbit (Homo floresiensis) fossils were found, are also said to use a base six system. 111122112_6 = 2,017,016.

sexarish	1	2	3	4	5	10	11	12	13	14	15
mates	4	5	6	7	8	44	45	46	47	48	49
broken-up	1	3	10	43	225	2,293	25,448	307,669	4,025,145	56,659,699	853,920,630
broken-down	1	2	7	30	157	1,600	17,757	214,684	2,808,649	39,535,770	595,845,199
broken	1	2	3	7	10	30	43	157	225	1,600	2,293
sum	1	3	5	9	14	24	35	67	80	94	109
super	1	2	6	24	120	1200	13,200	158,400	2,059,200	28,828,800	432,432,000

hypersum	1	12	123	1,234	12,345	1,234,510	123,451,011	12,345,101,112

56 Besides seven the Indo-European root septm̥- gives us the words and prefix: hepta-, heptad, heptomad, September, septennial, septentrion, septet, Septuagint, septuple, seventeen, seventy. ("Indo-European and the Indo-Europeans" by Calvert Watkins)

57 The Brzöibians use base 7. ("Brzöibian Magic Squares" by Victor Serebriakoff) *Rikchik Culture and Language* by Denis Moskowitz notes that the Rikchiks of alpha Centauri A II's base 7 system is simpler, only getting as high as 48. They are heptapodal cephalopodans, who call 49 "many" and 343 "very many".

The Tau of *Warhammer 40,000* also use base 7, wherein the year 40,000 AD would be 224,422. Tenplex would require twelve symbols/syllables and a googol one hundred nineteen. 111122112_7 = 6,728,297.

septarish	1	2	3	4	5	6	10	11	12	13	14
mates	3	4	5	6	7	8	33	34	35	36	37
broken-up	1	3	10	43	225	1,393	14,155	157,098	1,899,331	2,484,801	349,776,945
broken-down	1	2	7	30	157	972	9,877	109,619	1,325,305	17,338,584	244,065,481
broken	1	2	3	7	10	30	43	157	225	972	1,393

gnomon		1	1	1	1	1	4	1	1	1	1
sum	1	3	5	7	9	11	16	21	23	25	27
super	1	2	6	24	120	720	7,200	79,200	950,400	12,355,200	172,972,800

hypersum	1	12	123	1,234	12,345	123,456	12,345,610	1,234,561,011

One of the most interesting "magic squares" is this nested one whose rows and columns not only add to the same number, but each of the rows and columns of the smaller magic squares within do too, from the 13x13 to the 11x11 to the 9x9 to the 7x7 to the 5x5 all the way down to the 3x3: 8,101 + 955 + 4,537 = 967 + 4,531 + 8,095 = 4,525 + 8,107 + 961 = 8,101 + 967 + 4,525 = 955 + 4,531 + 8,107 = 4,537 + 8,095 + 961 = 13,593.

247	8017	187	8221	421	8371	157	8227	8755	787	85	7981	7447
9061	7255	2347	1951	5917	1237	3541	7915	7807	7411	2095	2365	1
925	7261	3187	6655	2725	2551	6667	3001	6505	3061	6427	1801	8137
9001	6781	6331	5461	3691	3817	5671	3607	3925	5365	7621	2215	9
817	6847	1441	3697	4621	4087	4735	5167	4045	5545	2731	2281	8297
8515	1387	5857	3757	3751	8101	955	4537	5311	5305	3205	7675	547
1105	1777	5965	5641	4321	967	4531	8095	4741	3421	3097	7285	7957
9403	7855	2971	3877	4945	4525	8107	961	4117	5185	6091	1207	565
625	2137	7177	6673	5923	5881	4327	4801	5347	4201	3697	8737	9343
8737	1345	6121	3517	5371	5245	3391	5455	5137	3601	2941	7717	325
877	1405	2635	2407	6337	6511	2395	6061	2557	6001	5875	7657	8185
8881	6697	6715	7111	3145	7825	5521	1147	1255	1651	6967	1807	181
1615	1045	8875	841	8641	691	8905	835	307	8275	8977	1081	8815

58 In "Unicorn Jelly" Jennifer Diane Reltz explained the Talcryl's zero-less base-7 number system in which niekyi means 172_7 or 1(49) + 7(7) + 2 = 100. Yutoyuek means 2626_7 = 1000. Ninisisisiniyu means 1133311_7 = million. In Talcryl there would be no limit to monoconsonantal numbers, unlike English's lone "one" and "nine". The gnomon shows that it is just the sum of the powers of seven minus one, $\Sigma 7^z - 1$.

Talcryl	ni	nini	ninini	ninini	nininini	niniininini	niniininininini	nininininininini
base 7	0	10	110	1110	11110	111110	1111110	111111110
untrimmed	1	21	221	2221	22221	222221	2222221	22222221
septarish	0	7	56	399	2,800	19,607	137,256	969,799
gnomon		7	49	343	2,401	16,807	117,649	823,543
ratio			7	7	7	7	7	7

hypersum	0	7	756	756,399	7,563,992,800	756,399,280,019,607
holiness	1	0	1	3	7	10
untrimmed	1	8	867	86,741,010	867,410,103,911	> seventeenplex

This polymonosyllabic yet infinite series is perhaps the origin of the Knights Who Say "Ni" ("Monty Python and the Holy Grail" by Monty Python (Graham Chapman, John Cleese, Terry Gillam, Eric Idle, Terry Jones and Michael Palin), "Spamalot" by Eric Idle), in contrast to "nīþ", a word shouted by Vikings as a gross insult and term of abuse or the Japanese "ní", meaning two, load, burden, package, luggage and freight. This last ambiguously could be in any base three or higher.

ní...ní	2	22	222	2222	22222	222222	2222222	22222222	222222222
base 3 to 10	2	8	26	80	242	728	2,186	6,560	19,682
base 4 to 10	2	10	42	170	682	2,730	10,922	43,690	174,762
base 5 to 10	2	12	62	312	1,562	7,812	39,062	195,312	976,562

There are also monosyllabic numbers in English, but again only a few, one, two, three, four, five, six, eight, nine, ten, and twelve, as well as the even rarer one-voweled numbers, seven, seventeen and googol.

The Forgiving sequence from Matthew 18:23 is similar, beginning with seven and then seventy times seven or four hundred ninety, both 7^z10^{z-1}.

forgiving	7	490	34,300	2,401,000	168,070,000	11,764,900,000	823,543,000,000
dusub	1	66	4,642	324,940	22,745,801	1,592,206,074	111,454,425,163
1st ratio		70	700	7,000	70,000	700,000	7,000,000
2nd ratio			10	10	10	10	10
mates	2	509	65,699	7,598,999	831,929,999	88,235,099,999	176,456,999,999

hypersum	7	7,490	749,034,300	7,490,343,002,401,000
dusub	1	1,014	101,370,769	10,137,707,691,769,320

The St. Ives Pilgrim asked about a husband with seven wives with forty-nine sacks, three hundred forty-three cats and twenty-four hundred one kittens, who may or may not also have been going to St. Ives. (*A Book of Nursery Rhymes* by Sabine Baring-Gould) The cat-lovers would likely have been traveling more slowly and the cats and kittens could be dead, alive or indeterminate like Schrödinger's cat.

59 Llewellyn "Loophole" Pohl is a German-Welch American, pre-law major, who is also a H. P. Lovecraft fan and so nicknamed "Cthu Lou".

60 Besides nine the Indo-European root newn̥- gives us the English words and prefixes: ennead, nineteen, ninety, ninth, November, novena, nona-, nonagenarian, nonagon, nonanonic acid, nones, and noon. ("Indo-European and the Indo-Europeans" by Calvert Watkins)

61 The Vipswarznee, described in *Time Warp #2* as like 4.5-meter rotten tomatoes with seven green tentacles and two long arms, use base nine.

nonarish	1	2	3	4	5	6	7	8	10	11	12	13

gnomon		1	1	1	1	1	1	1	2	1	1	1
sum	1	3	5	7	9	11	13	15	18	21	23	25
super	1	2	6	24	120	720	5,040	40,320	403,200	4,435,200	53,222,400	691,891,200

By backformation we have arish, the numbers not nonarish, those with nines.

arish	9	19	29	39	49	59	69
broken-up	9	172	4,989	194,743	9,547,396	563,491,107	38,890,433,779
dusub	1	23	675	26,356	1,292,100	76,260,229	5,263,247,871
broken-down	1	19	552	21,547	1,056,355	62,346,492	4,302,964,303
broken	1	9	19	172	552	4,989	21,547
gnomon	9	10	10	10	10	10	10

hypersum	9	919	91,929	9,192,939	919,293,949	91,929,394,959
dusub	1	124	12,441	1,244,129	124,412,907	12,441,290,705

62 The Telefol and Okasapmin use a zeroless base 27 for counting. (*Counting on Your Body in Papua New Guinea* by James V. Rauff) The nymber "one" would not mean the usual 24(729) + 23(27) + 14 = 18,131, but only the wordnum, 15(729) + 14(27) + 5 = 11,318. The nymber "eight" would qualify as what Joyce called a half-pryme, since 8 is not prime, while its wordnum, $EIGHT_{27}$ = 2,839,691 is.

The Odd Squad, since being made public by PBS, assigns badge numbers by the weight of the agents' names. AJAS did the same. Alberto "Al" Fuori was 13, Calvin "Cal" Pantof 16, though he was usually called by his jersey's "11", Ellen "El" Emeno 17, Daniel "Dan" Svetlost 19, Catherine "Cat" Katz 24, Patrick "Pat" Shaw 27, Lana Coperta 28, Ann Tabiat 29, Samson "Sam" Znowu 33, Kathleen "Kay" Emeno 36, Filip "Fil" Pinno 37, Maxine "Max" Miniver 38, though she was sometimes called 86, after Maxwell Smart of "Get Smart", Arthur "Art" Koenig 39, Llewellyn "Lew" Pohl 40, Phillip "Pip" Miniver 41, Clair Peau 43, Razilee "Raz" Purdue 45, Br. Kenneth "Bro. Ken" Kendrick 50, Thomas "Tom" Ficka 52, Robert "Bobby" Katz 56, Tavola "Taffy" Tafelberg 58, Robin Robb 60, Kamika Schmidt 64 and Victoria Emeno 97.

63 Higher than the Number of the Beast the sequence continues, but since different numbers can have the same value, eventually it becomes cyclical.

MXXXXVII	MMDCCCCLX	DCCXXVI	MDCXX	MCCXXXXVII	MMCCCCLXVI
1,047	2,860	726	1,620	1,247	2,466
MCXXXII	**MDCCCLXIII**	**DCCIII**	**XL**	**DCXXX**	
1,132	1,863	703	40	630	

After eighteen steps an eleven-step cycle begins.

MDCCCVII	CDLXXIII	MCCLXVII	MXCVI	MLIII	C	III	XXX	MDXXX	LIII	LX
1,807	473	1,267	1,096	1,053	100	3	30	1,800	53	60

Starting similar series with different Roman numerals also cycle unpredictably, except for not unexpected twins like IV and VI and IX and XI, rather like the (in)famous HOTPO (half or triple plus one) function of the Collatz (aka Kakutani's, Syracuse oe Ulam's) problem.

numeral	**I**	**II**	**III**	**IV**	**V**	**VI**	**VII**	**VIII**	**IX**	**X**	**XI**	**XII**	**XIII**
start	18	12	0	12	16	12	4	1	11	21	11	4	1
period	11	6	11	7	12	7	7	12	7	11	7	7	11

The repeated applications of the HOTPO function all seem to eventually reach one, via a "hailstone sequence", but in a seemingly unpredictable number of steps.

	1	**2**	**3**	**4**	**5**	**6**	**7**	**8**	**9**	**10**	**11**	**12**	**13**	**14**	**15**	**16**	**17**	**18**	**19**	**20**	**21**	**22**	**23**	**24**	**25**	**26**	**27**
steps	0	1	7	2	5	8	16	3	19	6	14	9	9	17	17	4	12	20	7	7	15	15	10	23	10	111	18

64 The Scrabble values are: A = 1, B = C = 3, D = 2, E = 1, F = 4, G = 2, H = 4, I = 1, J = 8, K = 5, L = 1, M = 3, N = O = 1, P = 3, Q = 10, R = S = T = U =1, V = W = 4, X = 8, Y = 4, Z = 10, Repeated evaluations all seem to repeat rather quickly with a cycle of 4, except odd twelve which just repeats. Eventually the sequences get higher as the arithmonym lengths do.

Scrabble values	**one**	**two**	**three**	**four**	**five**	**six**	**seven**	**eight**	**nine**	**ten**	**eleven**	**twelve**	**thirteen**
1st	3	6	8	7	10	10	8	9	4	3	9	12	11
2nd	8	10	9	8	3	3	9	4	7	8	4	12	9
3rd	9	3	4	9	8	8	4	7	8	9	7	12	4
4th	4	8	7	4	9	9	7	8	9	4	8	12	7
5th	7	9	8	7	4	4	8	9	4	7	9	12	8
repeats at step	2	4	1	0	3	3	1	0	0	2	4	0	5

In Superscrabble the values are multiplied and in Hyperscrabble they're hyperadded. In the 3-D versions words can be changed by stacking letters, like playing F and V on NINE to make FIVE. In the 4-D versions stacked letters can move anywhere on or off the board.

	one	**two**	**three**	**four**	**five**	**six**	**seven**	**eight**	**nine**	**ten**	**eleven**	**twelve**
hyper	111	141	14,111	4,111	4,141	118	11,411	1,141	1,111	111	111,411	141,141
super	1	4	4	4	16	8	4	4	1	1	4	16
2nd super	1	4	4	4	8	4	4	4	1	1	4	8
3rd super	1	4	4	4	4	4	4	4	1	1	4	4
repeats at step	1	1	1	1	3	2	1	1	1	1	1	3

This is similar to the hard-to-recognize pattern in the cyclical darts and roulette sequences. Taking double zero as infinity, we get high surreally fast after one cycle.

darts	1	18	4	13	6	10	15	2	17	3	19	7	16	8	11	14	9	12	5	1	18

sum	1	19	23	36	42	52	67	69	86	89	108	115	131	139	150	164	173	185	190	191	209

super	1	18	72	936	5,616	56,160	842,400	1,684,800	28,641,600	85,924,800	1,632,571,200

hypersum	1	118	1,184	118,413	1,184,136	118,413,610	1,1841,361,015	118,413,610,152

hypersuper	1	18	444,444,444,444,444,444	> seventeenduplex

roulette	1	13	**36**	**24**	3	**15**	**34**	**22**	5	17	**32**	**20**	7	11	**30**	**26**	**9**	**28**
sum	1	14	50	74	77	92	126	148	153	170	202	222	229	240	270	296	305	333

roulette	**0**	2	**14**	**35**	23	**4**	**16**	**33**	**21**	**6**	**18**	31	19	**8**	**12**	29
sum	333	335	349	384	407	411	427	460	481	487	505	536	555	563	575	604

roulette	25	10	27	00	1	13	36
sum	629	639	666	666 + 00	667 + 00	680 + 00	716 + 00

hypersum	1	113	11,336	1,133,624	11,336,243	1,133,624,322	11,336,243,225

66 At dogsquad.co.uk/tarot/gem is a program for converting Roman, Greek or Hebrew letters into nymbers. Iconian isopsephia had: A (alpha), B (beta), Γ (gamma), Δ (delta), E (eta), F (digamma), Z (zeta), H (heta), and Θ (theta) represent one through nine, I (iota), K (kappa), Λ (lambda), M (mu), N (nu), Ξ (xi), O (omicron), and Π (pi) ten through ninety, P (rho), Σ (sigma), T (tau), Y (upsilon), Φ (phi), X (chi), Ψ (psi), Ω (omega) one hundred through eight hundred. The number of the Beast (Revelation 13:18) 666, would therefore be Chi Xi Digamma, XΞF, a most evil fraternity, the archenemy of the Delta Schwa sorority. **36**

67 The Hebrews' gematria had: א (aleph), ב (beth), ג (gimel), ד (daleth), ה (he), ו (waw), ז (zayin), ח (heth), ט (teth), י (yodh) are one through ten. ך (kaph), ל (lamedh), ם (mem), ן (nun), ס (samekh), ע ('ayin), ף (pe), ץ (sadhe), ק (qoph) are twenty through a hundred. ר (resh), ש (shin), ת (taw) are two hundred through four hundred. The Masorites later added כ (final kaph), מ (final mem), נ (final nun), פ (final pe), צ (final sadhe) to complete the pattern to nine hundred. The number of the Beast, 666, would be, right-to-left, ם ס ז, zayin samekh mem. The 'ayin holds special significance as the "Vulcan peace sign", the initial Hebrew letter representing the whole of Deuteronomy 4:40, "And now, O Israel, hearken unto the statutes and unto the ordinances, which I teach you, to do them; that ye may live, and go in and possess the land which the LORD, the God of your fathers, giveth you.", usually abbreviated to just "Live long and prosper."

67 Yet other variants of obsolete numerologies are the so-called new merologies in which the number and its evaluation are equal, or as equal as possible. They were, at least, new when introduced by Lee Sallows in 1990. One popular version of such a system has: **a** = +936.5; **b** = +999,999,766; **d** = +296.5; **e** = +107; **f** = -95.5; **g** = -16.5; **h** = -251.5; **i** = +178; **l** = +3.5; **m** = +999,766; **n** = -129; **o** = +23; **q** = +999,999,999,998,510.5; **r** = +49.5; **s** = +106.5; **t** = -9; **u** = +27; **v** = -184.5; **w** = -12; **x** = -287.5; **y** = +72; **z** = -179.5. Thus zero = -179.5 + 107 + 49.5 + 23 = 0, one = 23 - 129 +107 = 1. Most High = 999,545.

Another version gives all the numbers from zero through twelve correctly, but fails with more of the higher numbers:
a = 947.5, **b** = 1000000010, **c** = g(2, 303, 10), **d** = 44.5, **e** = 3, **f** = 9, **g** = 6, **h** = 1, **i/j** = -4, **l** = 0, **m** = 1000010, **n** = 5, **o** = -7, **q** = 999999999999016, **r** = -6, **s** = -1, **t** = 2, **u** = 8, **v** = -3, **w** = 7, **x** = 11, **y** = 25, **z** = 10. In this Most High would be slightly higher 1,000,008.
Joyce's multi-operational merology is more complex, but also more often correct.
a = 1331/1008; **b** = 250; **c** = 33,078,375(10^{16}); **d** = 43,980,465,111,040/14,348,907; **e** = 1,089/4; **ec** = 48,721,035,264/175; **en** = **n** = 10; **ex** = 1,375g(2, 13, 10); **f** = 11/2; **fi** = 10/11; **fo** = 80/121; **g** = 8/45; **h** = 3; **i** = 40/121; **in** = -4; **j** = 307 = 55/1842; **l** = 128/5445; **ll** = 79,720,245; **lv** = 968/3; **m** = 1/4; **n** = **en** = 10; **ne** = 1,089/200; **nov** = 8g(2, 24, 10); **o** = 152,587,890,625,000/7,623; **p** = 5g(2, 30, 10)/693; **pl** = + 1 +; **q** = g(2, 8, 10); **re** = 8/1089; **s** = 7/55; **t** = 1; **te** = +; **ti** = 16/7,623; **tr** = 2,500; **u** = 1, **ua** = 189/13,456; **ui** = 25,000; **us** = - 1 +; v = 18/11; **w** = 1,089/100 ; **we** = 1/5; **x** = 3,993/14; **y** = 10 +; **z** = -605/2 +
So Most High = (1/4)(152,587,890,625,000/7,623)(7/55)1(3)(40/121)(8/45)3 > 6,000,000

68 François Fauvel-Gouraud's *Phreno-Mnemotechny or The Art of Memory* (1845). The digits zero through nine he represented by the struck-through letters ~~S~~, ~~T~~, ~~N~~, ~~M~~, ~~R~~, ~~L~~, ~~J~~, ~~K~~, ~~F~~, ~~P~~, pronounced iss, it, in, im, ir, ill, ij, ik, if, and ip. Allen Krill and others aptly renamed this pseudonumerology, which not coincidentally has the pseudonum. ~~STNMRLJ~~ or zero through six. Most High would just be ~~MST~~, pronouced im-iss-it aka 401. For more visit Krill's website, www.pseudonumerology.com.

69 The GOOD/BAD ratio is only 71/91 ≈ 78%. "Not half BAD" > 45. "Bad" originally came from *baeddling* meaning a homosexual. Four is interesting in that it retains its tail during pseudonumerological contraction and expansion.

4	84	1,184	2,102,122,141,184	11,5895,702,214,112,295,702,214,112,113,552,221,418,417,102,122,141,184
5	88	11	582	88,214,111
6	70	821	121,411,212	882,141,112,214,112,123,552,842,141,582,102,112,141,158
7	82	11	582	88,214,111
8	1	2	1	2
9	22	111	22,141,582	1,213,552,221,418,412,102,188,214,111
10	12	158	22,141,881	121,355,222,141,841,210,211,214,112

70 Nymber sequences generated from bases:

	A	B	C	D	E	F	G	H	I	J	K	L	M	N	O	P	Q	R	S	T	U	V	W	X	Y	Z
0	1	2	3	4	5	6	7	8	9	10	11	12	13	14	15	16	17	18	19	20	21	22	23	24	25	26

wordnums: 1, ONE_{27} = 34, THIRTY-FOUR_{27} = 160, ONE HUNDRED SIXTY_{27} = 895, EIGHT HUNDRED NINETY-FIVE_{27} > thirty-fourplex

0	1	2	3	4	5	6	7	8	9	A	B	C	D	E	F	G	H	I	J	K	L	M	N	O	P	Q	R	S	T	U	V	W	X	Y	Z
0	1	2	3	4	5	6	7	8	9	10	11	12	13	14	15	16	17	18	19	20	21	22	23	24	25	26	27	28	29	30	31	32	33	34	35

alphadecimal, base 36: 1, ONE_{36} = 31946, THIRTY-ONE THOUSAND, NINE HUNDRED

FORTY-SIX$_{36}$ > sixty-threeplex

	,	-	a	b	c	d	e	f	g	h	i	j	k	l
32	44	45	97	98	99	100	101	102	103	104	105	106	107	108
m	n	o	p	q	r	s	t	u	v	w	x	y	z	
109	110	111	112	113	114	115	116	117	118	129	120	121	122	

0	1	2	3	4	5	6	7	8	9
48	49	50	51	52	53	54	55	56	57

base 128: 1, ONE$_{128}$ = 1,832,805, ONE MILLION, EIGHT HUNDRED THIRTY-TWO THOUSAND, SIX HUNDRED FIVE$_{128}$ > hundred-thirty-fourplex

base 256: 1, ONE$_{256}$ = 7,302,757, SEVEN MILLION, THREE HUNDRED TWO THOUSAND, SEVEN HUNDRED FIFTY-SEVEN$_{256}$ > hundred-sixty-threeplex

71 Patrick "Rickshaw" Shaw is an economics major Irish-American.

72 Some digital possibilities can be eliminated by logic. According to the rules leading zeroes are not allowed, Two WRONGS do not make a RIGHT, though they could make ARIGHT, demonstrating both Benchley's First and Second Laws, "Most problems have either many answers or no answer, few a single answer." and "An answer may be wrong, right, both, or neither; most are both." S and U must both be the result of a carry-overs, so S is most likely = 1, U = 0 and G = 9. A, E and T must therefore all be 2 or greater and their sum greater than 10, so E too must be the result of a carry-over. Since both M result from different digits One or the other, but most likely both, are also the result of carry-overs and so H, R, A and E must all between 4 and 8. Eventually trial-and-error determines the final solution

HAVE + A + GREAT = SUMMER = 7685 + 6 + 94563 = 102254,
DO + YOU + FEEL = LUCKY = 57 + 870 + 9441 = 10368,
SEND + MORE = MONEY = 9567 +1085 = 10652,
APPLES + ORANGES = BANANAS = 855140 + 6983240 or 855240 + 6983140 = 7838380 ,
APPLE + APPLE = ORANGE = 63390 + 63390 = 126780,
THEN + JESUS + WENT = THENCE =1072 + 97464 + 8721 = 107257,
ONE + NINE + FIFTY + TWENTY = EIGHTY = 984 + 8584 + 75732 + 364833 = 450132,
JESUS + BIBLE = CHURCH = 94353 + 80864 = 175217,

These and other alphametics from bases 2 through 16 can be found at http://www.tkcs-collins.com/truman/alphamet/alpha_solve.shtml. You also can easily find solutions to such unique arithmetics problems as **TWO + TWO = FOUR** in base 7. **TWO + TWO = FIVE**, while having no unique solution in any base, does have an ever-higher number of solutions in bases 7 through 16: 3, 7, 18, 35, 72, 124, 178, 275, 376, and 537, for a total of 1675. To really get high there are exponentional arithmetics including:

- 3,245^012 ≤ **HEAD^TOE** ≤ 6,754^987 ≈ threeduplex
- 69,5,488,485^01234 ≤ **BEGINNING^END** ≤ 695,488,485^987 ≈ fourduplex
- 045,567^0123 ≤ **PILLAR^POST** ≤ 954,432^9,876 ≈ fiveduplex
- 567^01,234 ≤ **BAD^WORSE** ≤ 432^98,765 ≈ fiveduplex

- 512,674^01,234 ≤ **CRADLE^GRAVE** ≤ 487,325^98,765 ≈ sixduplex
- 35,678^012,134 ≤ **START^FINISH** ≤ 64,324^987,865 ≈ sevenduplex

This triple exponential however requires a higher than base ten solution, at least base eleven: **TINKERS^EVERS^CHANCE** ≤ 93X467^45467^012304 ≈ eightduplex

There's even never-ending alphametics using fractions like **EVE/DID** = 242/909 = **.TALKTALKTALK...**

Much more difficult is nimber alphametrics without carry-over. Nimbers are a subgroup of games, as are numbers, which obey the rules of the game Nim. Nim-summing however operates as if a bitwise exclusionary or xor, either but not both. Thus one nim-sum one equals not two but zero, since there is no carryover to the two's place. The symbol for nim-summing is an encircled plus, ⊕, approximated with a plus within parentheses, (+). The winning strategy in Nim requires finding the values for which z ⊕ z' ⊕ z'' = 0.

Nim-summing alternately gives higher and lower nimbers, yet all nim-sums do get higher in the long run just as with normal addition. It is easier to see what is happening when expressed in binary.

⊕	1	10	11	100	101	110	111	1000	1001	1010	1011	1100	1101	1110	1111
1	1	11	10	101	100	111	110	1001	1000	1011	1010	1101	1100	1111	1110
decimal	0	3	2	5	4	7	6	9	8	11	10	13	12	15	14

Similarly nim-products are symbolized by an encircled x, ⊗, approximated by (x). 2 ⊗ z = z ⊕ z = 0; 3 ⊗ z = z ⊕ z ⊕ z = z, etc. Nim-tetration and higher operations could be symbolized by multiple encircled x's.

73 Scott Kim's E, F, H, I, N, O, R, S, V, Z letter-numbers inspired extrapolating throughout the rest of the alphabet. Some were tempted to go on inspired by *On Beyond Zebra* by Dr. Seuss by adding the ampersand, &, Classical shorthand for the Latin *et*. The GOOD/BAD ratio for this transformation would be 6009/949 > 6. "Not half BAD" > 475. Googol gets reduced to a mere 600,607, while "six hundred thousand, six hundred seven" becomes a twenty-eight-place number. In Bowden's hexadecimal C is perhaps not coincidentally represented a mirrored two, though D is a mirrored three, E a mirrored four and F a mirrored five.

74 At http://www.contestcen.com/rithms.htm can be found many other challenging alphametics, like GOOD/A = OLD = 1440/3 = 480, and GOOD/T = TOO = 1996/4 = 499.

At http://www.gootar.com/folder/base.html you can convert any base between 2 and 36 into any other one. R2-D2 and C-3PO are the (in)famous rebel robots from the Jedi galaxy (IC 10) in "Star Wars" by George Lucas. R2-D2 would, of course be greater in any positive base because C-3PO, a single-place number minus a triple-place number would have to be less value. R2-D2 in base 36 is 974 - 470 = 504 in base 10, 10k (10 kilometers) in base 22. C-3PO in base 36 is 12 - 4812 = -4800 in base 10, -3pc (parsecs) in base 36.

75 The English number name sequence "one, two, three, ..." would be transformed to: ETHO, AWO, NGEE or 8, 1, 1303. Eight (181601) already becomes a 42-place number. The nymber sequence would be: 1. one, ETHO, 8, eight, IERPA, 181601, etc. Atbash is one of the languages used in T*he Rise and Fall of the Sheshak Empire*. Many others are also shifted alphabets that like Atbash preserve vowels as vowels, and consonants as similar consonants to form English cognates in a Playfair cipher. (See http://michaelhalm.tripod.com/Appendix_--_Languages.html)

	0	1	2	3	4
0	A	E	I	O	U
1	B	F	M	P	V
2	C	J	S	X	Z
3	D	G	K	Q	T
4	H	L	R	W	Y

With N for null, ONE in Krufoz, a descendant of Globish, would just encode as a rightward substitution code into UNI or 20540, so we'd get high like dyu, dlwii, muaw, mobi, xoz, xibin, iokld, noni, din, while the related Kruvos encodes bigrams like a Playfair cipher: uni, qyu, dylii, peiy, mefu, xoz, jifun, iodld, noni, gud, sharing the words for one, six and nine.

76 By analogy with Atbash this cipher should logically be called Azby, but the name of the librarian Atoz was made famous in the Star Trek episode, "All Our Yesterdays" by Jean Lisette Aroeste. His name also reminds of "The last shall be first and the first last.", which would apply to Atbash. For more on Atoz and the Sarpeidonians see "Who Mourns for Zarabeth?" addendum (http://michaelhalm.tripod.com/id100.htm) In Atoz GOOD/BAD ≈ 577.

number	1	2	3	4	5
name	one	two	three	four	five
Atoz	lmv	gdl	gsivv	ulfi	urev
Atozish	121,322	70,412	719,092,222	21,120,609	21,180,522
sum	121,322	191,734	719,283,956	740,404,565	761,585,087
Aybz	ymu	gdi	gsjuu	viej	vofu
Aybzish	251,321	70,409	719,102,121	22,090,510	22,150,621
sum	251,321	321,730	719,423,851	741,514,361	763,664,982

In Aybz GOOD/BAD = 20090923/262523_{26} = 400487/727 ≈ 550. "Good" comes from the Indo-European ghedh- meaning "unite", related to deghedh- "together". ("Indo-European and the Indo-Europeans" by Calvert Watkins)

"Abracadabra" in Aybz would become the even more cryptic "yzjyxywyzjy'.

Latin	uno	duo	tres	quattuor	quintus
+4	ars	has	zxiy	vaezzasx	vaerzay
nymber	151,213	221,513	18,160,517	11,151,414,150,912	11,151,908,141,513
sum	151,213	372,726	18,533,243	11,151,432,684,155	22,303,340,825,668

In this Latin code GOOD/BAD would be BONUS/MALUS = FSRAY/QEPVY = 2009081513/719061513 ≈ 3.

77 Other isograms are: adventurishly, bracketing, comedians, constipated, Democrats, dermatoglyphics, documentary, forensical, formidable, formulated, girandole, handsomely, measuringly, musicotherapy, mythopeic, nefariously, playgrounds, porcelain, profamily, profluency, questionably, quicksandy, Republicans, rudimental, schmaltzy, signature, slotmachine, solemnity, stenographic, subordinately, sympathizer, tenaciously, universal, valedictory, workmanship, xylophagic, zygomatic, zymogenic. Each give a scrambled alphabet with and without repeated letters and so many different GOOD/BAD ratios.

A polygraphic cipher can be even more cryptic using two or more keywords, for example, turning Democrats into Republicans or "Dinos". "Most High" could be the simple substitution PUAC FGDF or, playing less fair, one via the intermediary Playfair OCBS IKHJ to MARHZDNSQ or 130,118,082,604,141,917.

D	**E**	**M**	**O**	**C**		**R**	**E**	**P**	**U**	**B**
R	**A**	**T**	**S**	B		**L**	**I**	**C**	**A**	**N**
F	G	H	I/J	K		**S**	D	F	G	H
L	N	P	Q	U		J	K	M	O	Q
V	W	X	Y	Z		T	V	W	X	Y

The eleven most common letters are contained, though not in order, in the not-so-common isogram "threnodials". The average frequencies are in English, according to *Cipher Systems* by H. Beker and F. Piper: E 12.7%, T 9.1%, A 8.2%, O 7.5%, I 7%, N 6.7%, S 6.3%, H 6.1%, R 6%, D 4.3%, L 4%, U and C only 2.8%, M and W 2.4%, F 2.2%, G and Y 2%, P 1.9%, B 1.5%, J, K, Q, V, X, Z 1%, hence the traditional mnemonic for the twelve most common English letters, Etaoin Shrdlu. A reverse frequency cipher might be called ETOZ. "Most High" would be UQBX PKDP or 2,117,022,416,110,416.

Etoz	V	S	W	G	Z	L	D	P	K	N	I	F	U	J	Q	H	O	Y	B	X	M	A	C	T	R	E
nymber	22	19	23	7	26	12	4	16	11	14	9	6	21	10	17	8	15	25	2	24	13	1	3	20	18	5

At http://thinkzone.wlonk.com/Gibber/GibGen.htm a Gibberish generator can be found samples of Gibberish using monogrammic, trigrammic and hexagrammic frequencies. Using the three-letter frequencies generated an almost passable Old English text and the six-letter frequencies generatesd a passable Modern English text especially at lower reading levels..

Level 1 Gibbrish: Waf nnodg tfin faton lifs mhec nwaw'hd tho etso gw'ao otdvlh.

Level 3 Gibberish: But his recome withe thas of ming whout eved didn't and was it hesh.

Level 6 Gibberish: All things made were made through Him.

The noun "gibberish" and the verb "gibber" are said to have been influenced by association with Geber aka Abu Mūsā Jābir ibn Hayyān, who to avoid accusations of sorcery disguised his writings, rather like fellow polymath Leonardo "Obranoel" da Vinci did by mirror-writing. "Algebra" does not come from Geber, but from "al-jebr wa'lmuqabalah" meaning "the science of reuniting and equating".(*Dictionary of Word Origins* by Joseph T. Shipley)

78 Other pangrams are: "I, quartz pyx, who fling muck beds." by Augustus de Morgan in

1872, "Mr. Jock, TV quiz PhD, bags few lynx.", "New Job: Fix Mr. Gluck's hazy TV PDQ.", "J. Q. Vandz struck my big fox whelp." and "J. Q. Schwartz flung D. V. Pike my box," referring to John Quincy Schwartz and Dick Van Pike. The last two could have several variations depending on the initials of Vandz, Schwartz and Pike and who threw to whom. The Jock and Gluck pangrams could also be rephrased giving them the initials R. M. or putting Jock's title first.

I	Q	U	A	R	T	Z	P	Y	X	W	H	O	F	L	I	N	G	M	U	C	K	B	E	D	S
M	R	J	O	C	K	T	V	Q	U	I	Z	P	H	D	B	A	G	S	F	E	W	L	Y	N	X
N	E	W	J	O	B	F	I	X	M	R	G	L	U	C	K	S	H	A	Z	Y	T	V	P	D	Q
J	Q	V	A	N	D	Z	S	T	R	U	C	K	M	Y	B	I	G	F	O	X	W	H	E	L	P
J	Q	S	C	H	W	A	R	T	Z	F	L	U	N	G	D	V	P	I	K	E	M	Y	B	O	X
T	V	Q	U	I	Z	P	H	D	R	M	J	O	C	K	B	A	G	S	F	E	W	L	Y	N	X

Some imperfect pangrams include "Waltz, nymph, for quick, bad jigs vex.", "Quickly wafting zephyrs vex bold Jim." and "Jackdaws love my big sphinx of quartz." in 32 letters, "Pack my box with five dozen liquor jugs." and "The qualmish Afghan Jew packed over sixty fez with bees." In *The Valley of Fear* Watson suggests use of the *Bible* for such a cipher. Holmes however, discovers that his informant Porlock has rather used the previous years' *Whitaker's Almanac*. Watson likely had in mind the King James (1611) pangrams from 1 *Chronicles* 12:40 in 214 letters (MOREVTHYAWNIGUSCDZBLXKFPJQ) or *Ezra* 7:21 in only 172 letters (ANDIEVRTXSKGOMCLUWHBYZPFQYJ). Such pangrams can rather easily be found in any sufficiently large sample, such as William Shakespeare's "Sonnet 27" in 454 letters (WEARYITHOLSMBUDPFVNGJKXZCQ).

In 2000 Ovaltine, which had sponsored both "Orphan Annie" and "Capt. Midnight", had a decoder ring. (ASLWIMVHFKXDPOEJBTNQZGUYRC) and in 2005 Mr. Lobo's Sleepless Knights of Insomnia had a decode ring (RWADHKOEMQUFTBSPNIJCXZLGYV). They were nicknamed Wisconsin and Washington.

'00	one	two	three	four	five	six
wordnum	141915	180414	1808251515	9142325	9050715	20511
decimal	1,662	1,580	17,511	5,967	3,967	2,796
sum	1,662	3,242	20,753	24,740	27,536	27,844

'05	one	two	three	four	five	six
wordnum	71708	130207	1305010808	12071101	12182608	151821
decimal	1,034	1,114	11,726	3,781	5,084	1,722
sum	1,034	2,148	13,874	16,655	21,739	23,461

With fifty states plus possessions and just twenty-six letters there's more than a fifty-fifty chance of a scrambled alphabet containing at least one uspobigram: AK, AL, AR, AZ, CA, CO, CT, DC, DE, FL, GA, HI, IA, ID, IL, IN, KS, LA, MA, MD, ME, MI, MN, MO, MS, MT, NE, NH, NJ, NM, NV, NY, OH, OK, OR, PA, PR, RI, SC, SD, TN, TX, UT, VA, VT, WA, WI, WV, WY.

Since none begin with New Jersey's final J, Texas's final X, Wyoming's final Y or Arizona's final Z, no perfect panuspobigram is possible. Spoonerisms however give Jew Nersey's JN, Xetas's XT, Ywoming's YW and Zariona's ZA.

79 Typoglycemia is an intentional typo for hypoglycemia. ("The Significance of Letter Position in Word Recognition" PhD thesis, Nottingham University (1976), *New Scientist* (1999), *Aerospace and Electronic Systems Magazine* (2007)). Chris Cummings in "Scrambled Word Recognition: Implications for Position Coding" referenced W. A. Winkelgren's emphasis of trigram recognition and C. Whitney's bigram recognition. Others have confirmed that the initial letter and word length are more important for word recognition than the last letter and that swapping letters round the middle, and so breaking up the bigrams and trigrams, reduces word recognition.

typoglycemic	1,021	1,031	1,033	1,061	1,069
mates	2,036	2,048	2,050	2,132	2,158
broken-up	1,021	1,052,652	1,087,390,537	1,153,722,412,409	1,233,330,346,255,758
heptasub	1	960	991,572	1,052,058,661	1,124,651,700,322
broken-down	1	1,022	1,055,727	1,120,127,369	1,197,417,213,188
broken	1	1,021	1,022	1,052,652	1,055,727

hypersum	1,021	10,211,031	102,110,311,033	1,021,103,110,331,061
heptasub	1	9,311	93,112,551	931,125,511,283

80 The Navaho alphabet code had several alternatives for many letters. A could also be be-la-sana (apple) or tse-nill (axe), B na-hash-chid (badger), toish-jeh (barrel) or shush (bear), C ba-goshi (cow), tla-gin (coal) or moasi (cat), D lha-cha-eh (dog), chindi (devil) or be (deer), E ah-jah (ear), ah-nah (eye) or dzeh (elk), F tsa-e-donin-ee (fly), chuo (fir) or ma-e (fox), G klizzie (goat), ah-tad (girl) or heha (gum), H tse-gah (hair), cha (hat) or lin (horse), I yeh-hes (itch), tkin (ice) or a-chi (intestine), J ah-yatsinne (jaw), tkele-cho-g (jackass) or yil-doi (jerk), K ba-ah-ne-ditini (key), jad-ho-lani (kettle) or klizzie-yazzie (kid), L nash-doie-tso (lion), dibeh-yazzie (lamb) or ah-jad (leg), M tsin-tliti (match), be-tas-tni (mirror) or na-as-tso-si (mouse), N a-chin (nose) or tsah (needle), O ne-ahs-jah (owl), tlo-chin (onion) or a-kha (oil), P cla-gi-aih (pant), bi-so-dih (pig) or ne-zhoni (pretty), Q ca-yeilth (quiver), R dah-nes-tsa (ram), than-zie (turkey), ah-losz (rice) or gah (rabbit), S dibeh (sheep) or klesh (snake), T a-woh (tooth) or d-ah (tea), U no-da-ih (Ute), shi-da (uncle), V a-keh-di-glini (victor), W gloe-ih (weasel), X al-na-as-dzoh (cross), tsah-as-zih (yucca) and besh-do-tliz (zinc).

Because several letters have more than one substitution the nymbers can vary greatly. There are eighteen different combinations for “one”.

Royal Navy (1914): Apples, Butter, Charlie, Duff, Edward, Freddy, George, Harry, Ink, Johnnie, King, London, Monkey, Nuts, Orange, Pudding, Queenie, Robert, Suger, Tommy, Uncle, Vinegar, William, Xerxes, Yellow, Zebra.

ITU (1932): Amsterdam, Baltimore, Casablanca, Denmark, Edison, Florida, Gallipoli, Havana, Italia, Jerusalem, Kilogramme, Liverpool, Madagascar, New York, Oslo, Paris, Quebec, Roma, Santiago, Tripoli, Upsala, Valencia, Washington, Xanthippe, Yokohama, Zurich.

US (1941): Able, Baker, Charlie, Dog, Easy, Fox, George, How, Item, Jig, King, Love, Mike, Nan, Oboe, Peter, Queen, Roger, Sugar, Tare, Uncle, Victor, William, X-ray, Yoke, Zebra.
NATO (1955): Alfa, Bravo, Charlie, Delta, Echo, Foxtrot, Golf, Hotel, India, Juliett, Kilo, Lima, Mike, November, Oscar, Papa, Quebec, Romeo, Sierra, Tango, Uniform, Victor, Whiskey, X-ray, Yankee, Zulu. Victor Charlie was VC, which in turn stood for Viet Cong, the enemy in the Vietnam War.

81 "Crazy ABC's" sung by Barenaked Ladies has some good words, and even gets creative with argyle for R but misses other craziness, like words with silent initials, rather like initial zeros, aesthetic, euphemistic, oestrogen, and xylophone, perhaps because they were limited by their tune. Their alphabet had A as in aisle, B as in bdellium, C as in czar, D as in djinn, E as in Euphrates, F as in fohn, G as in gnarly, H as in hour, I as in irk, J as in jalapeño, K as in knicknack, L as in llama, M as in mnemonic, N as in ndomo, O as in ouija board, P as in pneumonia, Q as in qat, R as in argyle, S as in Szár, T as in tsunnami, U as in urn, V as in vraisemblance, W as in who, X as in Xian, Y as in yiperite, Z as in zed. Even crazier are alphabettors, which use puns like "A for ism" (aphorism), B for pork (Beef or pork), C for yourself (see for yourself), D for mation (deformation), etc.

82 Razilee "Raz" Purdue is the co-author of *Hierogamous Hymns* and translator of *The Wizard Who Couldn't and Other Basilian Tales* with Michael Halm and some of Joyce's writings. Razilee means "my secret" in Hebrew and Purdue "for God" in French. RAZ_{36} = $12HR_{32}$.

83 Pataphysicist Alfred Jarry wrote "Commentair pour servir ... la construction pratique de la machine ... explorer le temps" in 1911 (translated by Roger Shattuck as "How to Construct a Time Machine"). See reprint at http://michaelhalm.tripod.com/id92.htm.

84 Br. Kenneth "Bro. Ken" Kendrick is featured in the as yet unpublished *Uffda*, the story of a Vietnam War widow and Turkees groupee set in the Sixties. It's set mostly in Wells, the largest city in East Dakota, named for Henry Wells partner of William Fargo. The city was originally called Centralia. When Dakota territory was divided in 1889, between Minnesota and the Dakotas, East Dakota became the 38th state and West Dakota the 39th. The East Dakota Agricultural College, the future East Dakota State University, was founded in 1890. The Kendricks are one of the Piscatorian families in *The Rise and Fall of the Sheshak Empire*.

85 The reference is to the long-running "Dial H for HERO" series in which the dial-like H-dial transformed the dialer. The sequence of four symbols that Robert Reed likened with the letters H, E, R and O turned the user into a random superhero. Reversing the order, OREH, reversed the transformation. Daffy Dagan used five other positions, A, I, L, N and V, to become a villain. Victoria Grant dialed HEROINE to turned into a heroine. INE changed the hero to female and ENI changed the heroine to male. Christopher King dialed HORROR and became a monster, which RORROH would reverse. Robert "the Wizard" Reed dialed HIDE YOURSELF using five symbols previously not used and made the H-dial disappear. Other unrecorded possibilities might have been invisibility (HIDE ME) or animal (DEER, EEL, FISH, HORSE, HOUND, LION, SNAIL, VOLE), elemental (AIR, FIRE, LAND, SEA), or other (ALIEN, ELDERLY, ELF, HAIRY, IRISH, NORSE, RUSSIAN, UNDEAD) transformations.

Prof. John Frink, Jr., recreated an H-Dial-like cellphone which was (ab)used by Milhouse

Mussolini van Houten. ("Dial 'M' for Milhouse" by Ian Boothby)
The Society's compilation of KLondike-5 phony numbers can be found at michaelhalm.tripod.com/555 directory.html.

phony primes	5550011	5550019	5550029	5550059	5550073	5550113	5550121
inter	5550015	5550024	5550044	5550066	5550093	5550117	555127
mates	4440020	4440052	4440062	4440070	4440076	4440092	4449110

If you allow three-digit area codes you can get primes higher than the prime 5559979, like those between 0015550009 and 200555XXXX.

86 Ludlings vary in how much they add to the text, usually somewhat in synch with the original number name, unlike the phonetic alphabets. Some uninterestingly add exactly the same number of letters per word,
5 letters/word **Louchébem dialects** [from Fren.] move initial consonant(s) to the tail and add l- and -oche or -oque

sum	8	16	26	35	45	63	73	83	92	100	111	122	135	148	160	172	196	209	222

4 letters/word **Cha-Bhasa** [from Marachi] adds chan- prefix and moves initial syllable at end, **Gasó** [Rosarigasino] -gas- between doubled final stressed vowel, **Ke-an** [Malay] ke- before and -an after roots

sum	7	14	23	31	40	57	66	75	83	90	100	110	122	134	145	156	179	191	203

3 letters/word **Allspråket** [Swed.] moves initial consonant(s) to the tail and adds -all, **Izzle** adds -izz- after initial consonant(s), **Largonij** [from Fren.] moves initial consonant(s) to the tail and adds l- and -ij, **Louchébem** [from Fren.] moves initial consonant(s) to the tail and adds plus l- and -em. ("...the first shall be last." -- Mt 20:16b)

sum	6	12	20	27	35	41	59	67	74	80	89	98	109	120	130	140	162	173	184	193

2 or 3 letters/word **Mattenenglish** [from Bernese German] changes initial vowel to i, moves initial consonant(s) to the tail and adds -(h)ee, **Pig Latin dialects** initial consonants to the tail plus -ay, if none adds -way, -yay or -hay.

sum	6	11	18	24	30	35	42	50	56	61	70	78	88	98	107	116	127	135	146	157	170

Some more interesting ones add letters according to the arithmonym's duration.
6 letters/2 syllables **Firkonspråket** [from Swed.] puts fir- -kon around separated reversed syllables and so has words like "firteen-somethingkon" and "firty-somethingkon". **Utrovački** [from Serbian] puts u-za-kje around two swapped syllables.

sum	3	6	11	15	20	23	33	38	42	45	57	63	77	91	104	117	132	146	160	172	187

2 letters/syllable Kongarian [from Hung.] -ko- before syllable(s), **Pileshki** [from Bulg.] -pi- before syllable(s) in Cyrillic, which could be extrapolated to any of 126 other consonant-vowel combinations from **Bangarian** to **Zyngarian** and even more from **Chaleshki** to **Zhyleshki**.

	B	C	D	F	G	H	J	K	L	M	N	P	Q	R	S	T	V	W	X	Y	Z
A	ba	ca	da	fa	ga	ha	ja	ka	la	ma	na	pa	qa	ra	sa	ta	va	wa	xa	ya	za
E	be	ce	de	fe	ge	he	je	ke	le	me	ne	pe	qe	re	se	te	ve	we	xe	ye	ze
I	bi	ci	di	fi	gi	hi	ji	ki	li	mi	ni	pi	qi	ri	si	ti	vi	wi	xi	yi	zi
O	bo	co	do	fo	go	ho	jo	ko	lo	mo	no	po	qo	ro	so	to	vo	wo	xo	yo	zo
U	bu	cu	du	fu	gu	hu	ju	ku	lu	mu	nu	pu	qu	ru	su	tu	vu	wu	xu	yu	zu
Y	by	cy	dy	fy	gy	hy	jy	ky	ly	my	ny	py	qy	ry	sy	ty	vy	wy	xy	yy	zy

sum	5	9	16	22	27	32	39	46	52	57	65	73	83	93	102	111	122	130	141	152	165

Others add letters according to the number of vowels, which tends to be higher than the number of syllables.

5 letters/vowel **Jibberish dialects** add -atheb-, -uddag- or -uthug- before vowels.

sum	13	21	36	50	64	72	87	102	116	124	145	161	184	212	234	256	285	313	341

4 letters/vowel **Jargon** [from Fren.] adds -d-g- between tripled vowel(s), **Jibberish dialects** add -theb-, -idig-, -uvug- or -adag- before vowels, **Kinabayo** [Cebuano] on the other hand adds -g-d- between tripled vowel(s).

sum	11	18	31	43	50	57	70	83	95	102	116	130	160	179	200	204	228	252	260

3 letters/vowel **Egglish** adds -egg- before vowel(s), **Aigy Paigy** adds -aig- before vowel(s), **Jibberish dialects** add -dag- before vowels or -r-g-, -l-g- around doubled vowels, **Löffelsprache dialects** [Germ.] add -lef-,-lew- or -lev- between doubled vowel(s).

sum	9	15	26	36	46	52	63	74	84	90	105	117	134	154	175	191	212	229	249

2 letters/vowel **Turkey Irish**, the ludling used in the Turkee's lyrics, adds -ab- before vowels. **Af Jinni** adds any consonant between doubled vowel(s), and so includes **Jeringonza** [from Span.], **Păsărească** [from Romanian], **Pupiņvaloda** [from Latvian] with -p-, **Kuş Dili** [from Turk] with -g- or **Korakistika** [from Greek] with -k-, **Yägra** [from Amharic] or **Zargari** [from Persian] with -z- between doubled vowel(s), which could be extrapolated to the remaining consonants **Bargari, Cargari, Dargari, Fargari, Gargari, Hargari, Kargari, Largari, Margari, Nargari, Pargari, Qargari, Rargari, Sargari, Targari, Vargari, Wargari**, and **Yargari.**

sum	7	12	21	29	37	43	52	61	69	74	86	96	110	126	139	152	177	193	209	219

Still others add letters according to the number of consonants, which may be more than the number of vowels.

4 letters/consonant(s) **Madárnyelv -arg- dialect** [from Hung.] adds -arg- between doubled consonant(s).

sum	7	17	32	43	54	64	79	94	105	115	131	150	159	175	216	235	260	271

3 letters/consonant(s) Arpy Darpy adds -arp- after consonant(s), **Fuzzy Wuzzy dialects** -uz- or -ez- between doubled consonant(s), **Madárnyelv dialect** [from Hung.] -av- between doubled consonants(s).

letter sum	6	15	29	39	49	58	72	86	96	105	120	138	146	161	181	200	219	243	263

2 letters/consonant Javanais [from Fren.] adds -av- after consonant(s), **Madárnyelv dialect** [from Hung.] -av- between doubled consonants(s), **Ibbish** -ib- after consonant(s), **Obish** [Pig Greek dialect] -ob- after consonant(s), **Opish** [Pig Greek dialect] -op- after consonant(s), **Rövarspråket** -o- between doubled consonant(s), **Ubbi Dubbi** [aka Pig Greek] -ub- after consonant(s).

letter sum	5	12	21	29	37	44	55	66	74	81	93	107	125	141	156	171	187	203	217

letter(s)/consonant(s) Splantziana [from Grk.] adds following vowel(s) before consonant(s).

letter sum	4	8	15	21	27	31	38	43	49	53	61	69	80	91	101	110	123	133	145	153

One adds a variable syllable per word.

1 syllable/word Binaliktad [from Filipino, Tagalog] prefixes final syllable plus -s.

letter sum	7	12	20	28	36	42	50	61	69	75	84	95	107	119	130	141	154	166	178	186

Daniel Tammet in *Thinking in Numbers* tells of the **Veddas** of Sri Lanka who name one *ekkamai*, two *dekkamai* and three *otameekai* ("and another"), each one letter longer. Four would, if it were named, might therefore be *dotameekai* (and another two), five *dotameekai otameekai* (four plus one) and six *dotameekai dotameekai* (four plus two), with each additional number having more letters.

letter sum	7	8	9	10	19	20	29	30	39	40	49	50	59	60	69	70	79	80	89	90	99	100	109

The **Munduruku** of Brazil, he writes, get higher by adding syllables. One is *pug*, two *xepxep*, three *ebapug* and four *ebaxepxep*. If the prefix *eba-* means "(two)-plus-" and *-eba* means "-plus-(two)" and if they counted higher than four, five might be *ebapugeba* (three plus two), six *ebaxepxepeba* (four plus two), seven *ebaxepxepebapug* (four plus three), eight *ebapugebapugeba* (three plus three plus two) and so on.

Munduruku	3	6	6	9	9	12	15	15	18	18	21	24	24	27	27	30	33	33
letter sum	3	9	15	24	33	45	60	75	93	111	132	156	180	207	234	264	297	330

Some notable secret languages that do not change the word length at all and so are technically not ludlings, Joyce named with the portmanteau “nudlings”, from "non-ludlings": **A-keili** [from Fin.] transforms all vowels into a. This naturally can be extrapolated to other univocal dialects, **E-keili**, **I-keili**, **O-keili**, **U-keili** and **Y-keili** or more exotic ones like **Æ-keili or Ä-keili or Ø-keili**. **Jaredian** reverses words and transliterates the letters into Cyrillic. Assigning the values as follows according the position in the Cyrillic alphabet, with Ts transliterated for C, Dzh for J, Q and X as Kh, W as a double U and the absent H as ', Most High would be НГИН ТСОМ, 4090018171412.

A	B	Ch	D	E	F	G	H	I	K	Kh	L	M	N	O	P	R	S	Sh	Shch	T	Ts	U	V	Y	Z	Zh
А	В	Ч	Д	Е	Ф	Г	ь	И	К	Х	Л	М	Н	О	П	Р	С	Ш	Щ	Т	Ц	У	Б	Ы	З	Ж
1	2	23	5	6	20	4	0	9	10	21	11	12	13	14	15	16	17	24	25	18	22	19	3	26	8	7

Jaredian	ЕНО	ОУУТ	ЕЕРьТ	РУОФ
Jaredianish	61,314	14,191,918	606,160,018	15,191,420

Some nudlings are anagrammic. **Ngalskcab** (or Backslang) usually phonetically and sometimes creatively reverses the word, like Jaredian. **Šatra** [from Balkan Šatrovački], **Verlen** [from Fren. l'envers], **Verzin** [Hung.], and **Vésre** [from Span.] all reverse the syllable(s). **Dhochi** [from Luo] reverses the initial and final consonants, like amagrans, in monosyllable words and the last and next-to-last syllables in polysyllabic words.

Other linguistic transformations are pre-asperation; b-devoicing (B = D), d-voicing (D = T), g-devoicing (G = K), h-dropping (H = '), h-buccalization (H = TH), k-voicing (K = G), lallation (L = W), denasalization (N = '), ooglification (vowel = OO), yodding (OO = YOO), p-devoicing (P = B), rhoticism (R = W); lisping (S = TH), sibilation (S = SS), t-devoicing (T = D), t-glottalization (T = '), th-fronting (TH = F or V), th-alveolarizing (TH = T or D); sigmatism (TH = S), th-debuccalization (TH = H), assibilization (T(I) = S(H)), zazzification (consonant = Z), gemination (letter doubling) and degemination (letter undoubling). Zazzooglification would make everything "zoo".

In "The Goddess of Atvarabar" William Richard Bradshaw introduced the constructed language or conlang Atvarbarian in which substitutions sometimes depended upon the position in the word. One through nine would be ami, dy, dlei, foyl, faqui, caz, cequem, eejd, mami and Most High "Nasd Fejoh".

	A	B	C	D	E	F	G	H	I	J	K	L	M	N	O	P	Q	R	S	Sh	T	U	V	W	X	Y	Z
	O	P	S/K	T	E	F/V	J		A	G	C	R	N	M	A	B	V	L	S	H	D	Y	QU	S	Z	W	X
initial	O	P	W	T	D	F	J	F	E	G	C	R	N	M	A	B	V	L	C	H	D	Y	QU	C	Z	W	X
final	o	p	k	t	i	r	j	oh	e	g	c	r	n	m	a	b	v	l	e	h	d	y	q	y	z	w	x

An anagram with the letters in alphabetical order is called an "alphome", coined by Susan Thorpe from the beginnings of alpha and omega. Some worlds are natural alphomes like aegilops, beefily or billowy. Anagrams are usually designated by numbers indicating the rearrangement from the alphome. The alphome of "alphome" (1234567), for example, is "aehlmop" (1742653). Joyce named an anagram in reverse order a "margana", a reverse alphome an "emohpla" (7654321), one that swaps the first two letters a "naagram" (2134567), the second and third an "aangram" (1324567) and a word turned into palindrome by prefixing a "marg", or suffix of reversed letters, a "marganagram". An anagram with just swapped outer consonants is an "amagran" (aka Davrosism, used in "The Daleks" by Terry Nation and *The Rise and Fall of the Sheshak Empire* for devolved, killer mutants). Other consonant-swapping anagrams are the aganram, aragnam, anargam, anamrag and anagmar. A number in numerical order is called "sorted".

87 Count von Count is a character who lives on "Sesame Street" with an obsessive-compulsive disorder fixated on counting. A better translation of his name would be the French Comte de Comte. The title "count" comes from Old French "comte", from Latin "comes", meaning companion or fellow traveler from the Indo-European "comei-" meaning go with. The verb "count" comes from the Old French "comte" from the Latin "computare" meaning think from the Indo-European "computo-" meaning cut or strike with. ("Indo-European and the Indo-Europeans" by Calvert Watkins) The count seems however to be from Transylvania, so the Romanian Count de Numărare might be better.

88 Tutnese (Double Dutch) has a third of its substitutions regular, like tut for t, but the rest irregular: bub, cash, dud, fud, gug, hutch/hash, jug, kuck, lul, mum, nun, pub, quack, rug, sus, tut, vuv, wash/wack, xux, yuck/yub, zub/zug, and squa(t) (double). Some dialects also include verbalized punctuation: per (period), que (question mark) and com (comma). Most High would be the 25-place "Mumosustut Hutchigughutch".

number	number name	Tutnese name	length	sum
1	one	onune	5	5
2	two	tutwasho	8	13
3	three	tuthutchrugsquate	17	30
4	four	fudourug	8	38
5	five	fudivuve	8	46
6	six	susixux	7	53
7	seven	susevuvenun	11	64
8	eight	eigughutchtut	13	77
9	nine	nuninune	8	85
10	ten	tutenun	7	92

This gets high irregularly amplifying the natural number name length fluctuations. "Most High" would be a 25-place nymber in Tutnese and a 67-place nymber in Double Tutnese.

Double Tutnese	length	sum
onununune	9	
tutututwashasushutcho	21	30
tutututhutchututcashhutchrugugugsquackuatute	44	74
fudududourugugug	16	90
fudududivuvuvuve	16	106
susususixuxuxux	15	121
susususevuvuvuvenununun	23	144
eigugugughutchututcashhutchtututut	34	178
nunununinununune	16	194

tutututenununun	15	209

89 The prefixes **zetta-** or twenty-firstplex and **yotta-** or twenty-fourthplex were adopted in 1991 and obviously not from the Greek. Tenyottaplex would be to-the-twenty-fifth ten. Since then there have been several proposed extensions of the list to get even higher including the following.

power of 10	Morgan Burke's	Jeff K. Aronson's	Jim Blowers'	André Joyce's
27	harpo-	xenta-	xona-	xova-
30	groucho-	wekta-	weka-	wieca-
33		vendeka-	vunda-	vunda-
36		udeka-	uda-	uda-
39			treda-	treda-
42			sorta-	satta-

Since **zetta-** (z + (s)etta) and **yotta-** (y + otto) obviously, at least to the Joyceans, seemed formed from the reverse alphabet (as Aronson and Blowers noticed) and Italian numbers (nove, dieci, undici, dodici, tredici, quattordici, quindici, sedici, diciasette, diciotto, diciannove, venti, etc.) they extrapolated **xova-** and **weica-**. To keep other prefixes the traditional bisyllabic length they used just the initial syllables: **rinda-** = forty-fiveplex, **qeda-** = forty-eightplex, **pica-** = fifty-oneplex, **oca-** = fifty-fourplex, **nica-** = fifty-sevenplex, **menta-** = sixtyplex. See michaelhalm.tripod.com/prefixes. Until any of these are officially adopted, if ever, we can always continue to compound what is official.

90 Kilo- comes from the Indo-European root *gheslo-* meaning a thousand, as do chiliad, mil, mile, milfoil, millennium, millepore, milliary, millime, million, and millipede. **Mega-** comes from meg-, "great", as do acromegaly, almagest, maestoso, maestro, magisterial, magistral, magistrate, magnanimous, magnate, magnific, magnificent, magnifico, magnify, magniloquent, magnitude, magnum, maharajah, maharani, maharishi, mahatma, mahayana, majesty, major, majordomo, majority, majuscule, master, maxim, maximum, May, mayor, mega-, megalo-, mickle, mister, mistral, mistress, much, and omega. **Tera-** comes from kwer-ōr-, meaning "that which makes harm, monster", as do karma, peloria, teratogen, teratoid, teratoma. **Myria-** comes from the Greek murias, "countless", and **giga**- from the Greek gigas, "giant", as does giant, giantess, giantism and gigantic. ("Indo-European and the Indo-Europeans" by Calvert Watkins)

- quexa- = 2g(2, g(2, 18, 10), 2) > g(2, 3g(2, 17, 10), 10)
- quexbi- = g(2, g(2, 60, 10), 2) > g(3, 4, 10)
- quextre- = g(2, g(2, 60, 10), 2) > g(2, 4g(2, 28, 10), 10)
- quocto- = g(2, g(2, 60, 8), 2)
- quoctoginta- = g(2, g(2, 60, 80), 2)
- quoctingenti- = g(2, g(2, 60, 800), 2)

Richard Crandall in "The Challenge of Large Numbers" estimates it would take approximately sixduplex years before a parrot, pecking randomly at a keyboard, could reproduce by chance *The Hound of the Baskervilles*. "This time span, though enormous, pales" he says, "to the [thirty-threeplex] years before fundamental quantum fluctuations might topple a beer can on a level surface."

According to the Kardashev-Sagan classification of civilizations we are globally an H.7 one , though it varies greatly individually, with ten kilowatt-hours per capita and a hundred terabits of information per capita. At a growth rate of three to five percent it's estimated that by about AD 10,000 we will as citizens of an Intergalactic Empire have access to a hundred-sixty-sevenplex bits or a hundred-fifty-threeplex *Hamlets*. Well, at least that much information, if not that many masterpieces. nowhere near what Joseph S. Madachy called "rather large", to-the-fourth-nine.

91 In 1996 the International Electrotechnical Commision introduced a set of binary prefixes based on the decimal prefixes, to avoid the confusion between the two. **Mebi-** = g(2, 20, 2) = 1,048,576; **gebi-** = g(2, 30, 2) = 1,073,741,824; **tebi-** = g(2, 40, 2) = 1,073,741,824 > nineplex; **pebi-** = g(2, 50, 2) = 1,125,899,906,842,624 > fifteenplex,
exbi- = g(2, 60, 2) = 1,152,921,504,606,846,976 > eighteenplex,
zebi- = g(2, 70, 2) > 1,180,591,620,717,411,303,424 > twenty-oneplex,
yobi- = g(2, 80, 2) > 1,208,925,819,614,629,174,706,176 > twenty-fourplex.

Higher than this lies the region of the little googol, two-to-the-hundredth > thirtyplex. Blowers also refers in "*Hamlet* is a Big Number" to a range of high numbers he calls literary, those with approximately a book's worth of information. *Hamlet* downloaded as a file is actually a rather small literary number, less than sevenduplex.

The Joyceans not unexpectedly extrapolated these to ternary and higher with:
metri- = g(2, 20, 3) = 3,486,784,401 > nineplex,
getri- = g(2, 30, 3) = 205,891,132,094,649 > fourteenplex,
tetri- = g(2, 40, 3) = 12,157,665,459,056,928,801 > nineteenplex,
petri- = g(2, 50, 3) = 717,897,987,691,852,588,770,249 > twenty-threeplex,
extri- = g(2, 60, 3) = 42,391,158,275,216,203,514,294,433,201 > twenty-eightplex,
zetri- = g(2, 70, 2) = 2,503,155,504,993,241,601,315,571,986,085,849 > thirty-threepklex,
yotri- = g(2, 80, 3) = 147,808,829,414,345,923,316,083,210,206,383,297,601> thirty-eightplex,

By extension, we get new prefixes, compoundable with all the previous prefixes and themselves, **ki-** meaning -to-the-tenth-, **me-** -to-the-twentieth-, **ge-** to-the-thirtieth-, **te-** -to-the-fortieth-, **pe-** -to-the-fiftieth-, **ex-** -to-the-sixtieth-, **ze-** -to-the-seventieth- and **yo-** -to-the-eightieth-. The -bi- infix can be further extrapolated through the Latin prefixes: **yotri-** = three-to-the-eightieth > thirty-eightplex, **yocent-** = hundred-sixtyplex. The prefix **yoyo-** could be interpreted as Asimov did his **teratera-** to mean twenty-fourduplex. Yoyodyne, the Red Lectroids' front organization in *Buckaroo Banzai Across the Eighth Dimension* by Earl MacRauch, would refer to a force of that many dynes, far more than all of that in a merely four-dimensional cosmos.

92 "Umbu-Ungu" by Raymond G. Gordon, Jr., *Ethnologue: Languages of the World*, "Kaugel Valley systems of reckoning" by Nancy Bowers and Pundia Lepi, *Journal of the Polynesian Society* 84 (3): 309–324, "Rarities in Numeral Systems" by Harald Hammarström, *Proceedings of Rara and Rarrissima Conference*

Skoish

broken-up	1	3	10	43	225	2,293	25,448	307,669	4,025,145	56,659,699

broken-down	1	2	7	30	157	1,600	17,757	214,684	2,808,649	14,257,929
broken	1	2	3	7	10	30	43	157	225	1,600
sum	1	3	6	10	15	25	36	48	61	75
super	1	2	6	24	120	1,200	13,200	158,400	2,059,200	28,828,800

hypersum	1	12	123	1,234	12,345	1,234,510	123.451,011	12,345,101,112

A hybrid 24/7 base would soon diverge.

24/7	1	2	3	4	5	6	10	11	12	13	14	15	16	20	21	22	23	24	25	26	30	31	32	100	101	102

broken-up	1	3	10	43	225	1,393	14,155	157,098	1,899,331	24,848,401
broken-down	1	2	7	30	157	972	9,877	109,619	1,325,305	6,736,144
broken	1	2	3	7	10	30	43	157	225	1,600
sum	1	3	6	10	15	25	36	48	61	75
super	1	2	6	24	120	1,200	13,200	158,400	2,059,200	28,828,800

hypersum	1	12	123	1,234	12,345	123,456	12,345,610	1,234,561,011

Ngitish

broken-up	1	3	10	103	1,143	13,819	180,790	3,629,619	76,402,789
broken-down	1	2	7	72	799	9,660	126,379	2,537,240	53,408,419
broken	1	2	3	7	10	72	103	799	1,143
sum	1	3	6	16	27	39	52	72	93
super	1	2	6	60	660	7,920	108,960	217,920	653,760

hypersum	1	12	123	12,310	1231011	123,101,112	12,310,111,213

Supyire-ish

broken-up	1	3	10	43	440	4,883	59,036	772,351	10,871,950
broken-down	1	2	7	30	307	3,407	41,191	538,890	7,585,651
broken	1	2	3	7	10	30	43	307	440
sum	1	3	6	10	20	31	43	56	70
super	1	2	6	24	240	2,640	31,680	411,840	5,765,760

hypersum	1	12	123	1,234	123,410	12,341,011	1,234,101,112

93 *Biting the Wax Tadpole* by Elizabeth Little

94 *Sherlock Holmes in Babylon* ed. by Marlow Anderson

95 *A Place in England* by Melvyn Bragg

96 Richard Feynman pointed out the difference between a number and its representation in *Surely You're Joking, Mr. Feynman.* "Honest" number expressions are those with the same number of letters as the number they express. "Four" is honest in English. Roman numbers make one through three honest. Joseph DeVincentis gives five as "a five", six as "one six", seven as "one 'n' six", eight as "two cubed" and nine as "just a nine" and proved all numbers above twelve can be made honest.

Bertram Russell noted in what's become known as the Berry paradox that "the smallest positive integer not nameable in fewer than twelve words" is nameable with these eleven words. Googology demonstrates that any number can be named in just one word with the application of logic and wit.[152]

◆ The four-syllable 1,177 = two-ones-two-sevens or ${}_{2}1_{2}7$
- 1,777 = one-three-sevens or $1_{3}7$
- 27,777 = two-four-sevens or $2_{4}7$

◆ five-syllables: 11,777 = five-syllable two-ones-three-sevens or ${}_{2}1_{3}7$
- 111,777 = three-ones-three-sevens or ${}_{3}1_{3}7$
- 177,777 = one-one-five-sevens or ${}_{1}1_{5}7$
- 1,177,777 = five-syllable two-ones-five-sevens or ${}_{2}1_{5}7$
- 1,777,777 = five-syllable one-one-six-sevens or ${}_{1}1_{6}7$
- 2,777,777 = one-two-six-sevens or ${}_{1}2_{6}7$
- 11,777,777 = two-ones-six-sevens or ${}_{2}1_{6}7$
- 111,777,777 = three-ones-six-sevens or ${}_{3}1_{6}7$
- 177,777,777 = one-one-eight-sevens or ${}_{1}1_{8}7$

◆ six-syllables: 1,127 = two-ones-twenty-seven or ${}_{2}127$
- 27,777,777 = one-two-seven-sevens or $12_{7}7$

seven-syllables: 1,127,777 = two-ones-one-two-four-sevens or ${}_{2}1_{1}2_{4}7$

97 Paul "Cordwainer Smith" Linebarger wrote of the pentapaul in “Under Old Earth” attributing it to C'paul the mad cat-man minstrel's mimicking of the cadence of congohelium, that unstable mixture of matter and antimatter that blew up the underseas, killing "Sun-Boy" fka Yebayee and so breaking the link with the Douglas-Ouyang planets in the Fifteenth millennium.

98 Raymond Queneau's *100,000,000,000,000 Poems* is more modest and more entertaining than *Nepalese Limericks.* In it any of ten different choices of the fourteen lines can be read independently as fourteenplex sonnets. The first starts with "Don Pedro from his shirt has washed the fleas" and the last ends with the couplet "Poor reader smile before your lips go numb; The best of all things to an end must come." He estimated it would take at least 190,258,751 years to read them all, a short time compared to eternity, the time needed to finish the sequences in this book.

Based on the variations of traditional poem forms, André Joyce named quite a large number of other kinds of poetic numbers or arithmopoems.

Caharthamese: [J. A. Lindon, poet from Cahartham couldn't start 'em] without first line

Crewesque: [referring to "There was a young lady of Crewe Whose limericks stopped at line two. (She was not done After line one, But one more's all she could do.)"] with only first two lines

Ecuadorean: [limerick from Ecuador without line four] without fourth line

haikuized: [Raymond Queneau] abbreviated to only the haiku-like line-endings
Manx: [tailless like Manx cat] without last line
Moorean: [J. A. Lindon, limerick of Moore ended before four] with only three lines
Percyan: [J. A. Lindon, like the verses of Percy which are vice-versy] with lines reversed
Peruvian: [limerick from Peru without line two] without second line
Tuplettian: [J. A. Lindon, referred to in "There was a poet named Tuplett Who was so quick his limerick's just a couplet. When asked 'Why did you whittle Out the poem's middle?', He answered 'Because I am Tuplett.'"] with only first and last lines
Verdunese: [limerick from Verdun done at line one] with only one line
Wendhamese: [J. A. Lindon] without last line's rhyme

99: [Galician]

gaita-gallega	1,177	1,277	1,377	1,477
broken-up	1,177	1,503,030	2,069,673,487	3,056,909,243,329
heptasub	1	1,371	1,887,298	2,787,540,409
broken-down	1	1,178	1,622,107	2,395,853,217
broken	1	1,177	1,178	1,503,030

hypersum	1,177	11,771,277	117,712,771,377	1,177,127,713,771,477
heptasub	1	10,734	107,340,153	1,073,401,533,343

100 The hudibrastic with 2(6) = 12 syllables comes from *Hudibras* by Samuel Butler.

hudibrastic	27,111	27,121	27,122
broken-up	27,111	7,352,717,432	19,942,194,537,815
dekasub	1	33,381	90,537,431
broken-down	1	27,112	735,331,665
broken	1	27,111	27,112

hypersum	27,111	2,711,127,121	271,112,712,127,122
dekasub	1	123,085	12,308,498,088

The Caharthamese hudibrastic or "udibrastic" has one six-syllable line.

udibrastic	111	121	122	123	124	125
mates	100	101	102	103	104	105
broken-up	111	13,432	1,638,815	201,587,677	24,998,510,763	3,125,015,433,052
pentasub	1	90	11,042	1,358,287	168,438,641	21,056,188,359
broken-down	1	121	14,763	1,815,970	225,195,043	28,151,196,345
broken	1	111	121	13,432	14,763	1,638,815

hypersum	111	111,121	111,121,122	111,121,122,123	111,121,122,123,124
pentasub	1	749	748,728	748,728,231	7,448,728,231,344

101 [Japanese] with 5 + 7 + 5 = 17 syllables, comes from the hokku or introductory three lines of a longer poem. It contains three lines, two with five syllables, the other with seven for a total of seventeen, thus incorporating five of the first seven primes. Daniel Tammet in *Thinking in Numbers* connects this with Shichigosan, the feast of Seven-Five-Three, in which seven-year-old girls, five-year-old boys and three-year-olds are celebrated. This could be extrapolated through other non-repeating three-digit numbers from ichinisan (one-two-three) to kyuuhachishichi (nine-eight-seven), five hundred four feasts.(See endnotes 316-319)

haiku	11,077,077	11,077,107	11,077,113
broken-up	11,077,077	122,701,967,176,240	1,359,183,555,733,490,318,008
hexakaidekasub	1	13,808,288	152,955,958,920,013
broken-down	1	11,077,078	122,702,044,715,815
broken	1	1,077,077	11,077,078

hypersum	11,077,077	1,107,707,711,077,107	110,770,771,107,710,711,077,113
hexakaidekasub	1	124,656,081	1,246,560,808,039,400

102 villanelle: [French] with 5(3) + 4 = 19 syllables

villanelle	101,101,127	101,101,137	101,101,147
broken-up	101,101,127	10,221,438,891,681,400	1,033,399,195,939,398,197,464,654
oktakaidekasub	1	155,672,307	1,573,864,882,235,942
broken-down	1	101,101,137	1,022,144,000,379,382
broken	1	101,101,127	101,101,137

hypersum	101,101,127	101,101,127,101,101,137	101,101,127,101,101,137,101,101,147
18th sub	1	1,539,768,118	15,397,681,117,917,399,808

103 gayatri: [Sanskrit] with 24 syllables

gayatri	7,777,777	13,777,777	14,777,777
broken-up	7,777,777	107,160,477,061,730	1,583,593,633,231,846,396,434
hexakaidekasub	1	12,059,323	178,209,986,200,055
broken-down	1	7,777,778	114,938,268,839,507
broken	1	7,777,777	7,777,778

hypersum	7,777,777	777,777,713,777,777	77,777,771,377,777,714,777,777
hexakaidekasub	1	87,527,351	8,752,735,091,272,890

104 tanka or **haikai** [Japanese] with 5 + 7 + 5 + 2(7) = 31 syllables The second part of the tanka is called the shimo-no-ku, "the lower phrase", a shorter madrigal, the upper phrase being a haiku. It have five lines of two five-syllable lines and three seven-syllables lines for a total of thirty-one, incorporating seven primes, three of them Mersenne primes.

tanka	11,077,011,077,127	11,077,011,077,137
broken-up	11,077,011,077,127	122,700,174,402,905,030,853,345,400
30th sub	1	1,148,181,970,733
broken-down	1	11,077,011,077,128
broken	1	11,077,011,077,127

hypersum	11,077,011,077,127	1,107,701,107,712,711,077,011,077,137
30th sub	1	103,654,493,281,422

anka	77,011,077,127	77,011,077,137	77,011,077,147
broken-up	77,011,077,127	> twenty-oneplex	> thirty-twoplex
25th sub	1	82,365,312,022	> twenty-oneplex
broken-down	1	77,011,077,128	> twenty-oneplex
broken	1	11,077,011,077,127	> twenty-oneplex

hypersum	77,011,077,127	7,701,107,712,777,011,077,137
25th sub	1	106,952,551,613

105 a-gallega: [beheaded ta-gallega] with 7 syllables, 127, 137, 147, 157, 167, 187, ...
agne: [reverse enga] with 2(7) + 5 + 7 + 19(5 + 2(2(7) + 5 + 7) = 615 syllables
aiku: [beheaded haiku] with 4 + 7 + 5 = 16 syllables, 7077107, 7077113, 7077114, ...
aita-gallega: [beheaded gaita-gallega] with 10 syllables, aka quincuncigayatri, 1127, 1137, ...
allega: [beheaded gallega] with 5 syllables, 77, 107, 113, 114, 115, 116, 118, ...
atri: [beheaded yatri] with 21 syllables, 7177127, 7177137, 7177747, 7177157, ...
ayatri: [beheaded gayatri] with 23 syllables, 71177127, 71177137, 71177147, ...
begayatri: [two-third gayatri] with 16 syllables, 117127, 117137, 117147, 117157, ...
dibrastic: [beheaded udibrastic] with 4 + 6 = 10 syllables, 7117, 7127, 7137, 7147, ...
ega: [beheaded lega] with 2 syllables, 7, 13, 14, 15, 16, 18, ...
egayatri: [beheaded begayatri] with 15 syllables, 77127, 77137, 77147, 77157, ...
enga: [beheaded renga] with 19(7 + 5 + 2(7) + 5) + 7 + 5 + 2(7) = 615 syllables
eptuncigayatri: [beheaded septuncigayatri] with 13 syllables, 11127, 11137, 11147, ...
extragayatri: [beheaded dextragayatri] with 19 syllables, 777177, 177277, 177377, ...
[The extra- prefix here means 5n/12 - 1.]
gallega: [beheaded a-gallega] with 6 syllables, aka quartigayatri, 1071, 1072, 1072, ...
hai: [curtailed haik] with 5 + 7 + 3 = 15 syllables, 11077011, 11077027, 11077037, ...
haik: [curtailed haiku] with 5 + 7 + 4 = 16 syllables, 11077071, 11077072, 11077073, ...

iah: [reverse hai] 1 iku [beheaded aiku] with 3 + 7 + 5 = 15 syllables, 1077077, 1077107, ...
ibrastic: [beheaded dibrastic] with 3 + 6 = 9 syllables, 1111, 1117, 1121, 1122, ...
ita-gallega: [beheaded aita-gallega] with 9 syllables, 1111, 1117, 1121, 1122, ...
kaih: [reverse haik] with 4 + 7 + 5 = 16 syllables, 7077077, 7077107, 7077113, ...
lega: [beheaded llega] with 3 syllables, 11, 17, 21, 22, 23, 24, 25, 26, 28, ...
llega: [beheaded allega] with 4 syllables, 27, 37, 47, 57, 67, 71, ...
semigayatri: [half gayatri] with 12 syllables, 11111, 11117, 11121, 11137, 11147, ...
septuncigayatri: [seven-twelfth gayatri] with 14 syllables, 71111, 71117, 71121, ...
ta-gallega: [beheaded ita-gallega] with 8 syllables, aka tertigayatri, 727, 737, 747, ..
tri: [beheaded atri] with 20 syllables, aka dextragayatri, 777777, 1127777, 1137777, ...
udibrastic: [beheaded hudibrastic] with 5 + 6 = 11 syllables, 11111, 11117, 11121, ...
ukia: [reverse aiku] with 5 + 7 = 12 syllables, 11127, 11137, 11147, 11157, ...
yatri: [beheaded ayatri] with 22 syllables, 1177777, 1277777, 1377777, 1477777,...

106 naga-uta: [Japanese] with (5 + 7)n = 12n syllables and the **atu-agan** (7 + 5)n = 12n syllables. The **aga-uta** numbers would have 7 + (5 + 7)n = 7 + 12n syllables. The **atu-aga** numbers (7 + 5)n + 7 syllables With less than all of the three rightmost columns we can get as high as twenty-seven-sevens, $_{27}7$, eighty-two syllables. With just the next two columns we can get as a hundred-sixty-threeplex.

syllables	5	4	3	2	1
	twenty-sevenplex	twenty-oneplex	eleven	seven	one
	sixty-sevenplex	twenty-fourplex	seventy	twenty	two
	seventy-threeplex	thirty-threeplex	fifteenplex	thirty	three
	seventy-sixplex	thirty-sixplex	eighteenplex	forty	four
	seventy-nineplex	thirty-nineplex	thirtyplex	fifty	five
	hundred-sevenplex	forty-twoplex		sixty	six
	hundred-twentyplex	forty-fiveplex		eighty	eight
	hundred-sixtyplex	forty-eightplex		ninety	nine
		sixty-oneplex		hundred	ten
		sixty-fourplex		thousand	twelve
		seventyplex		million	
		eighty-oneplex		nineplex	
		eighty-threeplex		twelveplex	
		eighty-sixplex			
		eighty-nineplex			
		ninety-twoplex			
		ninety-fiveplex			
		ninety-eightplex			

		hundred-oneplex			
		hundred-fourplex			

107 madrigal: [French] with 2 or 3 lines with 7- or 11-syllables = 14, 21, 22 or 33 syllables. There would also be **adrigal** numbers with 2 or 3 lines with 6- or 10-syllables, **drigal** numbers with 2 or 3 lines of 5- or 9-syllables, **rigal** numbers with 2 or 3 lines of 4- or 8-syllables, **igal** numbers with 2 or 3 lines with 3- or 7-syllables. The **semimadrigal** has just one line of 7- or 11-syllables.

semimadrigal	127	137	147	157	167	171	172	173
madrigal	77,127	77,137	77,147	77,157	77,167	77,171	77,172	77,173
adrigal	71,117	71,121	71,122	71,123	71,124	71,125	71,126	71,128
drigal	11,077	11,107	11,113	11,114	11,115	11,116	11,118	11,119
rigal	7,027	7,037	7,047	7,057	7,067	7,071	7,072	7,073
igal	1,017	1,021	1,022	1,023	1,024	1,025	1,026	1,028

108 rondelet: [French] The **telednor** numbers would have 4 + 2(8) + 2(4 or 8) = 28 or 32 syllables, **elednor** numbers 3 + 2(8) + 2(4 or 8) = 27 or 31 syllables.

rondelet	1,001,111,111,027	1,001,111,111,037	1,001,111,111,047	1,001,111,111,057
telednor	7,117,117,727	7,117,117,737	7,117,117,747	7,117,117,157
elednor	1,117,117,727	1,117,117,737	1,117,117,747	1,117,117,757

109 choka: [Japanese] with (5 + 7)z + 2(7) = 12z + 14 syllables. The **hoka** [Hoka from *Earthman's Burden* by Poul Anderson and Gordon Dickman] would have 4 + 7 + (5 + 7)n + 2(7) = 12n + 25 syllables and the **oka** 3 + 7 + (5 + 7)n + 2(7) = 12n + 24 syllables.

choka	11,077,077,127	11,077,077,137	11,077,077,147	11,077,077,157	11,0077,077,167
hoka	7,077,077,127	7,077,077,137	7,077,077,147	7,077,077,157	7,0077,077,167
oka	1,077,077,127	1,077,077,137	1,077,077,147	1,077,077,157	1,0077,077,167

110 [Chinese] **tertilüshi** 2(5 or 7) = 10 or 14 syllables, **semilüshi** would have 3(5 or 7) = 15 or 21 syllables, **belüshi** 4(5 or 7) = 20 or 28 syllables. The **sesquilüshi** has 9(5 or 7) = 45 or 63 syllables.

lüshi	11,011,011,011,077	11,011,011,011,107	11,011,011,011,113
tertilüshi	11,077	11,107	11,113
semilüshi	11,011,077	11,011,107	11,011,107
belüshi	11,011,011,077	11,011,011,107	11,011,011,113

sesquilüshi	1,001,007,007,011,011,011,011,077	1,001,007,007,011,011,011,011,107

111: [French] **semibergette** would have 2 lines of 8- or 10-syllables or 5 lines of 4- or 5-

syllables for either 16, 20, 24 or 25 syllables. The **sesquibergette** has 6 lines of 8- or 10-syllables and so 48 or 60 syllables.

bergette	111,111,111,177	111,111,111,277	111,111,111,377	111,111,111,477
semibergette	111,177	111,277	111,377	111,477

sesquibergette	107,111,111, 111,111,177	107,111,111, 111,111,277	107,111,111, 111,111,377

112 dizain: [French] with 10 lines of 8- or 10-syllables for 80 or 100 syllables. The **semidizain** would have 5 lines for 40 or 50 syllables, the **quintidizain** 2 lines for 16- or 20-syllables, The **sesquidizain** would have 15 lines for 120 or 150 syllables.

semidizain	117,117,117,117,177	117,117,117,117,277	117,117,117,117,377

113 James Cooke Brown's Loglan is nearly as economical as Tevmekian.[123] The digits are just two letters each, though ironically the evens end in -o and the odds in -e, like one and two. Zero ends in -i and like one begins with n, two and three logically begin with t-, four and five with f- and six and seven with s-. Eight and nine however break the pattern by beginning with v-. Hundreds end with -ma, thousands with -mo and -sua- is the exponential operator. The tetrational operator could use the ordinal suffix, -ri, as an infix, so to-the-third-ten would be the six-syllable *nenisuarite* and to-the-ninth-ninth the five-syllable *vesuarive*. It is the most arithmopoetic language.

cielito: [Spanish "little heaven"] with 4(8) = 32 syllables rhyming abcb
cyhydedd hir: [Welsh] with 2(3(5) + 4) = 38 syllables rhyming aaabaaab
cyrch a chwta: [Welch] with 6(7) = 42 syllables rhyming aaaaba
nonet: with 9 + 8 + 7 + 6 + 5 + 4 + 3 + 2 + 1 = 45 syllables rhyming aabbccddd
cywydd llosgyrnog: [Welch] with 2(2(8) + 7) = 46 syllables rhyming aabccb with midrhymes of a in lines 3 and 5
décima: [Spanish] with 10(8) = 80 syllables rhyming abbaaccdde
diciottina: 18-verse quenina
droighneach: [Irish] with 6(13) = 78 syllables rhyming ababcdcd with crossrhyme and alliteration in each couplet
englyn: [Welsh] with 10 + 6 + 2(7) = 30 syllables rhyming abbb
espinella: [Vincent Espinel] 80-syllable poem with 10 8-syllabled lines rhyming abbaaccdde
huitain: [French] with 6(8 or 10) = 48 or 60 syllables rhyming ababbcbc or abbaacac
kyrielle: [French] with 4(7) = 28 syllables rhyming aabBccbB... where B is refrain
irrational sonnet: [Jacques Bens, 1965] 14 line poem rhyming aaabccccddddd, i. e., pi-like 3, 1, 4, 1, 5
limatherick: mathematical limerick, i. e., $(12 + 144 + 20 + 3\sqrt{4})/7 + 5 + 11 = 9^2 + 0$ (A dozen, a gross and a score plus three times the squareroot of four divided by seven plus five times eleven is nine squared and not a bit more.)
limeraiku: [limerick + haiku] with 5 + 7 + 5 = 17 syllables rhyming aba
muwashshah: [Arabic] with 5(4, 5 or 6) + 2 = 22, 27 or 32-syllable syllables rhyming abcd(c(d)) abef(e(f)) abgh(g(h)) abij(i(j) abkl(k(l)) ab
nonet: 45-syllable poem with 9 lines with 9 syllables in 1st line and one less in each succeeding line rhyming aabbccddd

nonina: 81-line quenina with spirally permutated pattern -- 123456789, 918273645, 594168327, 752934816, 671582493, 369741258, 835629174, 487315962, 246897531 [*Etoffe* by Jacques Rouband, 1974]
quatrain: 4 verses rhyming abab
quattordicina: 14-verse quenina
quenina: [*Subsidia Pataphysica* by Raymond Queneau] poem with a number of lines suitable for spirally permuting like in the sestina, numbers half of a prime minus one, like the tritina, quintina, ottina, nonina, undicina, quattordicina and so on.

p	7	11	13	17	19	23	29	31	37	41	47	53	59	61	67	71	73	79	83	89	97	101
(p - 1)/2	3	5	6	8	9	11	14	15	18	20	23	26	29	30	33	35	36	39	41	44	48	50

quinta: [10:5::decima:?] 40-syllable poem in 5 8-syllable lines each rhyming abbaa or aabba abcdef
quintina: 120-line poem with 60 stanzas with five permuting endwords
sestina: 39-line poem with endwords spirally permutated -- abcdef faebdc cfdabe ecbfad deacfb bdfeca with mid- and endwords of the three -- badfec
sonnet, Petrarchan: 14 line poem rhyming abbaabbacdecde
sonnet, Shakespearean: 14 line poem rhyming ababcdcdefefgg
sonnet, Spenserian: 14 line poem rhyming ababbcbaadcdee
tritina: 6-line poem with permuted endwords, abc acb bac bca cab cba
undecina: 11-verse quenina

Arithmopoems with reversed syllable and rhyme patterns can also be named by their ananyms and more by their ananym's beheadings, which may no longer rhyme..

114 The zth Fibonacci number is give by $F(z) = (\phi^z - \phi^{-z})/\surd 5$ with the first term getting closer and closer to the exact value as the second term gets smaller and smaller. H. E. Huntley notes in *The Divine Proportion* that the golden ratio is found in such diverse places as the atomic energy states, bee genealogy, leaf positions, multiple reflection optics, musical scales, the Parthenon, the pentangle, the pentagram, planetary orbits, seashell spirals, sunflower spirals, etc. The Fibonacci numbers are named for Leonardo Fibonacci, or filius (son of) Bonacci, who in his *Liber Abaci* in 1202 described them in connection with rabbit reproduction.

A converse number is a Fibonacci equivalent to a multiple of five times its index (modulo the index). ("Integer 4181" in *Lure of the Integers* by Joe Roberts) [NOTE: Counting from zero increases the index by two from that starting the count at one.]

converse	4,181	75,025	1,346,269	165,580,141	956,722,026,041	1,548,008,755,920

From Euler's equation we get F = E - V + 2, which can be misread as Fibonacci equals an even minus V plus two, so V is an indicator of the oddness of Fibonaccis.

F	1	2	3	5	8	13	21	34	55	89	144	233	377	610	987	1,597
V	1	0	1	1	0	1	1	0	1	1	0	1	1	0	1	1

115 Lucas also named the aforementioned Fibonacci series in 1877. The ratio of pairs of its numbers, and indeed any two numbers in such an additive sequence, interestingly gets ever closer and closer to phi as the numerators and denominators get higher and higher.

116 The famous poem by Gellett Burgess about the purple cow was inspired by a news story about a mineral-deficient bovine at the Florida State Nutrition Laboratory.

117 Some other common and not-so-common bases are: **6** (fathom-foot), **7** (week-day), **9** (span-inch, yard-hand), **22** (rod-yard), **42** (barrel-gallon), **49** (week-of-Sundays-day), **52** (year-week), **64** (square-mile-square-furlong), **66** (rod-foot), **72** (inches-points), **80** (mile-chain), **88** (mile-per-minute-foot-per-second), **144** (square-foot-square-inch), **168** (week-hour), **240** (league-rod), shilling-pence), **256** (pounds-drams), **288** (hand-point), **320** (mile-rod), **336** (fortnight-hour), 360 (revolution-degree), **480** (shilling-ha'penny), **512** (cubic-mile-cubic-furlong), **550** (horsepowers-foot-pound-per-second), **640** (square-mile-acre), **648** (span-point), **864** (foot-point), **1,440** (day-minute), **1,728** (cubic-foot-cubic-inch), **1,760** (mile-yard), **2,592** (yard-point), **2,640** (league-fathom), **2,655** (watt-hour-foot-pound), **3,410** (kilowatt-hour-BTU), **4,840** (acre-square-yard), **5,184** (fathom-point, square-inch-square-point), **5,280** (league-yard, mile-foot), **6,000** (league-link), **7,000** (pound-grain), **8,000** (mile-link), **10,080** (week-minute), **12,000** (foot-mil), **15,840** (league-foot), **20,160** (fortnight-minute), **21,120** (league-span), **21,600** (revolution-minute), **33,000** (horsepower-foot-pound-per-minute), **43,560** (acre-square-foot), **47,520** (league-hand), **57,024** (rod-point), **57,600** (square-league-square-rod), **63,360** (mile-inch), **190,080** (league-inch), **373,248** (cubic-inch-cubic-point), **570,240** (furlong-point), **604,800** (week-second), **746,496** (square-foot-square-point), **1,209,600** (fortnight-second), **1,296,000** (revolution-second),**4,561,920** (mile-point), **6,718,464** (square-yard-square-point), **6,969,600** (square-league-square-fathom), **13,685,760** (league-point), **13,824,000** (cubic-league-cubic-rod), **27,878,400** (square-league-square-yard), **36,000,000** (square-league-square-link), **63,360,000** (mile-mil), **79,833,600** (furlong-per-second-foot-per-fortnight), **190,080,000** (league-mil), **250,905,600** (square-league-square-foot), **446,054,400** (square-league-square-span), **2,258,150,400** (square-league-square-hand), **3,251,736,576** (square-furlong-square-point), **4,790,016,000** (furlong-per-second-inchper-week), **6,386,688,000** (mile-per-second-foot-per-fortnight), **9,580,032,000** (furlong-per-second-inch-per-fortnight) and, of course, all the possible hybrid bases.

Ovid W. Eshbach in *Handbook of Engineering Fundamentals* (3-16 III) lists the twelve comprehensive unit systems in use from Karl Freidrich Gauss (1833) to the present. James Clarke Maxwell's 1881 QES system is based on the quadrant, a billion centimeters. E. Bennet's 1917 CGSS system's 'gram-seven' is ten million grams or a hundred quadrillion of Maxwell's 'eleventh-grams'. From the suffixes -seven and -eleventh any number of other power-of-ten prefixes and suffixes could be similarly extrapolated.

Max Planck's so-called cG system uses the most extreme dimensionless scales, based on c, the speed of light, and G, the gravitational constant. The Planck unit of time is G/c^5, about forty-fiveminex seconds, the Planck length unit, G/c^3, about thirty-threeminex centimeters and the proton mass unit, $m_p = c/G$, about thirtyminex grams.

118 Kamika "the Mad Hawaiian" Schmidt is a German-American from Hawaii. Her first name is Hawaiian for smith as is her surname in German.

119 "The Hawaiian Number System" by Barnabas Hughes, Forty is a rare alphome number, one with all its letters in alphabetical order. To-the-second-ten would also be seven quadragintary bits and a googol sixty-eight. $COFFEE_{40}$ = 1,229,719,015.

120 After the Europeans came many Hawaiianized words were introduced like "Meli Kalikama" for "Merry Christmas" or "kamika" for smith. This Anglo-Hawaiian is more like a substitution code rather than a ludling since it only sometimes, but not always, the word length higher and in the case of the number two decreases it.

The Hawaiian alphabet has only twelve letters, A, E, H, I, K, L, M, N, O, P, U, W. With "Merry Christmas" the ER becomes transformed into an A, the R becomes an L, the Y, pronounced as EE, becomes a I, the Ch, pronounced as a K, becomes a K, an A is added between the CH and R, both S's become silent, unless initial, and the T another K. B and F both become P, D, G, Q, TH, X, and Z all become K. V becomes a W. The sound 'ay' becomes EI and 'eye' sound AI. The final vowel is repeated, if the word does not end with one. "Meli Kalikima!"

English	one	two	three	four	five	six	seven	eight	nine	ten	eleven
length	3	3	5	4	4	3	5	5	4	3	6
Anglo-Hawaiian	wunu	ku	kali	polo	paiwai	kiki	kewene	eikei	nainai	kene	elewene
length	4	2	4	4	6	4	6	5	6	4	7
gnomon	+1	-1	+1	0	+2	+1	+1	0	+2	+1	+1

121 Since a great hundred is one-zero-zero in duodecimal, 'great' could be used to describe any number of dozens. A great googol would be twelve hundredplex.

A glerint, the Ferengi equivalent for gargross, was popularized in "Infinite Bureaucracy" by Anne E. Clements in *Strange New Worlds VII*. Since Kieran Cockburn's prefix gar-, used in his coinage 'gargoogol', means "squared" and the adjective great means to-the-three-halves-power, we get higher like so:

- great gargross = twelve-to-the-sixth = 2,985,984
- great great gargross = second great gargross = twelve-to-the-nineth ≈ tenplex

The equivalence of great great and second great comes from yet another discipline, genealogy, where a great great grandparent can also be called a second great grandparent. Great great great would be third great and so on.

- gargreat gargross = g(2, 12, 12) ≈ thirteenplex
- garglerint = g(2, 16, 12) > seventeenplex
- great gargreat gargross = g(2, 18, 12) ≈ nineteenplex
- great garglerint = g(2, 24, 12) ≈ twenty-sixplex
- gargreat gargross = g(2, 36, 12) ≈ thirty-nineplex
- gargreat garglerint = g(2, 48, 12) ≈ fifty-twoplex
- gargrossth great gargross = g(2, 31104, 20736) > fiveduplex

In *Being Human* by Peter David is referenced the number system of the Selelvians, apparently a hybrid base 12/5, since the Selvelvians had trouble in base ten, confusing one-two, our seven; two-two, our twelve; one-two-two, our twenty-four; and two-two-zero, our thirty-four.

Selvelvianish	1	2	3	4	10	11	12	13	14	20
broken-up	1	3	10	43	440	4,883	59,036	772,351	1,0871,950	218,211,350

broken-down	1	2	7	30	307	3,407	41,191	538,890	7,585,651	152,251,910
broken	1	2	3	7	10	30	43	307	440	3,407
sum	1	3	6	10	20	31	43	56	70	90
super	1	2	6	24	240	2,640	31,680	411,840	5,765,760	115,315,200

122 Since Pandas have six fingers per hand for better climbing, while Koalas have two thumbs per hand, Pandans would count by twelves. By analogy with million, billion and trillion, two- and three-greats gross were nicknamed by Joyce "bo" and "tro", the trice great aka Hermetic for Hermes Trismegistus. Four-greats gross or mo squared could be called garmo, using Kieran Cockburn's prefix gar-, to avoid confusion with the momo, the skunkape-like Missouri monster (*Monster Spotter's Guide to North America* by Scott Francis). Using the double placeholder version avoids the differing notations of other systems.

duodecarish	...	10	11	100	101
broken-up	...	7,489,051	83,120,346	8,312,034,610	83,951,549,621
broken-down	...	5,225,670	57,999,271	5,805,152,770	586,378,429,041
broken	...	972	1,393	6,961	9,976

NOTE: Elevenmo is not eleven great gross, as it might seem -- that would be el mo (not to be confused with Elmo from "Sesame Street"). Rather it is Anglo-Italian for duodecimo, a page or book the size of a twelfth of a full sheet of paper, 5" by 7"9"', that is, five inches by 7.75 inches.

123 Homonyms and so homonums as well are vital in the universal language of Tevmekian of "Galaxy Quest" (see The *Unofficial Galaxy Quest Guide* by Michael Joseph Halm at michaelhalm.tripod.im4gq.htlm). The letter count gets higher very slowly.

- khax means zero, not, nothing, none
- vuk means one, first, won, win, victory
- thub means two, second, twoth, tooth, but greb means to
- means three, third
- shan means four, fourth, but khaz means for, avkav means fore
- phej means five, fifth
- lep means six, sixth
- rug mans seven, seventh
- khaz means eight, eighth, ate, eat, eaten
- haj means ten(-), tenth, -teen, teen(s)

	khax	**vuk**	**thub**	**zem**	**shan**	**phej**	**lep**	**rug**	**khaz**	**haj**	**huk**	**hub**	**hem**	**han**
letter sum	4	7	11	14	18	22	25	28	32	35	38	41	44	47

Since none of the numbers from zero to nine share the same vowel or final consonant and none of the numbers from one to ten share the same initial consonant(s), other numbers can be contracted to unique portmanteaux, for example twenty (two-tens, thub-haj) to thaj, rather like contracting five-eight to "fiveight", 'seven-nine to "sevenine" or one-eight-two to "oneightwo".

+	0	10	20	30	40	50	60	70	80	90	100
0	khax	haj	thaj	zaj	shaj	phaj	laj	raj	khaj	braj	drap
1	vuk	huk	thuk	zuk	shuk	phuk	luk	ruk	khuk	bruk	druk
2	thub	hub	thuth	zub	shub	phub	lub	rub	khub	brub	drub
3	zem	hem	them	zez	shem	phem	lem	rem	khem	brem	drem
4	shan	han	than	zan	shash	phan	lan	ran	khan	bran	dran
5	phej	hej	thej	zej	shej	pheph	lej	rej	khej	brej	drej
6	lep	hep	thep	zep	shep	phep	lel	rep	khep	brep	drep
7	rug	hug	thug	zug	shug	phug	lug	rur	khug	brug	drug
8	khaz	haz	thaz	zaz	shaz	phaz	laz	raz	khaz	braz	draz
9	brom	hom	thom	zom	shom	phom	lom	rom	khom	brob	drom

Higher number names can also be formed by using the word for the preposition "to", greb, not a homonum for thub. "Ten-to-", hej-greb-, could therefore be contracted to heb-, the equivalent of -plex. Googol would be translated as another two-syllable word, hebdrap. To-the-nine-hundredth-ten could be contracted from greb-drom-haj to gromhaj.

124 réi (zero) means soul, spirit, salutation

- gó (five) also means word, term, language
- shíchí (seven) also mean fatal position
- hachí (eight) also means bowl, basin

125 The fear of the number 666 is technically called hexakosiohexekontahexaphobia. This could also be symptomatic of a fear of sixes, hexaphobia. The Blue-Seven Effect identifies blue and seven as the most common responses to "Name a color and number." So not doing so, avoiding the number seven, would be a symptom of heptaphobia. Every digit or combination of digits theoretically could have a sequence and associated phobia. Individual numbers without a specific digit or combination of digits can be called a something-free number, as in one-free. A lipogram is a word without a particular letter, so a liponum is a number without a particular digit or string of digits.

monophobic	2	3	4	5	6	7	8	9	20	22	23	24	25	26	27	28	29	30	32	33	34
duphobic or Bantu	1	3	4	5	6	7	8	9	10	11	13	14	15	16	17	18	19	31	33	34	35
triphobic	1	2	4	5	6	7	8	9	10	11	12	14	15	16	17	18	19	20	21	22	24
tetraphobic	1	2	3	5	6	7	8	9	10	11	12	13	15	16	17	18	19	20	21	22	23
pentaphobic	1	2	3	4	6	7	8	9	10	11	12	13	14	16	17	18	19	20	21	22	23
hexaphobic	1	2	3	4	5	7	8	9	10	11	12	13	14	15	17	18	19	20	21	22	23
heptaphobic	1	2	3	4	5	6	8	9	10	11	12	13	14	15	16	18	19	20	21	22	23

oktaphobic	1	2	3	4	5	6	7	9	10	11	12	13	14	15	16	17	19	20	21	22	23
enneaphobic	1	2	3	4	5	6	7	8	10	11	12	13	14	15	16	17	18	20	21	22	23
dekaphobic	1	2	3	4	5	6	7	8	9	11	12	13	14	15	16	17	18	19	20	21	22

126 gar- added to great gross modifies not the great but squares the whole number. The connection between the garglerint and the powerful drink celebrated in *The Hitchhiker's Guide to the Galaxy* or the filksong, "The Pan-Galactic Gargle Blaster Blues" is purely co-incidental.

- great long gargross = g(2, 6, 13) = 4,826,809
- second-great long gargross = g(2, 9, 13) ≈ tenplex
- gargreat long gargross = g(2, 12, 13) > thirteenplex
- long garglerint = g(2, 16, 13) ≈ eighteenplex
- great gargreat long gargross = g(2, 18, 13) ≈ twentyplex
- great long garglerint = g(2, 24, 13) ≈ twenty-sevenplex
- gargreat long gargross = g(2, 36, 13) ≈ fortyplex
- gargreat long garglerint = g(2, 48, 13) ≈ fify-threeplex

triskaidekarish	...	10	11	12	100	101	102	103	104	105	106	107	108	109	110

Friday the thirteen is said to be connected to the curse on the thirteenth person at the Last Supper, Judas, (vampirism in "Dracula 2000") or alternatively the Friday 13, 1307, when the Knights Templar were declared illegal.

In Pyramid cards adding to thirteen are paired. Applying this to just digits we get the pairs four and nine, five and eight, and/or six and seven. A pyramid po or "poramid" is two pairs and a pyramid proil or "proilamid" all three.

pyramid	49	58	67	76	85	94	149	158	167	176	185	194	249	258	267	276	285	294	349

poramid	4,589	4,598	4,679	4,697	4,769	4,796	4,859	4,895	4,949	4,994	5,489	5,498

proilamid	456,789	456,798	456,879	456,897	456,978	456,987	457,689	457,698

Extrapolating further as with the three kinds of dozens, "pyralo" describes digit pairs adding to twelve, "poralo" two pairs, "proilalo" three pairs, "pyrahi" those adding to fourteen and "porahi" two pairs.

pyralo	39	48	57	75	84	93	139	148	157	175	184	193	239	248	257	284	293	339

poralo	3,489	3,498	3,579	3,597	3,759	3,795	3,849	3,894	3,948	3,984	4,389

proilalo	345,789	345,798	345,879	345,897	354,789	354,798	354,879	345,897

pyrahi	49	68	86	94	149	168	186	194	249	268	286	294	349	368	386	394	449

porahi	4,499	4,949	4,994	6,489	6,498	6,849	6,894	6,948	6,984	8,469	8,496

127 The knock-knock may trace its origin to *Macbeth* Act II, Scene iii. (*QPB Encyclopedia of*

Word and Phrase Origins by Robert Hendrickson)

Although a fortnight is a contraction for fourteen nights or two weeks, the prefix fort- from forty can also confusingly mean three hundred-twentyplex as in fortillion. A fortillion nights would be almost eighteenplex googol years.

Since the poulter's dozen is the second dozen with an extra two, a third dozen would by extrapolation have an extra three for a total of fifteen, a fourth sixteen, and so on.

	dozen	gross	great gross	gargross	great gargross	2nd great gargross
third	15	225	3,375	50,625	759,375	11,390,625
fourth	16	256	4,096	65,536	1,048,576	16,777,216
fifth	17	289	4,913	83,521	1,419,857	24,137,569
sixth	18	324	5,832	104,976	1,889,568	34,012,224
seventh	19	361	6,859	130,321	2,476,099	47,045,881
eighth	20	400	8,000	160,000	3,200,000	64,000,000
nineth	21	441	9,261	194,481	4,084,101	85,766,121
tenth	22	484	10,648	234,256	5,153,632	113,379,904

128 The Psychlos in L. Ron Hubbard's *Battlefield Earth* also used base eleven. The base eleven system of the Minbari originated, according to J. Michael Straczynski's "Babylon 5", from indicating the counted numbers on two five-fingered hands and then a headnod. Daniel Tammet in *Thinking in Numbers* speculates that Anne Boleyn, who had an extra finger, might have called ten with the French "dix" and so might have called a hundred "dixty" and a hundred ten, "dixty-dix" to rhyme with sixty-six.

129 According to Darren Doyles and the Kryptonese Project, based upon the Superman mythos, http://inventurous.net/kryptonese/.

130 *A Universal History of Numbers* by Georges Ifrah gives the Fifth Century B. C. numbers: **koti** = sevenplex, **ayuta** = nineplex, **niyuta** = elevenplex, **kankara** = thirteenplex, **pakoti** = fourteenplex, **vivara** = fifteenplex, **kshobhya** = seventeenplex, **vivaha** = nineteenplex, **kotippakoti** = twenty-oneplex, **bahula** = twenty-threeplex, **nagabala** = twenty-fiveplex, **nahuta** = twenty-eightplex, **titlambha** = twenty-nineplex, **vyavasthanapjnapati** = thirty-oneplex, **hetuhila** = thirty-threeplex, **ninnahuta** = thirty-fiveplex, **hetvindriya** = thirty-sevenplex, **samaptalambha** = thirty-nineplex, **gananagati** = forty-one-plex, **akkhobini** = forty-twoplex, **niravadya** = forty-fiveplex, **sarvabala** = forty-sevenplex, **bindu** = forty-nineplex, **sarvajna** = fifty-oneplex, **vibhutangama** = fifty-threeplex, **abbuda** = fifty-sixplex, **nirabbuda** = sixty-threeplex, **ahaha** = seventyplex, **ababa** = seventy-sevenplex, **atata** = eighty-fourple, **soganghika** = ninety-oneplex, **uppala** = ninety-eightplex, **kumuda** = hundred-fiveplex, **pundarika** = hundred-twelveplex, **paduma** = hundre-nineteenplex, **kathana** = hundred-twenty-sixplex, **mahakathana** = hundred-thirty-threeplex, **asankheya** = hundred-fortyplex, **dhvajagraishkamani** = four-hundred-twentyoneplex

By extrapolation from nahuta and ninnahuta comes the prefix nin- which seems to mean to-the-nine-sevenths, giving: **ninsonganghika** = hundred-seventeenplex, **ninkumuda** = hundred-thirty-fiveplex, **ninpundarka** = hundred-forty-fourplex, **ninpaduma**, = hundred-fifty-threeplex, and **ninkathana** = hundred-sixty-twoplex. Similarly extrapolating from koti- and pakoti-, comes the prefix pa- which seems to mean -squared, giving: **pasoganghika** = hundred-eighty-twoplex, **pakumuda** = two-hundred-tenplex, **papundarika** = hundred-twenty-fourplex, **papaduma** = two-hundred-thirty-eightplex, **pakathana** = two-hundred-fifty-twoplex, **pamahakathana** = two-hundred-sixty-sixplex, **padhvajagraishkamani** = eight-hundred-forty-twoplex.

From kathana and mahakathana comes the prefix maha- which seems to mean sevenplex more and so to **manapadhvajagraishkamani** = eight-hundred-forty-nineplex and **manapadhvajagraishkamani** = eight-hundred-fifty-sixplex.

131 Before Chinese and Korean diverged they were Sino-Korean and shared a myriad-based number system. **Baek** = hundred, **cheon** = thousand, **man** = myriad, **eok** = eightplex, **jo** = twelveplex, **gyeong** = sixteenplex, **hae** = twentyplex, **ja** = twenty-fourplex, **gan** = thirty-sixplex, **jeong** = fortyplex, **jae** = forty-fourplex, **geud** = forty-eightplex, **hanghasa** = fifty-twoplex, **nayuta** = sixtyplex, **bulgasaui** = sixty-fourplex, **muryang daesu** = sixty-eightplex. From ja and jae we can extrapolate the suffix -e to mean -to-the-twentieth, as does me-, and so get **baeke** = twenty-twoplex, **cheone** = twenty-threeplex, **eoke** = twenty-eightplex, **joe** = thirty-twoplex, **gane** = fifty-sixplex, and **hanghasae** = seventy-twoplex.

132 The Japanese number system has what appears to be some co-incidental similarities or borrowings from Sino-Korean. **Oku** = eightplex, **chou** = twelveplex, **gai** = twentyplex, **kou** = thirty-twoplex, **jo** = twenty-fourplex (the square of the Sino-Korean jo), **jou** = twenty-eightplex, **sei** = fortyplex, **sai** = forty-fourplex, **goku** = forty-eightplex, **kougasha** = fifty-twoplex, **asougi** = fifty-sixplex, **tukashigi** = sixty-fourplex, **muryou** = sixty-eightplex, **taisui** = seventy-twoplex. From kou and kougasha we get the suffix -gasha meaning -to-the-twentieth, as does -e and me-, and from jo and jou we get the suffix -u meaning -to-the-fourth and so **taisuigashau** = ninety-sixplex.

133 In his article "Kazilliard et Au-delá", Joyce explained his use of Mayan prefixes for indicating the nesting numbers in his more generalized exponential notation. Since zillion means g(2, 3z+3, 10) a kazillion would mean g(2, 2, 3z+3, 10) or g(2, 3z+3, g(2, 3z+3, 10)). Similar prefixes could be formed from other Mayan numbers. He also used the Mayan word for three, ox, as a suffix to replace the awkward triple I from the Roman numeral suffixes, so that the prefix three-to-the- would be the suffix **-plox**, rather than -pliii or even -pliij.

ox = 3, **kan** = 4, **ho** = 5, **uak** = 6, **uuk** = 7, **uaxak** = 8, **bolon** = 9, **lahun** = 10, and **bulak** = 11.

From ka- and kan- **-n** would seem to mean squared.

-ab = -to-the-fourth, **-nab** = -to-the-eighth, **alaun-** = twenty-to-the-twelfth ≈ sixteenplex, **hablan-** = twenty-to-the-fourteenth ≈ eighteenplex, **kinchilab-** = to-the-second-twenty > twenty-sixplex, **kinchilnab-** = twenty-to-the-fortieth > fifty-twoplex, **alaunab-** = twenty-to-the-forty-eighth > sixty-twoplex, **hablanab-** = twenty-to-the-fifty-fourth > seventyplex, **hablal-** = twenty-to-the-seventieth > ninety-oneplex.

An oxgoogol would be a mere g(3, 2, 50, 100), while a hablalgoogol would be g(g(2, 70, 20), 2, 50, 100)).

134 A D'ni number converter can be found at http://www.ookii.org/software/dni/default.aspx.

135 alternative #1 can be found at http://www.fadedtwilight.org/DniDictionary/numbers.htm, and alternative #2 at http://en.wikibooks.org/wiki/D%27ni/Numbers. The website http://english.stackexchange.com/questions/36938/correct-usage-of-replacing-cuss-words-with-symbols explains using the asterisk for the interior letters of offensive words as was done between 1785 and 1960 and why usually just four-letter words. Cubed rokubi and primes, same as primes until 27, are better called "parc" than c**p; Fibonacci-unindexed cubed kyuubi, better "swyve" than f**k (which it replaced in 1503); shichibi and gobi better "gash" than sh*g and shichibi-indexed ternarish better "unshut" than sh*t.

swyve	59,319	205,379	326,509	493,039	704,969	1,295,029	1,685,159
gash	5	7	15	17	25	27	35
unshut	21	122	1,000	1,101	10,010	101,011	101,120

In Stan Lee's Superhero Name Generator's alienation option @ was substituted for at, # for ll, % for oo, $ for sh, and + for t and in proto-Roman (for c, |) for d, (|) for m, \ / for v and >< for x and from Latin shorthand & for et. "Beetle Bailey" and *Let's get down to grawlixes* both by American cartoonist Mort Walker has done much to popularize ideographic substitutions

🕷☐☹☠👎☆✶💣✸✵✹🗡ϟ

The most famous typewriter cipher is the first, Edgar Allen Poe's "Gold Bug", many of the less frequent letter substitutions however are not revealed in the story, but some have posed solutions. Most High could become 9‡() 4634.

A	B	C	D	E	F	G	H	I	J	K	L	M	N	O	P	Q	R	S	T	U	V	W	X	Y	Z
5	2	-	†	8	1	3	4	6	,	7	0	9	*	‡	.	]	$	(	)	;	?	¶	¢	:	[

Transforming number names into numbers, like in pseudonumerology ignoring the letters not included in the system, we get an irregular Gold Bug sequence. "One" becomes E or 8, "eight" becomes eight thousand, six hundred thirty-four.

Gold Bug	1	8	8,634	863,445,648,461	> twenty-sevenplex

Although emoticons or emoji made with non-alphanumeric characters came much later than Andrew Bierce's "For Brevity and Clarity" or Poe, the Gold Bug code's YT, colon right parenthesis, could alread have been read as "smile"; his TT, double right parentheses, "kiss", his YL, "surprise", his YS, left parenthesis, 'frown'; his E, 8, "glasses"; his O, double dagger, ‡, "sealed eyes or lips"; his U, semicolon, "wink".

Some letters on a Qwerty keyboard become punctuation on a Dvorak keyboard and vice versa. Only the A remains the same in this base thirty.

	1	2	3	4	5	6	7	8	9	10	11	12	13	14	15	16	17	18	19	20	21	22	23	24	25	26	27	28	29
Qwerty	A	B	C	D	E	F	G	H	I	J	K	L	M	N	O	P	Q	R	S	T	U	V	W	X	Y	Z	,	;	.
Dvorak	A	X	J	E	.	U	I	D	C	T	N	L	W	M	R	L	I	P	O	Y	G	K	,	Q	B	;	V	-	Z

"Most High" in Dvorakish would be WROY DCID or 231815250004030904; "One", 181329; "two", 252718; "three", 2504162929; "four", 21180716; "five", 21031129.

136 $\pi \approx$ 196,314/625 or 3.141024

137 Dzu Tse was a contemporary of Sun-Tsi famous for the Chinese Remainder Theorem and

Diophantus of Alexandria famous for the Diophantus problem. Jootsy is alternatively said by folk etymology to come from the acronym for jumping-out-of-the-system. In any case it demonstrates Anderson's Law, "Any system or program, however complicated, if looked at in exactly the right way, will become even more complicated."

138 The answer is read as $10^{16} + (2 + 2)10^{15} + (3 + 4 + 3)10^{14} + (4 + 6 + 6 + 4)10^{13} + \ldots + (2 + 8 + 2)10 + 1 = 15{,}241{,}578{,}750{,}190{,}521$.

139 The Cheela have a metabolism based on nuclear reactions much faster than chemical ones. Their calendar counts the rotation period of their pulsar Ova Draconis, just 0.1992687 seconds. Their whole recorded history therefore from 0/0/0/1 to the prophesied closest approach of the starship *St. George* at 1/1/1/1 was therefore less than a week in AD 2050. The perihelion of Ova Draconis was not predicted to occur until 129/1/1/1/1 in AD 2403, 353 years later. (*Dragon's Egg* by Robert Forward).

140 The tenth of a cent, the mil, symbolized like the cent with a slashed letter, ₥, would be a hundred times larger than the millicent, m¢, said to be named for Millicent Garrett Fawcett, author of *Political Economy for Beginners*. The proper name "Millicent" means honest or diligent.

141 Dominissimo is Latin for "the Lord of lords" or "the domino game of all domino games". More complete rules and terminology are at http://michaelhalm.tripod.com/id76.html, gleaned from many other games including Authors, Baccarat, Bingo, Black Jack, Boule, Bowling, Bridge, Buzz, Canasta, Casino, Chish, Chiss, Chist, Chosh, Choshogi, Choss, Chost, Courier, Craps, Cribbage, Dice, Dominoes, Double Dutch, Fairy Chess, Faro, Football, Gin, Gleek, horseracing, Imperial, Jetan, Lowball, Mah Jongg, Michigan, Odd John, Panguingue, Pedro, Pinoche, Piquet, Poker, Pontoon, RennChess, Roulette, Rummikub, Rummy, Samba, Scrabble, See-Low, Shogi, Slogger, Snip Snap, Snoozer, Soccer, Solitaire, Tiddly-winks, Tombola, Tressette, Tien Gow, Tiu U, Xianggi, and Yacht. Dominissimo is André Joyce's similar game using nine-pip dominoes in which the object is to get rid of dominoes by correctly adding to the dealer's sequence. With the dominoes' paired digits a player can play one, two or more digits using whole or half-dominoes. If not, the player must leave that domino to the side to indicate what has been incorrectly played and redraw.

According to *Mathematical Circus* by Martin Gardner, quoting the *Encyclopedia Britannica*, the Eskimos played with bone sets between sixty and a hundred forty-eight pieces -- for wives. That would include twelve through sixteen pips. The Chinese zero-less Dominoes are called *kwat p'ai*. If also without nines, it would be called trimmed Dominoes.

142 Eleusis is a card game invented by Robert Abbott, and popularized by Martin Gardner in 1959 and then simplified by John Golden. See www.logicmazes.com/games/eleusis.

143 The *On-line Encyclopedia of Integral Sequences* at www.research.att.com
is an expansion of the original *Encyclopedia of Integral Sequences* by Neil Sloane and Simon Plouffe. If any reader of this book is free to add or add to a sequence in it on-line as long as a reference to the book is cited.

The OEIS sequence includes the nth term in nth numbered sequence.

OEIS	1	2	1	0	2	3	0	6	6

hypersum	1	12	121	1,210	12,102	121,023	1,210,230	12,102,306	121,023,066

144 *The Index of Forbidden Sequences*, of course, is a tongue-in-cheek reference to the *Index of Forbidden Books*, the list of books banned by the Church between 1559 and 1966.

145 There are several quite different ways of seeing "invisible" parentheses. The system known by the acronym BODMAS ranks operations with bracketed operations first, then division, multiplication, addition and subtraction, similar to PEDMAS, which puts exponentials between parentheses and division. The one known by the mnemonic My Dear Aunt Sally ranks multiplication first, then division, then addition and then subtraction. FORTRAN nests from the right, while Texas Instrument nests from the left. If you used Sally, My Dear Aunt you'd get something even more strange, even without ever including tetration or higher operations.

BODMAS and PEDMAS	(1+2)-(3*(4/5)) = -3/5
FORTRAN	((((1+2)-3)*4)/5) = 0
Sally, My Dear Aunt	1+(((2-3)*4)/5) = 1/5
My Dear Aunt Sally	(1+2)-((3*4)/5) = 3/5
TI	(1+(2-(3*(4/5)))) = 7/5

146 The digits in alphabetical order, 'digitalphabeticals', are eight, five, four, nine, one, seven, six, three, two, zero. This would not allow any leading zeroes or monodigitals.

digitalphabetical	85	84	89	81	87	86	83	82	80	58	54	59	51	57	56	53	52	50	45	49

broken-up	85	7,141	635,634	51,493,495	4,480,569,699	385,380,487,609
tetrasub	1	131	11,642	943,136	82,064,497	7,058,489,846
broken-down	1	86	7,655	620,141	53,959,922	385,271,354,861
broken	1	85	86	7,141	7.655	635,634

hypersum	10	1,012	101,213	10,121,316	1,012,131,617	101,213,161,720
dusub	1	137	13,698	1,369,771	136,977,119	13,697,711,909

The digits in reverse alphabetical or zeewyexical order are zero, two, three, six, seven, one, nine, four, five, eight.

zeewyexical	20	23	26	27	21	29	24	25	28	30	32	36	37	31	39	34	35	38	60	62

broken-up	20	461	12,006	324,623	6,829,089	198,368,204	4,767,665,985
tressub	1	23	598	16,162	340,000	9,876,171	237,368,112
broken-down	1	21	547	14,790	311,137	9,037,763	217,217,449
broken	1	20	21	461	547	12,006	14,790

hypersum	20	2,023	202,326	20,232,627	2,023,262,721	202,326,272,129

tressub	1	101	10,073	1,007,323	100,732,319	10,073,231,943

147 One of the most enduring such problems has been the Four Fours Problem which has occupied problem solvers since first published in *The Schoolmaster's Assistant: Being a Compendium of Arithmetic Both Practical and Theoretical* by Thomas Dilworth circa 1744. The problem is to express the highest number limited to just four fours and four usual operators of addition, subtraction, multiplication and division. He however only asked for solutions to a mere 100. With these limits the maximum would, of course, be two fifty-six.

David A. Wheeler gives solutions to the Four Fours Problem from 0 to 112 at his website, http://www.wheels.org/math/44s.html. Where more than one expression is possible, the one with the fewest and simplest symbols is usually preferred, which depends on which operators are considered simpler.

A subproblem to the Four-Fours is finding numbers expressible with just one four and building on them. The other three fours could be reduced to just adding 3 - 2 -1 = [!!!4] - 4 - [4]. In 1964 Donald A. Knuth found, albeit with the help of a computer, ways of expressing all the integers less than 208 using just a four, square root, √, factorial, !, and left and right brackets, [z], for the floor (rounding down) function, aka //z\\. Allowing the use of the primorial sign, #, simplifies some of them. Some allow lesser known symbols like the per mill symbol, ‰ = 1/1000, or \\z// the ceiling (rounding up) function or the binary logarithm, lgg, repeating decimal indicated with decimal point and overline, $.\overline{z}$, exponentiation can be indicated not by superscription but with a caret, ^, and tetration with a double caret, ^^, and so on for the higher operations.

Here are some creative candidates for Four Fours: **0** = 44%44, **1** = 44/44, **2** = 4/4 + 4/4, **3** = (4 + 4 + 4)/4, 4 = 4_{444}, **5** = 4 + 4^{4-4}, **6** = 4 + (4 + 4)/4, **7** = 4 + 4 - 4/4, **8** = (4 + 4)(4/4), **9** = 4 + 4 + 4/4, **10** = (44 - 4)/4, **11** = 4/.4 + 4/4, **12** = (44 + 4)/4, **13** = 4! - 44/4, **14** = 4 + 4 + 4 + √4, **15** = 44/4 + 4, **16** = 4 + 4 + 4 + 4, **17** = (44 + 4!)/4, **18** = 44(.4) + .4, **19** = 4! - 4!! + 4/4, **20** = 4!! + 4 + √4 + √4

The "fourth-four" sequence includes the other digits with four anchored in the fourth place, "out of the money". After getting as high as 9,994 it jumps to 10,040.

fourth-four	1,004	1,014	1,024	1,034	1,054	1,064	1,074	1,084	1,094	1,104

broken-up	1,004	1,018,057	1,042,491,372	1,077,937,096,705
heptasub	1	928	950,629	98,2951,399
broken-down	1	1,005	1,029,121	1,064,112,119
broken	1	1,004	1.005	1,018,057

hypersum	1,004	10,041,014	100,410,141,024	10,041,014,102,401,034
heptasub	1	9,156	91,562,197	9,156,219,675,858

There are, of course, many, many other "anchored digit" sequences "between" the "first one" (mono phoenix) and the "last one" (ichibi).

Peter Karsanow at http://www.geocities.com/hentaihelper/44sfaq.htm refers to "The 4444 Problem" to prove that all numbers between 0 and 4444 are expressible. He uses .4~ for a repeating decimal, rather than .4' .4 or .(4), but arbitrarily disallows log, ln, ^, #, $, ${}_zC_{z'}$ and ${}_zP_{z'}$.

Another sub-problem is finding just the primes expressible with four fours, for example, 73 = (4!/4)/4 + 4.

The Math Chat Book: Episode 19 by Frank Morgan, James F. Bredt has:

- 257 = 4^4 + 4/4 [William Foster]
- 577 = 4!(4!) + 4/4 = 577 [James Grimm and Michael Stern]
- 331,777 = $(4!)^4$ + 4/4 [Garrett Gray and Michael Eastep]
- 479,001,599 = (4!/4)! - 4/4 = 12! - 1 [Eric Brahinsky]
- 1,197,503,999 = ((4!/.4)! -.4)/.4 [William Foster wrote, "I doubt that a larger one will be found by publishing date." The next day he found (((4/.4)!)! - .4)/.4, higher than a hundred-ninety-nineplex.

148 Matt Hudelson's website is http://www.sci.wsu.edu/math/faculty/hudelson/moser.html. The wedge symbol indicates z nestings about the exponential of z| = g(3, 2, z) and is more easily understood in Joyce's More Generalized Exponential notation: z< = g(z, 1, 3, 2, z) = g(z - 1, 1, 3, g(3, 2, z), z). The triangle indicates z nestings of nestings, zΔ = g(z, 1, z, 3, 2, z) .The square indicates z nestings of triangles, z□ = g(g(2, z, z), 1, z, 3, 2, z). He gave special names to 2Δ or to-the-five-hundred-twelfth-two = zelda, "the infamous twenty-fifth letter of the Greek alphabet" [sic], > hundred-fifty-fourplex and to-the-zeldath-two = a-ooga, "the twenty-sixth letter of the Greek alphabet" [sic] > tentriplex. By extrapolation a pentagon would indicate yet another nesting, z⌂ = g(g(2, 1, 2, z, z) 1, z, 3, 2, z) and the hexagon yet another, z○ = g(g(3, 1, 2, z, z), 1, z, 3, 2, z).

Clifford Pickover's superfactorial for the product of all previous factorials, is indicated with an s before the exclamation mark, defined recursively as zs! = z!(z - 1)s! = g(z - 1, 2, (z- 1)!, z) or Πz!.

A. Berezin's supersuperfactorial, symbolized by a rightward dollar sign indicates the factorial tetrated-to-the-second, z$ = g(3, 2, z!). The hectosupersuperfactorial would be symbolized with a right ¢ and the kilosupersuperfactorial with a right ₥. the symbol for a mil. Mike Wilber's exfunctory he expressed with a superscripted exclamation mark, $z^{!}$ = g(4, 2, z!). His z hyperexfunctory he expressed with a superscripted h!, $z^{h!}$ = g(2, 4, 2, z) and his z superhyperexfunctory with a superscripted sh!, $z^{sh!}$ = g(3, 4, 2, z). Expressions analogous to left factorials would be the much higher $^{!}z$, $^{h!}z$ and $^{sh!}z$, to the multifactorials the not-so-high $z^{!}$, $z^{h!!}$ and $z^{sh!!}$ and the to the hybrid left multifactorials $^{!!}z$, $^{h!!}z$ and $^{sh!!}z$, etc.

There also dyadic operators that are placed between two numbers: the much lower remainder or modulo represented by a percent sign between, z%z' = z (mod z'). Aalbert Torius's much higher torian, the product of factorials from z to z' represented by a exclamation mark between, z!z' = z!((z - 1)!...(z')!)) and the analogous morian the product of primorials from z to z' represented with a hash mark between, z#z' = z#((z - 1)#...(z')#) and their multitorian, z!!z', and multimorian, z##z', variants.

Adam Clarkson's three-variable or triadic hyperfactorial function is a generalization of the factorial, multifactorial, exponential factorial, etc. The first variable is the multiple of a multifactorial, the second is the operational level of the factorial and the third is the number factorialized. In Joyce's notation, hypf(z, z', z") = g([(z" - 1)/z'], z + 1, [(z" - 1)/z'], z"). The Clarkson function is a four-variable or tetradic hyperfactorial function with the rightmost variable now the number of iterations, hypf(z, z', z", z''') = g(z, [(z" '- 1)/z"], z' + 1, [(z''' - 1)/z"], z''').

149 **1** = (√3√3)/3, **2** = 3 - 3/3, **3** = 3+ 3 - 3, **4** = 3 + 3/3, **5** = 3!/3 + 3, **6** = 3/.3 - 3, **7** = 3! + 3/3, **8** = [3√3 + 3], **9** = 3^3/3, **10** = [3(√3 + √3)], **11** = 33/3, **12** = 3(3) + 3, **13** = 3/.3 + 3, **14** = [3(√3 + 3)], **15** = 3(3!) - 3, **16** = [3(3!) - √3], **17** = 31/.3 - 3, **18** = 3(3 + 3), **19** = [$3!^{\sqrt{3}}$] - 3, **20** = (3 + 3)/.3, **21** = 3(3!) + 3, **22** = [[3! - √3]!! - √3], **23** = [3! - √3]!! - [√3], **24** = [3! - 3/√3]!!, **25** = [3! - √3]!! + [√3], **26** = [[3! - √3]!! + √3], **27** = 3(3)3, **28** = 3^3 + [√3], **29** = [(3!)!!/√3 + √3], **30** = 33 + 3, **31** = 3/.[√3] + [√3], **32** = 33 - [√3], **33** = 3^3 + 3!, **34** = 33 + [√3], **35** = 33 + [√3!], **36** = 33 + 3, **37** = [√[√((3!)!)]!/3 - 3, **38** = 33 + [√(√((3!)!)], **39** = 33 + 3!, **40** = (3!)!/3(3!), **41** = [√((3!)!)]/3 + [√3], **42** = [√((3!)!)]/3 + [√3!]

The third-three anchored numbers

third three	103	113	123	143	153	163	173	183	193	203	213	223	243	253	263

broken-up	103	11,640	1,431,823	204,762,329	31,330,068,160	51,070,058,172,409
pentasub	1	78	9,648	1,379,678	211,100,339	34,410,734,892
broken-down	1	104	12,793	1,829,503	279,926,752	45,629,890,079
broken	1	103	104	11,640	12,793	1,431,823

hypersub	103	103,113	103,113,123	103,113,123,143	103,113,123,143,153
pentasub	1	695	694,771	694,770,759	694,770,758,649

150 **0** = 1 - 1 in three symbols, **1** in one symbol, **2** in one symbol, **3** = S2 in two symbols, **4** = 2*2 in three symbols, **5** = S2*2 in four symbols, **6** = S2! in three symbols, **7** = SS2! in four symbols, **8** = 2^3 in three symbols, or 2^3 in two, **9** = 3^3 in three symbols or 3^3 in two.

151 Brian Boutel began by expressing ϕ with Four Fours. That prompted the answers with Five Fives by Jaroslaw Tomasc Wroblewski and Seth Breidbart's (5/5 + √5)/$\log_{\sqrt{5}}$5. David G. Caraballo expressed ϕ as (5 + 5(√5))/(5 + 5), while Phil Hanna expressed it as Eight Eights, 8 + 8√(√(8 + 8) + 8/8))/(8 + 8) and Nine Nines, (9 + 9√(√9 + 9/9 + 9/9))/(9 + 9).

fifth-five	10,005	10,015	10,025	10,035
broken-up	10,005	100,200,076	1,004,505,771,905	10,080,215,521,266,751
enneasub	1	12,366	123,965,861	1,243,997,422,631
broken-down	1	10,006	100,310,151	1,006,612,375,291
broken	1	10,005	10,006	100,200,076

fifth-five	10,005	1,000,510,015	100,051,001,510,025	10,005,100,151,002,510,035
enneasub	1	123,473	12,347,274,495	1,234,727,449,502,830

152 A third category, if it can be called such, are the amateur googologists, whose numbers aren't very helpful contributions to either syntactic or abstract googology, whose definitions are inelegant, and have no consistency or mathematical significance. Most importantly their googolisms violate the Gentleman's Rule of large number duels, ("On Salad Numbers" by Nathan Ho), which allows numerical tricks, but requires both logic and wit. ("Profs Duke It Out in Big Number Duel")

Douglas Hofstader promoted a Luring lottery in a series of essays on the Prisoner's Dilemma starting in the June 1983 *Scientific American* and reprinted in his book, *Metamagical Themas* with the alleged million dollar prize money to be split between all entries for the largest number definable on a postcard. Since so many entries, googolplex and higher, were received the prize money per entry came to less than a millicent each. The Largest Number Game however continues. In 2003 the British "Who Wants to Be a Millionaire?" scandal raised the prize to a million British pounds just for knowing how much a googol is.

153 0 = 111 - 111, **1** = 111/111, **2** = 1/1 + 11/11, **3** = 1/1 + 1/1 + 1/1, **6** = 1 + 1 + 1 + 1 + 1 + 1, **7** = 11 - 1 - 1 - 1 - 1, **8** = 11 - 1/1 - 1 - 1, **9** = 11 - 1/1 - 1/1, **10** = 11 - 11/11, **11** = 11(1(1(1(1)))), 111(1(1(1))) = **111**, 11(11(1(1))) = **121**, 1111(1(1)) = **1,111**, 11(11(11)) = 1,331, 111(111) = **12, 321**, 111(1(11!)) = **4,430,764,800**, 11!((11!)11!) **> 22plex**, 1111^{11} **> 31plex**, 111^{111} > **125plex**, 1(1(1(111!))) **> 180plex**, 111^{111} **> 227plex,** 111!(111!) **> 360plex**, 1(1(1111!)) **> 3191plex**, 11!(1111!) **> 3198plex**, 11^{1111} **> 21159plex**, 1(11111!) **> 40553plex**, 111111! **> 300000plex,** 11^(11!^(11!))! [Holt] ≈ **g(3, 6, 10),** ${}^{11}1111$ **> g(3, 33, 10)**, ${}^{111}111$ **> g(3, 227, 10)**, ${}^{1111}11$ **> g(3, 1157, 10)**

154 12345678^{9} **> 63plex**, 1234567^{89} **> 542plex**, 123456^{789} **> 4017plex**, 12345^{6789} **> 27777plex**, 1234^{56789} > **175552plex**, 123^{456789} > **954645plex**, $12^{3456789}$ **> 3730501plex**, ${}^{12}3456789$ **> 78plex**, ${}^{123}456789$ > **696plex,** ${}^{1234}56789$ **> 5866plex**, ${}^{12345}6789$ > **47303plex**, ${}^{123456}789$ **> 357661plex**, ${}^{1234567}89$ > **2406662plex**, ${}^{12345678}9$ **> 11780770plex**

155 In 1536 Hudalricus Regius proved that two-to-the-eleventh minus one, 2,047, was not prime and therefore 2,096,128 not perfect and so not all prime exponents of two produced perfect numbers. By 1603 Pietro Cataldi correctly verified that two-to-the-seventeenth minus one and two-to-the-nineteenth minus one indeed were both prime and so 8,589,869,056 and 137,438,691,328 were perfect. Mersenne suspected two-to-the-thirty-first minus one and two-to-the-hundred-twenty-seventh minus one were primes, but they were not proved to be until 1883 and 1876 by Pervouchine and Lucas respectively, giving perfects greater than eighteenplex and seventy-sixplex. The prime exponents which do give primes are called Mersenne.

Mersenne exponents	2	3	5	7	13	17	19	31	61	89	107	127	521	607	1,279	2,203

In 1811 Peter Barlow wrote that the former "is the greatest that will be discovered; for as they are merely curious, without being useful, it is not likely that any person will attempt to find one beyond it." He did not realize all numbers are interesting,[27] even those Mersenne numbers not prime. Prime hunting however has become much easier with the ever-growing power of the computer. Pierre Simon de LaPlace defined as 'extraordinary' a probability of at most a millionth, or one microkan. A kan is one hundred percent probability, backformed from a millikan, named for Robert Andrews Millikan, for a probability of a thousandth, ‰. The gaps between the primes is approximately root-z-times-e-to-root-z, so a probability of a thousandth is reached at about fifteenplex and a millionth at over four-hundredplex.

156 Mersenne listed the first such primes in his *Cognitata Physica-Mathematica* in 1644. The notations for Mersennes, M_n, and for perfects, P_n, sometimes are used in the Four Fours and other constructions.

hyperMersenne	2	23	235	2,357	235,713	23,571,317	2,357,131,731	235,713,173,167
sub	1	8	86	867	86,714	8,672,403	867,140,304	86,714,030,421

n	p_n	M_n length	P_n length	year
12	521	157	314	1952
13	607	183	366	1952
14	1,279	368	770	1952
15	2,203	664	1,327	1952
16	2,281	687	1,373	1952
17	3,217	969	1,937	1957
18	4,253	1,281	2,561	1961
19	4,423	1,332	2,663	1961
20	9,689	2,917	5,834	1963
21	9,941	2,993	5,985	1963
22	11,213	3,376	6,751	1963
23	19,937	6,002	12,003	1971
24	21,701	6,533	13,066	1978
25	23,209	6,987	13,973	1979
26	44,497	13,395	26,790	1979
27	86,243	25,962	51,924	1982

The search for Mersenne primes and their associated perfects got another boost with the networking of computers in the 1990's into the Great International Mersenne Prime Search (GIMPS), which has discovered many of the newest, highest ones, though still just twenty-two million-something digits.

157 The first titanic discovered was g(2, 44'497, 2) - 1 > g(2, 13'395, 10). The first Titan or titanic discoverer was Alexander Hurwitz, who also found g(2, 4'253, 2) - 1 > g(2, 12611, 10). By 1984 Samuel Yates listed a hundred Titans. By 1994 his titanic list was up to 1,426. By 1998 perfects were over a million digits. The first megaprime discovered was g(2, 697'254, 2) - 1 > g(2, 2'098'960, 10) by Nayan Hajratwada in 1999. A prime higher than nineplex has been called a bevaprime, from bev, billion electon volts, or perhaps Beverly, but would more consistently be called a gigaprime.

158 The base 3 sequence is called the Morris sequence after cryptologist Roger Morris and The audio-active elements ranked in ternary and decimal are: **3** = U, 12_3 = **5** = Ca, 13_3 = **6** = Pa, 22_3 = **8** = H, 132_3 = **14** = Pm, 312_3 = **32** = Zn, 1112_3 = **41** = K, 1113_3 = **42** = Th, 3112_3 = **95** = Ar, 3113 = **96** = Ac, 11131_3 = **127** = La, 11132_3 = **128** = Hf, 13211_3 = **184** = Sn, 31132_3 = **289** = Cr, 32112_3 = **311** = Co, 111312_3 = **385** = Nd, 131112_3 = **527** = Cu, 132112_3 = **554** = Cl, 132113_3 = **555** = Ra, 311311_3 = **868** = Ba, 311312_3 = **869** = Lu, 311332_3 = **875** = Sm, 1112133_3

= **1,128** = Y, 1113222_3 = **1,160** = Eu, 1321132_3 = **1,667** = Ho, 1322112_3 = **1,688** = Si, 1322113_3 = **1,689** = At, 3112112_3 = **2,579** = Sr, 3112221_3 = **2,590** = Sb, 11131221_3 = **3,454** = In, 11133112_3 = **3,497** = Ni, 13122112_3 = **4,847** = Fe, 13211312_3 = **5,000** = V, 13211321_3 = **5,002** = Cs, 31131112_3 = **7,817** = Pr, 123222112_3 = **13,838** = Na, 123222113_3 = **13,839** = Pb, 311311222_3 = **23,462** = **Er**, 1113122112_3 = **31,091** = S, 1113122113_3 = **31,092** = Fr, 1113222112_3 = **31,334** = Al, 1113222113_3 = **31,335** = Po, 1321122112_3 = **44,942** = Rb, 1321131112_3 = **44,996** = Yb, 1321133112_3 = **45,050** = Ce, 3113112211_3 = **70,384** = Cd, 3113322112_3 = **70,943** = Mg, 3113322113_3 = **70,944** = Bi, 13221133112_3 = **136,904** = Gd, 111213322112_3 = **274,334** = Ne, 111213322113_3 = **274,335** = Tl, 111311222112_3 = **279,194** = Mn, 111312211312_3 = **279,833** = Dy, 132113212221_3 = **405,241** = Ag, 132211331221_3 = **410,722** = Te, 11311222112_3 = **633,488** = P, 311311222113_3 = **633,489** = Rn, $1113122113111 2_3$ = **2,518,493** = Ti, 11131221131211_3 = **2,518,501** = Xe, 11131221133112_3 = **2,518,547** = Tm, 11131221222112_3 = **2,518,682** = Kr, 31121123222112_3 = **5,643,176** = F, 31121123222113_3 = **5,643,177** = Hg, 311322113212221_3 = **17,175,157** = Tc, 3113112211322112_3 = **51,310,895** = Br, 3113112221131112_3 = **51,312,650** =Tb,3113112221133112_3 = **51,312,704** =Sc, 1322113312211332_3 = **99,805,835** = Ga, $11131221131211321 1_3$ = **203,998,684** = Pd, 132112211213322112_3 = **294,869,795** = O, 132112211213322113_3 = **294,869,796** = Au, 311311222113111221_3 = **461,813,857** = I, $1321132221131211321 1_3$ = **2,659,256,104** = Mo, 13211321222113222112_3 = **2,658,791,291** = Se, $13221133122211311221 1_3$ = **8,084,272,144** = Ru, $12322211331222113112211_3$ = **22,074,456,469** = Zr, $311311222113111221132 22_3$ = **112,220,767,439** = Ge,$1113122122211211232221 12_3$ = **148,725,686,939** = N, $3113112221131112211312 21_3$ = **336,662,302,291** = Rh, $11131221131211322113322112_3$ = **1,338,435,390,740** = As, $3122113222122211211232221 12_3$ = **9,281,776,294,625** = Li, $31221132221222112112322211 3_3$ = **9,281,776,294,626** = W, $1113122113322113111221131221_3$ = **12,046,188,301,990** = Nb, $311311221132211221121332211 2_3$ = **27,268,713,801,176** = C, $3113112211322112211213322113_3$ = **27,268,713,801,177** = Ir, $1311222113321132211221121332211 2_3$ = **1,249,394,596,541,164** =He, $13112221133211322112211213322113_3$ = **1,249,394,596,541,165** = Ta, $13211321222113222122211211232221 12_3$ = **18,275,976,907,276,352** = B, $1321132122211322212221121123222113_3$ = **18,275,976,907,276,353** = Os, $1113122113121132211332113221122112133221 12_3$ = **57,615,254,840,843,428,793** = Be, $111312211312113221133211322112211213322113_3$ = **57,615,254,840,843,428,794 =** Re

Carl Sagan in *Broca's Brain* points out that even the most stable element, iron, has an estimated half-life of "only" five-hundredplex years.

159 13211321322113_3= $10{,}942{,}254_{10}$ ≠ 1321132_3 + 1322113_3= 5,001 + 5,067

160 HoAt is 13211321322113_3 while AtHo is slightly higher 13221131321132_3= 11,086,530.

161 "The Magnetic Monster" by Curt Siodmak, defeated by the A-men

162 Greek prefixes like that used to name protactinium, of which actinium is the decay product, can easily be extrapolated: proto- first, deutero- second, trito- third, tetaro- fourth, pempto- fifth, ekt- sixth, esdomo- seventh, ondo- eighth, enato- ninth, dekato- tenth, endekato- eleventh, duodekato- twelfth, eikosto- twentieth, triakosto- thirtieth, tessarakosto- fortieth, pentekosto- fiftieth, esdomekosto- seventieth, ondoeikosto- eightieth, ekatosto-

hundredth, and chilosto- thousandth.

The mixed Greco-Roman prefixes are from the rules of the International Union of Pure and Applied Chemisty's Commission on Nomenclature of Inorganic Chemistry.

163 Mendeleev's system for naming as yet unknown elements is based on the first three Sanskrit prefixes. From the so-called magic numbers of nuclear physics we can extrapolate much higher. From the atomic numbers for the first column of elements: hydrogen, lithium, sodium, potassium, rubidium, cesium and francium the other thirty-two known columns can be named. Not all elements however can be easily named because they are in completely new areas of the periodic chart. Higher than ekaradium are unbunium, unbibium, unbitrium, unbiquadium, unbipentium, unbihexium, unbiseptium, unboctium, unbenium, untrinilium, etc.

"magic"	1	3	11	19	37	55	87	119	175	231	321	411	547	683	879	1,135
gnomon		2	8	8	18	18	32	32	56	56	90	90	136	136	196	256
2nd gnomon			6	0	10	0	16	0	24	0	34	0	46	0	60	0

Alternatively "magic" could be the acronum for Mersenne-and-gobi-indexed cubes.

Magic	3	7	31	125	127	2,047	3,375	8,191	15,625	42,875	91,125	131,071	166,375

164 Higher than francium would be ekafrancium, etc., to kotifrancium, higher yet element twenty-oneplex; ayutafrancium, nearly twenty-sevenplex; niyutafrancium, about thirty-threeplex; kankafrancium, thirty-nineplex; pakotifrancium forty-twoplex; vivarafrancium; forty-fiveplex; kshobhyafrancium, fifty-oneplex; vivahafrancium, fifty-sevenplex; kotippakotifrancium, sixty-twoplex; bahulafrancium, sixty-nineplex; nagabalafrancium, seventy-fiveplex; nahutafrancium, eighty-fourplex.

Much simpler are the aptly named elementary numbers, with digits able to be broken up into element symbols, like one, "oxygen neonite", ONe, or nine, "nickel neonite", NiNe.

elementary	1	9	11	19	91	99	111	119	191	199	911	919	991	999	1,111	1,119

broken-up	1	10	111	2,119	192,940	19,103,179	172,121,551	1,912,440,240
broke-down	1	2	23	439	39,972	3,957,667	35,658,975	396,206,392
broken	1	2	10	23	111	439	2,119	39,972

hypersum	1	19	1,911	191,119	19,111,991	1,911,199,199	1,911,199,199,911

hypersuper	1	9	11,111,111,111,111,111,111,111,111,111,111,111,111	> eighteenplex

"Chemonums" are acronums using element symbols: Ac = actinium = -and-cube, Ag = silver = -and-gobi, Al = aluminim = -and-lucky, Am = americium = -and-Mersenne, Ar = Argon = -and-rokubi, As = arsenic = -and-sanbi, At = astinine = -and-ternarish, B = boron = binarish, Ba = barium = binarish-and-, Be = beryllium = binarish even, Bi = bismuth = binarish ichibi, C = carbon = cube, Ca = calcium = cube-and-, Ce = cesium = cube-even, Cf = californium = cubed Fibonacci, Cl = chlorine = cube lucky, Cm = curium = cubed Mersenne, Co = cobalt = cubed odd, Cr = chromium = cubed rokubi, Cu = copper = cube-unindexed, Er = erbium = even rokubi, Es = einsteinium = evens sorted, Eu = europium = even-unindexed, F = flourine

= Fibonacci, Fe = iron = Fibonacci even, Ga = gallium = gobi-and-, H = hydrogen = hachibi, Hf = hafnium = hachibi Fibonacci, I = iodine = -indexed-, In = indium = indexed-nibi, Ir = iridium = -indexed-rokybi, K = potassium = kyuubi, La = lanthanium = lucky-and-, Li = lithium = lucky ichibi, Lr = lawrencium = lucky rokubi, Lu = lutetium = lucky-unindexed-, Mo = molybdenium = Mersenne odd, N = nitrogen = nibi-, Na = sodium = nibi-and-, Nb = niobium = nibi binarish, Ni = nickel = nibi-indexed-, Np = neptunium = nibi prime = 2, O = oxygen = odd, P = phosphorus = prime, Pa = proactinium = -prime-and-, Pb = lead = prime binarish, Pm = promethium = prime Mersenne, Po = polonium = prime odd, Pt = platinum = prime ternarish, Pu = prime-unindexed-, Ra = radium = rokubi-and-, Re = rhenium = rokubi even, Ru = ruthenium = rokubi-unindexed-, S = sulfur = sanbi, Sb = antimony = sanbi binarish, Sc = scandium = sanbi cube, Se = selenium = sorted even, Si = silicon = sanbi-indexed-, Sm = samarium = sanbi Mersenne, Sn = tin = sorted nibi, Sr = strontium = sorted rokubi, Ta = tantalium = ternarish-and-, Tc = technetium = ternarish cubed, Te = tellurium = tenarish even, Ti -titanium = ternarish ichibi, Tl = thallium = ternarish lucky, U = uranium = -unindexed-, Y = yttrium = yongbi.

165 The chained arrows are more clearly expressed in Joyce's notation as z→z' = g(3, z', z) and z→z'→z'' = g(z', 1, g(3, g(1, z'', 2), z)), z→z'→z''→z''' = g(z''', 1, 1, z', 1, g(3, g(1, z'', 2), z)) = g(z''' - 1, 1, g(z', 1, g(3, g(1, z'', 2), z), z'')

Stephen Houban's website is http://www.win.tue.nl/~stephanh, where he shared his partial calculation of four-pentated to-the-fourth by calculating the number four-to-the-zth, where z = 3.407807929942597099574024998205846127479365820592393377723561443721764030073546976801874298166903427690031858186486050853753882811946569946433649006084096g(2, 152, 10), he wrote "We can safely say that computing all digits of [this number] will never happen."

To which Alistair Cockburn replied, "Never happen? Nonsense! Here they are: 0, 1, 2, 3, 4, 5, 6, 7, 8, and 9. (Some assembly required.)" In binary it would be even simpler with just two parts to assemble, and we already know that it is even and so ends in zero.

- 2 → 2 → 2 → 2 = 2 → 2 → (2 → 2 → 1 → 2) → 1 = 2 → 2 → (2 → 2) = g(3, g(2, 2, 2), 2) = 65,536
- 2 → 2 → 3 → 2 = 2 → 2 → (2 → 2 → 2 → 2) → 1 = 2 → 2 → (2 → 2 → 2 → 2) = 2 → 2 → (2 → 2 → (2 → 2)) = 2 → 2 → 65,536 = g(3, 2, g(3, g(2, 2, 2), 2), 2) = g(2, 3, 2, g(2, 2, 2)) = g(3, 65536, 2) > 2g(2, 19728, 10)
- 3 → 3 → 3 → 3 = 3 → 3 → (3 → 3 → 2 → 3) → 2 = 3 → 3 → (3 → 3 → (3 → 3 → 1 → 3) → 2) → 2 = 3 → 3 → (3 → 3 → (3 → 3) → 2) → 2 = 3 → 3 → (3 → 3 → 27 → 2) → 2 = g(g(3, 3, 3), 4, g(3, 3, 3), 3) = g(27, 4, 27, 3) = g(3, g(2, 27, 2), 2) = g(3, 134217728, 2)
- 2 → 2 → 2 → 3 = 2 → 2 → (2 → 2 → 1 → 3) → 2 = 2 → 2 → (2 → 2) → 2 = 2 → 2 → 4 → 2 = 2 → 2 → (2 → 2 → 3 → 2) → 1
- 4 → 4 → 4 → 4 = 4 → 4 → (4 → 4 → 3 → 4) → 3 = 4 → 4 → (4 → 4 → (4 → 4 → 2 → 4) → 4) → 3 = 4 → 4 → (4 → 4 → (4 → 4 → (4 → 4 → 1 → 4) → 4) → 4) → 3 = 4 → 4 → (4 → 4 → (4 → 4 → (4 → 4) → 4) → 4) → 3 = 4 → 4 → (4 → 4 → (4 → 4 → 256 → 4) → 4) → 3 = 4 → 4 → (4 → 4 → (4 → 4 → (4 → 4 → 255 → 4) → 3) → 4) → 3 = g(256, 5, 256, 4) = g(3, g(2, g(2, 256, 2), 2), 4)

166 What has been called the Conway-Guy function, from *The Book of Numbers* by John Horton Conway and Richard Guy, is a number linked by one less than that number of

Conway's chained arrows, a rather fast-growing function.

- CG(2) = 2→2 = 2↑2 = g(2, 2, 2) = 4
- CG(3) = 3→3→3 = 3↑↑↑3 = g(4, 3, 3) = g(3, g(3, 3, 3), 3) = g(3, 27, 3) = 7,625,597,484,986.

The recursive Conway-Guy function is even more absurdly fast-growing, the second term already inexpressible in even tetration's left-superscription.

- RCG(2, 2) = CG(CG(2)) = CG (4) = 4→4→4→4 = g(5, 4, 4) = g(4, g(4, 4, 4), 4) = g(4, g(3, g(3, 4, 4), 4), and since to-the-fourth-four is nearly a googolplex, and to-the-googolplexth-four is only slightly less than googolduplex, just the argument of the pentation has a height of googolduplex.
- RCG(3,2) = CG(CG(CG(2))) = CG(CG(4)) = g(2, 1, 1, 5, 4, 4)
- RCG(4, 2) = CG(CG(CG(CG(2)))) = CG(CG(CG(4))) = g(2, 1, 1, 2, 1, 1, 5, 4, 4)

167 To get higher Knuth did not use the usual Latin prefixes, nor the Greek, but the prefix latin- on the numbers that Latin had no name for. Thus latinbillionyllion was his name for ten-to-the-eight-nineplex-eighth. Knuth gave the special name, umptyllion, to an even higher number that otherwise would have been latinbyllionyllionyllionyllion, higher than tenpentaplex.

168 The googolisms from the Cockburns, Howell, Leach, Kreitzberg and Lavrov all came from the site, http://c2.com/cgi/wiki?ReallyBigNumbers.

169 Matt Leach's prefix ban- is said to be an acronym for "bad-a** number".

170 g(4, 2, 4) = g(3, g(3, 4, 2), 4) = g(3, 256, 4) ≈ g(3, 93, 10)

171 He also defined the expostfacto function recursively as z-to-the-factorial-of-the-z-minus-oneth-expostfacto. Since xpf(1) = 1!^0! = 1, xpf(2) = 2!^(1!^0!) = 2, xpf(3) = 3!^(2!^(1!^0!)) = 9, one would expect based on gnomonics something like 32 = 9 + (9 - 2) + (2 - 1) + (7 - 1), but instead we get xpf(4) = 4!^(3!^(2!^(1!^0!))) = g(2, 9, 24) = 2,641,807,540,224 while xpf(5) = 5!^(4!^(3!^(2!^(1!^0!)))) = g(2, 2'641'807'540'224, 120) > twelveduplex.

Dmytry Lavrov shared with Cockburn, etal. a very powerful notation for indicating repeated iteration of any function. In addition to the usual superscription to indicate a multiply iterated function, as in a double f(z) or f(f(z)), he uses square brackets for f(z) iterations of f(z). If for example the function is doubling, the iterations would double each time as well. f[1](1) = 2, f[2](2) = f(f(2)) = f(4) = 8, f[3](3) = f(f(f(3))) = f(f(6)) = f(12) = 24, etc., so the function gets higher exponentially. Lavrov however has even faster-rising higher functions. defined by the equation f[1]g(z, z', z'') = g(2, z, z', z'') in Joyce's more generalized exponential notation, through all the letters between f and z.

- f[z] = g(z + 1, z', z'', z''')
- f[z, z] = g(g(2, z, z + 1), z', z'', z'''),
- f[z,..., z]g(z', z'', z''') = g(g(z - 1, 2, z, z + 1), z', z'', z'''),
- g[1]g(z', z'', z''') = g(g(g(2, 2, z - 1), 2, z, z + 1), z', z'', z'''),
- g[z]g(z', z'', z''') = g(g(g(2, z + 1, z - 1), 2, z, z + 1), z', z'', z'''),
- g[z, ..., z]g(z', z'', z''') = g(g(g(2, g(3, z, z + 1), z - 1), 2, z, z + 1), z', z'', z'''),
- h[z, ..., z]g(z', z'', z''') = g(g(g(2, g(4, z, z + 1), z - 1), 2, z, z + 1)), z', z'', z'''),
- y[z, ..., z]g(z', z'', z''') = g(g(g(2, g(2, 1, z, z + 1), z - 1), 2, z, z + 1)), z', z'', z'''),
- z[1](z', z'', z''') = g(g(g(2, g(2, 2, g(2, 1, z, z + 1), z - 1), 2, z, z + 1)), z', z'', z''').

For z = googol we get Lavrov's number = g(g(g(2, g(2, 2, g(21, g(2, 100, 10), g(2, 100, 10) + 1), g(2, 100, 10) - 1), 2, g(2, 100, 10), g(2, 100, 10) + 1)), 2, 100, 10)

Jonathan "Polyhedron Dude" Bowers writes a googol as {10, 3, 100} in his array notation and coins many new terms:

- tripent = {5, 5, 5} = 5{5}5 = g(5, 2, 5) = g(4, 5, 5)
- trisex = {6, 6, 6} = 6{6}6 = g(6, 2, 6) = g(5, 6, 6)
- trisept = {7, 7, 7} = 7{7}7 = g(7, 2, 7) = g(6, 7, 7)
- tridecal = {10, 10, 10} = 10{10}10 = g(10, 2, 10) = g(9, 10, 10), his first so-called "infinity scraper"

The four-variable or tetradic version is defined by {z, z', z'', z'''} = {z, z', {z, z', z''}, z''' - 1}, a nesting about the operator rather than about the base as in iteration.

- tetratri = {3, 3, 3, 3} in Bowers's notation, that is, to-the-second three with the power tower doubled twice or g(3, g(2, 1, 1, 2, 2), 3) = g(3, 8, 3) = g(2, 8, 8) ≈ tentriplex
- corporal = {10, 100, 1, 2} = 10{{1}}100 = 10{10{...{10{10}10}...}10}10 with 100 tens from center outward or g(99, 1, g(8, 10, 10), 10) > tentetraplex.
- grand tridecal = {10, 10, 10, 2}. The operation of z on z' with double brackets, z{{10}}z' = g(2, 1, 1, 9, z', z), Bowers calls expandodecation, from expando- plus decation, <u>not</u> expan- plus dodecation, an expansion of z{10}z' = g(9, z', z) which is dekation.
- pentatri = {3, 3, 3, 3, 3} or to-the-second-three with the power tower doubled to-the-second-three-times or g(3, g(2, g(3, 2, 3), 2), 3) = g(3, 134217728, 3) > g(3, 64038130, 10)
- pentadecal = {10, 10, 10, 10, 10} or tenplex with the power tower doubled tenplex times or g(3, g(2, g(3, 2, 10), 2), 10) > g(3, g(2, 3010299956, 10), 10)
- iteral = {10, 10, 10, 10, 10, 10, 10, 10, 10, 10} or tenplex with the power tower doubled sevenplex-times or g(3, g(2, g(3, 7,10), 2), 10) From the analogy tera:iteral::peta:? comes **ipetal** for tenplex with the power tower doubled tenenneaplex times or g(3, g(2, g(3, 10,10), 2), 10) and **izettal** tenplex with the power tower doubled tendudekatplex times or g(3, g(2, g(3, 13,10), 2), 10) and iyottal tenplex with the power tower doubled tenpentakaidekaplex times or g(3, g(2, g(3, 16,10), 2), 10).
- ultatri = g(3, g(2, 24, 2), 3) In general ultaz = g(3, g(2, g(2, g(3, 2, z) - 3, z), 2), z), so that **ultadecal** would be g(3, g(2, g(2, 999'999'997, 10), 2), 3)
- emperal = g(3, g(2, g(2, g(3, 3, 10) - 3, 10), 2), 10)
- hyperal = g(3, g(2, g(2, g(3, 4, 10) - 3, 10), 2), 10)
- dutritri = g(3, g(2, g(2, g(3, g(2, 2, 3), 3) - 3, 3), 2), 3)
- dutridecal = g(3, g(2, g(2, g(3, g(2, 2, 10), 10) - 3, 10), 2), 10), so that in general dutriz = g(3, g(2, g(2, g(3, g(2, 2, z), z) - 3, z), 2), z)
- xappol = g(3, g(2, g(2, g(3, g(3, 2, 10), 10) - 3, 10), 2), 10) By extrapolation **cappol** = g(3, g(2, g(2, g(3, g(3, 2, 100), 100) - 3, 100), 2), 100) and **mappol** would be g(3, g(2, g(2, g(3, g(3, 2, 1'000), 1'000) - 3, 1'000), 2), 1'000), and in general **zappol** would be g(3, g(2, g(2, g(3, g(3, 2, z), z) - 3, z), 2), z)
- dimentri = g(3, g(2, g(2, g(3, g(2, 3, 3), 3) - 3, 3), 2), 3)
- colossal = g(3, g(2, g(2, g(3, g(2, 3, 10), 10) - 3, 10), 2), 10)
- dimendecal = g(3, g(2, g(2, g(3, g(2, 10, 10), 10) - 3, 10), 2), 10), so that in general dimenz = g(3, g(2, g(2, g(3, g(3, 2, z), z) - 3, z), 2), z)

- gongulus = X{10, 100, 3}, 2X in his exploding array notation or g(3, g(2, g(2, g(3, googol, 10) - 3, 10), 2), 10) = g(3, g(2, g(2, g(3, g(2, 100, 10), 10) - 3, 10), 2), 10)
- dulatri = g(3, g(2, g(2, g(3, g(2, 2, g(4, 2, 3)), 3) - 3, 3), 2), 3) and in general dulaz = g(3, g(2, g(2, g(3, g(2, 2, g(4, 2, z)), z) - 3, z), 2), z)
- trimentri = g(3, g(2, g(2, g(3, g(3, 3, 3), 3) - 3, 3), 2), 3), so that trimenz = g(3, g(2, g(2, g(3, g(4, 2, z), z) - 3, z), 2), z) and in general zmenz' = g(3, g(2, g(2, g(3, g(z + 1, 2, z'), z') - 3, z'), 2), z'). Since -women- is the complimentary infix to -men-, zwomenz' = g(2, [log(g(3, g(2, g(2, g(3, g(z + 1, 2, z'), z') - 3, z'), 2), z'))] + 1, 10)/9] - g(3, g(2, g(2, g(3, g(z + 1, 2, z'), z') - 3, z'), 2), z') - 1)
- goppatoth = g(3, g(2, g(2, g(3, g(3, 100, 10), 10) - 3, 10), 2), 10)
- tridecatrix = g(3, g(2, g(2, g(3, g(9, 100, 10), 10) - 3, 10), 2), 10)
- boogol = g(99, 10, 10) = g(100, 2, 100), which can be extrapolated to **troogol** = g(100, 3, 10), etc.
- golapulus = X{10, 100, 3}, 3X or g(3, g(2, g(2, g(3, gongulus, 10) - 3, 10), 2), 10) = g(3, g(2, g(2, g(3, g(3, g(2, g(2, g(3, g(2, 100, 10), 10) - 3, 10), 2), 10), 10) - 3, 10), 2), 10)
- golapulusplux = X{10, 100, 3} = g(3, g(2, g(2, g(3, g(2, 100, 10), 10) - 3, 10), g(3, g(2, g(2, g(3, g(2, 100, 10), 10) - 3, 10), 2), 10)), 10), so that **oneplux** = googol, **twoplux** = gongulus, **threeplux** = golapulus and in general zplux = X{10, 100, 3}, zX = g(3, g(2, g(2, g(3, g(3, g(2, g(2, g(3, z - 1, 10) - 3, 10), 2), 10) , 10) - 3, 10), 2), 10) Extrapolating we get zpluc = X{100, 100, 3}, zX = g(3, g(2, g(2, g(3, g(3, g(2, g(2, g(3, z - 1, 100) - 3, 100), 2), 100) , 100) - 3, 100), 2), 100) and zplum = X{1000, 100, 3}, zX = g(3, g(2, g(2, g(3, g(3, g(2, g(2, g(3, z - 1, 1'000) - 3, 1'000), 2), 1'000) , 1'000) - 3, 1'000), 2), 1'000).

172 The Joyceans further extrapolated these. "Muggle" according to the *Scotichronicon* by Abbott Walter Bower means "an Englishman's tail". In *A History of the Borders from Early Times* by Alistair Moffat notes that in the Thirteenth Century the Scots claimed that the English had tails. With J. K. Rowling's Harry Potter series it became associated with non-magic and so muggle got extrapolated to buggle, truggle, etc., comic to domic and momic, bomic, tromic, etc.

non-magic "muggle"	g(891, 890, 981, 2)
non-tragic "comic"	g(2'673, 2'672, 2'673, 2)
buggle	g(2, 891, 890, 981, 2)
truggle	g(3, 891, 890, 981, 2)
domic	g(13'473, 13'472, 13'473, 2)
momic	g(26'973, 26'972, 26'973, 2)
bomic	g(2, 26'973, 26'972, 26'973, 2)
tromic	g(3, 26'973, 26'972, 26'973, 2)

173 See http://home.earthlink.net/~mrob/pub/largenum-3.html.

174 See www.users.cs.york.ac.uk/%7Esusan/cyc/b/big.htm.

175 This would be familiar to any one who'd ever seen Curly in a Three Stooges film.

176 The term **googology** can be found correctly defined with other -ologies at the website, users.tinyonline.co.uk/gswithenbank/ologies.htm. A googol is, of course a composite number with two hundred prime factors, a hundred twos and a hundred fives. A **googolism** is any large number name or arithmonym. Googology should not to be confused with the related term **googlology**, though both seem to be much too often misspelled. **Google** has become a verb, though googol seems not to have yet. Googlology with an extra -lo- is the study of the origins, history, and structure of the popular search engine named Google. See www.googlology.info.

The word **googol** on the other hand was the name given by Milton Sirotta, nephew of mathematician Edward Kasner sometime before 1940, according to most reliable sources, but the origin of the word has become legendary. Milton in some references is Kasner's grandson or even an unnamed non-relative with an age varying anywhere between five and ten. Clarence Nelson even wrote that a teacher he had at Brooklyn Tech claimed he'd named it.

Daniel Tammet in *Thinking in Numbers* suggests that the word "googol" is related to other words suggesting bigness, "great", "grand", "gross", "gargantuan", "grow", and "gain", the elongation of "oo" and of "ll" as in "pull". "Great", "grand" and "gross" all come from words meaning "thick and coarse". "Gargantuan" comes from *Gargantua and Pantagruel* by François Rablais, from which gar- may also come. "Grow" goes back to the Indo-European root ghrē-. The final -l may go back to kel- from which comes "colonel", "excel", "hill". ("Indo-European and the Indo-Europeans" by Calvert Watkins)

177 In 1993 came Theoni Pappas' *Fractals, Googols and Other Mathematical Tales*. David M. Schwartz explained the googol to beginning readers in *G is for Googol* in 1998. Robert E. Wells proposed the question *Can You Count to a Googol?* in 2000 and Bob Frybarger did one better in "Want to get a feeling for a Googol?" at http://aaoj.homestead.com/files/googol.html where he calculates that a googol salt crystals at 75 per inch would fill nearly four tenplex known universes. If the early universe is taken to be inflationary, that could be thought of as the equivalent universe with a twenty-two trillion light year radius, with 99.999999997% of everything still unknown.

A much more modest attempt to imagine a googol can be found at the Megapenny Project, http://www.kokogiak.com/megapenny/nineteen.asp which uses pennies. One megapenny is a mile-high stack. Ten is an acre spread out. A hundred is two adult Blue whales' volume. Ten thousand is nearly a years' minting. A hundred thousand are half the world's supply. A trillion would cover the surface of the Earth twice.

178 The first computer-calculated prime was 180g(2, g(2, 127 - 1, 2), 2) + 1, nearly a hundred-eightyplex. It held the record until January 30, 1952 and computers and their programmers have been getting higher ever since.

179 Such usable Roman numerals would be:
-i = 1, -ij = 2, 3 = üj, -iv = 4, -v = 5, -vi = 6, -vij = 7, -vüj, -ix = 9, -xi = 11, -xij = 12, -xüj = 13, -xiv = 14, -xvi = 16, -xvij = 17, -xvüj = 18, -xix = 19, -xxi = 21, -xxij = 22, -xxüj = 23, -xxiv = 24, -xxix = 29, -xl = 40, -xli = 41, -xlij= 42, -xlüj = 43, -xliv = 44, -xlix = 49, -li = 51, -lij = 52, -lüj = 53, -liv = 54, -lv = 55, -lvi = 56, -lvij = 57, -lüj = 58, -lix = 59, -lx = 60, -lxi = 61, -lxij = 62, -lxüj = 63, -lxiv = 64, -lxix = 69, -lxxi = 71, -lxxij = 72, -lxxüj = 73, -xxiv = 74, -xxiv = 79, -c = 100, -ci = 101, -cij = 102, -cüj = 103, -civ = 104, -cvi = 106, -cvij = 107, =cvüj = 108, -cix = 109, -cx = 110, -cxi = 111, -cxij = 112, -cvüj = 113, -cxiv = 114, -cxix = 119, -cli = 151, -clij = 152, -clüj =

153, -cliv = 154, -clix = -159, -cci = 201, -ccij = 202, -ccüj = 203, -cciv = 204, -ccix = 209, -cdi = 401, -cdij = 402, -cdüj = 403, -cdix = 409, -d = 500, -di = 501, -dij = 502, -düj = 503, -div = 504, -dvi = 506, -dvij = 907, -dvüj = 908, -dxix = 909, -dxi = 511, -dxij = 512, -dxüj = 513, -dxiv = 514, -dxix = 519, -dli = 551, -dlij = 552, -dlüj = 553, -dliv = 554, -dlix = 559, -cmi = 901, -cmij = 902, -cmüj = 903, -cmiv = 904, -cmix = 909, -mi = 1001, -mij = 1002, -müj = 1003, -miv = 1004, -mvi = 1006, -mvij = 1007, -mvüj = 1008, -mvix = 1009, -mxi = 1011, -mxij = 1012, -mxüj = 1013, -mxix = 1019, -mli = 1051, -mlij = 1052, -mlüj = 1053, -mliv = 1054, -mlix = 1059, -mci = 1101, -mcij = 1102, -mcüj = 1103, -mciv = 1104, -mcvi = 1106, -mcvij = 1107, -mcüj = 1108, -mcix = 1109, -mcxi = 1111, -mcxij = 1112, -mcxüj = 1113, -mcxiv = 1114, -mcxix = 1119, -mcxli = 1141, -mcxlij = 1142, -mcxlüj = 1143, -mcxliv = 1144, -mcxlix = 1149, -mcli = 1151, -mclij = 1152, -mclüj = 1153, -mcliv = 1154, -mclix = 1159, -mcci = 1201, -mccij = 1202, -mccüj = 1203, -mcciv = 1204, -mccix = 1209, -mccli = 1251, -mcclij = 1252, -mcclüj = 1253, -mccliv = 1254, -mcclix = -1259, -mcdi = 1401, -mcdij = 1402, -mcdüj = 1403, -mcdix = 1409, -mcmi = 1901, -mcmij = 1902, -mcmüj = 1903, -mcmiv = 1904, -mcmix = 1909, mmi = 2001, -mmij = 2002, -mmüj = 2003, -mmiv = 2004, -mmix = 2009.

There could also be the rarer, but more pronounceable, ones, like: xxxxxix = -il = 49, -lxxxxxix = -ic = 99, clxxxxix = -cic = 199, -cxxxxix = -cil = 149, -ccclxxxxix = -cid = 399, -dccclxxxxix = -cim = 899, -cclxxxxix = -ccic = 299, -cclix = -ccil = 249, -cccclxxxxix = -id = 499, -dccccix = -dic = 599, -dclxxxxix = -dcic = 699, -dccccxxxxxix = -im = 999, -mlxxxxix = -mic = 1099, -mxxxxix = -mil = 1049, -mccccIxxxxix = -mid = 1499, -mdccccIxxxxix = -mim = 1099, -mcmic = 1999, -mcmix = 1909, -mcmil = 1949, -mmlxxxxix = -mmil = 2049, -mmlxxxxix = -mmic = 2099. -mmccccIxxxxix = -mmid = 2499, -mmdccccIxxxxix = -mmim = 2999.

Thus number names can be formed like:

- googoc = g(2, 100, 200) > g(2, 230, 10)
- googoci = g(2, 101, 202)

or the adjective-like

- googoccic = g(2, 299, 598) > g(2, 830, 10)

or the Italian-like

- googocci = g(2, 221, 442) > g(2, 584, 10)

or the rather high

- googommim = g(2, 2'999, 5'998) > g(2, 9'441, 10)

By writing the terminating Roman numerals V, C, and D phonetically as -vy [as in Ivy], -cy [as in Nancy] and -dy [as in Lady] still others are possible: -ccy = 200, -cdy = 400, -cvy = 105, -dcy = 600, -dvy = 505, -lvy = 55, -mcy = 1,100, -mdy = 1,500.

180

gogoz	1	4	27	256	3,125	46,656
gogooz	1	16	729	65,536	9,765,625	2,176,782,336
googoz	1	2	216	4,096	100,000	2,985,984
googooz	1	4	256	46,656	16,777,216	10,000,000,000

181

geegoz	3	9	27	81	243	729	2,187	6,561	19,683
sub	1	3	10	30	89	268	805	2,414	7,241

gorgoz	4	16	64	256	1,024	4,096	16,384	65,536	262,144
sub	1	6	24	94	377	1,507	6,027	24,109	96,437
giegoz	5	25	125	625	3,125	15,625	78,125	390,625	1,953,125
dusub	1	3	17	85	423	2,115	10,573	52,865	264,327
gigoz	6	36	216	1,296	7,776	46,656	279,936	1,679,616	10,077,696
dusub	1	5	29	175	1,053	6,314	37,885	227,311	1,363,868
gegoz	7	49	343	2,401	16,807	117,649	823,543	5,764,801	40,353,607
dusub	1	7	46	325	2,275	15,922	111,455	780,181	5,461,267
geigoz	8	64	512	4,096	32,768	262,144	2,097,152	16,777,216	134,217,728
dusub	1	9	69	554	4,435	35,477	283,819	2,270,549	18,164,394

182 In order higher than a googol we have:

- gigiexi = g(2, 5(11), 6(11)) > g(2, 55, 66)
- gorgiexij = g(2, 5(12), 4(12)) = g(2, 60, 48) > 7g(2, 100, 10)
- geigeexvi = g(2, 3(14), 5(14)) = g(2, 48, 128) > 3g(2, 101, 10)
- geigevüj = g(2, 7(8), 8(8)) = g(2, 56, 64) > g(2, 101, 10)
- gegooxxüj = g(2, 2(23), 7(23)) = g(2, 46, 161) > g(2, 101,10)
- gegorxüj = g(2, 7(13), 4(13)) = g(2, 52, 91) > 7g(2, 101, 10)
- giegix = g(2, 6(10), 5(10)) = g(2, 60, 50) > 8(2, 101, 10)
- gigeexvij = g(2, 3(17), 6(17) = g(2, 51, 102) > 2g(2, 102, 10)
- geegex = g(2, 7(10), 3(10)) = g(2, 70, 30) > 2g(2, 103,10)
- geegiexüj = g(2, 3(13), 5(13)) = g(2, 39, 65) > 2(2, 103, 10)
- geigeimump = g(2, 8(2'000), 8(2'000)) = g(3, 2, 16'000) > 8g(2, 87, 10)

Extrapolating the Joyceans got even higher:

- from eleven to gelgol = g(2, 50, 11(50)) = g(2, 50, 550) > g(2, 934, 10)
- from twelve to gwelgol = g(2, 50, 12(50)) = g(2, 50, 6(50)) >g(2, 951, 10)
- from hundred gugum = g(3, 2, 100'000) = g(2, 500'000, 10)
- from thousand to gougoum = g(3, 2, 1'000'000) = g(2, 6'000'000, 10)
- from trillion to grigrim = g(3, 2, g(2, 15, 10) > g(3, 2, 10)
- from quadrillion to guaguam = g(3, 2, g(2, 18, 10)) ≈ g(2, 2, g(3, 2, 10))
- from quintillion to guiguim = g(3, 2, g(2, 21, 10)) > g(2, 2, g(3, 2, 10))
- from duodecillion to guoguom = g(3, 2, g(2, 39, 10)) ≈ g(2, 4, g(3, 2, 10))
- from tredecillion to gregrem = g(3, 2, g(2, 45, 10) > g(2, 4, g(3, 2, 10))

Googologist Jonathan Bowers independently coined the names:

- gaggol for ten-pentated-to-the-hundredth, Joyce's g(4, 100, 10) or givungen
- gygol for ten-heptated-to-the-hundredth, Joyce's g(6, 100, 10) or gaviungen
- gagol for ten-enneated-to-the-hundredth, Joyce's g(8, 100, 10) or gavüjungen
- boogol for ten-oktakaidekated-to-the-hundredth, Joyce's g(17, 100, 10) or gaxvijungen

Matt Hudelson on the other hand used the indeterminate number koogol for the much lower numbers ten-to-the-tenth-to-the-kth, and so substituting Roman numerals for k gives **coogol** for googolplex, **doogol** for five-hundredduplex, and **moogol** for thousandduplex so that

Joyce's **boogol** was g(2, 2, g(2, 1000, 10), 10) and **troogol** g(3, 2, g(2, 1000, 10), 10).

183 Such numbers higher than a googol would be:

- googgol = g(3, 50, 2(50)) = g(3, 50, 100) = g(3, 100, 10)
- geeggol = g(3, 50, 3(50)) = g(3, 50, 150) > 6g(3, 108, 10)
- googgool = g(3, 2(50), 2(50)) = g(3, 100, 100) > g(3, 200, 10)
- geeggool = g(3, 3(50), 3(50) = g(3, 100, 150) > 4g(3, 217, 10)
- giggel = g(3, 7(50), 6(50)) = g(3, 350, 300) > 9g(3, 866, 10)
- geiggeil = g(3, 8(50), 8(50)) = g(4, 2, 400) > 6g(3, 1'040, 10)

184 In many languages the letter count begins small with just two, three or four letters.

Loglan	ne	to	te	fo	fe	so	se	vo	ve	neni
letter sum	2	4	6	8	10	12	14	16	18	22
Danish	en	to	tre	fire	fem	seks	syv	otte	ni	ti
sum	2	4	7	11	14	18	21	25	27	29
Hmong	ib	ob	peb	plaub	tsib	rau	xya	yim	cuaj	kaum
sum	2	4	7	12	16	19	22	25	29	33
Catalan	un	dos	tres	quatre	cinc	sis	set	vuit	nou	deu
sum	2	5	9	15	19	22	25	29	32	35
Galician	un	dous	tres	catro	cinco	seis	sete	oito	nove	dez
sum	2	6	10	15	20	24	28	32	36	39
French	un	deux	trois	quatre	cinq	six	sept	huit	neuf	dix
sum	2	6	14	20	24	27	30	34	38	41

Esperanto	unu	du	tri	kvar	kvin	ses	sep	ok	nau	dek
sum	3	5	8	12	16	19	22	24	27	30
Albanian	nje	dy	tre	kater	pese	gjashte	shtate	tete	nente	dhjete
sum	3	5	8	13	17	24	30	34	39	45
Basque	bat	bi	hiru	lau	bost	sei	zazpi	zortzi	bederatzi	hamar
sum	3	5	9	12	16	19	24	30	39	44
Italian	uno	due	tre	quattro	cinque	sei	sette	otto	nove	dieci
sum	3	6	9	16	22	25	30	34	38	43
Romanian	unu	doi	trei	patru	cinci	sase	sapte	opt	noua	zece
sum	3	6	10	15	20	24	29	32	36	40
English	one	two	three	four	five	six	seven	eight	nine	ten
sum	3	6	12	16	20	23	28	33	37	40

Igbo	utu	abuo	ato	ano	ise	isii	asaa	asato	itoolu	iri
sum	3	7	10	13	16	20	24	29	35	38
Afrikaan	een	twee	drie	vier	vyf	ses	sewe	agt	nege	tien
sum	3	7	11	15	18	21	25	28	32	36
Cebuano	usa	duka	talo	upat	lima	anom	pito	walo	siyam	napul
sum	3	7	11	15	19	23	27	31	36	41
Somali	mid	laba	saddex	afar	shan	lix	todoba	sideed	sagaal	toban
sum	3	7	13	17	21	24	30	36	42	47
Filipino	isa	dalawa	tatlo	apat	lima	anim	pitu	walo	siyam	sampa
sum	3	9	14	18	22	26	30	34	39	44

Samoa	tasi	lua	tolu	fa	lima	ono	fitu	valu	iva
sum	4	7	11	13	17	20	24	28	31
Sundanese	hiji	dua	tilu	upat	lima	genep	tujuh	dalapan	salpan
sum	4	7	11	15	19	24	29	36	42
Indonesian	satu	dua	tiga	empat	lima	enam	tujuh	delapan	sembilan
sum	4	7	11	16	20	24	29	36	44
Malagasy	iray	roa	telo	efatra	dimy	enina	fito	valo	sivy
sum	4	7	11	17	21	26	30	34	38
Hausa	daya	biyu	uku	hudu	biyar	shicha	bakwai	takwas	taru
sum	4	8	11	15	20	26	32	38	42
German	eins	zwei	drei	vier	fünf	sechs	sieben	acht	neun
sum	4	8	12	16	20	25	31	35	39
Javanese	siji	loro	telu	papat	lima	enem	pitu	wolu	sanga
sum	4	8	12	17	21	25	29	33	38
Swahili	moja	mbili	tatu	nne	tano	sita	saba	nane	tisa
sum	4	9	13	16	20	24	28	32	36
Finnish	yksi	kakai	kolme	nelja	viisi	kuusi	seilsemän	kahdekeksan	yhdeksan
sum	4	9	14	19	24	29	38	49	57

Some others are higher.

Shona	mumwe	vaviri	vatatu	vana	shanu	tanhatu	nomwe	vasere	pfumbamwe	gumi
sum	5	11	17	21	26	33	38	44	53	57

Moari	kotahi	e rua	e tora	e wha	e rima	e ono	e whitu	e waiu	e iwa	tekau
sum	6	10	15	19	24	28	33	38	42	47

Maltese	wieħed	tnejn	tlieta	erbgħa	ħamsa	sitta	sebgħa	tmienja	disgħa	għaxra
sum	6	11	17	23	28	33	39	46	52	57

Chichewa	chimodzi	ziwiri	zitatu	zinai	zisan	ndi chimodzi	seveni	eyiti	naini	teni
sum	8	14	20	25	30	41	47	52	57	61

185 T-is

broken-up	1	5	56	901	21,680	629,621	20,799,173	728,600,676
broken down	1	2	23	370	8,903	258,557	8,541,284	299,203,497
broken	1	2	5	23	56	370	901	8,903
sum	1	5	15	27	40	53	62	68
super	1	4	44	704	16,896	489,984	16,169,472	565,931,520

hypersum	1	14	1,411	141,116	14,111,624	1,411,162,429	141,116,242,933

hypersuper	1	4	11,111,111	> sevenduplex

T-est: "T est prima et quarta et undecima et sexima decima et nona decima et nona vicesima ... littera in hic sententiam."

T-est	1	4	11	16	29	33	42	56	70
broken-up	1	5	56	901	17,175	498,976	16,483,383	692,801,062	38,813,342,855
sub	1	2	21	331	6,318	183,563	6,063,898	254,867,268	14,278,630,879
broken-down	1	4	45	724	13,801	400,953	13,245,250	556,701,453	31,188,526,618
broken	1	4	5	45	56	724	901	13,801	17,175
sum	1	5	16	32	61	94	136	192	262
super	1	4	44	704	20,416	673,728	28,296,576	1,584,608,256	110,922,577,920

hypersum	1	14	1,411	141,116	14,111,629	1,411,162,933	141,116,293,342

B-est: "B est non secunda ... littera in hic sententiam." Since B is letter number one and no other, Latin having no word for billionth, this is the best sequence, numero uno.

186

R-is	1	30	62	74	96	132	166	186	208	248	289	330	371	416	460	507	550	592
E-is	1	6	13	16	25	28	31	34	41	44	53	56	59	62	68	69	74	77
F-is	1	7	19	49	98	116	151	190	213	215	218	230	269	312	338	340	352	395

I-is	1	2	8	22	51	65	80	115	129	134	148	162	167	181	195	199	213	234
L-is	1	12	38	42	77	111	148	179	186	227	269	310	356	399	439	484	530	575
N-is	1	26	29	38	60	63	72	75	77	94	97	129	132	158	161	190	193	203
D-is	1	9	17	45	79	93	106	141	168	175	197	210	232	237	242	249	271	290

H-is is, of course, a reference to the Most High.

broken-up	1	16	97	2,431	87,613	3,331,725	1,566,678,688	7,680,587,437
broken-down	1	5	81	2,030	73,161	2,782,148	130,834,117	6,413,653,881
broken	1	5	16	81	97	2,030	2,431	73,161
sum	1	6	22	47	83	130	179	236
super	1	5	80	2,000	72,000	2,736,000	128,592,000	6,301,008,000

hypersum	1	15	1,516	151,625	15,162,536	1,516,253,647	151,625,364,749

hypersuper	1	5	1,616,161,616	> nineduplex

S-is:

broken-up	1	4	41	947	22,769	1,162,166	60,455,401	4,777,138,845
broken-down	1	3	31	716	17,215	878,661	45,708,627	3,611,860,214
broken	1	3	4	31	41	716	947	17,215
sum	1	4	14	37	61	112	164	243
super	1	3	30	690	16,560	844,560	43,917,120	3,469,452,480

hypersum	1	13	1,310	131,023	13,102,324	1,310,232,451	131,023,245,152

hypersuper	1	3	101,010	> fiveduplex

187 S-ain'ts The word "saint" comes from the Indo-European sak- as do sacred, consecrate, sacerdotal, corposant, sacrosanct, sanctify. ("Indo-European and the Indo-Europeans" by Calvert Watkins)

broken-up	1	3	43	225	1,393	9,976	81,201	9,123,047	110,298,550
broken-down	1	2	30	157	972	6,961	56,660	6,365,831	76,963,533
broken	1	2	3	30	43	157	225	972	1,393

hypersum	2	23	234	2,345	23,456	234,567	2,345,678	234,567,810	23,456,781,011
trimmed	1	12	123	1,234	12,345	123,456	1,234,567	12,345,670	1,234,567,000
sub	1	8	86	863	8,629	86,292	862,927	86,292,675	8,629,267,490

188 Since zero is not countable, it can't be a s-ain't, so it must be included in the s-inners, if at all. The greatest of sinner would be Satan himself however, who called himself "Mr. Zero" in the Monkees episode, "The Devil and Davey Jones", in which he was defeated by the love of music. If zero is included, the supersinners would all become zeroes. Other Mr. Zeroes were associated with Bruce "Batman" Wayne (aka Dr. Victor Fries) and Clark “Superman” Kent's favorite Martian.

super	1	9	279	10,044	984,312	105,321,384	16,430,135,904
broken-up	1	10	311	11,206	1,098,499	18,338,991,943	3,007,712,229,251
broken-down	1	9	280	10,089	989,002	16,510,984,270	2,707,907,253,583
broken	1	9	10	280	311	10,089	11,206

hypersum	1	19	1,931	193,136	19,313,698	19,313,698,107

189 numbers with odd letter count ordinals, aka "olco numbers"

broken-up	1	5	36	401	5,650	90,801	1,730,869	36,439,050
broken-down	1	2	15	167	2,353	37,815	720,838	15,175,413
broken	1	2	5	15	36	167	401	2,353

hypersum	1	14	147	14,711	1,471,114	147,111,416	14,711,141,619

numbers with even letter count ordinals, aka "elco numbers"

broken-up	2	7	30	157	972	7,933	72,369	731,623	8,851,845	115,805,608
trimmed	1	6	2	46	861	6,822	61,258	620,512	7,740,734	47,457
sub	1	2	11	57	357	2,918	26,623	269,149	3,256,411	42,602,502
broken-down	1	3	13	68	421	3,436	31,345	316,886	3,833,977	50,158,587
broken	1	2	3	7	13	30	68	157	421	971

hypersum	2	23	234	2,345	23,456	234,568	2,345,689	23,456,891
trimmed	1	12	123	1,234	12,345	123,457	1,234,578	12,345,780
sub	1	8	86	863	8,629	86,293	862,931	8,629,308

The-first

broken-up	1	3	22	289	4,068	77,581	1,943,593	60,328,964	2,234,115,261
broken-down	1	2	15	197	2,773	52,884	1,324,873	41,123,947	1,522,910,912
broken	1	2	3	15	22	197	289	2,773	4,068
sum	1	3	5	12	25	39	58	83	114
super	1	2	14	182	2,548	48,412	1,210,300	37,519,300	1,38,214,100

hypersum	1	12	127	12,713	1,271,314	127,131,419	12,713,141,925

hypersuper	1	2	77	13,131,313,131,313	> thirteenplex

The-second	2	4	8	9	11	13	14	18	19
trimmed	1	3	7	8	0	2	3	7	8
broken-up	1	9	74	675	7,499	98,162	1,381,767	24,969,968	475,811,159
sub	1	3	27	248	2,759	36,112	508,324	9,185,938	175,041,143
broken-down	1	4	33	301	3,344	43,773	616,166	11,134,761	212,176,625
broken	1	4	9	33	74	301	675	3,344	43,773

hypersum	2	24	248	2,489	248,911	24,891,113	2,489,111,314	248,911,131,418
trimmed	1	13	137	1,378	137,800	13,780,002	1,378,000,203	13,780,020,307
sub	1	9	91	916	91,569	9,156,929	915,692,879	91,569,287,927

hypersuper	2	44	88,888,888,888,888,888,888,888,888,888,888,888,888,888,888
trimmed	1	33	77,777,777,777,777,777,777,777,777,777,777,777,777,777,777
sub	1	16	> forty-threeplex

190 Four-is

broken-up	1	11	144	2,315	44,129	973,153	24,372,954	683,415,865
broken-down	1	2	27	434	8,273	182,440	4,569,273	128,122,084
broken	1	2	11	27	144	434	2,315	8,273
sum	1	11	21	34	50	69	91	116
super	1	10	130	2,080	35,920	889,440	21,736,000	608,608,000

hypersum	1	110	11,013	1,101,316	110,131,619	11,013,161,922

hypersuper	1	10	131,313,131,313,131,313,131,313	> hundred-twentyplex

Other sequences that work because of the letter count in the phrase "is the number of letters in the" are two-is, three-is, six-is and seven-is.

two-is	2	5	7	10	13	16	19	22	25	28	31	34	37	40	43	46	49	52	55	58	61	64	67	70	73	76
trimmed	1	4	6	0	2	5	8	11	14	17	20	23	26	3	32	35	38	41	44	47	50	53	56	6	62	65

broken-up	2	11	79	801	10,492	168,673	3,215,279	70,904,811	1,775,835,554
trimmed	1	0	68	70	381	57,562	2,104,168	683,700	664,724,443
sub	1	4	29	294	3,859	62,051	1,182,835	26,084,422	653,293,391
broken-down	1	3	22	223	2,921	46,959	895,142	19,740,083	494,397,217
broken	1	2	3	11	22	79	223	801	2,921

hypersum	2	25	257	25,710	2,571,013	257,101,316	25,710,131,619
trimmed	1	14	146	1,460	146,002	14,600,205	1,460,020,508
sub	1	9	95	9,458	945823	94,582,288	9,458,228,852

three-is	3	6	8	10	12	14	16	18	20	22	24	26	28	30	32	34	36	38	40	42	44	46	48	50	52	54

broken-up	3	19	155	1,569	18,983	267,331	4,296,279	7,600,353	1,556,303,339
sub	1	6	57	577	6,983	98,345	1,580,512	28,547,574	572,32,002
broken-down	1	4	33	334	4,041	56,908	914,569	16,519,150	331,297.569
broken	1	3	4	19	33	155	334	1,569	4,041

hypersum	3	36	368	36,810	3,681,012	368,101,214	36,810,121,416
sub	1	13	135	13,542	1,354,169	135,416,869	13,541,686,896

six-is	4	9	11	18	25	32	39	46	53	60	67	74	91	98	105	112	129	136	143	150	157	164

sub	1	3	4	7	9	12	14	17
broken-up	4	37	411	7,435	186,286	5,968,587	232,961,179	1,072,218,221
sub	1	14	151	2,735	68,531	2,195,720	85,701,628	3,944,470624
broken-down	1	5	56	1,013	25,381	813,205	31,730,376	1,460,870,501
broken	1	4	5	37	56	411	1,013	7,435

hypersum	4	49	4,911	491,118	49,111,825	4,911,182,532	491,118,253,239
sub	1	18	1,807	180,672	18067231	1,806,723,085	180,672,308,551

seven-is	6	13	20	27	34	41	48	55	62	69	76	83	90	97	104	111	118	125	132	139	146	153
holiness	1	0	1	0	0	0	2	0	1	2	1	2	1	1	1	0	2	0	0	1	1	0

broken-up	6	79	1,586	42,901	1,460,220	59,911,921	2,877,232,428
holiness	1	1	3	2	3	3	4
dusub	1	10	214	5,805	197,618	8,108,196	389,391065
broken-down	1	7	141	3,814	129,817	5,326,311	255,792,745
broken	1	6	7	79	141	1,586	3,814

hypersum	6	613	61,320	6,132,027	613,202,734	61,320,273,441
holiness	1	1	2	2	2	1
dusub	1	83	8,299	829,880	82,987,966	8,298,796,574

mates	19	29	39	49	59	69	79	89	90	109	119	129	139	149	159	169	179	189	190	191
holiness	1	1	1	1	1	2	1	3	2	2	1	1	1	1	1	2	1	3	2	1

broken-up	0	1	20	601	24,060	12,003,601	72,240,120
holiness	1	0	1	2	3	4	2
untrimmed	1	2	31	712	35,171	23,114,712	83,351,231
broken-down	1	1	21	631	25,261	1,263,681	75,846,121
broken	0	1	20	21	601	631	24,060
trimmed	1	2	31	32	712	742	35,171

hypersum	0	10	1,020	102,030	10,203,040	1,020,304,050
holiness	1	1	2	3	4	5
untrimmed	1	21	2,131	213,141	21,314,151	2,131,415,161

one-ish

mates	8	18	28	38	48	58	68	78	80
holiness	2	2	2	2	2	2	3	2	3
broken-up	1	11	122	1,475	19,297	271,633	4,093,792	65,772,305	3,095,392.127
broken-down	1	10	111	1,342	17,557	247.140	3,724,657	205,103,275	12,720,127,707
broken	1	2	11	23	122	278	1,475	3,637	19,297

hypersum	1	110	11,011	11,012	110,111,213	11,011,121,314

twoish

mates	7	17	27	37	47	57	67	70
broken-up	2	25	502	10,567	23,976	5,369,015	129,089,336	3,232,602,415
trimmed	1	14	41	456	12,865	425,804	1,878,225	212,151,304
sub	1	9	185	3,887	85,707	1,975,150	47,489,313	1,189,207,970
broken-down	1	3	61	1,284	28,309	709,009	1,446,327	2,155,336
broken	1	2	3	25	61	502	1,284	10,567

hypersum	2	212	21,220	2,122,021	212,202,122	21,220,212,223
trimmed	1	101	1,011	101,110	10,111,011	1,011,101,112
sub	1	78	7,806	780,648	78,064,798	7,806,479,814

three-ish

mates	6	16	26	36	46	56	60	61
holiness	1	1	1	1	1	1	2	1

broken-up	3	40	1,203	37,333	1,195,859	39,500,680	1,344,218,979	47,087,164,945
sub	1	15	443	13,734	4389,932	14,531,488	494,510,527	17,322,399,926
broken-down	1	13	300	9,013	279,703	8,959,509	304,903,009	10,071,038,509
broken	1	3	13	40	300	91,203	9,013	37,333

hypersum	3	313	31,323	3,132,330	313,233,031	31,323,303,132
sub	1	115	11,523	1,152,320	115,231,992	11,523,199,252

fourish

mates	15	25	35	45	50	51	52
broken-up	4	57	1,372	46,705	1,916,277	80,530,339	3,464,720,854
dusub	1	8	186	6,321	253,020	10,380,103	43,621,833
broken-down	1	14	337	11,472	459,217	18,839,369	791,712,715
broken	1	4	8	14	337	1,372	11,472

hypersum	4	414	41,424	4,142,434	414,243,440	4,142,434,041
sub	1	152	15,239	1,523,916	152,391,645	1,523,916,320

five-ish

broken-up	5	76	1,905	66,751	3,005,700	150,351,751	7,670,945,001
dusub	1	10	258	9,034	406,777	20,347,897	1,038,149,514
broken-down	1	6	151	5,291	238,246	12,394,083	657,124,645
broken	1	5	6	76	151	1,905	5,291

hypersum	5	515	51,525	5,152,535	515,253,545	51,525,354,550
dusub	1	70	6,973	697,320	69,731,984	6,973,198,452

sixish

broken-up	6	97	2,528	91,105	4,193,358	234,919,153	14,099,342,538
holiness	1	1	2	2	3	2	5
dusub	1	13	342	12,330	567,509	31,792,850	1,908,138,516
broke-down	1	7	183	6,595	303,553	18,523,328	1,148,749,889
broken	1	6	7	97	183	2,528	6,595

hypersum	6	616	61,626	6,162,636	616,263,646	61,626,364,656
holiness	1	2	3	4	5	6
dusub	1	84	8,340	834,022	83,402,215	8,340,221,516

sevenish

broken-up	7	120	3,247	120,259	5,655,420	322,479,199	21,611,761,753

dusub	1	16	440	16,275	765,378	43,642,814	2,724,833,898
broken-down	1	8	217	8,037	377,956	26,464,957	1,879,389,903
broken	1	7	8	120	217	3,247	8,037

hypersum	7	717	71,727	7,172,737	717,273,747	71,727,374,757
dusub	1	97	9,707	970,724	97,072,446	9,707,244,578

eightish

broken-up	8	145	4,068	154,729	7,431,060	431,156,209	29,326,053,272
holiness	2	0	4	1	3	3	3
dusub	1	19	551	20,940	1,005,685	58,350,648	3,968,849,726
broken-down	1	9	253	9,623	462,157	36,057,869	2,885,091,677
broken	1	8	9	145	253	4,068	9,623

hypersum	8	818	81,828	8,182,838	818,283,848	81,828,384,858
dusub	1	111	11,074	1,107,427	110,742,676	1,174,267,642

nine-ish

broken-up	9	172	4,997	195,055	9,562,692	564,393,883	38,952,740,619
dusub	1	23	676	26,398	1,294,170	76,382,406	5,271,680,185
broken-down	1	10	291	11,359	556,882	44,005,037	3,917,005,175
broken	1	9	10	172	291	4,997	11,359

hypersum	9	919	91,929	9,192,939	919,293,949	91,929,394,959
holiness	1	2	3	4	5	6
dusub	1	124	12,441	1,244,129	124,412,907	12,441,290,705

192 The Inuit have been said to use same words for odd and even numbers, one and six, two and seven, three and eight, four and nine, five and ten, but while counting on the right and left hands respectively.

broken-up	1	3	10	43	225	2,518	3,044	398,251	5,605,955	84,487,576
broken-down	1	2	7	30	157	1,757	21,241	277,890	3,911,701	58,953,405
broken	1	2	3	7	10	30	43	157	225	1,757
sum	1	3	6	10	15	26	38	51	65	80
super	1	2	6	24	120	1,320	15,840	205,920	2,882,880	43,243,200

hypersum	1	12	123	1,234	12,345	1,234,511	123,451,112	1,2345,111,213

The neighbors of the left-handed or sinister numbers would be the downtown "left-behinds", a reference to the series by Tim LaHaye and Jerry B. Jenkins and the uptown would be "left-overs".

left-behinds	0	1	2	3	4	10	11	12	13
holiness	1	0	0	0	0	1	0	0	0
untrimmed	1	2	3	4	5	21	22	23	24
broken-up	0	1	2	7	30	307	3,407	41,191	538,890
holiness	1	0	0	0	1	1	1	0	6
broken-down	1	1	3	10	43	440	4,883	59,036	772,351
broken	0	1	2	3	7	10	30	43	307
holiness	1	0	0	0	0	1	1	0	1
untrimmed	1	2	3	4	8	21	42	54	418

hypersum	0	1	12	123	1,234	123,410	12,341,011	1,234,101,112	123,410,111,213
holiness	1	0	0	0	0	1	1	1	1
untrimmed	1	2	24	234	2,345	234,521	23,452,122	2,345,212,223	234,521,222,324

left-overs	2	3	4	5	6	22	23	24	25
trimmed	1	2	3	4	5	11	12	13	14
broken-up	2	7	30	157	972	21,541	49,6415	11,935,501	29,888,940
trimmed	1	6	2	46	861	10,430	385,304	82,440	1,877,783
broken-down	1	3	11	58	358	7,924	182,621	4,390,825	109,953,257
broken	1	2	3	7	11	30	58	157	358

hypersum	2	23	234	2,345	23,456	2,345,622	234,562,223	23,456,222,324
trimmed	1	12	123	1,234	12,345	1,234,511	123,451,112	12,345,111,213
dusub	1	8	86	863	8,629	862,908	86,290,620	8,629,061,961

right-handed	0	6	7	8	9	60	66	67	68
holiness	1	1	0	2	1	1	2	1	3
untrimmed	1	7	8	9	10	71	77	78	79

hypersum	0	6	67	678	6,789	678,960	67,896,066	6,789,606,667
holiness	1	1	1	3	4	6	8	9
untrimmed	1	7	78	789	78,910	7,891,071	789,107,177	78,910,717,778

Numbers with an equal number of both right-handed and left-handed digits would be "ambidextrous".

ambidextrous	10	16	17	18	19	20	26	27
holiness	1	1	0	2	1	1	1	0
broken-up	10	161	2,747	49,607	945,280	18,955,207	304,228,592	5,190,841,271
holiness	1	1	0	3	4	3	4	3
dusub	1	22	372	6,713	127,930	2,565,308	41,172,863	702,503,974
broken-down	1	11	188	3,395	64,693	1,297,255	20,820,773	355,250,396
broken	1	10	11	161	188	2,747	3,395	49,607

hypersum	10	1,016	101,617	10,161,718	1,016,171,819	101,617,181,920
holiness	1	2	2	4	5	6
dusub	1	138	13,752	1,375,239	137,523,901	13,752,390,097

Those with "marching orders", alternating left-right or right-left, are marching numbers, differing from ambidextrous at three places.

marching	...	95	101	102	103	104	105	161	162	163	164	165	171	172	173	174	175

The complimentary sequence to the right numbers is "wrong", with digits four through eight.

wrong	4	5	6	7	8	44	45	46	47	48	54	55	56	57	58	64	65	66	67	68	74	75	76	77	78	84	85

"Not-quite-right" numbers are downtown neighbors of the ichibi rights with a zero, that is, left tail. Almost-right numbers are quite different, being digitally expanded.

not-quite-right	10	20	30	40	50	110	120	130	140	150	210	220	230	240	250	310	320

o-ish

broken-up	0	1	2	9	128	2,697	59,462	1,429,785	44,382,797	1,421,679,289
holiness	1	0	0	1	2	2	2	3	3	5
untrimmed	1	2	3	10	239	27,108	610,573	25,310,896	554,938,108	253,278,103,910
broken-down	1	1	3	13	185	3,898	85,941	2,066,482	64,146,883	2,04,766,738
broken	0	1	2	3	9	13	128	185	2,697	3,898
holiness	1	0	0	0	1	0	2	2	2	3
untrimmed	1	2	3	4	10	24	239	296	37,108	49,109

hypersum	0	1	12	124	12,414	1,241,421	124,142,122	12,414,212,224
untrimmed	1	2	23	235	23,525	2,352,532	235,253,233	23,525,323,335

n-ish

broken-up	1	8	73	738	8,191	107,221	1,509,285	22,746,496	365,453,221

broken-down	1	7	64	647	7,181	94,000	1,323,181	19,941,715	32,039,062
broken	1	7	8	64	73	647	738	7,181	8,191
sum	1	8	17	27	38	51	65	80	96
super	1	7	63	630	6,930	90,090	1,261,260	18,918,900	302,702,400

hypersum	1	17	179	1,7910	1,791,011	179,101,113	17,910,111,314	1,791,011,131,415

hypersum	1	7	9,999,999	> sevenduplex

e-ish

broken-up	0	3	16	115	936	8,539	86,326	958,125	11,583,826
holiness	1	0	1	0	2	3	3	3	5
untrimmed	1	4	27	226	1,047	96,410	97,437	1,069,236	22,694,937
broken-down	1	4	21	151	1,229	11,212	113,349	1,258,051	15,209,961
broken	0	1	3	4	16	21	115	151	936
holiness	1	0	0	0	1	0	0	0	2
untrimmed	1	2	4	5	27	32	226	262	1,047

hypersum	0	1	13	135	1,357	13,578	135,789	13,578,910	1,357,891,011
holiness	1	0	0	0	0	2	3	4	4
untrimmed	1	2	24	246	2,468	24,689	2,468,910	246,891,021	24,689,102,122

t-ish

broken-up	2	7	58	587	7,102	92,913	1,307,884	19,711,173	316,686,652
trimmed	1	6	47	476	601	81,802	26,773	8,600,062	205,575,541
sub	1	3	21	216	2,613	34,181	481,144	7,251,335	116,502,509
broken-down	1	3	25	253	3,061	40,046	563,705	8,495,621	136,493,641
broken	1	2	3	7	25	58	253	587	3,061

hypersum	2	212	21,220	2,122,021	212,202,122	21,220,212,223	2,122,021,222324
trimmed	1	101	1,011	101,110	10,111,011	1,011,101,112	101,110,111,213
sub	1	78	7,806	780,648	78,064,798	7,806,479,814	780,647,961,422

193 h-ish

broken-up	3	13	172	2,421	55,855	1,342,941	40,344,085	1,252,009,576
sub	1	2	24	328	7,560	181,748	5,459,979	169,441,071
broken-down	1	4	53	746	17,211	413,810	12,431,511	12,357,732,343

broken	1	3	4	13	53	172	746	2,421

hypersum	3	38	3,813	381,318	38,131,823	3,813,182,328	381,318,232,830
sub	1	14	1,403	140,279	14,027,914	140,279,184	14,027,913,402

r-ish

broken-up	3	13	172	2,421	55,855	1,342,941	40,344,085	1,252,009,576
dusub	1	2	24	328	7,560	181,748	5,459,979	169,441,071
broken-down	1	4	53	746	17,211	413,810	12,431,511	385,790,651
broken	1	3	4	13	53	172	746	2,421

hypersum	3	34	3,413	341,314	34,131,423	3,413,142,324	341,314,232,430
sub	1	13	1,256	125,562	12,556,249	1,255,624,891	125,562,489,090

194 i-ish

broken-up	5	31	253	2,308	34,873	560,276	10,119,841	192,837,255
dusub	1	5	35	313	4,720	75,826	1,369,572	26,097,685
broken-down	1	6	49	447	6,754	108,511	1,959,952	37,347,599
broken	1	5	6	31	49	253	447	2,308

hypersum	5	56	568	5,689	568,915	56,891,516	5,689,151,618	568,915,161,819
dusub	1	8	77	770	76,994	7,699,429	769,942,946	76,994,294,562

f-ish

broken-up	4	21	298	4,491	108,082	2,706,541	92,130,476	3,227,273,201
dusub	1	3	41	698	14,628	366,291	12,468,505	436,763,933
broken-down	1	5	71	1,070	25,751	644,845	2,195,048	768,911,680
broken	1	4	5	21	71	298	1,070	4,491

hypersum	4	45	4,514	451,415	45,141,524	4,514,152,425	451,415,242,534
sub	1	17	1,661	166,066	16,606,639	1,660,663,871	166,066,387,160

u-ish

broken-up	4	57	1,372	46,705	2,056,392	711,936,264	526,394,375,409
dusub	1	8	186	6,321	278,303	96,350,096	71,239,731,891
broken-down	1	14	337	11,472	505,105	74,682,193	129,296,569,424
broken	1	4	14	57	337	1,372	11,472

hypersum	4	414	41,424	4,142,434	414,243,444	41,424,344,445	4,142,434,444,546
sub	1	152	15,239	1,523,916	152,391,647	15,239,164,685	1,523,916,468,549

v-ish

broken-up	5	36	437	6,591	112,484	2,818,691	76,217,141	2,670,418,626
dusub	1	5	60	893	15,224	381,469	10,314,869	982,392,112
broken-down	1	7	85	1,282	21,879	548,257	14,824,818	519,416,887
broken	1	5	7	36	85	437	1,282	6,591

hypersum	5	57	5,712	571,215	57,121,517	5,712,151,725	571,215,172,527
dusub	1	8	773	77,305	7,730,557	773,055,671	77,305,567,163

195 s-ish

broken-up	6	43	694	11,841	308,560	8,342,961	300,655,156	11,132,583,733
holiness	1	0	2	2	5	4	4	2
tressub	1	3	35	590	15,363	415,372	14,968,739	554,258,708
broken-down	1	7	113	1,928	50,241	1,358,435	48,953,901	1,812,652,772
broken	1	6	7	43	113	694	1,928	11,841

hypersum	6	67	6,716	671,617	67,161,726	6,716,172,627	671,617,262,736
dusub	1	9	909	90,893	9,089,351	908,935,125	90,893,512,479

186 l-ish

broken-up	11	133	14,774	1,654,821	349,182,005	23,023,131,784,996
tressub	1	7	735	82,389	17,384,749	1,146,254,236,223
broken-down	1	12	1,333	149,308	31,505,321	2,077,286,764,281
broken	1	11	12	133	1,333	14,774

hypersum	11	1,112	1,112,111	1,112,111,112	1,112,111,112,211	1,112,111,112,211,212
dusub	1	150	150,508	150,507,872	150,507,872,362	150,507,872,361,689

187 y-ish

broken-up	20	601	24,060	1,203,601	72,240,120	5,058,012,001
holiness	1	2	3	3	2	6
tetrasub	1	11	441	22,045	1,323,125	92,640,722
broken-down	1	30	1,201	60,080	3,606,001	252,480,150
broken	1	20	30	601	1,201	24,060

hypersum	20	2,021	202,122	20,212,223	2,021,222,324	202,122,232,425
holiness	1	1	1	1	1	1

dusub	1	100	10,063	1,006,307	100,630,734	10,063,073,405

d-ish	100	101	102	103	104	105
holiness	2	1	1	1	1	1
broken-up	100	10,101	1,030,402	106,141,507	11,039,747,130	159,279,590,157
holiness	2	2	3	2	3	4
hexasub	1	26	2,555	263,099	27,364,798	394,814,631
broken-down	1	101	10,303	1,061,310	110,386,543	11,591,648,325
broken	1	100	101	10,101	10,303	1,030,402

hypersum	100	100,101	100,101,102	100,101,102,103	100,101,102,103,104
holiness	2	3	4	5	6
pentasub	1	675	674,476	674,475,920	674,475,920,521

m-ish	1,000,000	1,000,001	1,000,002
holiness	6	5	5
broken-up	1,000,000	1,000,001,000,001	10,000,030,000,030,000,003
holiness	6	10	16
15th sub	1	305,903	3,059,032,382,098
broken-down	1	1,000,001	1,000,001,000,001
broken	1	1,000,000	1,000,001

hypersum	1,000,000	10,000,001,000,001	100,000,010,000,011,000,002
holiness	6	11	16
14th sub	1	8,315,288	83,152,880,225,654

p-ish	1,000,000,000	1,000,000,001	1,000,000,0002
holiness	9	8	9
broken-up	1,000,000,000	1,000,000,001,000,000,001	100,000,000,300,000,001,200,000,003
holiness	9	16	22
22nd sub	1	278,946,810	2,789,468,101,237,325
broken-down	1	1,000,000,001	100,000,001,200,000,003
broken	1	1,000,000,000	1,000,000,001

hypersum	1,000,000,000	10,000,000,001,000,000,001	> twenty-nineplex
holiness	9	17	

21st sub	1	7,582,560,429	75,825,611,862

198 Ir-ish

broken-up	13	391	12,122	388,295	12,825,857	436,467,433	15,289,186,012
tressub	1	20	604	19,333	638,562	21,730,435	761,203,750
broken-down	1	30	931	29,822	985,057	33,521,760	1,174,246,657
broken	1	13	30	391	931	12,134	29,822

hypersum	13	1,330	133,031	13,303,132	1,330,313,233	133,031,323,334
tressub	1	66	6,623	662,324	66,232,396	6,623,239,590

Harry Matthew's Prisoner's Restriction (*Oulipo Compendium* ed. by Harry Matthews and Alastair Brotchie) limits letters to those without extenders above or below the center line, so that the maximum number of lines of text can be written, just a, c, e, i, m, n, o, r, s, u, v, w, x and z. For the number names that leaves only zero, one, six, seven and nine. Higher numbers could only be represented recursively, six-ones, six-sevens, etc. or as homonums in bases seven and above.

prisoner	0	1	6	7	9	111,111
holiness	1	0	1	0	1	0
untrimmed	1	2	7	8	10	222,222
broken-up	0	1	7	50	457	50,777,777
holiness	1	0	0	1	0	1
untrimmed	1	2	8	61	568	61,888,888
broken-down	1	1	6	43	393	43,666,666
broken	0	1	6	7	43	50
holiness	1	0	1	0	0	1
untrimmed	1	2	7	8	54	61

hypersum	0	1	16	167	1,679	1,679,111,111	1,679,111,111,777,777
holiness	1	0	1	1	2	2	2
untrimmed	1	2	27	278	27,810	27,810,222,222	> sixteenplex

Those with extenders include two through five and eight, "free" numbers, which are NOT Fibonacci rokubi evenly evens. All of these have ascenders. Only eightish numbers also have descenders.

free	2	3	4	5	8	10	11	12	13
trimmed	1	2	3	4	7	0	0	1	2
broken-up	2	7	30	157	1,286	13,017	144,473	1,746,693	22,851,482

trimmed	1	6	2	46	175	206	33,362	635,572	11,740,371
sub	1	3	11	57	473	4,788	53,148	642,572	8,406,590
broken-down	1	3	13	68	557	5,638	62,575	756,538	9,897,569
broken	1	2	3	7	13	30	68	157	557

hypersum	2	23	234	2,345	23,458	2,345,810	234,581,011	2,348,101,112
trimmed	1	12	123	1,234	12,347	123,470	12,347,000	123,700,001
sub	1	8	86	863	8,630	862,975	86,297,531	863,818,125

ten-ish

ten-ish	10	10,000	10,010	10,000,000
holiness	1	4	3	7
broken-up	10	100,001	1,001,010,020	10,010,100,200,100,001
holiness	1	4	6	11
dusub	1	13,533	135,471,974	1,354,719,745,807,410
broken-down	1	11	110,111	1,101,110,000,011
broken	1	10	11	10,000

hypersum	10	1,010,000	101,000,010,010	10,100,001,001,010,000,000
holiness	1	5	8	15
dusub	1	136,689	13,668,864,962	126,688,649,616,176,000

199 The Middling, cheap and dear numbers come from a game in *The Games and Diversions of Argyleshire* by R. C. MacLagan. "Middling", of course, comes from "middle" which comes from Indo-European *medhyo-* from which comes amid, intermediate, mean, medial, mediate, medieval, mediocre, Mediterranean, medium, meridian, mid.

mates	3	4	5	13	14	15	23
broken-up	4	21	130	1,841	27,745	445,761	10,726,009
dusub	1	2	17	249	3,754	60,327	1,451,607
broken-down	1	5	31	439	6,616	106,295	2,557,696
broken	1	4	5	21	31	130	439

hypersum	4	45	456	45,614	4,561,415	456,141,516	4,561,411,624
sub	1	17	168	16,780	1,678,051	167,805,086	1,678,049,559

200 middlinger and middlingest are extrapolations of middling.

mates	34	35	43	45	53	54

broken-up	45	2,071	111,879	6,267,295	401,218,759	26,085,486,630
pentasub	1	15	755	42,229	2,703,392	175,762,627
broken-down	1	46	2,485	139,206	8,911,669	579,397,691
broken	1	45	46	2,071	2,485	111,879

hypersum	45	4,546	454,654	45,465,456	4,546,545,664	454,654,566,465
tetrasub	1	83	8,427	832,729	83,272,889	8,327,288,858

201 middlingest

mates	345	354	435	453	634
broken-up	456	212,041	115,774,842	65,297,222,929	42,116,824,564,047
holiness	1	1	2	3	5
heptasub	1	194	105,573	59,543,361	38,405,572,767
broken-down	1	465	253,891	143,194,989	92,361,021,796
broken	1	456	465	212,041	253,891

hypersum	456	456,465	456,465,546	456,465,546,564	4,564,655,465,641,456
holiness	1	2	3	4	5
hexasub	1	1,131	1,131,465	1,131,464,967	11,314,649,671,191

202 "Cheap" comes from the Old English "god chep" (good price or bargain), from Latin caupo (shopkeeper), which may be related to Indo-European kau- (cut) from which comes haggle, hay, hew and hoe. ("Indo-European and the Indo-Europeans" by Calvert Watkins)

broken-up	1	3	10	113	1,592	23,993	385,480	6,577,153	118,774,234
broken-down	1	2	7	79	1,113	16,774	269,497	4,598,223	83,037,511
broken	1	2	3	10	79	113	1,113	1,592	16,774
sum	1	3	6	16	27	41	56	72	89
super	1	2	6	60	840	12,600	201,600	3,427,200	61,689,600

hypersum	1	12	123	12,310	1,231,014	123,101,415	12,310,141,516

203 cheaper

broken-up	12	157	3,309	76,264	2,367,493	75,836,040	8,496,003,973
tressub	1	9	166	3,798	117,871	3,775,655	422,991,132
broken-down	1	13	274	6,315	196,039	6,279,563	703,507,095
broken	1	12	13	157	274	3,309	6,315

hypersum	12	1,213	121,321	12,132,123	1,213,212,331	121,321,233,132

tressub	1	164	16419	1,641,904	164,190,434	16,419,043,449

204 cheapest

broken-up	123	16,237	3,458,604	798,953,761	249,277,032,036	80,018,726,237,317
enneasub	1	2	427	98,600	307,632,230	9,875,095,329
broken-down	1	132	28,117	6,495,159	2,026,517,725	650,518,684,884
broken	1	123	132	16,237	28,117	3,458,604

hypersum	123	123,132	123,132,213	123,132,213,231	123,132,213,231,312
pentasub	1	830	829,658	829,658,327	829,658,326,633

205 "Dear" comes from the Old English *dēore*, perhaps from Indo-European *dher-* (hold firm).

broken-up	7	57	8,897	160,666	3,061,551	82,822,543	2,322,092,755
tressub	1	3	444	8,000	152,426	4,123,492	115,610,191
broken-down	1	8	1,249	22,555	429,794	11,626,993	325,985,598
broken	1	7	8	57	1,249	8,897	22,555

hypersum	7	78	789	78,917	7,891,718	789,171,819	78,917,181,927
dusub	1	11	107	10,680	1,068,028	106,802,792	10,680,279,168

206 dearer

broken-up	78	6,163	536,259	47,733,214	4,630,658,017	453,852,218,880
holiness	2	2	2	0	6	9
pentasub	1	42	3,614	321,625	31,201,129	3,058,032,197
broken-down	1	79	6,874	611,865	59,357,779	5,817,674,207
broken	1	78	79	6,163	6,874	536,259

hypersum	78	7,879	787,987	78,798,789	7,879,878,997	787,987,899,798
holiness	2	3	5	8	9	12
tetrasub	1	144	14,432	1,443,250	144,325,018	14,432,501,821

207 dearest

broken-up	789	829,623	553,439,406	496,435,776,805
holiness	3	4	3	6
heptasub	1	574	504,671	452,690,832
broken-down	1	798	701,443	629,195,169
broken	1	789	798	629,623

hypersum	789	789,798	789,798,879	789,798,879,897	789,798,879,897,978
holiness	3	6	9	12	15
heptasub	1	720	720,203	720,203,355	720,203,354,994

208 uglier is from Old Norse *uggr* (fear).

broken-up	6	61	738	201,096	4,033,051	96,994,320	291,382,651
holiness	1	1	2	4	2	4	4
tressub	1	4	37	10,013	200,795	4,829,063	14,507,089
broken-down	1	10	121	32,971	661,245	15,902,851	477,746,775
broken	1	6	10	61	121	738	32,971

hypersum	6	610	61,012	6,101,215	610,121,518	61,012,151,820	6,101,215,182,024
holiness	1	2	2	2	4	5	5
dusub	1	82	8,257	825,710	82,570,968	8,257,096,847	825,709,684,747

209 ugliest

tressub	1	3	4	6	7	9
broken-up	30	1,801	54,060	3,245,401	486,864,210	29,215,098,001
holiness	1	3	2	1	7	7
tressub	1	90	2,691	161,579	24,239,542	4,363,279,087
broken-down	1	31	931	55,891	8,384,581	1,509,280,471
broken	1	30	31	931	1,801	54,060

hypersum	30	3,060	306,090	306,090,120	306,090,120,150	306,090,120,150,180
holiness	1	3	5	6	7	10
tressub	1	152	15,239	15,239,330	15,238,329,738	15,239,329,738,645

Babylonian ugliest. The Ekari (aka *Ekagi*, *Kapauku* and *Mee*) from the Paniai Lake region of the Indonesian province of Paua, also use base sixty.

broken-up	30	31	4,060	8,151	1,878,790	5,644,521	1,864,570,720	7,463,927,401
holiness	1	0	2	0	6	1	5	3
broken-down	1	31	4,031	8,093	1,865,421	5,604,356	1,851,302,901	7,410,815,960
broken	1	30	31	4,031	4,060	8,093	8,151	1,865,421

hypersum	30	301	30,113	3,011,302	3,011,302,230	30,113,022,303	30,113,022,303,330
holiness	1	1	1	2	3	3	4

tressub	1	15	1,499	149,924	14,992,314,910	1,499,239,100	1,499,239,100,179

210 decimal bipolar

broken-up	90	81,001	73,629,999	72,893,780,011	656,044,093,728,999
holiness	2	4	5	7	10
tetrasub	1	1,484	1,348,581	1,335,096,152	12,015,866,715,827
broken-down	1	900	818,101	809,920,890	7,289,288,828,101
broken	1	90	900	81,001	818,101

hypersum	90	90,900	90,900,909	90,900,909,990	909,009,099,909,000
holiness	2	5	8	11	15
tetrasub	1	1,665	16,649,098	1,664,908,242	16,649,082,420,507

ternarish bipolar	20	200	202	220	2,000
holiness	1	2	1	1	3
broken-up	20	4,001	808,222	177,812,841	355,626,490,222
holiness	1	2	1	4	4
tressub	1	199	40,239	8,852,780	17,705,600,382
broken-down	1	21	4,243	933,481	1,866,966,243
broken	1	20	21	4,001	4,243

hypersum	20	20,200	20,200,202	20,200,202,220	202,002,022,202,000
holiness	1	3	4	5	8
tressub	1	1,006	1,005,709	1,005,708,849	10,057,088,489,818

quadary bipolar	30	300	303	330	3,000
holiness	1	2	1	1	3
broken-up	30	9,001	2,727,333	900,028,891	2,700,089,400,333
holiness	1	3	0	8	8
tressub	1	448	13,786	44,809,800	134,429,535,574
broken-down	1	31	9,394	3,100,051	9,300,162,394
broken	1	30	31	9,001	9,394

hypersum	30	30,300	30,300,303	30,300,303,330	303,003,033,303,000
holiness	1	3	4	5	8
tressub	1	1,509	1,508,563	1,508,563,273	15,085,632,734,727

quintary bipolar	40	400	404	440	4,000
holiness	1	2	1	1	3
broken-up	40	16,001	6,464,444	2,844,371,361	11,377,491,908,444
holiness	1	3	2	3	5
tetrasub	1	293	118,401	52,096,479	208,386,033,255
broken-down	1	41	16,565	7,288,641	29,154,580,565
broken	1	40	41	16,001	16,565

hypersum	40	40,400	40,400,404	40,400,404,440	404,004,044,404,000
holiness	1	3	4	5	8
tetrasub	1	740	739,959	739,959,219	7,399,592,186,892

senary bipolar	50	500	505	550	5,000
holiness	1	2	1	1	3
broken-up	50	25,001	12,625,555	6,944,080,251	34,720,413,880,555
holiness	1	2	1	6	6
tetrasub	1	458	231,245	127,185,266	635,926,562,704
broken-down	1	51	25,756	14,165,851	70,829,280,756
broken	1	50	51	25,001	25,756

hypersum	50	50,500	50,500,505	50,500,505,550	505,005,055,505,000
holiness	1	3	4	5	8
tetrasub	1	925	924,949	924,949,023	9,249,490,233,615

septary bipolar	60	600	606	660	6,000
holiness	2	3	3	3	4
broken-up	60	36,001	21,816,666	14,399,035,561	86,394,235,182,666
holiness	1	3	6	4	9
tetrasub	1	659	399,586	263,727,536	1,582,365,613,674
broken-down	1	61	363,601	269,977,260	1,439,863,923,601
broken	1	60	61	36,001	363,601

hypersum	60	60,600	60,600,606	60,600,606,660	606,006,066,606,000
holiness	2	5	8	11	15
tetrasub	1	1,110	1,109,939	1,109,938,828	1,109,388,280,338

octary bipolar	70	700	707	770	7,000
holiness	1	2	1	1	3
broken-up	70	49,001	34,643,777	26,675,757,291	186,730,335,680,777
holiness	1	3	1	3	8
pentasub	1	897	634,523	488,583,538	3,420,085,397,901
broken-down	1	71	50,198	38,652,531	270,567,767,198
broken	1	70	71	49,001	50,198

hypersum	70	70,700	70,700,707	70,700,707,770	707,007,077,707,000
holiness	1	3	4	5	8
tetrasub	1	1,295	1,294,929	1,294,928,633	12,949,286,327,061

nonary bipolar	80	800	808	880	8,000
holiness	3	4	5	5	5
broken-up	80	64,001	51,712,888	4,550,740,544	364,059,295,240,888
holiness	3	3	6	2	11
tetrasub	1	1,172	947,155	833,497,205	6,667,978,585,719
broken-down	1	81	65,449	57,595,201	460,761,673,449
broken	1	80	81	64,001	65,449

hypersum	80	80,800	80,800,808	80,800,808,880	808,008,088,808,000
holiness	3	7	12	17	22
tetrasub	1	1,480	1,479,918	1,479,918,437	14,799,184,373,784

Numbers with one added to zero and eight are, at least in some fonts, the same when reflected vertically, when seen through a swizzle stick as C. C. Bombaugh described them in *Oddities and Curiosities*. These digits also happen to all be cubes, "pancubic".

swizzled	0	1	8	10	11	18	80	81
holiness	1	0	2	1	0	2	1	2
untrimmed	1	2	9	21	22	29	91	92
broken-up	0	8	81	899	16,263	1,301,939	105,473,322	9,282,954,275
holiness	1	2	2	4	2	3	1	4
sub	1	3	30	331	5,983	478,957	38,801,467	3,415,008,031
broken-down	1	9	91	1,010	18,271	1,462,690	118,496,161	10,429,129,124,858
broken	0	1	8	9	81	91	899	1,010

holiness	1	0	2	1	2	1	4	2
untrimmed	1	2	9	10	92	102	91,010	2,121

hypersum	0	1	18	1,810	181,011	18,101,118	1,810,111,880	181,011,188,081
holiness	1	0	2	3	3	5	8	10
untrimmed	1	2	29	2,921	292,122	29,212,229	2,921,222,991	292,122,299,192

A subsequence of these reflectable numbers, the ones that are also palindromic, are also invariant when reflected horizontally as well.

pswizzled	0	1	8	11	88	101	111
holiness	1	0	2	0	4	1	0
untrimmed	1	2	9	22	99	212	222
broken-up	0	1	8	89	7,840	791,929	87,911,959
holiness	1	0	2	3	3	3	5
untrimmed	1	2	9	910	8,951	810,210,310	98,102,210,610
broken-down	1	1	9	100	8,809	889,809	98,777,608
broken	0	1	8	9	89	100	7,840
holiness	1	0	2	1	3	2	3
untrimmed	1	2	9	10	910	211	8951

hypersum	0	1	18	1,811	181,188	181,188,101	181,188,101,111
holiness	1	0	2	2	6	7	7
untrimmed	1	2	29	2,922	292,299	292,299,212	292,299,212,222

211 Bantu

untrimmed	1	4	5	6	7	8	9	10	21	22
broken-up	0	3	13	68	421	3,015	24,541	223,884	2,263,381	25,121,075
holiness	1	0	0	3	0	1	0	4	3	1
untrimmed	1	4	24	59	532	4126	35652	334995	3374492	36232186
broken-down	1	4	17	89	551	3,946	32,119	293,017	2,962,289	32,878,196
broken	0	1	3	4	13	17	68	89	421	551
holiness	1	0	0	0	0	0	3	3	0	0
untrimmed	1	2	4	5	24	28	79	910	532	662

hypersum	0	3	34	345	3,456	34,567	345,678	3,456,789	345,678,910
holiness	1	0	0	0	1	1	3	4	5

untrimmed	1	4	45	456	4,567	45,678	456,789	45,678,910	456,789,101

212 Caliban

broken-up	1	3	10	43	311	3,153	44,453	758,854	15,221,533	320,411,047
broken-down	1	2	7	30	217	2,200	31,017	529,489	10,620,797	223,566,226
broken	1	2	3	7	30	43	217	311	2,200	3,153
sum	1	3	6	10	17	27	41	58	78	99
super	1	2	6	24	168	1,680	23,520	399,840	7,996,800	167,932,800

hypersum	1	12	123	1,234	12,347	1,234,710	123,471,014	12,347,101,417

213 Suburbans are <u>not</u> the same as the sub of urbans.

broken-up	1	3	16	131	1,195	12,081	134,086	1,621,113	24,450,781
broken-down	1	2	11	90	821	8,300	92,121	1,113,752	16,798,401
broken	1	2	3	11	16	90	131	821	1,195
sum	1	3	8	16	25	35	46	58	73
super	1	2	10	80	720	7,200	79,200	950,400	14,256,000

hypersum	1	12	125	1,258	12,589	1,258,910	125,891,011	12589,101,112

214 turban

<u>broken-up</u>	1	6	37	265	2,422	26,907	26,907,002,422	26,907,029,329,002,423
<u>broken-down</u>	1	5	31	222	2,029	22,541	22,541,002,029	22,541,024,570,002,038
<u>broken</u>	1	2	13	93	850	9,443	9,443,000,850	9,443,010,293,010,290
sum	1	6	12	19	28	39	1,000,039	2,000,040
super	1	5	30	210	1,890	20,790	20,790,000,000	20,790,020,790,000,000

hypersum	1	15	156	1,567	15,679	1,567,911	15,679,111,000,000

hypersuper	1	5	66,666	> sixty-thousandplex

urban

broken-up	1	3	16	99	709	5,771	52,648	532,251	5,907,409	71,421,159
broken-down	1	2	11	68	487	3,964	36,163	365,594	4,057,958	49,057,958
broken	1	2	3	11	16	68	99	487	709	3,964
sum	1	3	8	14	21	29	48	59	71	86
super	1	2	10	60	420	3,360	302,400	3,326,400	39,916,800	598,752,000

hypersum	1	12	125	1,256	12,567	125,678	1,256,789	125,678,910

hypersuper	1	2	55	> fifty-fourplex

turban

broken-up	1	6	37	265	2,422	26,907	26,907,002,422	2,690,702,933,029,329
broken-down	1	2	13	93	850	9,443	9,443,000,850	9.443,010,293,010,290
broken	1	2	6	13	37	93	265	850
sum	1	6	12	19	27	36	46	57
super	1	5	30	210	1,680	15,120	151,200	1,663,200

hypersum	1	15	156	1,567	15,678	156,789	15,678,910	1,567,891,011

hypersuper	1	5	66,666	> six-thousandplex

215 Ariel

broken-up	5	31	253	2,308	25,641	310,000	4,055,641	4,055,645,365,641
dusub	1	4	34	312	3,469	41,953	548,870	548,871,914,266
broken-down	1	6	49	447	4,966	60,039	785,473	785,473,845,512
broken	1	5	6	31	49	253	447	2,308

hypersum	5	56	568	5,689	568,911	56,891,112	5,689,111,213
dusub	1	8	77	770	76,994	7,699,375	769,937,477

216 flawless

broken-up	1	4	25	179	1,457	13,292	134,377	1,760,193	28,297,465
broken-down	1	3	19	136	1,107	10,099	102,097	1,337,360	21,499,857
broken	1	3	4	19	25	136	179	1,107	1,457
sum	1	4	10	17	25	34	44	57	73
super	1	3	18	126	1,008	9,072	90,720	1,179,360	18,869,760

hypersum	1	13	136	1,367	13,678	136,789	13,678,910	1,367,891,013

hypersuper	1	3	666	> six-hundredplex

217 godless

broken-up	3	16	99	709	6,480	65,509	727,079	8,790,457	11,5003,020
broken-down	1	5	31	222	2,029	20,512	227,661	2,752,444	36,009,433
broken	1	3	5	16	31	99	222	2,029	6,480

hypersum	3	35	356	3,567	35,679	3,567,910	356,791,011	35,679,101,112
sub	1	13	131	1,312	13,126	1,312,561	131,256,078	13,125,607,779

218 harmless

broken-up	1	3	16	99	709	5,771	52,648	532,251	5,907,409	71,421,159
broken-down	1	2	11	68	487	3,964	36,136	365,594	4,057,697	49,057,958
broken	1	2	3	11	16	68	99	487	708	3,964
sum	1	3	8	14	21	29	48	59	71	86
super	1	2	10	60	420	3,360	302,400	3,326,400	39,916,800	598,752,000

hypersum	1	12	125	1,256	12,567	125,678	1,256,789	12,5678,910	12,567,891,011

hypersuper	1	2	55	> fifty-fourplex

useless

broken-up	2	81	3,404	170,281	15,328,694	1,410,410,129	282,097,354,494
trimmed	1	70	233	6,170	4,217,583	3,030,018	17,186,243,383
sub	1	30	1,252	62,643	5,639,111	518,860,890	103,777,817,127
broken-down	1	40	1,681	84,090	756,781	696,503,942	139,308,358,181
broken	1	2	40	81	1,681	3,404	84,090

hypersum	2	240	24,042	2,404,250	24,425,090	2,442,509,092	2,442,509,092,200
trimmed	1	13	1,331	13,314	133,148	13,314,881	133,148,811
sub	1	88	8,845	884,474	8,985,488	898,548,880	898,548,879,895

219 worthless

broken-up	5	31	222	2,029	22,541	1,252,267,284	83,483,650,777,685
dusub	1	4	30	274	3,050	169,475,948	11,298,283,523,625
broken-down	1	6	43	393	4,366	242,553,523	16,170,073,168,684
broken	1	5	6	31	43	222	393

hypersum	5	56	567	5,679	567,911	56,791,155,555	5,679,115,555,566,666
dusub	1	8	77	769	76,858	7,685,847,122	768,584,712,246,068

220 flawed

flawed	2	4	5	11	12	14	15	20	21
trimmed	1	3	4	0	1	3	4	1	10
broken-up	2	9	20	89	1,088	15,321	230,903	4,633,381	97,531,904
trimmed	1	8	1	78	77	4,210	1,282	3,522,270	8,642,083
sub	1	3	7	33	400	5,636	84,944	1,704,526	35,879,982
broken-down	1	3	7	31	379	5,337	80,434	1,614,017	33,974,791
broken	1	2	7	9	20	31	89	379	1,088

hypersum	2	24	245	24,511	2,451,112	245,111,214	24,511,121,415	2,451,112,141,520
trimmed	1	13	134	12,400	1,340,001	134,000,103	13,400,010,304	134,000,103,041
sub	1	9	90	9,017	901,714	90,171,376	9,017,137,649	901,713,764,871

221

godly	0	1	2	4	8	14	18	21	22	24
holiness	1	0	0	0	2	0	2	0	0	0
untrimmed	1	2	3	5	9	25	29	32	33	35
broken-up	0	2	9	74	1,045	18,884	397,609	87,66,282	210,788,377	5,910,840,838
holiness	1	0	1	0	1	6	4	6	5	9
untrimmed	1	3	10	85	2,156	29,995	41,087,110	9,877,393	321,899,488	61,021,951,949
broken-down	1	3	13	107	1,511	27,305	574,916	12,675,457	304,785,884	8,546,680,209
broken	0	1	2	3	9	13	74	107	1,045	1,511
holiness	1	0	0	0	1	0	0	1	1	0
untrimmed	1	2	3	4	10	24	85	218	2,156	2,622

hypersum	0	1	12	124	1,248	124,814	12,481,418	1,248,141,821
holiness	1	0	0	0	1	2	4	4
untrimmed	1	2	23	235	2,359	235,925	23,592,529	2,359,252,932

222

harmed	3	4	13	14	21	22	23	24
broken-up	3	13	172	2,421	51,013	1,124,707	25,919,274	623,187,283
sub	1	5	63	891	18,767	413,757	9,535,168	229,257,789
broken-down	1	4	53	746	15,719	346.564	7,986,691	192,027,148
broken	1	3	4	13	53	172	746	2,421

hypersum	3	34	3,413	341,314	34,131,421	3,413,142,122	341,314,212,223
sub	1	13	1,256	125,562	1,255,624,816	125,562,481,656	12,556,248,165,656

223 The used numbers include all those between three and twenty-nine. The over-used numbers would describe their uptown neighbors and the "cliché" their complimentary sequence.

over-used	1	2	4	5	6	7	8	9	10
broken-up	1	3	13	68	421	3,015	24,541	223,884	2,263,381
broken-down	1	2	9	47	291	2,084	16,963	154,751	1,564,473

broken	1	2	3	9	13	47	68	291	421
sum	1	3	7	12	18	25	33	42	52
super	1	2	8	40	240	1,680	13,440	120,960	1,209,600
hypersum	1	12	124	1,245	12,456	124,567	1,245,678	12,456,789	1,245,678,910

cliché	1	2	3	4	5	7	8	10	11
broken-up	1	3	10	43	225	1,618	13,169	133,308	1,479,557
broken-down	1	2	7	30	157	1,129	9,189	93,019	1,032,398
broken	1	2	3	7	10	30	43	157	225
sum	1	3	6	10	15	22	30	40	51
super	1	2	6	24	120	640	6,720	67,200	739,200
hypersum	1	12	123	1,234	12,345	123,457	1,234,578	123,457,810	12,345,781,011

224

worthy	0	1	2	3	4	8	10	12	13	14
holiness	1	0	0	0	0	2	1	0	0	0
untrimmed	1	2	3	4	5	9	21	23	24	25
broken-up	0	2	7	30	247	2,500	30,247	395,711	5,570,201	83,948,726
holiness	1	0	0	1	0	2	1	1	2	6
untrimmed	1	3	8	41	358	3,611	41,358	4,106,822	6,681,312	941,059,837
broken-down	1	3	10	43	354	3,583	43,350	567,133	7,983,212	120,315,313
broken	0	1	2	3	7	10	30	43	247	354
holiness	1	0	0	0	0	1	1	0	0	0
untrimmed	1	2	3	4	8	21	41	54	358	465

hypersum	0	1	12	123	1,234	12,348	1,234,810	123,481,012	12,348,101,213
holiness	1	0	0	0	0	2	3	3	3
untrimmed	1	2	23	234	2,345	23,459	2,345,921	234,592,123	23,459,212,324

225

	zero	**one**	**two**	**three**	**four**	**five**	**six**	**seven**	**eight**	**nine**	**ten**
flawlessness	1	1	3/4	1	3/4	3/4	1	1	1	1	1
godlessness	2/3	2/3	2/3	1	2/3	1	1	1	2/3	1	1
harmlessness	3/4	1	1	3/4	1	1	1	1	3/4	1	1
uselessness	2/3	2/3	1	2/3	2/3	2/3	1	1/3	2/3	2/3	2/3

worthlessness	2/5	4/5	2/5	2/5	2/5	1	1	1	3/5	1	3/5

226 All these mostlies were inspired by Doug Adam's *Mostly Harmless.*

mostly flawless	2	4	5	11	14	15	20	21	22
trimmed	1	3	4	0	3	4	1	10	11
broken-up	2	9	47	526	7,411	11,691	2,241,231	47,177,542	1,040,147,155
sub	1	3	17	194	2,726	4,301	824,503	17,355,648	382,648,754
broken-down	1	4	21	235	3,311	49,900	1,001,311	21,077,431	464,704,793
broken	1	2	4	9	21	47	235	526	3,311

hypersum	2	24	245	24,511	2,451,114	245,111,415	24,511,141,520	245,111,415,202
trimmed	1	13	134	13,400	1,340,003	134,000,304	1,230,003,041	13,400,030,411
sub	1	9	90	9,017	901,714	90,171,450	9,017,145,045	90,171,450,449

mostly godless	0	1	2	4	8	14	18	21	22
holiness	1	0	0	0	2	0	2	0	0
untrimmed	1	2	3	5	9	25	29	32	33
broken-up	0	2	9	74	1,045	18,884	397,609	8,766,282	210,788,377
holiness	1	0	1	0	1	6	4	6	5
untrimmed	1	3	10	85	2,156	29,995	41,087,110	9,877,393	321,899,488
broken-down	1	3	13	107	1,511	27,305	574,916	12,675,457	304,785,884
broken	0	1	2	3	9	13	74	107	1,045
holiness	1	0	0	0	1	0	0	1	1
untrimmed	1	2	3	4	10	24	85	218	2156

hypersum	0	1	12	124	1,248	124,814	12,481,418	1,248,141,821	124,814,182,122
holiness	1	0	0	0	2	2	4	4	4
untrimmed	1	2	23	235	2,359	235,925	23,592,529	2,359,252,932	235,925,293,233

mostly harmless	0	4	8	14	24	28	34	38	40
holiness	1	0	2	0	0	2	0	2	1
untrimmed	1	5	9	25	35	39	45	49	51
broken-up	0	1	8	113	2,720	76,273	2,596,002	98,724,349	39,5169,962
holiness	1	0	2	0	1	1	4	4	5
untrimmed	1	2	9	224	3,831	87,384	36,107,113	1,098,354,510	410,627,101,073

broken-down	1	4	33	466	11,217	314,542	10,705,645	407,129,052	16,295,867,725
broken	0	1	4	8	33	113	466	2,720	11,217
holiness	1	0	0	2	0	0	2	1	0
untrimmed	1	2	5	9	44	224	577	3,831	22,328

hypersum	0	4	48	4,814	481,424	48,142,428	4,814,242,834	481,424,283,438
holiness	1	0	2	2	2	4	4	6
untrimmed	1	5	59	5,925	592,535	59,253,539	5,925,353,945	592,535,394,549

mostly useless	0	1	3	4	5	8	9	10	11	12
holiness	1	0	0	0	0	2	1	1	0	0
untrimmed	1	2	4	5	6	9	10	21	22	23
broken-up	0	3	13	68	557	5,081	51,367	570,118	6,892,783	90,176,297
holiness	1	0	0	3	0	3	1	3	6	4
untrimmed	1	4	24	79	668	6,192	62,478	681,229	79,103,894	1,012,873,108
broken-down	1	4	17	89	729	6,650	67,229	746,169	9,021,257	118,022,510
broken	0	1	3	4	13	17	68	89	557	729
holiness	1	0	0	0	0	0	3	3	0	1
untrimmed	1	2	4	5	24	28	79	910	668	8,310

hypersum	0	1	13	134	1,345	13,458	134,589	13,458,910	1,345,891,011
holiness	1	0	0	0	0	2	3	4	4
untrimmed	1	2	24	245	2,456	24,569	2,456,910	245,691,021	24,569,102,122

mostly worthless	1	8	10	12	15	16	17	18	19
broken-up	1	9	91	1,101	16,606	266,797	4,552,155	82,205,587	1,566,458,308
broken-down	1	8	81	980	14,781	237,476	4,051,873	73,171,190	1,394,304,483
broken	1	8	9	81	91	980	1,101	14,781	16,606

hypersum	1	18	1,810	181,012	18,101,215	1,810,121,516	181,012,151,617

hypersuper	1	8	1,010,101,010,101,010	> fifteenduplex

mostly faulty	234	235	238	243	245
broken-up	234	54,991	13,088,092	3,180,461,347	779,226,118,107
broken-down	1	235	55,931	13,591,468	3,329,965,591
broken	1	234	235	54,991	55,931

hypersum	234	234,235	234,235,238	234,235,238,243	234,235,238,243,245
pentasub	1	1,578	1,578,265	1,578,264,620	1,578,264,620,601

There are also mostly numbers based not on number names, but numbers in a sequence, the rounded sum of a half-number plus a half. The 'mostly odd' are all the integers higher than zero and the 'mostly even' all those including zero. Many slowly-growing sequences however have repeated terms. The more interesting ones are those that don't stutter. Curiously the 'mostly primes' only have more primes than the 'not mostly primes' for the first five terms

mostly prime	1	2	3	4	6	7	9	10	12	15	16	19	21	22	24	27	30	31	34	36
prime count	0	1	2	2	2	3	3	3	3	3	3	4	4	4	4	4	4	5	5	5
not mosty prime	0	5	8	11	13	14	18	20	23	25	26	28	29	32	33	35	38	39	41	43
prime count	0	1	1	2	3	3	3	3	4	4	4	4	5	5	5	5	5	5	6	7

227 The useful number names, those with all three of E, S and U, would include a 'four', 'hundred' or 'thousand' for a U, a 'six', 'seven', 'sixteen', 'seventeen', 'sixty', 'seventy' or 'thousand' for an S and a 'one', 'three', 'five', 'seven', 'eight', 'nine', 'ten', 'eleven', 'twelve', '-teen', 'twenty', 'seventy', 'eighty', 'hundred' or the suffix -plex for an E.

useful	74	106	107	116	117	126
broken-up	74	7,845	839,489	97,388,569	11,395,302,062	1,435,905,448,381
tetrasub	1	143	15,375	1,783,733	208,712,237	26,299525,670
broken-down	1	106	11,343	1,315,894	153,970,941	19,401,654,460
broken	1	74	106	7,845	11,343	839,489

hypersum	74	74,106	74,106,107	74,106,107,116	74,106,107,116,117
tetrasub	1	1,357	1,357,301	1,357,300,697	1,357,300,697,389

The harmful number names would include a 'three', 'eight', 'hundred', or 'thousand' for an H, a 'thousand' for an A, a 'three', 'four', 'thirty', 'forty', or 'hundred' for an R and a 'million' for an M.

harmful	3,003,000	3,003,001	3,003,002
holiness	5	4	4
broken-up	3,003,000	9,018,012,003,001	27,081,108,081,033,006,027,081,108,081,033,006,001
holiness	5	9	29
15th sub	1	2,758,630	8,284,173,803,750
broken-down	1	3,003,001	9,018,018,009,003
broken	1	3,003,000	3,003,001

hypersum	3,003,000	30,030,003,003,001	300,300,030,030,013,003,002
holiness	5	9	13
15th sub	1	9,186,248	91,862,476,032,950

228 "Banned from Arrgho" is a filksong by Francis X. Purgator published in the fanzine *NSEA Bulletin* inspired by the Galaxy Quest episode "The Pirates of Arrgho", in which the *Protector* visited Arrgho in the Evri system. (See "The Unofficial Questarian Guide" at the webpage, michaelhalm.tripod.com/im4gq.htm)

229 emirps

broken-up	13	222	6,895	255,337	18,135,822	1,324,170,343	22,529,031,653
tressub	1	11	343	12,712	902,930	65,926,559	1,121,654,439
broken-down	1	14	433	16,109	1,144,174	83,540,811	1,421,337,961
broken	1	13	14	222	433	6,895	16,109

hypersum	13	1,317	131,731	13,173,137	1,317,313,771	131,731,377,173
tressub	1	65	6,559	655,852	65,585,191	6,558,519,082

Prime numberdromes are included in acronums containing the infix -pp-: "blipping", "chippie", "chipping", "clapping", "clipping", "crappie", "crappier", "crappies","crapping", "crippling", "eohippus", "flappable, "flappier", "flippant", "flipping", "floppier", "floppies", "floppily", "flopping", "frapping", "fripping", "hippie", "hippo", "hippos", "hiccupping", "Lapp", Lapps", "kippur", "misapplier", "nappie", "opp", "opposable", "opposing", "opposite", "oppugus", "Philippine", "pippin", "plopping", "popping", "reapplier", "recappable", "recapping", "rippling", "shipping", "slapping", "slipping", "sloppily", "snappier", "snappily", "snappish", "snipping", "stepping", "stopping", "yippie" and "yuppie".

Ananums, like the emirps, can be subdivided into three sequences: sorted, with digits in numerical order, like thirteen, unsorted like one hundred seven or "detros", in reverse numerical order, like thirty-one.

sorted	13	17	37	79	113	149	157	167	179
unsorted	107	701	709	733	739	743	769	907	937
detros	31	71	97	311	751	761	941	971	991

230 gnilddim

broken-up	45	2,071	111,879	6,267,295	401,218,759	26,085,486,630
tetrasub	1	38	2,049	114,789	7,348,578	477,772,353
broken-down	1	46	2,485	139,206	8911669	579,397,691
broken	1	45	46	2,071	2,485	111,879

hypersum	45	4,546	454,654	45,465,456	4,546,545,664	454,654,566,465
tetrasub	1	83	8,327	832,729	83,272,889	8,327,288,858

231 Leira

broken-up	12	157	2,367	38,029	686,889	13,088,920	275,554,209	6,350,835,727
tressub	1	7	117	1,893	34,198	651,658	13,719,035	316,189,492
broken-down	1	13	196	3,149	56,878	1,083,831	22,817,329	525,882,398
broken	1	12	13	157	196	2,367	3,149	38,029

hypersum	12	1,213	121,315	12,131,516	1,213,151,618	121,315,161,819
dusub	1	164	16,418	1,641,822	164,182,218	16,418,221,786

232 Nabilac

broken-up	14	239	5,750	155,489	6,380,799	268,149,047	12,609,386,008
tressub	1	11	286	7,741	317,680	13,350,34	627,784,362
broken-down	1	15	361	9,762	400,603	16,835,088	791,649,739
broken	1	14	15	239	361	5,750	9,762

hypersum	14	1,417	141,724	14,172,427	1,417,242,741	141,724,274,142
tressub	1	71	7,056	705,604	70,560,361	7,056,036,126

233 paehc

broken-up	12	157	3,309	76,264	2,367,493	75,836,040	8,496,003,973
dusub	1	21	448	10,321	320,405	10,263,292	1,149,809,104
broken-down	1	13	274	6,315	196,039	6,279,563	703,507,095
broken	12	13	157	274	3,309	6,315	76,264

hypersum	12	1,213	121,321	12,132,123	1,213,212,331	121,321,233,132
dusub	1	2	21	37	448	855

234 raed

broken-up	78	6,163	53,625	47,733,214	4,630,658,017	453,852,218,880
holiness	2	2	1	0	6	9
tetrasub	1	113	9,822	874,264	84,813,460	8,312,593,350
broken-down	1	79	6,874	611,865	59357779	5,817,674,207
broken	1	78	79	6,163	6,874	53,625

hypersum	78	7,879	787,987	78,798,789	7,879,878,997	787,987,899,798
holiness	2	3	5	8	9	12
tetrasub	1	144	14,432	1,443,250	144,325,018	14,432,501,821

235 Utnab

broken-up	10	131	1,844	27,791	446,500	7,618,291	137,575,738
holiness	1	0	2	1	3	3	2
sub	1	17	249	3,760	60,427	1,031,023	18,618,851
broken-down	1	11	155	2,336	37,531	640,363	11,564,065
broken	1	10	11	131	155	1,844	2,336

hypersum	10	1,013	101,314	10,131,415	1,013,141,516	101,314,151,617
holiness	1	1	1	1	2	2
dusub	1	137	13,711	1,371,138	137,113,794	13,711,379,405

236 "Live" comes from the Indo-European *leip-* (stick, adhere; fat) as do aliphsatic, delay, leave, lebensraum, life, lipo-, lively, liver, relay, synalepha.

broken-up	15	256	6,927	201,139	6041097	205,598,437	3,501,214,526
sub	1	13	345	10,014	300,768	10,236,144	174,315,207
broken-down	1	16	433	12,573	377,623	12,851,755	218,857,458
broken	1	15	16	256	433	6,927	12,573

hypersum	15	1,517	151,727	15,172,729	1,517,272,930	151,727,293,034
tressub	1	75	7,554	755,406	75,540,571	7,554,057,112

"Life is short, death is sure, sin the cause, Jesus the cure." (St. Anonymus)

"Life", on the other hand, describes lucky-indexed Fibonacci evens and "limb" their complimentary sequence, even though the original phrase was the synonym pair, "lith and limb".

life	2	34	10,946	196,418	63,245,986	1,134,903,170	6,557,470,319,842
trimmed	1	23	835	85,307	52,134,875	238,206	544,636,208,731
sub	1	13	4,027	72,258	23,266,898	41,750,744	2,412,358,516,762

broken-up	2	69	755,276	148,349,801,437	9,382,529,464,788,607,158
trimmed	1	58	644,165	3,723,870,326	827,141,835,367,756,047
sub	1	25	27,781	54,574,842,051	3,451,639,696,280,819,712
broken-down	1	3	32,839	6,450,170,706	4,079,470,610,672,960
broken	1	2	3	69	32,839

hypersum	2	234	23,410,946	23,410,946,196,418	2,341,094,619,641,863,245,986

trimmed	1	123	1,230,835	1,230,835,085,307	123,083,508,530,752,134,875
sub	1	86	8,612,406	8,612,405,804,033	861,240,580,403,318,882,304

Since they are all even we also have "half-life" numbers, some of which are prime.

half-life	1	17	5,473	83,209	31,822,993	567,451,585	3,278,735,159,921

237 The p- prefix or -p suffix when combined with h for hachibi, r for rokubi or y for yongbi cannot mean prime, but rather a palindromic number. "burp" describes binarish-unindexed rokubi palindromes, "burping" binarish-unindexed rokubi palindrome-indexed ng, "caliph" cube-and-lucky-indexed palindromic hachibi, "caliphal" cube-and-lucky-indexed palindromic-hichibi-and-luckies, "caliphate" cube-and-lucky-indexed palindromic-hachibi-and-ternarish-evens, "carp" cubes-and-rokubi-palindromes, "carpal" cubes-and-rokubi-palindromes-and-luckies, "carpel" cubes-and-rokubi-palindromic-even-luckies, "carping" cubes-and-rokubi-palindrome-indexed ng, "carpus" cubes-and-rokubi-palindrome-unindexed sanbi, "chirp" cubed hachibi-indexed rokubi palindromes, "chirpier" cubed hachibi-indexed even rokubi palindrome-indexed even rokubi, "chirping" cubed hachibi-indexed rokubi palindrome-indexed ng, "chirpiest" cubed hachibi-indexed rokubi palindrome-indexed even sorted ternarish, "crispy" cubed rokubi-indexed sorted palindromic yongbi, "earpiece" evens-and-rokubi-palindrome-indexed even cubed evens, "earplug" evens-and-rokubi-palindromic-lucky-unindexed gobi, "elephant" even-lucky-even-palindromic-hachibi-and-nibi-ternarish, "epitaph" even prime-indexed ternarish-and-palindromic-hachibi, "escarp" even-sorted-cubes-and-rokubi-palindromes, "escarping" even-sorted-cubes-and-rokubi-palindrome-indexed ng, "gapy" gobi-and-palindromic-yongbi, "gulpy" gobi-unindexed palindromic yongbi, "harp" hachibi-and-rokubi-palindromes, "harpies" hachibi-and-rokubi-palindrome-indexed evens sorted, "harping" hachibi-and-rokubi-palindrome-indexed ng, "harpist" hachibi-and-rokubi-palindrome-indexed sanbi ternarish, "inphase" palindromic-hachibi-and-sorted-evens downtown neighbors, "kaliph" kyuubi-and-lucky-indexed palindromic hachibi, "occupy" odd cubed cube-unindexed palindromic yongbi, "pericarp" palindromic even rokubi-indexed cubes-and-rokubi-palindromes, "phase" palindromic-hachibi-and-sorted-evens, "phalli" palindromic-hachibi-and-luckier-lucky-ichibi, "phaser" palindromic-hachibi-and-sorted-even-rokubi, "phasic" palindromic-hachibi-and-sanbi-indexed cubes, "phallic" palindromic-hachibi-and-luckier-lucky-indexed cubes, "phallus" palindromic-hachibi-and-luckier-lucky-unindexed sanbi, "phaseal" palindromic-hachibi-and-sorted-evens-and-luckies, "phasing" palindromic-hachibi-and-sanbi-indexed ng, "phallism" palindromic-hachibi-and-luckier-lucky-indexed sanbi Mersennes, "phallist" palindromic-hachibi-and-luckier-lucky-indexed sanbi ternarish, "phantasm" palindromic-hachibi-and-nibi-ternarish-and-sanbi-Mersennes, "phantast" palindromic-hachibi-and-nibi-ternarish-and-sanbi-ternarish, "phantastasies" palindromic-hachibi-and-nibi-ternarish-and-sanbi-ternarish-and-sanbi-indexed-evens-sorted, "phantasy" palindromic-hachibi-and-nibi-ternarish-and-sorted-yongbi, "pharisee" palindromic-hachibi-and-rokubi-indexed sorted evenly evens, "phase" palindromic-hachibi-and-sorted-evens, "phaseout" palindromic-hachibi-and-sorted-evens neighbors, "pharisaic" palindromic-hachibi-and-rokubi-indexed sanbi-and-ichibi cubes, "pheasant" palindromic-hachibi-evens-and-sanbi-and-nibi-ternarish, "philatelic" palindromic-hachibi-indexed-luckies-and-ternarish-even-luckies-indexed cubes, "philippic" palindromic hachibi-indexed lucky-indexed palindromic prime-indexed cubes, "Phillipine" palindromic hachibi-indexed luckier lucky-indexed prime-indexed nibi evens,"Philistine" palindromic hachibi-indexed lucky-indexed sanbi ternarish-indexed nibi evens, "pipy" prime-indexed palindromic yongbi, "practical" palindromic-rokubi-and-cubed-ternarish-indexed cubes-and-luckies, "practice" palindromic-rokubi-and-cube-ternarish-indexed cubed evens, "practicing" palindromic-rokubi-and-cubed-ternarish-indexed cube-indexed ng, "Prague" palindromic-rokubi-and-gobi-unidexed evens, "praise" palindromic-rokubi-and-ichibi-sorted-evens, "praline" palindromic-rokubi-and-lucky-indexed nibi evens, "pram" palindromic-rokubi-and-Mersennes, "prance" palindromic-rokubi-and-ninbi-cubes, "prancing" palindromic-rokubi-and-nibi-cube-indexed ng, "praos" palindromic-rokubi-and-odd-sanbi, "prat" palindromic-rokubi-and-ternarish", "prate" palindromic-rokubi-and-ternarish-evens, "pratfall" palindromic-rokubi-and-ternarish-Fibonacci-and-luckier-luckies, "prating" palindromic-rokubi-and-ternarish-indexed ng, "praus" palindromic-rokubi-and-unindexed-sanbi, "pray" palindromic-rokubi-and-yongbi, "praying" palindromic-rokubi-and-yongbi-indexed ng, "preach" palindromic-rokubi-evens-and-cubed-hachibi, "preachier" palindromic-rokubi-evens-and-cubed-hachibi-indexed even rokubi, "preachiest" palindromic-rokubi-evens-and-cubed-hachibi-indexed even sorted sanbi, "preaching" palindromic-rokubi-evens-and-cubed-hachibi-indexed ng, "preallot" palindromic-rokubi-evens-and-luckier-lucky-odd-ternarish, "preamp" palindromic-rokubi-evens-and-Mersenne-primes, "precarious" palindromic-rokubi-even-cubes-and-rokubi-indexed odd-unindexed sanbi, "precast" palindromic-rokubi-even-cubes-and-sanbi-ternarish, "preciously" palindromic rokubi even cube-indexed odd-unindexed slorted lucky yongbi, "precis" palindromic rokubi even cube-indexed sanbi, "precise" palindromic rokubi even cube-indexed sorted evens, "precision" palindromic-rokubi-even-cube-indexed-sorted-ichibi-and-nibi, "precising" palindromic rokubi even cube-indexed sanbi-indexed ng, "precut" palindromic rokubi even cube-unindexed ternarish, "prefab" palindromic-rokubi-even-Fibonacci-and-binarish, "preface" palindrome-rokubi-even-Fibonacci-and-cube-evns, "prefigure" palindromic rokubi even Fibonacci-indexed gobi-unindexed rokubi evens, "prelate" palindromic-rokubi-even-luckies-and-ternarish-evens, "prelatic" palindromic-rokubi-even-lucky-and-ternarish-indexed cubes, "prelim" palindromic rokubi even lucky-indexed Mersennes, "prelimit" palindromic rokubi even lucky-indexed Mersenne-indexed ternarish, "priceless" without palindromic rokubi-indexed cubed evens, "prickiest" palindromic rokubi-indexed cubed kyuubi-indexed even sorted ternarish, "prickling" palindromic rokubi-indexed cubed kyuubi lucky-indexed ng, "primacies" palindromic rokubi-indexed Mersenne-and-cube-indexed evens sorted, "primatial" palindromic rokubi-indexed Mersenne-and-ternarish-ichibi-and-luckies, "primaries" palindromic rokubi-indexed Mersenne-and-rokubi-indexed evens sorted, "primarily" palindromic rokubi-indexed Mersenne-and-rokubi-indexed lucky yongbi, "presift" palindromic rokubi even sorted-indexed Fibonacci ternarish, "price" palindromic rokubi-indexed cubed evens, "priceless" without palindromic rokubi-indexed cubed evens, "pricier" palindromic rokubi-indexed cube-indexed even rokubi, "pricier" palindromic rokubi-indexed cube-indexed even rokubi, "priciest" palindromic rokubi-indexed cube-indexed even sorted ternarish, "pricing" palindromic rokubi-indexed cube-indexed ng, "prick" palindromic rokubi-indexed cubed kyuubi, "prickier" palindromic rokubi-indexed cube kyuubi-indexed even rokubi, "prickiest" palindromic rokubi-indexed cubed kyuubi-indexed even sorted ternarish, "pricking" palindromic rokubi-indexed cubed kyuubi-indexed ng, "prickling" palindromic rokubi-indexed cubed kyuucbi-indexed ng, "pricy" palindromic rokubi-indexed cubed yongbi, "prier" palindromic rokubi-indexed even rokubi, "priest" palindromic rokubi-indexed even sorted ternarish, "prig" palindromic-rokubi-indexed gobi, "prim" palindromic rokubi-indexed Mersennes, "primacies" palindromic-rokubi-indexed-

Mersennes-and-cube-indexed evens sorted, "primacy" palindromic rokubi-indexed Mersenne-and-cubed-yongbi, "primal" palindromic rokubi-indexed Mersennes-and-luckies, "primaries" palindromic-rokubi-indexed Mersennes-and-rokubi-indexed evens sorted, "primarily" palindromic-rokubi-indexed-Mersennes-and-rokubi-indexed lucky yongbi, "primas" palindromic rokubi-indexed Mersennes-and-sanbi, "primate" palindromic rokubi-indexed Mersenne-and-ternarish-evens, "primatial" palindromic-rokubi-indexed-Mersennes-and-ternarish-ichibi-and-luckies, "prismatic" palindromic rokubi-indexed sanbi-Mersenne-and-ternarish-indexed cubes, "primo" palindromic rokubi-indexed Mersenne odds, "primp" palindromic rokubi-indexed Mersenne primes, "primping" palindromic rokubi-indexed Mersenne prime-indexed ng, "primulas" palindromic rokubi-indexed Mersenne-unindexed lucky-and-sanbi, "prince" palindromic rokubi-indexed nibi cubed evens, "principal" palindromic rokubi-indexed nibi cube-indexed primes-and-luckies, "print" palindromic rokubi-indexed nibi ternarish, "printing" palindromic rokubi-indexed nibi ternarish-indexed ng, "printout" palindromic rokubi-indexed nibi ternarish neighbors, "priss" palindromic rokubi-indexed sorted sanbi, "prissies" palindromic rokubi-indexed sorted sanbi-indexed evens sorted, "prissier" palindromic rokubi-indexed sorted sanbi-indexed even rokubi, "prissily" palindromic rokubi-indexed sorted sanbi-indexed lucky yongbi, "pristine" palindromic rokubi-indeed sanbi ternarish-indexed nibi evens, "prune" palindromic rokubi-unindexed nibi evens, "pruning" palindromic rokubi-unindexed nibi-indexed ng, "prurient" palindromic rokubi-unindexed rokubi-indexed even nibi ternarish, "Prussian" palindromic-rokubi-unindexed-sorted-sanbi-indexed-and-nibi, "prussic" palindromic rokubi-undexed sorted sanbi-indexed cubes, "puerperal" prime-unindexed even-rokubi-palindromic-even-rokubi-and-luckies, 'pulpy" prime-unindexed lucky palindromic yongbi, "purple" prime-unindexed rokubi palindromic lucky evens, "purpling" prime-unindexed rokubi palindromic lucky-indexed ng, "purplish" prime-indexed rokubi palindromic lucky-indexed shichibi, "Ralph" rokubi-and-lucky-palindromic-hachibi, "scarp" sanbi-cubes-and-rokubi-palindromes, "Raphael" rokubi-and-palindromic-hachibi-and-even-luckies, "raspy" rokubi-and-sorted-palindromic-yongbi, "scarping" sanbi-cubes-and-rokubi-palindrome-even-rokubi-indexed ng, "seraph" sorted-even-rokubi-and-palindromic-hachibi, "seraphic" sorted-even-rokubi-and-palindromic-hachibi-indexed cubes, "sharp" shichibi-and-rokubi-palindromes, "sharpie" shichibi-and-rokubi-palindromes-indexed evens, "sharpies" shichibi-and-rokubi-palindromes-indexed evens sorted, "sharping" shichibi-and-rokubi-palindrome-indexed ng, "sirupy" sanbi-indexed rokubi-unindexed palindromic yongbi, "slurp" sanbi lucky-unindexed rokubi palindromes, "slurping" sanbi lucky-unindexed rokubi palindrome-unindexed ng, "soapy" sanbi-odd-and-palindromic-yongbi, "soupy" sanbi-odd-unindexed palindromic yongbi, "sprang" sorted-palindromic-rokubi-and-ng, "sprat" sorted-palindromic-rokubi-and-ternarish, "spray" sorted-palindromic-rokubi-and-yongbi, "spraying" sorted-palindromic-rokubi-and-yongbi-indexed ng, "sprier" sorted palindromic rokubi-indexed even rokubi, "spriest" sorted palindromic rokubi-indexed even sorted ternarish, "sprig" sorted-palindromic-rokubi-indexed gobi, "spring" sorted palindromic rokubi-indexed ng, springiest" sorted palindromic rokubi-indexed ng-indexed even sorted ternarish, "springing" sorted palindromic rokubi-indexed ng-indexed ng, "sprint" sorted palindromic rokubi-indexed nibi ternarish, "sprinting" sorted palindromic rokubi-indexed nibi ternarish-indexed ng, "sprit" sorted palindromic rokubi-indexed ternarish, "sprite" sorted palindromic rokubi-indexed ternarish evens, "spruce" sorted palindromic rokubi-unindexed cubed evens, "sprucer" sorted palindromic rokubi-unindexed cubed even rokubi, "sprucing" sorted palindromic rokubi-unindexed cube-indexed ng, "sprucy" sorted palindromic rokubi-unindexed cubed yongbi, "sprung" sorted palindromic rokubi-unindexed ng, "spy" sorted palindromic yongbi, "staph" sanbi-ternarish-and-palindromic-hachibi, "sugarplum" sanbi-unindexed gobi-and-rokubi-palindrome-lucky-unindexed Mersennes, "sulphas" sanbi-unindexed palindromic-hachibi-and-sanbi, "sulphate" sanbi-unindexed lucky-palindromic-hachibi-and-ternarish-evens, "sulphur" sanbi-unindexed lucky palindromic hachibi-unindexed rokubi, "sulphuric" sanbi-unindexed lucky palindromic hachibi-unindexed rokubi-indexed cubes, "sulphuring" sanbi-unindexed lucky palindromic hachibi-unindexed rokubi-indexed ng, "surplice" sanbi-unindexed rokubi palindromic lucky-indexed cubed evens, "surpass" sanbi-unindexed rokubi-palindromes-and-sorted-sanbi, "surplus" sanbi-unindexed rokubi palindromic luckies-unindexed sanbi, "surprise" sanbi-unindexed rokubi palindromic rokubi-indexed sorted evens, "tarp" ternarish-and-rokubi-palindromes and "tarpaulin" ternarish-and-rokubi-palindomes-and-unindexed-lucky-indexed nibi.

238 cubans

broken-up	7	134	4,965	302,999	27,577,874	3,502,692,997	591,982,694,367
dusub	1	18	671	41,006	3,732,259	474,037,948	80,116,145,613
broken-down	1	8	297	18,125	1,649,672	209,526,469	35,411,622,933
broken	1	7	8	134	297	4,965	18,125

hypersum	7	719	71,937	7,193,761	719,376,191	719,376,191,127
dusub	1	97	9,736	973,570	97,356,981	97,356,980,580

239 The first three cubans are primes.

sub	1	3	5	17	29	45
broken-up	7	134	4,965	630,689	136,864,478	45,302,772,907
dusub	1	18	671	85,354	18,522,592	6,131,063,602
broken-down	1	8	297	37,727	8,187,056	2,709,953,263
broken	1	7	8	134	297	4,965

hypersum	7	719	71,937	71,937,127	71,937,127,217	71,937,127,217,331
dusub	1	97	9,736	9,735,631	9,735,631,487	9,735,631,487,186

240 The fortunate primes are not named because they bring good fortune, but for their discoverer Reo Fortunate. (*Unsolved Problems in Number Theory* by R. K. Guy)

sub	1	2	3	5	6	7	8	14
broken-up	3	16	115	1,511	25,802	491,749	11,336,029	419,924,822
sub	1	6	42	556	9,492	180,904	914,014	6,579,001
broken-down	1	5	36	473	8,077	153,936	3,548,605	131,452,321

broken	1	3	5	16	36	115	473	1,511

hypersum	3	35	357	35,713	3,571,317	357,131,719	35,713,171,923
sub	1	13	131	13,138	1,313,814	131,381,417	13,138,141,729

unfortunate primes	2	11	29	31	41	43	53
trimmed	1	0	18	20	30	32	42
mates	7	16	26	46	58	68	70
broken-up	2	23	669	20,762	851,911	36,652,935	1,943,457,466
sub	1	8	246	7,638	313,401	13,483,861	714,958,047
broken-down	1	11	320	9,931	407,491	17,532,044	929,605,823
broken	1	2	11	23	320	669	9,931

hypersum	2	211	21,129	2,112,931	211,293,141	21,129,314,143	2,112,931,414,353
trimmed	1	100	10,018	1,001,820	100,182,030	10,018,203,032	1,001,820,303,242
sub	1	78	7,773	777,304	77,730,403	7,773,040,279	777,304,027,946

etanurof primes	17	71	199	991	1,021	1,061
broken-up	17	1,208	240,409	238,246,527	243,249,944,476	258,088,429,335,563
tresssub	1	60	11,969	11,861,596	12,110,701,616	1,284,946,627,684
broken-down	1	18	3,583	3,550,771	3,625,340,774	3,846,490,111,985
broken	1	17	18	1,208	3,583	240,409

hypersum	17	1,771	1,771,199	1,771,199,991	17,119,911,021	171,199,110,211,061
tressub	1	88	88,183	88,182,855	852,350,180	8,523,501,804,596

241 The first prime not a Solinas is forty-three!

non-Solinas primes	43	53	83	89	101	103
broken-up	43	2,280	189,283	16,848,467	1,701,884,450	175,310,946,817
tetrasub	1	42	3,467	308,591	31,171,101	3,210,931,995
broken-down	1	44	3,653	325,161	32,844,914	3,383,351,303
broken	1	43	44	2,280	3,653	189,283

hypersum	43	4,353	435,383	43,538,389	43,538,389,101	43,538,389,101,103
tetrasub	1	80	7,975	797,434	797,433,413	797,433,412,573

Sanilos primes

broken-up	13	222	6,895	255,337	18,135,822	1,324,170,343	22,529,031,653
tressub	1	11	343	12,712	902,930	65,926,559	1,121,654,439
broken-down	1	14	435	16,109	1,144,174	1,421,337,961	44,145,017,602
broken	1	13	14	222	435	6,895	16,109

hypersum	13	1,317	131,731	13,173,137	1,317,313,771	131,731,377,173
tressub	1	65	6,559	655,852	65,585,191	6,558,519,082

242 beheadable primes

broken-up	13	222	5,119	189,625	8,158,994	383,662,343	20,342,263,173
tressub	1	11	255	9,441	406,212	19,101,423	325,130,408
broken-down	1	14	323	11,965	514,818	24,208,411	412,057,805
broken	1	13	14	222	323	5,119	11,965

hypersum	13	1,317	131,723	13,172,337	1,317,233,743	131,723,374,353
tressub	1	65	6,558	655,812	65,581,207	6,558,120,645

243 curtailable primes

broke-up	23	668	20,731	767,715	40,709,626	2,402,635,649	170,627,840,705
tressub	1	33	1,032	38,222	206,813	119,620,185	3,471,012,187
broken-down	1	24	745	27,589	1,462,962	8,634,347	2,505,391,025
broken	1	23	24	668	745	20,731	27,589

hypersum	23	2,329	232,931	23,293,137	2,329,313,753	232,931,375,371
tressub	1	116	11,597	1,159,697	115,969,703	11,596,970,311

Hyperadding a new tail, especially one more (modulo 10) than the original, is called "caudation", the opposite of curtailing and different from untrimming.

1	116	11,597	1,159,697	115,969,703	11,596,970,311

caudated	12	23	34	45	56	67	78	89	90	101	112	123	134	145	156	167	178	189	190
trimmed	1	12	23	34	45	56	67	78	8	0	1	12	23	34	45	56	67	78	8

244 two-sided primes

tressub	1	2	3	4	6	15
broken-up	23	852	45,179	3,298,919	372,823,026	116,696,906,057
tressub	1	42	2,249	164,243	18,561,766	5,809,996,840
broken-down	1	24	1,273	92,953	10,504,962	3,288,146,059

broken	1	23	24	852	1,273	45,179

hypersum	23	2,337	233,753	23,375,373	23,375,373,113	23,375,373,113,131
tressub	1	116	11,638	1,163,791	1,163,791,299	1,163,791,299,308

245 permutable primes

broken-up	13	222	6,895	255,337	18,135,822	1,324,170,343	22,529,031,653
tressub	1	11	343	12,712	902,930	65,926,559	1,121,654,439
broken-down	1	14	435	16,109	1,144,174	8,340,811	1,421,337,961
broken	1	13	14	222	435	6,895	16,109

hypersum	13	1,317	131,731	13,173,137	1,317,313,771	131,731,377,173
tressub	1	65	6559	655,852	65,585,191	6,558,519,082

246 odders (*Strength in Numbers* by Sherman K. Stein)

broken-up	2	7	37	266	2,165	24,081	291,137	3,808,862	65,041,791	1,174,561,100
trimmed	1	6	26	155	1,054	1,370	180,026	277,751	5,430,680	634,500
sub	1	3	14	98	796	8,859	107,103	1,401,202	23,927,538	432,096,881
broken-down	1	3	16	115	936	10,411	125,868	1,646,695	28,119,683	507,800,989
broken	1	2	3	7	16	37	115	266	936	2,165

hypersum	2	23	235	2,357	23,578	2,357,811	235,781,112	23,578,111,213
trimmed	1	12	124	1,246	12,467	1,246,700	124,670,001	12,467,000,102
sub	1	8	86	867	8,674	867,390	86,739,024	8,673,902,377

hypersuper	2	33	555,555,555,555,555,555,555,555,555,555,555	> thirtyduplex

247 eveners

broken-up	4	25	229	2,315	32,639	491,900	7,903,039	166,455,719	3,669,928,857
sub	1	9	84	852	12,007	180,960	1,097,767	10,060,859	101,706,360
broken-down	1	5	46	465	6,556	98,805	599,386	5493279	55,532,176
broken	1	4	5	25	46	229	465	2,315	6,556

hypersum	4	46	469	46,910	4,691,014	469,101,415	46,910,141,516	4,691,014,151,621
sub	1	17	173	17,257	1,725,728	172,572,766	17,257,276,646	1,725,727,664,626

248 urtailable odds

broken-up	11	144	2,171	37,051	706,140	21,927,391	724,310,043	25,372,778,896

dusub	1	19	294	5,014	95,566	2,967,550	38,673,711	583,073,221
broken-down	1	12	181	3,089	58,872	1,828,121	23,824,445	359,194,796
broken	1	11	12	144	181	2,171	3,089	37,051

hypersum	11	1,113	111,315	11,131,517	1,113,151,719	111,315,171,931
dusub	1	150	15,065	1,506,487	150,648,703	15,064,870,322

urtailable evens

broken-up	20	441	10,604	276,145	7,742,664	309,982,705	13,027,016,274
holiness	1	0	3	1	2	6	3
tetrasub	1	9	195	5,058	141,813	5,677,532	238,598,127
broken-down	1	22	529	13,776	386,257	15,464,056	649,876,609
broken	1	20	22	441	529	10,604	13,776

hypersum	20	2,022	202,224	20,222,426	2,022,242,628	202,224,262,840
holiness	1	1	1	2	4	5
tetrasub	1	101	10,068	1,006,815	100,681,532	10,068,153,200

249 non-curtailable odds

broken-up	21	484	12,121	327,751	9,516,900	390,520,651	16,801,904,893
tetrasub	1	10	223	6,004	174,309	7,152,636	307,737,623
broken-down	1	23	576	15,575	452,251	18,557,866	798,440,489
broken	1	21	23	484	576	12,121	15,575

hypersum	21	2,123	212,325	21,232,527	2,123,252,729	212,325,272,931
tressub	1	106	10,571	1,057,105	105,710,529	10,571,052,879

non-curtailable evens

mates	21	23	25	27	29	41	43	45
broken-up	10	121	1,704	27,385	494,634	14,866,405	476,219,594	16,206,332,601
holiness	1	0	1	2	2	5	3	5
tressub	1	7	85	1,364	24,627	740,156	23,709,578	806,865,790
broken-down	1	12	169	2,716	49,057	1,474,426	47,230,689	1,607,317,852
broken	1	10	12	121	169	1,704	2,716	27,385

hypersum	10	1,012	101,214	10,121,416	1,012,141,618	101,214,161,820
holiness	1	1	1	2	4	5

dusub	1	137	13,698	1,369,785	136,978,473	13,697,847,257

250 Werewives, aka werewomen (http://bigclosetr.us/topshelf/fiction/31012/new-werewoman-handbook), are men who are said to turn into wildwomen with the full moon, analogous to werebeasts, who are men who turn into wild beasts. Their opposite, wifeweres, would be women who turn into wildmen, or in the numberworld, evens which permute into odds.

...	110	112	114	116	118	130	132	134	136	138	150	152	154
inter	111	113	115	117	124	133	135	137	147	151	153	155	157

251 panodd

broken-up	11	144	2,171	37,051	708,140	21,927,391	285,762,223	4,308,360,736
dusub	1	19	294	5,014	95,566	2,967,550	38,673,711	583,973,221
broken-down	1	12	181	3,089	58,872	1,828,121	23,824,445	359,194,796
broken	1	11	12	144	181	2,171	3,089	37,051

hypersum	11	1,113	111,315	11,131,517	1,113,151,719	111,315,171,921
dusub	1	150	15,065	1,506,487	150,648,703	15,064,870,320

paneven	0	2	4	6	8	20	22	24	26
holiness	1	0	0	1	2	1	0	0	1
broken-up	0	1	4	25	204	4,105	90,514	2,176,441	56,677,980
holiness	1	0	0	0	1	1	2	1	6
broken-down	1	2	9	56	457	9,196	202,769	4,875,652	126,969,721
broken	0	1	2	4	9	25	56	204	457
holiness	1	0	0	0	1	0	1	1	0

hypersum	0	2	24	246	2,468	246,820	24,682,022	2,468,202,224	246,820,222,426
holiness	1	0	0	1	3	4	4	4	5
untrimmed	1	3	25	357	3,579	357,931	35,793,133	3,579,313,335	357,931,333,537
trimmed	1	13	135	1,357	13,571	1,357,111	135,711,113	13,571,111,315	1,357,111,131,517

252 Both Steven Todd sequences start out the same.

Steven Todd	2	1	6	1	10	3	14	1	18
broken-up	2	3	20	23	250	773	11,072	11,845	224,282
broken-down	1	1	7	8	87	269	3,853	4,122	78,049
broken	1	2	3	7	8	20	23	87	250

hypersum	2	21	216	2,161	216,110	2,161,103	21,610,314	216,103,141	21,610,314,118
trimmed	1	10	105	1,050	10,500	105,002	1,050,203	10,502,030	1,050,203,007
sub	1	8	79	795	79,502	79,500	7,949,990	79,499,903	7,949,990,281

Another hailstone sequence turning odds to evens and evens to odds, for example, applies the DOHPO (double-or-halve-plus-one) function. It can be applied to the whole number or more confusingly to each digit individually.

	1	**2**	**3**	**4**	**5**	**6**	**7**	**8**	**9**	**10**	**11**	**12**	**13**	**14**	**15**	**16**	**17**	**18**	**19**	**20**	**21**	**22**	**23**	**24**
DOHPO	2	1	6	3	10	4	14	5	18	6	22	7	26	8	30	9	34	10	38	11	42	12	46	13
digitally	2	1	6	3	10	4	14	5	18	21	22	22	26	23	210	24	214	25	218	21	22	22	26	23

253 oddly odd

broken-up	9	136	2,865	71,761	1,940,412	64,105,357	2,245,627,907
holiness	1	1	3	1	2	2	3
tressub	1	7	143	3,573	96,608	3,191,618	111,803,231
broken-down	1	15	316	7,915	214,021	7,070,608	247,685,301
broken	1	9	15	136	2,865	7,915	71,761

hypersum	9	915	91,521	9,152,125	915,212,527	91,521,252,733	9,152,125,273,335
holiness	1	1	1	1	1	1	1
dusub	1	124	12,386	1,238,606	123,860,547	12,386,054,661	1,238,605,466,084

evenly even

broken-up	4	33	400	6,433	129,060	3,103,873	87,037,504	2,788,304,001
sub	1	12	147	2367	47479	1141851	32,019,308	1,025,759,718
broken-down	1	8	97	1,560	31,297	752,688	21,106,561	676,162,640
broken	1	4	8	33	97	400	1,560	6,433

hypersum	4	48	4,812	481,216	48,121,620	4,812,162,024	481,216,202,428
sub	1	18	1,770	177029	17,702,955	1,770,295,476	177,029,547,632

254 oddly even

broken-up	6	61	860	15,541	342,762	8,927,353	268,163,352
holiness	1	1	4	0	1	3	4
dusub	1	8	116	2,103	46,388	1,208,186	36,291,963
broken-down	1	10	141	2,548	56,197	1,463,670	43,966,297
broken	1	6	10	61	141	860	2,548

hypersum	6	610	61,014	6,101,418	61,041,822	6,104,182,226	61,041,222,630
holiness	1	2	2	4	4	5	4
dusub	1	82	8,257	825,737	8,261,112	826,111,230	8,261,031,154

255 minimal primes

mates	2	4	6	7	38	58	80	88	118
broken-up	2	7	37	266	2,963	56,563	2,322,046	141,701,369	12,613,743,887
trimmed	1	6	26	155	1,852	45,452	121,135	3,060,258	1,502,632,776
sub	1	3	14	98	1,090	20,808	854,233	52,129,020	4,640,337,052
broken-down	1	3	16	115	1,281	24,454	1,003,895	51,262,049	5,453,326,256
broken	1	2	3	7	16	37	115	266	1,281

hypersum	2	23	235	2,357	235,711	23,571,119	237,111,941
trimmed	1	12	124	1,246	124,600	12,460,008	124,000,830
sub	1	8	86	867	86,713	8,671,330	87,228,608

256 minimal odds

broken-up	1	4	21	151	1,380	29,131	671,393	16,813,956	454,648,205
broken-down	1	3	16	115	1,051	22,186	511,329	12,805,411	346,257,426
broken	1	3	4	16	21	115	151	1,051	1,380
sum	1	4	9	16	25	46	69	94	121
super	1	3	15	105	945	19,845	456,435	11,410,875	308,093,625

hypersum	1	13	135	13,521	1,352,123	135,212,325	1,3521,232,527

257 dihedrals

hypersum	0	1	12	125	1,256	12,568	125,689	12,568,910	125,681,011
holiness	1	0	0	0	1	3	4	5	4
untrimmed	1	2	23	236	2,367	23,679	2,367,910	236,791,021	23,679,102,122

Dihedral digits can be subdivided into homodihedrals and heterodihedral, those whose segment count corresponds to the number represented, the middlings, and those whose don't. In Hex A, b, C, d, E, and F are added with six, five, four, five and four segments each.

heterodihedral	0	1	2	3	7	8	9	10	11	12	13
segments	6	2	5	5	4	7	6	8	4	7	7
holiness	1	0	0	0	0	2	1	1	0	0	0
untrimmed	1	2	3	4	8	9	10	21	22	23	24

broken-up	0	1	2	7	51	415	3,786	38,275	424,811	5,136,007	67,192,902
holiness	1	0	0	0	0	0	3	2	2	3	4
untrimmed	1	2	3	8	62	526	4,897	49,386	535,922	6,247,116	7,821,031,013
broken-down	1	1	3	10	73	594	5,419	608,043	7,351,300	96,174,943	21,454,363,589
broken	0	1	2	3	7	10	51	73	415	594	3,786
holiness	1	0	0	0	0	1	0	0	0	1	3
untrimmed	1	2	3	4	8	21	62	84	526	6,103	4,897

hypersum	0	1	12	123	1,237	12,378	123,789	12,378,910	1,237,891,011
holiness	1	0	0	0	0	2	3	4	4
untrimmed	1	2	23	234	2,348	23,489	2,348,910	234,891,021	23,489,102,122

Seven digits are homovocalic: "zero" and "one" both have an E and an O, "three" and "seven" both have two E's, and three digits, "five", "eight" and "nine" each have an E and an I. The remaining three, "two", 'four" and "six", are heterovocalic, all three having different vowels than the three homovocalic groups and from each other.

homovocalic	0	1	3	5	7	8	9	10	11	13	15	17	18	19	30	31	33	35	37	38
holiness	1	0	0	0	0	2	1	1	0	0	0	0	2	1	1	0	0	0	0	2
untrimmed	1	2	4	6	8	9	10	21	22	24	26	28	29	210	41	42	44	46	48	49
heterovocalic	2	4	6	22	24	26	42	44	46	62	64	66	222	224	226	242	244	246	262	264
trimmed	1	3	5	11	13	15	31	33	35	51	53	55	111	113	115	131	133	135	151	153

258 dihedral primes

broken-up	2	11	13	123	12,434	2,250,677	2,658,061,971
trimmed	1	0	2	12	12,323	114,566	154,750,860
sub	1	4	5	45	4,574	827,978	977,846,352
broken-down	1	5	6	56	5,661	1,024,697	1,210,172,818
broken	1	2	5	6	11	13	56

hypersum	2	25	2,511	2,511,101	2,511,101,181	25,111,011,811,811
trimmed	1	14	1,400	140,000	140,000,070	1,400,000,700,700
sub	1	9	924	923,782	923,782,499	9,237,824,992,579

259 numberdromic primes

dusub	1	14	18	21	25	26
broken-up	11	1,112	145,683	21,999,245	3,982,009,028	760,585,723,593
sub	1	150	19,716	2,977,274	538,906,320	102,934,064,326

broken-down	1	12	1,573	237,535	42,995,408	8,212,360,463
broken	1	11	12	1,112	1,573	145,683

hypersum	11	11,101	10,101,131	10,101,131,151	10,101,131,151,181
dusub	1	1,502	1,367,039	1,367,039,445	1,367,039,445,355

The Great International Math on Keys Book by Texas Instrument, describes the game Flipit in which numbers and letters are transformed into each other by flipping the calculator over.

0	**1**	**2**	**3**	**ч**	**5**	**6**	**7**	**8**	**9**
O	I	Z	E	h	S	g	L	B	G

There are few numbers that flip into words. Seven times two would, for example, yield the answer fourteen, flipped "hI and 0.773ч 'hello'. "The answer 5379908.3ч5 would flip into 'she boggles', 577385.773ч into 'hell's bells' and 3215.918 into 'big size' and 32157.17 to "li'l size" and ч91ч 'high'. There are fewer numbers that flip into number names. 709009 would be a flipped 'googol' and 1709009 a flipped 'googoli'.

260 numberrdromes

broken-up	1,001	1,222,222	1,626,778,483	2,344,189,016,225	363,583,790,943,458
holiness	2	0	6	5	8
oktasub	1	411	545,724	786,387,808	121,968,773,973
broken-down	1	1,221	1,625,152	2,341,845,253	3,632,203,612,555
broken	1	1,001	1,221	1,222,222	1,625,125

hypersum	1	11,001	110,011,221	11,100,112,211,331	111,001,122,113,311,441

261 Cancrine from the Latin refers to the crab's characteristic sideways walk. It took the name for word palindrome, while the Greek word kept the meaning for letter for letter reversables.

broken-up	101	20,403	6,182,210	2,497,633,243	1,261,310,969,925
holiness	1	2	4	2	6
hexasub	1	51	15,325	6,191,015	3,126,477,313
broken down	1	202	61,207	2,472,830	12,487,615,357
broken	1	101	202	20,403	61,207

hypersum	101	101,202	101,202,303	101,202,303,404	101,202,303,404,505
holiness	1	2	3	4	5
pentasub	1	682	681,896	681,895,757	681,895,756,525

From Queneleyev's Table we also have syllabic palindromes, like sixty-six and eighty-eight positioned between the letter and word numberdromes. The name Queneleyev itself is a portmanteau of those of Raymond Queneau and Dmitri Mendeleyev. Queneau organized literary devices by what they did horizontally and by what they did it on vertically, like Mendeleyev had organized the chemical elements, in "Classification des Travaux de l'Oulipo".

We can find more syllabic palindromes, what Joyce called by another portmanteau "sylindromes", repunits in bases higher than seven that otherwise would include zeros and sevens, particularly eight and nine. Those higher than decimal would depend upon the higher number names. Those below seven-seven would be ordinary repunits.

sylindrome	1	2	3	4	5	6	11	22	33	44
from octarish	1	2	3	4	5	6	9	18	27	36
broken-up	1	3	10	43	225	1,393	12,762	231,109	6,252,705	225,328,489
broken-down	1	2	7	30	157	972	8,905	161,262	4,362,979	157,228,508
broken	1	2	3	7	10	30	43	157	225	972
sum	1	3	5	7	9	11	15	27	45	63
super	1	2	6	12	20	30	54	162	486	972

hypersum	1	12	123	1,234	12,345	123,456	12,345,611	1,234,561,122

hypersuper	1	2	33	444,444,444,444,444,444,444,444,444,444

from nonarish	1	2	3	4	5	6	8	10	20	30
broken-up	1	3	10	43	225	1,393	11,369	115,083	2,313,029	69,505,953
broken-down	1	2	7	30	157	972	7,933	80,302	1,613,973	48,499,492
broken	1	2	3	7	10	30	43	157	225	972
sum	1	3	5	7	9	11	14	18	30	50
super	1	2	6	12	20	30	48	80	200	600

hypersum	1	12	123	1,234	12,345	123,456	123,468	12,346,810	1,234,681,020

262 Promanish

broken-up	1	3	10	53	540	10,313	516,190	51,629,313	9,810,085,660
broken-down	1	2	7	37	377	7,200	360,377	36,044,900	6,848,891,377
broken	1	2	3	7	10	53	377	540	7,200
sum	1	3	5	8	15	29	69	150	290
super	1	2	6	15	50	190	950	5,000	19,000

hypersum	1	12	123	1,235	123,510	12,351,019	1,235,101,950	1,235,101,950,100

263 Paramæanish

broken-up	1	3	10	43	354	4,291	69,010	1,384,491	33,296,794	1,000,288,311

broken-down	1	2	7	30	247	2,994	48,151	966,014	23,232,487	697,940,624
broken	1	2	3	7	10	30	43	247	354	2,994
sum	1	3	5	7	12	20	28	36	44	54
super	1	2	6	12	32	96	192	320	480	720

hypersum	1	12	123	1,234	12,348	1,234,812	123,481,216	12,348,121,620

264 unholey quinarish

broken-up	6	73	1,174	24,727	644,076	19,991,083	720,323,064	30,273,559,771
holiness	1	0	0	0	3	6	3	2
dusub	1	10	159	2,247	87,166	2,705,499	97,485,126	4,097,080,786
broken-down	1	7	113	2,380	61,993	1,924,163	69,331,861	2,913,862,325
broken	1	6	7	73	113	1,174	2,380	24,727

hypersum	6	612	61,216	6,121,621	612,162,126	61,216,216,246
holiness	1	1	2	2	3	4
dusub	1	83	8,285	828,471	82,847,135	8,284,713,964

265 psenarish

dusub	1	3	4	6	7	14	15
broken-up	11	243	8,030	353,363	19,453,995	1,965,207,058	218,157,437,433
dusub	1	33	1,087	47,850	2,632,812	265,961,854	29,524,398,585
broken-down	1	12	397	17,480	961,797	97,158,977	2,138,459,291
broken	1	11	12	243	397	8,030	17,480

hypersum	11	1,122	112,233	11,223,344	1,122,334,455	1,122,334,455,101
dusub	1	152	15189	1,518,915	151,891,451	1,518,914,151,367

pseptarish

dusub	...	7	9	14	15
broken-up	...	19,453,995	1,284,317,233	129,735,494,528	14,401,924,209,841
broken-down	...	961,797	63,496,082	98,911,880,936	9,991,384,291,769
broken	...	8,030	17,480	353,363	961,797

poctarish

dusub	...	9	10	14
broken-up	...	1,284,317,233	98,911,880,936	9,991,384,291,769

broken-down	...	63,496,082	4,890,160,111	493,969,667,293

peven

broken-up	2	9	56	457	10,110	445,297	1,791,298	11,193,085	91,335,978
trimmed	1	8	45	346	0	334,186	680,187	8,274	80224867
sub	1	3	21	168	3,719	163,816	658,982	4,117,706	33,600,629
broken-down	1	3	19	155	3,429	151,031	607,553	3,796,349	30,978,345
broken	1	2	3	9	19	56	155	457	3,429

hypersum	2	24	246	2,468	246,822	24,682,244	2,468,224,466	246,822,446,688
trimmed	1	13	135	1,357	135,711	13,571,133	1,357,113,355	135,611,335,577
sub	1	9	90	908	90,801	9,080,090	908,009,037	908,800,903,756

266 podd

broken-up	1	4	21	151	1,380	15,3321	507,303	27,916,996
broken-down	1	2	11	79	722	8,021	265,415	14,605,846
broken	1	2	4	11	21	79	151	722
sum	1	4	8	12	16	20	44	88
super	1	3	15	105	945	10,395	343,035	18,866,925

hypersum	1	13	135	1,357	13,579	1,357,911	135,791,133	13,579,113,355

268 pyranibish

broken-up	1	11	1,112	123,443	123,567,555	137,283,677,048
broken-down	1	12	1,213	134,655	134,790,868	149,752,789,003
broken	1	11	12	1,112	1,213	123,443
sum	1	12	112	212	1,112	2,112
super	1	11	1,111	123,321	1,233,444,321	1,370,356,640,631

hypersum	1	111	111,101	111,101,111	1,111,011,111,001	11,110,111,110,011,111

pyranirtish

broken-down	1	3	34	751	75,885	3,339,691	6,755,267
broken-up	1	2	23	508	51,331	2,259,072	4,569,475
broken	1	2	3	23	34	508	751
sum	1	3	14	36	137	248	359
super	1	2	22	484	48,884	5,426,124	10,852,248

hypersum	1	12	1,211	121,122	121,122,101	121,122,101,111	121,122,101,111,121

pyrardauqish

broken-up	1	2	7	79	1,745	57,664	5,825,809	646,722,463
broken-down	1	3	10	113	2,496	82,481	8,333,077	925,054,028
broken	1	2	3	7	10	79	113	1,745
sum	1	3	6	17	39	72	173	284
super	1	2	6	66	1,452	47,916	95,832	287,496

hypersum	1	12	123	12,311	1,231,122	123,112,233	123,112,233,101

pyratniuqish, "pyrat" for short

broken-up	1	3	10	43	483	10,669	21,821	76,132
broken-down	1	2	7	30	337	7,444	15,225	53,119
broken	1	2	3	7	10	30	43	337
sum	1	3	6	10	21	43	76	120
super	1	2	6	24	264	5,808	11,616	34,848

hypersum	1	12	123	1,234	123,411	12,341,122	1,234,112,233	123,411,223,344

270 plucky

plucky	1	3	7	9	33	99	111	141	151	171	303	717	727	777	787	979	999

broken-up	1	4	29	265	8,774	868,891	2,615,447	19,177,020	175,208,627
broken-down	1	2	15	137	4,536	449,201	1352139	9,914,174	90,579,705
broken	1	2	4	15	29	137	265	4,536	8,774
sum	1	4	11	20	53	152	263	404	555
super	1	3	21	189	6,237	617,463	1,852,389	12,966,723	116,700,507

hypersum	1	13	137	1,379	137,933	13,793,399	13,793,399,111	13,793,399,111,141

271 lucky

broken-up	1	4	29	265	3,474	52,375	160,599	1,176,568	10749711
broken-down	1	2	15	137	1,796	27,077	83,027	608,266	5,557,421
broken	1	2	4	15	29	137	265	1,796	3,474
sum	1	4	11	20	33	48	69	94	125
super	1	3	21	189	2,457	36,855	110,565	773,955	6,965,595

hypersum	1	13	137	1,379	137,913	13,791,315	1379131521	1,379,131,525

272

ykcul	13	15	31	37	51	73
broken-up	13	196	6,089	225,489	11,506,028	840,165,533
tressub	1	10	303	11,227	1,557,172	41,829,379
broken-down	1	14	435	18,109	821,992	60,021,671
broken	1	13	14	196	435	6,089

hypersum	13	1,315	131,531	13,153,137	1,315,313,751	131,531,375,173
tressub	1	65	6,549	654,856	6,548,616	6,548,561,568

273 deuterosurvivors

broken-up	1	4	29	265	3,474	52,375	998,599	21,033,954	526,572,449
broken-down	1	2	15	137	1,796	27,077	516,259	10,868,516	272,229,159
broken	1	2	4	15	29	137	265	1,796	3,474
sum	1	4	11	20	33	48	67	88	113
super	1	3	21	189	2,457	36,855	700,245	14,705,145	36,762,825

hypersum	1	13	137	1,379	137,913	13,791,315	1,379,131,519	137,913,151,921	

Broken tritosurvivors are the same as those of the deuterosurvivors to 52,375.

broken-up	1	4	29	265	3,474	52,375	1,103,349	27,636,100	747,278,049
broken-down	1	2	15	137	1,796	27,077	570,413	14,287,402	386,330,267
sum	1	4	11	20	33	48	69	94	121
super	1	3	21	189	2,457	36,855	773,955	1,934,875	522,419,625

hypersum	1	13	137	1,379	137,913	13,791,315	1,379,131,521	137,913,152,125	

274 lucky twins

broken-up	1	4	29	265	3,474	52,375	1,627,099	53,746,642	2,635,212,557
broken-down	1	2	15	137	1,796	27,077	841,183	27,786,116	1,362,360,867
broken	1	2	4	15	29	137	265	1,796	3,474
sum	1	4	10	16	22	28	46	64	82
super	1	3	21	189	2,457	36,855	1,142,505	37,702,665	1,847,430,585

hypersum	1	13	137	1,379	137,913	13,791,315	1,379,131,531	137,913,153,133

275 lucky cousins

broken-up	3	22	201	2,635	55,536	1,391,035	45,959,691	1,701,899,602
sub	1	8	74	969	20,430	511,733	16,907,625	626,093,874
broken-down	1	4	37	485	10,222	256,035	8,459,377	313,252,984
broken	1	3	4	22	37	201	485	2,635

hypersum	3	37	379	37,913	3,791,321	379,132,125	37,913,212,533	3,791,321,253,337
sub	1	14	139	13947	1,394,749	139,474,914	13,947,491,440	1,394,749,143,979

276 oblong

broken-up	2	13	158	3,173	95,348	4,007,789	224,531,532	16,170,278,093
trimmed	1	2	47	2,062	84,237	36,678	113,420,421	50,616,782
sub	1	5	58	1,167	35,076	1,474,383	82,600,534	5,948,712,868
broken-down	1	3	37	743	22,327	938,477	52,577,039	3,786,485,285
broken	1	2	3	13	37	158	743	3,173

hypersum	2	26	2,612	261,220	26,122,030	2,612,203,042	261,220,304,256
trimmed	1	15	1,501	15,011	150,112	15,011,231	1,501,123,145
sub	1	10	961	96,097	9,609,758	960,975,795	96,097,579,552

oblong neighbors

broken-up	1	4	21	109	1,220	15,969	304,631	6,413,220	186,288,011
broken-down	1	2	11	57	638	8,351	159,307	3,353,798	97,419,449
broken	1	2	4	11	21	57	109	638	1,220

hypersum	1	12	1211	121,157	121,157,638	1,211,576,388,351	1,211,576,388,351,159,307

277 poblong

broken-up	2	13	3,538	21,249,241	61,622,416,417,200
trimmed	1	2	2,427	10,138,130	505,113,053,061
sub	1	5	1,301	7,817,159	22,669,620,115,193
broken-down	1	3	817	4,906,905	14,229,936,176,527
broken	1	2	3	13	817

hypersum	2	26	26,272	262,726,006	2,627,260,062,899,982
trimmed	1	15	15,161	1,516,155	15,161,551,788,871
sub	1	10	9,665	96,651,496	966,514,963,751,694

278 aka Motzkin sums or Riordan numbers, (z - 1)(2f(z - 1) + 3f(z - 2))/(z + 1), ("Relations between hypersurface cross ratios and a combinatorial formula for partitions of a polygon, for permanent preponderance and for non-associative products" by T. S. Motzkin and "Enumeration of plane trees by branches and endpoints" by J. Riordan)

broken-up	1	4	25	379	13,669	1,244,258	288,681,525	174,076,203,833
broken-down	1	2	13	197	7,105	646,752	150,053,569	90,482,948,859
broken	1	2	4	13	25	197	379	7,105
sum	1	4	9	21	51	127	323	835
super	1	3	18	270	9,720	884,520	205,208,640	123,740,809,920

hypersum	1	13	136	13,615	1,361,536	136,153,691	136,153,691,232

279 *1 + 1 = 1 (A how to guide for Christian Growth and Development)* is by Wayne A. Mack, *One Plus One Equals One: Secrets to the Plan for a Successful Marriage* is by Dr. Gideon Adjel and Prof. Dominique Endrinal, *Divine Mathematics: How one plus one equals three in the Kingdom* is by Selvyn Hughes, *One Plus One Equals Three: A Master class in Creative Thinking* is by Dave Trott, *Networking: 1 + 1 = 4* is by David Stewart, *1 + 1 = 5 and other unlikely additions* is by David LaRochelle. The Ten Movement claims "1 + 1 = 14", that is two cats can multiply into a family of fourteen in a year. Cat multiplication therefore seems to refer to a one-one-five Multibonacci sequence, adding all its previous numbers.

cat multiplication	1	1	5	7	14	28	56	112	224	448	896	1,792	3,584	7,168	14,336	28,672

280 This "digitization" of a growing sequence may well have inspired Control Systems Engineering, Inc.'s digital conveyor, the matter transmitter that digitizes transported objects and then reassembles them, considered safer by the NSEA than the accident-prone UFP transporter [*Galaxy Quest Technical Manual*].

The Champernowne constant is made up of the almost-natural numbers as a decimal fraction. Similarly the Copeland-Erdős constant is made up of the almost-primes as a decimal fraction. Other almost-constant sequences have been named for the constants of other mathematicians.

Artin's	3	7	3	9	5	5	8	1	3	6	1	9	2	0	2	2	8	8	0
Catalan's	9	1	5	9	6	5	5	9	4	1	7	7	2	1	9	0	1	5	0
Gauss's	5	7	7	2	1	5	6	6	4	9	0	1	5	3	2	8	6	0	6
Giesesing's	1	6	1	4	9	4	1	6	0	6	4	0	9	6	5	3	6	2	5
Khintchnine's	2	6	8	5	4	5	2	0	0	1	0	6	5	3	0	6	4	4	5
Lehner's	1	1	7	6	2	8	0	8	1	8	2	5	9	9	1	7	5	0	6
Levy's	3	2	7	5	8	2	2	9	1	8	7	2	1	8	1	1	1	5	9
Niven's	1	7	0	5	2	1	1	1	4	0	1	0	5	3	6	7	7	6	4
Sierpiński's	2	5	8	4	9	8	1	9	8	1	7	5	9	5	7	9	2	5	3
Soldner's	1	4	5	1	3	6	9	2	3	4	8	8	3	3	8	1	0	5	0

The almost-squares sequence has zeros when it digitally expands a hundred, but this can be 'collapsed' into an equivalent continued fraction sequence by multiplying the zeros' neighbors, since z/(0 + 1/z') = z/(0 + 1/(0 + 1/z') = zz'.

collapsed almost-squares	1	4	9	1	6	2	5	3	6	4	9	6	4	8	1	1	1	2	1	1	4

broken-up	1	5	46	51	352	755	4,127	13,136	82,943	344,908	3,187,115
broken-down	1	2	19	21	145	311	1,700	5,411	34,166	142,075	1,312,841
broken	1	2	5	19	21	46	51	145	311	352	755
sub	1	5	14	15	21	23	28	31	37	41	50
super	1	4	36	36	216	432	2,160	6,480	38,880	155,520	1,399,680

hypersum	1	14	149	1,491	14,916	149,162	1,491,625	14,916,253	149,162,536

281

collapsed almost-evens	2	4	6	8	1	2	1	4	1	6	1	8	4	2	2	4

almost-cube

broken-up	1	9	19	142	871	3,626	4,497	12,620	67,597	147,814	215,411
broken-down	1	2	5	37	227	945	1,172	3,289	17,617	38,523	56,140
broken	1	2	5	9	19	37	142	227	871	946	1,172
sum	1	9	11	18	24	28	29	34	36	37	43
super	1	8	16	112	672	2,688	2,688	5,376	26,880	53,760	53,760

hypersum	1	18	182	1,827	18,276	182,764	1,827,641	18,276,412	182,764,125

282 almost-cheap

contracted	1	2	3	1	4	1	5	1	6	1	7	1	8
broken-up	1	3	10	13	62	75	437	512	3,509	4,021	31,656	35,677	281,395
broken-down	1	2	7	9	43	52	303	355	2,433	2,788	21,949	24,737	195,108
broken	1	2	3	7	9	10	13	43	62	75	303	355	437
sum	1	3	6	7	11	12	17	18	24	25	32	33	41

hypersum	1	12	123	1,231	12,314	123,141	1,231,415	12,314,151	123,141,516

283 almost-Bantu

contracted	1	3	4	5	6	7	8	9	1	1	1
broken-up	1	4	17	89	551	3,946	32,119	293,017	325,136	618,153	943,289
broken-down	1	2	9	47	291	2,084	16,963	154,751	171,714	1,185,035	1,356,749
broken	1	2	4	9	17	47	89	291	551	2,084	3,946

sum	1	4	8	13	19	26	34	43	44	45	46

hypersum	1	13	134	1,345	13,456	134,567	1,345,678	13,456,789	134,567,891

284 almost-Ir-ish

contracted	1	3	6	1	3	2	3	3	3	4	3	5
broken-up	1	4	25	29	112	253	871	2,866	9,469	40,742	131,695	699,217
broken-down	1	2	13	15	58	131	451	1,484	4,903	21,096	68,191	362,051
broken	1	2	4	13	15	25	29	58	112	131	253	451
sum	1	4	10	11	14	16	19	22	25	29	32	37

hypersum	1	3	133	1,330	13,303	133,031	1,330,313	13,303,132	133,031,323

285 almost-ugly

contracted	1	2	3	4	5	6	8	9	1	2	1
broken-up	1	3	10	43	225	1,393	11,369	103,714	115,083	333,880	448,963
broken-down	1	2	7	30	157	972	7,933	80,302	232,973	313,275	1,799,348
broken	1	2	3	7	10	30	43	157	225	972	1,393
sum	1	3	6	10	15	21	29	38	39	41	42

hypersum	1	12	123	1,234	12,345	123,456	1,234,568	12,345,689	123,456,891

286 **"Rock"**, aka Ca, Charta, Evvan, Gunting, Hick, Jan, Jo, Kamen, Kivi, Kamień, Muck, Ollő, Pā, Paa, Pierre, Pietra, Pihera, Reau, Ro, Rocca, Roche, Shnik, Steen, Sten, Yan, approximates a militant fist or zero. **"Paper"**, aka Cham, Chee, Chi, Choki, Ciseaux, Mi, Papel, Papelsh, Papier, Papir, Papyrus, Quem, Sakset, Shnak, Tijhera, approximates a national-socialist salute or one. **"Scissors"**, aka Baa, Bato, Beaux, Bo, Ciseaux, Cisoria, Forbici, Fou, Gũ, Hock, Kem, Ken, Kő, Misparayayim, Nożyczki, Niýar, Po(n), Pu(n), Quem, Sax, Schaar, Sciseaux, Shkarje, Shnuk, approximates a pacifist peace sign or two. The modulo operator can be symbolized by the percent sign between numbers.

The prime contest leaves rock far behind the nearly even paper and scissors, while the lucky contest leaves scissors far behind the nearly even rock and paper.

p	2	3	5	7	11	13	17	19	23	29	31	37	41	43	47	53	59	61	67	71	73	79	83	89	97
%3	2	0	2	1	2	1	2	1	2	2	1	1	2	1	2	2	2	1	1	2	1	1	2	2	1
rock	0	1	1	1	1	1	1	1	1	1	1	1	1	1	1	1	1	1	1	1	1	1	1	1	1
paper	0	0	0	1	1	2	2	3	3	3	4	5	5	6	6	6	6	7	8	8	9	10	10	10	11
scissors	1	1	2	2	3	3	4	4	5	6	6	6	7	7	8	9	10	10	10	11	11	11	12	13	13

luckies	1	3	7	9	13	15	21	25	31	33	37	43	49	51	63	67	69	73	75	79	87	93	99	105
z%3	1	0	1	0	1	0	0	1	1	0	1	1	1	0	0	1	0	1	0	1	2	0	0	0

rock	0	1	1	2	2	3	4	4	5	5	6	7	8	8	8	9	9	10	10	11	11	12	13	14
paper	1	1	2	2	3	3	3	4	5	5	6	7	8	8	8	9	9	10	10	11	11	11	11	11
scissors	0	0	0	0	0	0	0	0	0	0	0	0	0	0	0	0	0	0	0	0	1	1	1	1

Sam Kass and Karen Bryla's variant adds lizard and Spock between rock and scissors and David C. Lovelace's RPS-9 adds fire between rock and scissors, Human and sponge between scissors and paper and air, water and gun between paper and rock. With modular arithmetic however any number of numbers can play and combining it with hypermath getsbus into hypermodular arithmetic symbolized with the permil sign.

hypersum	1	12	123	1,234	12,345	123,456	1,234,567	12,345,678	123,456,789
z‰2	1	0	1	0	1	0	1	0	1
z‰3	1	0	0	1	2	0	1	2	0
z‰4	1	0	3	2	1	0	3	2	1
z‰5	1	2	3	4	0	1	2	3	4
z‰6	1	0	3	4	3	0	1	0	0
z‰7	1	5	4	2	4	4	5	2	1
z‰8	1	4	3	2	1	0	7	0	5
z‰9	1	3	6	1	6	3	1	0	0

287 interemirp

broken-up	15	361	12,289	627,100	45,163,489	3,433,052,264	302,153,762,721
tressub	1	18	612	31,221	2,248,557	170,921,607	15,043,350,042
broken-down	1	16	545	27,811	2,002,937	152,251,023	13,400,092,961
broken	1	16	18	545	612	27,811	31,221

hypersum	15	1,524	152,434	15,243,451	1,524,345,172	152,434,517,276
tressub	1	76	7,589	758,927	75,892,677	7,589,267,733

287 interddo

broken-up	2	9	56	457	4,626	55,969	788,192	12,667,041	228,794,930
tressub	1	3	20	168	1,702	20,590	289,960	4,659,944	84,168,951
broken-down	1	3	19	155	1,569	18,983	267,331	4,296,279	77,600,353
broken	1	2	3	9	19	56	155	457	1,569

hypersum	14	1,416	141,618	14,161,825	1,416,182,533	141,618,253,336
tressub	1	71	7,051	705,076	70,507,577	7,050,757,661

288 interpodd

broken-up	2	9	56	457	4,626	102,229	4,502,702	297,280,561	26,165,192,070
trimmed	1	8	45	346	3,515	1,118	34,161	18,617,450	150,540,816
sub	1	3	20	168	1,702	37,608	1,656,451	109,363,407	9,625,636,237
broken-down	1	3	19	155	1,569	34,673	1,527,181	37,092,776	3,264,726,107
broken	1	2	3	9	19	56	155	457	1,569

hypersum	2	24	246	2,468	246,810	24,681,022	2,468,102,244	246,810,224,466
trimmed	1	13	135	1357	13,570	1357011	135,701,133	13,570,113,355
sub	1	9	90	908	90,796	9,079,641	907,964,074	90,796,407,452

289 interneve

broken-up	25	676	23,009	944,045	41,560,989	1,954,310,528	10,557,439,501
tressub	1	34	1,146	47,001	2,069,200	97,299,392	5,256,236,361
broken-down	1	26	885	36,311	1,598,569	75,169,054	4,060,727,485
broken	1	25	26	676	885	23,009	36,311

hypersum	25	2,527	252,734	25,273,441	2,527,344,144	252,734,414,447
tressub	1	126	12,583	1,258,290	125,829,055	12,582,905,571

290 Morrisish

broken-up	1	12	253	306,395	34,077,558,548	10,639,388,632,136,023
broken-down	1	2	43	52,075	5,791,833,618	1,808,274,165,761,473
broken	1	2	12	43	253	52,075
sum	1	12	33	54	111,275	423486
super	1	11	231	4,851	539,533,071	168,448,159,629,981

hypersum	1	111	11,121	111,211,211	111,211,211,111,221	111,211,211,111,221,312,211

291 intermorrisish

broken-up	6	97	59,758	3,359,355,825	711,229,377,905,458
holiness	1	1	3	3	5
dusub	1	13	8,087	454,639,372	96,254,429,305,035
broken-down	1	7	4,313	242,459,615	51,332,579,853,653
broken	1	6	7	97	4,313

hypersum	6	616	616,616	61,661,656,216	61,661,656,216,211,716
holiness	1	2	4	6	7

dusub	1	84	83,450	8,344,997,709	8,344,997,708,859,670

292 Duintermorrish is as far as this goes however, since its second term is even.

broken-up	11	3,477	98,785,058	13,233,345,158,215
dusub	1	470	13,369,104	1,790,938,515,155
broken-down	1	12	340,933	45,671,725,625
broken	1	11	12	470

hypersum	11	11,316	1,131,628,411	1,131,628,411,133,961
dusub	1	1,531	153,149,251	15,314,925,151,539,413

293 interugliest

broken-up	45	3,376	354,525	47,864,251	7,897,955,940
tetrasub	1	62	6,493	876,664	144,656,108
broken-down	1	46	4,831	652,231	107,622,946
broken	1	45	46	3,376	4,831

hypersum	45	4,575	4,575,105	4,575,105,135	4,575,105,135,165
tetrasub	1	84	83,796	83795973	83,795,973,534

duinterugliest

broken-up	60	5,401	648,180	97,232,401	17,502,480,360
holiness	1	1	6	2	6
tetrasub	1	99	11,872	1,780,873	320,569,110
broken-down	1	61	7,321	1,098,211	197,685,301
broken	1	60	61	5,401	7,321

hypersum	60	6,090	6,090,120	6,090,120,150	6,090,120,150,180
holiness	2	4	5	6	9
tetrasub	1	111	111,544	111,544,441	111,544,441,460

294 interprimorial

sub	1	7	44	464	5,949	99,427
broken-up	4	73	8,764	11,042,713	178,560,677,974	4,825,959,444,707,573
sub	1	27	3,224	4,062,387	65,688,802,428	1,775,371,263,635,061
broken-down	1	5	601	757,265	12,244,975,651	3,309,449,569,953,035
broken	1	4	5	73	601	8,764

hypersum	4	418	418,120	4,181,201,260	418,120,126,016,170
sub	1	154	153,818	1,538,177,983	153,817,798,301,362

295 duinterprimorial

dusub	1	9	93	1,179	19,383
broken-up	11	760	524,411	4,570,242,625	654,550,149,276,911
dusub	1	103	70971	618,515,080	88,583,729,844,958
broken-down	1	12	8,281	72,168,927	10,336,033,733,221
broken	1	11	12	760	8,281

hypersum	11	1,169	1,169,690	11,696,908,715	11,696,908,715,143,220
dusub	1	158	158,300	1,583,004,454	1,583,004,453,956,710

296 interlucky

sub	1	4	33	368	5,189	93,783	2,161,727	60,622,116
broken-up	2	11	90	1,001	14,104	254,873	5,876,183	164,787,997
trimmed	1	0	8	0	303	143,762	4,765,072	53,676,886
sub	1	4	33	368	5,188	93,762	2,161,727	60,622,116
broken-down	1	3	25	278	3,917	70,784	1,631,949	45,765,356
broken	1	2	3	11	25	90	278	1,001

hypersum	2	25	258	25,811	2,581,114	258,111,418	2,581,141,823	258,114,182,328
trimmed	1	14	147	14,700	1,470,003	147,000,307	1,470,030,712	147,003,071,217
sub	1	9	95	9,495	949,539	94,953,884	949,549,011	94,954,901,153

297 eban

broken-up	2	9	56	1,129	24,894	598,585	15,588,104	468,241,705
trimmed	1	8	45	18	13,783	487,474	447,703	35,713,064
sub	1	3	20	415	9,158	220,207	5,734,543	172,256,497
broken-down	1	3	19	383	8,445	203,063	5,288,083	158,845,553
broken	1	2	3	9	19	56	383	1,129

hypersum	2	24	246	24,620	2,462,022	246,202,224	24,620,222,426	24,262,022,242,630
trimmed	1	13	135	1,351	135,111	13,511,113	131,111,215	2,321,112,152
sub	1	9	90	9,057	905,727	90,572,737	89,254,991,843	8,925,499,184,308

298 intereban

broken-up	3	16	211	4,447	102,492	2,566,747	71,971,408	2,377,623,211
sub	1	6	77	1,636	37,705	944,253	26,476,801	874,678,698
broken-down	1	4	53	1,117	25,744	644,717	18,077,820	597,212,777
broken	1	3	4	16	53	211	1,117	4,447

hypersum	3	35	3,513	351,321	35,132,123	3,513,212,325	35,132,132,528
sub	1	13	1,292	129,244	12,924,386	1,292,438,587	12,924,389,282

299 nabe

broken-up	20	481	12,526	501,521	21,076,408	970,016,289	58,222,053,748
holiness	1	2	1	1	5	7	5
tressub	1	24	623	24,969	1,049,332	48,294,267	2,898,705,370
broken-down	1	21	547	21,901	920,389	42,359,795	2,542,508,089
broken	1	20	21	481	547	12,526	21,901

hypersum	20	2,024	202,426	20,242,640	2,024,264,042	202,426,404,246
holiness	1	1	2	3	3	4
sub	1	101	10,078	1,007,822	100,782,172	10,078,217,228

300 internabe

broken-up	22	551	18,205	746,956	32,884,269	1,743,613,213	106,393,290,262
tressub	1	27	906	37,188	1,637,211	86,809,390	5,297,010,016
broken-down	1	23	760	31,183	1,372,812	72,790,219	4,441,576,171
broken	1	22	23	551	760	18,205	31,183

hypersum	22	2,225	222,533	22,253,341	2,225,334,144	222,533,414,453
tressub	1	111	110,79	1,107,929	110,792,863	11,079,286,320

The other lipovocal numbers, those with number names without a particular vowel, differ quite a bit. Ubans include every one until a hundred and abans until a thousand.

iban	1	2	3	4	7	10	11	12	14	17	20	21	22	23	24	27	40	41	42	43	44	47

broken-up	1	2	7	30	217	2,200	24,417	295,204	4,157,273	70,968,845
broken-down	1	3	10	43	311	3,153	34,994	423,081	5,958,128	101,711,257
broken	1	2	3	7	10	30	43	217	311	2,200
sum	1	3	6	10	17	27	38	50	64	81
super	1	2	6	24	168	1,680	18,480	221,760	3,104,640	21,732,480

hypersum	1	12	123	1,234	12,347	1,234,710	123,471,011	12,347,101,112

nabi	10	12	14	17	20	24	27	40	41	42	71	72	73	74	100	102	103	107	110
holiness	1	0	0	0	1	0	0	1	0	0	0	0	0	0	2	1	1	1	1

broken-up	10	121	1,704	29,089	583,484	14,032,705	379,466,519	15,192,693,465
holiness	1	0	1	5	4	2	4	4
dusub	1	16	231	3,937	78,966	1,899,120	51,355,209	2,056,107,473
broken-down	1	11	155	2,646	53,075	1,276,446	34,517,117	1,381,961,126
broken	1	10	11	121	155	1,704	2,646	29,089

hypersum	10	1,012	101,214	10,121,417	1,012,141,720	101,214,172,024
holiness	1	1	1	1	2	2
dusub	1	137	137,698	1,369,785	136,978,486	13,697,848,638

oban	3	5	6	7	8	9	10	11	12	13	15	16	17	18	19	20	23	24	25	26	27	28	29	30

broken-up	3	16	99	709	5,771	52,648	532,251	5,907,409	71,421,159	934,382,476
sub	1	6	36	261	2,123	19,368	195,804	2,173,214	26,274,376	343,740,103
broken-down	1	4	25	179	1,457	13,292	134,377	1,491,439	18,031,645	235,902,824
broken	1	3	4	16	25	99	179	709	1,457	5,771

hypersum	3	35	356	3,567	35,678	356,789	35,678,910	356,7891,011
sub	1	13	131	1,312	13,125	131,255	13,125,537	1,312,553,751

nabo	30	35	36	37	38	39	50	53	56	57	58	59	60	63	65	67	68	69	70	73	75	79	80
holiness	1	0	1	0	2	1	1	0	1	0	2	1	2	1	1	1	3	2	1	0	0	1	3

broken-up	30	1,051	37,866	1,402,093	53,317,400	2,080,780,693	104,092,352,050
holiness	1	1	4	3	2	9	5
tressub	1	52	1,885	69,806	2,654,517	103,595,970	5,182,453,048
broken-down	1	31	1,117	41,360	1,572,797	61,380,443	3,070,594,947
broken	1	30	31	1,051	1,117	37,866	41,360

hypersum	30	3,035	303,536	30,353,637	3,035,363,738	303,536,373,839
holiness	1	1	2	2	4	5
sub	1	151	15,112	1,511,219	151,121,862	15,112,186,196

The ybans don't get interesting until they accelerate, getting higher faster after nineteen.

The naby do so a bit sooner.

yban	1	2	3	4	5	6	7	8	9	10	11	12	13	14	15	16	17	18	19	100	101	102	103	104

broken-up	...	183,586,751,854,827,751	18,368,309,566,828,627,750
broken-down	...	263,103,209,266,016,890	27,623,918,451,319,637,951

naby	10	100	102	103	104	105	106	107	108	109	110	112	113	114	115	116	117	118
holiness	1	2	1	1	1	1	1	1	1	1	1	0	0	0	0	1	0	2

broken-up	10	1,001	102,112	10,518,537	1,094,029,960	114,883,664,337
holiness	1	2	1	3	7	6
dusub	1	135	13,819	1,423,529	148,060,854	15,547,813,252
broken-down	1	11	1,123	115,680	12,031,843	1,263,459,195
broken	1	10	11	1,001	1,123	102,112

hypersum	10	10,100	10,100,102	10,100,102,103	10,100,102,103,104
holiness	1	3	4	5	6
dusub	1	1,367	1,366,900	1,366,900,179	1,366,900,178,842

301 strong primes

mates	20	28	32	40	58	62	70
holiness	1	2	0	1	2	1	1
dusub	1	2	4	5	6	8	9
broken-up	11	188	5,463	202,319	8,300,542	489,934,297	32,833,898,441
dusub	1	25	739	27,381	1,123,356	66,305,397	4,443,584,945
broken-down	1	12	349	12,925	530,274	31,299,091	2,097,569,371
broken	1	11	12	25	188	349	5,463

hypersum	11	1,117	1,111,729	11,172,937	1,117,293,741	111,729,374,159
dusub	1	151	150,456	1,512,093	151,209,265	15,120,926,498

302 weak primes

mates	10	16	26	38	52	56	68	76
holiness	1	1	1	2	0	1	3	1
sub	1	3	5	7	8	11	16	17
broken-up	3	22	289	5,513	127,088	3,945,241	169,772,451	7,983,250,438
sub	1	8	106	2,028	46,753	1,451,373	62,455,794	2,936,873,710

broken-down	1	4	53	1,011	23,306	723,497	31,133,677	1,464,006,316
broken	1	3	4	22	53	289	1,011	5,513

hypersum	3	37	3,713	371,319	37,131,923	3,713,192,331	371,319,233,143
sub	1	14	1,366	136,601	13,660,071	1,366,007,120	136,600,711,985

303 weaker primes

sub	1	3	42	192	4,167	5,769
broken-up	3	22	2,489	1,301,769	1,727,449,962	16,498,875,793,321
sub	1	8	916	478,894	635,493,323	6,069,597,206,804
broken-down	1	4	453	236,923	314,397,274	3,002,808,600,897
broken	1	3	4	22	453	2,489

hypersum	3	37	37,113	37,113,523	3,711,352,311,327	371,135,231,132,715,683
sub	1	14	13,653	13,653,302	1,365,330,214,281	136,533,021,428,138,000

304 strong curvaceous

broken-up	6	49	1,476	56,137	3,369,696	222,456,073	15,130,382,660	1,210,653,068,873
holiness	1	1	1	1	6	2	5	8
dusub	1	7	200	7,597	456,039	30,106,156	2,047,674,623	163,844,075,977
broken-down	1	7	211	8,025	481,711	31,800,951	2,162,946,379	173,067,511,271
broken	1	6	7	49	211	1,476	8,025	56,137

hypersum	6	68	6,830	683,038	68,303,860	6,830,386,066	683,038,606,668
holiness	1	3	4	6	8	10	13
dusub	1	9	924	92,439	9,243,922	924,392,233	92,439,223,295

305 strong curvilinear

tressub	1	3	11	13	26	28
broken-up	22	1,145	254,212	64,062,569	33,440,915,230	18,459,449,269,529
tressub	1	57	12,656	3,189,487	1,664,925,133	919,041,862,815
broken-down	1	23	5,107	1,286,987	671,812,321	370,841,688,179
broken	1	22	23	1,145	5,107	254,212

hypersum	22	2,252	2,252,222	2,252,222,252	225,222,252,522
tressub	1	112	112,131	112,131,543	112,131,155,684

306 strong linear

dusub	1	6	10	15	19	23
broken-up	11	452	32,103	3,563,885	502,539,888	206,547,457,853
dusub	1	61	4,345	482,319	68,011,378	27,953,158,710
broken-down	1	12	853	94,695	13,352,848	5,488,115,223
broken	1	11	12	452	853	32,103

hypersum	11	1,141	114,171	114,171,111	114,171,111,141	114,171,111,141,171
dusub	1	155	15,451	1,451,380	15,451,379,664	15,451,379,663,729

307 strong lucky

dusub	1	2	3	4	7	8	9
broken-up	7	92	1,939	60,201	2,951,788	186,022,845	12,466,482,403
dusub	1	12	262	8,147	399,481	25,175,454	1,687,154,927
broken-down	1	8	169	5,247	257,272	16,213,383	1,086,553,933
broken	1	7	8	92	169	1,939	5,247

hypersum	7	713	71,321	7,132,131	713,213,149	71,321,314,963
sub	1	96	9,652	965,229	96,522,904	9,652,290,361

308 Zenophobia is a homonym for xenophobia, the irrational fear of strangers. The zeno- prefix is a reference to the Zeno paradox having to do with converging fractional sums.

spell-and-say starting at zero

broken-up	0	1	8	105	2,213	66,495	2,396,033	107,887,980	5,828,346,953
holiness	1	0	2	1	0	3	3	9	6
untrimmed	1	2	9	216	3,323	775,106	34,105,144	2,189,981,091	69,394,571,064
broken-down	1	1	9	118	2,487	74,728	2,692,695	121,246,003	6,549,976,857
broken	0	1	8	9	105	118	2,213	2,487	66,495
holiness	1	0	2	1	1	0	0	2	3
untrimmed	1	2	9	10	216	229	3,324	3,598	775,106

hypersum	0	4	48	4,813	481,321	48,132,130	48,132,130,936	4,813,213,093,645
holiness	1	0	2	2	2	3	5	5
untrimmed	1	5	59	5,924	592,432	59,243,241	592,432,411,047	59,243,241,104,756

309 spell-and-count starting at one

broken-up	1	4	33	433	9,126	274,213	9,880,794	44,909,943	24,035,017,716

broken-down	1	2	17	223	4,700	141,223	5,088,728	229,133,983	12,378,323,810
broken	1	2	4	17	33	223	433	4,700	9,126
sum	1	4	12	25	46	76	112	157	211
super	1	3	24	312	6,552	196,560	7,076,160	318,427,200	17,195,068,800

hypersum	1	13	138	13,813	1381321	138,132,130	13,813,213,036	1,381,321,303,645

310 strong spell-and-count

broken-up	30	2,551	403,088	80,620,151	16,930,634,798	4,232,739,319,651
holiness	1	0	6	5	7	3
dusub	1	127	20,068	4,013,841	842,926,672	210,735,681,890
broken-down	1	31	4,899	979,831	205,769,409	51,443,332,081
broken	1	30	31	2,551	4,899	403,088

hypersum	30	3,085	3,085,158	3,085,158,200	3,085,158,200,210	3,085,158,200,210,250
holiness	1	3	5	7	8	9
tressub	1	154	153601	153,600,982	153,600,982,239	153,600,982,239,545

311 Siamese primes

mates	2	16	20	88	25,468	25,472
trimmed	1	5	1	77	14,357	14,361
broken-up	7	78	6,169	512,105	114,205,584	25,925,179,673
dusub	1	11	835	69,306	15,456,045	3,508,591,534
broken-down	1	8	633	52,547	11,718,614	2,660,177,925
broken	1	7	8	78	633	6,169

hypersum	7	711	71,179	7,117,983	7,117,983,223	7,117,983,223,227
dusub	1	96	9,633	963,314	963,314,275	963,314,275,589

strong Siamese primes	79	223	439	1,087	13,687
mates	20	560	776	8,912	...
holiness	1	2	1	3	
tetrasub	1	4	8	20	251
broken-up	79	17,618	7,734,381	8,407,289,765	115,070,582,747,936
holiness	1	3	2	7	6
tetrasub	1	322	141,660	153,984,883	2,107,591,240,327

broken-down	1	80	35,121	38,176,607	522,523,255,130
broken	1	79	80	17,618	35,121

hypersum	79	79,223	79,223,439	792,234,391,087	79,223,439,108,713,687
holiness	1	1	2	5	8
tetrasub	1	1,451	1,451,028	14,510,279,023	1,451,027,902,238,820

312 Siamese

broken-up	2	7	44	315	3,509	49,441	893,447	20,598,722	557,058,941
trimmed	1	6	33	204	248	38,330	782,336	1,487,611	44,647,830
sub	1	3	16	116	1,291	18,188	328,681	7,577,846	204,930,532
broken-down	1	3	19	136	1,515	21,346	385,743	8,893,435	240,508,488
broken	1	2	3	7	19	44	136	315	1,515

hypersum	2	23	236	2,367	236,711	23,671,114	2,367,111,418	236,711,141,823
trimmed	1	12	125	1,256	125,600	12,560,003	1,256,000,307	125,600,030,712
sub	1	8	87	871	87,081	8,708,116	870,811,626	87,081,162,573

313 strong Siamese

broken-up	6	67	944	21,779	828,546	38,963,441	2,416,561,888	190,947,352,593
holiness	1	1	1	1	5	4	8	4
dusub	1	9	128	2,947	112,131	5,273,128	327,046,088	25,841,914,046
broken-down	1	7	99	2,284	86,891	4,086,161	253,428,873	20,024,967,128
broken	1	6	7	67	99	944	2,284	21,779

hypersum	6	611	61,114	6,111,423	611,142,338	61,114,233,847	6,111,423,384,762
holiness	1	1	1	1	3	3	4
dusub	1	83	8,271	827,091	82,709,121	8,270,912,147	827,091,214,756

314 Siamese males

sub	1	3	4	8	10	17	19	29
broken-up	3	22	245	5,657	152,984	7,195,905	367,144,139	29,011,582,886
sub	1	8	90	2,081	56,280	2,647,226	135,064,781	10,672,764,900
broken-down	1	4	45	1,039	28,098	1,321,645	67,431,993	5,328,449,092
broken	1	3	4	22	45	245	1,039	5,657

hypersum	3	37	3,711	371,123	37,112,327	3,711,232,747	371,123,274,751

sub	1	14	1,365	136529	13,652,862	1,365,286,229	136,528,622,921

strong male Siamese

tressub	1	2	4	6	8	11
broken-up	23	1,082	85,501	9,919,198	1,656,591,567	369,429,838,639
tressub	1	54	4,257	493,847	82,476,837	18,392,828,633
broken-down	1	24	1,897	220,076	36,754,589	8,196,493,423
broken	1	23	24	1,082	1,897	85,501

hypersum	23	2,347	234,779	234,779,116	234,779,116,167	234,779,116,167,223
tressub	1	117	11,689	11,688,964	11,688,963,908	11,688,963,907,964

315 female Siamese

sub	1	2	5	7	13	14	23	24
broken-up	2	13	184	3,325	113,234	4,306,217	267,098,688	17,632,819,625
trimmed	1	2	73	2,214	2,123	325,106	15,687,577	6,521,708,514
sub	1	5	68	1,223	41,656	1,584,169	98,260,116	6,486,751,830
broken-down	1	3	43	777	26,461	1,006,295	62,416,751	4,120,511,861
broken	1	2	3	13	43	184	777	3,325

hypersum	2	26	2,614	261,418	26,141,834	2,614,183,438	261,418,343,862
trimmed	1	15	1,503	150,307	15,030,723	150,307,237	15,030,723,751
sub	1	10	962	96,170	9,617,043	961,704,342	96,170,434,252

strong female Siamese

broken-up	14	477	18,140	1,197,717	170,093,954	32,999,424,793
tressub	1	24	903	59,631	8,468,480	1,642,944,618
broken-down	1	15	571	37,701	5,354,113	1,038,735,623
broken	1	14	15	477	571	18,140

hypersum	14	1,434	143,438	14,343,866	14,343,866,142	14,343,866,142,194
tressub	1	71	7,141	714,139	714,139,044	714,139,044,281

316 The ichibi is a one-tailed tanuki (raccoon-dog).

ichibi	11	21	31	41	51	61	71
dusub	1	3	4	6	7	8	10
broken-up	11	232	7,203	295,555	15,080,508	920,206,543	65,349,745,061

dusub	1	31	975	39,999	2,040,925	124,536,413	8,844,126,257
broken-down	1	12	373	15,305	780,925	47,651,913	3,384,066,751
broken	1	11	12	232	373	7,203	15,305

hypersum	11	1,121	112,131	11,213,141	1,121,314,151	112,131,415,161
dusub	1	152	15,175	1,517,534	151,753,368	15,175,336,830

The Langford numbers are ones whose pairs of shepherd digits count the digits between them, the "sheep". Ichibi can have nearly any number of shepherds and sheep.

Langford	312,132	41,312,432	17,126,425,374,635	3,181,375,264,285,746
shepherds	12	20	56	72
sheep	136	1,482	1,730,444	194,750,939

The shepherded numbers start out looking like ambisæna. All have pairs shepherds even if they have no sheep.

shepherded	101	121	131	141	151	161	171	181	191	2,101	2,121	2,131	2,141
shepherds	2	2	2	2	2	2	2	2	2	2	2	2	2
sheep	0	2	3	4	5	6	7	8	9	0	2	3	4

ichibi primes

mates	28	38	58	68	88	118	178
holiness	2	2	2	3	4	2	2
broken-up	11	342	14,033	856,355	60,815,238	6,143,195,393	804,819,411,721
dusub	1	48	1,899	115,895	8,230,447	831,391,088	108,920,463,040
broken-down	1	12	493	30,085	2,136,528	215,819,413	28,274,479,631
broken	1	11	12	342	493	14,033	30,085

hypersum	11	1,131	113,141	11,314,161	1,131,416,171	1,131,416,171,101
dusub	1	153	15,312	1,531,205	153,120,528	153,120,527,974

The ibihci or ikunat would be those which still remain ichibi when reversed.

ikunat	1,011	1,021	1,031	1,041	1,051
broken-up	1,011	1,032,232	1,064,232,203	1,107,866,755,555	1,164,369,024,320,508
holiness	1	1	3	5	8
heptasub	1	6,019	970,454	1,010,243,715	1,061,767,114,528
broken-down	1	1,012	1,043,373	1,086,152,305	1,141,547,115,928

broken	1	1,011	1,012	1,032,232	1,043,373

hypersum	1,011	10,111,021	101,110,121,031	1,011,101,210,311,041
holiness	1	2	3	4
heptasub	1	9,220	92,200,496	922,004,959,033

ikunat primes	1,021	1,031	1,051	1,061	1,091
holiness	1	1	1	2	2
mates	8,048	8,068	8,098	8,128	8,138
holiness	5	6	6	4	4
broken-up	1,021	1,052,652	1,106,338,273	1,173,825,960,305	1,280,645,229,031,028
holiness	1	2	2	6	9
heptasub	1	6,019	1,008,850	1,070,390,724	1167797288627
broken-down	1	1,022	1,074,123	1,139,645,525	1,243,354,341,898
broken	1	1,021	1,022	1,052,652	1,074,123

hypersum	1,021	10,211,031	102,110,311,051	1,021,103,110,511,061
holiness	1	2	3	5
heptasub	1	9,311	93,112,551	931,125,511,447

strong ikunat primes	1,051	1,151	1,171	1,291
holiness	1	0	0	1
broken-up	1,051	1,209,702	1,416,562,093	1,828,782,871,765
holiness	1	2	4	9
heptasub	1	6,019	1,291,737	1,667,634,120
broken-down	1	1,052	1,231,893	1,590,374,915
broken	1	1,051	1,052	1,209,702

hypersum	1,051	10,511,151	105,111,511,171	1,051,115,111,711,291
holiness	1	1	1	2
heptasub	1	9,585	95,849,291	958,492,914,091

317 The sanbi is a three-tailed turtle, aka umibouzu ("green turtle"), but it could also be called a bashaw from the name for a Turkish official's hat, which also had three tails.

sanbi	13	23	33	43	53	63
broken-up	13	300	12,913	684,689	49,995,210	4,150,287,119

tressub	1	194	2,507,955	1,717,169,311,834	> 19plex	> 29plex
broken-down	1	14	603	31,973	2,334,632	193,806,429
broken	1	13	14	300	603	12,913

hypersum	13	1,323	132,333	13,233,343	1,323,334,353	132,333,345,363
tressub	1	66	6,588	658,849	65,884,938	6,588,489,313

sanbi primes

mates	16	26	46	56	76	86	116
holiness	1	1	1	1	1	3	1
broken-up	13	300	12,913	684,689	49,995,210	4,150,287,119	427,529,568,467
tressub	1	15	643	34,088	2,489,115	206,630,628	21,285,443,854
broken-down	1	14	603	31,973	2,334,632	193,806,429	19,964,396,819
broken	1	13	14	300	603	12,913	31,973

hypersum	13	1,323	132,343	13,234,353	1,323,435,373	132,343,537,383
tressub	1	66	6589	658,900	65,889,967	6,588,996,744

strong sanbi primes

broken-up	43	3,140	323,463	52,727,609	11,758,580,270	3,092,559,338,619
tetrasub	1	57	5,924	965,740	215,365,910	56,642,200,088
broken-down	1	44	4,533	738,923	164,784,362	43,339,026,129
broken	1	43	44	3,140	4,533	323,463

hypersum	43	4,373	4373103	4,373,103,163	4,373,103,163,223	4,373,103,163,223,263
tetrasub	1	80	80,096	80,096,178	80,096,178,361	80,096,178,360,779

shepherded	30,003	30,013	30,023	30,043	30,053	30,063	30,073	30,083	30,093
shepherds	6	6	6	6	6	6	6	6	6
sheep	0	1	2	4	5	6	7	8	9

318 The shichibi is a beetle with six wings and seven tails.

shichibi	17	27	37	47	57	67	77
broken-up	17	460	17,037	801,199	45,685,380	3,061,721,659	235,798,253,123
tressub	1	23	848	39,889	2,274,541	152,434,145	11,739,703,749
broken-down	1	18	667	31,367	1,788,586	119,966,629	9,231,519,019

broken	1	17	18	460	667	17,037	31,367

hypersum	17	1,737	173,747	17,374,767	1,737,476,797	1,737,476,797,107
tressub	1	86	8,650	865,039	86,503,876	86,503,876,085

ibihicihs	7,017	7,027	7,037	7,047
holiness	1	1	1	1
broken-up	7,017	49,308,460	346,983,640,037	2,445,193,760,649,199
holiness	1	6	7	7
enneasub	1	814	42,821,183	301,760,882,956
broken-down	1	7,018	49,385,667	348,020,802,367
broken	1	7,017	7,018	49,308,460

hypersum	7,017	70,177,027	701,770,277,037	7,017,702,770,377,047
holiness	1	2	3	4
enneasub	1	8,661	86,605,332	866,053,324,031

shepherded	700,000,007	700,000,017	700,000,027	700,000,037	700,000,047
shepherds	14	14	14	14	14
sheep	0	1	2	3	4

"Shack" describes shichibi and cubed kyuubi combined, which of course doesn't differ from pure shichibi until nine cubed or seven twenty-nine, but its complimentary sequence "grass" differs from nibi much sooner at a mere two seventy. ("My Little Grass Shack in Kealakekua, Hawai'i" by Tommy Harrison, Bill Cogswell, and Johnny Noble (1933))

A number containing the frag "007" is called a Bond number from the James "007" Bond series by Ian Flemming, many of the early ones are shichibi. The first Bond prime is four double-O seven and the first strong Bond number ten double-O seven.

Bonds	1,007	2,007	3,007	4,007
holiness	2	2	2	2
broken-up	1,007	2,021,050	6,077,298,357	24,351,736,537,549
holiness	2	3	5	2
heptasub	1	1,843	5,541,778	22,205,909,378
broken-down	1	1,008	3,031,057	12,145,446,407
broken	1	1,007	1,008	2,021,050

hypersum	1,007	10,072,007	100,720,073,007	1,007,200,730,074,007
holiness	2	4	6	8
heptasub	1	9,184	91,844,818	918,448,181,448

strong Bonds	10,007	10,070	20,070	30,070
holiness	3	3	3	3
broken-up	10,007	100,770,491	2,022,463,764,377	60,815,485,495,586,881
holiness	3	4	3	14
enneasub	1	814	249,591,857	7,505,227,150,447
broken-down	1	10,008	200,860,561	6,039,877,079,278
broken	1	10,007	10,008	100,770,491

hypersum	10,007	1,000,710,070	100,071,007,020,070	10,007,100,702,007,030,070
holiness	3	6	9	12
enneasub	1	123,497	12,349,743,371	1,234,974,337,110,360

Bond primes	4,007	6,007	9,007	10,007
holiness	2	3	3	3
mates	3,992	5,992	11,992	17,992
holiness	2	2	2	2
broken-up	4,007	24,070,050	216,798,944,357	2,169,507,060,250,549
holiness	2	4	5	8
oktasub	1	8,074	72,727,943	727,788,539,685
broken-down	1	4,008	36,100,057	361,253,274,407
broken	1	4,007	4,008	24,070,050

hypersum	4,007	40,076,007	400,760,079,007	40,076,007,900,710,007
holiness	2	5	8	11
oktasub	1	13,444	134,440,029	13,444,002,926,214

Similarly a number containing the frag "86" is a Smart number from Maxwell "86" Smart of "Get Smart". The first strong Smart is seventeen eighty-six.

Smarts	86	186	286	386	486
holiness	3	3	3	3	3
broken-up	86	15,997	4,575,228	1,766,054,005	858,306,821,658

holiness	3	2	2	5	11
tetrasub	1	293	83,798	32,346,407	15,720,437,801
broken-down	1	87	24,883	9,604,925	4,668,018,433
broken	1	86	87	15,997	24,883

hypersum	86	86,186	86,186,286	86,186,286,386	86,186,286,386,486
holiness	3	6	9	12	15
tetrasub	1	1,579	1,578,557	1,578,556,899	1,578,556,898,616

The smart primes, perhaps not coincidentally, contain 8699, the hypersum of 86 and 99, aka M/M Smart.

Smart primes	863	1,861	1,867	2,861	3,863
holiness	3	3	3	3	3
mates	136	1,132	1,136	1,138	1,300
holiness	1	0	1	2	2
broken-up	863	1,606,044	2,998,485,011	8,578,667,222,515	33,139,394,479,060,456
holiness	3	4	7	6	7
heptasub	1	1,464	2,734,264	7,822,731,929	30,219,216,174,852
broken-down	1	864	1,613,089	4,615,048,493	17,827,933,941,548
broken	1	863	864	1,606,044	1,613,089

hypersum	863	8,631,861	86,318,611,867	863,186,118,672,861
holiness	3	6	9	12
heptasub	1	7,871	78,712,385	787,123,854,535

shichibi primes

mates	32	52	62	82	112	122	142
broken-up	17	630	29,627	1,985,639	192,636,610	20,614,102,909	2,618,183,706,053
tressub	1	31	1,475	98,859	9,590,812	1,026,315,751	130,351,691,173
broken-down	1	18	847	56,767	5,507,246	589,332,089	589,332,089
broken	1	17	18	630	847	29,627	56,767

hypersum	17	1,737	173,747	17,374,767	1,737,476,797	1,737,476,797,107
tressub	1	86	8,650	865,039	86,503,876	86,503,876,085

strong shichibi primes

broken-up	37	3,590	455,967	127,187,109	35,976,898,430	11,404,793,989,419
tetrasub	1	66	8,351	2,146,356	658,939,880	208,886,088,310
broken-down	1	38	4,827	1,240,577	380,861,866	120,734,483,799
broken	1	37	38	3,590	4,827	455,967

hypersum	37	3,797	3,797,127	3,797,127,257	3,797,127,257,307
tetrasub	1	70	69,547	69,546,812	69,546,811,659

319 The kyuubi is a nine-tailed fox.

kyuubi	19	29	39	49	59	69	79
holiness	1	1	1	1	1	2	1
broken-up	19	552	21,547	1,056,355	62,346,492	4,302,964,303	339,996,526,429
holiness	1	0	0	1	3	4	6
tressub	1	27	1,073	52,593	3,104,049	214,231,978	16,927,430,306
broken-down	1	20	781	38,289	2,259,832	155,966,697	12,323,628,895
broken	1	19	20	552	781	21,547	52,593

hypersum	19	1,929	192,939	19,293,949	1,929,394,959	192,939,495,969
holiness	1	2	3	4	5	7
tressub	1	96	9,606	960,589	96,058,919	9,605,891,877

kyuubi prime

mates	10	20	40	70	80	140	160
holiness	1	1	1	1	2	1	2
broken-up	19	52	32,587	2,574,925	229,200,912	24,985,474,333	3,473,210,133,199
holiness	1	0	2	1	4	3	3
tressub	1	27	1,622	128,198	11,411,241	1,243,953,519	172,920,950,357
broken-down	1	20	1,181	93,319	8,306,572	905,509,667	125,874,150,285
broken	1	19	20	52	1,181	32,587	93,319

hypersum	19	1,929	192,959	19,295,979	1,929,597,989	192,959,475,989,109
holiness	1	2	3	4	7	9
tressub	1	96	9,607	960,690	96,069,027	9,606,886,623,297

strong kyuubi prime

broken-up	59	4,662	648,077	128,971,985	29,535,232,642	10,307,925,164,043
holiness	1	2	4	6	2	5
tetrasub	1	85	11,869	2,362,204	540,956,656	188,796,234,997
broken-down	1	60	8,341	1,659,919	380,129,792	132,666,957,327
broken	1	59	60	4,662	8,341	648,077

hypersum	59	5,979	5,979,139	5,979,139,199	5,979,139,199,229
holiness	1	2	3	5	6
tetrasub	1	110	109,512	109,511,754	109,511,754,439

ibuuyk	9,019	9,029	9,039	9,049
holiness	3	3	3	3
mate	90,010	90,020	90,030	90,040
holiness	4	4	4	4
broken-up	9,019	81,432,552	736,068,846,547	6,660,687,073,836,355
holiness	3	2	8	11
enneasub	1	814	90,838,112	821,994,086,865
broken-down	1	9,020	81,531,781	737,781,095,289
broken	1	9,019	9,020	81,432,552

hypersum	9,019	90,199,029	901,990,299,039	9,019,109,290,399,049
holiness	3	6	9	11
enneasub	1	11,131	111,314,446	1,113,046,510,564

ibuuyk primes	9,029	9,049	9,059	9,109
holiness	3	3	3	3
mates	90,050	90,070	90,140	90,160
holiness	4	4	3	4
broken-up	9,029	81,703,422	740,151,308,927	6,742,038,354,719,465
holiness	3	3	5	6
enneasub	1	10,083	91,341,928	832,033,632,500
broken-down	1	9,030	81,802,771	745,141,450,069
broken	1	9,029	9,030	81,703,422

hypersum	9,029	90,299,049	902,990,499,059	9,029,904,990,599,109

holiness	3	6	9	11
enneasub	1	11,144	111,437,881	1,114,378,805,811

shepherded	90,000,000,009	90,000,000,019	90,000,000,029	90,000,000,039
shepherds	18	18	18	18
sheep	0	1	2	3

The nibi is a two-tailed cat aka neko-mata ("forked-cat"). There is only one nibi prime, two, and no yongbi, gobi, rokubi, hachibi or love primes or strong yongbi, gobi, hachbi or love numbers.

nibi	12	22	32	42	52	62	72
mates	17	27	37	47	57	67	77
dusub	1	3	4	6	7	8	10
broken-up	12	265	8,492	356,929	18,568,800	1,151,622,529	82,935,390,888
dusub	1	36	1,149	48,305	2,513,014	155,855,161	11,224,084,616
broken-down	1	13	417	17,527	911,821	56,550,429	4,072,542,709
broken	1	12	13	265	417	8,492	17,527

hypersum	12	1,222	122,232	12,223,242	1,222,324,252	122,232,425,262
dusub	1	166	16,542	1,654,236	165,423,599	16,542,359,893

ibin	2,012	2,022	2,032	2,042
holiness	1	1	1	1
mates	7,017	7,027	7,037	7,047
broken-up	2,012	4,068,265	8,266,716,492	16,880,639,144,929
holiness	1	5	6	10
oktasub	1	2,214	3,773,175	5,662,823,568
broken-down	1	2,013	4,090,417	8,352,633,527
broken	1	2,012	2,013	4,068,265

hypersum	2,012	20,122,022	201,220,222,032	2,012,202,220,322,042
holiness	1	2	3	4
oktasub	1	6,750	67,501,864	675,018,644,700

The yongbi is a four-tailed monkey.

yongbi	14	24	34	44	54	64	74

mates	15	25	35	45	55	65	75
broken-up	14	337	11,472	505,105	27,287,142	1,746,882,193	129,296,569,424
tressub	1	16	571	25,147	1,358,547	86,972,143	6,437,297,141
broken-down	1	15	511	22,499	1,215,457	77,811,747	5,759,284,735
broken	1	14	15	337	511	11,472	22,499

hypersum	14	1,424	142,434	14,243,444	1,424,344,454	142,434,445,464
tressub	1	71	7,092	709,139	70,913,935	7,091,393,474

shepherded	400,004	400,014	400,024	400,034	400,054	400,064	400,074
shepherds	8	8	8	8	8	8	8
sheep	0	1	2	3	5	6	7

The gobi is a five-tailed whale-horse, incredibly their mates are four-tailed monkeys.

gobi	15	25	35	45	55	65	75
broken-up	15	376	13,175	593,251	32,641,980	2,122,321,951	159,206,788,305
tressub	1	19	656	29,536	1,625,148	105,664,188	7,926,439,254
broken-down	1	16	561	25,261	1,389,916	90,369,801	6,779,124,991
broken	1	15	16	376	561	13,175	25,261

hypersum	15	1,525	152,535	15,253,545	1,525,354,555	152,535,455,565
tressub	1	76	7,594	759429	75,942,932	7,594,293,155

igob	5,015	5,025	5,035	5,045
holiness	1	1	1	1
mates	4,014	4,024	4,034	4,044
broken-up	5,105	25,200,376	126,883,898,175	640,129,291,493,251
holiness	1	3	10	5
enneasub	1	814	15,658,717	78,998,230,453
broken-down	1	5,016	25,255,561	127,414,310,261
broken	1	5,015	5,016	25,200,376

hypersum	5,015	50,155,025	501,550,255,035	5,015,502,550,355,045
holiness	1	2	3	4
enneasub	1	6,190	61,896,219	618,962,187,136

shepherded	5,000,005	5,000,015	5,000,025	5,000,035	5,000,045	5,000,055
shepherds	10	10	10	10	10	10
sheep	0	1	2	3	4	5

The rokubi is a six-tailed slug with a horse-head and branching nettles. Rokubi numbers are called "rare" from the redundant "rokubi-and-rokubi-evens" acronum.

rokubi	16	26	36	46	56	66	76
holiness	1	1	1	1	1	2	1
mates	13	23	33	43	53	63	73
broken-up	16	417	15,028	691,705	38,750,508	2,556,225,233	194,463,868,216
holiness	1	0	3	3	6	1	8
dusub	1	20	748	34,438	1,929,274	127,366,534	9,681,785,902
broken-down	1	17	613	28,215	1,580,653	104,351,313	7,932,280,441
broken	1	16	17	417	613	15,028	28,215

hypersum	16	1,626	162,636	16,263,646	1,626,364,656	162,636,465,666
holiness	1	2	3	4	5	7
tressub	1	81	8,097	809,719	80,971,938	8,097,193,835

shepherded	60,000,006	60,000,016	60,000,026	60,000,036	60,000,046
shepherds	12	12	12	12	12
sheep	0	1	2	3	4

The hachibi is an eight-tailed, horned cephalopod.

hachibi	18	28	38	48	58	68	78
holiness	2	2	2	2	2	3	2
mates	11	21	31	41	51	61	71
broken-up	18	505	19,208	922,489	53,523,570	3,640,525,249	284,014,492,992
holiness	1	1	4	4	1	3	6
tressub	1	25	956	45,928	2,664,781	181,251,079	14,140,248,980
broken-down	1	19	723	34,723	2,014,657	137,031,399	10,690,463,779
broken	1	18	19	505	723	19,208	34,723

shepherded	8,000,000,008	8,000,000,018	8,000,000,028	8,000,000,038
shepherds	16	16	16	16

sheep	0	1	2	3

Loved numbers are those ending in a zero, aka a goose egg or "love" from the French *l'oef*, multiples of ten.

loved	0	10	20	30	40	50	60
holiness	1	1	1	1	1	1	2
untrimmed	1	21	31	41	51	61	71
mates	19	29	39	49	59	69	79
holiness	1	1	1	1	1	2	1

hypersum	0	10	1,020	102,030	10,203,040	1,020,304,050	102,030,405,060
holiness	1	1	2	3	4	5	7
untrimmed	1	21	2,131	213,141	21,314,151	2,131,415,161	213,141,516,171

shepherded	100	200	300	400	500	600	700	800	900	1,100	1,200	1,300	1,400
shepherds	2	2	2	2	2	2	2	2	2	2	2	2	2
sheep	0	0	0	0	0	0	0	0	0	0	0	0	0

The opposite of "love" is not hate, but fear, for "There is no fear in love, but perfect love casts out all fear." (1 Jn 4:18) The 'loved' numbers could therefore also be called 'fearless'.

feared	1	2	3	4	5	6	7	8	9	11	12	13	14	15	16	17	18	19	21	22	23

broken-up	...	740,785	8,229,836	99,498,817	1,301,714,457	18,323,501,215
broken- down	...	516,901	2,641,165	32,210,881	421,382,618	5,931,567,533

These names for the untailed and tailed numbers can be remembered with the mnemonic, referring to the never-to-be-forgotten Gothic romance by Hope Anthony, *Love in Syugarshhak* [pronounced sugarshack].

320 alliteratives

mates	11	22	33	44	55	66	77
broken-up	22	727	32,010	1,761,277	116,276,292	8,965,035,761	788,159,423,260
tressub	1	36	1,594	87,689	5,789,056	445,844,977	39,240,147,090
broken-down	1	23	1,013	55,738	3,679,721	283,394,255	24,942,374,161
broken	1	22	23	727	1,013	32,010	55,738

hypersum	22	2,233	223,344	22,334,455	2,233,445,566	223,344,556,677

tressub	1	111	11,120	1,111,967	111,196,707	11,119,670,713

321 strong alliteratives

broken-up	202	42,825	9,421,702	285,481,531	942,099,536,932
holiness	1	2	2	4	6
pentasub	1	44,473	63,483	19,235,616	6,347,816,748
broken-down	1	203	44,661	13,532,486	4,465,765,041
broken	1	202	203	42,825	44,661

hypersum	202	202,212	202,212,220	202,212,220,303	202,212,220,303,330
holiness	1	1	2	3	4
pentasub	1	1,363	1,362,495	1,362,495,223	1,362,495,222,971

322 spooneristic

mates	11	12	13	15	20	21
broken-up	46	2,163	103,870	5,091,793	325,978,622	21,845,659,467
holiness	1	1	4	3	4	5
tetrasub	1	40	1,902	93,259	5,970,507	400,117,210
broken-down	1	47	2,257	110,640	7,083,217	474,686,179
broken	1	46	47	2,163	2,257	103,870

hypersum	46	4,647	464,748	46,474,849	4,647,484,964	464,748,496,467
holiness	1	1	3	4	5	6
tetrasub	1	85	8,512	851,217	85,121,656	8,512,165,635

323 strong spooneristic

broken-up	64	4,289	296,005	21,908,659	1,709,171,407	143,592,306,847
holiness	1	3	4	6	3	5
tetrasub	1	78	5,421	401,271	31,304,566	2,629,984,839
broken-down	1	65	4,486	332,029	25,902,748	2,176,162,861
broken	1	64	65	4,289	4,489	296,005

hypersum	64	6,467	646,769	64,676,974	6,467,697,478	646,769,747,884
holiness	1	2	4	4	6	8
tetrasub	1	118	11,846	1,184,600	118,460,011	11,846,001,146

324 stronger primes

mates	70	88	458	872	8,638	8,848
holiness	1	4	2	2	4	6
broken-up	11	320	40,651	21,992,511	19,947,248,128	22,959,304,587,839
dusub	1	43	5,502	2,976,363	2,699,566,475	3,107,203,989,311
broken-down	1	12	1,525	825,037	748,310,084	861,305,731,721
broken	1	11	12	320	1,525	40,651

hypersum	11	1,129	1,129,127	1,129,127,541	1,129,127,541,907
dusub	1	153	152,811	152,810,796	152,810,795,694

stronger prime strengths

broken-up	1	3	16	99	808	8,179	115,314	2,314,459	74,178,002	3,043,612,541
broken-down	1	2	11	68	555	5,618	79,207	1,589,758	50,951,463	2,090,599,741
broken	1	2	3	11	16	68	99	555	808	5,618

hypersum	1	12	125	1,256	12,568	1,256,810	125,681,014	1,256,101,420

325 stronger lucky

broken-up	7	218	13,741	1,745,325	329,880,166	85,440,708,319	59,552,503,578,509
dusub	1	29	1,859	236,204	44,644,426	11,563,142,460	8,059,554,939,247
broken-down	1	8	505	64,143	12,123,532	3,140,058,931	2,188,633,198,439
broken	1	7	8	505	13,741	64,143	1,745,325

stronger lucky strengths

broken-up	1	3	13	68	489	3,980
broken-down	1	2	9	47	338	2751
broken	1	2	3	9	13	47
sum	1	3	7	12	19	27
super	1	2	8	40	280	2,240
hypersum	1	12	124	1,245	12,457	124,578

327 stronger curvaceous

mates	3	69	699	6,999	69,999
broken-up	6	181	54,306	162,918,181	4,887,545,484,306
holiness	1	2	2	6	8
dusub	1	24	7,349	22,048,578	661,457,352,450

broken-down	1	72,101	6,303,007	189,090,212,101	56,727,063,636,603,007
broken	1	6	181	54,306	72,101

hypersum	6	630	630,300	6,303,003,000	630,300,300,030,000
holiness	1	2	4	7	11
dusub	1	85	85,302	853,018,696	85,301,869,628,682

328 stronger curvilinear

tressub	1	3	11	26	111	260
broken-up	22	1,145	254,212	132,699,809	294,859,229,810	1,539,755,030,767,629
tressub	1	57	12,657	6,606,734	14,680,176,633	76,659,888,986,591
broken-down	1	23	5,107	2,665,877	5,923,583,801	30,932,957,274,699
broken	1	22	23	1,145	5,107	254,212

hypersum	22	2,252	2,252,222	225,222,522	22,522,225,222,222
tressub	1	112	112,131	11,213,169	1,121,315,566,935

stronger curvilinear strength

mates	2	17	87	167	867	1,667
trimmed	1	6	76	56	756	556
broken-up	7	85	6,977	921,049	766,319,745	1,020,738,821,389
dusub	1	11	944	124,650	103,710,100	138,141,977,503
broken-down	1	8	657	86,732	72161681	96,119,445,824
broken	1	7	8	85	657	6,977

hypersum	7	712	71,282	711,282,132	71,282,132,832	712,821,328,321,332
dusub	1	96	9,647	96,261,569	9,646,987,637	96,469,876,365,466

329 stronger linear

dusub	1	6	15	56	150	556
broken-up	11	452	50,163	20,625,656	22,915,163,996	94,204,259,821,443
dusub	1	61	6,791	2,791,380	3,101,230,210	12,749,160,185,030
broken-down	1	12	1,333	547,875	606,690,458	2,502,327,020,713
broken	1	11	12	452	1,333	50,163

hypersum	11	1,141	1,141,111	1,141,111,411	11,411,114,111,111
dusub	1	155	154,432	154,432,636	1,544,326,360,272

330 strongest

mates	4	5	8	30	59	78	89	279
broken-up	1	5	26	265	5,591	223,905	15,455,036	1,700,277,865
broken-down	1	2	11	112	2,363	94,632	6,531,971	718,611,442
broken	1	2	5	11	26	265	2,363	5,591
sum	1	5	10	20	41	81	150	260
super	1	4	20	200	4,200	168,000	11,592,000	210,394,800,000

hypersum	1	14	145	14,510	1,451,021	145,102,140	14,510,214,069	14,510,214,069,110

331 strongest male or Samson numbers

broken-up	1	6	127	8,769	1,447,012	470,267,669	265,713,979,997
broken-down	1	2	43	2,969	489,928	159,229,569	89,965,196,413
broken	1	2	6	43	127	2,969	8,769
sum	1	6	27	96	261	586	1,151
super	1	5	105	7,245	1,195,425	388,513,125	219,509,915,625

hypersum	1	15	1521	152169	152,169,165	152,169,165,325

332 strongest even

mates	5	59	89	279	565	763	889
sub	1	4	15	40	87	160	265
broken-up	4	41	1,644	180,881	42,689,560	18,527,449,921	13,339,806,632,680
sub	1	15	605	66,542	15,704,611	6,815,867,923	4,907,440,609,365
broken-down	1	5	201	22,115	5,219,341	2,265,216,109	1,630,960,817,821
broken	1	4	5	201	1,644	22,115	180,881

hypersum	4	410	41,040	41,040,110	41,040,110,236	41,040,110,236,434
sub	1	151	15,098	15,097,813	15,097,812,819	15,097,812,819,394

333 strongest ichibi

tressub	1	45	228	650	1,411
broken-up	21	18,922	86,681,703	1,132,149,741,805	32,086,255,919,177,208
tressub	1	942	4,315,628	56,366,416,598	1,597,480,617,116,860
broken-down	1	22	100,783	1,316,326,785	37,306,017,514,468
broken	1	21	22	18,922	100,783

hypersum	21	21,901	219,014,581	21,901,458,113,061	2,190,145,811,306,128,341
tressub	1	1,090	10,904,094	1,090,409,392,431	109,040,939,243,088,992

334 strongest gobi

pentasub	1	2	4	13	18
broken-up	165	53,626	30,296,855	58,325,349,501	154,270,579,729,000
holiness	1	2	5	4	6
pentasub	1	361	204,152	392,993,113	1,039,466,989,732
broken-down	1	166	93,791	180,547,841	477,549,133,236
broken	1	165	166	53,626	93,791

hypersum	165	165,325	165,325,565	1,653,255,651,925
holiness	1	1	2	3
pentasub	1	1,114	1,113,955	11,139,548,958

335 strongest kyuubi

tetrasub	1	25	107	284	593
broken-up	69	93,082	542,575,047	8,414,796,497,005	272,546,844,284,069,992
holiness	2	4	1	7	10
tetrasub	1	1,705	9,937,609	154,122,373,961	4,991,869,580,171,110
broken-down	1	70	408,031	6,328,152,849	204,962,543,034,292
broken	1	69	70	93,082	408,031

hypersum	69	691,349	6,913,495,829	691,349,582,915,509
holiness	2	3	6	8
tetrasub	1	12,662	126,625,093	12,662,509,306,558

336 strongest loved

dusub	1	6	15	97	150	219
broken-up	10	40	44,120	31,766,801	35,261,193,230	57,123,164,790,401
holiness	1	1	1	5	3	4
dusub	1	54	5,971	4,299,169	4,772,083,573	7,730,779,687,499
broken-down	1	11	1,211	871,931	967,844,621	1,567,909,157,951
broken	1	10	11	40	1,211	44,120

hypersum	10	1,040	1,040,110	1,040,110,720	10,401,107,201,110
holiness	1	2	3	4	5

dusub	1	141	140,764	140,763,679	1,407,636,789,037

337 strongest yonbi

hexasub	1	8	25	57
broken-up	434	1,329,777	13,156,814,072	301,606,807,116,305
hexasub	1	3,296	32,612,481	747,608,529,637
broken-down	1	435	4,303,891	98,662,397,719
broken	1	434	435	1,329,777

hypersum	434	4,343,064	43,430,649,894	4,343,064,989,422,924
hexasub	1	10,765	107,653,818	10,765,381,795,936

338 strongest rokubi

hexasub	1	15	54	133
broken-up	236	534,777	4,329,554,828	85,404,799,071,905
holiness	1	0	5	6
hexasub	1	3,603	29,172,311	575,453,009,614
broken-down	1	237	1,918,753	37,849,321,915
broken	1	236	237	534,777

hypersum	236	2,362,266	2,322,668,096	232,266,809,619,726
holiness	1	3	7	9
pentasub	1	15,917	15,650,014	1,565,001,452,864

339 second strongest

broken-up	1	6	43	565	14,166	638,125	47,873,543	5,601,842,656	969,66,653,031
broken-down	1	2	15	197	4,940	222,497	16,692,215	1,953,211,652	337,922,308,011
broken	1	2	6	15	43	197	565	4,940	14,166
sum	1	6	13	26	51	96	171	288	461
super	1	5	35	455	11,375	511,875	38,390,625	4,491,703,125	777,064,640,625

hypersum	1	15	157	15,713	1,571,325	157,132,545	15,713,254,575	15,713,254,575,117

340 third strongest

broken-up	1	7	64	1,031	29,963	1,499,181	121,463,624	15,062, 988,557
broken-down	1	2	19	306	8,893	444,956	36,050,329	4,470,685,752
broken	1	2	7	19	64	306	1,031	8,893

sum	1	7	16	32	61	111	192	316
super	1	6	54	864	25,056	1,252,800	101,476,800	12,583,123,200

hypersum	1	16	169	16,916	1,691,629	169,162,950	16,916,295,081	16,916,295,081,124

341 fourth strongest

broken-up	1	8	89	1,699	56,156	3,090,279	268,910,429	35,230,356,478
broken-down	1	2	23	439	14,510	798,489	69,483,053	9,103,078,432
broken	1	2	8	23	89	439	1,699	14,510
sum	1	8	19	38	71	126	213	344
super	1	7	77	1,463	48,279	2,655,345	231,015,015	30,262,966,965

hypersum	1	17	1,711	171,119	17,111,933	1,711,193,355	171,119,335,587

342 fifth strongest

broken-up	1	9	118	2,605	96,503	5,792,785	538,825,508	74,363,712,889
broken-down	1	2	27	596	22,079	1,325,336	123,278,327	17,013,734,462
broken	1	2	9	27	118	596	2,605	22,079
sum	1	9	22	44	81	141	234	372
super	1	8	104	2,288	84,656	5,079,360	472,380,480	6,518,850,640

hypersum	1	18	1,813	181,322	18,132,237	1,813,223,760	181,322,376,093

343 sixth strongest

broken-up	1	10	151	3,785	155,336	10,100,625	1,000,117,211	145,027,096,220
broken-down	1	2	31	777	31,888	2,073,497	205,308,091	29,771,746,692
broken	1	2	10	31	151	777	3,785	31,888
sum	1	10	25	50	91	156	255	400
super	1	9	135	3,375	138,375	8,994,375	890,443,125	129,114,253,125

hypersum	1	19	1,915	191,525	19,152,541	1,915,254,165	191,525,416,599

344 dustrongest

broken-up	1	7	22	359	8,997	486,197	2,926,179	9,264,734	151,161,923
broken-down	1	2	7	114	2,857	154,392	929,209	2,942,019	48,001,513
broken	1	2	7	22	114	2,857	8,997	154,392	486,197
sum	1	7	10	26	51	105	188	324	517
super	1	6	18	288	7,200	388,800	32,270,400	4,388,774,400	847,033,459,200

hypersum	1	16	163	16,316	1,631,625	163,162,554	16,316,255,483	16,316,255,483,136

345 superfactorial

dusub	1	39	4677	3,367,575
broken-up	12	3,457	119,473,932	2,972,893,744,745,860
dusub	1	468	16,169,038	402,337,416,977,534
broken-down	1	13	449,281	11,179,548,979,213
broken	1	12	13	3,457

hypersum	12	12,288	1,228,834,560	12,283,456,024,883,200
dusub	1	1,663	166,304,673	1,662,385,000,252,050

intersuperfactorial	150	17,424	12,458,880	62,718,105,600
holiness	1	0	7	7
pentasub	1	117	83,947	422,591,271
broken-up	150	2,613,601	32,562,541,227,030	> twenty-fourplex
holiness	1	3	3	
pentasub	1	17,610	219,404,676,943	> twenty-threeplex
broken-down	1	151	1,881,290,881	> twentyplex
broken	1	150	151	2,613,601

duintersuperfactorial	8,787	6,238,152	31,365,282,240
holiness	4	3	4
enneasub	1	770	3,870,783
broken-up	24	54,814,641,625	> twenty-oneplex
enneasub	1	6,764,664	> seventeenplex
broken-down	1	8,788	275,636,100,325,121
broken	1	24	8,788

346 superodd

sub	1	6	39	348	3,824	49,713
broken-up	3	46	4,833	4,567,231	47,476,371,078	6,415,719,410,192,760
sub	1	17	1,778	1,680,190	17,465,861	2,360,211,271,334,400
broken-down	1	4	421	397,849	4,135,640,776	58,869,816,662,609
broken	1	3	4	421	4,833	397,847

hypersum	3	315	315,105	315,105,945	31,510,594,510,395

sub	1	116	115,921	115,920,999	11,592,099,899,464

intersuperodd	9	60	525	5,670	72,765	1,081,080	18,243,225	344,594,250
holiness	1	2	0	2	1	7	2	2

dusub	1	8	71	767	9,848
broken-up	9	541	284,034	1,610,473,321	117,186,091,486,599
holiness	1	0	3	2	10
dusub	1	73	38,440	217,953,863	15,859,412,882,730
broken-down	1	10	5,251	29,773,180	2,166,445,447,951
broken	1	9	10	541	5,251

347 The superevens are neighbors to the intersuperodds, the intersuperodds neighbors to the superevens.

dusub	1	4	26	208	2,079	24,945
broken-up	8	257	49,352	75,804,929	1,164,363,758,792	214,61528,096,346,369
holiness	1	0	1	5	5	9
sub	1	34	6,679	10,259,081	157,579,499,087	29,045,053,281,894,300
broken-down	1	9	1,729	2,655,753	40,762,367,809	7,518,849,237,210,633
broken	1	8	9	257	1,729	49,352

The superevens, unlike the superodds, can withstand a couple repeated averagings before getting an odd gnomon.

inter	20	112	864	8,448	99,840	1,382,400	21,934,080	392,232,960
duinter		66	488	4,656	54,144	741,120	11,658,240	20,708,320
tresinter			277	2,572	29,400	397,632	6,199,680	16,183,280
gnomon				2,295	26,828	368,232	5,802,048	9,983,600

348 superevil

broken-up	15	451	24,369	2,193,661	263,263,689	47,389,657,681
tressub	1	22	1,213	109,216	13107127	2,359,392,127
broken-down	1	16	865	77,866	9,344,785	1,682,139,166
broken	1	15	16	451	865	24,369

hypersum	15	1,530	153,054	15,305,490	15,305,490,120	15,305,490,120,180
tressub	1	76	7,620	762,016	762,015,483	762,015,483,017

349 superodious

sub	1	3	10	21	32	53	67
broken-up	2	17	478	26,785	2,357,558	337,157,579	61,365,036,936
trimmed	1	6	367	1,674	1,246,447	226,046,468	5,025,425,825
sub	1	6	176	9,854	867,297	124,033,342	22,574,935,495
broken-down	1	3	85	4,763	419,229	59,954,510	10,912,140,049
broken	1	2	3	17	85	478	4,763

hypersum	2	28	2,828	282,856	28,285,688	28,285,688,143	28,285,688,143,182
trimmed	1	17	1,717	171,745	17,174,577	17,174,577,032	17,174,577,032,071
sub	1	10	1,040	104,057	10,405,723	10,405,723,147	10,405,723,147,264

350 superbinarish

broken-up	1	11	1,211	13,321,011	1,479,964,222,211	1,808,664,398,186,419,847,168
broken-down	1	2	221	2,431,002	2,700,843,222,221	330,070,050,187,630,804,992
broken	1	2	11	221	1,211	2,431,002
sum	1	11	121	11,121	1,122,121	123,332,121
super	1	10	1,100	12,100,000	13,443,100,000,000	1,642,881,250,000,000,000,000

hypersum	1	110	110,110	11,011,011,000	110,110,110,001,111,000

351 superternarish

broken-up	1	3	61	13,423	35,436,781	9,355,310,197,423
broken-down	1	2	41	9,022	2,318,121	6,287,983,953,022
broken	1	2	3	41	61	9,022
sum	1	3	23	243	2883	266883
super	1	2	40	8,800	23,232,000	6,133,248,000,000
hypersum	1	12	1,220	1,220,220	12,202,202,640	12,202,202,640,264,000

352 superquadrarish

broken-up	1	3	19	1,143	754,399	5,974,841,223
broken-down	1	2	13	782	516,133	4,087,774,142
broken	1	2	3	13	19	782
sum	1	3	9	69	729	8,649
super	1	2	12	720	475,200	3,763,584,000

hypersum	1	12	126	12,660	12,660,660	126,606,607,920

353 superquintarish

broken-up	1	3	19	459	110,179	290,873,019	9,214,857,352,099
broken-down	1	2	13	314	75,373	198,985,034	6,303,845,952,493
broken	1	2	3	13	19	314	459
sum	1	3	9	33	273	2,913	34,593
super	1	2	12	288	69,120	182,476,800	5,780,865,024,000

hypersum	1	12	126	12,624	12,624,240	126,242,402,640	12,624,240,264,031,680

354 supersenarish

broken-up	1	3	19	459	55,099	66,119,259	872,774,273,899
broken-down	1	2	13	314	37,696	45,231,914	597,061,302,493
sum	1	3	9	33	153	1,353	14,553
super	1	2	12	288	34,560	41,472,000	547,430,400,000

hypersum	1	12	126	12624	12,624,120	126,241,201,200	12,624,120,120,013,200

355 supercomposites

sub	1	9	71	636	6,357	76,283
broken-up	4	97	18,628	32,189,281	556,230,794,308	115,340,017,539,896,160
sub	1	36	6,853	11,841,775	204,625,873,772	14,721,630,214,055
broken-down	1	5	961	1660613	28,695,393,601	5,950,276,818,763,973
broken	1	4	5	97	961	18,628

hypersum	4	424	424,192	4,241,921,728	424,192,172,817,280
sub	1	156	156,052	1,560,515,795	156,051,579,485,321

356 balanced binarish

broken-up	10	10,011	10,111,120	11,122,242,011	111,234,655,587,324,241
holiness	1	2	2	1	3
tressub	1	499	503,404	553,743,824	55,380,474,025,020
broken-down	1	11	1,111	12,222,111	1,222,345,554,332
broken	1	10	11	1,111	10,011

hypersum	10	101,001	1,010,011,010	10,100,110,101,100	10,100,110,101,100,100,011
holiness	1	3	5	7	10
dusub	1	13,559	136,690,126	1,366,901,261,253	> eighteenplex

superb binarish

dusub	1	1,355	1,368,253	1,505,078,572
broken-up	10	100,101	10,120,331,120,110	11,254,909,411,179,453,220,110
holiness	1	3	4	6
dusub	1	13,547	136,963,518,284	> twenty-oneplex
broken-down	1	11	1,112,111,101	1,236,790,887,442,110,011
broken	1	10	11	100,101

357 balanced ternarish

broken-up	102	12,241	2,460,543	516,726,271	51,735,670,165,605	> eighteenplex
holiness	1	0	2	2	5	
pentasub	1	82	16,579	3,481,674	348,592,203,538	34,156,236,615,438,600
broken-down	1	103	20,704	4,347,943	435,324,769,750	43,624,765,830,534,944
broken	1	102	103	20,704	2,460,543	4,347,943

superb ternarish

pentasub	1	82	16,577	3,481,165
broken-up	102	1,248,481	3,071,562,895,542	1,586,924,198,606,933,983,200
holiness	1	4	5	15
pentasub	1	8,412	20,696,027,995	163,058,930,972,882
broken-down	1	102	103	1,248,681

hypersum	102	10,212,240	102,122,402,460,240	102,122,402,460,240,516,650,400
holiness	1	2	5	10
pentasub	1	68,810	688,095,335,196	> twentyplex

357 superbs

21st sub	1	794,245,762	> seventeenplex
broken-up	1,023,456,789	10,234,567,891,047,463,808,161,301,622	> fifty-oneplex
holiness	5	15	
21st sub	1	> twenty-oneplex	> forty-oneplex
broken-down	1	1,023,456,790	> twenty-threeplex
broken	1	1,023,456,789	1,023,456,790

hypersum	1,023,456,789	10,234,567,891,023,456,798
holiness	5	10
21st sub	1	> eighteenplex

359 superperfect

pentasub	1	561	4,563,544	153,108,435,060,865
broken-up	168	13,999,105	9,481,453,601,453,464,490	> thirty-twoplex
holiness	3	4	8	
pentasub	1	94,325	63,885,531,840,956	> thirtyplex
broken-down	1	169	114,462,007,297	> twenty-sevenplex
broken	1	168	169	13,999,105

hypersum	168	1,681,399,105	168,139,991,059,481,453,601,453,464,490
holiness	3	6	15
pentasub	1	11,329,178	1,132,918,348,185,489,861,557,485,568

360 dusuperfactorials

dusub	1	468	1,617,170	39,522,434,102,816
broken-up	12	41,473	495,575,807,292	144,724,728,971,116,760,215,436,736
dusub	1	5,613	670,668,892,245	19,586,362,186,648,097,680,064,512
broken-down	1	13	155,341,681	45,364,972,118,980,181,901,312
broken	1	12	13	42,473

hypersum	12	123,456	12,345,611,949,360	12,345,611,949,360,292,033,482,752,000
dusub	1	16,708	1,670,796,889,896	167,079,688,989,599,864,470,216,704

361 dusuperperfects

pentasub	1	94,325	> twenty-plex
broken-up	169	2,365,848,577	> thirty-twoplex
holiness	2	5	
pentasub	1	15,940,962	> thirtyplex
broken-down	1	170	> twenty-fiveplex
broken	1	169	170

hypersum	169	16,913,999,104	16,913,999,104,132,731,845,556,620,643,794,944
holiness	2	6	14
pentasub	1	113,965,630	> thirty-twoplex

362 1,858,126,910,059,070,281,032,375,730,176 to be exact, with a sub depth of eighty-five!

363

broken-up	1	8	121	2,791	72,687	2,110,714	65,504,821	2,556,798,733

broken-down	1	2	31	715	18,621	540,724	16,781,065	655,002,259
broken	1	2	8	31	121	715	2,791	18,621

hypersum	1	17	1,715	171,523	17,152,326	1,715,232,629	151,523,262,931

364 Hebrew 2:10

365 *Amusements in Mathematics* by Henry Ernest Dudeney

curious	48	1,680	57,120	1,940,448	65,918,160	2,239,277,040
holiness	1	4	1	4	6	3

supercurious	48	80,640	4,606,156,800	893,800,770,246,400
holiness	2	5	8	11

Twenty-four curiously is not only the downtown neighbor of five squared and half the downtown neighbor of seven squared, it's also three times the downtown neighbor of three squared. Three sixty is also both a downtown neighbor of nineteen squared and three times the downtown neighbor of eleven squared, though not itself the downtown neighbor of a square.

Dudeney also named the torn numbers, squares which can be torn into two frags that sum to their square root, like the eight and one of eighty-one sum to nine.

torns	81	100	2,025	3,025	9,801	10,000	88,209	494,209	998,001	1,000,000
holiness	2	2	1	1	4	4	6	3	6	6

tetrasub	1	2	37	55	180
broken-up	81	8,101	16,404,606	49,623,941,251	486,364,264,605,657
holiness	2	3	5	3	8
tetrasub	1	148	300,461	908,894,188	8,908,072,238,902
broken-down	1	82	166,051	502,304,357	4,923,085,169,008
broken	1	81	82	8,101	166,051

hypersum	81	81,100	811,002,025	8,110,020,253,025	81,100,202,530,259,801
holiness	2	4	5	6	10

Half-torn numbers, of course, would be the halves of those torn numbers that can be torn exactly in half.

half-torns	1	8	20	25	30	98	209	494	998	1,000	1,729	1,984

broken-up	1	9	181	4,534	136,201	13,352,232	2,790,752,689
broken-down	1	2	41	1,027	30,851	3,024,425	632,135,676
broken	1	2	9	41	181	1,027	4,534
sum	1	9	29	54	84	182	391
super	1	8	160	4,000	129,000	11,760,000	2,457,840,000

hypersum	1	18	1,820	182,025	18,202,530	1,820,253,098	1,820,253,098,209

The automorphic numbers confusingly have also been called 'curious', those whose squares end in the same digit(s) as their squares. Higher than the trivial zero and one they're all gobi or rokubi.

automorphic	1	5	6	25	76	376	625	9,376	90,625	109,376	890,625	2,890,625

broken-up	1	6	37	931	70,793	26,619,099	16,637,007,668
broken-down	1	2	13	327	24,865	9,349,567	5,843,504,240
broken	1	2	6	13	37	327	931
sum	1	6	12	37	113	489	1,114
super	1	5	30	750	57,000	21,432,000	13,395,000,000

hypersum	1	15	156	15,625	1,562,576	1,562,576,625	15,625,766,259,376

hypersuper	1	5	666,666	> millionplex

Those not automorphic would be heteromorphic, which are much more prime-prone.

heteromorphic	2	3	4	7	8	9	10	11	12	13	14	15	16	17	18	19	20	21	22	23	24	26
trimmed	1	2	3	6	7	8	0	0	1	2	3	4	5	6	7	8	1	10	11	12	13	15

broken-up	2	7	30	217	1,766	16,111	162,876	1,807,747	21,855,840	264,077,827
trimmed	1	6	2	106	1,655	5,000	51,765	76,636	1,074,473	15,366,716
sub	1	3	11	80	650	5,927	59,919	665,033	8,040,314	97148803
broken-down	1	3	13	94	765	6,979	70,555	783,084	9,467,563	76,523,588

hypersum	2	23	234	2,347	23,478	234,789	23,478,910	2,347,891,011
trimmed	1	12	123	1,236	12,367	123,678	1,236,780	123,678,000
sub	1	8	86	863	8,637	86,374	8,637,408	863,740,833

The suffix -morphic can also be applied to tails that have the same form as any named sequence, so there are, for example, supersequences like the carmorphic, not the same as the automorphic, numbers with cube and/or rokubi tails, the cubes already including love, ichibi and hachibi.

cars	0	1	6	8	10	11	16	18	20	21	26	27	28	30	31	36	38	40	41	46	48	50	51	56	58	60
holiness	1	0	1	2	1	0	1	2	1	0	1	0	2	1	0	1	2	1	0	1	2	1	0	1	2	2

carmorphic	...	100	101	106	108	110	111	116	118	120	121	125	126	127	128	130	131
holiess	...	2	1	1	1	1	0	1	2	1	0	0	1	0	2	1	0

The primemorphic numbers would be those with prime tails, whether or not they are primes themselves.

primemorphic	13	15	17	22	23	25	27	32	33	35	37	42	43	45	47	52	53	62	63
primes	13	17	23	37	43	47	53	67	73	83	97	103	107	111	113	117	119	123	131

366 supercheap

dusubinter	1	4	61	909	14,494	245,552	4,406,300	8,348,780
sub	1	2	22	309	4,635	74,164	1,260,796	22,694,336

broken-up	2	13	782	656,893	8,276,852,582	1,668,613,581,188,093
trimmed	1	2	671	545,782	7,165,741,471	55,750,247,007,782
sub	1	5	288	241,657	3,044,883,903	613,848,594,990,611
broken-down	1	3	181	152,043	1,915,741,981	386,213,583,521,643
broken	1	2	3	13	181	782

hypersum	2	26	2,660	2,660,840	266,084,012,600	266,084,012,600,201,600
trimmed	1	15	155	15,573	15,573,015	15,573,015,105
sub	1	10	979	978,868	97,886,837,860	≈ seventeenplex

357 superdear

pentasubinter	1	3	58	1,039	19,744
tetrasub	1	9	157	2,825	53,670
broken-up	56	28,225	24,1831,856	37,296,276,187,969	> twentyplex
holiness	1	2	5	8	
tetrasub	1	517	4,429,305	683,105,126,553	> eighteenplex
broken-down	1	57	488,377	75,319,454,505	> seventeenplex
broken	1	56	57	28,225	488,377

hypersum	56	56,504	565,048,568	565,048,568,154,224	5,650,485,681,542,242,930,256
holiness	1	2	6	7	10
tetrasub	1	1,035	10,349,226	10,349,225,528,909	> twentyplex

358 superuglies

subinter	1	6	26	155	1,192	10,595	104,890
sub	1	2	9	44	265	2,119	19,071
broken-up	2	13	314	376,933	27,139,274	156,322,255,933	8,103,745,774,705,994
trimmed	1	2	203	265,822	16,028,163	45,211,144,822	70,263,466,364,883
sub	1	5	116	13,866	9,983,981	57,507,755,155	2,981,201,466,994,280
broken-down	1	3	73	8,763	6,309,433	36,342,342,843	1,883,987,059,290,553
broken	1	2	3	13	73	314	8,763

hypersum	2	26	2,624	2,624,120	2,624,120,720	26,241,207,205,760
trimmed	1	15	1,513	151,301	15,130,161	15,130,161,465
sub	1	10	965	965,360	965,360,064	9,653,600,642,519

369 superemirps

dodekasubinter	1	2	152	11,122	877,723
pentasub	1	46	1,708	121,267	8,852,471

broken-up	221	1,514,072	383,797,569,285	> eighteenplex
pentasub	1	10,202	2,586,007,680	> sixteenplex
broken-down	1	222	56,274,115	1,012,797,717,819,577
broken	1	221	222	1,514,072

hypersum	221	2,216,851	2,216,851,253,487	221,685,125,348,717,997,577
pentasub	1	14,937	14,937,026,251	> eighteenplex

370 supersquares

sub	1	13	212	5,297	190,709	9,344,726
broken-up	4	145	83,524	1,202,745,745	623,503,394,291,524	> twenty-twoplex
sub	1	3	30,727	442,465,433	229,374,080,260,464	> twenty-oneplex
broken-down	1	5	2,881	41,486,405	21,506,552,354,881	> twentyplex
broken	1	4	5	145	2,881	83,524

hypersum	4	436	436,576	43,657,614,400	43,657,614,400,518,400
sub	1	160	160,607	16,060,738,788	> sixteenplex

The complimentary sequence to squares would be described as 'fair', the two being mated by Sir Francis Bacon in 1604 and the products of the fairs 'superfair'.

fair	5	8	18	35	50	63	74	83	90	158	215	270	323	374	423	470	515	558	599

broken-up	5	41	743	26,046	1,303,043	82,117,755	6,078,016,913	504,557,521,534
dusub	1	6	100	3,525	176,348	11,113,430	822,570,140	68,284,435,086
broken-down	1	6	109	3,821	191,159	12,046,838	891,657,171	74,019,592,031
broken	1	5	6	41	109	743	3,821	26,046

hypersum	5	58	5,818	581,835	58,183,550	5,818,355,063	581,835,506,374
dusub	1	8	787	78,743	7,874,287	787,428,730	78,742,873,052

superfair	5	40	720	25,200	1,260,000	79,380,000	635,040,000	11,430,720,000

tetrasub	1	107	10,657	1,065,669	106,566,890
broken-up	5	337,445	196,337,311,633	11,423,601,788,264,574,595	> 27plex
dusub	1	6,180	3,596,043,300	> eighteeenplex	> 27plex
broken-down	1	59	34,328,266	1,997,340,381,224,359	> 25plex
broken	1	5	59	337,445	34,328,266

hypersum	5	540	540,720	54,072,025,200	540,720,252,001,260,000
dusub	1	73	73,179	7,317,852,846	> sixteenplex

371 supercubes

dusub	1	29	1,871	233,859	50,513,624
broken-up	8	1,729	23,901,704	41,302,144,513,729	> twenty-twoplex
holiness	2	1	3	2	
dusub	1	234	3,234,744	5,589,637,426,045	> twenty-oneplex
broken-down	1	9	124,417	214,992,576,009	> nineteenplex
broken	1	8	9	1,729	124,417

hypersum	8	8,216	821,613,824	8,216,138,241,728,000
holiness	2	3	5	10
dusub	1	1,112	111,193,339	1,111,933,396,055,420

372 supercubans

pentasub	1	33	2,023	184,057
broken-up	133	654,494	196,466,663,547	5,366,775,917,249,037,131
pentasub	1	4,410	1,323,781,966	> sixteenplex
broken-down	1	134	40,224,255	1,098,784,695,204,239
broken	1	133	134	654,494

hypersum	133	1,334,921	1,334,921,300,181	133,492,130,018,127,316,471
pentasub	1	8,995	8,994,628,969	> seventeenplex

373 superoblongs

dusub	1	19	390	11,693	491,105
broken-up	12	1,729	4,979,532	430,231,566,529	1,561,224,308,625,414,732
dusub	1	234	673,906	58,225,510,914	> seventeenplex
broken-down	1	13	37,441	3,234,902,413	11,738,813,876,331,841
broken	1	12	13	1,729	37,441

hypersum	12	12,144	121,442,880	12,144,288,086,400	> thirty-fourplex
dusub	1	1,644	16,435,507	1,643,550,667,880	> thirty-threeplex

374 streaky Catherine Katz, aka Mrs. Robert "Bobby" Katz, is a zoology major.

broken-up	1	9	1,945	19,325,629	24,002,307,019,945	3,756,169,030,165,252,265,529
broken-down	1	2	433	4,302,290	5,343,444,180,433	83,620,626,684,325,338,290
broken	1	2	9	433	1,945	4,302,290
sum	1	9	225	10,161	1,252,161	157,744,161
super	1	8	1,728	17,169,408	21,324,404,736,000	> twenty-oneplex

hypersum	1	18	18,216	182,169,936	1,821,699,361,242,000	> twenty-twoplex

375 CATS

broken-up	1	9	244	11,233	1,404,369	176,961,727	59,106,621,187
broken-down	1	2	55	2,532	316,555	3,988,462	13,323,062,863
broken	1	2	9	55	244	2,532	11,233
sum	1	9	36	82	207	333	667
super	1	8	216	9,936	1,242,000	156,492,000	52,268,328,000

hypersum	1	18	1,827	182,746	182,746,125	182,746,125,126	182,746,125,126,334

376 The subfactorial was so named by William Allen Whitworth in 1867, because of its relation to the factorial, but was previously called derangement and written about by Pierre Raymond de Montmort number in 1713 as the number of permutations in which no object appears in its natural place.

broken-up	1	3	28	1,235	327,303	604,529,876	8,931,324,715,327
broken-down	1	2	19	838	222,069	410,199,221	6,060,283,513,143
broken	1	2	3	19	28	838	1,235

sum	1	3	12	56	321	2,168	16,942
super	1	2	18	792	209,880	387,648,360	5,727,116,870,640

hypersum	1	12	129	12,944	12,944,265	1,294,426,521,847	129,442,652,184,714,774

377 By mere observation, aka the Vedic sub-sutra Vilokanam subsuperfactorials

broken-up	4	421	5,352,177	4,899,391,9245,430	2,260,395,670,580,463,520,479,407
sub	1	155	1,968,956	18,023,855,632,808	> twenty-threeplex
broken-down	1	5	63,566	581,884,244,627	26,845,956,551,159,972,034,213
broken	1	4	5	421	63,566

hypersum	4	4,105	410,512,713	4,105,127,139,154,017	> twenty-sixplex
sub	1	1,510	151,019,187	1,510,191,877,889,700	> twenty-sixplex

378 supersubfactorial

tressub	1	35	9,253	15,304,835	227,001,309,400
broken-up	16	11,265	2,093,667,856	643,605,692,456,004,609	> thirtyplex
holiness	1	1	7	11	
tressub	1	561	104,237,585	32,043,240,612,253,500	> twenty-nineplex
broken-down	1	17	3,159,553	971,264,993,436,689	> twenty-sevenplex
broken	1	16	17	11,265	3,159,553

hypersum	16	16,704	16,704,185,856	16,704,185,856,307,405,824
holiness	1	2	7	11
sub	1	832	831,652,443	> seventeenplex

379

subgogoz	0	1	10	94	1,150	17,164
holiness	1	0	1	1	1	1
subgogooz	0	6	268	24,109	3,592,573	800,793,469
holiness	1	1	3	2	1	7
subgoogooz	0	1	94	17,164	6,171,993	3,678,794,412
holiness	1	0	1	1	3	4

The acronum "sub" describes sanbi-unindexed binarish.

sub	1	10	100	101	110	111	1,000	1,001	1,010	1,011	1,100	1,110	1,111	10,000

380 subcubans

broken-up	3	22	311	6,864	226,823	10,667,545	661,614,613	52,939,836,585
sub	1	8	114	2,525	83,444	3,924,370	243,394,414	19,475,477,499
broken-down	1	4	57	1,258	41,571	1,955,095	121,257,461	9,702,551,975
broken	1	3	4	22	57	311	1,258	6,864

hypersum	3	37	3,714	371,422	37,142,233	3,714,223,347	371,422,334,762
sub	1	14	1,366	13,669	1,366,864	1,366,385,409	136,638,640,951

381 dusubcubans

broken-up	1	4	21	172	2,085	35,617	821,276	23,852,621	1,074,189,221
broken-down	1	2	11	90	1,091	18,637	429,742	12,481,155	562,081,717
broken	1	2	4	11	21	90	172	1,091	2,085
sum	1	4	9	17	29	46	69	98	143
super	1	3	15	120	1,440	24,480	563,040	16,328,160	734,767,200

hypersum	1	13	135	1,358	135,812	13,581,217	1,358,121,723	135,812,172,329

382 three-pseudonumerological

broken-up	3	43	293,005	322,945,454,856,569,999,833	> nineteenplex
sub	1	16	107,791	118,805,004,793,369,964	> nineteenplex
broken-down	1	4	27,257	300,422,351,241,543,610	> eighteenplex
broken	1	3	4	43	27,257

broken	3	314	3,146,814	314,681,411,021,842,141,158
sub	1	116	1,157,648	> twentyplex

383 four-pseudonumerological

broken-up	4	337	62,012	1,373,019,102,545	> thirty-nineplex
sub	1	124	22,813	505,105,500,162	> thirty-eightplex
broken-down	1	5	921	20,392,030,469	> thirty-sevenplex
broken	1	4	5	337	921

hypersum	4	484	484,184	48,418,422,141,184	> fortyplex
sub	1	178	178,121	17,812,142,079,702	> fortyplex

The five- and seven-pseudonumerological sequences dip at the third term and then converge.

five	5	88	22	222	121,411,211	2,214,112,113,552,842,141,582,102,112,141,582

dusub	1	12	3	30	16,431,220	> thirty-twoplex

broken-up	5	441	9,707	2,155,395	261,689,117,143,052	> forty-sevenplex
dusub	1	60	1,314	291,701	35,415,770,788,494	> forty-sixplex
broken-down	1	6	133	29,532	3,585,515,883,385	> forty-fiveplex
broken	1	5	6	133	441	9,707

hypersum	5	588	58,822	58,822,222	58,822,222,121,411,211	> fifty-oneplex
dusub	1	79	7,961	7,960,722	7,960,722,091,408,130	> fiftyplex

six	6	70	821	121,411,212	2,214,123,552,842,141,582,102,112,141,158
holiness	1	1	2	0	7

dusub	1	10	111	16,431,221	> twenty-nineplex
broken-up	6	421	345,647	41,965,421,194,585	> forty-threeplex
holiness	1	0	1	5	
dusub	1	57	46,778	5,679,402,163,513	> forty-threeplex
broken-down	1	7	5,748	697,871,646,583	> forty-twoplex

hypersum	6	670	670,821	670,821,121,411,212	> forty-fiveplex
holiness	1	2	4	4	

seven	7	82	22	222	121,411,211	2,214,112,113,552,842,141,582,102,112,141,582
sub	1	11	3	30	16,431,220	> thirty-twoplex

broken-up	7	575	12,657	2,810,429	341,217,588,332,176	> forty-sevenplex
dusub	1	78	1,713	380,350	46,178,778,962,249	> forty-sixplex
broken-down	1	8	177	39,302	4,771,703,414,899	> fifty-nineplex
broken	1	7	8	177	575	12,657

hypersum	7	782	78,222	78,222,222	78,222,222,121,411,211
dusub	1	106	10,586	10,586,226	> sixteenplex

Eight quickly alternates one and two.

eight	8	1	2	1	2	1	2	1	2	1	2	1	2	1
holiness	2	0	0	0	0	0	0	0	0	0	0	0	0	0

dusub	1	0	0	0	0	0	0	0	0	0	0	0	0	0
broken-up	8	9	26	35	96	131	358	489	1,336	3,161	7,658	10,819	29,296	40,115
holiness	2	1	1	0	2	0	2	3	1	1	3	4	3	1
broken-down	1	9	19	28	75	103	281	384	1,049	2,482	6,013	8,495	23,003	31,498
broken	1	8	9	19	26	28	35	75	96	103	131	281	358	384

hypersum	8	81	812	8,121	81,212	812,121	8,121,212	8,121,212	8,1212,121
holiness	2	2	2	2	2	2	2	2	2
dusub	1	11	110	1,099	10,991	109,909	1,099,087	10,990,866	1,099,086,654

384

nine	9	22	222	121,411,211	2,214,112,113,552,842,141,582,102,112,141,582
holiness	1	0	0	0	7
dusub	1	3	30	16,431,220	> thirty-twoplex

broken-up	9	199	44,187	536,479,780,656	> forty-sixplex
holiness	1	2	2	7	
dusub	1	27	5,980	726,046,345,951	> forty-fiveplex
broken-down	1	10	2,221	269,654,299,641	> forty-fourplex
broken	1	9	10	199	2,221

hypersum	9	922	922,222	922,222,121,411,211	> forty-nineplex
holiness	1	1	1	1	
dusub	1	125	124,809	124,809,192,008,257	> forty-eightplex

Ten converges with twelve.

ten	10	12	158	21,418,811	121,355,284,214,112,102,112,141,582
holiness	1	0	2	4	5

broken-up	10	121	19,128	409,699,016,929	> thirty-sevenplex
holiness	1	0	3	9	
dusub	1	17	2,589	55,446,732,498	> thirty-sixplex
broken-down	1	11	1,739	37,247,312,340	> thirty-sixplex
broken	1	10	11	121	1,739

hypersum	10	1,012	10,158	101,215,821,418,811	>forty-twoplex

holiness	1	1	3	7	
dusub	1	137	1,375	13,698,071,859,741	> forty-oneple

twelve	12	158	21,418,811	121,355,284,214,112,102,112,141,582
dusub	1	21	2,898,721	> twenty-fiveplex
broken-up	12	1,897	40,631,484,479	> thirty-sixplex
broken-down	1	257	5,798,873,460	> thirty-fiveplex

hypersum	12	12,158	1,215,821,418,811	> thirty-nineplex
dusub	1	1,646	164,543,536,080	> thirty-eight

385 thirteen

broken-up	13	18,357	385,897,956,185,237,419	> seventy-threeplex
broken-down	1	14	294,305,789,976,213	> seventyplex
broken	1	13	14	18,357

hypersum	13	131,412	13,141,221,021,842,141,158	> eighty-twoplex
tressub	1	6,543	654,262,869,451,667,968	> eightyplex

thirty-four

broken-up	34	482,257	4,056,795,128,294,870,972,322	> ninety-oneplex
tetrasub	1	8,833	> nineteenplex	> eighty-ninetyplex
broken-down	1	35	294,423,573,924,941,441	> eighty-sevenplex
broken	1	34	35	482,257

hypersum	34	3,414,184	34,141,848,412,102,112,141,184	> ninety-threeplex
tetrasub	1	62,533	> twentyplex	> ninety-oneplex

ninety-two

broken-up	92	203,413	224,183,937,055,619,459	> ninety-nineplex
holiness	1	1	7	
pentasum	1	1,371	1,510,539,485,927,070	> ninety-sevenplex
broken-down	1	93	102,496,429,167,127	> ninety-sixplex
broken	1	92	93	203,413

hypersum	92	922,211	9,222,111,102,112,141,582	> eighty-twoplex
holiness	1	1	4	
pentasub	1	6,214	62,138,095,825,709,496	> seventy-nineplex

386 look-and-say-10

broken-up	10	1,011,101	1,022,327,365,624,120	> thirtyplex
holiness	1	2	4	
dusub	1	136,838	1,383,556,963,587,281	> twenty-nineplex
broken-down	1	11	11,122,134,211	> twenty-fiveplex
broken	1	10	11	1,011,101

hypersum	10	101,110	1,011,103,110	1,011,103,110,132,110	> forty-twoplex
holiness	1	2	3	4	
dusub	1	13,684	136,837,926	136,837,925,791,150	> forty-oneplex

387 look-and-say-100

broken-up	100	112,001	23,655,731,310	2,888,651,286,291,886,301	> thirtyplex
holiness	2	2	2	17	
pentasub	1	755	159,391,064	194,635,792,741,020,992	> 28plex
broken-down	1	101	21,332,211	26,049,213,175,074,311	> 28plex
broken	1	100	101	112,001	21,332,211

hypersum	100	1,001,120	1,001,120,211,210	10,011,202,112,101,221,121,110
holiness	2	3	4	5
pentasub	1	6,745	674,594,923	67,454,949,228,470,198,272

388 spell-and-count from 3

sub	1	2	3	5	8	11	13	17
broken-up	3	16	147	1,927	40,614	1,220,347	43,973,106	1,980,010,117
sub	1	6	54	709	14,941	448,941	16,176,802	728,405,015
broken-down	1	4	37	485	10,222	307,145	11,067,442	498,342,035
broken	1	3	4	16	37	147	485	1,927

hypersum	3	35	359	35,913	3,591,321	359,132,130	35,913,213,036
sub	1	13	132	13,212	1,321,173	132,117,327	13,211,732,742

389

from 4	4	8	13	21	30	36	45	54
sub	1	3	5	8	11	13	17	20
broken-up	4	33	433	9,126	274,213	9,880,794	444,909,943	24,035,017,716
sub	1	12	159	3,357	100,877	3,634,941	163,673,221	8,841,988,886

broken-down	1	5	66	1,391	41,796	1,506,047	67,813,911	3,663,457,241
broken	1	4	5	33	66	433	1,391	9,126

hypersum	4	48	4,813	481,321	48,132,130	4,813,213,036	481,321,303,645
sub	1	18	1,771	177,068	17,706,821	1.770.682.122	177.068.212.209

from 5	5	8	13	21	30	36	45
broken-up	5	41	538	11,339	340,708	12,276,827	552,797,923
dusub	1	6	73	1,534	46,110	1,661,488	74,813,063
broken-down	1	6	79	1,665	50,029	1,802,709	81,171,934
broken	1	5	6	41	79	538	1,665

hypersum	5	58	5,813	581,321	58,132,130	5,813,213,036	581,321,303,645
dusub	1	8	787	78,673	7,867,328	786,732,833	78,673,283,280

390 In his article "The Killion" in the *New Yorker* and reprinted in his *Dating Your Mother* (1986), he reported some accidents involving such mind-boggling numbers, for example, Marcie Chang's check error that reportedly killed her and eleven other people including two scientists at the Institute for Catastrophe Control, Princeton, N. J., August 6, 1982.

391 To be exact **Shirham's number** = g(2, 64, 2) - 1 = 18,446,744,073,709,551,616; **its mate**, 81,553,255,926,290,448,383; **their viva la difference**, 63,106,511,852,580,496,767. and **Mahrihs' number** would be its ananum, 61,615,590,737,044,764,481.

Adam Clarkson's number, called "the lynz" also started as a doubling. He was assigned writing the line "I must always tuck my shirt in whilst participating in a Chemistry Lesson." a hundred times on February 26, 1998. He didn't so the next day the number doubled and kept doubling until September 17th, when the Chemistry teacher added squaring the number of lines, already higher than sixty-threeplex. On the next day it broke the googol barrier.

392 Rubik's number = g(2, 6, 2)g(2, 5, 3)g(2, 2, 7)(11)13(17)19(23)31(37)41(43)47 = 278,914,005,382,139,703,576; **its mate**, 821,085,994,617,860,296,423; **their viva la difference**, 542,171,989,235,720,592,847; **its ananum, Kibur's number**, 675,307,931,283,500,419,872.

393 *mahakalpa* = g(2, 27, 2)g(2, 3, 3)g(2, 23, 5) = 424,068,126,192,000,000,000; **it's mate**, 575,931,873,807,999,999,999; **their viva la difference**, 151,863,747,615,999,999,999; **it's ananum, *aplakaham***, only 291,621,860,424.

394 The number is named for Count Anedeo Avogadro and is actually a measure of how exactly one can count atoms and/or molecules. It's approximately 6.022140857 twenty-threeplex, rather like guesstimating the number of gumballs in twentyplex gumball machines.

395 91,409,924,241,424,243,424,241,924,242,500. **It's mate** is almost ten times higher, 908,590,075,758,575,756,575,758,075,757,499, **their viva la difference**, 817,180,151,517,151,513,151,517,051,514,999; **its ananum only**

524,242,914,242,414,242,990,419.

396 FHUGE = g(2, 6, 5)g(2, 104, 2)+1 = 316,912,650,057,057,350,374,175,801,344,000,001; **its mate**, 683,087,349,942,942,649,625,824,198,655,999,998; **their viva la difference**, 366,123,699,885,885,299,251,648,397.
its ananum EGUHF = 100,000,443,108,571,473,053,750,750,056,219,613.

397 g(2, 256, 2) + 1 = 340,282,366,920,938,463,463,374,607,431,768,211,457.
Its mate is 659,717,633,079,061,536,536,625,392,568,231,788,542;
their viva la difference, 319,435,266,158,123,073,073,250,775,136,463,577,085;
its ananum, the Scamorg number,
754,112,867,134,706,473,364,364,839,029,663,282,043.

398 64!/(31!)(8!)2(2!)424 = 120,108,745,018,747,840,297,009,996,995,758,129,152; **its mate,** 879,891,254,981,252,159,702,990,003,004,241,870,847; **their viva la difference**, 759,782,509,962,504,319,405,980,096,998,483,741,795; **its ananum, the ssehc number**, 251,921,857,599,699,900,792,048,747,810,547,801,021.

399 g(2, 56, 7) = 211,587,613,802,425,391,637,729,361,787,678,676,290,060,193,601; **its mate**, 788,412,386,197,574,608,362,270,638,212,321,323,709,939,806,398; **their viva la difference**, 556,823,680,395,149,216,704,541,276,430,642,647; '**nevesagaf**', 106,391,060,092,676,876,787,1363,927,736,193,524,208,316,785,112.

400 52! = 374,144,419,156,711,147,060,143,317,175,368,453,031,918,731,001,856; **its mate** 625,855,580,843,288,852,939,856,682,824,631,546,968,081,268,998,143; **their viva la difference**, 251,711,161,686,577,705,879,713,365,649,263,093,936,762,537,996,287; **its ananum**, 341,899,862,180,869,645,136,428,286,658,939,258,882,348,085,558,526.

401 http://www.gamersnook.com/blog/ gives the number as the quantity 9.460536 times twentyplex cubed times a fifth of ten-to-the-twentieth, higher than sixty-twoplex.

402 80,658,175,170,943,878,571,660,636,856,403,766,975,289,505,440,883,277,824;
its mate, 19,341,824,829,056,121,428,339,363,143,596,233,024,710,494,559,116,722,175;
their viva la difference,
61,315,350,341,887,757,143,321,273,712,807,533,905,579,010,881,766,555,639; **its ananum**, 42,877,238,804,450,598,257,966,739,465,863,606,617,587,834,907,157,185,608

403 1,151,322,190,187,639,925,650,955,597,973,971,522,401g(2, 36, 10) The mates would be the aptly named **'ridiculous' numbers**, beginning with 87 and then catapulting to 8,848,677,809,812,360,074,349,044,402,026,028,477,599g(2, 36, 10) - 1. By backformation the **lime numbers** would be e times higher than this, beginning with 32 and catapulting to 3,129,618,188,335,882,615g(2, 57, 10).

404 6,086,555,670,238,378,989,670,371,734,243,169,622,657,830,773,351,885,970,528,-324,860,512,791,691,264; **its mate**, 3,913,444,329,761,621,010,329,628,265,756,830,377,-342,169,226,648,114,029,471,675,139,487,208,308,735; **their viva la difference**, 2,173,-111,340,476,757,979,340,743,468,486,339,245,315,661,546,703,771,941,056,649,721,025,-583,382,529; **its ananum,** 4,621,961,972,150,684,238,250,795,881,533,770,387,562,269,-613,424,371,730,769,898,738,320,765,556,806.

405 Eddington's number = 17g(2, 260, 2) = 31,495,448,272,550,005,155,211,307,922,363,-

110,936,089,435,829,054,233,418,732,462,850,152,371,262,062,592. **Its mate** is higher, 68,504,551,727,449,994,844,788,692,077,636,889,063,910,564,170,945,766,581,267,537,-149,847,628,737,937,407; and its ananum, "not **Gnidde's number"** ≠ 29,526,026,217,325,-105,826,423,781,433,245,092,853,498,063,901,136,322,970,311,255,150,005,527,284,459,-413.

406 g(2, 28, 8400000) = 758,263,253,073,010,241,157,973,569,975,696,406,218,966,848,-080,183,296 eightyplex; **it's mate**, 241,736,746,926,989,758,842,026,430,024,303,593,781,-033,151,919,816,704 eightyplex -1; **its ananum**, 692,381,080,848,669,812,604,696,579,-569,379,751,142,010,370,352,362,857.

407 191,082,339,774,398,580,771,809,339,633,875,494,367,179,645,716,206,190,592 eighty-sevenplex; **its mate**, 808,917,660,225,601,419,228,190,660,366,124,505,632,820,-354,283,793,809,503 eighty-sevenplex -1; **its ananum just** 295,091,602,617,546,971,763,-494,578,336,933,908,177,085,893,477,933,280,191.

408 M. Beeler estimates it as four hundred sixty-three hundred-fiveplex.

409 2,835 two-hundred-fifteenplex.

410 6,030,058,771,879,169,382,471,003,705,612,628,383,934,925,076,850,273,779,688,-931,328 156plex

411 g(4, 2, g(2, 3, 3) > 9,391plex. Higher than this may be the sixth Busy Beaver number, what the Society called the second Busier Beaver number, the Busy Beaver number of the second Busy Beaver number. Tibor Radő in his 1962 paper, "On Non-Computable Functions" defined Tibor Radő in his 1962 paper, "On Non-Computable Functions" defined the Busy Beavers and gave the first two. They are the number of steps that a certain state of Turning machine reaches the maximum number of ones. The third Busier Beaver number would be the twenty-first Busy Beaver number and the fourth would be the hundred seventh. The Recursive Busy Beaver number get higher very much faster. Shen Lin found the value of the third Busy Beaver in 1964 and Allen Brady the fourth in 1983. The exact number for states five and six are not yet known, but have been given minimum values of at least 47,176,870 and 18,267plex.

state	**1**	**2**	**3**	**4**	**5**	**6**
BB(z)	1	6	21	107	BB(5)	BB(6)
BB(BB(z))	1	BB(6)	BB(21)	BB(107)	BB(BB(5))	BB(BB(6))
zBB(z)	1	BB(BB(2))	BB(BB(BB(3))	4BB(4)	5BB(5)	6BB(6)

sum	1	7	28	135
super	1	6	126	13,482
broken-up	1	7	148	15,843
broken-down	1	2	43	4,603
broken	1	2	7	43
hypersum	1	17	17,148	1,714,815,843

hypersuper	1	7	148,148,148,148,148,148,148	> twentyduplex

412 mega = g(5, 2, 2) = g(2, 65'536, 2) > 19,720plex

413 g(2, (3!)!, 4) = g(3, 4, 362'880) > 37,849plex

414 Adams' number = g(2, 261'199, 2) > 78319plex and Adam's radio number = g(2, 260'709, 2) > 83,289plex. If we included Klondike-5 phone numbers they'd be between 167,076plex and 1,673,727plex. The maximum two-to-a-seven-digit-number would be g(2, 999'999, 2) over three-millionplex.

415 Archimedes' cattle problem, aka *bovinum problema*, solution was finally calculated by H. C. Williams, etal., in 1965. It is the total number of Helios' white, black, spotted, and brown bulls, and white, black, spotted and brown cows, W + X + Y + Z + w + x + y + z, such that W = (1/2 + 1/3)(X - Z), X = (1/4 + 1/5)(Y - Z), Y = (1/4 + 1/5)(Y - Z), w = (1/3 + 1/4)(X + x), x = (1/4 + 1/5)(Y + y), y = (1/5 + 1/6)(Z + z), z = (1/6 + 1/7)(W + w) and W + X is square and Y + Z is triangular, namely: [(251,941/184,119,152)(109,931,986,732,829,734,979,866,232,821,433,-543,901,088,049 + 5,054,944,823,431,503,307,447,781,973,554,040,898,634(√472,994)))] > 202,544plex.

416 [g(3, 4, e) + 1/2] > 1,656,513plex

417 g(3, 3, 9) = g(2, 387'420'489, 9) > 369,723,452plex

418 aka Carl Friedrich Gaussens meßbare unendlichkeit

419 g(2, 9, 8)! = 134,217,728! The ananum, **Yksniharb's number**, would be the same length and nearly the same height.

420 *Ramanujan: Twelve Lectures on Subjects Suggested by His Life and Work* by G. H. Hardy = g(2, 2.5g(2, 70, 10), 10) > to-the-third-ten

421 g(2, g(2, 23, 2), 2) > g(2, 2'525'222, 10)

422 by Wolfram Research, 1.4403971939817846g(2, 323'228'010, 10)

423 g(3, 4, 9) = g(2, g(2, 387'420'489, 9), 9)) > g(2, 352'776'871, 10)

424 9.265 times to-the-third-ten. This was below the four surreal "infinity minus one" responses. Most were much lower. Of the hundred numbers lower than 99 most were below twelve. Most common was three. Of all one hundred forty-six responses 69% were only one or two digits.

425 Ivan Matveevich Vinogradov's number, connected to his theorem of three-prime sums, = g(2, 41.96, g(3, 4, e)) > eighteenduplex. Its ananum would be Vodargoniv's number.

426 g(2, g(2, 41, 3), 2) = g(2, 36'472'996'377'170'786'403, 2) > nineteenplex

427 g(2, (4!)!, 5) = g(2, 620'448'401'733'239'439'360'000, 5) > twenty-threeduplex. The ananum, Grebztierk's number, is the same length, but smaller beginning with five rather than seven.

428 g(2, 42, g(3, 2, 10)) = forty-twoduplex. The ananum, "not Elttil's number", is one, so Elttil's numbers are everything else except one.

Elttil	0	2	3	4	5	6	7	8	9	10	11
holiness	1	0	0	0	0	1	0	2	1	1	0
broken-up	0	1	3	13	68	421	3,015	24,541	223,884	2,263,381	25,121,075
holiness	1	0	0	0	3	0	1	0	4	3	0
broken-down	1	2	7	30	157	972	6,961	56,660	516,901	5,225,670	57,999,271
broken	0	1	2	3	7	13	30	68	157	421	972
holiness	1	0	0	0	0	0	1	3	0	0	1

hypersum	0	2	23	234	2,345	23,456	234,567	2,345,678	23,456,789	2,345,678,910
holiness	1	0	0	0	0	1	1	3	4	4

429 This, of course, can be more simply written in just two fours if left superscription is allowed for tetration. g(3, 4, 4) = g(2, g(2, g(2, 4, 4), 4), 4) = g(2, g(2, 256, 4), 4) > ninety-twoplex.

430 The inflation factor of Dr. Thanu Padmanabhan's inflationary cosmology model is higher than tentetraplex. g(2, 13, g(2, 10, g(3, 2, e))) = g(2, 1.45g(2, 4342944819032, 10), 10) > g(3, 5.04205, 10)

- Knapowski's number is a little higher than that, g(2, 35, g(3, 5, e)) > g(3, 4, 10)(6.8879689056405g(2, 14, 10)) > g(3, 5.176, 10). It's ananum would be Ikswopank's number.
- Folkman's number is triple two-tetrated-to-the-two-tetrated-to-the-nine-hundred-first. g(3, 3, g(3, g(2, 901, 2), 2)) > g(2, 2, g(3, 5, 10)) > g(3, 5.301, 10). It's ananum is Namklof's number.
- Dingleus' number is googol-to-the-myriadplex-to-the-googolth. (2, g(2, g(2, 100, 10), g(2, 10000, 10)), g(2, 100, 10)) > to-the-fifth-ten. Its ananum, Suelgnid's is the same as Elttil's.
- Three superfactorial, 3$, is three-factorial-pentated-to-the-second or to-the-sixth-six. g(4, 2, 3!) = g(2, 6, 6) > g(3, 3, 10)(2.0691973765g(2, 36'305, 10))
- Ten exponential factorial is ten-to-the-ninth-to-the-eighth-to-to-the-seven-to-the-sixth-to-the-fifth-to-the-fourth-to-the-third-to-the-second-to-the-first. 10^! = 10g(2, 1, g(2, 2, g(2, 3, g(2, 4, g(2, 5, g(2, 6, g(2, 7, g(2, 8, 9)))))))) > g(2, 4.829261048g(2, 183'230, 10), g(3, 5, 10)) > tenpentaplex

431 Joyce preferred the eleven-word French version, "Moi, j'aime a faire connaitre un nombre utile aux sages." ("Me, I like to teach a number useful to wisemen.") "Poe, E.: Near A Raven" by M. Keith, a paraphrase of Edgar Allan Poe's "The Raven", encodes pi to 740 digits with ten representing zero and words longer than ten letters representing two digits.

broken-up	3	4	19	23	134	1,229	2,592	16,781	86,497	276,272	1,467,857
broken-down	1	4	17	21	122	1,119	2,360	15,279	78,755	409,054	2,124,025
broken	1	3	4	17	19	21	23	122	134	1,119	1,229

hypersum	3	31	314	3,141	31,415	314,159	3,141,592	31,415,926	314,159,265
sub	1	11	116	1,156	11,557	115,573	1,155,727	11,557,273	115,572,735

Just as integers can have mates, so can irrationals, like pi's mate, "wopi".

hypersum	6	68	685	6,858	68,584	685,840	6,858,407	68,584,073	685,840,734
holiness	1	3	3	5	5	6	6	6	6
dusub	1	9	93	928	9,282	92,818	928,184	9,281,845	92,818,450

432 John "Jack" Horner was first mentioned in *The History of Jack Horner, Containing the Witty Pranks he play'd from his Youth to his Riper Years, Being pleasant for Winter Evenings* in 1764. He has been identified as a descendant of Thomas Horner, steward for the last abbot of Glastonbury abbey, who was given the secret mission to deliver a pie with deeds to King Henry VII. The Horners however claim that Thomas did not steal, but bought, the title to Mells Manor, Somerset. In any case all the Church's property in England including the abbey was ~~stolen~~ owned by Henry by 1541.

broken-up	2	15	17	66	347	2,842	40,135	484,462	2,946,907	147,829,812
trimmed	1	4	6	55	236	1,731	3,024	373,351	183,586	36,718,701
broken-down	1	3	4	15	79	647	9,137	110,291	670,883	3,464,706
broken	1	2	3	4	15	17	66	79	347	647

hypersum	2	27	271	2,713	27,135	271,358	27,135,814	2,713,581,412
trimmed	1	16	160	1,602	16,024	160,247	16,024,703	1,602,470,301
sub	1	10	100	998	9,982	99,827	9,982,708	998,270,813

433 sliced pi

sub	1	5	6	34	240	2,169	3,430
broken-up	3	43	648	59,659	38,957,975	229,735,238,234	2,141,821,626,113,557
sub	1	16	238	21,947	14,331,838	84,514,871,059	787,932,142,903,566
broken-down	1	4	61	5,616	3,667,309	21,626,126,789	201,620,383,721,156
broken	1	3	4	43	61	648	5,616

hypersum	3	314	31,415	3,141,592	3,141,592,653	31,415,926,535,897
sub	1	116	11,557	1,155,727	1,155,727,350	11,557,273,497,909

434 diced pi

sub	1	5	58	976	216,973
broken-up	3	43	6,840	18,146,563	10,702,715,838,299
sub	1	16	2,516	6,675,747	3,937,309,121,610

broken-down	1	4	637	1,689,965	996,729,527,882
broken	1	3	4	43	637

hypersum	3	314	314,159	3,141,592,653	3,141,592,653,589,793
sub	1	116	115,573	1,155,727,350	1,155,727,349,790,920

435 pancake

broken-up	2	9	65	724	11,649	257,002	74,364,707	276,451,161
trimmed		1	8	54	613	538	1,461	6,325,366
sub		1	3	24	266	4,285	94,546	2,746,112	101,700,699
broken-down		1	3	22	245	3,942	86,969	2,526,043	935,505,560
broken		1	2	3	9	22	245	724	3,942

hypersum	2	24	247	24,711	2,471,116	247,111,622	24,711,162,229
sub	1	9	91	9,091	909,073	90,907,285	9,090,728,552

hypersuper	2	44	77,777,777,777,777,777,777,777,777,777,777,777,777,777,777,777

The uptown neighbors atop the pancakes we call 'syrup'.

syrup	3	5	8	12	17	23	30	38	47	57	68	80	93	107	122	138	155

sub	1	2	3	4	6	8	11	14
broken-up	3	16	131	1,588	27,127	625,509	18,792,397	714,736,595
sub	1	6	48	584	9,979	230,112	6,913,337	262,936,899
broken-down	1	4	33	400	6,833	157,559	4,733,603	180,034,473
broken	1	3	4	16	33	131	400	1,588

hypersum	3	35	358	35,812	3,581,217	358,121,723	35,812,172,330
sub	1	13	132	13,174	1,317,456	131,745,619	13,174,561,944

436 cakes

broken-up	2	9	74	1,119	29,168	1,226,175	78,504,368	7,302,132,399
trimmed	1	8	63	8	18,057	445,064	674,257	621,021,288
sub	1	3	27	412	10,730	451,085	28,880,143	2,686,304,386
broken-down	1	4	33	499	13,007	546,793	35,007,759	3,256,268,380
broken	1	2	4	9	33	74	499	1,119

hypersum	2	24	248	24,815	2,481,526	248,152,642	24,815,264,264
trimmed	1	13	137	13,704	1,370,415	137,041,531	13,704,153,153

sub	1	9	91	9,129	912,902	91,290,255	9,129,025,550

hypersuper	2	44	88,888,888,888,888,888,888,888,888,888,888,888,888,888,888

The uptown neighbors atop the cakes, we call "frosting".

frosting	3	5	9	17	27	43	65	94	131	177	233	300	379	471	577	698

sub	1	2	3	6	10	16	24	35
broken-up	3	16	147	2,515	68,052	2,928,751	190,436,867	17,903,994,249
sub	1	6	54	925	25,035	1,077,427	70,057,808	6,586,511,399
broken-down	1	4	37	633	17,128	737,137	47,931,033	4,506,254,239
broken	1	3	4	16	37	147	633	2,515

hypersum	3	35	359	35,917	3,591,727	359,172,743	35,917,274,365
sub	1	13	132	13,213	1,321,323	132,132,268	13,213,226,822

437 andigital

broken-up	23,456,789	50,221,161,301,623	12,906,471,203,891,676,671,406
holiness	4	4	11
17th sub	1	22,778,814	534,319,869,534,067
broken-down	1	23,456,790	550,223,084,758,411
broken	1	23,456,789	23,456,790

hypersum	23,456,789	2,345,678,923,456,798	234,567,892,345,679,823,456,879
holiness	4	8	12
17th sub	1	971,096,467	9,710,964,649,296

438 Marie Antoinette

broken-up	2	7	30	157	972	17,653	530,562	16,995,637	578,382,220	1,9681,991,117
trimmed	1	6	2	46	861	6,542	42,451	5,884,526	46,727,111	8,570,880,006
sub	1	3	11	58	358	6,494	195,183	6,252,345	212,774,928	7,240,599,893
broken-down	1	3	13	68	421	7,646	229,801	7,361,278	250,513,253	151,044,796
broken	1	2	3	7	13	30	68	157	421	972

hypersum	2	23	234	2,345	23,456	2,345,618	234,561,830	23,456,183,032
trimmed	1	12	123	1,234	12,345	1,234,507	12,345,072	1,234,507,221
sub	1	8	86	863	8,629	862,905	86,290,475	8,629,047,506

439 doughnuts of the form z cubed plus three z squared plus eight z all over six

sub	1	2	5	9	15	23	33	47
broken-up	2	13	171	4,117	164,851	10,224,879	930,628,840	119,130,716,399
trimmed	1	2	60	3,006	53,760	113,768	8,251,773	802,605,288
sub	1	5	63	1,515	60,645	3,761,523	342,359,218	43,825,741,375
broken-down	1	3	40	963	38,560	2,391,683	217,681,713	27,865,650,947
broken	1	2	3	13	40	171	963	4,117

hypersum	2	26	2,613	261,324	26,132,440	2,613,244,062	261,324,406,291
trimmed	1	15	1,502	150,213	1,502,133	150,213,351	15,021,335,180
sub	1	10	961	96,136	9,613,587	961,358,765	96,135,876,551

The penholodigital numbers are those without zeros.

zeroless	1	2	3	4	5	6	7	8	9	11	12	13	14	15	16	17	18	19	21	22	23	24	25	26	27

Doughnuts without 'holes', that is, without zero, or any of the other "holey" digits, six, eight or nine, we call "doughnut wholes" rather than "unholy doughnuts", while those that don't we call "doughnut holes", rather than "holy doughnuts".

doughnut wholes	2	13	24	174	574	2,323	4,524	7,174	14,233	15,224	17,342	23,477
trimmed	1	2	13	63	463	1,212	3,413	6,063	3,122	4,113	6,231	12,366
sub	1	5	9	64	211	855	1,664	2,639	5,236	5,601	6,380	8,637

broken-up	2	27	650	113,127	64,935,548	150,845,391,131	682,424,614,412,192
trimmed	1	16	54	2,016	53,824,437	4,734,280,020	571,313,503,301,081
sub	1	10	239	41,617	23,888,453	55,492,918,193	251,049,985,791,595
broken-down	1	3	73	12,705	7,292,743	16,941,054,694	76,641,338,728,399
broken	1	2	3	27	73	650	12,705

hypersum	2	213	21,324	21,324,174	21,324,174,574	213,241,745,742,323
trimmed	1	102	10,213	10,213,063	10,213,063,463	102,130,634,631,212
sub	1	78	7,845	7,844,725	7,844,725,426	78,447,254,258,109

doughnut holes	6	40	62	91	128	230	297	376	468	695	832	986	1,158	1,349	1,560
holiness	1	1	1	1	2	1	1	1	3	3	2	4	2	1	2
dusub	1	6	8	12	17	31	40	51	63	94	113	134	157	182	211

broken-up	6	241	14,948	1,360,509	174,160,100	40,058,183,509	11,897,454,662,273
holiness	1	0	3	4	4	8	5

dusub	1	33	2,023	184,125	23,570,006	5,421,285,611	1,610,145,396,514
broken-down	1	7	435	39,592	5,068,211	1,165,728,122	346,226,320,445
broken	1	6	7	241	435	14,948	39,592

hypersum	6	640	64,062	6,406,291	6,406,291,128	6,406,291,128,230
holiness	1	2	3	4	6	7
dusub	1	86	8,670	866,997	866,997,224	866,997,224,335

440 E Day commemorates February 7, 1828.
almost-e

broken-up	2	15	17	151	319	2,703	3,022	26,879	56,780	140,439	116,902,028
trimmed	1	4	6	40	208	162	211	15,768	4,567	3,328	58,117
broken-down	1	3	4	35	74	627	701	6,235	13,171	32,577	27,117,235
broken	1	2	3	4	15	17	35	74	151	319	627

hypersum	2	27	271	2,718	27,182	271,828	2,718,281	27,182,818	271,828,182
trimmed	1	16	160	1,607	16,071	160,717	1,607,170	16,071,707	160,717,071
sub	1	10	100	1,000	10,000	100,000	1,000,000	10,000,000	100,000,000

sliced e

sub	1	3	7	10	67	311	1,926
broken-up	2	15	272	7,631	1,389,114	1,173,808,961	1,061,124,689,858
trimmed	1	4	161	6,520	278,003	6,277,850	50,013,578,767
sub	1	6	100	2,807	511,026	431,820,185	390,365,957,918
broken-down	1	3	55	1,543	280,881	237,345,988	214,561,054,033
broken	1	2	3	15	55	272	1,542

hypersum	2	27	2,718	271,828	271,828,182	271,828,182,845	271,828,182,845,904
trimmed	1	16	1,607	160,717	160,717,071	160,717,071,734	16,071,707,173,483
sub	1	10	1,000	100,000	100,000,000	100,000,000,000	100,000,000,000,000

diced e dips at the seventh term because of a leading zero.

sub	1	26	305	672	16,887	192,598	105,751	12,973,728	285,373,172

broken-up	2	143	118,406	216,446,311	9,935,751,578,550	> eighteeenplex
trimmed	1	32	735	105,335,200	882,464,046,744	
sub	1	53	43,559	79,626,148	3,655,158,738,335	> eighteenplex

broken-down	1	3	2,485	4,542,583	208,522,732,517	> seventeenplex
broken	1	2	3	143	2,485	118,406

hypersum	2	271	271,828	2,718,281,828	271,628,182,845,904	271,628,182,845,904,523,536
trimmed	1	160	160,717	16,071,717	160,717,173,483	160,717,173,483,412,425
sub	1	100	100,000	1,000,000,000	99,926,424,111,766	99,926,424,111,765,667,840

441 Phi Day commemorates January 6, 1803.
sliced phi

broken-up	1	17	275,061	4,450,580,225,696	> twenty-twoplex
broken-down	1	2	32,361	523,611,950,381	> twenty-oneplex
broken	1	2	17	32,361	275,061
sum	1	17	16,197	16,196,536	16,196,536,423
super	1	16	25,880	4,188,766,160,320	> twenty-twoplex

hypersum	1	16	16,180	16,180,339	16,180,339,887	161,803,398,874,989

diced phi

broken-up	1	62	49,787	198,550,618	14,889,112,342,989	> eighteenplex
broken-down	1	2	1,607	6,408,718	480,583,355,709	> seventeenplex
broken	1	2	62	1,607	49,787	6,408,718
sum	1	62	865	4,853	79,842	564,666
super	1	61	48,983	195,344,204	1,468,666,513,756	> eighteenplex

hypersum	1	161	161,803	1,618,033,988	161,803,398,874,989	> twentyplex

442

John Wallis named them 'continued fractions' in 1653 in *Arithmetica Infinitorum.* Seemingly random numbers can be re-interpreted to show a hidden pattern. http://www.maths.surrey.ac.uk/hosted-sites/R.Knott/Fibonacci/cfINTRO.html#sqrtCalc can calculate continuing fractions of the form the quantity z plus/minus the square root of z prime divided by z double-prime or z as well as z-to-z-prime, even when z equals e. The notation used there is fraction list, with a semicolon indicating the decimal point and overlining indicating repeated numbers, all within square brackets. Phi would therefore be simply [1;1].

443 almost-root-two Root Two's Day commemorates January 4, 1421. The first root-two prime is higher than fifty-fourplex.

sum	1	5	6	10	12	13	16	21	27
hypersum	1	14	141	1,412	14,121	141,213	1,412,135	14,121,356	141,213,562

sliced

broken-up	1	5	71	1,496	47,943	52,431	3,252,218	1,213,129,745

broken-down	1	2	29	611	21,414	1,328,279	495,469,481	2,478,675,684
broken	1	2	5	29	71	611	1,496	21,414
sum	1	5	19	40	72	137	510	515
super	1	4	56	1,176	41,160	2,551,920	951,866,160	4,759,330,800

hypersum	1	14	1,414	141,421	14,142,132	1,414,213,256	1,414,213,256,237

diced

sub	1	15	155	1,310	13,725	185.735
broken-up	1	42	17,683	23,978,190	1,495,591,662,553	> seventeenplex
broken-down	1	2	843	3,002,768	112,030,272,155	> sixteenplex
broken	1	2	42	843	17,683	3,002,768
sum	1	42	463	4,025	41,334	546,214
super	1	41	17,261	61,483,682	2,293,894,691,738	> eighteenplex

hypersum	1	141	141,421	1,414,213,562	141,421,356,237,309	> twenty-sevenplex

fracked

broken-up	1	3	7	17	41	99	239	577	1,393	3,363	8,119	19,601	47,321
broken-down	1	2	5	12	29	70	169	408	985	2,378	5,741	13.860	33,461
broken	1	2	3	5	7	12	17	29	41	70	99	169	239
sum	1	3	5	7	9	11	13	15	17	19	21	23	25
super	1	2	4	8	16	32	64	128	256	512	10,24	2,048	4,096

hypersum	1	12	122	1,222	12,222	122,222	1,222,222	12,222,222	122,222,222

444 almost-root-three Root Three Day commemorates January 7, 320.

hypersum	1	17	173	1,732	17,320	173,205	1,732,058	17,320,580

collapsed	1	7	3	7	15	5	6	8	8	7
sum	1	8	11	18	33	38	44	52	60	75
super	1	7	21	147	2,205	11,035	66,150	529200	4233600	63504000

broken-up	1	8	25	183	2,770	14,033	86,968	709,777	5,765,184	46831249
broken-down	1	2	7	51	772	3,911	24,238	197,815	1,606,758	24,299,185

hypersum	1	17	173	1,737	173,715	1,737,156	17,371,568	173,715,688

sliced

broken-up	1	8	2,561	13,009,888	98,458,834,945	863,680,913,147,428
broken-down	1	2	641	3,256,262	24,643,542,817	216,173,160,847,006

broken	1	2	8	641	2,561	3,256,262
sum	1	8	328	5,408	12,976	21,748
super	1	7	2,240	11,379,200	86,117,785,600	755,425,215,283,200

hypersum	1	17	17,320	173,205,080	1,732,050,807,568	17,320,508,075,688,772

diced

broken-up	1	74	15,171	12,243,071	696,471,595,148	> seventeenplex
broken-down	1	2	411	331,679	18,868,223,684	> sixteenplex
broken	1	2	74	411	15,171	331,679
sum	1	74	279	1,086	57,973	787,325
super	1	73	14,965	12,076,755	687,010,361,685	> seventeenplex

hypersum	1	173	173,205	173,205,807	17,320,580,756,887	17,320,580,756,887,729,352

445 almost-root-five Root Five Day commemorates February 23, 606.

hypersum	2	22	223	2,236	22,360	223,606	2,236,067	22,360,679	223,606,797
trimmed	1	11	112	1,125	1,125	11,255	112,556	1,125,568	11,255,686
sub	1	8	82	823	8,226	82,260	822,603	8,226,034	82,260,344

collapsed	2	2	3	12	7	9	7	9	7
trimmed	1	1	2	1	6	8	6	8	6
broken-up	2	5	17	209	1,480	13,529	96,183	879,176	6,250,415
trimmed	1	4	6	18	37	2,418	85,072	768,065	514,304
sub	1	2	6	77	544	4,977	35,384	323,431	2,299,399
broken-down	1	3	10	123	871	7,962	56,605	517,407	368,454
broken	1	2	3	5	10	17	123	209	871

hypersum	2	22	223	2,236	22,312	223,127	2,231,279	22,312,797	223,127,977
trimmed	1	11	112	1,125	11201	112,016	1,120,168	11,201,686	112,016,866
sub	1	8	82	823	8,208	82,084	820,842	8,208,419	82,084,195

sliced

sub	1	8	22	25	36	285	367
broken-up	2	47	2,822	189,121	18,347,559	14,201,199,787	14,158,614,535,198
trimmed	1	36	1,711	78,010	7,236,448	310,088,676	3,047,503,424,087
sub	1	17	1,038	69,574	6,749,690	5,224,329,442	5,208,663,202,971
broken-down	1	3	181	12,130	1,176,791	910,848,364	908,116,995,699

broken	1	2	3	47	181	2,822	12,130

hypersum	2	223	22,360	2,236,067	223,606,797	223,606,797,774	223,606,797,774,997
trimmed	1	112	1,125	112,556	11,255,686	11,255,686,663	11,255,686,663,886
sub	1	82	8,226	822,603	82,260,344	82,260,343,807	82,260,343,807,602

diced

sub	1	8	223	2,935	18,386	356,711
broken-up	2	47	28,484	227,216,915	11,355,847,006,354	> nineteenplex
trimmed	1	36	17,373	116,105,804	2,447,365,243	
sub	1	17	10,479	83,588,432	4,177,582,650,726	> eighteenplex
broken-down	1	3	1,819	14,510,166	725,189,078,167	> seventeenplex
broken	1	2	3	47	1,819	28,484

hypersum	2	223	223,606	2,236,067,977	223,606,797,749,978	> twentyplex
trimmed	1	112	11,255	112,556,866	11,255,686,658,867	
sub	1	82	82,260	822,603,438	82,260,343,798,398	>nineteenplex

446 almost-root-six Root Six Day commemorates February 9, 383.

broken-up	2	9	38	351	1,442	11,887	108,425	770,862	1,650,149
trimmed	1	8	27	240	331	776	7,314	66,751	54,038
sub	1	3	14	129	530	4,373	39,887	283,584	607,056
broken-down	1	3	13	120	493	4,064	37,069	263,547	564,163

hypersum	2	24	244	2,449	24,494	244,948	2,449,489	24,494,897	244,948,972
trimmed	1	13	133	1,338	13,383	133,837	1,338,378	13,383,786	133,837,861
sub	1	9	90	901	9,011	90,111	901,117	9,011,169	90,111,691

broken-up	2	9	443	216,636	160,744,355	125,863,046,601	224,036,383,694,135
trimmed	1	8	332	105,525	5,633,244	147,523,550	11,325,272,583,024
sub	1	3	163	79,696	59,34,543	46,302,427,248	82,418,379,635,469
broken-down	1	3	148	72,375	53,702,398	42,049,050,009	74,847,362,718,418
broken	1	2	3	9	148	443	72,375

hypersum	2	24	2,449	2,449,489	2,449,489,742	2,449,489,742,783	24,494,897,427,831,780
sub	1	9	901	901,117	901,116,917	901,116,917,730	9,011,169,177,302,560

Almost root-seven

sub	1	16	349	3,584	28,811	297,916
broken-up	2	89	84,374	821,971,597	64,374,346,623	52,131,571,457,300,561,920
sub	1	33	31,039	302,386,452	23,681,999,773,775	19,178,133,375,100,796,928
broken-down	1	3	2,845	27,715,993	2,170,633,426,626	1,757,820,190,944,556,800
broken	1	2	3	89	2,845	84,374

hypersum	2	244	244,948	2,449,489,742	244,948,974,278,317	244,948,974,278,317,809,819
trimmed	1	133	133,837	1,338,378,651	133,837,865,167,206	13,383,786,516,720,678,708
sub	1	90	90,111	901,116,917	90,111,691,773,025	90,111,691,773,025,599,488

447 almost-cube root two Cube Root Two's Day commemorates December 5, 992.

hypersum	1	12	125	1,259	12,599	125,992	1,259,921	12,599,210	125,992,104
collapsed	1	2	5	9	9	2	5	9	4
broken-up	1	3	16	147	1,339	2,825	15,464	142,001	583,468
broken-down	1	2	11	101	920	1,941	10,625	97,566	400,889
broken	1	2	3	11	16	101	147	920	1,339
sum	1	3	8	17	26	28	33	42	46
super	1	2	10	90	810	1,620	8,100	72,900	291,600

hypersum	1	12	125	1,259	12,599	125,992	1,259,925	12,599,259	125,992,594

448 almost-cube root three Cube Root Three Day commemorates January 4, 422.

hypersum	1	14	144	1,442	14,422	144,224	1,442,249	14,422,495	144,224,957

collapsed	1	4	4	2	2	4	9	17	4	1
broken-up	1	5	21	47	115	507	4,678	80,033	324,810	1,379,273
broken-down	1	2	9	20	49	216	1,993	34,097	138,381	310,859
broken	1	2	5	9	20	21	47	49	115	216
sum	1	5	9	11	13	17	26	43	47	49
super	1	4	16	32	64	256	2,304	3,916	156,672	313,344

hypersum	1	14	144	1,442	14,422	144,224	1,442,249	14,422,495	1,442,249,517

449 almost-cube root four Cube Root Four Day commemorates January 5, 874.

hypersum	1	15	158	1,587	15,874	158,740	1,587,401	15,874,010	158,740,105

collapsed	1	5	8	7	10	1	9	6	8	2

broken-up	1	6	49	349	3,539	3,888	38,531	235,074	1,919,123	15,588,058
broken-down	1	2	17	121	1,227	1,348	13,359	81,502	665,375	6,735,252
broken	1	2	6	17	49	121	349	1,227	1,348	3,539
sum	1	6	14	21	31	32	41	47	55	65

hypersum	1	15	158	1,587	158,710	1,587,101	15,871,019	158,710,196	1,587,101,968

450 almost-cube root five Cube Root Five commemorates January 7.

hypersum	1	17	170	1,709	17,099	170,997	1,709,975	17,099,759	170,997,594

collapsed	1	16	9	9	7	5	9	4	6
broken-up	1	17	154	1,403	9,975	51,278	471,477	1,937,186	12094593
broken-down	1	2	19	173	1,230	6,323	58,137	238,871	1,491,363
broken	1	2	17	19	154	173	1,230	1,403	6,323
sum	1	17	26	35	42	47	56	60	66
super	1	16	144	12,96	9,072	45,360	408,240	1,632,960	9,797,760

hypersum	1	116	1,169	11,699	116,997	1,169,975	11,699,759	116,997,594	1,169,975,946

451 almost-cube root six Cube Root Six commemorates January 8, 1712.

hypersum	1	18	181	1,817	18171	181,712	1,817,120	18,171,205	181,712,059

collapsed	1	8	1	7	1	7	9	2	8	3	2
broken-up	1	9	10	79	89	702	6,407	13,516	114,535	929,796	1,974,127
broken-down	1	2	3	23	26	205	1,871	3,947	33,447	37,394	108,235
broken	1	2	3	9	10	23	26	79	89	205	702
sum	1	9	10	17	18	25	34	44	45	47	48

hypersum	1	18	181	1,817	18,171	181,717	1,817,179	18,171,792	181,717,921
weight	1	9	10	17	18	25	34	36	37

452 fracked pi

hypersum	3	37	3,715	37,151	37,151,292	371,512,921
sub	1	14	1,367	13,667	13,667,197	136,671,966

fracked half-pi

hypersum	1	11	111	1,113	111,331	1,113,311	1,113,311,145
weight	1	2	3	6	37	38	18

Two over pi is the probability of a randomly dropped stick crossing lines its length apart.

hypersum	6	63	636	6,366	63,661	636,619	6,366,197	63,661,977	636,619,772
holiness	1	1	2	3	3	4	4	4	4
tetrasub	1	8	86	862	8,616	86,157	861,571	8,615,712	86,157,117

453 fracked e

hypersum	3	37	3,715	37,151	37,151,292	371,512,921
sub	1	14	1,367	13,667	13,667,197	136,671,966

fracked half-e

hypersum	1	12	121	1,213	12,131	121,311	1,213,111	12,131,113	121,311,133
sum	1	3	4	7	9	10	11	14	17

Logarithms also can be digitally expanded.

almost-ln 4	1	3	8	6	2	9	4	3	6
hypersum	1	13	138	1,386	13,862	138,629	1,386,294	13,862,943	138,629,436

almost-ln 5	1	6	0	9	4	3	7	9
hypersum	1	16	160	1,609	16,094	160,943	1,609,437	16,094,379
collapsed	1	17	4	3	7	9	1	2
hypersum	1	117	1,174	11,743	117,437	1,174,379	11,743,791	117,437,912

almost-log 13	1	1	1	3	9	4	3	3
hypersum	1	11	111	1,113	11,139	111,394	1,113,943	11,139,433

almost-log 14	1	1	4	6	1	2	8	0	3
hypersum	1	11	114	1,146	11,461	114,612	1,146,128	11,461,280	114,612,803
collapsed	1	1	4	6	1	2	11	5	7
hypersum	1	11	114	1,146	11,461	114,612	11,461,211	114,612,115	1,146,121,157

The first four e-primes are one-digit, two-digits, six digits and eighty-four digits.

454 Inflationary English was popularized by the Clown Prince of Denmark, Børge Rosenbaum aka "Victor Borge". He also introduced phonetic punctuation. Loglan also has phonetic punctuation but many different words for each depending upon the context.

455 two-place almost-phi

hypersum	16	1,618	16,183	1,618,339	161,833,988	16,183,398,874
holiness	1	3	3	4	8	8
tressub	1	81	806	80,572	8,057,240	805,723,986

456 three-place almost-phi

hypersum	161	161,803	161,803,398	161,803,398,874	161,803,398,874,989
holiness	1	4	7	9	13
pentasub	1	1,090	1,090,223	109,222,726	1,090,222,725,892

457 one/two-place almost-phi

hypersum	1	161	1,618	16,183	1,618,398	16,183,988	1,618,398,874
sum	1	62	70	73	171	179	253

458 two/one-place almost-phi

hypersum	16	161	16,183	161,839	16,183,988	161,839,887
holiness	1	1	3	4	8	8
tressub	1	8	806	8,057	805,753	8,057,534

459 three powers

sub	1	3	10	30	89	268	805
broken-up	3	28	759	61,507	1,496,960	10,896,395,347	23,830,431,570,849
sub	1	10	279	22,627	5,498,679	4,008,559,831	8,766,725,849,158
broken-down	1	4	109	8,833	2,146,528	1,564,827,745	3,422,280,424,843
broken	1	3	4	28	109	759	8,833

hypersum	3	39	3,927	392,781	392,781,243	392,781,243,729	3,927,812,437,292,187
sub	1	14	1,445	144,496	144,496,144	144,496,144,446	1,444,961,444,457,290

460 broken-up

broken-up	1	4	41	1,767	397,616	553,880,855	6,297,071,838,111	> seventeenplex
broken-down	1	2	21	905	203,646	283,679,783	3,225,155,656,573	> seventeenplex
broken	1	2	4	21	41	905	1,767	203,646

broken-down

broken-up	1	3	22	663	104,113	101,198,499	802,807,796,680	> sixteenplex
broken-down	1	2	15	452	70,979	68,992,040	547,313,924,299	> sixteenplex
broken	1	2	3	15	22	452	663	70,979

broken

broken-up	1	3	10	73	740	22,273	958,479	150,503,476	33,864,240,579
broken-down	1	2	7	51	517	15,561	669,640	105,149,041	23,659,203,865
broken	1	2	3	7	10	51	73	517	740

461 broken-up odd

broken-up	1	5	6	29	40,026	61,363,835	123,146,842,227,731	> twentyplex
broken-down	1	2	3	14	19,323	296,240,927	59,450,517,972,464	> twentyplex
broken	1	2	3	5	6	14	29	19,323

broken-down odd

broken-up	1	4	5	19	19,974	233,216,443	35,644,567,951,651	> nineteenplex
broken-down	1	2	3	11	11,564	135,021,275	20,636,516,661,289	> nineteenplex
broken	1	2	3	4	5	11	19	11,564

broken odd

broken-up	1	4	5	19	404	46,479	7,018,733	7,376,734,862	10,179,901,128,293
broken-down	1	2	3	11	234	26,921	4,065,305	4,272,662,476	5,896,278,282,185
broken	1	2	3	4	5	11	19	234	404

462 broken-down even

broken-up	1	3	4	11	5,031	23,273,417	1,302,589,881,104	> eighteenplex
broken-down	1	2	3	8	3,659	16,926,542	947,361,632,857	> seventeenplex
broken	1	2	3	4	8	11	3,659	5,031

463 beans

broken-up	5	36	185	1,331	33,460	904751	31,699,745	1,173,795,316
dusub	1	5	25	180	4,528	122,445	4,290,094	158,855,922
broken-down	1	6	31	223	5,606	151,585	5311081	196,661,582
broken	1	5	6	31	36	185	223	1,331

hypersum	5	57	5,715	571,517	57,151,725	5,715,172,527	571,517,252,735
dusub	1	8	773	77,346	7,734,645	773,464,493	77,346,449,274

"Counterbeans" describes a number name with a word with five or seven letters, that is, three or seven.

counterbeans	3	7	103	107	203	207	301	302	303	304	305	306	307	308	309	310

464 The Heinz cannery used the slogan "57 varieties" from 1892 to 1969. Since Heinz numbers includes five-seven frag, their mates include four-two frag

mates	42	142	242	342	420	421	422	423	424	425	426	427	428	429	542

Znieh numbers include 575 or 757 frags but exclude numberdromes.

Znieh	1,575	1,757	2,575	2,757	3,575	3,757	4,575	4,757	5,575	5,757	5,758

465 bingo

mates	199	299	399	499	599	699	799	899	1,991	1,992	1,993	1,994	1,995	1,996
holiness	2	2	2	2	2	3	2	4	2	2	2	2	2	3

ognib	1,002	1,003	1,004	1,005	1,006	1,007	1,008	1,009	2,001	2,003	2,004	2,005
holiness	2	2	2	2	3	2	4	2	2	2	2	2

466 po

mates	11	22	33	44	55	66	77	88	100	101	112	113	114	115	116	117	118	119

op	112	113	114	115	116	117	118	119	122	133	144	155	166	177	188	199	220

467 Pos include at least two different doubled digits, also spelled 'poes' after Edgar Allen Poe. Ironically "po"could also stand for the acronum for prime odds, which excludes the only even prime, two.

broken-up	1,001	1,011,011	1,012,023,012	1,022,144,253,131	1,158,090,450,820,435
holiness	2	2	3	1	9
heptasub	1	922	922,845	932,074,911	1,056,041,796,584
broken-down	1	1,002	1,003,003	1,013,034,032	1,147,768,561,259
broken	1	1,001	1,002	1,011,011	1,003,003

hypersum	1,001	10,011,010	100,110,101,100	1,001,101,011,001,122
holiness	2	4	6	6
heptasub	1	9,129	91,288,596	912,885,957,630

"Pubo" or Tutnese po describes a pair with two other digits between, which start out like four-place numberdromes.

468 Wonks can be alternatively spelled 'wonx' with an X, like how Jonas Bronck's land became 'the Bronx' and then superpluralized.

wonxes	1,111	10,111	11,110	11111	11,112	11,113	11,114	11,115	11,116

469 Dux also can be superpluralized to 'duxes', describing at least four twos.

duxes	2,222	12,222	20,222	21,222	22,122	22,212	22,221	22,222	22,223

"Duduxux or Tutnese dux describes two twos with two other digits between.

duduxux	2,002	2,012	2,032	2,042	2,052	2,062	2,072	2,082	2,092	2,102	2,132
holiness	2	1	1	1	1	2	1	3	2	1	0

470 Ferty-free describes numbers with a thirty-three frag, while all others would be called by backformation 'ferty'.

firty-free	33	133	233	333	433	533	633	733	833	933	1,033	1,133	1,233	1,333
tressub	1	7	12	17	21	26	32	36	42	46	52	56	61	66

"Countercrabs" describe number names with a word with a letter count of three, that is, one, two, six or ten.

countercrab	1	2	6	10	101	102	103	104	105	106	107	108	109	110	111	112

471 "Doorman" would describe a number with three fours.

doorman	444	1,444	2,444	3,444	4,044	4,144	4,244	4,344	4,414	4,424	4,434	4,440

"Counterman" describes a number name with two words with letter counts of four, that is four, five or nine.

counterman	404	405	409	504	505	509	904	905	909	1,404	1,405	1,409	1,504
holiness	1	1	2	1	1	2	2	2	3	1	1	2	1

"Mumanun" or Tutnese man describes a number with two fours with two other digits between.

mumanun	4,004	4,014	4,024	4,034	4,054	4,064	4,074	4,084	4,094	4,104	4,114
holiness	2	1	1	1	1	2	1	3	2	1	0

Alternatively "man" is the acronum for Mersenne-and-nibi and "superman" their ever higher products.

man	2	3	7	12	22	31	32	42	52
trimmed	1	2	6	1	11	20	21	31	41
superman	2	6	42	504	11,088	343,728	10,999,296	461,970,432	241,970,432
trimmed	1	5	31	43	77	232,617	888,185	35,086,321	13,086,321

472 "Doormen" describes a number with five fours.

doormen	44,444	404,444	414,444	424,444	434,444	440,444	441,444	442,444

"Countermen" describes a number with four words with letter counts of four.

countermen	404,404	404,405	404,409	404,504	404,505	404,509	404,904	404,905
holiness	2	2	3	2	2	3	3	3

473 Since five-five is 'rootbeer', its square would be "beer" numbers with the frag three-oh-two-five in bases six and higher.

beerish	3,025	13,025	23,025	30,250	30,251	30,252	30,253	30,254	30,255	30,256
holiness	1	1	1	2	1	1	1	1	1	2

"Counternickle", not "kupfernickel", describes a number name with a word with a letter count of five, that is, three, seven, eight, forty, fifty or sixty.

counternickle	3	7	8	40	50	60	103	107	108	140	150	160	203	207	208	240	250	260

"Eery" describes beheaded beerish numbers.

eery	...	259	3,025	13,025	23,025	25,000	25,001	25,002	25,003	25,004
holiness	...	1	1	1	1	3	2	2	2	2

474 "Counterwoman", analogous to counterman, describes a number name with two words with letter counts of five and "counterwomen" with four.

counterwoman	303	307	308	340	350	360	703	707	708	740	750	760	1,303	1,307	1,308
holiness	1	1	3	1	1	2	1	1	3	1	1	1	1	1	1

counterwomen	303,303	303,307	303,308	303,340	303,350	303,360	303,703	303,707
holiness	2	2	4	2	2	3	2	2

475 "Yak" describes the mate of buzz, numbers with a four (yongbi) or nine tail (kyuubi).

yak	4	9	14	19	24	29	34	39	44	49	54	59	64	69	74	79	84	89	94	99	104	109	204
sub	1	3	5	7	9	11	13	14	16	18	20	22	24	25	27	29	31	33	35	36	38	40	75

pyak	404	414	424	434	444	454	464	474	484	494	909	919	929	939	949	959
holiness	1	0	0	0	0	0	1	0	2	1	3	2	2	2	2	2

"Counterbuzz" describes number names with a word with a letter count of a multiple of five, that is, seven, eight, twenty-, thirty- or eighty-four, -five or -nine, etc.

counterbuzz	7	8	24	25	29	34	35	39	84	85	89	107	108	124	125	129	134	135	139	184

476 "Hutcheavuvenun", "tutomums" or Tutnese toms all describe a number with two sixes with two other digits between.

tutomums	6,006	6,016	6,026	6,036	6,046	6,056	6,076	6,086	6,096	6,106	6,116	6,126
holiness	4	3	3	3	3	3	3	5	4	3	2	2

"Countertom" describes number names with a word with a letter count of six, that is, eleven, twelve, twenty, thirty or eighty.

countertom	11	12	20	30	80	111	112	120	130	180	211	212	220	230	230	280	311	312

The ananym of tom is "Mot", the name of the genius Martian infant in "Rocket-bye Baby" by Michael Maltese. Tom numberdromes would be Mot not.

mots	166	266	366	466	566	660	661	662	663	664	665	667	668	669	766	866	966

holiness	2	2	2	2	2	3	2	2	2	2	2	2	4	3	2	4	3

477 beastly

primes	6,661	16,661	26,669	46.663	56,663	60,661	63,667	64,661	64,667	66,601
holiness	3	3	4	3	3	4	3	3	3	4

"Counterbeast" describes number names with three words with letter counts of six. The first prime is twenty million, eleven thousand, eight hundred eleven.

counterbeast	11,011,011	11,011,012	11,011,020	11,011,030	11,011,080	11,011,111
holiness	2	2	3	3	5	1

478 Tomtoms is aka second heaven, so "seventh heaven" would have fourteen sixes.

prime	166,667	166,669	266,663	606,661	616,669	626,663	646,669	66,0661
holiness	4	5	4	5	5	4	5	5

479 "Quackeenuns" or Tutnese queens describes numbers with two fours with two other numbers between.

quackeenuns	4,004	4,014	4,024	4,034	4,054	4,064	4,074	4,084	4,094	4,104	4,114
holiness	2	1	1	1	1	2	1	3	2	1	0

primes	40,241	40,343	40,543	40,949	41,047	41,143	41,147	41,341	41,543	41,641
holiness	1	1	1	3	1	0	0	0	0	1

"Counterqueens" describes a number name with two words with letter counts of four.

counterqueens	404	405	409	504	505	509	904	905	909	4,004	4,005	4,009	4,104	4,105
holiness	1	1	1	1	1	2	2	1	3	2	2	3	1	1

prime merries also spelled "primaries"

prime	233	313	331	337	353	373	383	433	733	1,033	1,303	1,373	1,433	1,733	1,933	2339

480 "Three Little Kittens" was included in a collection by Eliza Lee Cabot Follen in 1833, attributed to "Mother Goose". Kittens and mittens overlap so we have the supersequence "kittens with mittens" and the subsequence "kittens without mittens".

kittens with mittens	3	6	13	16	23	26	30	31	32	33	34	35	36	37	38	39	43
kittens without mittens	3	13	23	30	31	32	33	34	35	37	38	39	43	53	66	73	83

481 The OWAD (Our Werebeasts Are Different) trope also includes many other Human bimorphs who are also werecars, werebots, wereplants, weredemons, etc. The term for an animal able to transform into a Human is "beastwere", the product of werebeastiality or transmutation.

wereprimes	2,333	3,133	3,323	3,331	3,343	3,373	3,433	3,533	3,733	3,833	5,333	7,333

"Wuweruge", aka Tutnese were, describes numbers with three threes counting every third number.

wuweruge	3,003,003	3,003,013	3,003,023	3,003,043	3,003,053	3,003,063
holiness	4	3	3	3	3	4

"Underwear" is a punning reference to the were's downtown neighbor, like the motto "Semper ubi sub ubi," mistranslated as "Always wear underwear."

underwear	332	1,332	2,332	3,032	3,132	3,232	3,302	3,312	3,322	3,329	3,330	3,331

"Sububi", of course, is the acronym for sanbi-unindexed-binarish-unidexed binarish ichibi.

binarish ichibi	1	11	101	111	1,001	1,011	1,101	1,111	10,001	10,011	10,101	10,111

sububi	11	101	111	1,001	1,011	1,101	1,111	10,001	10,101	10,111	11,001

482 "Countercrutch" describes number names with a word with a letter count of seven, that is, fifteen, sixteen, seventy, hundred or million.

countercrutch	15	16	70	101	102	103	104	105	106	107	108	109	110	111	112	113	114

483 "Wuwaafuds" aka Tutnese waafs describes a number with two eights with two other digits between.

wuwaafuds	8,008	8,018	8,028	8,038	8,048	8,058	8,068	8,078	8,089	8,108	8,118	8,128
holiness	6	5	5	5	5	5	6	5	6	5	4	4

primes	80,287	80,387	80,489	80,681	80,683	80,687	80,783	80,789	80,963	80,989
holiness	5	5	6	6	6	6	5	6	5	7

"Counterwaaf" describes number names with a word with a letter count of eight, that is, thirteen, eighteen, nineteen, thousand or nineplex.

counterwaafs	13	18	19	113	118	119	213	218	219	313	318	319	413	418	419

484 "Pothooks" is another term ripe for superpluralization, such that 'pothooxes' would mean with at least four nines.

pothooxes	9,999	19,999	29,999	39,999	49,999	59,999	69,999	79,999	89,999	90,999
holiness	4	4	4	4	4	4	5	4	6	5

primes	49,999	59,999	79,999	94,999	98,999	99,991	139,999	179,999	190,999
holiness	4	4	4	4	6	4	4	4	5

485 cockeyed

primes	127	211	241	251	281	421	521	821	1,021	1,123	1,129	1,201	1,213	1,217	1,223

486 One-duzz does include primes that don't include a "dozen".

primes	127	1,123	1,201	1,213	1,217	1,223	1,229	1,231	1,231	1,237	1,239	1,259	1,277

"Counterduzz" describes a number name with a word with a letter count of a dozen, that is, seventy-seven.

counterduzz	77	177	277	377	477	577	677	777	877	977	1,077	1,177	1,277	1,377	1,477

487 Similarly sams include all two-one frags.

primes	211	421	521	821	1,021	1,213	1,217	1,321	1,721	2,111	2,113	2,129	2,131	2,137

488 Two-duzz primes include all two-four frags.

primes	241	1,249	2,243	2,411	2,417	2,423	2,437	2,447	2,457	2,467	2,473	2,477

489 three-duzz

primes	367	1,361	1,367	3,361	3,607	3,613	3,617	3,623	3,631	3,637	3,643	3,659
holiness	1	1	1	1	2	1	1	1	1	1	1	2

490 four-duzz

primes	487	1,481	1,483	1,487	1,489	4,481	4,483	4,801	4,813	4,817	4,831	4,861
holiness	2	2	2	2	3	2	2	3	2	2	2	3

491 Does-too's mates naturally are "does-nots" with five-seven frags.

primes	421	1,423	1,427	1,429	2,423	4,201	4,211	4,217	4,219	4,229	4,231	4,241

does-nots	57	157	257	357	457	557	570	571	572	573	574	575	576	577	578	579	657

492 does-three

primes	163	263	463	563	631	863	1,063	1,163	1,637	1,663	2,063	2,633	2,663	2,963
holiness	1	1	1	1	1	3	2	1	1	2	2	1	2	2

493 does-for aka altruistic

paltruistic	484	848	4,884	8,448	48,084	48,184	48,284	48,384	48,484	48,584
holiness	2	4	4	4	5	4	4	4	4	4

primes	1,847	2,843	3,847	5,843	5,849	6,841	7,841	8,419	8,423	8,429	8,431	8,443
holiness	2	2	2	2	3	3	2	3	2	3	2	2

494 six-o

primes	601	607	1,601	1,607	1,609	2,609	3,607	4,603	6,007	6,011	6,029	6,037
holiness	2	2	2	2	3	3	3	2	3	2	3	2

495 o-six

primes	1,061	1,063	1,069	2,063	2,069	3,061	3,067	6,067	7,069	8,069	9,067	10,061
holiness	2	2	3	2	3	2	2	3	3	5	3	3

496 In 1362 "dime" meant a tithe, a tenth; by 1786 a tenth of a dollar. We take it to meant "one-zero".

primes	101	103	107	109	1,009	1,013	1,019	1,021	1,031	1,033	1,039	1,049	1,051
holiness	1	1	1	1	2	1	1	1	1	1	2	2	1

"Counterdime" describes a number name with a word with a letter count of ten, that is, twenty-, thirty-, eighty-four, -five, -nine, forty-, fifty-, sixty-seven, or -eight.

counterdime	24	25	29	34	35	39	47	48	57	58	67	68	84	85	89	124	125	129	134	135

The "five-and-dimes", aka Woolworths, would be a subsequence and "fives-and-dimes" a subsequence of that. All numbers not Woolworths would be "Woolworthless".

five-and-dimes	105	510	1,005	1,015	1,025	1,035	1,045	1,055	1,065	1,075	1,085
holiness	1	1	2	1	1	1	1	1	2	1	3

fives-and-dimes	1,055	5,105	5,510	10,055	10,155	10,255	10,355	10,455	10,550
holiness	1	1	1	2	1	1	1	1	2

497 o-one

primes	101	401	601	701	1,013	1,019	1,201	1,301	1,601	1,801	1,901	2,011
holiness	1	1	2	1	1	2	1	1	2	3	2	1

498 score

primes	1,201	2,003	2,011	2,017	2,027	2,029	2,039	2,063	2,069	2,081	2,083
holiness	1	2	1	1	1	2	2	2	3	3	3

499 o-two

primes	1,021	2,027	2,029	3,023	4,021	4,027	5,021	5,023	6,029	7,027	9,029
holiness	1	1	2	1	1	1	1	1	3	1	3

500 half-past

primes	307	1,301	1,303	1,307	2,309	3,001	3,011	3,079	3,083	3,089	3,301
holiness	1	1	1	1	2	2	1	2	3	4	1

501 o-three

primes	103	503	1,103	1,303	2,003	2,203	2,503	2,803	2,903	3,203	3,803
holiness	1	1	1	1	2	1	1	3	2	1	3

502 four-o

primes	401	409	1,409	3,407	4,001	4,003	4,007	4,013	4,019	4,031	4,027
holiness	1	2	2	1	2	2	2	1	2	1	1

503 o-four

primes	1,049	3,041	3,049	4,049	6,043	6,047	7,043	9,041	9,043	9,049	10,427
holiness	2	1	2	2	2	2	1	2	2	3	1

504 five-o from "Hawaii Five-O" by Leonard Freeman

primes	503	509	2,503	4,507	5,003	5,009	5,011	5,021	5,023	5,039	5,057
holiness	1	2	1	1	2	3	1	1	1	2	1

Four pieces-of-eight would include five-o, along with fives without an "o", but we also have other frags: one piece-of-eight or one-two-five, two pieces-of-eight or two-five, three pieces-of-eight or three-seven-five, five pieces-of-eight or six-two-five, six pieces-of-eight or seven-five, seven pieces-of-eight or eight-seven-five.

piece(s)-of-eight

one	125	1,125	1,250	2,125	3,125	4,125	5,125	6,125	7,125	8,125	9,125
two	25	125	225	250	325	425	525	625	725	825	925
three	375	1,375	2,375	3,375	3,750	4,375	5,375	6,375	7,375	8,375	9,375
five	625	1,625	2,625	3,625	4,625	5,625	6,250	6,625	7,625	8,625	9,625
six	75	175	275	375	475	575	675	750	775	875	975
seven	875	1,875	2,875	3,875	4,875	5,875	6,875	7,875	8,750	8,875	9,875

505 o-five

primes	1,051	2,053	4,051	4,057	5,051	5,059	6,053	7,057	8,053	8,059	9,059
holiness	1	1	1	1	1	2	2	1	3	4	3

506 seven-o

primes	701	709	1,709	2,707	3,701	3,709	4,703	5,701	6,701	6,703	6,709
holiness	1	2	2	1	1	2	1	1	2	2	3

507 o-seven seems to break the stuttering habit, but eventually succumbs in its seventeenth term.

primes	107	307	607	907	1,307	1,607	1,907	2,207	2,707	3,079	3,307
holiness	1	1	2	2	1	2	2	1	1	2	1

508 The eight-o numbers commemorate the famous racehorse Potoooooooo [read as pohtayt-ohs], famous not for its races but for its univocalicity.

primes	809	1,801	2,801	2,803	4,801	5,801	5,807	6,803	8,009	8,011	8,017
holiness	3	3	3	3	3	3	3	4	5	3	3

509 o-eight

primes	1,087	2,081	2,083	2,087	2,089	3,083	3,089	5,081	5,087	6,089	8,081
holiness	3	3	3	3	4	2	4	3	3	5	5

510 nine-o

primes	907	1,901	1,907	2,903	2,909	3,907	4,903	4,909	5,903	6,907	7,901
holiness	2	2	2	2	3	2	2	3	2	3	2

511 o-nine

primes	109	409	509	709	809	1,009	1,091	1,093	1,097	1,109	1,409	1,609	1,709
holiness	2	2	2	2	4	3	2	2	2	2	2	3	2

512 harmonious

pharmonious	131	313	1,331	3,113	13,031	13,131	13,231	13,331	13,431	13,531	13,631

primes	13	31	103	113	131	137	139	163	173	193	311	313	317	331	431	613	631	1,013

513 Fewteen numbers include interior as well as tail fewteens.

pfewteens	313	414	515	3,113	4,114	5,115	13,031	13,131	13,231	13,331	13,431	13,531

primes	13	113	213	613	1013	1213	1301	1393	1307	1319	1321	1327	1361	1367

514 Primes cannot have a midteen tail, so they can only be interior at hundreds, etc.

primes	1601	1607	1609	1613	1619	1621	1623	1637	1657	1663	1667	1669	1693
holiness	2	2	3	1	2	1	1	1	1	2	2	3	2

515 moreteens

primes	17	19	317	419	617	619	919	1019	1117	1217	1319	1619	1709	1721	1723

What the society's extrapolation "scoreteens", thirty-three through thirty-nine, has not become as popularized as "thirty-something" for the more inclusive thirty-one through twenty-nine.

scoreteens	33	34	35	36	37	38	39	133	134	135	136	137	138	139	233	234	235	236

516 fiffish primes contain a one-five frag, palindromic fiffish numbers both one-five and/or five-one.

primes	1,511	1,523	1,543	1,549	1,553	1,559	1,567	1,571	1,579	1,583	1,597

pfiffish	1,551	5,115	15,051	15,151	15,251	15,351	15,451	15,551	15,651	15,751

"Counterfiff' describes a number name with a word with a letter count of fifteen, that is, twenty-, thirty- or eighty-sevenplex! Beheaded fiffish is "iffish".

iffish	5	15	50	51	52	53	54	55	56	57	58	59	150	151	152	152	153	154	155	156	157

517 Fewty primes are those with a twenty, thirty or forty.

primes	2,011	2,017	2,027	2,029	2,039	2,053	2,063	2,069	2,081	2,083	2,087
holiness	1	1	1	2	2	1	2	3	3	3	3

pfewty	2,002	20,002	20,102	20,202	20,302	20,402	20,502	20,602	20,702	20,802
holiness	2	3	2	2	2	2	2	3	2	4

518 morety primes are those with a six-, seven-, eight- or nine-zero

primes	6,007	6,011	6,029	6,037	6,043	6,047	6,067	6,073	6,079	6,089	6,091
holiness	3	2	3	2	2	2	3	2	3	5	3

519 Fewplex is between ten and the ten thousands, midplex in the hundred thousands and moreplex between a million and nineplex. Beyond that is fewduplex, midduplex, moreduplex, fewtriplex, midtriplex, moretriplex, etc.

520 Step-prime contain thirty-nine.

step-primes	139	239	439	739	839	1,039	1,439	2,039	2,239	2,539	2,939	3,539
holiness	1	1	1	1	3	2	1	2	1	1	2	1

521 Half-step-primes contain either a three or a nine.

half-step-primes	3	13	19	23	31	37	43	53	73	79	83	89	97	103	109	113	131	137

Adding a four to the either of the half-steps, but not both, would make the 'Step-foured wives', an allusion to both the movie "The Stepford Wives" by William Goldman and the book by Ira Levin and the amended step four of the Twelve Steps, "Made a searching and fearless moral inventory of our [wives]."

wives	34	43	49	134	143	149	234	243	249	304	314	324	334	340	341	342	343

inventory	1,314	1,413	1,419	111,314	111,413	111,419	121,314	121,413	121,419

522 valles

primes	1,753	2,357	3,257	3,457	3,517	3,527	5,237	5,347	5,387	5,437	5,5,73	5,737	5,743

"Countervalles" describes number names with words with a letter count of three, five or seven, that is, one, two, three, six, seven, eight, fifteen, sixteen, seventy, hundred and million.

countervalles	1	2	3	6	7	8	15	16	70	101	102	103	104	105	106	107	108	109	110	111	112

The complimentary sequence of the valles, the wovalles, contain a two, four and six.

wovalles	246	264	426	462	624	642	1,246	1,264	1,426	1,462	1,624	1,642	2,046	2,064

holiness	1	1	1	1	1	1	1	1	1	1	1	1	2	2

523 tertivalles

primes	3	5	7	17	23	31	43	47	59	71	79	103	107	113	131	139	151	179	193	197	223

524 vo

primes	1,753	2,357	2,753	3,571	4,357	5,573	5,737	7,351	7,537	7,573	7,753	8,573	8,753

"Vuvo" or Tutnese valles po describes two of three valles with two other digits between. The first vuvo prime is three thousand two hundred fifty-seven.

vuvo	3,005	3,007	3,015	3,017	3,025	3,027	3,045	3,047	3,065	3,067	3,085	3,087	3,095
holiness	2	2	1	1	1	1	1	1	2	2	3	3	2

"Wovo" describes numbers containing two of two, four or six.

wovo	24	26	42	46	62	64	124	126	142	146	162	164	204	206	214	216	240	241	242	243

primes	241	263	269	421	461	463	467	641	643	647	1,249	1,423	1,427	1,429	1,621	1,627

525 lok

primes	421	1,423	1,427	1,429	2.423	4,201	4,211	4,217	4,219	4,229	4,231	4,241	4,243

kol	1,424	2,424	3,424	4,241	4,243	4,244	4245	4,246	4,247	4,248	4,249	4,424	5,424

"Lulokuck" or Tutnese lok describes a four and two with two other digits between, while "wolok" is another name for a Heinz.

lulokuck	4,002	4,012	4,032	4,052	4,062	4,072	4,082	4,092	4,102	4,112	4,132	4,152	4,162
holiness	2	1	1	1	2	1	3	2	1	0	0	0	1

primes	40,123	40,127	40,129	40,423	40,427	40,429	40,529	40,627	40,823	40,829
holiness	1	1	2	1	1	2	2	2	3	4

526 semilok

primes	23	29	41	43	127	149	211	223	227	229	233	239	251	257	263	269	271	277

527 Ticketty-boo means "as it should be, correct or satisfactory".

primes	1,621	1,627	2,621	3,623	4,621	5,623	6,203	6,211	6,217	6,221	6,229	6,247
holiness	1	1	1	1	1	1	2	1	1	1	2	1

"Boo-boo" means a minor injury or mistake (1937). "Boo Boo" refers to William Hannah and Joseph Barbera's character, Yogi Bear's sidekick. (1961)

528 partition

primes	641	643	647	2,647	3,643	4,643	4,649	5,641	5,647	6,421	6,427	6,449	6,451

holiness	1	1	1	1	1	1	2	1	1	1	1	2	1

529 semipartition

primes	41	43	47	61	67	149	163	167	241	263	269	347	349	367	401	409	419	421	431

530 trombones

primes	7	17	37	47	61	67	71	73	79	107	127	137	153	167	173	179	227	263	269	271

531 proil

liorp	1,011	1,101	1,211	1,222	1,311	1,333	1,411	1,444	1,511	1,555	1,611	1,666
holiness	1	1	0	0	0	0	0	0	0	0	1	3

"Pugrugoilul" or Tutnese proil is a number with three of the same digit with two other digits between.

pugrugoilul	1,001,001	1,001,021	1,001,031	1,001,041	1,001,051	1,001,061	1,001,071
holiness	4	3	3	3	3	4	3

532 gleek

keelg	1,444	1,555	1,666	1,777	1,888	1,999	2,444	2,555	2,666	2,777	2,888	2,999

The first gugluleekuck prime is forty million, forty thousand, forty-one.

gugluleekuck	4,004,004	4,004,014	4,004,024	4,004,034	4,004,054	4,004,064
holiness	4	3	3	3	3	4

533 raffle

elffar	1,011	1,011	1,110	1,112	1,113	1,114	1,115	1,116	1,117	1,118	1,119	1,121
holiness	1	1	1	0	0	0	0	1	0	2	1	0

534 The first sliorp prime is one million, one hundred seventeen.

sliorp	100,011	100,101	101,001	110,001	111,222	111,333	111,444	111,555	111,666
holiness	3	3	3	3	0	0	0	0	3

535 The first gleex prime is four million, four hundred forty-five thousand, five hundred fifty-seven. Gleex can be superpluralized to "gleexes" or at least four gleeks.

gleexes	444,555,666,777	445,455,666,777	445,545,666,777	445,554,666,777
holiness	3	3	3	3

536 "Tuhc" does not differ from "chut" until it skips over two hundred fifty-two, the first numberdrome. "Cashhashtut" or Tutnese chut describes a two and five, three and four, in either order, with two other digits between, with its first prime four thousand three.

cashhashtut	2,005	2,015	2,035	2,045	2,065	2,075	2,085	2,095	2,105	2,115	2,135

holiness	2	1	1	1	2	1	3	2	1	0	0

"Wochut" describes numbers with a four and seven or a five and six.

wochut	47	56	65	74	147	156	165	174	247	256	265	274	347	356	365	374	407	417

537 "Sesquichut" describes a chut and a half, a number with three of two through five, with its first prime fourteen twenty-three.

sesquichut	234	235	243	245	253	254	1,234	1,235	1,243	1,245	1,253	1,254	1,324

538 The first ertauq prime is ten thousand, two hundred twenty-three.

ertauq	1,023	1,032	1,203	1,302	2,013	2,031	2,103	2,301	3,012	3,021	3,102	3,201
holiness	1	1	1	1	1	1	1	1	1	1	1	1

539 Before and after semiquatre are numbers with just one or three of zero through three.

quartiquartre	0	1	2	3	14	15	16	17	18	19	24	25	26	27	28	29	34	35	36	37	39	40	41	42
holiness	1	0	0	0	0	0	1	0	2	1	0	0	1	0	2	1	0	0	1	0	1	1	0	0

dodriquartre	102	103	120	123	130	132	201	203	210	213	301	302	310	312	1,024
holiness	1	1	1	0	1	0	1	1	1	0	1	1	1	0	1

540 The first turugoisus prime is thirteen thousand, four hundred twenty-one.

tutrugoisus	1,340	1,342	1,350	1,352	1,360	1,362	1,370	1,372	1,380	1,382	1,390	1,392
holiness	1	0	1	0	2	1	1	0	3	2	2	

541 betrois

siorteb	12	21	103	104	105	106	107	108	109	112	121	122	123	124	125	126	127	128

542 The first Tutnese deux prime, "the dude", is twelve thousand, two hundred three.

dudeuxux	1,220	1,230	1,240	1,250	1,260	1,270	1,280	1,310	1,320	1,340	1,350	1,360
holiness	1	1	1	1	2	1	3	1	1	1	1	2

semideux	0	1	11	12	13	14	15	16	17	18	19	20	21	30	31	40	41	50	51	60	61	70	71	80
holiness	1	0	0	0	0	0	0	1	0	2	1	1	0	1	0	1	0	1	0	2	1	1	0	3
untrimmed	1	2	22	23	24	25	26	27	28	29	210	31	32	41	42	51	52	61	62	71	72	81	82	91

xued	102	103	104	105	106	107	108	109	201	301	401	501	601	701	801	901	1,002
holiness	1	1	1	1	2	1	3	2	1	1	1	1	2	1	3	2	2

The numbers including seven, eight and nine, rather than being called "woduex", are called "seven-up". The drink originally called by Charles L. Grigg "Bib-Label Lithiated Lemon-Lime Soda" in 1929 was re-named for its seven ingredients (carbonated water, sugar, citrus oils, citric acid, sodium citrate and lithium citrate) and its lithium "uppers". Lithium drinks were banned in 1948.

seven-up	789	798	879	897	978	987	7,789	7,798	7,879	7,889	7,897	7,898	7,899	7,978
holiness	3	3	3	3	3	3	3	3	3	5	3	5	4	3

543 "Un" is just French for zeroish, but it gives us a simple name for Tutnese uns or "ununsus", a number with zeroes with two other digits between. The first ununsus prime is 101,107, so the primes below that would by backformation all be unsus primes.

ununsus	10,110	10,120	10,130	10,140	10,150	10,160	10,170	10,180	10,190	10,210
holiness	2	2	2	2	2	3	2	4	3	2

544 Extrapolating from bicycle we can call even bicycles divided by two "unicycles" rather than semicinqs and odd bicycles not divisible by two, "tricycles", rather than terticinqs.

unicycles	5,117	5,162	5,171	5,216	6,017	6,512	6,521	7,016	7,115	7,151	7,160

tricycles	10,243	10,423	12,043	14,023	14,203	20,143	20,341	20,413	20,431	21,043
holiness	1	1	1	1	1	1	1	1	1	1

545 "Xis" would not differ from "six" until they skip over twelve thousand, three hundred forty and are not a reference to other letters of the Greek alphabet, like the transfinite epsilons through the omegas. Six mate or "wosix" describes numbers with four through nine.

susixux	1,660,662,663,664,665	1,660,662,663,664,675	1,660,662,663,664,685
holiness	11	10	12

wosix	456,789	456,798	456,879	456,897	456,978	456,987	465,789	465,879	465,897
holiness	4	4	4	4	4	4	4	4	4

"Counterwosix" describes number names that include words with letter counts of four though nine and therefore not one, two, six, ten, seven-one, etc.

counterwosix	3	4	5	7	8	9	11	12	13	14	15	16	17	18	19	20	21	22	23	24	25	26	27	28	29

546 tertisix

hypersum	10	1,012	101,213	10,121,314	1,012,131,415	101,213,141,520
holiness	1	1	1	1	1	1
dusub	1	137	13,698	1,369,771	136,977,092	13,697,709,175

547 semisix

hypersum	102	102,103	102,103,104	102,103,104,105	102,103,104,105,120
holiness	1	2	3	4	5
pentasub	1	688	687,965	687,965,304	687,965,303,902

548 besix

hypersum	1,023	10,231,024	102,310,241,025	1,023,102,410,251,032
holiness	1	2	3	4
heptasub	1	9,329	93,294,863	932,948,636,823

549 Tutnese sept or "susepugtut" describes a number with zero through six, with two other digits between. Sept mate or "wosept" describes a number with three through nine.

susepugtut	1,002,003,004,005,006,000	1,002,003,004,005,006,010	1,002,003,004,005,006,020
holiness	13	12	12

wosept	3,456,789	3,456,798	3,456,879	3,456,897	3,456,978	3,456,987	3,457,689
holiness	4	4	4	4	4	4	4

550 aka "seven-down", the antonym of seven-up; Tutnese huit or "hutchuitut" describes a number with seven through zero, with two other digits between.

hutchuitut	100,200,300,400,500,600,700	100,200,300,400,500,600,710	100,200,300,400,500,600,720
holiness	15	14	14

The "pepper" numbers, also named for a soda pop, are named for Dr. Pepper time, "ten, two and four". The first pepper prime is 10,243.

pepper	1,024	1,042	2,410	4,210	10,204	10,214	10,224	10,234	10,240	10,241	10,242
holiness	1	1	1	1	2	1	1	1	2	1	1

551 Three, two and one of the huit digits are call semdodri-, quart- and sesuncihuit.

semdodrihuit	102	103	104	105	106	107	120	123	124	125	126	127	130	132	134	135
holiness	1	1	1	1	2	1	1	0	0	0	1	0	1	0	0	0

quartihuit	10	12	13	14	15	16	17	20	21	23	24	25	26	27	30	31	32	34	35	36	37	40	41	42
holiness	1	0	0	0	0	1	0	1	0	0	0	0	1	0	1	0	0	0	0	1	0	1	0	0

sesuncihuit	0	1	2	3	4	5	6	7	18	19	28	29	38	39	48	49	58	59	68	69	78	79	188	189	288
holiness	1	0	0	0	0	0	1	0	2	1	2	1	2	1	2	1	2	1	2	1	2	1	4	3	4

552 Neuf mate or "woneuf" describe numbers with digits one through nine.

woneuf	123,456,789	123,456,798	123,456,879	123,456,897	123,456,978
holiness	4	4	4	4	4

Tutnese neuf or "nuneufud" describes a number with zero through eight with two other digits between.

nuneufud	1,000,002,003,004,005,006,007,008	1,000,002,003,004,005,006,007,018
holiness	20	19

Tertineuf describes a number with one-third of the numbers between zero and eight and bebeuf two-thirds.

tertineuf	102	103	104	105	106	107	108	120	123	124	125	126	127	128	130	132
holiness	1	1	1	1	2	1	3	1	0	0	0	1	0	2	1	0

beneuf	102,345	102,346	102,347	102,345	102,346	102,347	102,348	102,354
holiness	1	2	1	1	2	1	3	1

The Hansen numbers are named for Annika Hansen, aka "Seven of Nine", from Star Trek: Voyager: "Scorpion (Part 2)" by Brannon Braga and Joe Menosky, including seven of the digits one through nine. By beheading six of nine would be "Ansen numbers" and four of nine "En numbers". The first Ansen prime is 123,547. Other similar numbers can be described with Borg designations from One of One (repdigits) to Ten of Ten (pandigitals).

Hansen	1,234,567	1,234,576	1,234,657	1,234,675	1,234,756	1,234,765	1,235,467
holiness	1	1	1	1	1	1	1

Ansen	123,456	123,457	123,458	123,459	123,465	123,467	123,468	123,469	123,475
holiness	1	0	2	1	1	1	3	2	0

En	1,234	1,235	1,236	1,237	1,238	1,239	1,245	1,246	1,247	1,248	1,249	1,253	1,254

Tutnese En or "enun" describes numbers with four of one through nine, with two other digits between.

Enun	1,002,003,004	1,002, 003,054	1,002,003,064	1,002,003,074	1,002,003,084
holiness	6	5	6	5	7

Between the two extremes, the repunits and pandigitals, are the Malcolm numbers, named for "Malcolm in the Middle" by Linwood Boomer, Michael Glouberman and Andy Bobrow, describing those with two of the digits one through three; three of one through five; four of one through seven or five of one through nine.

Malcolm	12	13	21	23	31	32	102	103	112	113	120	121	122	124	125	126	127	128

554 Besides meaning five hundred nine in Roman numerals, "dix" may be where we Anglos get our verb "dicker" for haggling over price and "Dixie" from the nickname for Louisiana's ten-dollar bill. "Decidix" would be another name for repunits.

555 Tutnese book or "bubookuck" describes four of one digit, with two other digits between.

bubookuck	1,001,001,001	1,001,001,021	1,001,001,031	1,001,001,041	1,001,001,051
holiness	6	5	5	5	5

556 Tutnese boox or "bubooxux" describes eight of one digit, with two other digits between.

bubooxux	1,001,001,001,001,001,001,001	1,001,001,001,001,001,001,021
holiness	14	13

The first hyperpluralized **bubooxuxes** number is higher than forty-threeplex.

557 "Sesdextriwonxes" creatively describes in one word numbers with five ones.

sesdextriwonxes	11,111	101,111	112,111	113,111	115,111	116,111	117,111	118,111

558 Tutnese bust or "bubusustut" describes a number summing to at least twenty-two with two other digits between. The first bubusustut prime is seven million, eight thousand, thirty-seven.

bubusustut	7,007,008	7,007,018	7,007,028	7,007,038	7,007,048	7,007,058
holiness	6	5	5	5	5	5

"Counterbust" describes a number name with a letter count higher than twenty-one, that is, at least one hundred seventy-three".

counterbust	173	177	178	273	277	278	324	325	327	328	329	447	457

559 "Sequibusts" describes a number with a weight higher than sixty-three, "sesdextribusts" over a hundred five.

sesquibusts	19,999,999	29,999,999	39,999,999	49,999,999	59,999,999	69,999,999
holiness	7	7	7	7	7	8

sesdextribusts	799,999,999,999	899,999,999,998	899,999,999,999	999,999,999,997
holiness	11	12	13	11

560 "Buckets" describes both the right and left, two through six plus eight.

buckets	1,234,568	1,234,586	1,234,658	1,234,685	1,234,856	1,234,865	1,235,468
holiness	3	3	3	3	3	3	3

"Buck" on the other hand is the acronym that describes binarish-unindexed cubed kyuubi and "doe" its complimentary sequence. It can be pluralized to "bux" and hyperpluralized to "buxes". The first Tutnese buck or "bubuckuk" would be 6,008,005,009.

buck	6,859	24,389	59,319	117,649	205,379	328,509	493,039	704,969	970,299
holiness	4	3	2	2	2	4	3	4	4
doe	3,140	40,680	75,610	295,030	506,960	671,490	794,620	882,350	2,119,400

bux	243,896,859	593,196,859	685,924,389	685,959,319	1,176,496,859
holiness	7	6	7	6	6

does	40,603,140	756,103,140	314,040,680	314,075,710	2,953,029,700
holiness	4	3	6	2	5

buxes	24,389,685,959,396,859	243,896,859,685,959,319	5,924,389,243,896,859
holiness	13	12	11

Many other acronums end in -uck: "chuck" (cubed hachibi-unindexed cubed kyuubi), "guck" (gobi-unindexed cubed kyuubi), "muck" (Mersenne-unindexed cubed kyuubi), "puck" (prime-unindexed cubed kyuubi), "shuck" (shichibi-unindexed cubed kyuubi), "snuck" (sorted nibi-unindexed cubed kyuubi) and "suck" (sanbi-unindexed cubed kyuubi). Mates such as "cheese" for "chucky", "hockey" for "puck" and "blow" for "suck" are easily matched.

"Oe" describes oddly evens and so modifies sequences like Fibonacci, Fibonacci lucky, hachibi, rokubi, sanbi lucky, shichibi, ternarish. "Overshoe" describes shichibi uptown neighbors, i. e., hachibi. "Eo" however describes even/odds and so does nothing: "cameo" cubes-and-Mersennes, "eocene" cubed even nibi evens, "eohippus" hachibi-indexed palindromic prime-unindexed sanbi, "eoline" lucky-indexed nibe evens, "eon" nibi, "eons" nibi sorted, "Galileo" describes gobi-and-lucky-indexed luckies, "Leo" luckies, "mimeo" Mersenne-indexed Mersennes, "oleo" odd luckies and "paseo" prime-and-sanbi.

561 The singular of "craps" describes just one one, two or six and "becrap" one six with a one or two, but not both.

crap	1	2	6	10	13	14	15	17	18	19	20	23	24	25	27	28	29	31	32	36	41	42	46	51	52	56

becrap	16	26	61	62	106	136	146	156	160	163	164	165	167	168	169	206	236	246
holiness	1	1	1	1	2	1	1	1	2	1	1	1	1	3	2	2	1	1

Tutnese craps or "cashrugapugsus" describes craps with two other digits between.

cashrugapugsus	1,001	1,021	1,031	1,032	1,041	1,042	1,051	1,052	1,061	1,062	1,071
holiness	2	1	1	1	1	1	1	1	2	2	1

562 Tutnese quad or "quackuadud" describes four of one digit with two other digits between.

quackuadud	1,001,001,001	1,001,001,021	1,001,001,031	1,001,001,041	1,001,001,051
holiness	6	5	5	5	5

"Sesquiquad" describes six-of-a-kind. The next higher than a million is the non-repunit, a million, eleven thousand, one hundred eleven. The first sesquiquad prime is 1,111,151.

sesquiquad	111,111	222,222	333,333	444,444	555,555	666,666	777,777	888,888	999,999

563 The word redux means "bringing back" as in Fortuna Redux, the goddess of safe returns, or in John Dryden's *Astraea Redux*, Anthony Trollope's *Phineas Redux* or John Updike's *Rabbit Redux*. Tutnese redux or "rugeduduxux" describes four twos with two other digits between, which can be superpluralized to "rugeduxexes".

rugeduduxux	2,002,002,002	2,002,002,012	2,002,002,032	2,002,002,042
holiness	6	5	5	5

564 Tutnese quint or "quackuinuntut" describes five of one digit with two other digits between.

quackuinuntut	1,001,001,001,001	1,001,001,001,021	1,001,001,001,031
holiness	8	7	7

565 Tutnese go or "gugo" describes a number with a weight of thirty with two other digits between.

gugo	3,009,009,009	3,009,009,019	3,009,009,029	3,009,009,039	3,009,009,049
holiness	9	8	8	8	8

"Countergo" describes a number name with a word with a letter count of thirty, that is, at least fifteen-hundred-seventy-sevenplex.

566 "Semigugo" describes a number with a weight of fifteen with two other digits between. The first semigugo prime is six thousand, twenty-nine.

semigugo	6,009	6,019	6,029	6,039	6,049	6,059	6,069	6,079	6,089	6,099	6,109	6,119
holiness	4	3	3	3	3	3	4	3	5	4	3	2

567 "Sesquigugo" describes a number with a weight of forty-five with two other digits between.

semigugo	9,009,009,009,009	9,009,009,009,019	9,009,009,009,029
holiness	13	12	12

568 "Bubegugo" describes a number with a weigh of twenty with two other digits between.

bubegugo	2,009,009	2,009,019	2,009,029	2,009,039	2,009,049	2,009,059	2,009,069
holiness	6	5	5	5	5	5	6

569 "Tertigugo" describes a number with a weight of ten with two other digits between.

tertigugo	1,009	1,019	1,029	1,039	1,049	1,059	1,069	1,079	1,089	1,099	1,109
holiness	3	2	2	2	2	2	3	2	4	3	2

570 hatch

broken-up	56	3,641	203,952	1,326,021	3,394,897,328	899,661,052,441
holiness	1	1	2	2	6	7
tetrasub	1	67	3,735	242,875	62,179,713	16,477,866,959
broken-down	1	57	3,193	207,602	53,149,305	14,084,773,427
broken	1	56	57	3,193	3,641	203,952

hypersum	56	5,665	5,665,156	5,665,156,165	5,665,156,165,256
holiness	1	2	3	4	5
tetrasub	1	104	103,761	103,760,954	10,376,954,571

hatchling

broken-up	5	31	160	991	24,935	649,301	22,750,470	819,666,221
dusub	1	4	22	134	3,375	87,873	3,078,941	110,929,760
broken-down	1	6	31	192	4,831	125,798	4,407,761	158,805,194
broken	1	5	6	31	160	192	991	4,831

hypersum	5	56	5,615	561,516	56,151,625	5,615,162,526	561,516,252,635
dusub	1	8	760	75,993	7,599,296	759,929,611	75,992,961,092

571 Technically according to *Crisis on Cloud City* by Michael Stern, etal., it's the absolute value of twenty-three with cards valued at minus fifteen to plus fifteen. A sub-sabacc we contract to "subacc" for a number with a weight of twenty-two so that a "semisubacc" has a weight of eleven. "Countersabacc" describes a number name with a letter count of twenty-three, that is, at least a hundred-seventy-sevenplex.

subacc	499	589	598	688	778	787	877	958	985	1,399	1,489	1,498	1,588	1,678
holiness	2	3	3	5	2	2	2	3	3	2	3	3	4	3

semisubacc	29	38	47	56	65	74	83	92	119	128	137	146	155	164	173	182	191	209

572 The mate of manque is "womanque".

womanque	5,678	5,687	6,587	7,568	7,586	8,567	8,576	1,5678	1,5687
holiness	3	3	3	3	3	3	3	3	3

573 "Semimanque" describes a number with two of one, two, three or four.

semimanque	12	13	14	21	23	24	31	32	34	41	42	43	102	103	104	120	125	126	127

broken-up	12	157	1,896	24,805	572,411	13,762,669	427,215,150	13,684,647,469
dusub	1	21	256	3,357	77,467	1,862,575	57,817,283	185,201,5641
broken-down	1	13	157	2,054	47,399	1,139,630	35,375,929	1,133,169,358
broken	1	12	13	157	1,896	24,805	47,399	572,411

hypersum	12	1,213	121,314	12,131,421	1,213,142,123	121,314,212,324
dusub	1	164	16,418	1,641,809	164,180,933	16,418,093,286

574 semipasse

broken-up	67	4,557	305,386	20,770,805	1,620,428,176	128,034,596,709
holiness	1	0	4	5	5	7
tetrasub	1	84	5,593	380,431	29,679,177	2,345,035,439
broken-down	1	68	4,557	309,944	2,480,189	1,910,544,875

hypersum	67	6,768	676,869	67,686,976	6,768,697,678	676,869,767,879
holiness	1	4	6	7	9	10
tetrasub	1	124	12,397	1,239,730	123,973,022	12,397,302,243

575 quartipasse

broken-up	6	43	264	1,891	30,520	520,731	9,403,678	179,190,613	4,668,359,616
holiness	1	0	1	3	2	1	5	4	7
dusub	1	6	36	256	4,131	70,473	1,272,649	24250813	631,793,771
broken-down	1	7	43	308	4,971	84,815	1531641	29,185,994	76,036,7485
broken	1	6	7	43	264	308	1,891	4,971	30,520

hypersum	6	67	678	6,789	678,916	67,891,617	6,789,161,718	678,916,171,819
holiness	1	1	3	4	5	5	7	8
dusub	1	9	92	919	91,881	9,188,131	918,813,124	91,881,312,407

576 "Quarting" is a somewhat more memorable name for 'quartimanque'."

quarting	1	2	3	4	10	15	16	17	18	19	20	25	26	27	28	29	30	35	36	37	38	39	40	45

The mate of ng is "wong".

wong	58	67	76	85	158	167	176	185	258	267	276	285	358	367	376	385	458	467	476
holiness	2	1	1	2	2	1	1	2	2	1	1	2	2	1	1	2	2	1	1

The first Tutnese ng or "nungug" prime is two thousand three.

nungug	1,004	1,024	1,034	1,054	1,064	1,074	1,084	1,094	1,204	1,224
holiness	2	1	1	1	2	1	3	2	1	0

"King" describes kyuubi-indexed ng, "ping" prime-indexed, "sing" sanbi-indexed ng. Many, many other subsequences end in -ing, e. g., "singing', sing-indexed ng.

king	141	154	230	239	312	414	614	823	1,023	1,045	1,140	1,203	1,236	1,245	1,283
ping	23	32	104	123	143	145	149	164	203	232	234	241	293	320	326
sing	32	114	141	144	147	154	184	213	230	233	236	239	253	283	312

577 "Counterstiff" describes a number with between twelve and sixteen letters.

counterstiff	77	101	102	103	104	105	106	107	108	109	110	111	112	120	130	140	150

578 "Counterhalf-stiff" describes a number between six and eight letters. "Hundred" or "twoplex", "thousand" and "million" with a leading one understood count. So too would "fourplex", "fiveplex", "nineplex", "tenplex" and "googol".

counterhalf-stiff	11	12	20	30	70	80	90	100	1,000	10,000	100,000	1,000,000

Using Joyce's Roman numeral suffixes, as we do, many more numbers are included in the six-to-eight-letter limit: oneplev (5), twoplev (25), fourplev (625), fiveplev (3,125), sixplev (15,625), nineplev (1,953,125), tenplev (9,765,625), oneplel (50), twoplel (2,500), fourplel (6,250,000), fiveplel (312,500,000), sixplel (15,625,000,000), etc.

579 blind

hypersum	0	1	12	123	1,234	12,347	1,234,710	123,471,011	12,347,101,112
holiness	1	0	0	0	0	0	1	1	1
untrimmed	1	2	23	234	2,345	23,458	2,345,821	234,582,122	23,458,212,223

The triops is a kind of tadpole shrimp with two close-set compound eyes with a third eye between. The tetrops were chiropteran mutants with four eyes equally spaced around their heads in "Time and the Rani" by Pip and Jane Baker.

580 cyclops

broken-up	5	31	160	991	15,025	241,391	4,360,063	83,082,588	2,081,424,763
dusub	1	4	22	134	2,033	32,669	590,071	11,244,006	281,690,210
broken-down	1	6	31	192	2,911	46,768	844,735	16,096,733	403,263,060
broken	1	5	6	31	160	192	991	2,911	15,025

hypersum	5	56	568	5,689	568,915	56,891,516	5,689,151,618	568,915,161,819
dusub	1	8	77	770	76,994	7,699,429	769,942,946	76,994,294,562

581 monophemous

broken-up	56	3,249	182,000	10,559,249	718,210,932	49,567,113,557
holiness	1	1	5	3	4	2
tetrasub	1	60	3,333	193,399	13,154,492	907,853,353
broken-down	1	57	3,193	185,251	12,600,261	869,603,260
broken	1	56	57	3,193	3,249	182,000

hypersum	56	5,658	565,859	56,585,965	5,658,596,568	565,859,656,869
holiness	1	3	4	5	8	10
tetrasub	1	104	10,364	1,036,408	103,640,811	10,364,081,137

582 half-cockeyed

broken-up	1	3	4	11	158	2,381	38,254	652,699	11,786,836	212,815,747
broken-down	1	2	3	8	115	1,733	27,843	475,064	8,578,995	120,580,994
broken	1	2	3	4	8	11	115	158	1,733	2,381

hypersum	1	12	1,210	121,013	12,101,314	1,210,131,415	121,013,141,516

583 "Looped" subsequences can be formed by moving the tail digit to the head, for example, looping the primes yields the "eprim primes". The "looped" and "half-looped" numbers do not differ from each other or from "rebmuns" until they get as high as three digits.

eprim	13	17	31	37	71	73	79	97	113	131	197	199	271	311	313	337	373	379	397

Re-looping or "plooing" the primes yields the "mepri primes".

mepri	...	103	107	113	131	149	173	181	191	197	199	307	311	317	331	337	347	359
holiness	...	1	1	0	0	1	0	2	1	1	2	1	0	0	0	0	0	1

Re-looping once again yields the "oploed primes".

oploed	...	1,117	1,123	1,129	1,151	1,153	1,181	1,187	1,193	1,213	1,231	1,237	1,259

Re-looping yet again or yields the "rimep primes".

rimep	...	11,113	11,117	11,119	11,131	11,149	11,173	11,197	11,239	11,251	11,261

584 Tutnese yacht or "yuckatut" describes five of the same digit higher than five with two other digits between.

yuckatut	6,006,006,006,006	6,006,006,006,016	6,006,006,006,026
holiness	13	12	12

585 The mate of napoletana is "wonapoletana", a number with six, seven and eight. The first such prime is 2,687.

wonapoletana	678	687	768	786	867	876	1,678	1,687	1,768	1,786	1,867	1,876	2,678
holiness	3	3	3	3	3	3	3	3	3	3	3	3	3

586 The mate of benapoletana could be called either "bewonapoletana" or "wobenapoletana" and that of tritonapoletana either "tritowonapoletana" or "wotritonapoletana", so they're called "two of six, seven, eight" and "one of six, seven, eight" instead.

2 of 6, 7, 8	67	68	76	78	86	87	167	168	176	178	186	187	267	268	276	278	286	287
holiness	1	3	1	2	3	2	1	3	1	2	3	2	1	3	1	2	3	2
1 of 6, 7, 8	6	7	8	16	17	18	26	27	28	36	37	38	46	47	48	56	57	58
holiness	1	0	2	1	0	2	1	0	2	1	0	2	1	0	2	1	0	2

587 "Semibart" is just one of the four and "sesquibart" three of the four.

semibart	2	3	5	6	12	13	15	16	20	21	24	27	28	29	30	31	34	37	38	39	42	43	45	46
trimmed	1	2	4	5	1	2	4	5	1	10	13	16	17	18	2	20	23	26	27	28	31	32	34	35

sesquibart	235	236	253	256	325	326	356	365	523	526	532	536	623	625	632

The mate of bart or "wobart" is three and seven or four and six.

wobart	37	46	64	73	137	146	164	173	237	246	264	273	307	317	327	337	347	357	367

588 bubarugtut

primes	5,003	5,023	5,113	5,153	5,233	5,273	5,303	5,323	5,353	5,393	5,413	5,443
holiness	2	1	0	0	0	0	1	0	0	1	0	0

589 The first bubinungugo prime is 101,107. The mate of bubinungugo is "wobubingugo", third nines.

wobubinungugo	9,009	9,019	9,029	9,039	9,049	9,059	9,069	9,079	9,089	9,109	9,119
holiness	4	3	3	3	3	3	4	3	5	3	2

590 The mate of susamum or "wosusamum" is seven and eight with two other digits between. The first such prime is 170,081.

wosusamum	7,008	7,018	7,028	7,038	7,048	7,058	7,068	7,098	7,108	7,118	7,128
holiness	4	3	3	3	3	3	4	4	3	2	2

591 "Counterpontoon" describes a number name with twenty-one letters.

counterpontoon	127	137	174	175	179	187	197	227	237	274	275	279	287	297	321

592 Tutnese pat or "pugatut" describes a number with a letter count of seventeen to twenty-one, with two other digits between.

pugatut	8,009	8,019	8,029	8,039	8,049	8,059	8,069	8,079	8,089	8,099	8,109	8,119
holiness	5	4	4	4	4	4	5	4	4	5	4	3

The mate to pat is not wopat, but "mike" from Patricia Pemberton and Michael Conovan who mated in "Pat and Mike" (1952) by Ruth Gordon and Garson Kanin.

mike	10	105	106	107	108	109	118	127	136	147	158	163	172	181	190	204	205	206
holiness	1	1	1	1	3	2	0	0	1	0	2	1	0	2	3	1	1	2

593 "Semipat" describes a number weighing nine or ten.

semipat	9	18	19	27	28	36	37	45	46	54	55	63	64	72	73	81	82	90	108	109	117	118
holiness	1	2	1	0	2	1	0	0	1	0	0	1	1	0	0	2	2	2	3	2	0	2

"Tertipat" describes a number weighing six or seven.

tertipat	6	7	15	16	24	25	33	34	42	43	51	52	60	61	105	106	114	115	123	124	132
holiness	1	0	0	1	0	0	0	0	0	0	0	0	1	1	1	1	0	0	0	0	0

"Sesquipat" describes a number weighing twenty-seven to thirty.

sesquipat	999	1,999	2,999	3,999	4,899	4,989	4,998	5,799	5,889	5,898	5,988
holiness	3	3	3	3	4	4	4	2	5	5	5

594 Tutnese point or "pugoinuntut" describes a number with a weight between four and ten with two other digits between.

pugoinuntut	1,003	1,004	1,005	1,006	1,007	1,008	1,009	1,012	1,013	1,014	1,015
holiness	2	2	2	2	2	4	3	1	1	1	1

"Semipoint" describes a number weighting between two and five.

semipoint	2	3	4	5	11	12	13	14	20	21	22	23	30	31	32	40	41	50	101	102	103	104	110

"Sesquipoint" describes a number weighting between six and fifteen.

sesquipoint	6	7	8	9	15	16	17	18	19	24	25	26	27	28	29	33	34	35	36	37	38	39	42	43
holiness	1	0	2	1	0	1	0	2	1	0	0	1	0	2	1	0	0	0	1	0	2	1	0	0

"Counterpoint" describes a number name with a letter count of four through six or eight through ten.

counterpoint	4	5	7	8	9	11	12	13	14	18	19	20	24	25	29	30	34	35	39	40	44	45	49	50

595 'Farpoint' is an allusion to Star Trek: The Next Generation: "Encounter at Farpoint" by D. C. Fontana and Gene Roddenberry. "Farties", by extrapolation, describe numbers weighing twenty or ninety.

farties	299	389	398	479	488	497	569	578	587	596	659	668	677	686	695	767	776	866
holiness	2	3	3	1	4	1	2	2	2	2	2	4	1	4	3	1	1	4

"Farteen" describes numbers weighing either thirteen or nineteen.

farteens	49	58	67	76	85	94	139	148	157	166	175	184	193	229	238	247	256	265	274
holiness	1	2	1	1	2	1	1	2	0	2	0	2	1	1	2	0	1	1	0

596 "Semibaccarat" describes numbers with weights divisible by five, "quintibaccarat" those with weights divisible by two and "sesquibaccarat" those divisible by fifteen.

semibaccarat	5	14	19	23	28	32	37	41	46	50	55	64	69	73	78	83	87	91	96	104	113	122

quintibaccarat	2	4	6	8	11	13	15	17	19	20	22	24	26	28	31	33	35	37	39	40	42	44	46	48

sesquibaccarat	69	78	87	96	159	168	177	186	195	249	258	267	276	285	294	339	348
holiness	2	2	2	2	1	3	0	3	1	1	2	1	1	2	1	1	2

"Counterbaccarat" describes a number name with a word with a letter count divisible by ten, that is, twenty-, thirty-, eighty-four, -five, -nine, seventy-one, -two or -six, one-hundred-twenty-, -thirty, eighty-four, -five, -nine, etc.

counterbaccarat	24	25	29	34	35	39	71	72	76	124	125	129	134	135	139	184	185	189

597 "Sesquimournival" describes a number with six digits the same four or higher. The first such prime is 2,999,999.

sesquimournival	444,444	555,555	666,666	777,777	888,888	999,999	1,000,000

"Semimournival" describes a number with two digits the same four or higher.

semimournival	44	55	66	77	88	99	144	155	166	177	188	199	244	255	266	277	288

598 Combo

prime	23	43	67	89	103	107	109	127	167	239	367	389	431	439	457	541	547	653	659

599 catchamumbubo

prime	2,003	2,053	2,063	2,083	2,113	2,143	2,153	2,473	2.483	2,503	2,593	2,663
holiness	2	1	2	3	0	0	0	0	2	1	1	2

600 pugo

prime	1,021	1,031	1,051	1,061	1,091	1,201	1,231	1,291	1,301	1,321	1,361	1,381
holiness	1	1	1	2	2	1	0	1	1	0	1	2

601

pugrugoilul	1,001,041	1,001,081	1,001,291	1,001,321	1,001,381	1,001,401	1,001,431
holiness	3	5	3	2	4	3	2

602 "Run" alternatively can mean the acronum rokubi-unindexed nibi.

run	...	42	62	72	82	92	102	112	122	132	142	162	172	182	192	202	212
holiness	...	0	1	0	2	1	1	0	0	0	0	1	0	2	1	1	0

"Run-off" describes the neighbors of runs and "overrun" a run with a non-neighbor kicker. The first such prime 1,207.

run-off	101	103	119	121	233	235	242	244	323	325	344	346	353	355	455	457	464	466
holiness	1	1	1	0	0	0	0	0	0	0	0	1	0	0	0	0	1	2

overrun	1,024	1,025	1,026	1,027	1,028	1,029	1,042	1,052	1,062	1,072	1,082	1,092
holiness	1	1	2	1	3	2	1	1	2	1	3	2

603 rugunun

prime	2,440,481	2,440,541	2,440,561	2,440,591	2,440,621	2,440,681	2,440,831
holiness	3	1	2	2	2	4	3

604 Variations on straight numbers are "four-shortened", without a four; "indeuced", with an interior two; "ontwo", with a two on either end, "open-ender", with four neighboring digits; "timbuk", open-ender with a two far from six through nine.

four-shortened	1,023	1,032	1,203	1,230	1,302	1,320	1,235	1,253	1,325	1,352	1,523
holiness	1	1	1	1	1	1	0	0	0	0	0

indeuced	10,234	10,243	10,324	10,423	12,034	12,043	12,304	12,340	12,403	12,430
holiness	1	1	1	1	1	1	1	1	1	1

ontwo	10,342	10,432	13,042	13,402	14,032	14,302	20,134	20,143	20,314	20,341
holiness	1	1	1	1	1	1	1	1	1	1

open-ended	1,234	1,324	1,342	1,423	1,432	2,314	2,341	2,413	2,431	2,345	2,354

605 "Sesquisexette" decribes a number with nine neighboring digits, not quite the same as "andigital" in which they may not all be neighbors or "penholodigital", which include all except zero. Sexette excluding zero is "penholosexette".

sesquisexette	102,345,678	102,345,687	102,345,768	102,345,786	102,345,867
holiness	4	4	4	4	4

penholosexette	123,456	123,465	123,546	123,564	123,645	123,654	124,356	124,365
holiness	1	1	1	1	1	1	1	1

606 Septet excluding zero is "penholoseptet".

penholoseptet	1,234,567	1,234,576	1,234,657	1,234,675	1,234,756	1,234,765
holiness	1	1	1	1	1	1

607 Tutnese john or "jugohutchnun" describes five neighboring odds with two other digits between.

jugohutchnun	1,003,005,007,009	1,003,005,007,029	1,003,005,007,049
holiness	9	8	8

608 Similarly Tutnese steven or "sustutevuvenun" describes five neighboring evens with two other digits between.

sustutevuvenun	2,110,114,116,118	2,110,114,116,138	2,110,114,116,158
holiness	4	4	4

609 "Poh-tac" describes a non-numberdrome cat-hop that remains a cat-hop when reversed.

poh-tac	112	122	133	144	155	166	177	188	199	211	221	223	224	225	226	227	228
holiness	0	0	0	0	0	2	0	4	2	0	0	0	0	0	1	0	2

610 crag

prime	229	337	373	661	733	2,029	3,037	3,307	5,503	9,029	9,209	20,029	30,307
holiness	1	0	0	2	0	2	1	1	1	3	3	3	2

611 cashrugagug

prime	2,002,009	2,002,019	2,002,079	2,002,159	2,002,199	2,002,339
holiness	5	4	4	3	4	3

612 A Tutnese gow or "gugowash" is a gow with two other digits between.

gugowash	3,006	3,016	3,026	3,046	3,056	3,076	3,086	3,096	3,106	3,116	3,126
holiness	3	2	2	2	2	2	4	3	2	1	1

613 "Double gow" describes a number with three through six, while "gow po" describes pairs of threes and sixes or fours and fives.

double gow	3,456	3,546	3,564	3,645	3,654	4,356	4,365	4,536	4,563	4,635	4,653
holiness	1	1	1	1	1	1	1	1	1	1	1
gow po	3,366	3,636	3,663	4,455	4,545	4,554	5,445	5,454	5,544	6,336	6,363
holiness	2	2	2	0	0	0	0	0	0	2	2

614 "Double gow po" describes pairs of threes through sixes.

double gow po	33,445,566	33,445,656	33,445,665	33,446,556	33,446,565	33,446,655
holiness	2	2	2	2	2	2

615 "Double chut" describes a number with two through five, "chut po" pairs of twos and fives or threes and fours and "double chut po" pairs of twos through fives.

double chut	2,345	2,354	2,435	2,453	3,245	3,254	3,425	3,452	3,524	3,542	4,235	4,253
chut po	2,255	2,525	2,552	3,344	3,434	3,443	4,334	4,343	4,433	5,225	5,252	5,522

double chut po	22,334,455	22,334,545	22,334,554	22,335,445	22,335,454	22,335,544

616 In *World Wide Words* it's said to come from the confusion over whether the Merchant Taylors or the Skinners ranked sixth or seventh. Since 1484 they have alternate in the two rankings. Alternatively it may come from the much older "troubles" in Job 5:19. Rather than Taylor-Skinner or Skinner-Taylor primes, we call them Job primes.

Job primes	36,677	36,767	56,767	57,667	67,679	76,667	76,673	76,679	86,677	86,767
holiness	2	2	2	2	3	3	2	3	4	4

617 luan qi ba zao

primes	487	587	787	877	887	1,087	1,187	1,487	1,783	1,787	1,789	1,871	1,873	1,879
holiness	2	2	2	2	4	3	2	2	2	2	3	2	2	3

618 cinque et sice

primes	563	569	653	659	1,567	1,657	2,657	2,659	4,561	4,567	4,651	4,657	5,167	5,261
holiness	1	2	1	2	1	1	1	2	1	1	1	1	1	1

619 al tres y al cuatro

primes	33,449	34,403	34,439	34,483	34,543	34,843	39,443	40,343	40,433	42,433
holiness	1	1	1	2	0	2	1	1	1	0

620 sandwich

primes	241	1,249	1,423	1,427	1,429	2,141	2,143	2,243	2,341	2,347	2,411	2,417	2,423

half-sandwich

primes	23	29	41	43	47	127	149	211	233	239	251	257	263	269	271	277	281	283	293

621 half-odd

primes	23	29	43	47	61	67	83	89	1,009	1,021	1,049	1,061	1,063	1,069	1,087	1,201

622 quarter-odd and -even

quarter-odd	2,001	2,003	2,005	2,007	2,009	2,021	2,023	2,025	2,027	2,029	2,041	2,043
holiness	2	2	2	2	3	1	1	1	1	2	1	1
quarter-even	1,110	1,112	1,114	1,116	1,118	1,130	1,132	1,134	1,136	1,138	1,150	1,152
holiness	1	0	0	1	2	1	0	0	1	2	1	0

623 alternating

primes	23	29	41	43	47	61	67	83	89	103	107	109	127	149	163	167	181	307	347	349	359

624 two-third-odd

primes	101	103	107	109	127	149	163	167	181	307	347	349	367	383	389	503	509	521
holiness	1	1	1	2	0	1	1	1	2	1	0	1	1	2	3	1	2	0

625 The direct numbers, of course, are named for the opposite of alternating current, direct current, and include both the direct even and the direct odd. The direct odd subsequence is called the Tesla numbers after the odd Nikola Tesla associated with direct current.

primes	211	223	227	229	233	239	241	251	257	263	269	271	277	281	283	293	311	131

Those that remain Tesla when reversed are Alset (pronounced all-set).

primes	13	15	17	19	31	35	37	39	51	53	57	59	71	73	75	79	91	93	95	97	113	115	117	119

626 "Neider" is a backformation from 'Schneider', deleting the Yiddish prefix sch- perhaps from the ancient password "shibboleth" (Judges 12:1-16).

627

holy Mongolian primes	101	103	107	109	307	401	409	503	509	601	607	701	709	809
holiness	1	1	1	1	1	1	2	1	2	2	2	1	2	3

holier Mongolian primes	1,009	2,003	3,001	4,001	4,003	4,007	5,007	5,009	6,007	7,001
holiness	3	3	2	2	2	2	2	3	3	2

628 Some prefer the term mulatto, either because they think it an obsolete description for people. The mate of semiblank, a subsequence of half-nine, can be called by the new term "womulatto".

primes	89	1,399	1,499	1,699	1,949	1,979	1,993	1,997	2,099	2,399	2,699	2,909	2,939
holiness	3	2	2	3	2	2	2	2	3	2	3	3	2

half-nines	19	29	39	49	59	69	79	89	90	91	92	93	94	95	96	97	98	1,009	1,099	1,199
holiness	1	1	1	1	1	2	1	3	2	1	1	1	1	1	2	1	3	3	3	2

On the other hand semiblank can be pluralized to semiblanx and superpluralized to semiblanxes.

semiblanx	1,929	1,939	1,949	1,959	1,969	1,979	1,989	1,990	1,991	1,992	1,993	1,994
holiness	2	2	2	2	3	2	4	3	2	2	2	2

semiblanxes	19,293,949	19,293,959	19,293,969	19,293,979	19,293,989	19,293,990
holiness	4	4	5	4	6	5

629 Tritoblanks could be called triroons and their mates, wotriroons.

wotriroon	109	119	129	139	149	159	169	179	189	191	192	193	194	195	196	197	198
holiness	2	1	1	1	1	1	2	1	3	1	1	1	1	1	2	1	3

tritoblanx	109,119	109,129	109,139	109,149	109,159	109,169	109,179	109,189
holiness	3	3	3	3	3	4	3	5

tritoblanxes	109,119,129,139	109,119,129,149	109,119,129,159	109,119,129,169

holiness	5	5	5	6

630 Quartiblanks could be called quadroons, their mates woquadroons.

primes	1,013	1,019	1,021	1,031	1,033	1,049	1,051	1,061	1,063	1,069	1,087	1,091
holiness	1	2	1	1	1	2	1	2	2	3	3	2

woquadroon	1,009	1,019	1,029	1,039	1,049	1,059	1,069	1,079	1,089	1,091	1,092	1,093
holiness	3	2	2	2	2	2	3	2	4	2	2	2

quartiblanx	10,011,012	10,011,013	10,011,014	10,011015	10,011,016	10,011,017
holiness	3	3	3	3	4	3

quartiblanxes	1,011,101,210,131,014	1,011,101,210,131,015	1,011,101,210,131,016
holiness	4	4	5

631 Quintiblanks could be called quintoons, or 'toons for short, and their mates woquintoons.

primes	10,111	10,133	10,139	10,141	10,151	10,159	10,169	10,177	10,181	10,193
holiness	1	1	2	1	1	2	3	1	3	2

woquintoon	10,009	10,019	10,029	10,039	10,049	10,059	10,069	10,079	10,089	10,091
holiness	4	3	3	3	3	3	4	3	5	3

quintiblanx	1,011,110,112	1,011,110,112	1,011,110,113	1,011,110,114	1,011,110,115

quintiblanxes	10,111,101,121,011,310,114	10,111,101,121,011,310,115	10,111,101,121,011,310,116
holiness	4	4	5

632 Sextiblanks could be called sextroons and their mates wosextroons.

primes	101,111	101,113	101,117	101,119	101,141	101,149	101,159	101,161	101,173
holiness	1	1	1	2	1	2	2	2	1

wosextroon	10,009	10,019	10,029	10,039	10,049	10,059	10,069	10,079	10,089	10,091
holiness	4	3	3	3	3	3	4	3	5	3

sextiblanx	101,111,101,112	101,111,101,113	101,111,101,114	101,111,101,115
holiness	2	2	2	2

sextiblanxes	1,011,111,011,121,101,113,101,114	1,011,111,011,121,101,113,101,115
holiness	4	4

633 Septiblanks could be called septroons and their mates woseptroons.

primes	1,011,137	1,011,139	1,011,163	1,011,167	1,011,191	1,011,217	1,011,221	1,011,229
holiness	1	2	2	2	2	1	1	2

woseptroon	10,009	10,019	10,029	10,039	10,049	10,059	10,069	10,079	10,089	10,091
holiness	4	3	3	3	3	3	4	3	5	3

septiblanx	10,111,111,011,112	10,111,111,011,113	10,111,111,011,114
holiness	2	2	2

The septiblanxes start at higher than twenty-sixplex.

634 Octiblanks could be called octroons and their mates woctroons.

primes	1,011,117	10,111,121	10,111,139	10,111,147	10,111,187	10,111,229	10,111,247
holiness	1	1	2	1	3	2	1

woctroon	10,000,009	10,000,019	10,000,029	10,000,039	1,000,049	10,000,059
holiness	7	6	6	6	6	6

octiblanx	1,011,111,110,111,112	1,011,111,110,111,113	1,011,111,110,111,114
holiness	2	2	2

The octiblanxes start at higher than thirtyplex.

635 holy Devangari

hypersum	0	4	410	41,014	4,101,420	410,142,024	41,014,202,430	410,142,024,334
holiness	1	0	1	1	2	2	3	2

636 The downtown neighbors of the holiest Devangari are sometimes prime.

not-quite holiest	39	399	439	3,999	4,003	4,039	3,043	3,399	4,403	4,439	40,003
holiness	1	2	1	3	2	2	1	2	1	1	3

637 unholy Devangari

primes	2	3	5	7	11	13	17	19	23	29	31	37	53	59	61	67	71	73	79	83	89	97	113	127	131

638 Gurmukhi

unholy	2	3	5	6	7	8	9	10	11	14	22	23	25	26	27	28	29	32	33	35	36	37	38	39

639 Some holiest Gurmukhi numbers are neighbors, while some other neighbors are not-quite.

not-quite holiest	103	139	1,003	1,039	1,043	1,103	1,139	1,399	1,409	1,419	1,429
holiness	1	1	2	2	1	1	1	2	2	1	1

640 holier Gurmukhi

primes	41	101	103	107	109	149	409	419	421	431	461	491	541	601	641	701	719

641 Gujarati

unholy	2	3	5	6	8	9	22	23	25	26	28	29	32	33	35	36	38	39	52	53	55	56	58	59	62	63	65

642 holiest Gujarati

not-quite	1,046	1,073	1,406	1,469	1,703	1,739	10,046	10,073	10,146	10,173

643 holier Gujarati

primes	11	17	41	47	71	101	107	127	137	149	157	167	173	179	401	409	419	421	431

644 Telugu

unholy	1	3	6	7	8	9	11	13	16	17	18	19	20	24	25	31	33	36	37	38	39	61	63	66	67	68	69

645 holiest Telugu

not-quite	2,044	2,053	2,404	2,449	2,503	2,539	4,024	4,051	4,204	4,249	4,501
holiness	1	1	1	1	1	1	1	1	1	1	1

646 Telugu

unholier	1	2	3	4	5	6	7	8	9	10	11	12	13	14	15	16	17	18	19	21	23	26	27	28	29	30	31

647 Khmer

unholy	1	4	6	8	9	11	14	16	18	19	20	22	23	25	27	30	32	33	35	37	41	44	46	48	49	61	64

648 holiest Khmer

not-quite	20,356	20,376	20,536	20,572	20,734	20,752	23,056	23,074	23,506
holiness	2	2	2	1	1	1	2	1	2

649 holier Khmer

primes	23	37	53	73	103	107	109	113	223	227	233	283	307	317	337	347	353	359

650 Mayalam

unholy	2	3	5	6	8	10	22	23	28	32	33	35	36	38	52	53	55	56	58	62	63	65	66	68	82	83
trimmed	1	2	4	5	7	0	11	12	17	21	22	24	25	27	41	42	44	45	47	51	52	54	55	57	71	72

651 holiest Mayalam

not-quite	10,478	10,498	10,748	10,793	10,946	10,973	14,078	14,096

652 holier Mayalam

prime	17	19	41	47	71	79	91	97	101	103	107	109	127	137	139	149	157	167	173	191

653 Bengali

unholy	1	2	3	6	9	11	12	13	16	19	21	22	23	26	29	31	32	33	36	39	61	62	63	66	69	91

654 Bengali holiest

not-quite	3	43	443	4,443	44,443	444,443	4,444,443	44,444,443	444,444.443

655 holier Bengali

primes	47	107	157	257	307	347	457	547	557	577	587	607	647	743	751	757	787

656 Odia

unholy	3	6	8	9	33	36	38	39	63	66	68	69	83	86	88	89	93	96	98	99	333	336	338	339

657 holiest Odia

not-quite	4	54	554	5,554	55,554	555,554	555,554	5,555,554	55,555,554

658 holier Odia

primes	17	41	47	71	101	103	107	127	137	149	151	157	167	173	179	197	211

659 Vai

unholy	1	4	5	11	13	16	17	18	19	20	22	23	26	27	28	29	31	33	36	37	38	39	41	44	45

660 holiest V*ai*

not-quite	8	98	998	9,998	99,998	999,998	9,999,998	99,999,998	999,999,998
holiness	2	3	4	5	6	7	8	9	10

661 holier Vai

primes	23	29	37	67	73	79	83	89	103	107	109	127	137	139	163	167	173	179	197	223

662 Kannada

unholy	1	2	7	11	12	17	21	22	27	71	72	77	111	112	117	121	122	127	177	211	212

663 holiest Kannada

not-quite	3	4	43	44	53	54	443	444	453	454	543	544	553	554	4,443	4,444	4,453

664 holier Kannada

primes	43	53	59	83	89	349	353	359	383	389	409	433	439	443	449	463	499	503

665 holier Thai

primes	2	3	7	11	13	17	19	23	29	31	37	41	43	47	61	67	71	73	79	83	89	97	101	103
trimmed	1	2	6	0	2	6	8	12	18	20	26	30	32	36	50	56	60	62	68	72	78	86	0	2

666 Tamil

unholy	3	5	33	35	53	55	333	335	353	355	533	535	553	555	3,333	3,335	3,353	3,555

667 holier Tamil

primes	17	19	29	41	47	61	67	71	79	83	89	97	101	107	109	127	149	167	173	179	181

668 Holier Lao numbers are those with digits with "only" two holes, four, five, seven, eight or nine.

holier Lao	4	5	7	8	9	44	45	47	48	49	54	55	57	58	59	74	75	77	78	79	84	85	87	88	89	94

669 The curtain of the holy of holies was broken down when Jesus died on Good Friday.

broken-up	1	3	4	11	81	821	9,933	139,883	2,248,061	36,108,859	688,316,382
broken-down	1	2	3	8	59	88	1,064	14,984	240,808	1,700,640	32,552,968
broken	1	2	3	4	8	11	59	81	88	821	1,064

670 Arabic

holy-four	0	4	6	8	9	10	14	16	18	19	20	24	26	28	29	30	34	36	38	39	40	41	42
holiness	1	0	1	2	1	1	0	1	2	1	1	0	1	2	1	1	0	1	2	1	1	0	0

671 Arabic holiest

not-quite	7	87	887	8,887	88,887	888,887	8,888,887	88,888,887	888,888,887

672 holier Arabic

primes	89	109	199	269	499	509	569	599	601	609	661	683	691	709	769	809
holiness	3	2	2	2	2	2	2	2	2	3	2	3	2	2	2	4

673 unholy Arabic

primes	2	3	5	7	11	17	23	31	37	41	43	53	71	73	113	127	131	137	151	157	173	211
trimmed	1	2	4	6	0	6	12	20	26	30	32	42	60	62	2	16	20	26	40	46	62	100

674 The unholy binarish numbers are also known as the repunits.

holy binarish primes	101	10,111	101,111	11,110,111	11,111,101	101,111,111
holiness	1	1	1	1	1	1

675 The not-quite holiest binarish are nine times the repunits.

676 The first holier binarish prime is one million, one hundred ten thousand, one.

677 holy ternarish

primes	101	1,021	1,201	2,011	10,111	10,211	12,011	12,101	21,011	21,101
holiness	1	1	1	1	1	1	1	1	1	1

678 holiest ternarish

not-quite	99	199	999	1,999	9,999	19,999	99,999	199,999	999,999	1,999,999
holiness	2	2	3	3	4	4	5	5	6	6

679 holier ternarish

broken-up	1,001	1,003,003	1,004,007,004	100,601,6021,011	1,106,618,627,119,104
holiness	2	4	6	7	8
heptasub	1	915	915,536	917,367,867	1,009,105,568,817
broken-down	1	1,002	1,003,003	1,005,010,008	1,105,512,011,803
broken	1	1,001	1,002	1,003,003	1,004,007,004

680 unholy ternarish

broken-up	1	3	4	11	235	5,181	575,326	64,441,693	7,798,020,179	943,624,883,352
broken-down	1	2	3	8	171	184	20,432	2,288,568	276,937,160	5,817,968,928
broken	1	2	3	4	8	11	171	184	235	5,181

681 holy quadarish

hypersum	0	10	1,020	102,030	102,030,101	102,030,101,102	102,030,101,102,103
holiness	1	1	2	3	4	4	6

682 holier quadarish

broken-up	1,001	1,003,003	1,004,007,004	1,006,016,021,011	1,026,137,345,438,224
holiness	2	4	4	7	4
heptasub	1	915	915,536	917,367,867	935,716,139,487
broken-down	1	1,002	1,003,003	1,005,010,008	1,025,111,211,163

683 unholy quadrarish include the cheap, cheaper and cheapest numbers.

primes	2	3	11	13	23	31	113	131	211	223	233	311	313	331	1,123	1,213	1,223	1,231
trimmed	1	2	0	2	12	20	2	20	100	112	122	200	202	220	12	102	112	120

684 holiest quintarish

not-quite	99	199	299	399	999	1,999	2,999	3,999	9,999	19,999	29,999	39,999
holiness	2	2	2	2	3	3	3	3	4	4	4	4

685 holier quintarish

hypersum	1,001	10,011,002	100,110,021,003	1,001,100,210,031,004
holiness	2	4	6	8
heptasub	1	9,129	91,288,523	912,885,227,240

686 unholy quintarish

primes	2	3	11	13	23	31	41	43	113	131	211	223	241	311	313	331	1,123	1,213	1,223
trimmed	1	2	0	2	12	20	30	32	2	20	100	112	130	200	202	220	12	102	112

687 holy senarish

primes	101	103	1,013	1,021	1,031	1,033	1,051	1,103	1,301	1,303	2,003	2,011	2,053
holiness	1	1	1	1	1	1	1	1	1	1	2	1	1

688 holiest senarish

not-quite	99	199	299	399	499	599	999	1,999	2,999	3,999	4,999	5,999	9,999

689 holier senarish

hypersum	...	1,001,100,210,031,004	10,011,002,100,310,041,005	100,110,021,003,100,410,051,010
heptasub	...	~ fifteenplex	~ nineteenplex	~ twenty-twoplex

690 unholy senarish

primes	2	3	5	11	13	23	31	41	43	53	113	131	151	211	223	233	241	251	311	313	331	353
trimmed	1	2	4	0	2	12	20	30	32	42	2	20	40	100	112	122	130	140	200	202	220	242

691 holy septarish

primes	101	103	1,013	1,021	1,031	1,033	1,051	1,061	1,063	1,103	1,123	1,513	1,153
holiness	1	1	1	1	1	1	1	2	2	1	0	0	0

692 holiest septarish

not-quite	99	199	299	399	499	599	699	999	1,999	2,999	3,999	4,999	5,999	6,999
holiness	2	2	2	2	2	2	3	3	3	3	3	3	3	4

693 holier septarish

hypersum	...	10,011,002,100,310,041,005	100,110,021,003,100,410,051,006
holiness	...	10	11

694 unholy septarish

primes	2	3	5	11	13	23	31	41	43	53	61	67	113	131	151	163	211	223	233	241	251	263
trimmed	1	2	4	0	2	12	20	30	32	42	50	56	2	20	40	52	100	112	122	130	140	152

695 holy octarish

primes	101	103	107	1,013	1,021	1,031	1,033	1,051	1,061	1,063	1,103	1,117	1,123
holiness	1	1	1	1	1	1	1	1	2	2	1	0	0

696 holiest octarish

not-quite	99	199	299	399	499	599	699	799	999	1,999	2,999	3,999	4,999	5,999	6,999

697 holier octarish

hypersum	...	100,110,021,003,100,410,051,006	1,001,100,210,031,004,100,510,061,007

698 unholy octarish

primes	2	3	5	11	13	23	31	41	43	53	61	67	71	73	113	131	151	163	173	211	223	233	241
trimmed	1	2	4	0	2	12	20	30	32	42	50	56	60	62	2	20	40	52	62	100	112	122	130

699 See *Fractals, Chaos and Power Laws* by Manfred Schoeder for more on noble, near-noble and silver mean numbers.

almost-tau	6	1	8	0	3	3	9	8	8	7	4	9	8	9	4	8	4	8	2	0	4	5	8	6	8	3	4	3	6	5	6	3	8	1	1
holiness	1	0	2	1	0	0	1	2	2	0	0	1	2	1	0	2	0	2	0	1	0	0	2	1	2	0	0	0	1	0	1	1	2	0	0
collapsed	6	1	11	3	9	8	8	7	4	9	8	9	4	8	4	8	6	5	8	6	8	3	4	3	6	5	6	3	8	1	1	7	7	14	1
holiness	1	0	0	0	1	2	2	0	0	1	2	1	0	2	0	2	1	0	2	1	2	0	0	0	1	0	1	0	2	0	0	0	0	0	0

hypersum	6	61	6,111	61,113	611,139	6,111,369	61,113,698	611,136,988	6,111,369,887
holiness	1	1	1	1	2	3	5	7	7
dusub	1	8	827	8271	82,708	827,084	8,270,840	82,708,398	827,083,975

700 second noble number (5 + √5)/10

almost-2nd noble	7	2	3	6	0	6	7	9	7	7	4	9	9	7	8	9	6	9	6	4	0	9	1	7	3	6	6	8	7	3	1	2	7	6	2	3
collapsed	7	2	3	16	7	9	7	7	4	9	9	7	8	9	6	9	6	13	1	7	3	6	6	8	7	3	1	2	7	6	2	3	5	4	10	1

hypersum	7	72	723	72,316	723,167	7,231,679	72,316,797	723,167,974	7,231,679,749
dusub	1	10	98	9,787	97,870	978,701	9,787,014	97,870,143	978,701,427

third noble number (15 + √5)/22

almost-3rd noble	7	8	3	4	5	7	6	3	5	3	4	0	8	9	9	5	3	1	6	5	4	1	6	9	8	4	9	4	2	3	3
collapsed	7	8	3	4	5	7	6	3	5	3	12	9	9	5	3	1	6	5	4	1	6	9	8	4	9	4	2	3	3	4	6

hypersum	7	78	783	7,834	78,345	783,457	7,834,576	78,345,763	783,457,635
dusub	1	11	106	1,060	10,603	106,029	1,060,294	10,602,946	106,029,461

fourth noble number (29 + √5)/38

almost-4th noble	8	2	2	0	0	1	7	8	8	8	8	8	1	5	7	3	4	1	3	0	6	3	3	9	9	3	0	7	0	7	1	8	7	5
holiness	2	0	0	1	1	0	0	2	2	2	2	2	0	0	0	0	0	0	0	1	1	0	0	1	1	0	1	0	1	0	0	2	1	1
collapsed	8	2	3	1	7	8	8	8	8	1	5	7	3	4	1	9	3	3	9	9	17	1	8	7	5	6	19	3	38	11	10	11	3	8
holiness	2	0	0	0	0	2	2	2	2	0	0	0	0	0	0	1	0	0	1	1	0	0	2	0	0	1	1	0	2	0	1	0	0	2

hypersum	8	82	823	8,231	82,317	823,178	8,231,788	82,317,888	823,178,888
holiness	2	2	2	2	2	4	6	6	8
dusub	1	11	111	1,114	11,140	111,405	1,114,052	11,140,515	111,405,148

fifth noble number (47 + √5)/58

almost-5th noble	8	4	8	8	9	7	7	2	3	7	4	9	9	9	6	3	7	4	0	7	6	0	2	0	2	3	5	6	6	7	7	8	0	6
holiness	2	0	2	2	1	0	0	0	0	0	0	1	1	1	1	0	0	0	1	0	1	1	0	1	0	0	0	1	1	0	0	2	1	1
collapsed	8	4	8	8	9	7	7	2	3	7	4	9	9	9	6	3	7	11	10	3	5	6	6	7	7	14	2	4	7	4	8	97	6	1
holiness	2	0	2	2	1	0	0	0	0	0	0	1	1	1	1	0	0	0	1	0	0	1	1	0	0	0	0	0	0	0	2	1	1	0

hypersum	8	84	848	8,488	84,889	848,897	8,488,972	84,889,723	848,897,237
holiness	2	2	4	6	7	7	7	7	7
dusub	1	11	115	1,149	11,489	114,886	1,148,857	11,488,575	114,885,748

702 sixth noble number (69 + √5)/82

almost-6th noble	8	6	8	7	3	2	5	3	6	3	1	0	9	7	3	0	4	5	0	7	8	1	6	0	6	5	4	4	9	6	5	4	4	9
holiness	2	1	2	0	0	0	0	0	1	0	0	1	1	0	0	1	0	0	1	0	2	0	1	1	1	0	0	0	1	1	0	0	0	1
collapsed	8	6	8	7	3	2	5	3	6	3	10	7	7	12	8	1	12	5	4	4	9	6	5	4	4	9	6	7	2	2	8	17	9	5
holiness	2	1	2	0	0	0	0	0	1	0	1	0	0	0	2	0	0	0	0	0	1	1	0	0	0	1	1	0	0	0	2	0	1	0

hypersum	8	86	868	8,687	86,873	868,732	8,687,325	86,873,253	868,732,536
holiness	2	3	5	5	5	5	5	5	6
dusub	1	12	117	1,176	11,757	117,570	1,175,701	11,757,016	117,570,164

seventh noble (95 + √5)/110

almost-7th noble	8	8	3	9	6	4	2	5	4	3	4	0	9	0	7	1	7	9	0	5	8	2	6	5	2	1	5	1	7	0	2	8	4	3
holiness	2	2	0	1	1	0	0	0	0	0	0	1	1	1	0	0	0	1	1	0	2	0	1	0	0	0	0	0	0	1	0	2	1	1
collapsed	8	8	3	9	6	4	2	5	4	3	20	1	7	14	8	2	6	5	2	1	5	1	9	8	4	3	2	9	4	1	12	6	5	8
holiness	2	2	0	1	1	0	0	0	0	0	1	0	0	0	2	0	1	0	0	0	0	0	1	2	0	0	0	1	0	0	0	1	0	2

hypersum	8	88	883	8,839	88,396	883,964	8,839,642	88,396,425	883,964,254
holiness	2	4	4	5	6	6	6	6	6
dusub	1	12	120	1,196	11963	119,631	1,196,316	1,193,155	119,631,553

eighth noble (125 + √5)/142

almost-8th noble	8	9	6	0	2	8	6	4	7	7	2	8	8	7	1	7	5	8	4	2	5	4	1	6	7	1	5	9	7	6	9	8	0	8

holiness	2	1	1	1	0	2	1	0	0	0	0	2	2	0	0	0	0	2	0	0	0	0	0	1	0	0	0	1	0	1	1	2	1	2
collapsed	8	9	8	8	6	4	7	7	2	8	8	7	1	7	5	8	4	2	5	4	1	6	7	1	5	9	7	6	9	16	1	8	5	5
holiness	2	1	2	2	1	0	0	0	0	2	2	0	0	0	0	2	0	0	0	0	0	1	0	0	0	1	0	1	1	1	0	2	0	0

hypersum	8	89	898	8,988	89,886	898,864	8,988,647	89,886,477	898,864,772
holiness	2	3	5	7	8	8	8	8	8
dusub	1	12	121	1217	12,165	121,648	1,216,481	12,164,812	121,648,119

ninth noble (159 + √5)/178

almost-9th noble	9	0	5	8	2	0	6	0	6	6	1	5	1	6	7	3	5	7	8	4	4	9	9	5	3	5	7	6	8	9	6	5	5	1
holiness	1	1	0	2	0	1	1	1	1	1	0	0	0	1	0	0	0	0	2	0	0	1	1	0	0	0	0	1	2	1	1	0	0	
collapsed	14	8	14	6	1	5	1	6	7	3	5	7	8	4	4	9	9	5	3	5	7	6	8	9	6	5	5	1	8	8	4	4	9	7

hypersum	14	148	14,814	148,146	1,481,461	14,814,615	148,146,151	1,481,461,516

tenth noble (197 + √5)/218

almost-10th noble	9	1	3	9	2	6	9	1	7	3	2	7	9	8	0	6	8	6	6	8	0	7	7	6	0	2	6	0	0	3	3	5	4	4
holiness	1	0	0	1	0	1	1	0	0	0	0	0	1	2	1	1	2	1	1	2	1	0	0	1	1	0	1	1	1	0	0	0	0	0
collapsed	9	1	3	9	2	6	9	1	7	3	2	7	9	14	8	6	6	15	7	8	9	3	5	4	4	2	2	4	9	9	9	8	2	1
holiness	1	0	0	1	0	1	1	0	0	0	0	0	1	0	2	1	1	0	0	2	1	0	0	0	0	0	0	0	1	1	1	2	0	0

hypersum	9	91	913	9,132	91,326	913,269	9,132,691	91,326,917	913,269,173
holiness	1	1	1	1	2	3	3	3	3
dusub	1	12	124	1,236	12,360	123,598	1,235,975	12,359,754	123,597,542

703 (z + √5)/z'

z	5	15	29	47	69	95	125	159	197	239	277	327	381	449
gnomon		10	14	18	22	26	30	34	38	42	46	50	54	58
2nd gnomon			4	4	4	4	4	4	4	4	4	4	4	4
z'	10	22	38	58	82	110	132	178	218	262	310	362	398	458
gnomon		12	16	20	24	28	32	36	40	44	48	52	56	60
2nd gnomon			4	4	4	4	4	4	4	4	4	4	4	4

704 more almost-nobles

almost-1-2 noble	5	8	0	1	7	8	7	2	8	2	9	5	4	6	4	1
collapsed	5	9	7	8	7	2	8	2	9	5	4	6	4	1	...	

hypersum	5	59	597	5,978	59,787	597,872	5,978,728	59,787,282	597,872,829
dusub	1	8	81	809	8,091	80,913	809,133	8,091,329	80,913,289

705 first near-noble (√2)/2

almost-half root two	7	0	7	1	0	6	7	8	1	1	8	6	5	4	7	5	2	4	4	0	0	8	4	3
collapsed	14	7	8	1	1	8	6	5	4	7	5	2	4	12	3	6	2	5	8	4	12	9	2	8

hypersum	14	1,421	142,129	14,212,930	1,421,293,031	142,129,303,139
tressub	1	71	7,076	707,620	70,762,013	7,076,301,333

706 second near-noble (1 + √13)/6

almost-2nd near-noble	7	6	7	5	9	1	8	7	9	2	4	3	9	9	8	1

hypersum	7	76	767	7,675	76,759	767,591	7,675,918	76,759,187
dusub	1	10	104	1,039	10,388	103,882	1,038,822	10,388,226

third near-noble (1 + √5)/4

almost-3rd near-noble	8	0	9	0	1	6	9	9	4	3	7	4	9	4	7	4	2	4	1	0	2	2	9	3	4	1	7	1
holiness	2	1	1	1	0	1	1	1	0	0	0	0	1	0	0	0	0	0	0	1	0	0	1	0	0	0	0	0
collapsed	18	6	9	9	4	3	7	4	9	4	7	4	2	4	3	2	9	3	4	1	7	1	8	2	8	9	8	8
holiness	2	1	1	1	0	0	0	0	1	0	0	0	0	0	0	0	1	0	0	0	0	0	2	0	2	1	2	2

hypersum	18	186	1,869	18,699	186,994	1,869,943	18,699,437	186,994,374
holiness	2	3	4	5	5	5	5	5
tressub	1	9	93	931	9,310	93,099	930,990	9,309,902

707 [2; 2] = 2 + √2, [3;3] = (3 + √13)/2, [4; 4] = 2 + √5, [5; 5] = (5+√29)/2, [6; 6] = 3 + √10, [7; 7] = (7 + √53)/2; [8; 8] = 4 + √17, [9; 9] = (9 + √85)/2, [10; 10] = 5 + √26

708 The silver mean numbers are all of the form z' = (z + g(2, 0.5, (g(2, 2, z) + 4))/2.

z'	8	13	20	29	40	53	68	85
holiness	2	0	1	1	1	0	3	2
sub	1	2	3	4	6	7	9	11
broken-up	8	105	2,108	61,237	2,451,588	129,995,401	8,842,138,856	751,711,798,161
holiness	2	1	3	1	4	4	9	4
dusub	1	14	285	8,288	331,786	17,592,964	1,196,653,366	101,733,129,116
broken-down	11	13	261	7,582	303,541	16,095,255	1,094,780,881	93,072,470,140

709 almost-naturals

broken-up	1	3	10	43	225	1,393	9,976	81,201	740,785	6,748,266	7,489,051	14,237,317
broken-down	1	2	7	30	157	972	6,961	56,660	516,901	2,641,165	3,158,066	5799231
broken	1	2	3	7	10	30	43	157	225	972	1,393	6,961

710 zero in pi

broken-up	33	1,684	92,653	6,116,782	440,500,957	34,365,191,426	295,584,696,63,765
tressub	1	84	4,613	304,536	21,931,251	1,710,942,135	147,162,954,870
broken-down	1	51	2,806	185,247	13,340,590	1,040,751,267	89,517,949,552
broken	1	33	51	1,684	2,806	93,653	185,247

711 zero in e

broken-up	14	309	13,610	925,789	67,596,207	7,639,297,180	870,947,474,727
tressub	1	15	678	46,092	3,365,417	380,338,211	43,361,921,469
broken-down	1	22	969	65,914	4,812,691	54,399,997	62,009,412,349
broken	1	14	22	309	969	13,610	65,914

712 zero in phi

broken-up	5	106	4,245	203,866	13,459,401	942,361,936	73,517,690,409
tressub	1	14	575	27,590	1,821,532	127,534,819	9,949,537,454
broken-down	1	21	841	40,389	2,666,515	186,696,439	14,564,988,757
broken	1	5	21	106	841	4,245	40,389

713 collapsed almost-phi

broken-up	1	7	8	95	293	2,732	22,149	179,924	1,281,617	5,306,392	49,039,145
broken-down	1	6	7	83	256	2,387	19,352	157,203	1,119,773	4,636,295	42,846,428
broken	1	6	7	8	83	95	256	293	2387	2732	19,352

714 fully expanded

almost-squares	1	1	0	1	0	1	0	1	1	0	1	0	1	0	1	0	1	0	1	0	1	0	1	1	0	1	1	0	1	0	1	0	1	0	1	0
almost-cubes	1	1	0	1	0	1	0	1	0	1	0	1	0	1	0	1	1	0	1	0	1	0	1	0	1	0	1	0	1	0	1	0	1	0	1	0

715 recollapsed pseudoternarish

hypersum	2	23	233	2,334	23,343	233,434	2,334,344	23,343,445	233,434,454
trimmed	1	12	122	1,223	12,232	122,323	1,223,233	12,232,334	122,323,343

716 recollapsed pseudoquadrarish

hypersum	2	23	234	2,343	23,434	234,345	2,343,454	23,434,545	234,345,456
trimmed	1	12	123	1,232	12,323	123,234	1,232,343	12,323,434	123,234,345

717 recollapsed pseudoquintarish

hypersum	2	23	234	2,345	23,453	234,534	2,345,345	23,453,456	234,534,564
trimmed	1	12	123	1,234	12,342	123,423	1,234,234	12,342,345	123,423,453

718 recollapsed pseudosenarish

hypersum	2	23	234	2,345	23,456	234,563	2,345,634	23,456,345	234,563,456
trimmed	1	12	123	1,234	12,345	123,452	1,234,523	12,345,234	123,452,345

719 recollapsed pseudoseptarish

hypersum	2	23	234	2,345	23,456	234,567	2,345,673	23,456,734	234,567.345
trimmed	1	12	123	1,234	12,345	123,456	1,234,562	12,345,623	123,456,234

recollapsed pseudoseptarish

hypersum	2	23	234	2,345	23,456	234,567	2,345,678	23,456,783	234,567.834
trimmed	1	12	123	1,234	12,345	123,456	1,234,567	12,345,672	123,456,723

720 Johannes Widman introduced the minus sign for negative numbers in 1489. "In the red" however was not recorded until the *Wise-crack Dictionary* by George H. Maines and Bruce Grant in 1926, although red for negative was used long before by Dzu Tse, etal. Its older anagram "nithered" from the Seventeenth Century, perhaps not co-incidentally, meant "shrivelled with cold".

In his column "Bogglers" in *Discovery* Scott Kim introduced negative placeholders for five through nine.

Kimese	1	2	3	4	5	4	3	2	1	10	11	12	13	14	15	14	13	12	11	20
pseudosum	1	3	4	10	25	31	42	54	55	45	24	22	101	115	130	144	143	141	121	201

721 These seemingly different representations of continuing fractions all represent the same irrational, 0.7912878475...

broken-up	1	4	5	19	24	91	115	436	551	2,089	2,640	10,009	12,649	47,956	60,605
broken-down	1	3	4	15	19	72	91	345	436	1,653	2,089	7,920	10,009	37947	47,956
broken	1	3	4	5	15	19	24	72	91	115	345	436	551	1,653	2,089
sum	1	4	5	8	9	12	13	16	17	20	21	24	25	28	29

722 [1, 4] and [1, 11]

broken-up	1	5	6	29	35	169	204	985	1,189	5,741	6930	33,461	40,391	195,025
broken-down	1	4	5	24	29	140	169	816	985	4,756	5,741	27.720	33,461	161,564
broken	1	4	5	6	24	29	35	140	169	816	985	1,189	4,756	5,711
sum	1	5	6	10	11	15	16	20	21	25	26	30	31	35

hypersum	1	14	141	1,414	14,141	141,414	1,414,141	14,141,414

broken-up	1	12	13	155	168	2,003	2,171	25,884	28,055	334,489	362,544
broken-down	1	11	12	143	155	1,848	2,003	23,881	25,884	308,605	334,489
broken	1	11	12	13	143	155	168	1,848	2,003	2,171	23,881

723 Trimming followed by beheading Joyce called "rimming", giving us not only "trimmed", but also "rimmed" primes that remain prime after these transformations, also all sanbi primes, though the trimmed is a subsequence of the rimmed.

rimmed	13	103	113	163	173	193	1,013	1,033	1.063	1,093	1,103	1,123	1,153	1,163

If the binarish sequence is trimmed we get the null sequence. If we trim that we get nothing, the null set. The untrimmed almost-repunit sequence is much more interesting, a subsequence of ternarish. An untrimmed nine can be represented by a one-zero rather than a Roman X or other symbol for ten.

untrimmed	2	12	20	102	112	120	200	1,002	1,012	1,112	1,200	2,000	10,002	10,012
z%3	2	0	2	0	1	0	2	0	1	2	0	2	0	1
rock	0	1	1	2	2	3	3	4	4	4	5	5	6	6
paper	0	0	0	0	1	1	1	1	2	2	2	2	2	3
scissors	1	1	2	2	2	2	2	3	3	4	4	5	5	5

724 The primes that the safe primes become when decreased by one and halved are called Sophie Germain primes after their discoverer, Marie-Sophie Germain.

Sophie Germain primes	2	3	5	11	23	29	41	53	83	89	113	131	173	179	191	233	239
trimmed	1	2	4	0	12	18	30	42	72	78	2	20	62	68	80	122	128

Some are both safe and Sophie Germain.

both	5	11	23	83	179	359	719	1,019	1,439	2,039	2,063	2,459	2,819	2,903	2,963	3,023

The mates of the safe primes are "sound", of which only one, that is two, is prime. Safe and sound were mated by Thomas Lupset in 1529, long before William Shakespeare in 1592.

sound prime mates	2	4	16	40	52	76	88	112	136	160	280	412	456	496	520	532	616
trimmed	1	3	5	3	41	65	77	1	25	5	17	301	345	385	41	421	505

725 Some call the unsafe primes the Nader primes after *Unsafe at Any Speed: The Designed-in Dangers of the American Automobile by Ralph Nader.*

broken-up	7	92	651	8,555	248,746	7,719,681	285,876,943	11,728,674,344
dusub	1	13	88	1,158	33,664	1,044,745	38,689,237	1,587,303,464
broken-down	1	8	57	749	21,778	23,968	887,565	36,414,133

broken	1	7	8	57	92	61	749	8,555

hypersum	7	713	71,317	7,131,719	713,171,929	71,317,192,931	7,131,719,293,137

726 safe luckies

broken-up	1	4	5	19	594	25,561	1,304,205	82,190,476	5,508,066,097
broken-down	1	2	3	11	344	484	24,695	1,556,269	104,294,718
broken	1	2	3	4	5	11	19	344	594

hypersum	1	13	137	13,715	1,371,531	137,153,143	13,715,314,351

727 unsafe luckies

broken-up	1	10	11	109	2,736	90,397	3,347,425	164,114,222	11,327,228,743
broken-down	1	2	3	29	728	986	36,511	1,790,025	123,548,236
broken	1	2	3	10	11	29	109	728	986

hypersum	1	19	1,913	191,321	19,132,125	1,913,212,533	191,321,253,337

728 safe unluckies

broken-up	23	1,220	28,083	1,489,619	168,355,030	2,070,915,309	2,961,577,993,217
tressub	1	61	1,398	74,164	8,381,904	1,031,048,280	147,448,286,025
broken-down	1	24	553	29,333	3,315,182	3,637,292	520,162,089
broken	1	23	24	553	1,220	28,083	29,333

hypersum	23	2,353	235,383	235,383,103	235,383,103,113	23,583,103,113,123
tressub	1	117	11,719	11,719,035	11,719,034,647	11,719,034,647,020

safe unluckies

broken-up	5	116	585	13,571	638,422	35,126,781	2,073,118,501	147226540352
dusub	1	16	79	1,836	86,401	4,753,893	280,566,080	19,924,945,538
broken-down	1	6	31	716	33,824	40,264	2,376,295	168,757,209
broken	1	5	6	31	116	585	716	13,571

hypersum	5	523	52,335	5,233,539	523,353,947	52,335,394,755	5,233,539,475,559
dusub	1	71	7,083	708,282	70,828,255	7,082,825,472	708,282,547,255

729 unsafe unluckies

hypersum	0	2	24	246	2,468	246,810	24,681,011	2,468,101,112	246,810,111,213
holiness	1	0	0	1	3	4	4	4	4

untrimmed	1	3	35	357	3,579	357,921	35,792,122	3,579,212,221	357,921,222,124

730 unsafe squares

broken-up	1	5	6	29	1,050	51,479	3,295,706	267,003,665	26,703,662,206
broken-down	1	2	3	14	507	700	44,814	3,630,634	363,108,214
broken	1	2	3	5	6	14	29	507	700

hypersum	1	14	1,416	141,625	14,162,536	1,416,253,649	141,625,364,964

731 unsafe curvilinears

sub	1	8	9	19	20	82	83	93	94	192	193	203	204	817	819	828	830	928	929

broken-up	2	45	92	2,069	113,887	25,284,983	5,689,235,062	1,433,712,520,607
trimmed	1	34	81	158	2,776	14,173,872	457,812,451	3,226,014,156
sub	1	17	34	761	41,897	9,301,825	2,092,952,615	527,433,360,881
broken-down	1	3	7	157	8,642	35,011	7,877,632	1,985,198,275
broken	1	2	3	7	45	92	157	2,069

hypersum	2	222	22,225	2,222,552	222,255,255	222,255,255,222	222,255,255,222,225
trimmed	1	111	11,114	1,111,441	111,144,144	111,144,144,111	111,144,144,111,114

732 safe ternarish

tressub	1	10	100	996	9,957	99,574	995741	9,957,414	99,574,137	995,741,367

broken-up	21	4,222	88,683	17,829,505	3,565,918,918,188	> eighteenplex
tressub	1	210	4,415	887,679	177,536,648,974	>seventeenplex
broken-down	1	22	463	93,085	18,617,093,548	186,170,186,170
broken	1	21	22	463	4,222	88,683

hypersum	21	21,201	212,012,001	212,012,001,200,001	2,120,120,012,000,012,000,001

733 unsafe ternarish

broken-up	1	3	4	11	136	2,731	60,218	6,024,531	608,537,849	61,468,347,280
broken-down	1	2	3	8	99	168	3,704	370,568	37,431,072	449,543,432
broken	1	2	3	4	8	11	99	136	2,731	3,704

hypersum	1	12	1,210	121,011	12,101,112	1,210,111,220	121,011,122,022

734 safe quadarish

broken-up	21	484	10,185	234,739	469,722,924	940,855,251,511
tressub	1	24	507	11,687	23,386,127	46,842,424,731
broken-down	1	22	463	10,671	21,353,134	21,384,684
broken	1	21	22	463	484	10,185

hypersum	21	2,123	2,123,201	212,320,1203	21,232,012,032,001

unsafe quadarish

broken-up	1	3	4	11	125	1,511	19,768	396,871	8,750,930	192,917,331
broken-down	1	2	3	8	91	104	1,360	27,304	602,048	6,649,832
broken	1	2	3	4	8	11	91	104	125	1,360

hypersum	1	12	123	12,310	1,231,011	123,101,112	12,310,111,213	1,231,011,121,320

safe quinarish

broken-up	3	64	195	4,159	179,032	35,989,591	7,306,066,005	2,929,768,457,596
sub	1	24	72	1,530	65,862	13,239,831	2,687,751,479	1,077,801,582,942
broken-down	1	4	13	277	11,924	55,954	11,358,939	4,554,990,493
broken	1	3	4	13	64	195	277	4,159

hypersum	3	321	32,123	3,212,341	321,234,143	321,234,143,201	321,234,143,201,203	

unsafe quinarish

broken-up	...	19,768	278,263	5,585,028	111,978,823	2,693,076,780	80,904,282,223
broken-down	...	1,360	19,144	384,240	4,245,784	102,283,056	3,072,737,464
broken	...	91	104	125	1,360	1,511	19,768

hypersum	1	12	124	12,410	1,241,011	124,101,112	12,410,111,213	1,241,011,121,314

Nine of the first thirteen safe senarish are prime.

broken-up	3	16	51	271	6,284	157,371	4,884,785	200,433,556	8,623,527,693
sub	1	6	19	100	2,312	57,894	1,797,012	73,735,385	3,172,418,549
broken-down	1	4	13	69	1,600	1,794	55,683	2,284,797	98,301,954
broken	1	3	4	13	16	51	69	271	1,600

hypersum	3	35	3,511	351,121	35,112,123	3,511,212,325	351,121,232,531

unsafe senarish

broken-up	1	3	4	11	136	1,779	25,042	377,409	7,573,222	151,841,849	3,651,777,598

broken-down	1	2	3	8	99	112	1,576	23,752	476,616	5,743,144	138,312,072
broken	1	2	3	4	8	11	99	112	136	1,576	1,779

hypersum	1	12	124	12,411	1,241,112	124,111,213	12,411,121,314	1,241,112,131,415

Eight of the first eleven safe septarish are prime.

sub	3	5	11	13	21	23	25	31	35
broken-up	3	16	51	271	5,742	132,337	3,314,167	102,871,514	3,398,074,129
sub	1	6	19	100	2,112	48,684	1,219,214	37,844,315	1,250,081,612
broken-down	1	4	13	69	1,462	33,695	843,837	26,192,642	917,586,307
broken	1	3	4	13	16	51	69	271	1,462

hypersum	3	35	3,511	351,113	35,111,321	3,511,132,123	351,113,212,325

unsafe septarish

broken-up	1	3	4	11	114	1,379	19,420	292,679	4,702,284	75,529,223
broken-down	1	2	3	8	83	104	1,464	22,064	354,488	3,566,944
broken	1	2	3	4	8	11	83	104	114	1,379

hypersum	1	12	124	1,246	124,610	12,461,012	1,246,101,214	124,610,121,415

735 safe octarish

broken-up	3	16	51	271	3,574	75,325	1,736,049	43,476,550	1,175,602,899
sub	1	6	19	100	1,315	27,711	638,657	15,994,129	432,480,138
broken-down	1	4	13	69	910	1,518	34,983	876,093	23,689,494
broken	1	3	4	13	16	51	69	271	910

hypersum	3	35	357	35,711	3,571,113	357,111,321	35,711,132,123	3,571,113,212,325

736 unsafe octarish

broken-up	...	75,529,223	1,515,286,744	33,411,837,591	803,399,388,928
broken-down	...	3,566,944	71,693,368	1,580,821,040	38,011,398,328
broken	...	1,379	1,464	19,420	22,064

hypersum	...	1,246,101,214,1516	124,6101,214,151,617	124,610,121,415,161,720

737 Samuel Beatty published an article on what became known as Beatty sequences in *American Mathematical Monthly* in 1924. However, even earlier, such sequences were briefly mentioned by John W. Strutt, the third Baron Rayleigh, in the second edition of his book *The Theory of Sound* in 1894.

738 phi Beatty

broken-up	1	4	5	19	157	1,432	15,909	192,340	2,708,669	38,113,706	650,641,671
broken-down	1	2	3	11	91	110	1,221	14,762	207,889	1,677,874	28,731,747
broken	1	2	3	4	5	11	19	91	110	157	1,221

hypersum	1	13	134	1,346	13,468	134,689	13,468,911	1,346,891,112	134,689,111,213

non-phi Beatty

broken-up	2	11	24	131	1,727	26,036	470,375	9,433,536	217,441,703	5,010,592,705
trimmed	1	0	13	20	616	1,525	36,264	8,322,425	10,633,062	4,048,164
sub	1	4	9	48	635	9,578	173,041	3,470,404	79,992,332	1,843,294,044
broken-down	1	3	7	38	501	608	10,982	220,248	5,076,686	66,217,166
broken	1	2	3	7	11	24	38	131	501	608

hypersum	2	25	257	25,710	2,571,013	257,101,315	25,710,131,518	2,571,013,151,820
trimmed	1	14	146	1,460	146,002	14,600,204	1,460,020,407	14,600,204,071

739 duphi Beatty

broken-up	1	3	4	11	70	571	5,209	52,661	637,141	7,698,353	108,414,083	1,742,323,681
broken-down	1	2	3	8	51	72	656	6,632	80,240	488,072	6,913,248	11,110,040
broken	1	2	3	4	8	11	51	70	72	571	656	5,209

hypersum	1	12	124	1,245	12,456	124,568	1,245,689	124,568,910	12,456,891,012

non-duphi Beatty

sub	1	3	4	6	7	8	10	11	13
broken-up	3	22	69	505	9,664	222,777	6024643	186,986,710	6,563,572,783
sub	1	8	25	186	3,555	81,955	2,216,342	68,788,566	2,341,027,599
broken-down	1	4	13	95	1,818	2,280	61,655	1,913,585	65,123,545
broken	1	3	4	13	22	69	95	505	1,818

hypersum	3	37	3,711	371,115	37,111,519	3,711,151,923	371,115,192,327

phi-non-phi Beatty

broken-up	2	11	24	131	1,596	24,071	410,803	8,240,131	189,933,816
trimmed	1	0	13	20	485	1,360	3,072	713,020	78,822,705
sub	1	4	9	48	587	8,855	151,126	3,031,375	69,872,746

broken-down	1	3	7	38	463	608	10,374	208,088	4,796,398
broken	1	2	3	7	11	24	38	131	463

hypersum	2	25	257	25,710	2,571,012	257,101,215	25,710,121,517
trimmed	1	14	146	1,460	146,001	14,600,104	1,460,010,406

dunon-phi Beatty

broken-up	1	4	5	19	157	1,432	15,909	208,249	2,931,395	41,247,779
broken-down	1	2	3	11	91	110	1,221	15,983	1,815,847	32,910,229
broken	1	2	3	4	5	11	19	91	110	157

hypersum	1	13	134	1,346	13,468	134,689	13,468,911	1,346,891,113	134,689,111,315

740 [z(phi + 1)] - 1 generates the Wythoff AA numbers.

broken-up	1	5	6	29	354	4,985	85,099	1,621,866	35,766,151	788,477,188
broken-down	1	2	3	14	171	210	3,584	68,306	1,506,316	18,144,098
broken	1	2	3	5	6	14	29	171	210	354

hypersum	1	14	146	1,469	146,912	14,691,214	1,469,121,417	146,912,141,719

[z(phi +1)] generates the Upper Wythoff numbers.

broken-up	2	11	24	131	1,727	26,036	470,375	9,433,536	217,441,703
trimmed	1	0	13	20	616	1,525	36,264	8,322,425	10,633,062
sub	1	4	9	48	635	9,578	173,041	3,470,404	79,992,332
broken-down	1	3	7	38	501	608	10,982	220,248	5,076,686
broken	1	2	3	7	11	24	38	131	501

hypersum	2	25	257	25,710	2,571,013	257,101,315	25,710,131,518
trimmed	1	14	146	1,460	146,002	14,600,204	1,460,020,407
sub	1	9	95	9,458	945,823	94,582,288	9,458,228,815

[[z(phi + 1)]phi] generates the Wythoff AB numbers.

sub	1	3	4	6	8	9	11	12	14
broken-up	3	25	78	649	13,707	329,617	9,572,600	306,652,817	11,355,726,829
sub	1	9	29	239	5,043	121,259	3,521,563	112,811,267	4,177,538,440
broken-down	1	4	13	108	2,281	2,700	78,408	2,511,756	93,013,380
broken	1	3	4	13	25	78	108	649	2,281

hypersum	3	38	3,811	381,116	38,111,621	3,811,162,124	381,116,212,429
sub	1	14	1,402	140,205	14,020,482	1,402,048,192	140,204,819,250

741

4th protosubdivision	1	6	11	16	21	26	31	36	41	46	51	56	61	66	71	76
5th	1	7	13	19	25	31	37	43	49	55	61	67	73	79	85	91
6th	1	8	15	22	29	36	43	50	57	64	71	78	85	92	99	106
7th	1	9	17	25	33	41	49	57	65	73	81	89	97	105	114	121

1st protodiagonal	1	4	9	16	25	36	49	64	81	100	121	144	169	196	225
2nd	1	5	11	19	29	41	55	71	89	109	131	156	182	210	240
3rd	1	6	13	22	33	46	61	78	97	118	141	168	195	234	255
4th	1	7	15	25	37	51	67	85	105	127	151	180	208	248	270

742 Thue-Morsean sequences can be generated in higher bases by hyperadding mates.

base 3	1	2	2	1	2	1	1	2	2	1	1	2	1	2	2	1	2	1	1	2	1	2	2	1
base 4	1	3	3	1	3	1	1	3	3	1	1	3	1	3	3	1	3	1	1	3	1	3	3	1
base 5	1	4	4	1	4	1	1	4	4	1	1	4	1	4	4	1	4	1	1	4	1	4	4	1
base 6	1	5	5	1	5	1	1	5	5	1	1	5	1	5	5	1	5	1	1	5	1	5	5	1
base 7	1	6	6	1	6	1	1	6	6	1	1	6	1	6	6	1	6	1	1	6	1	6	6	1
base 8	1	7	7	1	7	1	1	7	7	1	1	7	1	7	7	1	7	1	1	7	1	7	7	1
base 9	1	8	8	1	8	1	1	8	8	1	1	8	1	8	8	1	8	1	1	8	1	8	8	1

743 sliced Thue-Morse

dusub	1	1,490	1,490,043	1,490,042,957	1,490,042,957,260
broken-up	11	121,111	1,332,232	14,667,995,431	> twenty-threeplex
dusub	1	16,391	180,298	1,985,097,316	> twenty-twoplex
broken-down	1	12	133	1,464,342	> nineteenplex
broken	1	11	12	133	1,490

hypersum	11	1111010	111101011010011	11110101101001111010011001

744 diced Thue-Morse

hypersum	0	1	111	111110	1111101101	11110110111010	11110110111010110100
holiness	1	0	0	1	2	4	7

745 fragged Thue-Morse

hypersum	0	1	1110	11101101001	11101101001110100110010110

holiness	1	0	1	4	10
untrimmed	1	2	2221	22212212112	> twenty-fiveplex

746 rock-paper-scissors

z%3	0	1	2	0	1	2	1	2	0	2	0	1	1	2	1	2	0	2	0	1	1	2	0	2	0	1
rock	1	1	1	2	2	2	2	2	3	3	4	4	4	4	4	4	5	5	6	6	6	6	7	7	8	8
paper	0	1	1	1	2	2	3	3	3	3	3	4	5	5	6	6	6	6	6	7	8	8	8	8	8	9
scissors	0	0	1	1	1	2	2	3	3	4	4	4	4	5	5	6	6	7	7	7	7	8	8	9	9	9

747 1st subsequence

broken-up	1	3	4	11	81	740	7,481	90,512	1,274,649	17,935,598
broken-down	1	2	3	8	59	539	5,449	65927	928427	6,564,916
broken	1	2	3	4	8	11	59	81	539	740

hypersum	1	12	124	1,245	12,457	124,579	12,457,910	1,245,791,012

748 Although wo- seems to be a prefix transforming a man into a woman, a portmaneau for "wombed man", and fe- a prefix to turn a male into a female, a "femme male", they are not, neither can the prefixes he-, she- or trans-. "Woman" comes from 'wif-man" meaning a female Human, with "wer-man" meaning "male Human". "Female" and "feminine" and "male" and "masculine" come from the Latin "femina" and "masculus". The suffix -mate therefore is usually preferred for the complimentary sequences. Mating is just one of countless examples of 'modulation', in base ten, $10^{\log(z + 1)}(9) - z$ (modulo 9), useful for en/decryption with nim keyexpressions using nim operations rather than keywords.

z	0	1	2	3	4	5	6	7	8	9	10	11	12	13	14	15	16	17	18	19	20	21	22	23	24	25	26
mates = 9 - z	9	8	7	6	5	4	3	2	1	0	89	88	87	86	85	84	83	82	81	80	79	78	77	76	75	74	73
z ⊗⊗ 3 ⊕ 1	1	2	9	8	5	6	7	4	3	0	21	22	29	28	25	26	27	24	23	20	91	92	99	98	95	96	97
z ⊘ 3 ⊕ 1	1	8	5	2	9	6	3	0	7	4	81	88	85	82	89	86	83	80	87	84	51	58	55	52	59	56	53
z ⊖ 9 = z ⊕ 1	1	2	3	4	5	6	7	8	9	0	21	22	23	24	25	26	27	28	29	20	31	32	33	34	35	36	37
3 ⊗ z ⊕ 1	1	4	7	0	3	6	9	2	5	8	41	44	47	40	43	46	49	42	45	48	71	74	77	70	73	76	79
3 ⊗⊗ z ⊕ 1	1	2	9	8	5	6	7	4	3	0	21	22	29	28	25	26	27	24	23	20	91	92	99	98	95	96	97

"Woe" describes the mates of evens, "woeful" that of even Fibonacci-unindexed luckies, "woefully" that of even Fibonacci-unindexed luckier lucky yongbi, "woful" that of Fibonacci-unindexed luckies, "wok" the mates of kyuubi, "wolf" that of lucky Fibonacci, "wolfbane" that if lucky-Fibonacci-binarish-and-nibi-evens, "wolfer" that of lucky Fibonacci even rokubi, "wolfram" that of lucky-Fibonacci rokubi-and-Mersennes, "won" that of the nibi, "wont" that of the nibi ternarish, "woo" that of the odds, "woof" that of the odd Fibonacci, "wool" that of odd luckies, "woolier" that of odd lucky-indexed even rokubi, "wooliest" odd lucky-indexed sorted ternarish, "woolskin" that of odd lucky sorted kyuubi-indexed nibi, "woop" that of odd primes, "woosh" that of odd shichibi, "wop" that of primes and "wore" that of rokubi evens.

749 Since all the luckies are necessarily odd, they can also be abbreviated as 'lo' and their complimentary sequence called "behold". The two were mated by Charles Dickens in *David Copperfield* in 1850. Twelve of the first fifteen (80%) untrimmed beholds are prime.

beholds	0	2	6	8	12	20	24	26	30	32	36	48	50	56	62	66	68	74	78	84
holiness	1	0	1	2	0	1	0	1	1	0	1	2	1	1	1	2	3	0	2	2
untrimmed	1	3	7	9	23	31	35	37	41	43	47	59	61	67	73	77	79	85	89	95

Lucky modifying another sequence is usually taken as meaning those luckies within the sequence. Lucky primes, for example, means the same as prime luckies. When modifying the lucky sequence however the additional lucky would be meaningless unless it meant applying the lucky operation on the lucky sequence, analogous to the oddly odds and the evenly evens. Many of the already rare lucky primes don't make the cut, like forty-three, sixty-seven, seventy-three, seventy-nine and a hundred twenty-seven.

lucky	1	3	7	9	13	15	21	25	31	33	37	43	49	51	63	67	69	73	75	79	87	93	99	105	111	127

1st survivors	1	7	13	21	31	37	49	63	69	75	87	99	111	133	141	159	169	189	195
2nd survivors	1	7	13	21	31	37	63	69	75	87	99	111	141	159	169	189	195	219	237
3rd survivors	1	7	13	21	31	37	63	69	75	87	99	111	159	169	189	195	219	237	259

All of these luckier luckies numbers are included in the ball, call, fall, gall, hall, mall, pall and tall numbers. Some are included in the bill, boll, bull, cull, fill, full, gill, gull, hill, hull, kill, lull, mill, moll, null, pill, poll, pull, rill, sill, shill, till numbers.

ball	1	7	10	11	13	21	31	37	63	69	75	87	99	100	101	110	111	141	159	169	189	195	219
call	1	7	8	13	21	27	31	37	63	64	69	75	87	99	125	141	159	169	189	195	216	219	237
fall	1	2	3	5	7	8	13	21	31	34	37	55	63	69	75	87	89	99	111	144	159	169	189
gall	1	5	7	13	15	21	25	31	35	37	45	55	63	65	69	75	85	87	95	99	105	111	115
hall	1	7	8	13	18	21	28	31	37	38	48	58	63	68	69	75	78	87	88	98	99	108	111
mall	1	3	7	13	21	31	37	63	69	75	87	99	111	127	159	169	189	195	219	237	259	261	267
pall	1	2	3	5	7	11	13	17	19	21	23	29	31	37	41	43	47	53	59	61	63	67	69
tall	1	2	7	10	11	12	13	20	21	22	31	37	63	69	75	87	99	100	101	102	110	111	112

750 Many curvaceous mates are curvaceous themselves, not so many ugly mates are also ugly.

curvaceous curvaceous mates	0	3	6	9	13	16	19	30	31	33	36	39	60	61	63	66	69

holiness	1	0	1	1	0	1	1	1	0	0	1	1	1	1	1	2	2

ugly mates	1	2	3	4	5	6	8	9	18	24	27	39	51	54

751 Nine is the first square mate that is also square.

752 class

sorted sanbi	13	23	33	34	35	36	37	38	39	113	123	133	134	135	136	137	138	139

cluckies	1	27	343	729	2,197	3,375	9,261	15,625	29,791	35,937	507,653	79,507

classy	1	13	23	27	33	34	35	36	37	38	39	113	123	133	134	135	136	137	138	139

"Lass", describing luckies and sorted sanbis, is mated with "lad".

lass	1	3	7	9	13	15	21	23	25	31	33	34	35	36	37	38	39	43	49	51	63
lad	2	6	8	12	20	24	26	30	32	36	48	50	56	62	66	68	74	78	84	86	90

"Bass", describing binarish and sorted sanbi, comes from the Middle English *bas* ("low") from the Greek *batos* from the Indo-European *gwm̥yo-* and its mate is from the Italian alto ("high") from the Latin *altus* from the Indo-European *al-*. Some basses are female and some altos male.

bass	1	10	11	13	23	33	34	35	36	37	38	39	100	101	110	111	113	123	133	134	135	136	137
alto	8	60	61	62	63	64	65	66	76	86	88	89	102	104	108	110	114	126	128	138	144	156	158

"Glass" describes gobi luckies and sorted sanbis.

glass	13	23	33	34	35	36	37	38	39	113	123	133	134	135	136	137	138	139

The mate of "mass", describing Mersennes-and-sorted-sanbis, is NOT "energy" since Albert Einstein proved they are equivalent with $E = mc^2$.

mass	2	3	5	7	13	17	19	23	31	33	34	35	36	37	38	39	61	89	107	113	123	127	133	134	135

The Mersennes include not only the primes, but the Mersenne composites, those with non-Mersenne, though still prime, exponents.

non-Mersenne exponents	11	23	29	41	43	47	53	59	71	73	79	83	97	101	103	109	113	131	137	139

753 "Bay" describes binarish and yongbi, "cay" cubes and yongbi, "fay" Fibonacci and yongbi and "lay" luckies and yongbi.

prime luckies	3	7	13	31	37	43	67	73	79	127	151	163	193	211	223	241	283	307	331

play	3	4	7	13	14	24	31	37	43	44	54	64	67	73	74	79	84	94	104	114	124	127	134

bay	1	4	10	11	14	24	34	44	54	64	74	84	94	100	101	104	110	111	124	134	144	154	164
cay	1	4	8	14	24	27	34	44	54	64	74	84	94	104	114	124	125	134	144	154	164	174	184
fay	1	2	3	4	5	8	13	14	21	24	34	44	54	55	64	74	84	89	94	104	114	124	134
lay	1	3	4	7	9	13	15	21	24	25	31	33	34	37	43	44	49	51	54	63	64	67	69

754 Since there are no even luckies these are plain sorted ternarish.

stale	1	11	12	22	111	112	122	222	1,111	1,112	1,122	1,222	2,222	11,112	11,122

755 "Team" describes ternarish evens and Mersenne, "beam" binarish-even-and-Mersenne, "cream" cubed-rokubi-and-Mersenne, "gleam" gobi-luckies-and-Mersenne, "ream" rokubi-and-Mersenne.

team	0	2	3	5	7	10	12	13	17	19	20	22	31	67	100	102	110	112	120	122	200	202

beam	0	3	7	10	100	110	127	1,000	1,010	1,100	1,110	2,027	10,000	10,010	10,100

cream	3	7	31	127	216	2,047	4,096	8,191	17,576	97,336	131,071	175,616	287,496

ream	3	6	7	16	26	31	46	56	76	86	96	106	116	126	127	136	146	156	166	176	186

756 "Global warming" was renamed "climate change" when it became obvious to everyone that it was neither global nor everywhere warming, thus also renaming the climate mates.

change	9,261	29,791	132,651	1,367,631	2,803,221	3,442,951	5,000,211	8,120,601
holiness	2	2	1	2	3	1	3	5

757

quadarish mates	6	7	8	9	66	67	68	69	76	77	78	79	80
quintarish mates	5	6	7	8	9	55	56	57	58	59	65	66	67
senarish mates	4	5	6	7	8	9	44	45	46	47	48	49	54
septarish mates	3	4	5	6	7	8	9	33	34	35	36	37	38
octarish mates	2	3	4	5	6	7	8	9	22	23	24	25	26

758

polygamous	0	9	90	900	9,000	90,000	900,000	9,000,000
holiness	1	1	2	3	4	5	6	7
untrimmed	1	10	101	1,011	10,111	101,111	1,011,111	10,111,111

759 "Old MacDonald Had a Farm" is similar to the variant with the opening line, "Old MacDougal had a farm in Ohio-i-o." from *Tommy's Tunes* by F. T. Nettleingham (1917). Since no hachibi can be odd however neither "ohio" or "ohio-i-o" can describe any acronums. Eieio can be read as even-indexed-evens indexed odds or even indexed even-indexed-odds, which happen to be the same, but some other doubly indexed or "binocular" acronums may not be. "Monocular" acronums are therefore to be preferred.

broken-up	7	106	749	11,341	443,048	20,834,597	1,146,345,883	72,240,625,226
dusub	1	14	102	1,535	59,960	2,819,656	155,141,045	9,776,705,476
broken-down	1	8	57	863	33,714	1,585,421	87,231,869	5,497,193,168
broken	1	7	8	57	106	863	11,341	33,714

hypersum	7	715	71,523	7,152,331	715,233,139	71,523,313,947	7,152,331,394,755
dusub	1	97	9,680	967,963	96,796,279	9,679,627,951	967,962,795,111

760 Phillip "Pip" Pirrip was chronicled in *Great Expectations* by Charles Dickens and "Pip" Phillips in Rod Serling's memorable "In Praise of Pip" in which his life is saved by his repentant and praying father Max. Maxine "Max" and Phillip "Pip" Miniver are married English majors ("Mrs. Miniver" by William Wyler.)

"Pups" describes prime-unidexed primes, "sups" sanb-unindexed primes and "sips" sanbi-indexed primes.

pups	2	7	13	19	23	29	43	47	53	61	71	73	79	89	97	101	103	107
sups	2	3	7	11	13	17	19	23	29	31	37	43	47	53	59	61	67	71
sips	5	41	83	137	191	241	307	367	431	487	563	617	739	859	937	1,009	1,063	1,129

761 "Fib" describes Fibonacci-indexed binarish,

fib	1	10	11	101	1,000	1,101	10,101	100,010
broken-up	1	11	12	131	131,012	144,244,343	1,457,012,239,655	> seventeenplex
broken-down	1	2	3	32	32,003	35,235,335	355,912,150,838	> sixteenplex
broken	1	2	3	11	12	32	131	32,003

hypersum	1	110	11,011	11,011,101	110,111,011,000	1,101,110,110,001,101

"Fibber" describes non-repunit Fibonacci-indexed binarish, an allusion to Fibber McGee of "Fibber McGee and Molly" by Jim and Marian Jordan, aka McGee or Jordan numbers.

fibber	10	101	1,000	1,101	10,101	100,010	110,111	1,011,001	10,010,000	11,101,001
holiness	1	1	3	1	2	4	1	3	6	3

762 Alternatively a "lie", rather than coming from LI, could be an acronum for lucky-indexed evens, a "half-lie" just the luckies themselves and the "truth" the lucky-unindexed evens.

lie	2	6	14	18	26	30	42	50	62

sub	1	2	5	7	10	11	15	18	23
truth	1	3	4	5	7	8	9	10	11
broken-up	2	13	28	181	4,734	142,201	5,977,176	299,001,001	18,544,039,238
trimmed	1	2	17	70	3,623	3,110	4,866,065	18,800	743,328,127
sub	1	5	10	67	1,742	52,313	2,198,880	109,996,321	6,821,970,792
broken-down	1	3	7	45	1,177	35,355	1,486,087	74,339,705	4,610,547,797
broken	1	2	3	7	13	28	45	181	1,177

hypersum	2	26	2,614	261,418	26,141,826	2,614,182,630	261,418,263,042
trimmed	1	15	1,503	150,307	15,030,715	150,307,152	15,030,715,231

763 Besides being a nickname for Lily, "lil" also meant book in George Henry Borrow's *Lavengro: the scholar -- the gypsy -- the priest* (1851).

broken-up	1	8	9	71	3,488	219,815	19,127,393	2,123,360,438	286,673,786,523
broken-down	1	2	3	23	1,130	71,213	6,196,661	687,900,584	92,872,775,501
broken	1	2	3	8	9	23	71	1,130	3,488

hypersum	1	17	1,721	172,131	17,213,149	17,212,314,963	172,131,496,387

764 "Pie" may also be an acronym for Proto-Indo-European aka Aryan. "Easy as pie" was used by Zane Grey in "The Young Forester" (1910), referring to eating, not making one.

sub	1	2	4	5	8	10	13	14	17
broken-up	4	25	104	649	14,382	374,581	12,750,136	484,879,749	22,317,218,590
sub	1	9	38	239	5,291	137,801	4,690,513	178,377,291	8,210,045,903
broken-down	1	5	21	131	2,903	75,609	2,573,609	97,872,751	4,504,720,155
broken	1	4	5	21	25	104	131	649	2,903

hypersum	4	46	4,610	461,014	46,101,422	4,610,142,226	461014222634

765 Oil comes from the Norman French euille.

broken-up	1	8	9	71	2,210	81,841	4,012,419	252,864,238	17,451,644,841
broken-down	1	2	3	23	716	26,515	1,299,951	81,923,428	5,654,016,483
broken	1	2	3	8	9	23	71	716	2,210

hypersum	1	17	1,713	171,321	17,132,131	1,713,213,137	171321313749

766 According to Ric Pashley of the Australian National University, Canberra, oil and water DO mix, if the water has no gases in it.

hypersum	0	2	23	234	2,345	23,456	234,568	2,345,689	234,568,910
holiness	1	0	0	0	0	1	3	4	5
trimmed	1	12	123	1,234	12,345	123,457	1234578	12,345,780	1234578011

767 binarish odd

broken-up	1	12	13	155	155,168	156,875,003	172,719,533,471	191,891,558,561,284
broken-down	1	2	3	35	35,038	35,423,453	39,001,256,791	43,330,431,718,254
broken	1	2	3	12	13	35	155	35,038

hypersum	1	111	111,101	111,101,111	1,111,011,111,001	11,110,111,110,011,011

768 ternarish odd

broken-up	1	12	13	156	17,218	2,083,533	418,807,351	88,370,434,594
broken-down	1	2	3	35	3,888	470,483	94,570,971	1,995,494,364
broken	1	2	3	12	13	35	156	3,888

hypersum	1	111	11,121	11,121,101	11,121,101,111	11,121,101,111,121

769 Some abbreviate Fibonacci odd as "fodd", the same ones who abbreviate prime odds as "podds", binarish odds as "bodds", and ternary odds "todds".

foils	1	7	13	49	87	349	559
broken-up	1	8	9	71	6,186	2,158,985	1,206,878,801
broken-down	1	2	3	23	2,004	699,419	390,977,225
broken	1	2	3	7	13	23	49

hypersum	1	17	1,713	171,349	17,134,987	17,134,987,349	17134987349559

770 "Silo" comes from the Spanish, which comes from the Latin sirus.

sub	3	18	36	58	81	122	150
broken-up	7	344	2,415	118,679	25,993,116	8,603,840,075	3,518,996,583,791
dusub	1	47	327	16,062	3,517,786	1,164,403,134	476,244,399,376
broken-down	1	8	57	2,801	613,476	203,063,357	83,053,526,489
broken	1	7	8	57	344	2,425	2,801

hypersum	7	749	74,999	74,999,159	74,999,159,219	74,999,159,219,331

771 "Pig" is from the Old English picga. The plural is spelled as pigges in Chaucer's *Canterbury Tales.*

broken-up	15	376	5,655	141,751	14,889,510	1,861,330,501	307,134,422,175

tressub	1	19	281	7,057	741,305	92,670,189	15,291,322,475
broken-down	1	16	241	6,041	634,546	79,324,291	13,089,142,561
broken	1	15	16	241	376	5,655	6,041

hypersum	15	1,525	152,545	15,254,565	15,254,565,105	15,254,565,105,125
tressub	1	76	7,595	759,480	759,480,076	759,480,075,811

772 "Pig-headed" was used by Benjamin Jonson in *Newes from New World* (c. 1637).

broken-up	150	22,651	3,397,800	513,090,451	79,019,327,254	12,248,508,814,821
holiness	1	1	5	3	3	9
pentasub	1	153	22,894	3,457,176	532,428,039	82,529,803,212
broken-down	1	151	22,651	3,420,452	526,772,259	81,653,120,597
broken	1	150	151	22,651	3,397,800	3,420,452

hypersum	150	150,151	150,151,152	150,151,152,153	150,151,152,153,154
holiness	1	1	1	1	1

773 "Padre Pio" is formally called St. Pius of Pietrelcina.

sub	1	2	3	5	8	9	12	14	17
broken-up	3	16	51	271	5,742	143,821	4,751,835	175,961,716	7,923,029,055
sub	1	6	19	100	2,112	52,909	1,748,102	64,732,698	2,914,719,501
broken-down	1	4	13	69	1,462	36,619	1,209,889	44,802,512	2,017,322,929
broken	1	3	4	13	16	51	69	271	1,462

hypersum	3	35	359	35,913	3,591,321	359,132,125	35,913,212,533
sub	1	13	132	13,212	1,321,173	132,117,325	13,211,732,557

774 Rio is the common contraction for Rio de Janeiro ("River of January"), Brazil.

dusub	1	4	7	10	12	15	18
broken-up	11	342	3,773	117,305	10,678,528	1,185,433,913	155,302,521,131
dusub	1	46	511	15,875	1,445,182	160,431,034	21,017,910,685
broken-down	1	12	133	4,135	376,418	41,786,533	5,474,412,241
broken	1	11	12	133	342	3,773	4,135

hypersum	11	1,131	113,151	11,315,171	1,131,517,191	1,131,517,191,111

775 "Lip" comes from the Latin labrum from the Indo-European leb-.

sub	1	2	6	11	15	17	27	36	47

broken-up	2	11	24	131	5,395	253,696	18,525,203	1,797,198,387	228,262,720,352
trimmed	1	0	13	20	4,284	142,585	741,412	686,087,276	11,715,161,241
sub	1	4	9	48	1,985	93,330	6,815,041	661,152,338	83,973,162,003
broken-down	1	3	7	38	1,565	73,593	5,373,854	521,337,431	66,215,227,591
broken	1	2	3	7	11	24	38	131	1,565

hypersum	2	25	2,517	251,731	25,173,141	2,517,314,147	251731414773
trimmed	1	14	1,406	140,620	14,062,030	1,406,203,036	140620303662

776 slips

dusub	1	6	18	26	42	50	66	93	123	149	193

broken-up	5	206	1,035	42,641	13,091,822	480,4741,315	2,339,922,112,227
dusub	1	28	140	5,771	1,771,785	650,251,027	316,674,021,810
broken-down	1	6	31	1,277	392,070	143,890,967	70,075,292,999
broken	1	5	6	31	206	1,035	1,277

hypersum	5	541	541,137	541,137,191	541,137,137,191	541,137,191307

777 sanbi lucky

sub	1	5	12	16	23	27	34	49
broken-up	3	40	123	1,639	103,380	7,548,379	702,102,627	93,387,197,770
sub	1	15	45	603	38,031	2,776,893	258,289,122	34,355,230,128
broken-down	1	4	13	173	10,912	796,749	74,108,569	9,857,236,426
broken	1	3	4	13	40	123	173	1,639

hypersum	3	313	31,333	3,133,343	313,334,363	31,333,436,373	3133343637393

778 "Non-slip" was used by Gerald Briggs in the *Journal of Political Economics* in 1901.

hypersum	0	1	12	123	1,234	12,346	123,467	1,234,678	12,346,789	1,234,678,910
holiness	1	0	0	0	0	1	1	3	4	5
untrimmed	1	2	23	234	2,345	23,457	234,578	2,345,789	234,578,910	2,345,7891,021

779 "Gip" is said to come from gypsy which in turn came from 'Gyptian.

dusub	1	6	13	20	26	35	42
broken-up	11	518	5,709	268,841	52,967,386	13,612,887,043	4,260,886,611,845
dusub	1	70	773	36,384	7,168,356	1,842,303,924	576,648,296,453
broken-down	1	12	133	6,263	1,233,944	3,172,987`	99,262,883,567

broken	1	11	12	133	518	5,709	6,363

hypersum	11	1,147	114,797	114,797,149	114,797,149,197	114,797,147,197,257
dusub	1	155	15,536	15,536,105	15,536,104,701	15,536,104,430,696

"Gipsy" describes numbers with gobi-indexed prime frags.

gipsy	1,147	1,197	4,711	4,797	9,711	9,747	11,147	11,197	11,257	11,313	11,379

780 "Hip" and "hippy" are said to come from the W. African *hipi* meaning "aware".

tressub	1	3	5	8	11	14	17
broken-up	19	1,160	22,059	1,346,759	30,034,916	81,396,011,395	27,430,756,189,431
holiness	1	2	2	2	4	6	5
tressub	1	58	1,098	67,051	14,953,512	4,052,468,784	135,696,933,785
broken-down	1	20	381	23,261	5,187,584	1,405,858,525	473,779,510,509
broken	1	19	20	381	1,160	22,059	23,261

hypersum	19	1,961	1,961,107	1,961,107,163	1,961,107,163,223	196,161,107,163,223,271
holiness	1	2	3	4	4	5
tressub	1	97	97,638	97,637,776	97,637,776,412	> fifteenplex

781 "Kip" is also an acronym for "Knowledge is power" from "scientia potestas est" in Sir Francis Bacon's *Meditationes Sacrae* (1597).

tressub	1	3	6	8	11	14
broken-up	23	1,542	35,489	2,379,305	540,137,724	149,620,528,853
tressub	1	77	1,767	118,459	26,891,874	7,449,167,499
broken-down	1	24	553	37,075	8,416,578	2,331,429,181
broken	1	23	24	553	1,542	35,489

hypersum	23	2,367	2,367,109	2,367,109,167	2,367,109,167,227	2,367,109,167,227,347
tressub	1	118	117851	117,851,426	117,851,425,943	117,851,425,942,946

782 Nip comes from niperkin meaning a half-pint of liquor, so it'd be 3/8 of the eight-letter word or 3/16 pint. Naps are nibi-and-primes and snaps are square-nibi-and-primes. The first snip, square-nibi-indexed prime, is the nineth prime, twenty-three.

naps	2	3	5	7	11	13	17	19	23	29	31	33	37	41	43	47	53	59	61	63	67	71	73	79	83	89	97
snaps	2	3	5	7	9	11	13	17	19	23	29	31	37	41	43	47	53	59	61	67	71	73	79	83	89	97	101

square nibi	9	169	529	1,089	1,849	2,809	3,969	5,329	6.889	8,649	10,609	12,769

783 "Rip" may also be an acronym for the English "rest in peace" from the earlier Latin "requiem in pacem" as well as "research in progress", which can mean much the same thing.

tressub	1	3	5	8	10	13	16
broken-up	13	690	8,983	476,789	94,889,994	24,956,545,211	7,911,319,721,881
tressub	1	34	447	23,738	4,724,295	1,242,513,223	39,388,145,873
broken-down	1	14	183	9,713	1,933,070	508,407,123	161,166,991,061
broken	1	13	14	183	690	8,893	9,713

hypersum	13	1,353	1,353,101	1,352,101,151	1,353,101,151,199	1,353,101,151,199,263
tressub	1	67	67,367	67,317,152	67,366,939,523	67,366,939,523,394

784 Alas, the fourth ship hypersum is not prime, since its weight is divisible by three.

tressub	1	3	5	8	11	13	17
broken-up	17	1,004	17,085	1,009,019	212,920,094	57,276,514,305	18,958,739,155,049
tressub	1	50	851	50,236	10,600,667	2,851,629,733	943,900,042,481
broken-down	1	18	307	18,131	3,825,948	1,029,198,143	340,668,411,281
broken	1	17	18	307	1,004	17,085	18,131

hypersum	17	1,759	1,759,103	1,759,103,157	1,759,103,157,211	1,759,103,157,211,269
tressub	1	88	87,581	87,580,589	87,580,589,154	87,580,589,154,203

785 shipmates

trimmed	3	71	35	115	18	245	301	38	44	50	557	62	677	731	785

786 "Tip" is also an acronym for "to insure promptness" and "The Infinity Project",

broken-up	2	7	16	55	2,051	1,216,298	728,564,553	437,868,512,651	283,301,656,249,750
trimmed	1	6	5	44	140	105,187	617,453,442	326,757,401,540	1,722,054,513,864
sub	1	3	6	20	755	447,451	268,023,921	161,082,823,741	283,301,656,249,750
broken-down	1	3	7	24	895	530,759	317,925,536	191,073,777,895	123,625,052,223,601
broken	1	2	3	7	16	24	55	895	2,051

hypersum	2	23	2,329	232,931	13,293,137	13,293,137,593	13,293,137,593,599
holiness	1	12	1,218	121,820	12,182,026	12,182,026,482	12,182,026,482,488
sub	1	8	857	85,691	4,890,272	4,890,272,029	4,890,272,029,348

787 Yips may also refer to a people living in Winkie Country to the west of the Emerald City,

Oz.

sub	1	6	12	19	26	34	42
broken-up	7	302	2,121	91,505	17,662,586	443,340,091	1,378,805,246,387
dusub	1	41	287	12,384	2,390,371	599,995,525	186,600,998,548
broken-down	1	8	57	2,459	474,644	119,138,103	3,705,242,677
broken	1	7	8	57	302	2,121	2,459

hypersum	7	743	74,389	74,389,139	74,389,139,193	74,389,139,193,251

788 Filip's surname Pinno in Italian means "tourist settlement", so he's not likely of Filippino descent.

sub	1	2	6	15	36	84	165	323
broken-up	2	11	24	131	12,731	2,890,068	1,297,653,263	1,138,044,801,719
trimmed	1	0	13	20	1,620	17,857	186,542,152	273,370,608
sub	1	4	9	48	4,683	1,063,197	477,379,957	418,663,285,684
broken-down	1	3	7	38	3,693	838349	376,422,394	330,123,277,887
broken	1	2	3	7	11	24	38	131

hypersum	2	25	2,517	251,741	25,174,197	25,174,197,227	25174197227449
trimmed	1	14	1,406	140,630	140,603,086	140,603,086,116	140,603,086,116,338
sub	1	9	926	92,610	9,261,070	9,261,069,608	9261069607974

789 The word "fils" is the French suffix used since 1786 as the equivalent to the English junior. Its antonym, the equivalent of senior, is peré.

broken-up	1	4	5	19	480	23,539	2,048,373	325,714,846	92,830,779,483
broken-down	1	2	3	11	278	13,633	1,186,349	188,643,124	53,764,476,689
broken	1	2	3	4	5	11	19	278	480

hypersum	1	13	137	13,713	1,371,325	13,7132,549	13713254987

790 "Lisp" may also mean "list processing language" or its acronymous nickname "lots of irritating superfluous parentheses".

sub	1	8	19	31	42	82	97
broken-up	3	70	213	4,969	561,710	12566299	32,945,598,347
sub	1	26	78	1,828	206,642	46,082,896	12,120,008,309

broken-down	1	4	13	303	34,252	7,638,499	2,008,959,489
broken	1	3	4	13	70	213	303

hypersum	3	323	32,353	3,235,383	3,235,383,113	3,235,383,113,223
sub	1	119	11,902	1,190,231	1,190,230,932	1,190,230,931,668

791 Since the binarish luckies are only those with ones or zeroes, one and one hundred eleven are the first two. The first prime is, of course, two and the hundred eleventh prime is six hundred seven. The first Mersenne prime is two, the hundred eleventh Mersenne prime is still unknown.

792 "Flip" reminds some of Clerow "Flip" Wilson, Jr. aka "Geraldine 'the devil made me do it' Jones".

sub	1	2	15	27	2,898
broken-up	2	11	24	131	1,031,911
trimmed	1	0	13	20	20,800
sub	1	4	9	48	379,619
broken-down	1	3	7	38	299,333
broken	1	2	3	7	11
hypersum	2	25	2,541	254,173	2,541,737,877
trimmed	1	14	1,430	143,062	1,430,626,766
sub	1	9	935	93,505	93,503,110

793 Broken fluckies are a very forbidden sequence.

broken-up	1	4	5	19	18,758
broken-down	1	2	3	11	10,660
broken	1	2	3	4	5
hypersum	1	13	1,313	131,321	131,321,987

794 "Limit" comes from the Latin limes, limit-. The "limitless" numbers, of course, are those with no "limit", one, two and those between four and a thousand ten, etc.

sub	1	372	8,138,233,450	...
broken-up	3	3,034	9,105	9,208,189
sub	1	1,116	3,350	3,387,503
broken-down	1	4	13	13,147
broken	1	3	4	13
hypersum	3	31,011	3,101,122,122,012,102	...

sub	1	11,408	1,140,839,073,250,210	...

795 Besides being the abbreviation for limit, lim is the name for a Bhutanese flute.

sub	1	11	192,878	> seventeenplex	...
broken-up	3	94	285	8,929	70,334,018
sub	1	35	105	3,285	25,874,439
broken-down	1	4	13	407	3,205,952
broken	1	3	4	13	94

hypersum	3	331	331,524,297	3,315,242,972,305,843,009,213,693,951
sub	1	122	121,960,973	> twenty-sevenplex

"Slim" describes sanbi lucky-indexed Mersennes. Since the first three sanbi luckies are three, thirteen and thirty-three, the slim numbers get higher even faster.

sluckies	3	13	33	43	63	73	93	133	163	193	223	273	283	303	423	463	483	513	553

slim	31	2,199,023,255,551	17,422,457,186,352,049,329,247,799,005,065,324,265,471
tressub	1	109,482,921,167	> thirty-nineplex

796 Piback takes an early lead with three, but then eback immediately gets higher than both piback and phiback. Phiback then catapults ahead because of and early zero, apparently maintaining the lead.

zero in phi	5	21	40	42	48	66	70	78	84	86	90	107	110	145	146	147
zero in e	14	22	44	68	73	111	112	114	132	139	140	149	155	173	195	205
zero in pi	33	51	55	66	72	78	86	98	107	117	122	129	133	147	152	155

Although "piback", "phiback" and "eback" are not acronums, many acronums do end in -ack, including "back" (binarish-and-cubed-kyuubi): "clack" (cubed-luckies-and-cubed-kyuubi), "crack" (cubed-rokubi-and-cubed-kyuubi), "flack" (Fibonacci-luckies-and-cubed-kyuubi), "hack" (hachibi-and-cubed-kyuubi), "lack" (lucky-and-cubed-kyuubi), "mack" (Mersenne--and-cubed-kyuubi), "manpack" (Mersenne-and-nibi-prime-and-cubed-kyuubi), "pack" (prime-and-cubed-kyuubi), "plack" (prime-luckies-and-cubed-kyuubi), "rack" (rokubi-and-cubed-kyuubi), "slack" (sanbi-luckies-and-cubed-kyuubi), "smack" (sanbi-Mersennes-and-cubed-kyuubi), "snack" (sorted-nibi-and-cubed-kyuubi), "stack" (sorted-ternarish-and-cubed-kyuubi), "tack" (ternarish-and-cubed-kyuubi), "tamarack" (ternarish-and-Mersennes-and-rokubi-and-cubed-kyuubi"yack" (yongbi-and-cubed-kyuubi). Some mates are easily named like "wise" for "crack", "clothes" for "rack", "thumb" for "tack", "yackity" for "yack".

797 rootback diagonals

1st	1	41	371	60,322	49,442	575,462	7,248,282	67,722,613
2nd	1	71	322	9,442	75,462	248,282	7,722,613	667,722,613

3rd	1	22	442	5,462	48,282	722,613	4,266,133	61,014,643

798 This is, of course, quite different than the trimmed back numbers.

trimmed root fiveback	1	11	211	5,211	65,211	865,211	6,865,211	66,865,211
trimmed root six back	1	31	331	8,331	38,331	738,331	8,378,331	68,378,331
trimmed root sevenback	1	51	351	4,351	64,351	464,351	20,464,351	...
trimmed eback	1	61	7,061	17,061	717,061	70,717,061	170,717,061	7,170,717,061
trimmed root eightback	1	71	171	7,71	37,171	137,171	6,137,171	...

799 quarterbacks

quarter root seventeenback	1	301	70301	770,301	6,770,301	6,046,770,301
quarter root eighteenback	1	601	60,601	660,601	106,606,601	7,106,606,601
quarter root nineteenback	1	801	9,801	79,801	279,801	4,279,801
quarter root twenty-fourback	1	21	221	4,221	74,221	474,221
quarter root twenty-sixback	1	21	721	4,721	74,721	574,721

quarter root twenty-sevenback	1	21	921	9,921	309,921	8,309,921
quarter root twenty-eightback	1	31	231	2,231	82,231	782,231
quarter root twentynineback	1	31	431	6,431	26,431	926,431
quarter root thirtyback	1	31	631	9,631	39,631	6,039,631
quarter root thirty-oneback	1	31	931	1,931	91,931	491,931

quarter root thirty-twoback	1	41	141	4,141	24,141	124,141
quarter root thirty-threeback	1	41	341	6,341	16,341	416,341
quarter root thirty-fourback	1	41	541	7,541	77,541	377,541
quarter root thirty-fiveback	1	41	741	9,741	109,741	9,109,741
quarter root thirty-sevenback	1	51	251	60,251	960,251	60,960,251
quarter root thirty-eightback	1	51	451	1,451	11,451	3,011,451
quarter root thirty-nineback	1	51	651	1,651	21,651	421,651

quarter root fortyback	1	51	851	1,851	11,851	311,851
quarter root forty-oneback	1	61	70,061	870,061	1,870,061	501,870,061
quarter root forty-twoback	1	61	261	10,261	810,261	5,810,261
quarter root forty-threeback	1	61	361	9,361	39,361	539,361
quarter root forty-fourback	1	61	561	8,561	38,561	138,561

quarter root forty-fiveback	1	61	761	7,761	507,761	90,507,761
quarter root forty-sixback	1	61	961	5,961	55,961	855,961

quarter root forty-sevenback	1	71	171	3,171	93,171	193,171
quarter root forty-eightback	1	71	371	2,371	502,371	80,502,371
quarter root fiftyback	1	71	671	7,671	77,671	677,671
quarter root fifty-oneback	1	71	871	5,871	35,871	535,871
quarter root fifty-twoback	1	81	2,081	72,081	772,081	5,772,081
quarter root fifty-threeback	1	81	281	200,281	7,200,281	47,200,281
quarter root fifty-fourback	1	81	381	7,381	17,381	117,381
quarter root fifty-fiveback	1	81	581	4,581	404,581	9,404,581
quarter root fifty-sixback	1	81	781	80,781	280,781	8,280,781
quarter root fifty-sevenback	1	81	881	7,881	47,881	547,881
quarter root fifty-eightback	1	91	3,091	93,091	493,091	3,493,091
quarter root sixtyback	1	91	391	6,391	46,391	946,391
quarter root sixty-oneback	1	91	591	2,591	52,591	652,591
quarter root sixty-twoback	1	91	691	8,691	58,691	1,058,691
quarter root sixty-threeback	1	91	891	4,891	34,891	134,891

800 almost-easy

almost-3e trimmed	7	0	4	3	7	3	4	3	7	4	2	6	6	0	2	4	1	2
collapsed	11	3	7	3	4	3	7	4	2	6	8	4	6	5	7	7	5	1
almost-5e collapsed	1	3	5	9	1	13	1	4	2	2	9	5	2	2	6	1	7	6
almost-7e collapsed	1	11	7	9	7	2	7	9	9	2	1	3	3	1	6	6	4	7
almost-9e trimmed	1	3	3	5	3	4	2	5	3	4	5	0	2	0	3	6	0	0
collapsed	1	3	3	5	3	4	2	5	3	4	10	13	1	3	1	4	7	6
almost-9e collapsed	2	4	4	6	4	5	3	6	4	5	6	1	11	1	1	8	2	4
trimmed	1	3	3	5	3	4	2	5	3	4	5	0	0	0	0	7	1	3

hypersums

almost-7e collapsed	1	111	1,117	11,179	111,797	1,117,972	11,179,727	111,797,279
almost-9e trimmed/collapsed	1	13	133	1,335	13,353	133,534	1,335,342	13,353,425

801 almost-too-easy

almost-2e trimmed	4	3	2	5	4	5	2	5	4	5	8	0	7	8	4	6	6	1	4
collapsed	4	3	2	5	4	5	2	5	4	5	15	8	4	6	6	1	4	6	3

almost-2e collapsed	5	4	3	6	5	6	3	6	5	6	9	1	22	14	7	7	4	9	4

almost-4e collapsed	9	7	3	1	2	7	3	1	3	8	3	6	1	17	4	1	4	13	8
holiness	1	0	0	0	0	0	0	0	0	2	0	1	0	0	0	0	0	0	2

almost-6e collapsed	1	6	12	6	18	14	5	4	2	7	1	4	1	2	1	6	1	7	3
almost-8e trimmed	1	0	1	3	5	1	4	3	1	6	5	6	1	2	5	0	7	7	1
collapsed	2	3	5	1	4	3	1	6	5	6	1	2	12	7	1	7	7	1	1
untrimmed	1	2	6	2	5	4	2	7	6	7	2	3	13	8	2	8	8	2	2

hypersums

almost-6e collapsed	1	16	1,612	16,126	1,612,618	161,261,814	1,612,618,145
almost-8e trimmed	1	10	106	1,063	10,635	106,351	1,063,514

802 almost-half-easy

almost-3e/2 trimmed	3	6	6	3	1	1	6	3	1	5	7	7	4	5	6	7	4	2	3
almost-3e/2 collapsed	11	7	7	4	2	2	7	4	2	6	8	8	5	6	7	8	5	11	3
almost-5e/2 trimmed	5	6	7	4	6	3	4	6	0	0	3	6	5	0	2	7	7	3	6
collapsed	5	6	7	4	6	3	4	9	6	7	7	7	2	6	7	7	3	6	0
almost-5e/2 collapsed	6	7	8	5	11	5	7	1	1	4	7	6	1	11	8	11	1	8	6
holiness	1	0	2	0	0	0	0	0	0	0	0	1	0	0	2	0	0	2	1
almost-7e/2 trimmed	8	4	0	2	8	7	5	2	8	8	5	5	5	4	7	2	1	2	6
collapsed	8	6	8	7	5	2	8	8	5	5	4	7	2	1	2	6	5	7	1
holiness	2	1	2	0	0	0	2	2	0	0	0	0	0	0	0	1	0	0	0
almost-7e/2 collapsed	9	5	1	3	9	8	6	3	9	9	12	6	5	8	3	2	3	7	6
holiness	1	0	0	0	1	2	1	0	1	1	0	1	0	2	0	0	0	0	1
almost-9e/2 collapsed	1	2	2	3	2	2	6	8	2	2	14	5	10	5	5	9	1	2	1

hypersums

almost-9e/2 collapsed	1	12	122	1,223	12,232	122,322	1,223,226	12,232,268

803 collapsed almost-decimals

almost-two-seventeenths	1	1	7	6	4	12	8	8	2	3	5	2	9
almost-three-seventeenths	1	7	6	4	12	8	8	2	3	5	2	9	1
almost-two-nineteenths	6	2	6	3	1	5	7	8	9	4	7	3	6
holiness	1	0	1	0	0	0	0	2	1	0	0	0	1

almost-three-nineteenths	1	5	7	8	9	4	7	3	6	8	4	2	6

trimmed almost-decimals

almost-two-sevenths	1	7	4	6	0	3	1	7	6
almost-three-elevenths	1	6	1	6	1	6	1	6	1
almost-three-thirteenths	1	2	6	5	8	1	2	6	5
almost-three-fourteenths	1	0	3	1	7	4	6	0	3
almost-four-fifteenths	1	5	5	5	5	5	5	5	5
almost-four-seventeenths	1	2	4	1	8	3	0	0	6
almost-four-nineteenths	1	0	4	1	5	2	0	4	7

804 collapsed almost-roots

almost-fourth root 2	1	1	8	9	9	1	1	5	2	7	2
almost-fourth root 3	1	3	1	13	5	2	9	5	2	4	9
almost-fourth-root 6	1	5	6	13	4	5	8	4	5	8	7
almost-fourth-root 8	1	6	8	1	7	9	2	8	15	4	2

There are also almost-transcendental easy sequences, including those almost-easy as pi-e.

almost-pi-e	8	5	3	9	7	3	4	2	2	2
holiness	2	0	0	1	0	0	0	0	0	0
almost-phi-e	1	6	7	9	9	9	0	5	6	1

805 *Partridge's Concise Dictionary of Slang and Unconventional English* by Eric Partridge, the inspiration for many of the terms in this section, refers to "Down by the Liffeyside" for "one-and-one".

dusub	1	14	17	18	19	21
broken-up	11	1,112	12,243	1,237,655	174,521,598	26,353,998,953
dusub	1	150	1657	167,498	23,618,930	3,566,625,913
broken-down	1	12	133	13,445	1,895,878	286,291,023
broken	1	11	12	133	1,112	12,243

hypersum	11	11,101	1,110,121	11,101,121,131	11,101,121,131,141
dusub	1	1,502	150239	1,502,373,372	1,502,373,372,527

806 Alternatively "fish" describes Fibonacci-indexed shichibi and "fudisushutch" Tutese fish.

fish	7	17	27	47	77	127	207	337	547	887	1,437	2,327	3,767	6,097	9,867	15,967	25,837

fudisushutch	1,007	1,027	1,037	1,047	1,057	1,067	1,087	1,097	1,207	1,227	1,237
holiness	2	1	1	1	1	2	3	2	1	0	0

"Chips" is not only the abbreviation for California Highway Patrolmen, but cubed-hachibi-indexed primes. The first is three thousand, six hundred seventy-one.

cubed hachibi	512	5,832	21,952	54,872	110,592	195,112	314,432	454,552

807 Tutnese money or "mumonuneyud" has two other digits between the one and four. The first prime would be ten thousand, two hundred forty-three.

mumonuneyud	1,004	1,024	1,034	1,054	1,064	1,074	1,084	1,094	1,204	1,224	1,234
holiness	2	1	1	1	2	1	3	2	1	0	0

808 Tutnese show or "sushutchowuw" has two other digits between the two and the four. The first prime is twenty thousand, forty-seven.

sushutchowuw	2,004	2,014	2,034	2,054	2,064	2,074	2,084	2,094	2,104	2,114
holiness	2	1	1	1	2	1	3	2	1	0

809 Tutnese ready or 'rugeadudy' has two other digits between the two and three.

rugeadudy	2,003	2,013	2,033	2,043	2,053	2,063	2,073	2,083	2,093	2,103	2,113	2,143
holiness	2	1	1	1	1	2	1	3	2	1	0	0

810 Tutnese poor or "pugoorug" numbers have ones and/or fours between two other numbers.

pugoorug	2,110	2,112	2,113	2,115	2,116	2,117	2,118	2,119	2,140	2,142	2,143	2,145
holiness	1	0	0	0	1	0	2	1	1	0	0	0

811 no-show

hypersum	0	1	13	135	1,356	13,567	135,678	1,356,789	135,678,910	13,567,891,011
holiness	1	0	0	0	1	1	3	4	5	5
untrimmed	1	2	24	246	2,467	2,4678	246,789	24,678,910	2,467,891,021	246,789,102,122

812 not-quite-ready

trimmed	1	2	1	2	1	10	11	13	14	15	16	17	18	2	20	22	23	24	25	26	27	28	31	32

hypersum	2	23	2,312	231,213	23,121,320	2,312,132,021	231,213,202,122
trimmed	1	12	1,201	120,102	1,201,021	120,102,110	12,010,211,011

813 Half-plate numbers would have either at least one one or one eight, but not both.

half-plate	1	8	10	11	12	13	14	15	16	17	19	21	28	31	38	41	48	51	58	61

Eightless numbers, those without an eight, are like those on the Classic Dvorak keyboard, which had an eight where Qwerty had another symbol and it had another symbol where Qwerty had a one.

Qwerty	1	2	3	4	5	6	7	8	9	0	
Dvorak		7	5	3	1	9	0	2	4	6	8

Only half-plateless numbers, those with neither ones nor eights, would be able to be both Dvorak and Qwerty and so Karovd-Ytrewq.

half-plateless	0	2	3	4	5	6	7	9	20	22	23	24	25	26	27	29	30	32	33	34
holiness	1	0	0	0	0	1	0	1	1	0	0	0	0	1	0	1	1	0	0	0
untrimmed	1	3	4	5	6	7	8	10	31	33	34	35	36	37	38	310	41	43	44	45
Dvorak	6	7	5	3	1	9	0	4	76	77	75	73	71	79	72	74	56	57	55	53
holiness	1	0	0	0	0	1	1	0	1	0	0	0	0	1	0	0	1	0	0	0
Karovd-Ytrewq	0	2	3	4	5	6	7	9	20	23	24	25	26	27	29	30	32	34	35	36
holiness	1	0	0	0	0	1	0	1	1	0	0	0	1	0	1	1	0	0	0	1
untrimmed	1	3	4	5	6	7	8	10	31	34	35	36	37	38	310	41	43	45	46	47

Alternatively "plate" could describe "prime luckies and ternarish evens".

plate	2	3	7	10	12	13	20	22	31	37	43	67	73	79	83	89	97	100	101	102	103	107	109
trimmed	1	2	6	0	1	2	1	11	20	26	32	56	62	68	72	78	86	0	0	1	2	6	8

814 stately

trimmed	17	71	17	17	107	117
broken-up	28	2,297	64,344	5,278,505	1,150,778,434	262,382,761,457
tressub	1	114	3,203	262,801	57,293,884	13,063,268,483
broken-down	1	29	813	66,695	14,540,323	3,315,260,339
broken	1	28	29	813	2,297	64,344

hypersum	28	2,882	2,882,128	2,882,128,208	2,882,128,08,218	2,882,128,208,218,228
trimmed	17	1771	1,771,017	177,101,717	17,710,177,107	17,710,177,107,117
holiness	2	4	6	9	11	13
tressub	1	143	143,493	143,492,714	14,349,270,787	143,492,714,147,511

"Status" describes sorted-ternarish-and-ternarish-unindexed-sanbi.

sternarish	1	11	12	22	111	112	122	222	1,111	1,112	1,122	1,222
tis	3	13	93	103	113	193	203	213	993	1,003	1,013	1,023
tus	23	33	43	53	63	73	83	123	133	143	153	163
status	1	11	12	22	23	33	43	53	63	73	83	111

815 "Interstates" appropriately begin with "double nickles".

broken-up	55	5,776	317,735	33,367,951	7,775,050,318	1,889,370,595,225
tetrasub	1	106	5,819	611,155	142,405,014	34,605,029,549
broken-down	1	56	3,081	323,561	75,392,794	18,320,772,503
broken	1	55	56	3,081	5,776	317,735

hypersum	55	55,105	55,105,168	55,105,168,213	55,105,168,213,233
tetrasub	1	1,009	1,009,286	1,009,286,362	1,009,286,361,897

816

unshod	...	2,344	2,346	2,347	2,348	2,349	2,350	2,351	2,352	2,353	2,355	...

The first Tutnese shod or sushutchdud prime would be higher than 2,004,005,002.

817 The shoeless numbers are called by some the Jackson numbers for Joseph Jefferson Jackson, the ballplayer featured in *Shoeless Joe* by William Patrick Kinsella (1982) on which "Field of Dreams" (1989) was based.

818 The term 'sabotage' comes from the French word 'sabot' or wooden shoe or sandal thrown into machinery.

trimmed	1	2	3	4	1	2	4	1	10	11	12	13	14	15	16	17	18	2	20	21	22	23	24	25	26

broken-up	2	7	16	55	676	8,843	124,478	1,876,013	37,644,738	754,770,773
trimmed	1	6	5	44	565	7,732	13,367	76,502	26,533,627	64,366,662
sub	1	3	6	20	249	3,253	45,793	690,147	13,848,725	277,664,650
broken-down	1	3	7	24	295	3,859	54,321	818,674	16,427,801	197,952,286
broken	1	2	3	7	16	24	55	295	676	3,859

hypersum	2	23	234	2,345	234,512	23,451,213	2,345,121,314
trimmed	1	12	123	1,234	123,401	12,340,102	1,234,010,203
sub	1	8	86	863	86,272	8,627,219	862,721,918

819 half-straight

broken-up	6	49	300	2,449	63,974	1,793,721	64,637,930	2,458,035,061

holiness	1	1	2	1	2	1	4	5
dusub	1	7	40	331	8,658	242,754	874,793	332,658,871
broken-down	1	7	43	351	9,169	257,083	9,264,157	352,295,049
broken	1	6	7	43	49	300	351	2,449

hypersum	6	68	6,816	681,618	68,161,826	6,816,182,628	681,618,262,836
holiness	1	3	4	6	7	9	10
dusub	1	9	922	92,247	9,224,700	922,470,007	9,227,000,660

820 appreciated

trimmed	1	3	5	7	1	3	5	7	1	10	11	12	13	14	15	16	17	18	3	30	31	42	33	34

broken-up	2	9	20	89	1,088	15,321	246,224	4,447,353
trimmed	1	8	1	78	77	4,210	135,113	3,336,242
sub	1	3	7	33	400	5,636	90,581	1,636,090
broken-down	1	3	7	31	379	5,337	85,771	1,549,215
broken	1	2	3	7	9	20	31	89

hypersum	2	24	246	2,468	246,812	24,681,214	2,468,121,416
trimmed	1	13	135	1,357	135,701	13,570,103	1,357,010,305
sub	1	9	90	908	90,797	9,079,711	90,791,127

821 unappreciated

untrimmed	1	4	6	8	10	21	22	24	26	28	210	41	42	44	46	48	49	410	61	64	64	66	68	610	81

822 One-two punch from boxing (1811) came from one-two in fensing. Punch, fka Punchinella, was mated with Judy in 1662. "Judy" is, of course, a number with a seven and eight.

broken-up	12	253	3,048	64,261	7,714,368	933,502,789	113,895,054,626
dusub	1	34	412	8,697	1,044,026	126,335,864	15,414,019,477
broken-down	1	13	157	3,310	397,357	48,083,507	5,866,585,211
broken	1	12	13	157	253	3,048	3,310

hypersum	12	1221	1,221,102	1,221,102,112	1,221,102,112,112,120
dusub	1	165	165,258	165,258,200	165,258,200,203,520

Judy	78	87	178	187	278	287	378	387	478	487	578	587	678	687	778	787	878	887

holiness	2	2	2	2	2	2	2	2	2	2	2	2	3	3	2	2	4	4

broken-up	78	6,787	529,464	46,070,155	12,608,032,554	3,675,951,413,153
holiness	2	3	2	3	5	2
tetrasub	1	124	9,697	843,804	234,587,299	67,327,398,656
broken-down	1	79	6,163	536,260	149,086,443	42,788,345,401
broken	1	78	79	6,163	6,787	529,464

hypersum	78	7887	7,887,178	7,887,178,187	7,887,178,187,278
holiness	2	4	6	8	10
tetrasub	1	145	144,459	144,458,708	144,458,707,529

Tutnese judy or jugududy

jugududy	7,008	7,018	7,028	7,038
holiness	2	3	3	3
broken-up	7,008	49,182,145	344,668,479,168	2,418,883,435,983,169
holiness	4	3	8	11
enneasub	1	6,070	42,535,470	298,513,930,943
broken-down	1	7,009	49,119,073	344,717,661,323
broken	1	7,008	7,009	49,119,973

hypersum	7,008	70,087,018	700,870,187,028	7,008,701,870,287,038
holiness	4	7	10	13
enneasub	1	8,650	86,494,253	864,942,524,714

823 false

broken-up	51	7,651	390,252	58,545,451	8,957,844,255	1,379,566,560,721
tetrasub	1	140	7,148	1,072,297	164,068,641	25,267,642,949
broken-down	1	52	2,653	398,002	60,896,959	9,378,529,688
broken	1	51	52	2,653	7,651	390,252

hypersum	51	51,150	51,150,151	51,150,151,152	51,150,151,152,153
tetrasub	1	937	936,848	936,847,698	936,847,697,607

824 half-true

broken-up	1	111	112	12,431	1,368,792,488,915	> twenty-twoplex
broken-down	1	2	3	332	36,556,922,719	> twentyplex

broken	1	2	3	111	112	332

hypersum	1	110	11,011	1,101,112	110,111,213	11,011,121,314	1,101,112,131,415

825 true

hypersum	0	2	23	234	2,346	23,467	234,678	2,346,789	234,678,920
holiness	1	0	0	0	1	1	3	4	5
untrimmed	1	12	123	1,235	12,356	123,567	1,235,678	12,356,781	1,235,678,111

tutrugue	2,110	2,130	2,140	2,150	2,160	2,170	2,180	2,190	2,320	2,330	2,340
holiness	1	1	1	1	2	1	3	2	1	1	1

broken-up	2,110	4,494,301	948,977,220	20,198,745,972,901	> sixteenplex
holiness	1	2	5	7	
oktasub	1	1,508	3,181,184	6,775,924,404	14,635,999,894,699
broken-down	1	2,111	4,454,211	9,487,471,541	20,492,942,982,771
broken	1	2110	2,111	4,454,211	4,494,301

hypersum	2,110	21,102,130	211,021,302,140	2,110,213,021,402,150
holiness	1	2	3	4
oktasub	1	7,079	70,789,761	707,897,605,594

826 dyed

broken-up	501	255,511	128,011,512	65,286,126,631	33,557,197,099,846
holiness	1	0	3	6	7
hexasub	1	633	317,309	161,828,128	83,179,975,354
broken-down	1	502	251,503	128,267,032	65,929,505,951
broken	1	501	502	251,503	255,511

hypersum	501	501,510	501,510,512	501,510,512,513	501,510,512,513, 514
holiness	1	2	2	2	2
hexasub	1	1,243	1,243,120	1,243,120,275	1,243,120,274,514

827 my

broken-up	1,001	1,013,013	1,014,027,014	1,026,196,351,181	1,041,590,310,475,729
holiness	2	2	3	6	5
heptasub	1	924	924,673	935,769,946	949,807,419,619
broken-down	1	1,002	1,003,003	1,015,040,038	1,030,266,641,573

broken	1	1001	1,002	1,003,003	1,013,013

hypersum	1,001	10,011,012	100,110,121,013	1,001,101,210,131,014
holiness	2	3	3	5
heptasub	1	9,129	91,288,614	912,886,139,213

828 thieves

broken-up	40	1,641	65,680	2,694,521	118,624,604	5,340,801,701
holiness	1	1	5	2	5	5
tetrasub	1	30	1,203	49,352	2,172,685	97,820,195
broken-down	1	41	1,641	67,322	2,963,809	133,438,727
broken	1	40	41	1,641	65,680	67,322

hypersum	40	4,041	404,142	40,414,243	4,041,424,344	404,142,434,445
holiness	1	1	1	1	1	1
tetrasub	1	74	7,402	740,213	74,021,269	740,216,889

829 dark

broken-up	2,359	5,849,806	13,327,894,713	31,920,313,487,441	> sixteenplex
holiness	1	7	3	4	
oktasub	1	1,895	4,471,011	10,708,072,246	31,428,196,512,929
broken-down	1	2,360	5,567,241	13,333,544,555	39,133,958,836,166
broken	1	2,359	2,360	5,567,241	5,849,806

hypersum	2,359	23,592,395	235,923,952,539	2,359,239,525,392,593
holiness	1	2	3	4
oktasub	1	7,915	79,143,669	791,436,691,040

dudarugkuck	2,003,005,009	2,003,005,019	2,003,005,049	2,003,005,069
holiness	6	6	6	7
trimmed	1,248	12,408	12,438	12,478

Alternatively our chocoholic members take "dark" as the mate to "milk" Mersenne-indexed lucky kyuubi.

milk	69	159	...
holiness	3	1	...
dark	30	840	...

holiness	1	3	...

830 twilight

broken-up	23	576	13,271	332,351	11,645,556	454,509,035	23,646,115,376
tressub	1	29	661	16,547	579,798	22,628,672	1,177,270,763
broken-down	1	24	553	13,849	18,939,301	985,328,920	52,241,372,061
broken	1	23	24	553	576	13,271	13,849

hypersum	23	2,325	232,529	23,252,932	2,325,293,235	23,229,323,539
tressub	1	116	11,577	1,157,695	115,769,533	1,156,519,919

831 light

untrimmed	1	5	7	8	9	21	22	25	27	28	29	51	52	55	57	58	71	72	75	77	78	79	81

hypersum	0	1	14	146	1,467	14,678	146,710	14,671,011	14,678,101,112
holiness	1	0	0	1	1	3	2	2	4
untrimmed	1	2	25	257	2,578	25,789	257,821	25,782,122	2,578,212,223

832 half-past

tressub	1	7	11	17	21	26
broken-up	30	3,901	117,060	15,221,701	6,545,448,490	3,469,102,921,401
holiness	1	2	3	1	5	5
tressub	1	194	5,828	757,844	325,878,691	172,716,464,323
broken-down	1	31	931	121,061	52,057,161	27,590,416,391
broken	1	30	31	931	3,901	117,060

hypersum	30	30,130	30,130,230	30,130,230,330	30,130,230,330,430
holiness	1	2	3	4	5
tressub	1	1,500	1,500,096	1,500,095,837	1,500,095,837,401

half-future	69	169	269	369	469	569	669	769	869	969	1,069	1,169	1,269	1,369
holiness	2	2	2	2	2	2	3	2	4	3	3	2	2	2

broken-up	69	11,662	804,747	136,013,905	63,791,326,192
holiness	2	2	3	4	4
tetrasub	1	214	14,739	2,491,181	1,168,378,895
broken-down	1	70	4,831	816,509	382,947,552
broken	1	69	70	4,831	11,662

hypersum	69	69,169	69,169,269	69,169,269,369	69,169,269,369,469
holiness	2	4	6	8	10
tetrasub	1	1,267	1,266,879	1,266,879,360	126,687,879,359,969

833 quarter-to

tetrasub	1	3	4	6	8	10
broken-up	45	6,526	293,715	42,595,201	18,955,158,160	10,330,603,792,401
tetrasub	1	120	5,380	780,158	347,175,832	189,211,608,564
broken-down	1	46	2,071	300,341	133,653,816	72,841,630,061
broken	1	45	46	2,071	6,526	293,715

hypersum	45	45,145	45,145,245	45,145,245,345	45,145,245,345,545
tetrasub	1	827	826,864	826,864,011	826,864,011,292

834 canteen

tressub	1	6	11	16	21	26
broken-up	19	2,262	42,887	5,118,905	2,144,864,192	1,113,189,634,553
holiness	1	1	4	4	4	4
tressub	1	113	2,141	254,855	106,786,500	55,422,448,442
broken-down	1	20	381	45,359	19,005,802	9,864,056,597
broken	1	19	20	381	2,262	42,887

hypersum	19	19,119	19,119,219	19,119,219,319	19,119,219,319,419
holiness	1	2	3	4	5
tetrasub	1	952	951,890	951,889,879	951,889,879,396

835 The phrase "nine-to-five" was first recorded in 1955, but works and faith were mated long before in "You have faith and I have works. Show me your faith without works." (James 2:18)

broken-up	5	31	160	991	9,079	500,336	28,027,895	1,598,090,351
dusub	1	4	22	134	1,229	67,713	3,793,163	216,278,010
broken-down	1	6	31	192	1,759	96,937	5,430,231	309,620,104
broken	1	5	6	31	160	192	991	1,759

hypersum	5	56	567	5,678	56,789	5,678,955	567,895,556	56,789,555,657
dusub	1	8	77	769	7,686	768,563	76,856,306	7,685,630,600

faith

broken-up	1	3	4	11	114	1,265	15,294	200,087	2,816,512	39,631,255	835,072,867

broken-down	1	2	3	8	83	921	11,135	145,676	2,050,599	20,651,666	435,735,585
broken	1	2	3	4	8	11	83	114	921	1,265	11,135

hypersum	1	12	123	1,234	123,410	12,341,011	1,234,101,112	123,410,111,213

The faithful or faith-filled numbers would contain all five of the digits zero through four, the unfaithful less.

faithful	10, 234	10,243	10,324	10,342	10,423	10,432	12,304	12,340	12,403	12,430
holiness	1	1	1	1	1	1	1	1	1	1

broken-up	10,234	104,826,863	1,072,798,126,176	> sixteenplex
holiness	1	7	6	
enneasub	1	12,937	132,393,807	1,356,109,773,694
broken-down	1	10,235	104,744,991	1,072,902,953,048
broken	1	10,234	10,235	104,744,991

hypersum	10,234	1,023,410,243	102,341,024,310,324	10,234,102,431,032,410,342
holiness	1	2	3	4
enneasub	1	126,299	12,629,885,760	1,262,988,576,016,730

836 Chronograms are words that contain Roman numbers within them, usually counting years, hence the chrono- prefix. If higher Roman numerals, L, C, D and M, are included however there can be ambiguity. "Million" could be read either as M + IL + LI = 1,000 + 49 + 51 = 1100 or as MI + L + LI = 1,001 + 50 + 51 = 1102.

broken-up	1	5	6	29	325	4,579	69,010	1,315,769	54,015,539	2,215,952,868
broken-down	1	2	3	14	157	2,212	33,337	635,615	26,093,552	287,664,687
broken	1	2	3	5	6	14	29	157	325	2,212

hypersum	1	14	145	1,459	145,911	14,591,114	1,459,111,415	145,911,141,519

837 "Pascal's triangle" was described by Zhu Shijie in 1303, Yang Hui in 1206 and al-Karkhī Abū Bakr ibn Muḥammad ibn al-Ḥusayn Al-Karajī in 990. Pascal is more accurately noted for Pascal's wager for Christ's promise of eternal life, which if the probability is greater than or equal to zero, the reward is infinitely great, while the probability of losing that reward if not making the wager is certain.

838 *Ancient Puzzles: Classic Brainteasers and Other Timeless Mathematical Games of the Last 10 Centuries* by Dominic Olivastro

839 tetrahedrals

broken-up	1	5	6	29	1,021	46,995	3,478,651	382,698,605
broken-down	1	2	3	14	493	22,692	1,679,701	184,789,802
broken	1	2	3	5	6	14	29	493

hypersum	1	14	1,410	141,020	14,102,035	1,410,203,546	141,020,354,674

840 pentatope

broken-up	1	6	7	41	2,877	333,773	63,419,747	19,026,257,873
broken-down	1	2	3	17	1,193	138,405	26,298,143	7,889,581,305
broken	1	2	3	6	7	17	41	1,193

hypersum	1	15	1,515	151,535	15,153,570	15,153,570,116	15,153,570,116,190

841 simplex diagonal

broken-up	1	3	4	11	774	179,579	151,565,450	474,703,168,979
broken-down	1	2	3	8	563	130,624	110,247,219	345,294,420,532
broken	1	2	3	4	8	11	563	774

hypersum	1	12	126	12,620	1,262,070	1,262,070,232	1,262,070,232,844

842 deuterohypersimplex

broken-up	1	4	5	19	2,209	932,217	145,985,031	3,956,205,356,227
broken-down	1	2	3	11	1,279	539,749	845,248,213	2,290,623,196,979
broken	1	2	3	4	5	11	19	1,279

hypersum	1	13	1,310	131,035	131,035,116	131,035,116,422	13,101,164,221,566

843 hypersimplex diagonal

broken-up	1	4	5	19	8,650	13,433,469
broken-down	1	2	3	11	5,008	7,777,435

broken	1	2	3	4	5	11	19	5,008	8,650	7,777,435	13,433,469

hypersum	1	13	1,315	131,574	131,574,455	1,315,744,551,553

844 duhypersimplex diagonal

broken-up	1	5	6	29	42,056	389,396,533
broken-down	1	2	3	14	20,303	187,985,491

broken	1	2	3	5	6	14	29	20,303	42,056	187,985,491	389,396,533

hypersum	1	14	1,428	1,428,210	14,282,101,450	142,821,014,509,259

845 treshypersimplex diagonal

broken-up	1	6	7	41	181,842	8,131,974,281
broken-down	1	2	3	17	75,398	3,371,798,577

broken	1	2	3	6	7	17	41	75,398	181,842	337,198,577	8,131,974,281

hypersum	1	15	1,545	1,545,445	15,454,454,435	1,545,445,443,544,720

846 Al-Karajī-Lucas

2nd	1	4	15	57	218	838	2,813	10,836	29,357	122,524
3rd	1	5	21	85	339	1,343	4,882	5,877	43,502	194,733
4th	1	6	28	121	505	2,069	3,808	14,145	72,209	318,900
5th	1	7	36	166	726	3,082	8,268	28,707	124,167	530,041
6th	1	8	45	221	1,013	4,460	14,562	51,958	211,141	878,830
7th	1	9	54	275	1,288	5,748	20,310	72,268	283,409	1,545,601

diagonal	1	4	21	121	726	4,460	20,310

1-3 Al-Karajī-Lucas	1	4	13	47	174	654	1,913	7,777	30,735	103,780
2nd	1	5	19	73	281	1,085	3,632	14,029	38,120	159,590
3rd	1	6	26	107	431	1,719	6,252	7,385	55,810	251,145
4th	1	7	34	150	634	2,620	4,765	17,690	91,555	407,110
5th	1	8	43	203	901	3,864	10,305	35,745	155,965	670,401
6th	1	9	53	267	1,244	5,540	18,055	64,410	263,291	1,102,754

1-3 diagonal	1	4	13	47	174	654	1,913	7,777	30,735	103,780

847 The diagonal is the centered square numbers, 2z(z - 1).

broken-up	1	6	7	41	2,508	213,221	24,096,481	3,494,202,966	632,474,833,327
broken-down	1	2	3	17	1,040	88,417	9,992,161	1,448,951,762	262,270,261,083
broken	1	2	3	6	7	17	41	1,040	2,508

hypersum	1	15	1,513	151,341	15,134,161	1,513,416,185	1,513,416,185,113

848 "squared" numbers

1	3	6	10	15	21	28	36	45	55	66	78	91
2	5	9	14	20	27	35	44	54	65	77	90	104
4	8	13	19	26	34	43	53	64	76	89	103	118

7	12	18	25	33	42	52	63	75	88	102	117	133
11	17	24	32	41	51	62	74	87	101	116	132	149
16	23	31	40	50	61	73	86	100	115	131	148	166
22	30	39	49	60	72	85	99	114	130	147	165	184

849 "squared" primes

1st column	2	3	7	17	31	53
2nd	5	11	19	37	59	83
3rd	13	23	41	61	89	127
4th	29	43	67	97	131	173
5th	47	71	101	137	179	229
6th	73	103	139	181	241	307

850 diagonal primes

trimmed	1	0	30	86	68	26	430
sub	1	4	15	36	66	113	199
broken-up	2	23	48	551	98,677	30,294,390	16,389,363,667
trimmed	1	12	37	440	87,566	218,328	5,278,252,556
sub	1	8	18	203	36,301	11,144,683	6,029,309,947
broken-down	1	3	7	80	14,327	4,398,469	2,379,586,056
broken	1	2	3	7	23	48	80

hypersum	2	211	21,141	2,114,197	2,114,197,179	2,114,197,179,307
trimmed	1	100	10,030	1,003,086	1,003,086,068	100,308,606,826
sub	1	78	7,777	777,770	777,769,677	777,769,676,850

851 boustrophedonic

1st column	1	3	4	10	11	21	22
2nd	2	5	9	12	20	23	35
3rd	6	8	13	19	24	34	39
4th	7	14	18	25	33	40	52
5th	15	17	26	32	41	51	60
6th	16	27	31	42	50	61	73
7th	28	30	43	49	62	72	85

852 boustrophedonic diagonal

broken-up	1	6	7	41	1,688	103,009	8,757,453

broken-down	1	2	3	17	700	42,717	3,531,645
broken	1	2	3	6	7	17	41

hypersum	1	15	1,513	151,325	15,132,541	1,513,254,161	151,325,416,185

Besides these ways of subdividing a sequence we also have the "bent diagonal". The first horizontal sequence is the squares just as in the "squared" isoscoles and the first vertical their uptown neighbors. The diagonal is the central polygonals.

vertical	1	2	5	10	17	26	37	50	65	82	101	122	145	170	197	226	257	290
diagonal	1	3	7	13	21	31	43	57	73	91	111	133	157	183	211	241	273	307

1	4	9	16	25	36	49
2	3	8	15	24	35	48
5	6	7	14	23	34	47
10	11	12	13	22	33	46
17	18	19	20	21	32	45
26	27	28	29	30	31	44
37	38	39	40	41	42	43

853 "squared" isosceles columns

1st column	1	2	3	5	6	10	11
2nd	4	7	8	12	13	19	20
3rd	9	14	15	21	22	30	31
4th	16	23	24	32	33	43	44
5th	25	34	35	45	46	58	59
6th	36	47	48	60	61	75	76
7th	49	62	63	77	78	94	95

Just squaring the odd isosceles greatly improves the odds of finding primes.

1		7		17		31		49
	5		15		29		47	
3		13		27		45		67
	11		25		43		65	
9		23		41		63		89
	21		39		61		87	

19		37		59		85		115
	35		57		83		113	

854 "squared" isosceles diagonal

diagonal	1	3	7	13	21	31	43	57	73	91	111	133	157

broken-up	1	4	5	19	404	12,543	539,753	30,778,464	2,247,367,625
broken-down	1	2	3	11	234	7,265	312,629	17,827,118	1,301,692,243
broken	1	2	3	4	5	11	19	234	404

hypersum	1	13	137	13,713	1,371,321	137,132,131	13,713,213,143

855 Egyptian mathematics is best known from the so-called Rhind papyrus, *Directions for Attaining Knowledge into All Obscure Secrets* copied by A'h-mose from an 1850 BC manuscript.

856 Egyptian pi

3	3.0	3.0
+1/8	0.125	3.125
+1/61	0.0163934426	3.1413934426
+1/5,020	0.0001992032	3.1415926458
+1/139,085,963	0.000000007189798	3.1415926529

sub	1	3	22	1,847	51,166,866
broken-up	3	25	78	649	90,266,790,065
sub	1	9	29	239	33,207,296,285
broken-down	1	4	13	108	15,021,284,017
broken	1	3	4	13	25
hypersum	3	38	3,861	38,615,020	3,861,502,013,905,963
sub	1	14	1,420	14,205,672	1,420,567,202,958,130

857 Egyptian half-pi

broken-up	1	3	4	11	762,128	5,112,197,114,244,123
broken-down	1	2	3	8	554,275	3,717,962,147,431,483
broken	1	2	3	4	8	11

hypersum	1	12	1,215	1,215,243	121,524,369,284	1,215,243,692,846,707,793,329

858 Egyptian phi

broken-up	1	13	14	181	2,337,531,375,480	> thirty-twoplex
broken-down	1	2	3	38	490752443471	> thirty-oneplex
broken	1	2	3	13	14	38

hypersum	1	12	129	129,145	12,914,537,986	129,145,379,862,345,721,554

859 Egyptian e

broken-up	2	5	12	29	289,983	1,049,800,386,449,645
trimmed	1	4	1	18	178,872	387,275,338,534
sub	1	2	4	11	106,679	386,199,979,508,660
broken-down	1	3	7	17	169,990	615,400,101,704,497
broken	1	2	3	5	7	12

hypersum	2	22	225	22,555	225,559,999	2,255,599,993,620,213,552
trimmed	1	11	114	11,444	114,448,888	114,448,888,251,102,441
sub	1	8	83	8,298	82,978,886	> seventeenplex

860 Egyptian half-e

broken-up	1	4	5	19	694,400,263
broken-down	1	2	3	11	402,021,205

broken	1	2	3	4	5	11	19	402,021,205

hypersum	1	13	1,339	13,396,006	1,339,600,636,547,382

861 Egyptian root-two

broken-up	1	4	5	19	4,145,824	> seventeenplex
broken-down	1	2	3	11	2,400,214	> seventeenplex

broken	1	2	3	4	5	11	19	2,400,214

hypersum	1	13	1,313	1,313,243	1,313,243,218,201	131,324,321,820,161,323,125,725

Egyptian root-three

broken-up	1	3	4	11	13,743	82,265,598,011	> twenty-fiveplex
broken-down	1	2	3	8	9,995	59,830,070,008	> twenty-fiveplex

broken	1	2	3	4	8	11	9,995	13,743	59,830,070,008

hypersum	1	12	125	12,532	125,321,249	1,253,212,495,986,000

Egyptian root-five

sub	1	2	10	1,040	4,328,184
broken-up	2	11	24	131	1,541,244,499
trimmed	1	0	13	20	430,133,388
sub	1	4	9	48	56,699,265
broken-down	1	3	7	38	447,078,557

broken	1	2	3	7	11	24	38	131	447,078,557

hypersum	2	25	2,528	25,282,828	2,528,282,811,765,225
trimmed	1	14	1,417	14,171,717	1,417,171,700,654,114
sub	1	9	930	9,301,033	930,103,267,915,554

Egyptian half-root-five

broken-up	1	10	11	109	255,683,649,397
broken-down	1	2	3	29	68,025,925,069

broken	1	2	3	10	11	29	109	68,025,925,069

hypersum	1	19	19,145	1,914,537,986	19,145,379,862,345,721,554

Egyptian root-six

broken-up	2	7	16	55	2,726,476
trimmed	1	6	5	44	1,615,365
sub	1	3	6	20	1,003,014
broken-down	1	3	7	24	1,189,735

broken	1	2	3	7	16	24	55	0	2,726,476

hypersum	2	23	239	239,199	23,919,949,572
trimmed	1	12	128	128,088	12,808,838,461
sub	1	8	88	87,996	8,799,657,681

862 Egyptian expansion

[3]	4	13	157	24,493	599,882,557	> seventeenplex
[4]	5	21	421	176,821	31,265,489,221	> twentyplex
[5]	6	31	931	865,831	749,662,454,731	> twenty-threeplex
[6]	7	43	1,807	3,263,443	10,650,056,950,807	> twenty-sixplex
[7]	8	57	3,193	10,192,057	103,878,015,699,193	> twenty-eightplex
[8]	9	73	5,257	27,630,793	763,460,694,178,057	> twenty-nineplex

863 Egyptian expansion diagonal

diagonal	3	13	421	865,831	10,650,056,950,807	> twenty-eightplex

864 Martin Gardner called these numbers magic in *Mathematical Circus.* Joyce called them prepdigits because they are the prerequisite factors of repdigits.

multiplier	3	6	9	12	15	18	21
repdigit	111	222	333	444	555	666	777
multipier	7	14	21	28	35	42	49
repdigit	111,111	222,222	333,333	444,444	555,555	666,666	777,777

865

101	1,010,101	10,101,010,101	101,010,101,010,101
8,547	854,708,547	85,470,854,708,547	8,547,085,470,854,708,547
15,873	15,873,015,873	15,873,015,873,015,873	15,873,015,873,015,873,015,873

866 prepdigits

x	3	**6**	**9**	11	**12**	**15**	**18**	**21**	**22**	**24**	**27**	**33**	**44**
37	111	222	333	~~407~~	444	555	666	777	~~814~~	888	999	~~1,221~~	~~1,628~~
101	~~303~~	~~606~~	~~909~~	1,111	~~1,212~~	~~1,515~~	~~1,818~~	~~2,121~~	2,222	~~2,424~~	~~2,727~~	3,333	4,444

867 Seminacci are half the Fibonacci divisible by two.

broken-up	1	5	6	29	8,851	11,435,521	625,866,073,181
broken-down	1	2	3	14	4,273	5,520,730	302,149,557,173

broken	1	2	3	5	6	14	29	4,373	8,851	5,520,730

hypersum	1	14	1,417	141,772	141,772,305	1,417,723,051,292

868 Tertinacci are a third the Fibonacci divisible by three.

broken-up	1	8	9	71	1,097,385	116,253,674,816	> sixteenplex
broken-down	1	2	3	23	355,491	37,659,650,090	> sixteenplex

broken	1	2	3	8	9	23	71	355,491	1,097,385	3,769,650,090

hypersum	1	17	1,748	17,482,255	1,748,225,515,456	1,748,225,515,456,105,937

869 Quartinacci are a fourth the Fibonacci divisible by four.

trimmed	1	25	535	481	170	2,621,477
sub	1	13	238	4,264	76,523	1,373,142
broken-up	2	73	148	5,401	1,123,462,158	4,193,421,369,410,305

trimmed	1	62	37	430	12,351,047	30,823,102,583,024
sub	1	27	54	1,987	413,298,631	1,542,673,509,975,050
broken-down	1	3	7	255	53,042,557	197,986,011,747,771

broken	1	2	3	7	73	148	255	5,401	53,042,557	1,123,462,158

hypersum	2	236	236,646	23,664,611,592	23,664,611,592,208,010
trimmed	1	125	125,535	12,553,500,481	12,553,500,481,170
sub	1	87	87,057	8,705,724,088	8,705,724,088,080,720

870 Quintinacci are a fifth the Fibonacci divisible by five.

broken-up	1	112	113	12,655	> eighteenplex
broken-down	1	2	3	335	> sixteenplex

broken	1	2	3	112	113	335	12,655	> sixteenplex

hypersum	1	111	111,122	1,111,221,353	111,122,135,315,005

Sextinacci are a sixth the Fibonacci divisible by six.

tressub	1	385	123,890	39,892,112	12,845,136,294
broken-up	24	185,473	4,451,376	34,400,419,201	> twenty-oneplex
tressub	1	9,234	221,621	1,712,696,023	> twentyplex
broken-down	1	25	601	4,644,553	> eighteenplex

broken	1	24	25	385	601	185,473	4,451,376	4,644,553	34,400,419,201

hypersum	24	247,728	2,477,282,488,392	2,477,282,488,392,801,254,496
trimmed	13	136,617	1,366,171,377,281	136,617,137,728,170,143,385
tressub	1	12,334	123,336,632,616	> twentyplex

Septinacci are a seventh the Fibonacci divisible by seven.

sub	1	52	2,437	114,479	5,378,090	252,655,762
broken-up	3	424	1,275	180,199	2,634,358,915,110	> twenty-oneplex
sub	1	156	469	66,292	969,126,485,536	> twentyplex
broken-down	1	4	13	1,837	26,855,406,118	> nineteenplex

broken	1	3	4	13	424	1,275	1,837	180,199	26,348,915,110	26,855,406,118

hypersum	3	3,141	31,416,624	31,416,624,311,187	3,141,662,431,118,714,619,165
sub	1	1,156	11,557,530	11,557,530,195,093	> twenty-oneplex

Octinacci are an eighth of the Fibonacci divisible by eight.

broken-up	1	19	20	379	39,434,970	73,597,247,901,559
broken-down	1	2	3	56	5,826,803	10,874,527,478,138

broken	1	2	3	19	20	56	379	5,826,803	39,434,970	10,874,527,478,138

hypersum	1	118	118,323	1,183,235,796	1,183,235,796,104,050

This subdivision can, of course, be applied to other composites as well. Semiprime, etc., however mean composites with two or more prime factors. Those with just two, no matter how many times they are multiplied, can be separated into a little and a not-so-little part like the bit and the googol.

There are no semiluckies since they are all odd, but there are tertiluckies, a third the luckies divisible by three, etc.

lucky = 0%3	3	9	15	21	33	51	63	69	75	87	93	99	105	111	129	135
tertilucky	1	3	5	7	11	17	21	23	25	29	31	33	35	37	43	45

871 Tribonacci is a variation on Fibonacci, replacing the tri- prefix for the fi- prefix.

broken-up	1	14	15	209	28,406,668	386,095,790,062,517
broken-down	1	2	3	41	5,572,600	75,741,280,170,641

broken	1	2	3	14	15	41	209	5,572,600	28,406,668

hypersum	1	13	135	1,359	135,917	13,591,731	1,359,173.157

872 Tetranacci is a contraction of the more logical extrapolation Tetrabonacci from Tribonacci that preserves the syllable count and rhythm.

broken-up	1	5	6	29	731	35,848	3,370,443	610,086,031	212,923,395,262
broken-down	1	2	3	14	353	17,311	1,627,587	294,610,558	10,820,712,329

broken	1	2	3	5	6	14	29	353	731	17,311	35,848	1,627,587

hypersum	1	14	147	14,713	1,471,325	147,132,549	14,713,254,994

873 Pentanacci

broken-up	1	6	7	41	1,360	88,441	11,410,249	2,886,881,438
broken-down	1	2	3	17	564	36,677	4,731,897	1,197,206,618

broken	1	2	3	6	7	17	41	564	1,360	36,677	4,731,897	11,410,249

hypersum	1	15	159	15,917	1,591,733	159,173,365	159,173,365,129

874 Hexanacci

broken-up	1	7	8	55	2,263	183,358	29,522,901	9,477,034,579
broken-down	1	2	3	20	823	66,683	10,736,786	3,446,574,989

broken	1	2	3	7	8	20	55	823	2,263	66,683	183,358

hypersum	1	16	1,611	161,121	16,112,141	1,611,214,181

875 Heptanacci

broken-up	1	8	9	71	3,488	338,407	65,316,039	25,147,013,422
broken-down	1	2	3	23	1,130	109,633	21,160,299	8,146,824,748

broken	1	2	3	8	9	23	71	1,130	3,488	109,633	338,407	21,160,299

hypersum	1	17	1,713	171,325	17,132,549	1,713,254,997	1,713,254,997,193

876 Oktanacci

broken-up	1	9	10	89	5,083	574,468	129,260,383	58,038,486,435
broken-down	1	2	3	26	1,485	167,831	37,763,460	16,955,961,371
broken	1	2	3	9	10	26	89	1,485

hypersum	1	18	1,815	181,529	18,152,957	18,152,957,113

hectonacci	1	199	298	497	993
kilonacci	1	1,999	2,998	4,997	9,993
myrianacci	1	19,999	29,998	49,997	99,993
meganacci	1	1,999,999	2,999,998	4,999,997	9,999,993

877 Stanislaw Ulam, who perhaps not coincidentally participated in the infamous Manhattan project, spiraled counterclockwise or widdershins, meaning backward, so his directions other than north and south are backward from our preferred designations.

878

E	1	2	11	28	53	86	127	176	233	298	371	452	541	638	743	856	977	1,106
SE	1	3	13	31	57	91	133	183	241	307	381	463	553	651	757	871	993	1,123
S	1	4	15	34	61	96	139	190	249	316	391	474	565	664	771	886	1,009	1,140
SW	1	5	17	37	65	101	145	197	257	325	401	485	577	677	785	901	1,025	1,157
W	1	6	19	40	69	106	151	204	265	334	411	496	589	690	799	916	1,041	1,174
NW	1	7	21	43	73	111	157	211	273	343	421	507	601	703	813	931	1,057	1,191
N	1	8	23	46	77	116	163	218	281	352	431	518	613	716	827	946	1,073	1,208
NE	1	9	25	49	81	121	169	225	289	361	441	529	625	729	841	961	1,089	1,225

879 Many luckies are also sexy; some both sexy and prime.

sexy luckies	1	3	7	9	13	15	21	25	31	37	43	49	63	67	69	73	75	87	93	99	105	111

880 prime twins

broken-up	3	16	51	271	3,574	61,029	1,163,125	33,791,654	1,048,704,399
sub	1	6	19	100	1,315	22,451	427,890	12,431,255	385,796,788
broken-down	1	4	13	69	910	15,539	296,151	8,603,918	267,017,609

hypersum	3	35	357	35,711	3,571,113	357,111,317	35,711,131,719
sub	1	13	131	13,137	1,313,739	131,373,912	13,137,391,180

Pythagoras is known for the most famous of triplets, two numbers whose squares sum to the square of the third. Pythagorean dates in any century would be those less than twelve for the month, less than thirty-one for the day of the month and less than ninety-nine for the last two digits of the year, in that order: 3/4/05, 3/5/08, 5/12/13, 6/8/10, 5/12/13, 9/12/15, 6/10/16, 8/15/17, 12/16/20, 9/15/24, 7/24/25, 10/24/26, 12/20/40, 7/25/48. Allowing permutations, we also have: 4/5/03, 5/4/03, 3/5/04, 5/3/04, 4/3/05, 6/10/08, 10/6/08, 6/16/10, 8/6/10, 5/13/12, 9/15/12, 12/5/13, 8/17/15, 9/24/15, 12/9/15, 10/6/16, 12/20/16, 8/15/17, 7/25/24, 10/26/24, an average of every three years. [NOTE: The use of two slashes for dates has in many cases been replaced by the use of two periods, since slashes can be confused with the already confusing double division, z/z'/z" = (z/z')/z" or z/(z'/z").]

1st triplet	3	3	5	6	7	7	8	9	9	10	11	11	12	13	13	14	15	15	16	17
2nd triplet	4	5	12	8	24	25	15	40	41	24	60	61	35	84	85	48	112	113	63	144
3rd triplet	5	8	13	10	25	48	17	41	80	26	61	120	37	85	168	50	113	224	65	145

Since some of these numbers are in overlapping sets of triplets, the smaller set is sometimes called Platonic rather than Pythagorean.

Palindromic dates in a century are similarly limited between 1/01/01 and 12/22/21. So-called doomsdays are those which occur on the same day of the week with equal month number and day of the month: 3/3, 5/5, 7/7 and 4/4, 6/6, 8/8, 10/10, 12/12 three days of the week earlier. 9/9 and 11/11 are two days earlier.

881 uptown twin primes

broken-up	5	36	185	1,331	41,446	1,783,509	108,835,495	7,946,774,644
dusub	1	5	25	180	5,609	241,372	14,729,283	1,075,478,997
broken-down	1	6	31	223	6,944	18,234,659	1,331,428,922	137,155,413,625

hypersum	5	57	5,713	571,319	57,131,931	5,713,193,143	571,319,314,361
dusub	1	8	773	77,320	7,731,966	773,196,612	77,319,661,228

882 downtown twin primes

sub	1	2	4	6	11	15	22	26	37
broken-up	3	16	51	271	7,910	324,581	19,158,189	1,360,556,000	137,435,314,189
sub	1	6	19	100	2,910	119,407	7,047,904	500,520,581	50,559,626,581
broken-down	1	4	13	69	2014	82,643	4,877,951	346,417,164	34,993,011,515
broken	1	3	4	13	16	51	69	271	2,014

hypersum	3	35	3,511	351,117	3,111,729	311,172,941	31,117,294,159
sub	1	13	1,292	129,169	1,144,741	114,474,128	11,447,412,786

883 uptown prime neighbors

broken-up	3	13	42	181	2,214	31,177	563,400	11,299,177	271,743,648
sub	1	5	15	67	814	11,469	207,263	4,156,735	99,968,901
broken-down	1	4	13	56	685	9,646	174,313	345,906	84,076,057

hypersum	3	34	346	3,468	346,812	34,681,216	3,468,121,618
sub	1	13	127	1,276	127,585	12,758,506	1,275,850,643

downtown prime neighbors

broken-up	1	3	4	11	114	1,379	22,178	400,583	8,835,004	194,770,671
broken-down	1	2	3	8	83	1,004	16,147	291,650	6,432,447	64,616,120
broken	1	2	3	4	8	11	83	114	1,004	1,379

hypersum	1	12	124	1,246	124,610	12,461,012	1,246,101,216	124,610,121,618

884 midtown prime neighbors

sub	1	2	4	7	11	15	22	26
broken-up	4	25	104	649	19,574	822,757	49,384,994	3,556,542,325
sub	1	9	38	239	7,201	302,675	18,167,724	1,308,378,803
broken-down	1	4	25	38	104	239	649	7,201

hypersum	4	46	4,612	461,218	46,121,830	4,612,183,042	461,218,304,260
trimmed	3	35	3,501	35,017	350,172	35,017,231	350,172,315
sub	1	17	1,697	169,673	16,967,273	1,696,727,320	169,672,732,029

885 spiraled odds

East	1	3	21	57	107	173	255
Southeast	1	5	25	63	105	181	265
South	1	7	29	69	123	193	277
Southwest	1	9	33	75	129	201	289
West	1	11	39	89	139	213	303
Northwest	1	13	43	87	147	223	315
North	1	15	47	93	155	233	327
Northeast	1	17	51	99	163	243	343

888 Wider and still wider spirals can be traced in the grid.

wider	1	3	15	37	69	111	163	176	241	316	401	496	601
still wider	1	4	19	46	96	151	218	316	411	518	664	799	946

889 Ordered clockwise they would be ESWN and give a spiral.

east	1	2	6	10	14	18	22	26	30	34	38	42	46	50	54	58	62	66	70	74	78	82	86	90
south	1	3	7	11	15	19	23	27	31	35	39	43	47	51	55	59	63	67	71	75	79	83	87	91
west	1	4	8	12	16	20	24	28	32	36	40	44	48	52	56	60	64	68	72	76	80	84	88	92
north	1	5	9	13	17	21	25	29	33	37	41	45	49	53	57	61	65	69	73	77	81	85	89	93
spiral	1	3	8	13	14	19	24	29	30	35	40	45	46	51	56	61	62	67	72	77	78	83	88	93

890 If we use the Latin form, we get 'crux', which can be superpluralized to 'cruxes'.

cruxes	12	14	16	18	21	24	26	28	41	42	46	48	61	62	64	68	81	82	84	86
trimmed	1	3	5	7	10	13	15	17	30	31	35	37	50	51	53	57	70	71	73	75

891 tau

broken-up	1	3	4	11	125	1,886	35,959	1,008,738	34,333,051	1,168,332,472
broken-down	1	2	3	8	91	1,373	26,178	734,357	24,994,316	275,671,833
broken	1	2	3	4	8	11	91	125	1,373	1,886

hypersum	1	12	124	1,246	124,611	12,461,115	1,246,111,519	124,611,151,928

Tutnese tau or tutau

tutau	1,001	1,021	1,031	1,041	1,051	1,061	1,071	1,081	1,081	1,091	1,201
holiness	2	1	1	1	1	2	1	3	3	2	1

892 St. Andrew cross

broken-up	1	4	5	19	176	2,307	39,395	829,602	20,779,445	520,315,727
broken-down	1	2	3	11	102	1,337	22,831	480,788	12,042,531	108,863,567
broken	1	2	3	4	5	11	19	102	176	1,337

hypersum	1	13	135	1,357	13,579	1,357,913	135,791,317	13,579,131,721

893 why

broken-up	1	5	6	29	441	9,290	232,691	7,920,784	340,826,403	14,663,456,113
broken-down	1	2	3	14	213	4,487	112,388	3,825,679	164,616,585	2,473,074,454
broken	1	2	3	5	6	14	29	213	441	4,487

hypersum	1	14	147	1,479	147,915	14,791,521	1,479,152,125	147,915,212,534

why-not	0	2	3	5	6	8	10	11	12	13	14	16	17	18	19	20	22	23	24	26	27	28	29	30	31	32
holiness	1	0	0	0	1	2	1	0	0	0	0	1	0	2	1	1	0	0	0	1	0	2	1	1	0	0
trimmed	1	2	4	5	7	0	0	1	2	3	5	6	7	8	1	11	12	13	15	16	17	18	2	20	21	22
untrimmed	1	3	4	6	7	9	21	22	23	24	25	27	28	29	210	31	33	34	35	37	38	39	310	41	42	43

894 Victoria

broken-up	1	8	9	71	1,784	76,783	3,764,151	274,859,806	22,267,408,437
broken-down	1	2	3	23	578	24,877	1,219,551	89,052,100	7,214,439,651
broken	1	2	3	8	9	23	71	578	1,784

hypersum	1	17	179	17,921	1,792,125	179,212,543	17,921,254,349	1,792,125,434,973

Kay

broken-up	1	6	7	41	704	17,641	512,293	24,095,412	1,229,378,305	62,722,388,967
broken-down	1	2	3	17	292	7,317	212,485	9,994,112	509,912,197	8,678,501,461
broken	1	2	3	7	17	41	292	704	7,317	17,641

hypersum	1	15	157	15,715	1,571,517	157,151,725	15,715,172,529

El, besides being a nickname for Ellen, is part of the names of the archangels Michael, Raphael, Gabriel and Kal-el aka Superman or Clark Kent or Kara Zor-el aka Supergirl or "Linda Lee Danvers".

broken-up	1	4	5	19	898	51,205	4,762,963	509,688,246	79,006,441,093
broken-down	1	2	3	11	520	29,651	2,758,063	295,142,392	45,749,828,823
broken	1	2	3	4	5	11	19	520	898

hypersum	1	13	1,315	131,521	13,152,147	1,315,214,757	131,521,475,793

895 The three sisters, Victoria, Ellen and Kathleen, who share the Greek surname Emeno shared the Greek letter subsequences, gamma and lambda.

broken-up	1	4	5	19	556	31,711	2,188,615	234,213,516	28,810,451,083
broken-down	1	2	3	11	322	18,365	1,267,507	135,641,614	16,685,186,029
broken	1	2	3	4	5	11	19	322	556

hypersum	1	13	137	13,721	1,372,129	137,212,957	13,721,295,769

broken-up	1	4	5	19	328	10,187	377,247	21,513,266	1,398,739,537	90,939,583,171
broken-down	1	2	3	11	190	5,901	218,527	12,461,940	810,244,627	13,786,620,599
broken	1	2	3	4	5	11	19	190	328	5,901

hypersum	1	13	135	13,513	1,351,317	135,131,731	13,513,173,137	1,351,317,313,757

896 Above the eyes, of course, are the brows, their uptown neighbors.

brows	2	5	9	16	24	35	47	62	78	97	117	140	164	191	219	250	282	317	392	432	475
trimmed	1	4	8	7	13	24	36	51	67	86	6	3	53	70	108	14	171	206	281	321	364

897 sedge

broken-up	1	5	6	29	267	4,034	84,981	1,958,597	49,049,906	1,228,206,247
broken-down	1	2	3	14	129	1,949	41,058	946,283	23,698,133	214,229,480
broken	1	2	3	5	6	14	29	129	267	1,949

hypersum	1	14	147	1,478	14,789	1,478,915	147,891,521	14,789,152,123

898 glorious

broken-up	1	4	5	19	252	3,799	64,835	2,013,684	68,530,091	2,332,036,778
broken-down	1	2	3	11	146	2,201	37,563	1,166,654	39,703,799	517,316,041
broken	1	2	3	4	5	11	19	146	252	2,201

hypersum	1	13	134	1,345	134,513	13,451,315	1,345,131,517	134,513,151,731

899 The dash is fallaciously called the fastest punctuation.

broken-up	1	3	4	11	213	5,975	239,213	12,684,264	875,453,429	60,418,970,865
broken-down	1	2	3	8	155	4,348	174,075	9,230,323	637,066,362	12,113,491,201
broken	1	2	3	4	8	11	155	213	4,348	5,975

hypersum	1	12	126	12,611	1,261,119	126,111,928	12,611,192,840	1,261,119,284,053

Inspired by Samuel Morse the mates to the dash numbers are called the dots.

dots	3	7	8	13	30	46	59	71	80	88

900 asterisk

broken-up	1	4	5	19	138	1,123	10,245	103,573	1,253,121	15,141,025	212227471
broken-down	1	2	3	11	80	651	5,939	60,041	726,431	5,145,058	72,757,243
broken	1	2	3	4	5	11	19	80	138	651	1,123

hypersum	1	13	134	1,345	13,457	134,578	1,345,789	134,578,910	13,457,891,012

The asterisk can be superplurialized to asterisxes.

asterisxes	13	14	15	17	18	19	31	34	35	37	38	39	41	43	45	47	48	49	51	53	54	55	57	58	59

The "fork" numbers can found in the truncated east and west spokes, the southern handle and four vertical "prongs".

forks	1	2	4	6	7	9	11	15	19	22	24	26	27	28	34	40	41	42	43	45	47	49	61	74	76	78	80

With just three prongs we have the "threek" numbers. Both can be superpluralized.

threeks	1	2	4	8	10	11	15	20	21	23	25	34	44	46	48	61	75	77	79	96	114	116	118	139

forxes	11	12	14	16	17	19	21	22	24	26	27	29	41	42	44	46	47	49	61	62	64	66	67	69	71	72

threexes	11	12	14	18	21	22	24	28	41	42	44	48	101	102	104	108	111	112	114	115	118	201

901 semiprime

broken-down	4	25	229	2,315	32,639	491,900	10,362,539	228,467,758
sub	1	9	84	852	12,007	180,960	3,812,165	84,048,591
broken-down	1	5	46	465	6,556	98,805	2,081,461	45,890,947
broken	1	4	5	25	46	229	465	2,315

hypersum	4	46	469	46,910	4,691,014	469,101,415	46,910,141,521

A composite with the same digits as at least one of its factorizations is called a vampire.

vampires	126	153	688	1,206	1,255	1,260	1,395	1,433	1,503	1,530
factorization	6x21	3x51	8x86	6x201	5x251	21x60	15x93	35x41	3x501	3x510
holiness	1	0	5	2	0	1	1	0	1	1

tertiprime

broken-up	8	97	1,754	35,177	951,533	26,678,101
holiness	2	1	0	0	1	5
dusub	1	13	237	4,761	128,776	3,610,488

broken-down	1	9	163	3,269	88,426	2,479,197
broken	1	8	9	97	163	1,754
hypersum	8	812	81,218	8,121,820	812,182,027	81,218,202,728
holiness	2	2	4	5	5	7
trimmed	7	701	70,107	701,071	70,107,117	7,010,711,617
dusub	1	110	10,992	1,099,169	109,916,885	10,991,688,470

quartiprime

broken-up	32	1,537	110,696	8,857,217	956,690,132	107,158,152,001
tressub	1	76	5,511	440,975	47,630,797	5,335,090,240
broken-down	1	33	2,377	190,193	20,543,221	2,301,030,945
broken	1	32	33	1,537	2,377	110,696
hypersum	16	1,624	162,436	16,243,640	1,624,364,054	162,436,405,456
holiness	1	1	2	3	3	4
tressub	1	81	8,087	808,723	80,872,324	8,087,232,424

quintiprimes	32	48	72	80	108	112	120	162	168	176
sextiprimes	64	144	160	216	240	324	336	400	528	540
septiprimes	128	192	288	480	648	672	800	1,008	1,056	1,080
octiprimes	256	576	896	960	1,296	1,344	1,440	1,600	1,944	2,112
noniprimes	512	768	1,152	1,280	1,728	1,792	1,920	2,592	2,688	2,816
deciprimes	1,024	1,536	2,304	2,560	3,456	3,584	3,840	5,184	5,384	5,632
undeciprimes	2,048	3,072	4,608	5,120	6,912	7,168	7,680	10,368	10,752	11,264
unciprimes	4,096	6,144	9,216	10,240	13,824	14,336	15,360	20,736	21,504	22,528
diagonal	4	12	36	80	240	672	1,440	2,592	5,384	11,264

903

4th vertical	10	20	40	80	216	480	960	1,280	2,560	3,120	10,240
5th vertical	14	27	54	108	240	648	1,296	1,728	3,456	6,912	13,824
6th vertical	15	28	56	112	324	672	1,344	1,792	3,584	7,168	14,336
7th vertical	21	30	60	120	336	800	1,440	1,920	3,840	7,680	15,360
8th vertical	22	42	81	162	400	1,008	1,600	2,592	5,184	10,368	20,736
9th vertical	25	44	84	168	528	1,056	1,944	2,688	5,384	10,752	21,504
10th vertical	26	45	88	176	540	1,080	2,112	2,816	5,632	11,264	22,528

904 emirpimes

broken-up	15	391	5,880	153,271	7,822,701	453,869,929
tressub	1	19	293	7,631	389,470	22,596,853
broken-down	1	16	241	6,282	320,623	18,602,416
broken	1	15	16	241	391	5,880
hypersum	15	1,526	152,639	15,263,949	1,526,394,951	152,639,495,158
tressub	1	76	7,600	759,947	75,994,730	7,599,472,981

905 quares

untrimmed	1	2	5	6	10	32	35	36	310	52	55	67	72	710	87	95	910	107	2	227

hypersum	0	1	14	145	1,459	145,921	14,592,124	1,459,212,425
holiness	1	0	0	0	1	1	1	1
untrimmed	1	2	25	256	25,610	2,561,032	256,103,235	25,610,323,536

906 ubes

untrimmed	1	5	8	23	27	36	310	54	107	2,108	322	372	442	486	487	759	764	839	855

hypersum	0	1	15	158	15,823	1,582,327	158,232,736	158,232,736,310
holiness	1	0	0	2	2	2	3	4
untrimmed	1	2	26	269	26,912	2,693,438	269,343,847	269,343,847,421

907 blongs

trimmed	1	13	45	61	10	45
dusub	1	3	8	10	12	21
broken-up	12	289	3,480	83,809	7,546,290	1,177,305,049
trimmed	1	178	237	7,278	643,518	662,438
dusub	1	39	471	11,342	1,021,279	159,330,912
broken-down	1	13	157	3,781	340,447	53,113,513
broken	1	12	13	157	289	3,480
hypersum	12	1,224	122,456	12,245,672	1,224,567,290	1,224,567,290,156
dusub	1	166	16,573	1,657,271	165,727,161	165,727,161,055

908 interblongs

trimmed	7	3	23	70	12	123
tressub	1	2	3	4	6	12
broken-up	18	721	12,996	520,561	64,041,999	14,986,348,327
holiness	2	0	3	2	5	6

tressub	1	36	647	25,917	3,188,463	746,126,349
broken-down	1	19	343	13,739	1,690,240	395,529,899
broken	1	18	19	343	721	12,996
hypersum	18	1,840	184,064	18,406,481	18,406,481,123	18,406,481,123,234
holiness	2	3	4	6	6	6
trimmed	7	73	7,353	735,370	735,370,012	73,539,912,123
tressub	1	92	9,164	916,405	916,404,734	916,404,734,094

909 rimorials

untrimmed	1	0	2	20	40	78,878
hypersum	0	10	1,030	1,030,310	103,031,010,510	103,031,010,510,699,690
holiness	1	1	2	3	5	8
untrimmed	1	21	2,141	2,141,421	214,142,121,621	214,142,121,621,710,107,101

910 ings

broken-up	1	4	5	19	613	130,588	70,126,569
broken-down	1	2	3	11	355	75,626	40,611,517
broken	1	2	3	4	5	11	19
hypersum	1	13	135	1,356	135,632	135,632,213	135,632,213,537

911 "Hut" can also be the acronum "hachibi-unindexed ternarish".

hut	1	2	10	11	20	21	100	101	102	110	111	201	202	210	211	212	220	221	222	1,000

912 "Ept" can also be the acronum "even prime ternarish", but that would just be two.

trimmed	123	124	125	132
broken-up	1,234	1,523,991	1,880,606,128	2,322,550,092,071
trimmed	123	1,412,880	7,755,017	1,211,448,160
heptasub	1	1,390	1,714,891	2,117,891,543
broken-down	1	1,235	1,523,991	1,882,130,120
broken	1	1,234	1,235	1,523,991
hypersum	1,234	12,341,235	123,412,351,236	1,234,123,512,361,243
trimmed	123	1,230,124	12,301,240,125	123,012,401,250,132
heptasub	1	11,254	112,537,497	1,125,374,974,189

913 "Rap" could also be the acronum for "rokubi and primes".

rap	2	3	5	6	7	11	13	16	17	19	23	26	29	31	36	37	41	43	46	47	53	56	59	61	66	67
trimmed	1	2	4	5	6	0	2	5	6	8	12	15	18	20	25	26	30	32	35	36	42	45	48	50	55	56

914 leek

broken-up	44	2,421	106,568	5,863,661	516,108,736	51,100,628,525
tetrasub	1	45	1,952	107,397	9,452,861	93,940,659
broken-down	1	45	1,981	109,000	9,593,981	949,913,119
broken	1	44	45	1,981	2,421	106,568
hypersum	44	4,455	445,566	44,556,677	4,455,667,788	445,566,778,899
tetrasub	1	82	8,161	816,084	81,608,402	8,160,840,223

915 Arthur Koenig was an art major.

broken-up	2	7	16	55	676	8,843	133,321	2,141,979
trimmed	1	6	5	44	565	7,732	22,210	1,030,868
sub	1	3	6	20	249	3,253	49,046	787,990
broken-down	1	3	7	24	295	3,859	58,180	934,739
broken	1	2	3	7	16	24	55	295
hypersum	2	23	235	2,356	235,612	23,561,213	2,356,121,315	235,612,131,516
trimmed	1	12	124	1,245	124,501	12,450,102	1,245,010,204	124,501,020,405
sub	1	8	86	867	86,677	8,667,686	866,768,593	86,676,859,275

Tutnese art

arugtut	1,002	1,003	1,005	1,006	1,023	1,025	1,026	1,032	1,035	1,036	1,042
holiness	2	2	2	3	1	1	2	1	1	2	1
trimmed	1	2	4	5	12	14	15	21	24	25	31

916 Tutnese ego

egugo	2,009	2,019	2,039	2,049	2,059	2,069	2,079	2,089	2,109	2,139	2,149
holiness	3	2	2	2	2	3	2	4	2	1	1
trimmed	18	108	128	138	148	158	168	178	108	1,028	1,038

917 From Sigmund Freud's id, ego and superego comes the name for the numbers with a weight below fourteen, "ids".

id	...	58	60	61	62	63	64	65	66	67	70	71	72	73	74	75	76	80	81	82	83	84	85	90
holiness	...	2	2	1	1	1	1	1	2	1	1	0	0	0	0	0	1	2	2	2	2	2	2	2

918 superego is the products of egos.

superego	29	1,102	31,958	1,214,404	58,291,392	2,856,278,208	159,951,579,648
holiness	1	1	3	1	4	8	6
tressub	1	55	1,591	60,462	2,902,158	142,205,718	7,963,520,232

919 "Ow" includes both both gow and semigow.

ow	3	4	5	6	13	14	15	16	23	24	25	26	30	31	32	36	37	38	39	40	41	42	45	47	0

920 *Brain Builder Numbers: "Number by Number"* and *Brain Builder Patterns* by Charles Phillips

921

25	13457	13467	2346	12467	1234567	136	1234567	12346	25123567	2525	2513457

922

...	12346	123456	24567	1257	34567	1245	1245	12567	2456	5	367	23456	257
holiness	1	1	1	0	1	0	0	1	1	0	1	1	0

923 All of these number generators were inspired by puzzles reprinted in *Brain Builder Numbers* by Charles Phillips.
Agent 8

broken-up	3	13	68	421	3,015	24,541	223,884	2,263,381
sub	1	5	25	155	1,109	9,028	82,362	832,651
broken-down	1	4	13	56	405	3,296	30,069	303,986
broken	1	3	4	13	56	68	405	421
hypersum	3	34	345	3,456	34,567	345,678	3,456,789	345,678,910
sub	1	13	127	1,271	12,716	127,168	1,271,682	127,168,164

924 Agent 8.1

broken-up	7	57	406	3,305	40,066	524,163	7,378,348
dusub	1	8	55	447	5,422	70,938	998,551
broken-down	1	8	57	464	5,625	73,589	1,035,871
hypersum	7	78	789	78,910	7,891,012	789,101,213	78,910,121,314
dusub	1	11	107	10,679	1,067,932	106,793,236	1,067,932,618

925 Agent 9

broken-up	7	64	455	4,159	54,522	767,467	11,566,527
dusub	1	9	61	563	7,379	103,865	1,565,359
broken-down	1	8	57	521	6,830	96,141	1,448,945

broken	1	7	8	57	455	521	4,159
hypersum	7	79	7,910	791,011	79,101,113	7,910,111,314	791,011,131,415
dusub	1	11	1,071	107,052	10,705,171	1,070,517,155	107,051,715,513

926 Agent 9.1

broken-up	7	64	455	4,159	54,522	767,467	11,566,527
dusub	1	9	61	563	7,379	103,865	1,565,359
broken-down	1	8	57	521	6,830	96,141	1,448,945
broken	1	7	8	57	455	521	4,159
hypersum	7	79	7,910	791,011	79,101,113	7,910,111,314	791,011,131,415
dusub	1	11	1,071	107,052	10,705,171	1,070,517,155	107,051,715,513

927 time-in numbers

broken-up	19,681	387,361,443	7,624,435,302,250
holiness	4	3	3
dekasub	1	17,586	346,148,827
broken-down	1	19,682	387,400,807
broken	1	19,681	19,682

hypersum	19,681	1,968,119,682	196,811,968,219,683	19,681,196,821,968,319,684
holiness	4	8	12	16
dekasub	1	89,352	8,935,249,534	893,524,953,359,006

The complimentary numbers to the time-in numbers would be the 'time-out' numbers, starting with ten thousand and jumping after eighty-nine thousand, nine hundred ninety-nine to a hundred thousand.

time-out	10,000	10,001	10,002	10,003	10,004	10,005	10,006	10,007	10,008	10,009
holiness	4	3	3	3	3	3	4	3	3	4

Another notable number generator that involves historical years past and future is that for the Umbugio sequence: 1492^z - 1770^z - 1863^z + 2141^z named for Euclide Paracelso Bombasto Umbugio, author of *Cheaper by the Googol, A Short Table of Even Primes, 1,000,000 Random Numbers in Ascending Order* and *22/7 to A Million Decimal Places.* He also noted that these numbers are all divisible by 1946, and so by two and seven as well.

Umbugio	206,276	1,124,101,062	4,106,026,092,896	12,565,214,785,548,390
holiness	3	3	10	6
1/1946	106	577,647	2,109,982,576	6,456,944,905,215

holiness	2	1	6	5
semi-Umbugio	103,138	562,050,531	2,053,013,046,448	6,232,607,392,774,195
holiness	3	3	6	5
septi-Umbugio	29,468	160,585,866	586,575,156,128	1,795,030,683,649,770
holiness	4	8	6	9

928 Antarctic numbers

broken-up	23	645	14,858	416,669	17,931,625	861,134,669
tressub	1	32	740	20,745	892,763	42,873,371
broken-down	1	24	553	15,508	667397	32,050,564
broken	1	23	24	553	645	14,858
hypersum	23	2,328	232,833	23,283,338	2,328,333,843	232,833,384,348
tressub	1	116	11,592	1,159,209	115,920,916	11592091625

929 new Antarctic numbers

sub	1	3	5	7	10
broken-up	18	1,117	118,420	17,764,117	3,446,357,118
holiness	2	0	3	1	3
tressub	1	58	5,896	884,423	171,584,018
broken-down	1	62	6,573	986,012	191,292,901
broken	1	18	62	1,117	6,573

hypersum	18	1,862	1,862,106	1,862,106,150	1,862,106,150,194
holiness	2	3	5	6	7
tressub	1	93	92,709	92,708,806	92,708,806,208

930 Arctic numbers

broken-up	25	676	16,925	457,651	15,119,408	529,636,931
tressub	1	34	842	22,785	752,751	26,369,070
broken-down	1	26	651	17,603	581,550	20,371,853
broken	1	25	26	651	676	16,925
hypersum	25	2,527	252,729	25,272,931	2,527,293,133	252,729,313,335
tressub	1	126	12,583	1,258,265	125,826,516	12,582,651,601

931

even Arctic	36	38	40	42	44	46	48	50	52	54	56	58	60	62	124	126	128	130	132
holiness	1	2	1	0	0	1	2	1	0	0	1	2	2	1	0	1	2	1	0

trimmed	25	27	3	31	33	35	37	4	41	43	45	47	5	51	13	15	17	2	21

932 N. Polar

broken-up	43	2,065	88,838	4,266,289	268,865,045	18,287,089,349
tetrasub	1	38	1,627	78,140	4,924,435	334,939,725
broken-down	1	44	1,893	90,908	5,729,097	389,669,504
broken	1	43	44	1,893	2,065	88,838
hypersum	43	4,348	434,853	43,485,358	4,348,535,863	434,853,586,368
tetrasub	1	80	7,965	796,462	79,646,212	7,964,621,257

933 S. Polar

broken-up	33	110,056	3,631,881	12,112,433,191	> nineteenplex	> thirty-oneplex
tressub	1	5,479	180,821	603,042,540	> eighteenplex	> twenty-nineplex
broken-down	1	34	1,123	3,745,239	> sixteenplex	> twenty-sevenplex
broken	1	33	34	1,123	110,056	3,631,881
hypersum	33	3,335	333,536	33,353,638	3,335,363,840	333,536,384,042
tressub	1	166	16,606	1,660,580	188,057,988	16,605,798,755

934

S. Polar even	36	38	40	42	44	46	48	50	52	54	56	58	60	62	64	66	68
holiness	1	2	1	0	0	1	2	1	0	0	1	2	2	1	1	2	3

broken-up	36	1,369	49,320	1,875,529	82,572,596	3,800,214,945
holiness	1	2	2	3	4	5
tetrasub	1	25	904	34,251	1,512,370	69,603,365
broken-down	1	37	1,333	50,691	2,231,737	102,710,593
broken	1	36	37	1,333	1,369	49,320
hypersum	36	3,638	363,840	36,384,042	3,638,404,244	363,840,424,446
holiness	1	3	4	4	4	5
tetrasub	1	67	6,664	666,397	66,639,698	6,663,969,827

935 even linears

sub	1	5	27	42	53	64
broken-up	4	57	232	3,305	476,152	82,853,753
sub	1	21	85	1,216	175,167	30,480,192
broken-down	1	5	21	299	43,077	7,495,697

broken	1	4	5	21	57	232
hypersum	4	414	41,474	41,474,114	41,474,114,144	41,474,114,144,174
trimmed	3	303	30,363	30,363	30,363,003,033	30,363,003,033,063
sub	1	152	15,257	15,257,474	15,257,473,934	15,257,473,934,439

936 odd linears

broken-up	1	8	9	71	2,920	137,311	9,752,001
broken-down	1	2	3	23	946	3,159,381	243,316,822
broken	1	2	3	8	9	23	71
hypersum	1	17	1,711	171,117	171,111,741	1,711,174,147	171,117,414,771

937 odd unluckies

broken-up	5	56	285	3,191	67,296	1,550,999	41,944,269
dusub	1	8	39	432	9,108	209,905	5,676,539
broken-down	1	6	31	347	7,318	4,561,165	132,442,446
broken	1	5	6	31	56	285	347
hypersum	5	511	51,117	5,111,719	511,171,921	51,117,192,123	5,111,719,212,327
dusub	1	69	6,918	691,796	69,179,597	6,917,959,674	691,795,967,426

938 safe kyuubi

tressub	1	2	3	4	5	6	7	8	9	10	11	12	13	14	15	16	17	18	19	24	55	56

broken-up	19	742	14,117	551,305	54,593,312	6,497,155,433
holiness	1	0	0	1	1	2
tressub	1	37	703	27,448	2,718,041	323,474,322
broken-down	1	20	381	14,879	1,473,402	175,349,717
broken	1	19	20	381	742	14,117
hypersum	19	1,939	193,959	19,395,979	1,939,597,999	1,939,597,999,119
holiness	1	2	3	4	6	7
tressub	1	96	9,657	965,669	96,566,898	96,566,898,188

939 curvaceous kyuubi

broken-up	39	2,692	105,027	7,249,555	2,457,704,172	906,900,089,023
holiness	1	2	2	1	1	11
tetrasub	1	49	1,924	132,780	4,514,422	16,610,454,539
broken-down	1	40	1,561	107,749	36,528,472	13,479,113,917

broken	1	39	40	1,561	2,692	105,027
hypersum	39	3,969	396,999	396,999,309	396,999,309,339	396,999,309,339,369
holiness	1	3	5	7	8	10
tetrasub	1	73	7,271	7,271,296	7,271,295,989	7,271,295,988,937

940 curvaceous safe kyuubi

tetrasub	1	2	6	7	12
broken-up	39	3,862	150,657	14,918,905	9,533,330,952
holiness	1	3	2	5	3
tetrasub	1	71	2,759	273,249	174,609,047
broken-down	1	40	1,561	154,579	98,777,542
broken	1	39	40	1,561	3,862
hypersum	39	3,999	3,999,339	3,999,339,399	3,999,339,399,639
holiness	1	3	4	6	8
tetrasub	1	73	73,250	73,250,456	73,250,456,237

941 curvaceous sanbi

sub	1	12	23	34	111
broken-up	3	100	303	10,099	3,060,300
sub	1	37	111	3,715	1,125,821
broken-down	1	4	13	433	131,212
broken	1	3	4	12	13
hypersum	3	333	33,363	3,336,393	3,336,393,303
sub	1	123	12,274	1,227,390	1,227,390,504

942 unsafe Fibonacci

broken-up	1	3	4	11	147	3,098	105,479	5,804,443
broken-down	1	2	3	8	107	2,255	76,777	4,224,990
broken	1	2	3	4	8	11	107	147
hypersum	1	12	123	1,238	123,813	12,381,321	1,238,132,134	123,813,213,455

943 "Fee-fi-fo-fum" appeared in "Jack and the Beanstalk" by Joseph Jacobs in 1890 and "Fie, fo, fum" in King Lear by William Shakespeare in 1605 and "fy, fa and fum" in "Haue with ypou to Saffron-Walden" by Thomas Nashe in 1596. Charles Mackay proposes in *The Gaelic Etymology of the Languages of Western Europe* that "Fa, fie, fi, fo, fum" comes from the Anglicized Gaelic "Faich fidh fiú fogh feum!" (Behold food good to eat, sufficient [for my] hunger.)

"Bee" describes binarish evenly even, "coatee" cubed-odds-and-ternarish-evenly-evens, "coulee" cubed odd-unindexed lucky evenly evens, "Cree" cubed rokubi evenly evens, "enlistee" even nibi lucky-indexed sorted ternarish evenly evens, "fellatee" Fibonacci-even-luckier-luckies-and-ternarish-evenly-evens, "fiancee" Fibonacci-indexed-and-nibi-cubed-evenly-evens, "filaree" Fibonacci-indexed-luckies-and-rokubi-evenly-evens, "flee" Fibonacci lucky evenly evens, "fusee" Fibonacci-unindexed sorted evenly evens, "Galilee" gobi-and-lucky-indexed-lucky-evenly-evens, "goatee" gobi-odds-and-ternarish-evenly-evens, "hee" hachibi evenly evens, "imitatee" ichibi-Mersenne-indexed-ternarish-and-ternarish-evenly-evens, "lee" lucky evenly evens, "lessee" lucky even sorted sanbi evenly evens, "lichee" lucky-indexed cubed hachibi evenly evens, "maharanee" Mersenne-and-hachibi-and-nibi-evenly-evens, "manatee" Mersenne-and-nibi-and-ternarish-evenly-evens, "matinee" mersenne-and-ternarish-indexed nibi evenly evens, "nee" nibi evenly evens, "passee" prime-and-sorted-sanbi-evenly-evens, "payee" prime-and-yongbi-evenly-evens, "puree" prime-unindexed rokubi evenly evens, "ranee" rokubi-and-nibi-evenly-evens, "saree" sanbi-and-rokubi-evenly-evens, "sayee" sanbi-and-yongbi-evenly-evens, "see" sorted evenly evens, "siree" sanbi-indexed rokubi evenly evens, "soiree" sanbi odd-indexed rokubi evenly evens, "tee" ternarish evenly evens and "tutee" ternarish-unindexed ternarish evenly evens,

"B. O." describes binarish odds, "ballo" binarish-and-luckier-lucky-odds, "basso" binarish-and-sorted-sanbi-odds, "Bilbo" binarish-indexed lucky binarish odds, "bolo" binarish odd lucky odds, "bubo" binarish-unindexed binarish odds, "bucko" binarish-unindexed cubed kyuubi odds, "cacao" cubes-and-cubes-and-odds, "calico" cubes-and-lucky-indexed cubed odds, "campo" cubes-and-Mersenne-prime-odds, "carabao" cubes-and-rokubi-and-binarish-and-odds, "chamiso" cubed-hachibi-and-Mersenne-indexed-sanbi-odds, "Chicago" cubed-hachibi-indexed-cubes-and-gobi-odds, "Chico" cubed hachibi-indexed cubed odds, "cigarillo" cube-indexed-gobi-and-rokubi-indexed-luckier-lucky-odds, "co" cubed odds, "cogito" cubed odd gobi-indexed ternarish odds, "curacao" cube-unindexed-rokubi-and-cubes-and-odds, "Eskimo" even sorted kyuubi-indexed Mersenne odds, "facto" Fibonacci-and-cubed-ternarish-odds, "fatso" Fibonacci-and-ternarish-ichibi-odds, "fiasco" Fibonacci-ichibi-and-sanbi-cubed-odds, "folio" Fibonacci odd lucky-indexed odds, "furioso" Fibonacci-unindexed rokubi-indexed odd sanbi odds, "gigolo" gobi-indexed gobi odd lucky odds, "gismo" gobi-indexed sanbi Mersenne odds, "go" gobi odds, "guiro" gobi-unindexed/indexed rokubi odds, "gusto" gobi-unindexed sanbi ternarish odds, "hallo" hachibi-and-luckier-lucky-odds, "halo" hachibi-and-lucky-odds, "hippo" hachibi-indexed palindromic prime odds, "hullo" hachibi-unindexed luckier lucky odds, "imago" ichibi-Mersennes-and-gobi-odds, "impasto" ichibi-Mersenne-prime-and-sanbi-ternarish-odds, "kilo" kyuubi-indexed lucky odds, "lasso" lucky-and-sorted-sanbi-odds, "limo" lucky-indexed Mersenne odds, "loco" lucky odd cubed odds, "logo" lucky odd gobi odds, "machismo" Mersenne-and-cubed-hachibi-indexed-sanbi-Mersenne, "mafioso" Mersenne-and-Fibonacci-indexed odd sanbi odds, "manifesto" Mersenne-and-nibi-indexed Fibonacci even sorted ternarish odds, "Mao" Mersenne-and-odds, "miso" Mersenne-indexed sanbi odds, "mo" Mersenne odds, "multo" Mersenne-unindexed lucky ternarish odds, "Nato" nibi-and-ternarish-odds, "nuncio" nibi-unindexed nibi cube-indexed odds, "nullo" nibi-unindexed luckier lucky odds, "olio" odd lucky-indexed odds, "Oslo" odd sanbi lucky odds, "outgo" gobi odd uptown neighbor, "paso" prime-and-sanbi-odds, "pianissimo" prime-ichibi-and-nibi-indexed-sorted-sanbi-indexed-Mersenne-odds, "Picasso" prime-indexed-cubes-and-sorted-sanbi-odds, "Plato" prime-luckies-and-ternarish-odds, "Pluto" prime lucky-unindexed ternarish odds, "po" prime odds, "politico" prime odd lucky-indexed ternarish-indexed cubed odds, "polo" prime odd lucky odds, "potato" prime-odd-ternarish-and-ternarish-odds, "rubato" rokubi-unindexed-binarish-and-ternarish-odds, "sago" sanbi-and-gobi-odds, "Santigo" sanbi-and-nibi-ternarish-indexed gobi odds, "shako" shichibi-and-kyuubi odds, "simpatico" sanbi-indexed-Mersenne-primes-and-ternarish-indexed odds, "so" sanbi or sorted odds, "staccato" sanbi-ternarish-and-cubed-cubes-and-ternarish-odds, "stucco" sanbi ternarish-unindexed cubed cubed odds, "sumo" sanbi-unindexed Mersenne odds, "tabasco" ternarish-and-binarish-and sanbi-cubed-odds, "taco" ternarish-and-cubed-odds, "tao" ternarish-and-odds, "to" ternarish odds, "tobacco" ternarish-odd-binarish-and-cubed-cubed-odds, "unco" non-cubed odds, "undergo" gobi downtown neighbors and "unto" non-ternarish odds.

"Fa" describes the note between mi and so, "Fy" Fibonacci yongbi, "basify" binarish-and-sanbi-indexed Fibonacci yongbi, "beatify" binarish-evens-and-ternarish-indexed Fibonacci yongbi, "calcify" cubes-and-lucky-cube-indexed Fibonacci yongbi, "citify" cube-indexed ternarish-indexed Fibonacci yongbi, "clarify" cubed-luckies-and-rokubi-indexed Fibonacci yongbi, "classify" cubed-luckies-and-sorted-sanbi-indexed Fibonacci yongbi, "coalify" cubed-odds-and-lucky-indexed Fibonacci yongbi, "crucify" cubed-rokubi-unindexed cube-indexed Fibonacci yongbi, "falsify" Fibonacci-and-lucky-sanbi-indexed Fibonacci yongbi, "gasify" gobi-and-sanbi-indexed Fibonacci yongbi, "gulfy" gobi-unindexed lucky Fibonacci yongbi, "leafy" lucky-evens-and-Fibonacci-yongbi, "minify" Mersenne-indexed nibi-indexed Fibonacci yongbi, "nullify" nibi-unindexed luckier lucky-indexed Fibonacci yongbi, "opacify" odd-prime-and-cube-indexed Fibonacci yongbi, "ossify" odd sorted sanbi-indexed Fibonacci yongbi, "pacify" prime-and-cube-indexed Fibonacci yongbi, "purify" prime-unindexed rokubi-indexed Fibonacci yonbi, "ramify" rokubi-and-Mersenne-indexed Fibonacci yongbi, "rarify" rokubi-and-rokubi-indexed Fibonacci yongbi, "reclassify" rokubi-even-cubed-luckies-and-sorted-sanbi-indexed Fibonacci yongi, "russify" rokubi-unindexed sorted sanbi-indexed Fibonacci yongbi, "salsify" sanbi-and-luckies-sorted-indexed Fibonacci yongbi, "satisfy" sanbi-and-ternarish-indexed sorted Fibonacci yongbi, "scarify" sanbi-cubes-and-rokubi-indexed Fibonacci yongbi, "simplify" sanbi-indexed Mersenne prime lucky-indexed Fibonacci yongbi, "stultify" sanbi-ternarish-unindexed lucky ternarish-indexed Fibonacci yongbi and "tackify" ternarish-and-cubed-kyuubi-indexed Fibonacci yongbi.

944 "Antebellum" describes binarish even luckier lucky-unindexed Mersenne downtown neighbors, "barium" binarish-and-rokubi/ichibi-unindexed Mersennes, "bum" describes binarish-unindexed Mersennes, "caecum" cubes-and-even-cube-unindexed Mersenne, "caesium" cubes-and-even-sorted-ichibi-unindexed Mersennes, "calcium" cubes-and-lucky-cubed-ichibi-unindexed Mersennes, "capsicum" cubes-and-prime-sanbi-indexed cube-unindexed Mersennes, "cerium" cubed even rokubi/ichibi-unindexed Mersennes, "cesium" cubes even sorted-ichibi-unindexed Mersennes, "chillum" cubed hachibi-indexed luckier lucky-unindexed Mersennes, "chum" cubed hachibi-unindexed Mersennes, "cilium" cube-indexed lucky ichibi-unindexed Mersennes, "citatum" cube-indexed-ternarish-and-ternarish-unindexed Mersennes, "coagulum" cubed-odds-and-gobi-unindexed lucky-unindexed Mersennes, "cranium" cubed-rokubi-and-nibi-indexed/unindexed Mersennes, "cum" cubed-unindexed Mersenne, "curium" cubed-unindexed rokubi/ichibi-unindex Mersenne, "elasticum" even-lucky-and-sanbi-ternarish-indexed cube-unindexed Mersennes, "elysium" even lucky yongbi sorted ichibi-unindexed Mersennes, "factotum" Fibonacci-and-cubed-ternarish-odd-ternarish-unindexed Mersennes, "francium" Fibonacci-rokubi-and-nibi-cube/ichibi-unindexed

Mersennes, "frustum" Fibonacci rokubi-unindexed sanbi ternarish-unindexed Mersennes, "fulcrum" Fibonacci-unindexed lucky cubed rokubi-unindexed Mersennes, "gallium" gobi-and-luckier-lucky-ichibi-unindexed Mersennes, "gum" gobi-unindexed Mersennes, "hafnium" hachibi-and-Fibonacci-nibi-indexed/unindexed Mersennes, "helium" hachibi even lucky ichibi-unindexed Mersennes, "hum" hachibi-unindexed Mersennes, "ilium" ichibi lucky ichibi-unindexed Mersennes, "kalium" kyuubi-and-lucky-ichibi-unindexed Mersennes, "labium" lucky-and-binarish-ichibi-unindexed Mersennes, "libitum" lucky-indexed binarish-indexed ternarish-uninddexed Mersennes, "linum" lucky-indexed nibi-unindexed, "lutetium" lucky-unindexed ternarish even ternarish/ichibi-unindexed Mersennes, "luteum" lucky-unindexed ternarish even-unindexed Mersennes, "lyceum" lucky yongbi cubed even-unindexed Mersennes, "marsupium" Mersenne-and-rokubi-sorted-unindexed prime ichibi-unindexed Mersennes, "minimum" Mersenne-indexed nibi-indexed Mersenne-unindexed Mersennes, "mum" Mersenne-unindexed Mersennes, "museum" Mersenne-unindexed sorted even-unindexed Mersennes, "oakum" odd-and-kyuubi-unindexed Mersenne, "opium" odd prime ichibi-unindexed Mersenne, "optimum" odd prime ternarish-indexed Mersenne-unindexed Mersennes, "osmium" odd sorted Mersenne ichibi-unindexed Mersennes, "ostium" odd sorted ternarish ichibi-unindexed Mersennes, "pablum" prime-and-binarish-lucky-unindexed Mersennes, "pabulum" prime-and-binarish-unindexed lucky-unindexed Mersennes, "platinum" prime-luckies-and-ternarish-indexed nibi-unindexed Mersennes, "plum" prime lucky-unindexed Mersennes, "possum" prime odd sorted sanbi-unindexed Mersennes, "postbellum" binarish even luckier lucky-unindexed Mersenne uptown neighbors, "potassium" prime-odd-ternarish-and-sorted-sanbi/ichibi-unindexed Mersennes, "rum" rokubi-unidexed Mersenne, "sacrum" sanbi-and-cubed-rokubi-unindexed Mersennes, "samarium" sanbi-and-Mersenne-and-rokubi-chibi-unindexed Mersennes, "sanatarium" sanbi-and-nibiand-ternarish-and-rokubi/ichibi-unindexed Mersennes, "sanctum" sanbi-and-nibi-cubed-ternarish-unindexed Mersennes, "scum" sanbi cube-unindexed Mersennes, "slum" sanbi lucky-unindexed Mersennes, "solarium" sanbi-odd-lucky-and-robubi/ichibi-unindexed Mersennes, "sputum" sanbi prime-unindexed ternarish-unindexed Mersennes, "stibium" sanbi ternarish-indexed binarish ichibi-unindexed Mersennes, "subgum" subgobi-unindexed-Mersennes, "sum" sanbi-unindexed Mersennes, "talcum" ternarish-and-lucky-cube-unindexed Mersennes, "tantalum" ternaris-and-nibi-ternarish-and-lucky-unindexed Mersennes, "timpanum" ternarish-indexed Mersenne prime-and-nibi-unindexed Mersennes and "titanium" ternarish-indexed ternarish-and-nibi/ichibi-unindexed Mersennes.

945 semidecimal

broken-up	10,234	104,744,991	1,071,960,248,128	> sixteenplex
holiness	1	3	8	
enneasub	1	12,927	132,290,404	1,353,992,300,018
broken-down	1	10,235	104,744,991	1,072,064,993,120
hypersum	10,234	1,023,410,235	102,341,023,510,236	10,234,102,351,023,610,237
holiness	1	2	4	5
broken	10,234	10,235	104,744,991	1,071,960,248,128
holiness	1	1	3	8

enneasub	1	126,199	12,629,885,661	1,262,988,566,142,850

946 quintidecimal

broken-up	10	121	1,220	14,761	222,635	3,576,921
holiness	1	0	1	1	1	2
dusub	1	17	165	1,998	30,130	484,084
broken-down	1	11	111	1,343	20,256	325,439
broken	1	10	11	111	121	1,220
hypersum	10	1,012	101,213	10,121,314	1,012,131,415	101,213,141,516
holiness	1	1	1	1	1	2
dusub	1	137	13,698	1,369,771	136,977,092	13,697,709,174

947 tertiternarish

broken-up	1	3	4	11	1,225	271,961	302,149,896
broken-down	1	2	3	8	891	197,810	21,976,801
broken	1	2	3	4	8	11	891
hypersum	1	12	1,211	121,122	121,122,111	121,122,111,222	1,211,221,112,221,111
weight	1	3	5	9	12	18	22

beternarish

broken-up	10	121	1,220	14,761	1,477,420	149,225,081
holiness	1	0	1	1	1	4
dusub	1	17	165	1,998	199,934	20,195,283
broken-down	1	11	111	1,343	134,411	13,576,854
broken	1	10	11	111	121	1,220
hypersum	10	1,012	101,220	10,122,021	10,122,021,100	10,122,021,100,101
holiness	1	1	2	2	4	5

948 tertisenarish

broken-up	10	121	1,220	14,761	222,635	4,467,461
holiness	1	0	1	1	1	2
dusub	1	17	165	1,998	30,130	604,605
broken-down	1	11	111	1,343	20,256	406,463
broken	1	10	11	111	121	1,220
hypersum	10	1,012	101,213	10,121,314	1,012,131,415	101,213,141,520
holiness	1	1	1	1	1	2

dusub	1	137	13,698	1,369,771	136,977,092	13,697,709,175

besenarish

broken-up	1,023	1,047,553	1,071,647,742	1,097,368,335,361
holiness	1	1	2	6
heptasub	1	955	977,216	1,000,670,395
broken-down	1	1,024	1,047,553	1,072,695,296
broken	1	1,023	1,024	1,047,553
hypersum	1,023	10,231,024	102,310,241,025	1,023,102,410,251,026
holiness	1	2	3	5
heptasub	1	9,329	93,294,863	932,948,636,823

949 tertinonarish

broken-up	102	10,507	1,071,816	110,407,555	11,704,272,646
trimmed	1	46	60,705	36,444	63,161,535
holiness	1	2	4	2	3
pentasub	1	71	7,222	743,920	78,862,768
broken-down	1	103	10,507	1,082,324	114,736,851
broken	1	102	103	10,507	1,071,816
hypersum	102	102,103	102,103,104	102,103,104,105	102,103,104,105,106
trimmed	1	102	10,203	1,020,304	102,030,405
holiness	1	2	3	4	6
pentasub	1	688	687,965	687,965,304	687,965,303,902

benonarish

broken-up	102,345	10,474,601,371	1,072,023,077,417,340
holiness	1	3	4
dodekasub	1	64,358	6,586,737,435
broken-down	1	102,346	10,474,601,371
hypersum	102,345	102,345,102,346	102,345,102,346,102,347
broken	1	102,345	
holiness	1	3	3
dodekasub	1	628,830	628,830,042,138

950 semiquadrarish

broken-up	10	111	1,120	12,431	249,740	5,356,971

holiness	1	0	1	0	1	2
dusub	1	15	152	1,682	33,799	711,454
broken-down	1	11	111	1,232	24,751	521,003
broken	1	10	11	111	1,120	1,132
hypersum	10	1,011	101,112	10,111,213	1,011,121,320	101,112,132,021
holiness	1	1	1	1	2	2
dusub	1	137	13,684	1,368,404	13,640,390	1,384,039,026

semoctarish

broken-up	1,023	1,047,553	1,071,647,742	1,097,368,335,361
holiness	1	1	2	5
heptasub	1	955	977,216	1,000,670,395
broken-down	1	1,024	1,047,553	1,072,695,296
broken	1	1,023	1,024	1,047,553
hypersum	1,023	10,231,024	102,310,241,025	1,023,102,410,251,026
holiness	1	2	3	5
heptasub	1	9,238	93,294,863	932,948,636,823

dodriquadrarish

broken-up	102	10,507	1,071,816	110,407,555	14,354,053,966
holiness	1	2	4	2	4
pentasub	1	71	7,222	743,920	96,716,855
broken-down	1	103	10,507	1,082,324	140,712,627
broken	1	102	103	10,507	1,071,816
hypersum	102	102,103	102,103,120	102,103,120,123	102,103,120,123,130
holiness	1	2	3	3	4
pentasub	1	688	687,965	687,965,412	68,795,411,831

dodroctarish

broken-up	102,345	10,474,601,371	1,072,023,077,417,340
holiness	1	3	4
dodekasub	1	64,358	6,586,737,435
broken-down	1	102,346	10,474,601,371
broken	1	102,345	102,346
hypersum	102,345	102,345,102,346	102,345,102,346,102,347

holiness	1	3	4
dodekasub	1	628,830	628,830,042,138

951 quartiquadrarish

broken-up	1	3	4	11	246	8,129	902,565
broken-down	1	2	3	8	179	5,915	656,744
broken	1	2	3	4	8	11	179
hypersum	1	12	123	12,311	1,231,122	123,112,233	123,112,233,111
trimmed	1	12	1,200	120,011	12,001,122	12,001,122,000	12,001,122,000,111

quartoctarish

holiness	1	0	0	0	0	1
broken-up	10	121	1,220	14,761	222,635	3576921
holiness	1	0	1	1	1	2
dusub	1	17	165	1,998	30,103	484,084
broken-down	1	11	111	1,343	20,256	325,439
broken	1	10	11	111	121	1,220
hypersum	10	1,012	101,213	10,121,314	1,012,131,415	101,213,141,516
holiness	1	1	1	1	1	2
dusub	1	137	13,698	1,369,771	136,977,092	13,697,709,174

952 dextrisenarish

broken-up	10,234	104,744,991	1,071,960,248,128
holiness	1	3	8
enneasub	1	12,927	132,290,404
broken-down	1	10,235	104,744,991
broken	1	10,234	10,235
hypersum	10,234	1,023,410,235	102,341,023,510,243
holiness	1	2	3
enneasub	1	126,299	12,629,885,661

953 penholodecimals

holiness	4	4	4
broken-up	123,456,789	> sixteenplex	> twentyfourplex
19th sub	1	85,395,470	> sixteenplex

broken-down	1	123,456,790	> sixteenplex
broken	1	123,456,789	123,456,790
hypersum	123,456,789	123,456,789,123,456,798	123,456,789,123,456,798,123,456,879
holiness	4	8	12
19th sub	1	691,703,258	> seventeenplex

954 There are no holy penholodigitals until six in base seven.

penholoseptarish	...	6	11	12	13	14	15	16	21	22	23	24	25	26	31	32	33	34
holiness	...	1	0	0	0	0	0	1	0	0	0	0	0	1	0	0	0	0
penholoctarish	...	6	7	11	12	13	14	15	16	17	21	22	23	24	25	26	27	31
holiness	...	1	0	0	0	0	0	0	1	0	0	0	0	0	0	1	0	0
penholononarish	...	6	7	8	11	12	13	14	15	16	17	18	21	22	23	24	25	26
holiness	...	1	0	2	0	0	0	0	0	1	0	2	0	0	0	0	0	1

955 pups

sub	1	3	5	7	8	11	14
broken-up	2	15	32	239	5,529	160,580	5,946,989
trimmed	1	4	21	128	4,418	547	4,835,878
sub	1	6	12	88	2,034	59,074	2187,775
broken-down	1	3	7	52	1,203	34,939	1,293,946
broken	1	2	3	7	15	32	52
hypersum	2	27	2,713	271,319	27,131,923	2,713,192,329	271,319,232,937
trimmed	1	16	1,602	160,208	16,020,812	1,602,081,218	160,208,121,826
sub	1	10	998	99,813	9,981,277	998,127,678	99,812,767,792

956 bishops

dusub	1	21	22	366	375
broken-up	7	1,100	7,707	1,211,099	3,351,118,640
dusub	1	149	1,043	163,904	453,524,590
broken-down	1	8	57	8,957	24,784,076
broken	1	7	8	57	1,100
hypersum	7	7,157	7,157,167	71,571,672,707	715,716,727,072,767
dusub	1	969	968,617	9,686,172,597	96,861,725,975,575

Over the bishops are the archbishops, the next higher primes than bishops.

archbishops	11	163	173	2,711	2,777	2,963	3,047

dusub	1	22	24	367	376	401	412

broken-up	11	1,794	19,745	3,220,229	8,942,595,678
dusub	1	243	2,672	435,811	1,210,248,719
broken-down	1	12	133	21,691	60,236,040
broken	1	11	12	133	1,794
hypersum	11	11,163	11,163,173	11,1631,732,711	1,116,317,327,112,777

Under the bishops are the priests, the next lower primes than the bishops.

priests	5	151	163	2,699	2,753	2,953	3,019
dusub	1	21	22	365	373	400	409

broken-up	5	25,756	128,785	663,397,291	> twenty-threeplex
dusub	1	3,486	17,429	89,781,060	> twenty-twoplex
broken-down	1	6	31	159,687	> nineteenplex
broken	1	5	6	31	25,756
hypersum	5	5,151	5,151,163	51,511,632,699	515,116,326,992,753
dusub	1	697	697,134	6,971,341,401	69,713,414,013,368

957 bumpy

dusub	1	4	17	1,108	17,738
broken-up	7	218	1,533	47,741	6,257,462,144
dusub	1	29	207	6,461	846,855,412
broken-down	1	8	57	1,775	232,651,082
broken	1	7	8	57	218
hypersum	7	731	731,127	7,311,278,191	7,311,278,191,131,071
dusub	1	99	98,947	989,473,905	989,473,904,818,393

958 clumps

dusub	1	17	1,108	17,738	70,954
broken-up	7	890	6,237	792,989	415,753,830,080
dusub	1	120	844	107,319	56,266,162,351
broken-down	1	8	57	7,247	3,799,507,946
broken	1	7	8	57	890
hypersum	7	7,127	71,278,191	71,278,191,131,071	71,278,191,131,071,524,287
dusub	1	965	9,646,454	9,646,454,185,317	> eighteenplex

"Mumps" describes Mersenne-unindexed Mersenne primes, so it excludes 31 and 524,287 and the other "mims". "Mums" do not. "Mimsy", from "'Twas Brillig" by Charles Ludwidge "Lewis Carrol" Dodgson, are numbers with "mim" frags. "Brillig" is not an acronum because no rokubi are binarish.

mumps	3	7	127	8,191	131,071	2,147,483,647	137,438,953,471
sub	1	3	47	3013	48,218	790,015,084	50,560,965,398

broken-up	3	22	69	505	66,190,924
sub	1	8	25	186	24,350,280
broken-down	1	4	13	95	12,451,758
broken	1	3	4	13	22
hypersum	3	37	37,127	371,278,191	371,278,191,131,071
sub	1	14	13,658	136,585,613	136,585,613,472,443

mimsy	31,524,287	52,428,731	> eightduplex

959 From the Italo-Spanish suits, as opposed to the French suits[985-8], we get the complimentary sequence "swords", the non-cup odd "discs" ("coins" being cubed odd-indexed nibi) and the non-sword even "wands" or "batons".

sword	4	6	10	16	20	26	28	32	38	40	52	56	58	62	68	70	80	82	86	88
trimmed	3	5	0	5	1	15	17	21	27	3	41	45	47	51	57	6	7	71	75	77

discs	1	7	9	15	21	23	25	27	33	35	39	45	49	51	53	55	57	63	65	69	75	77	81	85	87	91	93

coins	2	262	1,242	3,422	5,112	7,282	9,992	13,302	17,272	21,962	27,432	33,742
trimmed	1	151	131	1,311	4,001	6,171	8,881	221	6,161	10,851	16,321	22,631

wands	2	8	12	14	18	22	24	30	34	36	42	44	46	48	50	54	60	64	66	72	74	76	78	84
trimmed	1	7	1	3	7	11	13	2	23	25	31	33	35	37	4	43	5	53	55	61	63	65	67	73

960 Cut and dried were mated in 1710 by an unknown clergyman referring to his prepared sermon. The ternarish subsequence not including cuts, the cube-indexed ternarish, are "uncut".

dried	7	77	78	79	87	89	777	779	787	788	789	797	798	799	7,777	7,778	7,779	7,797

uncut	1	11	100	121	221	1,100	1,211	20,101	10,000	10,110	11,111	12,020	20,021

The sequence not including the cut numbers would be "cutless" or as a malapropism, "cutlass", while "cutback" is just the downward neighbors and "cutoff" both neighbors of the cuts (excluding those which are plain cuts).

cutlass	1	3	4	5	6	7	8	9	11	13	14	15	16	17	18	19	23
cutback	1	9	11	19	100	109	119	121	199	209	219	221	1,000	1,009	1,019	1,199	1,209
cutoff	1	3	9	11	13	19	23	100	103	109	113	119	121	123	199	203	209

961 All the "balls" are "foul balls" until 127 and 133. There's also "fubah", fouled-up beyond all hope or Fibonacci-unindexed binarish and hachibi; "fubar", fouled-up beyond all recognition or Fibonacci-unindexed binarish and rokubi. The "fubs" wouldn't intrude on the hachibi or rokubi until a hundred.

fubs	100	110	111	1,001	1,010	1,011	1,101	1,110	1,111	10,000	10,001	10,010	10,011

962 The "chumps", an allusion to Laurel and Hardy's "A Chump at Oxford" (1940), exclude the cubed hachibi-indexed Mersenne primes, the chimps, alluding to "The Chimp" (1932) and Cincinnati's chapter of the Sons of the Desert, the Chimp Tent. Both are subsequences of the "champs", cubed-hachibi-and-Mersenne-primes, an allusion to the lake snake of Lake Champlain, Ontario.

cubed hachibi	512	5,832	21,952	110,592	195,112	314,432	474,552	681,472	941,192

champs	3	7	31	127	512	5,832	8,191	21,952	110,592	131,071	195,112	314,432

963 Combining the two kinds of camels we get the generic "camels".

camels	3	7	13	17	23	27	30	31	32	33	34	35	36	37	38	39	43	47	53	57	63	67	70	71	72	73

Extrapolating from the bactrian camel with its two-humped m, we can get the three-humped "camnel" and the four-humped "cammel". (*The Xoo Book: A Guide to Exozoology*)

camnels	1,337	1,373	1,737	1,773	2,337	2,373	2,737	2,773	3,037	3,073	3,137	3,173	3,237

cammels	131,737	131,773	132,337	132,373	132,737	132,773	133,037	133,073	133,137

964 "Lump" originally meant to make a gloomy face as in "like it or lump it", hence the complimentary sequence.

likes	2	872	1,808	868,928	7,852,516,352	7,694,156,990,786,306,048
trimmed	1	761	77	757,817	6,741,405,241	6,583,045,886,752,537

965 limp

sub	1	11	192,874	50,560,965,398
broken-up	3	94	285	8,929
sub	1	35	105	3,285
broken-down	1	4	13	407
broken	1	3	4	13
hypersum	3	331	331,524,287	331,524,287,137,438,953,471

sub	1	122	121,960,969	> twentyplex

"Blimp" describes binarish lucky-indexed Mersenne primes, "crimp" cubed rokubi-indexed Mersenne primes, "pimp" prime-indexed Mersenne primes, "scrimp" sorted cubed rokubi-indexed Mersenne primes, "skimp" sorted kyuubi-indexed Mersenne primes and "simp" sanbi-indexed Mersenne primes.

986 "Pick-ups" describes the pick uptown neighbors and "trucks" their complimentary sequence, which also can be superpluralized to "truxes".

pick-ups	6,860	24,390	117,650	328,510	1,295,030	2,146,690
holiness	5	2	2	3	3	4

trucks	3,139	75,609	671,489	882,349
holiness	1	3	4	5
truxes	313,975,609	756,093,139	3,139,671,489	3,139,882,349
holiness	4	4	5	6

Picks also can be superpluralized to "pyxes", containers for the Blessed Sacrament.

pyxes	68,596,859	243,896,859	685,924,389	11,764,968,596
holiness	7	7	7	7

The mate of "pick" would be "choose", mated by Sir Thomas Herbert, aka "chews".

chews	3,140	75,610	671,490	882,350
holiness	1	2	3	5

"Nitpicky" describes nibi-indexed ternarish prime-indexed cubed kyuubi, the first being twenty-four thousand, three hundred eighty-nine and the second higher than fifteenplex.

nit	1	102	210	1,020	1,111	1,212	10,000	10,101	10,202	11,010	11,111	11,212

ternarish primes	2	11	211	1,021	1,201	2,011	2,111	2,221	10,111	10,211	12,011	12,101
trimmed	1	0	100	10	10	100	1,000	1,110	0	100	100	100

nitpicky	24,389	1,771,956,336,403,729
holiness	3	5

968 The sumps describe the sanbi-unindexed Mersenne primes, excluding the sanbi-indexed Mersenne primes, the "simps", beginning with thirty-one. The rumps describe the rokubi-unindexed Mersenne primes.

sumps	3	7	127	8,191	131,071	524,287	2,147,483,647	137,438,953,471
rumps	3	7	31	127	8,191	131,071	2,147,483,647	137,438,953,471

969 trimmed pushups and pulleys

pushups	1	2	4	6	0	2	8	12	18	20	26	30	32	36	42	48	50	56	60	62	68
pulleys	1	1	11	31	51	11	21	31	41	51	71	81	11	111	121	131	141	151	161	171	181

970 Sit-downs describe the sit's downtown neighbors, while "tight" describes the complimentary sequence, mated in print by Violet Hunt in 1897. "Up-tight" describes the uptown neighbors.

sit-downs	9	110	211	1,019	1,111	1,219	2,020	2,121	9,999	10,100	10,201	11,009	11,110
holiness	1	1	0	2	0	1	2	0	4	3	2	3	1

tight	89	787	888	7,877	7,978	8,779	8,887	8,979
holiness	3	2	6	2	3	3	6	4
up-tight	90	788	889	7,878	7,979	8,780	8,888	8,980
holiness	2	4	5	4	2	5	8	6

971 The hyphenym "slip-up" describes the uptown neighbors of slips. "Slip-up" was used in the New York *Evening Post* in 1909. "Slipup", on the other hand, describes slip-unindexed primes.

slip-up	6	42	138	192	308	368	488	692	908	1,104	1,427
holiness	1	0	2	1	3	3	4	2	4	1	0
slipup	2	3	5	7	13	17	19	23	29	31	37
trimmed	1	2	4	6	2	6	8	12	18	20	26

972 The mate of a "slump" is a "streak" from sports jargon.

streak	2	6	68	1,808	868,928	475,702	7,852,516,352	7,694,156,990,786,306,048
trimmed	1	5	57	77	757,817	36,461	6,741,405,241	2,305,843,009,213,693,951

973 The mate of "sunny" is "cloudy" from meteorology.

trimmed sunny	1	1	21	31	41	51	61	71	81	1	1	21	31	41	51	61	71	81
cloudy	7	17	27	37	47	57	67	87	107	117	127	137	147	157	167	187	207	217

"Sundowns" describe the downtown neighbors of the sanbi-unindexed nibi.

sundowns	1	11	31	41	51	61	71	81	91	101	111	131	141	151	161	171	181	191	201

974 The mate of "sin" is "grace".

graced	77	277	377	477	577	677	877	1,077	1,177	1,277	1,377	1,477	1,577	1,677	1,777

"Yin"

975 tipsy

broken-up	23	668	15,387	446,891	103,247,208	24,470,035,187
tressub	1	33	766	22,249	5,140,376	1,218,291,315
broken-down	1	24	553	16,061	3,710,644	879,438,689
broken	1	23	24	553	668	15,387
hypersum	23	2,329	232,932	232,932,229	232,932,229,231	232,932,229,231,237
tressub	1	116	11,597	11,597,013	11,597,012,822	11,597,012,821,815

976 "Cig" describes cube-indexed gobi, "fig" Fibonacci-indexed gobi, "gig" gobi-indexed gobi and "rig" rokubi-indexed gobi.

cig	5	75	265	635	1,245	2,155	3,425	5,115	7,285	9,995	13,305	17,275	21,965
fig	5	15	25	45	75	125	205	335	545	885	1,435	2,325	3,765
gig	45	145	245	345	445	545	645	745	845	945	1,045	1,145	1,245
rig	55	155	255	355	455	555	655	755	855	955	1,055	1,155	1,255

977 "Bib" describes binarish-indexed binarish, "crib" cubed rokubi-indexed binarish, "lib" lucky-indexed binarish, "nib" nibi-indexed binarish and "sib" sanbi-indexed binarish.

crib	11,010,110	1,000,000,000,000	100,010,010,101,000
holiness	3	12	10

bib	1	1,010	1,011	1,100,100	1,100,101	1,101,110	1,101,111
lib	1	11	111	1,001	1,101	1,111	10,011
nib	10	1,100	10,110	100,000	101,010	110,100	111,110
holiness	1	2	2	5	3	3	1
sib	11	1,101	10,111	100,001	101,011	110,101	111,110

"Libber" describes non-repunit lucky-indexed binarish.

libber	1,001	1,101	1,101	10,011	10,111	100,001	100,101	101,111
holiness	2	1	1	2	1	4	3	1

978 Biorhythmic numbers are those associated with so-called biorhythms, the lunar or menstrual cycle of 28, the "male" cycle of 23 promoted by William Fliess and "intellectual" cycle of 33 promoted by Hermann Swoboda. All three synchronize in a cycle of twenty-one thousand, two hundred fifty-two days (fifty-eight years and two months). The Mayan calendars synchronized in a cycle of 5,125 years.

biorhythmic	23	28	33	46	56	66	69	84	92	99	112	115	132	138	140	161	165	168
Fliess	23	46	69	92	115	138	161	184	207	230	243	266	289	312	335	358	381	404
Swoboda	33	66	99	132	165	198	231	264	297	330	363	396	429	462	495	528	561	594

979 "Bitter" describes those binarish-indexed ternarish with just two of the three "trits" and "bitterer" just one of the three trits, either a repunit or twice a repunit, one being the first one.

Rather than revenge, it's the mate to "bitter" that's "sweet" and their supersequence is "bittersweet". "Semi-sweet" is a euphemism for half-bitter when bitter is even and sweet is not.

bitter	101	1,101	1,101,001	1,101,110	111,212,111
holiness	1	1	3	2	0
sweet	898	8,898	8,898,998	8,898,889	888,787,888
holiness	5	7	11	12	14
semi-sweet	449	4,449	4,449,499	550,550	444,393,944
holiness	1	1	3	2	2
bittersweet	101	898	1,101	8,898	1,101,001
holiness	1	5	1	7	3

"Chit" describes cubed hachibi-indexed ternarish, starting at two hundred thousand, two hundred twenty-two. "Chat" describes cubed hachibi and ternarish; "chit-chat" comes from Samuel Palmer's *Moral essays on some of the most curious English, Scotch, and foreign proverbs* (1710).

cubed hachibi	512	5,832	21,952	110,592	195,112	314,432	474,552	681,472

980 "Bugs" alludes to Adelard "George 'Bugs' Moran" Cunin and/or Bugs Bunny. "Lugs" lucky-unindexed gobi and "mugs" Mersenne-unindexed gobi, avoiding the "migs", Mersenne-indexed gobi as if they were MiGs (Russian fighter planes named for Mikoyan and Gurevich).

lugs	15	35	45	55	75	95	105	115	165	175	185	195	215	225	235	255	265	275	285
mugs	5	15	35	45	55	75	85	95	105	115	125	135	145	155	165	175	185	195	205

migs	25	65	305	1,265	20,465	81,905	1,310,705	5,242,865	83,886,065	5,368,709,105

981 "Pun" describes prime-unindexed nibi and punish pun-indexed shichibi and "spun" sanbi prime-unindexed nibi.

puns	2	32	52	72	82	92	112	132	142	152	172	192	202	212	232	242	252	262	272
trimmed	1	21	31	61	71	81	1	21	31	41	61	81	11	101	121	131	141	151	161
sub	1	12	19	26	30	34	41	49	52	56	63	71	74	78	85	89	93	96	100
spun	2	12	32	42	52	62	72	82	92	102	112	132	142	152	162	172	182	192	202
trimmed	1	1	21	31	41	51	61	71	81	1	1	21	31	41	51	61	71	81	11
sub	1	4	12	15	19	23	26	30	34	38	41	49	52	56	60	63	67	71	74

982 "Mush" and "mushy" describe Mersenne-unindexed shichibi. "Mish" describes the Mersenne-indexed shichibi and "mash", not M*A*S*H (Mobile Army Surgical Hospital), but

Mersenne-and-shichibi, so "mishmash" is exactly the same as plain "mash". [4077 is a mash number.]

mushy	7	17	37	47	57	77	87	97	107
mish	27	67	307	1,267	81,907	1,310,707	5,242,867	83,886,067	5,368,709,107
mash	2	3	7	17	27	31	37	47	57

"Mishap" describes Mersenne-indexed shichibi-and-primes. 99.9% of the first thousand mishaps are prime!

mishap	2	3	5	7	11	13	17	19	23	27	31	37	41	43	47	53	59	61	67	71	73	79	83	89
trimmed	1	2	4	6	0	2	6	8	12	16	20	26	30	32	36	42	48	50	56	60	62	68	72	78

983 Re-spelling clicks as "clix", not to be mistaken for CLIX, they can be superpluralized to 'clixes', the first of which is greater than thirty-threeplex.

clix	19,465,109,729	72,919,465,109	40,318,322,589,729	72,940,318,322,589
holiness	5	5	7	7
24th sub	1	3	1522	2754

clixes	1,946,510,938,726,107,856,940,318,322,589,729
holiness	19
104th sub	1

"Buick" describes binarish-unindexed/indexed cubed kyuubi, "crick" cubed rokubi-indexed cubed kyuubi, "hick" hachibi-indexed cubed kyuubi, and "nick" nibi-indexed cubed kyuubi

984 Re-spelling cluck, not to be mistaken for base four CLUCK, as "clux", which could be superpluralized to "cluxes"

clux	243,896,859	593,196,859	685,924,389	685,959,319
holiness	7	6	7	6

cluxes	24,389,685,968,596,859	59,319,685,968,596,859
holiness	15	14

985 In the Jesus deck (1972) by Rev. Ralph M. Moore the suits are the four Gospels. The Clubs equivalent is Luke, symbolized by a winged ox and focusing on Jesus' birth and infancy.

986 Spades are John, symbolized by an eagle and focusing on Jesus' resurrection.

987 Diamonds are Matthew, symbolized by a winged man and focusing on Jesus' ministry and teachings.

988 Hearts are Mark, symbolized by a winged lion and focus on Jesus' passion and crucifixion. The game of Marks is similar to the game of Hearts, similarly the game of War becomes Peace, I Doubt It becomes I Believe, Gin Rummy becomes Witness, and Go Fish becomes Go Seek.

989 Jokers are symbolized with all four Gospels together (Ezek 10:14, Rev 4:7) with "Here we are fools for the sake of Christ." (1 Cor 4:10)

990 "Clued-in" describes the downtown neighbors of the "clued".

clued-in	4	6	8	12	14	16	18	20	22	24	26	28	30	32	34	36	38	40	42	44	46	48

991 "Glut" describes gobi lucky-unindexed ternarish and "glitter" those gobi lucky-indexed ternarish with all three 'trits'.

gobi lucky	15	25	75	105	135	195	205	235	285
glut	1	2	10	11	12	20	21	22	100
glitter	120	2,210	10,220	11,021	21,020	22,201	101,120	112,021	120,101
holiness	1	1	2	1	2	1	2	1	2

"Flit" describes Fibonacci lucky-indexed ternarish and "flut" Fibonacci unlucky-indexed ternarish, "flutter" those with two of three "trits" and "flutterer" with just one, beginning with two and twenty-two.

Fibonacci lucky	1	3	7	13	25	49	87	163	289	519
flit	1	10	21	111	221	1,211	10,020	20,001	101,201	201,020
flut	2	12	20	22	100	101	102	110	112	120
trimmed	1	1	1	11	0	0	1	0	1	1
flutter	12	20	100	101	110	112	121	122	200	211
trimmed	1	1	0	0	0	1	1	11	1	100

992 "Bub" describes binarish-unindexed binarish, "hub" hachibi-unindexed binarish, "lub" lucky-unindexed binarish, "pub" prime-unindexed binarish and "sub" sanbi-indexed binarish.

bub	10	11	100	101	110	111	1,000	1,001	1,100	1,101	1,110	1,111	10,000
holiness	1	0	2	1	1	0	3	2	2	1	1	0	4
hub	1	10	11	100	101	110	111	1,001	1,010	1,011	1,100	1,101	1,110
lub	10	100	101	110	1,000	1,010	1,011	1,100	1,110	10,000	10,001	10,010	10,011
holiness	1	2	1	1	3	2	1	2	1	4	3	3	2
pub	1	100	110	1,000	1,001	1,010	1,100	1,110	1,111	10,000	10,010	10,100	10,101
sub	1	10	100	101	110	111	1,000	1,001	1,010	1,011	1,100	1,110	1,111

"Flub" describes Fibonacci lucky-unindexed binarish and those non-repunit "flubber", after

Ned Brainard's serendipitously discovered flying rubber aka the Brainard numbers ("The Absent Minded Professor" by Bill Walsh, based on "A Situation of Gravity" by Samuel W. Taylor).

flub	10	100	101	110	1,000	1,001	1,010	1,011	1,100	1,110	1,111	10,000
holiness	1	2	1	1	3	2	2	1	2	1	0	4
flubber	10	100	101	110	1,000	1,001	1,010	1,011	1,100	1,110	10,000	10,001
holiness	1	2	1	1	3	2	2	1	2	1	4	3

993 "Blue" describes binarish lucky-unindexed evens, and alludes to "Think Blue, Count Two" by Paul "Cordwainer Smith" Linebarger. "Glue" gobi lucky-unindexed evens, "hue" hachibi lucky-unindexed evens and "sue" sanbi lucky-unindexed evens, a sequence named for "A Boy Named Sue" by Shel Silverstein and popularized by Johnny Cash.

blue	2	20	22	200	202	220	222	2,000	2,002	2,020	2,022
trimmed	1	1	11	1	11	11	111	1	11	11	111
glue	10	30	50	70	90	110	130	150	170	190	230
holiness	1	1	1	1	2	1	1	1	1	2	1
hue	16	36	56	76	96	116	136	156	176	196	216
holiness	1	1	1	1	2	1	1	1	1	2	1
sue	66	126	186	246	266	286	306	366	406	426	486
holiness	2	1	3	1	2	3	2	2	2	1	3

994 Fig newton cookies were NOT invented by Sir Isaac Newton after a fig fell on his head, nor by his brother Figbert. They were named for Newton, Massachusetts, which was a "new town" in 1638, and first made and mated in 1891 with Charles Rose's fig roll-making machine. Since all newtons are divisible by four we also have half-newtons (sticking with English prefixes).

half-newton	2	12	27	37	42	57	227	332	347	437	1,952	3,117	3,837	4,282	16,127

Dandy is said be a pet form of Andrew. Fine and Dandy were mated in 1671 in "My love is blithe and bucksome and sweet and fine as can be, fresh and gay as the flowers in May, and looks like Jack-a-dandy.", back before "gay" meant gobi-and-yongbi.

gay	4	5	14	15	24	25	34	35	44	45	54	55	64	65	74	75	84	85	94	95	104	105

995 "Hine" describes hachibi-indexed nibi, "line" lucky-indexed nibi, "nine" or better "non-nun" nibi-indexed nibi or evenly even.

hine	72	172	272	372	472	572	672	772	872	972	1,072	1,172	1,272	1,372	1,372
line	2	22	62	82	122	142	202	242	302	322	362	422	482	502	622

996 "Two-fisted" describes numbers with two hypersummed Fibonacci-indexed sorted ternarish.

two-fisted	111	112	121	1,111	1,112	1,121	1,122	1,221	1,222	11,111	11,112	11,121

997 "Hire" and "hired" describe hachibi-indexed rokubi evens, "hireling" their downtown neighbors, "mire" and "mired" Mersenne-indexed rokubi evens, "sir" and "sired" sanbi-indexed rokubi (evens) and "tire" and "tired" ternarish-indexed rokubi evens.

hired	76	176	276	376	476	576	676	776	876	976	1,076	1,176	1,276	1,376
holiness	1	1	1	1	1	1	2	1	3	2	2	1	1	1
hireling	75	175	275	375	475	575	675	775	875	875	1,075	1,175	1,275	1,325
sired	26	126	226	326	426	526	626	726	826	926	1,026	1,126	1,326	1,426
holiness	1	1	1	1	1	1	2	1	3	2	2	1	1	1
tired	6	16	96	106	116	196	206	216	996	1,006	1,016	1,096	1,106	1,116
holiness	1	1	2	2	1	2	2	1	3	3	2	3	2	1

mired	26	66	306	1,266	20,466	81,906	1,310,706	5,242,866	83,886,066	5,368,709,106
holiness	1	2	2	2	3	5	3	4	10	7

"Gunfire"describes gobi-unindexed nibi Fibonacci-indexed rokubi even, "misfire" Mersenne-indexed sanbi Finonacci-indexed rokubi even, "saltire" sanbi-and-lucky-ternarish-indexed-rokubi-evens, "satire" sanbi-and-ternarish-inedexed-rokubi-evens, "spire" sanbi prime-indexed rokubi evens, "spitfire" sanbi prime-indexed ternarish Fibonacci-indexed rokubi evens and "suspire" sanbi-unindexed sanbi prime-indexed rokubi evens.

997 "Baptist" describes binarish-and-prime-ternarish-indexed-sanbi-ternarish, "bassist" binarish-and-sorted-sanbi-indexed-sanbi-ternarish, "bigamist" binarish-indexed-gobi-and-Mersenne-indexed-sanbi-ternarish, "capitalist" cube-and-prime-indexed-ternarish-and-lucky-indexed-sanbi-ternarish, "casuist" cube-and-sanbi-unindexed/indexed sanbi ternarish, "cist" cube-indexed sanbi ternarish, "cubist" cube-unindexed binarish-indexed sanbi ternarish, "essayist" even-sorted-sanbi-and-yongbi-indexed-sanbi-ternarish, "fabulist" Fibonacci-and-binarish-unindexed-lucky-indexed-sanbi-ternarish, "fatalist" Fibonacci-and-ternarish-and-lucky-indexed-sanbi-ternarish, "fist" Fibonacci-indexed sorted ternarish, "fuguist" Fibonacci-unindexed gobi-unindexed/indexed sanbi ternarish, "guitarist" gobi-unindexed/indexed-ternarish-and-rokubi-indexed-sanbi-ternarish, "humanist" hochibi-unindexed-Mersenne-and-nibi-indexed-sanbi-ternarish, "Latinist" lucky-and-ternarish-indexed-nibi-indexed-sanbi-ternarish, "linguist" lucky-indexed ng-unindexed/indexed sanbi ternarish, "list" lucky-indexed sorted ternarish, "machinist" Mersenne-and-cubed-hachibi-indexed-hachibi-indexed-sanbi-ternarist, "minimalist" Mersenne-indexed-nibi-indexed-Mersenne-and-lucky-indexed-sanbi-ternarish "mist" and "misty" Mersenne-indexed sorted ternarish, "Maoist" Mersenne-and-odd-indexed-sanbi-ternish, "natualist" nibi-andternarish-unindexed-rokubi-and-lucky-indexed-sanbi-ternarish, "oculist" odd cube-unidexed lucky-indexed sanbi ternarish, "pacifist" prime-and-cube-indexed-Fibonacci-indexed-sanbi-ternarish, "papist" prime-and-prime-indexed-sanbi-ternarish, "parachutist" prime-and-rokubi-and-cubed-hachibi-unindexed-ternarish-indexed-sanbi-ternarish, "pianist" prime-ichibi-and-nibi-indexed-sanbi-ternarish, "populist" prime odd prime-unindexed lucky-indexed sanbi ternarish, "pugilist" prime-unindexed gobi-indexed lucky-indexed sanbi ternarish, "rapist" rokubi-and-prime-indexed-sanbi-ternarish, "realist" rokubi-even-and-lucky-indexed-sanbi-ternarish, "subsist" sanbi-unindexed binarish sanbi-indexed sanbi ternarish, "timpanist" ternarish-indexed-Mersenne-prime-and-nibi-indexed-sanbi-ternarish and "tsarist" ternarish-sanbi-and-rokubi-indexed-sanbi-ternarish.

998 "Fitter" describes those Fibonacci-indexed ternarish with any two of the three and 'fittest", those with any one of one or two, of which the first two are two and twenty-two.

fitter	10	12	22,122	111,222	21,1121	...

999 "Bahai" describes binarish-and-hachibi-and-ichibi, "Kali" kyuubi-and-lucky ichibi, "okapi" odd kyuubi-and-prime-ichibi and "Sufi" sanbi-unindexed lucky ichibi.

Bahai	1	8	10	11	18	21	28	31	38	41	48	51	58	61	68	71	78	81	88	91
Kali	1	9	19	21	29	31	39	49	51	59	69	79	89	99	109	111	119	129	139	141
okapi	9	19	29	31	39	49	59	69	79	89	99	109	119	129	139	149	159	169	179	189
holiness	1	1	1	0	1	1	1	2	1	3	2	2	1	1	1	1	1	2	1	3

Sufi	1	21	17,711	317,811	165,580,141	956,722,026,041	2,504,730,781,961

"Chili" describes cubed hachibi-indexed lucky ichibi, starting rather high with the five hundred twelfth lucky ichibi.

1000 "Gun" describes gobi-unindexed nibi and "nun" nibi-unindexed nibi.

gun	2	12	22	32	52	62	72	82	92	102	112	122	132	152	162	172	182	192	202
trimmed	1	1	11	21	41	51	61	71	81	1	1	11	21	41	51	61	71	81	11
nun	2	22	32	42	52	62	72	82	92	102	112	132	142	152	162	172	182	192	202
trimmed	1	11	21	31	41	51	61	71	81	1	1	21	31	41	51	61	71	81	11

1001 Since rokubi are all even, both "cur" and "cure" describe cube-unindexed rokubi, "lure" lucky-unindexed rokubi, "pure" prime-unindexed rokubi, "impure" prime-indexed rokubi, "sure" sanbi-unindexed rokubi and "unsure" sanbi-indexed rokubi.

cure	16	26	36	46	56	66	86	96	106	116	126	136	146	156	166	176
lure	16	36	46	56	76	106	116	136	166	176	186	196	216	226	236	256
pure	6	36	56	76	86	96	116	136	146	156	166	176	196	206	216	236
impure	16	26	46	66	106	126	186	226	286	306	366	406	426	466	526	586
sure	6	16	36	46	56	66	76	86	96	106	116	136	146	156	166	176
unsure	26	126	226	326	426	526	626	726	826	926	1,026	1,126	1,226	1,326	1,426	1,526

"Capture" describes cubes-and-prime-ternarish-unindexed-rokubi-evens, "caricature" cubes-and-rokubi-indexed-cubes-and-ternarish-unindexed-rokubi-evens, "cincture" cube-indexed nibi cubed tenarish-unindexed rokubi evens, "creature" cubed-rokubi-evens-and-tenarish-unindexed-rokubi-evens, "cubiture" cube-unindexed binarish-indexed ternarish-unindexed rokubi evens, "culture" cube-unindexed lucky ternarish-unindexed rokubi evens, "fracture" Fibonacci-rokubi-and-cubed-ternarish-unindexed-rokubi-evens, "future" Fibonacci-unindexed ternarish-unindexed rokubi evens, "hachure" hachibi-and-cubed-hachibi-unindexed-rokubi-evens, "langur" lucky-and-ng-unindexed-rokubi, "ligure" lucky-indexed gobi-unindexed rokubi evens, "ligature" lucky-indexed-gobi-and-ternarish-unindexed-rokubi-evens, "manicure" Mersenne-and-nibi-indexed cube-unindexed rokubi evens, "manufacture" Mersenne-and-nibi-unindexed-Fibonacci-and-cubed-ternarish-unindexed-rokubi-evens, "manure" Mersenne-and-nibi-unindexed-rokubi-evens, "mature" Mersenne-and-ternarish-unindexed-rokubi-evens, "nature" nibi-and-ternarish-unindexed-rokubi-evens, "musculature" Mersenne-unindexed-sanbi-cubed-unindexed-lucky-and-ternarish-unindexed-rokubi-evens, "parure" prime-and-rokubi-unindexed-rokubi-evens, "picture" prime-indexed cubed ternarish-unindexed rokubi evens, "puncture" prime unindexed nibi cubed ternarish-unindexed rokubi evens, "rapture" rokubi-and-prime-ternarish-unindexed-rokubi-evens, "sericulture" sorted even rokubi-indexed cube-unindexed lucky ternatish-unindexed rokubi evens, "sinecure" sinbi-indexed nibi even cube-unindexed rokubi evens, "stature" sanbi-ternarish-and-ternarish-unindexed-rokubi-evens, "suture" sanbi-unindexed ternarish-unindexed rokubi evens and "tincture" ternarish-indexed nibi cubed ternarish-unindexed rokubi evens.

1002 "Basin" describes binarish-and-sanbi-indexed-nibi, "biotin" binarish-indexed odd ternarish-indexed nibi, "bluefin" binarish lucky-unindexed even Fibonacci-indexed nibi, "buskin" binarish-unindexed sorted kyuubi-indexed nibi, "cabin" cube-and-binarish-indexed nibi, "capuchin" cube-and-prime-unindexed cubed hachibi-indexed nibi, "chaplin" cubed hachibi-and-prime-lucky-indexed nibi, "chin" describes cubed hachibi-indexed nibi, "chitlin" cubed-hachibi-indexed ternarish lucky-indexed nibi, "coalbin" cubed- odd-and-lucky-binarish-indexed nibi, "cousin" cubed odd-unindexed sanbi-indexed nibi, "cumin" cube-unindexed Mersenne-indexed nibi, "cutin" cube-unindexed ternarish-indexed nibi, "cyanin" cubed-yongbi-and-nibi-indexed-nibi, "elastin" even-lucky-and-sanbi-yernarish-indexed nibi, "gamin" gobi-and-Mersenne-indexed nibi, "hatpin" hachibi-and-ternarish-prime-indexed nibi, "histamin" hachibi-indexed-sanbi-ternarish-and-Mersenne-indexed-nibi, "kaolin" kyuubi-and-odd-lucky-indexed nibi, "lapin" lucky-and-prime-indexed nibi, "latin" lucky-and-ternarish-indexed nibi, "loin" lucky odd-indexed nibi, "lupin" lucky-unindexed cubed kyuubi-indexed nibi, "manikin" Mersenne-and-nibi-indexed-kyuubi-indexed-nibi, "matin" Mersenne-and-ternarish-indexed-nibi, "minikin" Mersenne-indexed nibi-indexed kyuubi-indexed nibi, "moulin" Mersenne odd-unindexed lucky-indexed nibi, "mullein" Mersenne-unindexed luckier lucky even-indexed nibi, "muslin" Mersenne-unindexed sanbi lucky-indexed nibi, "nuclein" nibi-unindexed cubed lucky even-indexed nibi, "pushpin" prime-unindexed shichibi prime-indexed nibi, "rein" rokubi-unindexed/indexed nibi, "saccharin" sanbi-and-cubed-cubed-hachibi-and-rokubi-indexed nibi, "satin" sanbi-and-ternarish-indexed nibi, "shin" shichibi-indexed nibi, "stalin" sanbi-ternarish-and-lucky-indexed nibi and "tin" ternarish-indexed nibi. "Fever" is the complimentary sequence to "cabin" "deer" to "rein".

1003 More than one tonic could be re-spelled 'tonix' and superpluralized to 'tonixes'.

tonix	15,757	25,757	35,757	45,757	55,757	57,557	57,657	57,757	57,857	57,957	65,757

tonixes	1,575,725,757	1,575,735,757	1,575,745,757	1,575,755,757	1,575,765,757

1004 guns

trimmed	1	21	41	51	61	71	81
sub	1	4	12	19	23	26	30
broken-up	2	25	52	649	40,290	2,901,529	237,965,668
trimmed	1	14	41	538	318	180,418	126,854,557
sub	1	9	19	239	14,822	1,067,413	87,542,677
broken-down	1	3	7	87	5,401	388,959	31,900,039
broken	1	2	3	7	25	52	87
hypersum	2	212	21,232	2,123,252	212,325,262	21,232,526,272	2,123,252,627,282
trimmed	1	101	10,121	1,012,141	101,214,151	10,121,415,161	1,012,141,516,171
sub	1	78	7,811	781,101	78,110,099	7,811,009,900	781,100,989,990

1005 Guns have been mated with roses at least since the hard rock band "Guns 'n' Roses" formed in 1985. By 2006 the TV series "Numb3rs" featured the episode "Guns and Roses". Some however prefer to call the sequence "bullets".

1006 "Crush" describes cubed rokubi-unindexed shichibi.

crokubi	216	4,096	17,576	46,656	97,336	175,616	287,496	438,976	636,056
crush	...	2,137	2,147	2,167	2,177	2,187	2,197	2,207	2,217

1007 "Guys and Dolls" refers to the 1950 Broadway play based on the short stories "The Idyll of Miss Sara Brown", "Blood Pressure" and "Pick the Winner" by Damon Runyon. "Buy" refers to binarish-unindexed yongbi.

buy	14	24	34	44	54	64	74	84	114	124	134	144	154	164	174	184	194	204	214	224

1008 "Guilt" and "guilty" describe gobi-unindexed/indexed lucky ternarish, i. e., all the lucky ternarish. "Gilt" and "silt" describe the rare gobi-indexed and sanbi-indexed lucky ternarish, "salt" the more common sanbi-and-lucky ternarish.

guilty	1	21	111	211	...
silt	111	...			

salt	1	3	13	21	23	33	43	53	63	73	83	93	103	111	113	123	133	143	153	163

1009 The complimentary sequence to "hits" is "misses". A "hit" with two of three "trits" is a "hitter" and its complementary sequence is "pitcher".

misses	77	799	7,791	7,887	7,988	8,789	8,897	8,998	9,008	9,018	9,028
hitters	22	200	1,001	1,010	2,112	2,220	11,000	11,101	12,111	12,212	20,020

pitchers	77	799	7,779	7,887	8,989	8,998	77,787	77,888	79,777	79,979	87.787

1010 "Buts" describes binarish-unindexed trinarish and "ifs" the complimentary sequence and both together are, of course, "ands". "No-ifs-ands-or-buts" is everything else.

buts	2	101	102	10,201	10,202	11,010	11,011	1,101,001	1,101,002	1,101,110
ifs	7	897	898	88,988	88,989	89,797	89,798	8,887,997	8,887,998	8,888,789
ands	2	7	10	101	102	897	898	10,201	10,202	88,988

1011 "Huggy" has since 2007 come to be particularly associated with Becky "Wordgirl" Botsford's sidekick, Capt. Huggy Face, both from the planet Lexicon. "Kiss" describes kyuubi-indexed sorted sanbi, rather than hug's complimentary sequence. The kiss, first symbolized with an X (two pairs of touching lips) by Gilbert White in 1763 (*Oxford English Dictionary*) and the hug, symbolized by an O (two pairs of enveloping arms) are associated with the much older, though perhaps less popular, game of tic-tac-toe. (*The Oxford History of Board Games* by David Parlett) The holiness of a kiss and a "kiss" both vary.

kissed	38	138	238	348	358	368	378	388	389	39	139	239	339	349	359	369	379
holiness	2	2	2	2	2	3	2	4	3	1	1	1	1	1	1	2	1

1012 Some prefer the more familiar "hunny" or "honey" mated with the complimentary sequence "bees".

bees	27	127	227	327	427	527	627	727	827	927	1,027	1,127	1,227	1,327	1,427

1013 The hyphenym "hush-up" describes the uptown neighbor of "hush", while "hushup" describes hachibi-unindexed shichibi-unindexed primes."Shush" and "shushed" describe shichibi-unindexed shichibi.

hush-up	...	77	79	80	81	82	83	84	85	86	87	88	89	90	91	92	93	94	95
shushed	7	17	27	37	47	57	77	87	97	107	117	127	137	147	157	177	187	197	207

1014 "Kicks" could be re-spelled 'kix' like the breakfast cereal introduced by General Mills in 1937 and superpluralized to 'kixes'. In 1954 they introduced a sugar-coated variation called Trix with with a silly rabbit trying to steal it from kids, which led to the shaggy dog story ending with "Silly rabbi, kicks are for Trids." "Trids" are therefore the complimentary sequence for "kix".

kix	6,751,269,704,969	7,049,696,751,269	24,137,569,704,969
holiness	7	7	6

kixes	67,512,697,047,049,696,751,269	704,996,967,512,696,751,269,704,969
holiness	10	15

trids	3,248,730,295,030	2,950,303,248,730	75,862,430,295,030

holiness	6	6	7

1015 "Kith and kin" comes from Middle English, kith meaning friends, neighbors and acquaintances and kin meaning relatives.

kith	17	117	217	317	417	517	617	717	817	1,017	1,117	1,217	1,317	1,417	1,517	1,817

1016 The phrase "kit and caboodle" is an Americanism, caboodle coming from the Dutch *boedal* meaning property and kit being personal property one can carry. (*QPB Encyclopedia of Word and Phrase Origins* by Robert Hendrickson)

caboodle	798	899	7,778	7,879	7,987	8,788	8,889	8,997	77,778	77,879	77,998

1017 "Skittish" describes numbers with sorted kyuubi-indexed ternarish frags.

skittish	100	1,001	1,002	1,003	1,004	1,005	1,006	1,007	1,008	1,009	1,100
holiness	2	2	2	2	2	2	3	2	4	3	2

1018 "Flick" and "flicked" describe Fibonacci lucky-indexed cubed kyuubi and "slick" sanbi lucky-indexed cubed kyuubi.

flicked	729	24,389	328,509	2,146,689	15,438,249	90,518,849
holiness	1	3	4	7	3	7
slick	24,389	2,146,689	35,611,289	78,953,589	248,858,189	387,420,489
holiness	3	5	4	6	9	6

1019 "Lite" describes lucky-indexed ternarish evens, "biotite" binarish-indexed odd ternarish-indexed ternarish evens, "bisulfite" binarish-indexed sanbi-unindexed lucky Fibonacci-indexed ternarish evens, "bite" binarish-indexed ternarish evens, "calcite" cubes-and-lucky cubes-indexed ternarish evens, "catamite" cubes-and-ternarish-and-Mersennes-indexed ternarish evens, "cite" cubed-indexed ternarish evens, "elite" even lucky-indexed ternarish evens, "finite" Fibonacci-indexed-nibi-indexed ternarish evens, "fleabite" Fibonacci-luckies-and-binarish-indexed ternarish evens, "halite" hachibi-and-lucky-indexed ternarish evens, "kite" kyuubi-indexed ternarish evens, "lucite" lucky-undexed cube-ndexed ternarish evens, "malachite" Mersennes-and-luckies-and-cubed-hachibi-indexed ternarish evens, "mite" Mersenne-indexed ternarish evens, "paracite" primes-and-rokubi-and-cubes-indexed ternarish evens, "reunite" rokubi evens-unindexed nibi-indexed ternarish evens, "ruralite" rokubi-unindexed-rokubi-and-lucky-indexed ternarish evens, "samite" sanbi-and-Mersennes-indexed ternarish evens, "site" sanbi-indexed ternarish evens, "spite" sanbi prime-indexed ternarish even, "stalactite" sorted-ternarish-and-luckies-and-cubed-ternarish-indexed ternarish evens and "steatite" sorted-ternarish-evens-and-ternarish-indexed-ternarish evens.

ternarish evens	10	12	100	102	110	112	120	122	1,000	1,010	1,012	1,100	1,102	1,110
holiness	1	0	2	1	1	0	1	0	3	2	1	2	1	1

"Blackout" binarish-kuckies-and-cube-kyuubi-odd-unindexed-ternarish, "caput" cubed-and-prime-unindexed-ternarish, "clout" cubed lucky odd-unindexed ternarish, "fallout" Fibonacci-

and-luckier-lucky-odd-unindexed-ternarish, "flameout" Fibonacci-luckies-and-Mersenne-evens/odds-unindexed-ternarish, "gamut" gobi-and-Mersenne-unindexed-ternarish, "halibut" hachibi-and-lucky-indexed-binarish-unindexed-ternarish, "kaput" kyuubi-and-prime-unindexed-ternarish, "layabout" lucky-and-yongbi-and-binarish-odd-unindexed-ternarish, "lilliput" lucky-indexed-luckier-lucky-indexed-prime-unindexed-ternarish, "miscut" Mersenne-indexed sanbi-unindexed ternarish, "peanut" prime-even-and-nibi-unindexed-ternarish, "ragout" rokubi-and-gobi-odd-unindexed-ternarish, "sacbut" sanbi-and-cubed-binarish-unindexed-ternarish, "scout" sanbi cubed odds-unindexed ternarish, "shout" shichibi odds-unindexed ternarish, "spout" sanbi prime odds-unindexed ternarish and "stout" sanbi ternarish odds-unindexed ternarish.

"Lute" describes lucky-unindexed ternarish evens, "chute" cubed hachibi-unindexed ternarish evens, "cute" cubed-unindexed ternarish evens, "flute" Fibonacci luckies-unindexed ternarish evens, "impute" ichibi Mersenne prime-unindexed ternarish evens, "minute" Mersenne-indexed nibi-unindexed evens, "mute" Mersenne-unindexed ternarish evens, "parachute" prime-and-rokubi-and-cubed-hachibi-unindexed-ternarish-evens, "salute" sanbi-and-lucky-unindexed ternarish evens, "statute" sanbi-ternarish-and-ternarish-unindexed ternarish evens, "subacute" sanbi-unindexed-binarish-and-cube-unindexed-ternarish-evens and "pluto" prime lucky-unindexed ternarish odds.

1020 "Shun" describes shichibi-unindexed nibi, "stun" sanbi ternarish-unindexed nibi and "yamun" yongbi-and-Mersenne-unindexed-nibi.

1021 Jesus Christ referred to Himself as the Bridegroom of Christians in Mark 2:19.

1022 The phrase "soup to nuts" comes from the mid-1800s referring to the first and last course of a formal meal. It was also the title of the first Stooges film with the Horwitz brothers, Moses "Harry Howard" and Samuel "Shemp Howard" and Louis "Larry Fine" Feinberg in 1930.

"Shells" describes the numbers surrounding the nuts, their neighbors, and "sea" their compliment sequence.

shells	0	2	9	11	12	13	19	21	22	23	99	101	102	110	112	119	121	122	123	199	200	203

sea	7	76	77	78	80	86	87	88	90	776	778	779	788	797	796						

1023 "Beaucoup" describes binarish-evens-and-cubed-odd-unindexed-primes, "coup" describes cubed odd-unindexed primes, "croup" cubed rokubi odd-unindexed primes, "loup" lucky odd-unindexed primes, "stoup" sanbi ternarish odd-unindexed primes.

1024 "Clamour" describes cubed-luckies-and-Mersenne-odd-unindexed-rokubi, "flour" Fibonacci lucky odd-unindexed rokubi, "four" describes Fibonacci odd-unindexed rokubi, "lour" lucky odd-unindexed rokubi, "paramour" primes-and-rokubi-and-Mersenn-odd-unindexed-rokubi, "pour" prime odd-unindexed rokubi, "scour" sanbi cubed odd-unindexed rokubi, "sour" sanbi odd-unindexed rokubi, "succour" sanbi-unindexed cubed cube odd-unindexed rokubi, and "tour" ternarish odd-unindexed rokubi.

"Masseur" describes Mersenne-and-sorted-sanbi-even-unindexed-rokubi, "pasteur" prime-and-sorted-ternarish-even-unindexed-rokubi, "sieur" sanbi-indexed even-unindexed rokubi and "coir" and "memoir" cubed odd-indexed rokubi and Mersenne even-/Mersenne odd-indexed rokubi.

1025 "Heirs" describes hachibi evens-indexed rokubi, "heiress" its complimentary sequence. "Hur" describes habichi-unindexed rokubi and "Ben" its complimentary sequence, alluding to *Ben-Hur* by Lewis Wallace (1880).

heirs	76	176	276	376	476	576	676	776	876	976	1,076	1,176	1,276	1,376
holiness	1	1	1	1	1	1	2	1	3	2	2	1	1	1
heiress	23	123	223	323	423	523	623	723	823	1,023	1,123	1,223	1,323	1,423
hur	6	16	26	36	46	56	66	86	96	106	116	126	136	146
holiness	1	1	1	1	1	1	2	3	2	2	1	1	1	1
ben	3	13	33	43	53	63	73	83	103	113	143	153	163	173

1026 The complimentary sequence of "mine" is "yours". "Yours, Mine and Ours" is a screenplay by Mort Lachman and Melville Shavekson.

yours	37	77	697	8,737	18,097	79,537	4,757,137	8,689,297	16,113,937	4,631,290,893

1027 "Cutout" describes cube-unindexed ternarish odds-unindexed ternarish, "freakout" Fibonacci-rokubi-evens-and-kyuubi-odd-unindexed-ternarish, "output" odd-unindexed ternarish prime odds-unindexed ternarish, "phaseout" palindromic-hachibi-and-sanbi-even/odd-unindexed-ternarish, "pullout" prime-unindexed luckier lucky odd-unindexed ternarish, "putout" prime-unindexed ternarish odd-unindexed ternarish, "shutout" shichibi-unindexed ternarish odd-unindexed ternarish, "sickout" sanbi-indexed cubed kyuubi odd-unindexed ternarish and "stakeout" sanbi-ternarish-and-kyuubi-even/odd-unindexed-ternarish.

1028 "Birdie" or "eagle" describes one below "par", "bogie" one above "par" and "albatross" two below "par".

eagle	1	2	4	5	6	10	12	15	16	18	22	25	28	30	35	36	40	42	45	51	54	57	59	64	66	69
bogie	3	4	6	7	12	14	17	20	24	27	30	32	38	42	44	47	48	54	57	60	65	67	70	74	77	80
albatross	0	1	3	4	5	9	11	14	15	17	21	24	37	29	34	35	39	41	44	45	48	49	55	58	61	66

1029 "Picnics" can be re-spelled picnix and superpluralized to "picnixes".

picnixes	12,072,692,177,925,159,780,352	51,597,803,521,207,269,217,792
holiness	6	6

1030 The Hebrew word for basket is the same as that for ark, connecting Moses (Ex 2:3) with Noah (Gn 6-8) and so also with Atrahasis of Shuruppak, Utnapishtim of Akkadia, Ziusudra of Sumeria and St. Brendan of Ireland's basket-like ships.

1031 "Spin" or "spine" describes sanbi prime-indexed nibi (evens)."

spin	22	122	222	422	522	722	922	1,022	1,122	1,622	1,722	1,922	2,222	2,322

1032 The condition of "pins-and-needles" is technically called paresthesia, so that the supersequence could be called the '.paresthesians". The non-palindromic ones could be called "aresthesians".

paresthesians	12	22	37	42	57	62	77	87	102	112	117	162	177	182	217	222
trimmed	1	11	26	31	46	51	66	76	1	1	6	51	66	71	106	111
aresthesians	12	37	42	57	62	87	102	112	117	162	177	182	217	277	282	297

1033 "The Pit and the Pendulum" by Edgar Allen Poe (1843).

1034 "Pitiful" describes prime-indexed ternarish-indexed Fibonacci-unindexed luckies.

pitif	2	89	233	17,711	> twenty-oneplex
pitifil	3	519	...		
pitiful	1	3	9	13	15

"Piteous" describes prime-indexed ternarish even/odd-unindexed sanbi, "bulbaceous" binarish-unindexed-lucky-binarish-and-cube-even/odd-unindexed-sanbi, "courageous" cubed-odd-unindexed-rokubi-and-gobi-even/odd-unindexed-sanbi, "farinaceous" Fibonacci-and-rokubi-indexed-nibi-and-cube-even/odd-unindexed-sanbi, "fugaceous" Fibonacci-and-gobi-and-cube-even/odd-unindexed-sanbi, "gallanaceous" gobi-and-luckier-lucky-indexed-nibi-and-cube-even/odd-unindexed-sanbi, "gaseous" gobi-and-sanbi-even/odd-unindexed-sanbi, "liliaceous" lucky-indexed-lucky-ichibi-and-cube-unindexed-even/odd-unindexed-sanbi, "nacreous" nibi-and-cubed-rokubi-even/odd-unindexed-sanbi, "pileous" prime-indexed lucky even/odd-unindexed sanbi, "ranunculaceous" rokubi-and-nibi-unindexed-nibi-cubed-unindexed-lucky-and-cube-even/odd-unindexed-sanbi, "siliceous" sanbi-indexed-lucky-indexed-cube-even/odd-unindexed-sanbi, "tiliaceous" ternarish-indexed-lucky-ichibi-and-cube-even/odd-unindexed-sanbi and "timeous" ternarish-indexed-Mersenne-even/odd-unindexed-sanbi.

1035 "Plucky", on the other hand, is a portmanteau for palindromic luckies.

plucky	1	3	7	9	33	111	141	151	171	303	535	717	727	777	787	979	999

1036 "Pluckable" describes the supersequence including both the formerly feathered "plucked" and the not-yet-plucked "feathered".

pluckable	270	729	3,140	6,859	40,680	59,319	117,649	295,030	493,039	506,960

1037

unplugged	1	2	3	4	6	7	8	9	10	11	12	13	14	16	17	18	19	20	21	22	23	24	25	26

1038 "Bibulous" describes binarish-indexed binary-unindexed lucky odd-unindexed sanbi, "bifarious" binarish-indexed-Fibonacci-and-rokubi-indexed-odd-unindexed-sanbi, "bigamous" binarish-indexed-gobi-and-Mersenne-odd-unindexed-sanbi, "bilious" binarish-indexed lucky-indexed odd-unindexed sanbi, "bus" describes binarish-unindexed sanbi, "cactus" cubes-and-cubed-ternarish-unindexed-sanbi, "calamitous" cubes-and-luckies-amd-Mersennes-indexed ternarish odd-unindexed sanbi, "calamus" cube-and-lucky-and-Mersenne-unindexed-sanbi, "calcaneus" cube-and-lucky-cube-and-nibi-even-unindexed-sanbi, "calculus" cube-and-lucky-cube-unindexed-lucky-unindexed-sanbi, "callous" cubes-and-luckier-lucky-odd-unindexed-sanbi, "callus" cubes-and-luckier-lucky-unindexed-sanbi, "campus" cube-and-Mersenne-prime-unindexed-sanbi, "canaliculus" cube-and-nibi-and-lucky-indexed-cube-unindexed-lucky-unindexed-sanbi, "coitus" cubed odd-indexed ternarish-unindexed sanbi, "cumulus" cube-unindexed Mersenne-unindexed lucky-unindexed sanbi, "curious" cube-unindexed-rokubi-indexed-odd-unindexed-sanbi, "fabulous" Fibonacci-and-binarish-unindexed-lucky-odd-unindexed-sanbi, "factious" Fibonacci-and-cubed-ternarish-ichibi-odd-unindexed-sanbi, "famous" Fibonacci-and-Mersenne-odd-unindexed-sanbi, "famulous" Fibonacci-and-Mersenne-unindexed-lucky-odd-unindexed-sanbi, "fatuous" Fibonacci-and-ternarish-unindexed-odd-unindexed-sanbi, "fetus" Fibonacci/even/ternarish-unindexed-sanbi, "fractious" Fibonacci-rokubi-and-cube-and-cubed-ternarish-ichibi-odd-unindexed-sanbi, "fungus" Fibonacci-unindexed ng-unindexed sanbi, "funiculous" Fibonacci-unindexed nibi-indexed cube-unindexed lucky odd-unindexed sanbi, "furious" Fibonacci-unindexed rokubi-indexed odd-unindexed sanbi, "gallous" gobi-and-luckier-lucky-odd-unindexed-sanbi, "habitus" hachibi-and-binarish-indexed-ternarish-unindexed-sanbi, "hamulus" hachibi-and-Mersenne-unindexed-lucky-unindexed-sanbi, "hibiscus" hachibi-indexed binarish-indexed sanbi cube-unindexed sanbi, "hilarious" hachibi-indexed-luckies-and-rokubi-indexed-odd-unindexed-sanbi, "humus" hachibi-unindexed-Mersenne-unindexed-sanbi, "impious" ichibi Mersenne prime ichibi odd-unindexed sanbi, "lapillus" lucky-and-prime-indexed-luckier-lucky-unindexed-sanbi, "lapsus" lucky-and-prime-sanbi-unindexed-sanbi, "limulus" lucky-indexed Mersenne-unindexed lucky-unindexed sanbi, "litigious" lucky-indexed ternarish-indexed gobi-indexed odd-unindexed sanbi, "lupus" lucky-unindexed prime-unindexed sanbi, "luscious" lucky-unindexed sanbi cubed ichibi odd-unindexed sanbi, "magus" Mersenne-and-gobi-unindexed-sanbi, "malleus" Mersenne-and-lucklier-lucky-unindexed-sanbi, "minibus" Mersenne-indexed nibi-indexed binarish-unindexed sanbi, "minimus" Mersenne-indexed nibi-indexed Mersenne-unindexed sanbi, "miraculous" Mersenne-indexed-rokubi-and-cube-unindexed-lucky-odd-unindexed-sanbi, "mucus" Mersenne-unindexed cube-unindexed sanbi, "multifarious" Mersenne-unindexed-lucky-ternarish-indexed-Fibonacci-and-rokubi-indexed-odd-unindexed-sanbi, "muticous" Mersenne-unindexed ternarish-indexed cubed odd-unindexed sanbi, "nucleus" nibi-unindexed cubed lucky even-unindexed sanbi, "opus" odd prime-unindexed sanbi, "passus" prime-and-sorted-sanbi-unindexed-sanbi, "patulous" prime-and-ternarish-unindexed-lucky-odd-unindexed-sanbi, "pious" prime ichibi odd-unindexed sanbi, "pompus" prime odd Mersenne prime-unindexed sanbi, "pus" prime-unindexed sanbi, "pusillanimous" prime-unindexed-sanbi-indexed-luckier-luckies-and-nibi-indexed-Mersenne-odd-unindexed-sanbi, "ranunculus" rokubi-and-nibi-unindexed-nubi-cubed-unindexed-lucky-unindexed-sanbi, "rubious" rokubi-unindexed binarish ichibi odd-unindexed sanbi, "rufous" rokubi-unindexed Fibonacci-odd-unindexed sanbi, "rumpus" rokubi-unindexed-Mersenne-prime-unindexed-sanbi, "sabulous" sanbi-and-binarish-unindexed-lucky-odd-unindexed-sanbi, "sagacious" sanbi-and-gobi-and-cube-indexed-odd-unindexed-sanbi, "scabious" sanbi-cube-and-binarish-ichibi-odd-unindexed-sanbi, "sinuous" sanbi-indexed nibi-unindexed odd-unindexed sanbi, "situs" sanbi-indexed ternarish-unindexed sanbi, "spirituous" sanbi prime-indexed rokubi-indexed ternarish-unindexed odd-unindexed sanbi, "spurious" sanbi prime-unindexed rokubi-indexed odd-unindexed sanbi, "stimulus" sanbi ternarish-indexed Mersenne-unindexed lucky-unindexed sanbi, "succubus" sanbi-unindexed-cubed-cube-unindexed-binarish-unindexed-sanbi, "talus" ternarish-and-lucky-unindexed-sanbi, "tubulous" ternarish-unindexed binarish-unindexed lucky-unindexed odd-unindexed sanbi, "tumultuous" ternarish-unindexed Mersenne-unindexed lucky ternarish-unindexed odd-unindexed sanbi, "tumulus" ternarish-unindexed Mersenne-unindexedblucky-unindexed sanbi.

"Cuss" describes cube-unindexed sorted sanbi, "fuss" Fibonacci-unindexed sorted sanbi, "huss" hachibi-unindexed sorted sanbi, "puss" prime-unindexed sorted sanbi and "schuss" sorted cubed hachibi-unindexed sorted sanbi.

1039 "Cutaneous" describes cube-unindexed-ternarish-and-nibi-even/odd-unindexed-sanbi, "luminous" lucky-unindexed Mersenne-indexed nibi odd-unindexed sanbi, "manus" Mersenne-and-nibi-unindexed-sanbi, "numinous" nibi-unindexed Mersenne-indexed nibi odd-unindexed sanbi, "rubiginous" rokubi-unindexed binarish-indexed gobi-indexed nibi-odd-unindexed sanbi, "sanguineous" sanbi-and-ng-unindexed/indexed-nibi-even/odd-unindexed-sangbi, "simultaneous" sanbi-indexed-Mersenne-unindexed-lucky-ternarish-and-nibi-even/odd-unindexed-sanbi and "sinus" sanbi-indexed nibi-unindexed sanbi.

1040 "Lush" describes the lucky-unindexed shichibi, "blush" the binarish-unindexed shichibi (without one thousand one hundred seven), "flush" Fibonacci-unindexed shichibi, "Slush" and "slushy" sanbi lucky-unindexed shichibi.

lush	17	37	47	57	77	97	107	117	137	157	177	187	197	217
flush	17	47	77	147	247	307	327	367	427	487	507	627	657	687
slush	17	207	307	507	1,107	1,407	1,507	1,707	2,007	2,107	2,307	2,407	3,307	3,507

1041 "Cosh" describes cubed odd shichibi and "slosh" sorted lucky odd shichibi.

cosh	343	4,913	19,683	50,653	103,823	185,193	300,763	456,533	658,503
shlucky	7	37	67	87	127	237	267	297	307
slosh	7	37	67	78	127	237	267	279	37

1042 "Shop" describes shichibi odd primes, "bop" binarish odd primes, "fop" Fibonacci oodd primes, "mop" Mersenne odd primes and "sop" sanbi odd primes.

shop	7	17	37	47	67	97	107	127	137
bop	11	101	10,111	101,111	1,011,001	1,100,101	10,010,101	10,011,101	10,101,101
cop	27	125	343	1,331	2,187	4,913	6.859	12,167	24,389
fop	3	5	13	89	233	28,657	514,229	24,157,817	...
mop	3	7	31	127	8,191	131,071	524,287	2,147,483,647	137,438,953,471
sop	11	31	41	61	71	101	131	151	18

1043 "Fro" is the complimentary sequence to "to" or ternarish odds.

fro	8	78	88	778	788	798	878	888	898	7,778	7,788	7,878	7,888	8,778	8,788	8,878	8,888
to	1	11	21	101	111	121	201	211	221	1,001	1,011	1,021	1,101	1,111	1,121	1,201	1,211

1044 "Jr" is the complimentary sequence to "sr".

Jr	3	30	31	32	33	43	63	73	83	311	321	331	431	531	631	731	831

1045 "Sr" describes sorted rokubi.

Sr	6	16	26	36	46	56	66	67	68	69	116	126	136	146	156	166	167	168	169
holiness	1	1	1	1	1	1	2	1	3	2	1	1	1	1	1	2	1	3	2

1046 An unintended pun doesn't count, while an intentional one does.
"How many sides does an Oregon have?", "If I pour my rootbeer into a square glass can I get beer?", "Parallel lines have so much in common that it's too bad that they'll never meet." "To make your newborn more mathematical give him or her quadratic formula."

sub	1	12	19	26	30	34	41
broken-up	2	65	132	4,289	351,830	32,372,649	3,626,088,518
trimmed	1	54	21	3,178	24,072	21,261,538	251,577,407
sub	1	24	49	1,578	129,431	11,909,232	1,333,963,418
broken-down	1	3	7	227	18,621	1,713,359	191,914,829
broken	1	2	3	7	65	132	227
hypersum	2	232	23,252	2,325,272	232,527,282	23,252,728,292	23,252,728,292,112
trimmed	1	121	12,141	1,214,161	121,416,171	12,141,617,181	12,141,617,181,001
sub	1	85	8,554	855,420	85,542,007	8,554,200,690	8,554,200,689,814

1047 "Pure and simple" were mated by Mary Ann "George Elliott" Cross in 1860.

1048 The hyphen helps distinguish the hyphenyms put-up and shut-up from the harder to calculate prime-unindexed-ternarish-unindexed primes and shichibi-unindexed-ternarish-unindexed primes. F. H. Hart abbreviated "Put up or shut up" to "P. U. or S. U." in *Sazerne Lying Club* in 1878.

"Put-off" is the supersequence of both neighbors, "putter" the subsequence with two "trits" and "putterer" the one with just one, starting with one, eleven, twenty-two and one eleven.

put-off	2	10	12	19	21	23	99	100	101	102	109	111	113
trimmed	1	0	1	8	1	12	88	0	0	1	8	0	2
putters	20	100	101	110	122	200	202	211	220	221	1,000	1,001	1,010
holiness	1	2	1	1	0	2	1	0	1	0	3	2	2

1049 "Shut-off" similarly is the supersequence of 'shut-up" and "shut-down". "Shutup" describes shut-unindexed primes.

shut-off	0	2	3	9	10	11	12	13	19	21	23	99	101	109	110	111	112	113	119	120
trimmed	1	2	8	0	0	1	2	2	8	10	12	88	0	8	0	0	1	2	8	1
holiness	1	0	0	1	1	0	0	0	1	0	0	2	1	1	1	0	0	0	1	1

1050 The complimentary sequence to the 'ribs' is 'Adams' from "Adam's Rib" by Jeane Macpherson (1923) or by Ruth Gordon and Garson Kanin (1949) and a subsequence of "rubs" is "rubes".

Adams	889	88,989	89,999	888,999	898,889	899,899	88,889,889	88,898,899
holiness	5	8	6	9	10	8	14	13
rubs	1	10	11	100	101	111	1,000	1,001
rubes	10	100	1,000	1,100	1,110	10,010	10,100	10,110
holiness	1	2	3	2	1	3	3	2

1051 "Rugs" describes rokubi-unindexed gobi.

rugs	5	15	25	35	45	65	75	85	95	105	115	125	135	145	165	175	185	195	205	215

1052 "Rush-in" or "Russian" describes the downtown neighbors.

rush-in	6	16	26	36	46	66	76	86	96	106	116	126	136	146	166	176	186	196	206	216

1053 "Offbeat" describes the neighbors of "beat" binarish-even-and-ternarish, "offer" those of even rokubi and "offset" those of sorted even ternarish.

offbeat	0	1	2	3	9	11	12	13	19	20	21	22	23
offer	5	7	15	17	25	27	35	37	45	47	55	57	65
set	2	12	22	112	122	222	1,112	1,122	1,222	2,222	11,112	11,122	11,222
offset	1	3	11	13	111	113	121	123	221	223	1,111	1,113	1,121

1054 "Beirut" describes binarish-even-indexed-rokubi-unindexed-ternarish, "2be" and "not-2be" come from "Hamlet" by William Shakespeare.

be	10	100	110	1,000	1,010	1,100	1,110	10,000	10,010	10,100
2be	210	2,100	2,110	21,000	21,010	21,100	21,110	210,000	210,010	210,100
not-2be	...	209	211	212	213	214	215	216	217	218
beir	96	996	1,096	9,996	10,096	10,996	11,096	99,996	100,096	100,996
Beirut	...	1,119	1,121	1,222	1,223	1,224	1,225	1,226	1,227	1,228

1055 "Barite" described binarish-and-rokubi-indexed-ternarish-evens,

barite		1	10	11	20	100	101	110	111	222	1,000

1056 Pvt. Sad Sack's misadventures were chronicled by Sgt. George Baker. "Coalsack" describes cubed-odds-and-lucky-sanbi-and-cubed-kyuubi), "packsack" primes-and-cubed-kyuubi-sorted-and-cubed-kyuubi, "ransack" rokubi-and-nibi-sorted-and-cubed-kyuubi, "rucksack" and "woolsack" the mates of odd-lucky-sanbi-and-cubed-kyuubi aka "slacks".

sad	106	116	126	136	146	156	166	176	186	196	206	216	226	236	246	256	266	270
holiness	2	1	1	1	1	1	2	1	3	2	2	1	1	1	1	1	2	1

1057 The complimentary sequence of "sick" is "well."

well	75,610	7,853,310	21,046,410	64,388,710	87,991,010	198,234,910	430,277,210
holiness	2	3	3	6	6	5	2

1058 The complimentary sequence of "sticky" is "wicket", mated in 1882.

wicket	270	8,704,970	10,870,670	1,891,563,270	8,232,827,670	8,636,061,970
holiness	1	5	6	5	6	8

1059 The hyphenym "stuck-up" describes the uptown neighbors of "stuck", while "stuckup" describes sanbi ternarish-unindexed cubed kyuubi-unindexed primes.

stuck-up	6,860	970,300	1,685,160	7,880,600	10,503,460	997,0023,000
holiness	5	4	5	8	4	7

1060 "Heads" describes the complimentary sequence of "tails".

heads	7	8	48	68	78	79	87	88	89	108	128	138	158	168	198	218	258	348	378	398	458

"Bail" describes binarish-and-ichibi-luckies, "camail" cubes-and-Mersenne-and-ichibi-luckies, "fail" Fibonacci-and-ichibi-luckies, "fantail" Fibonacci-and-nibi-and-ichibi-luckies, "flail" Fibonacci-luckies-and-ichibi-luckies, "frail" Fibonacci-rokubi-and-ichibi-luckies, "galinggale" gobi-and-lucky-indexed-ng-and-ichibi-luckies, "hail" hachibi-and-ichibi-luckies, "mail" Mersenne-and-ichibi-luckies, "nail" nibi-and-ichibi-luckies, "pail" primes-and-ichibi-luckies, "pintail" primes-indexed-nibi-ternarish-and-ichibi-luckies, "rail" rokubi-and-ichibi-luckies, "sail" sanbi-and-ichibi-luckies, and "snail" sorted-nibi-and-ichibi-luckies, while "tale" describes ternarish-and-lucky-evens, "bale" binarish-and-lucky-evens, "gale" gobi-and-lucky-evens, "gaol" gobi-and-odd-luckies, "hale" hachibi-and-lucky-evens, "kale" kyuubi-and-lucky-evens, "male" Mersenne-and-lucky-evens, "pale" primes-and-lucky-evens, "sale" sanbi-and-lucky-evens, "scale" sanbi-cubes-and-lucky-evens, "shale" shichibi-and-lucky-evens and "stale" sanbi-ternarish-and-lucky-evens.

1061 "Uptick" describes the uptown neighbor of "tick", "tick-off" both neighbors and "tock" the complimentary sequence.

uptick	730	6,860	970,900	1,295,030	1,685,160	7,880,600	9,129,330
holiness	1	5	5	3	5	8	3
tock	270	3,140	2,119,400	8,119,499	8,314,840	8,704,970	9,029,100
holiness	1	1	3	5	5	5	5
tick-off	728	730	6,858	6,860	970,898	97,900	1,295,028
holiness	2	1	5	6	7	4	4

1062 "Basic" describes binarish-and-sanbi-indexed cubes, "clinic" cubed lucky-indexed nibi-indexed cubes, "cubic" cubic-unindexed binarish-indexed cubes, "cyanic" cubed-yongbi-and-nibi-indexed-cubes, "hic" hachibi-indexed cubes, "italic" ichibi-ternarish-and-lucky-indexed-cubes, "manic" Mersenne-and-nibi-indexed cubes, "mimic" Mersenne-indexed Mersenne-indexed cubes, "music" Mersenne-unindexed sanbi-indexed cubes, "panic" prime-and-nibi-indexed cubes, "runic" rokubi-unindexed nibi-indexed cubes, "rustic" rokubi unindexed sanbi ternarish-indexed cubes, "sic" sanbi-indexed cubes, and "tactic" ternarish-and-cubed-ternarish-indexed-cubes.

1063 "Beanie" describes binarish-even-and-nibi-indexed-evens, "beastie" binarish-even-and-sanbi-ternarish-indexed-evens, "billie" binarish-indexed-luckier-lucky-indexed-evens, "challie" cubed-hachibi-and-luckier-lucky-indexed-evens, "cheapie" cubed-hachibi-evens-and-prime-indexed-evens, "collie" cubed odd luckier lucy-indexed evens, "cosie" cubed odd sanbi-indexed evens, "curie" cube-unindexed-rokubi-indexed-evens, "cutie" cube-unindexed-ternarish-indexed-evens, "eerie" evenly even rokubi-indexed evens, "Erie" even rokubi-indexed evens, "facie" Fibonacci-and-cube-indexed evens, "faerie" Fibonacci-and-even-rokubi-indexed-evens, "fantasie" Fibonacci-and-nibi-ternarish-and-sanbi-indexed-evens, "fogie" Fibonacci odd-indexed evens, "gillie" gobi-indexed-luckier-lucky-indexed-evens, "glassie" gobi-lucky-and-sorted-sanbi-indexed-evens, "goalie" gobi-odds-and-lucky-indexed-evens, "hackie" hachibi-and-cubed-kyuubi-indexed-evens, "heinie" hachibi-evens-indexed-nibi-indexed-evens, "kiltie" kyuubi-indexed lucky ternarish-indexed evens, "lassie" lucky-and-sorted-sanbi-indexed-evens, "mamie" Mersenne-and-Mersenne-indexed-evens, "mashie" Mersenne-and-shichibi-indexed-evens, "muskie" Mersenne-unindexed-sorted-kyuubi-indexed-evens, "okie" odd kyuubi-indexed evens, "outlie" odd-unindexed ternarish lucky-indexed evens, "pantie" prime-and-nibi-ternarish-indexed-evens, "talkie" ternarish-and-lucky-kyuubi-indexed-evens. The complimentary sequence of "eerie" is "indiana" from "Eerie, Indiana" by José Rivera and Karl Schaefer, of "pantie" is "hose" and of "talkie" is "silent".

1064

tied-up	3	5	21	23	25	41	43	45	201	203	205	221	223	225	241	243	245	401	403	405
tied-down	2	3	19	21	23	39	41	43	199	201	203	219	221	223	239	241	243	399	401	403
trimmed	1	2	8	10	12	28	30	32	88	10	12	108	110	112	128	130	132	288	30	32

1065 "Blastoff" describes binarish-lucky-and-sanbi-ternarish neighbors, "boiloff" binarish odd-indexed lucky neighbors, "castoff" cube-and-sanbi-ternarish neighbors, "cutoff" cube-unindexed ternarish neighbors, "falloff" Fibonacci-and-luckier-luckies neighbors, "kickoff" kyuubi-unindexed cubed kyuubi neighbors, "layoff" lucky-and-yongbi neighbors, "liftoff" lucky-indexed Fibonacci ternarish neighbors, "payoff" prime-and-yongbi neighbors, "pickoff" prime-indexed cubed kyuubi neighbors, "putoff" prime-unindexed ternarish neighbors, "ripoff" rokubi-indexed prime neighbors, "runoff" rokubi-unindexed nibi neighbors, "scoff" sanbi cube neighbors, "setoff" sorted even ternarish neighbors, "shutoff" shichibi-unindexed ternarish neighbors, "spinoff" sanbi prime-indexed nibi neighbors, "tipoff" ternarish-indexed prime neighbors and "toff" ternarish neighbors.

1066 "Tune" describes ternarish-unindexed nibi even "lune" lucky-unindexed nibi evens, picayune" prime-indexed-cubes-and-yongbi-unindexed-nibi-evens, "rune" rokubi-unindexed nibi evens and "teen" ternarish evenly even nibi. "fifteen" Fibonacci-indexed Fibonacci ternarish evenly even nibi, "lateen" lucky-and-ternarish-evenly-even-nibi, "sateen" sanbi-and-ternarish-evenly-even-nibi.

1067 Tit and tat were mated in 1556 by John Heywood in *The Spider and the Flie*. "Tut" describes ternarish-unindexed ternatish, alluding to King "Tut" Tutankamen discovered in 1923 by Earl of Carnarvon, etal. and to Steve Martin's classic song and dance from 1978 soon called "tutting".

tut	11	12	20	21	22	100	110	111	112	120	121	122	200	201	202	210	211	212	220	221

1068 Thomas "Tubby" Thompkins appeared with Lulu Moppet in "Little Lulu" by John Stanley.

1069 The tautonym "Lulu" suggests GRUsome "tautonums" or "repfrags" with repeated frags that are not merely repdigits.

tautonum	1,010	1,212	1,313	1,414	1,515	1,616	1,717	1,818	1,919	2,020	2,121
holiness	2	0	0	0	0	2	0	4	2	2	0

1070 "Tucks" re-spelled "tux" can be superpularized as "tuxes".

tux	2,438,959,319	5,931,924,389	11,764,924,289	10,537,959,319	105,379,242,389
holiness	5	5	5	4	5

tuxes	24,389,593,195,931,924,389	59,319,243,892,438,959,319	1,176,492,428,924,398,959,319
holes	9	10	11

1071 Friar Tuck refers to a character in the play "Robin Hood and the Sheriff" (bef. 1475), etc.

1072 "Tuckin", without a hyphen, on the other hand, would be ternarish-unindexed cubed kyuubi-indexed nibi.

tuckin	243,882	593,182	1,176,482	2,053,782	3,285,082	4,930,382	7,049,682
holiness	4	3	3	3	5	4	5

1073 "Boat" is NOT the complimentary sequence to "tug", since it describes binarish-odd-and-ternarish, "coat" cubes-odd-and-ternarish, "bloat" binary-luckies-odd-and-ternarish, "float' Fibonacci-lucky-odds-and-ternarish, "gloat" gobi-lucky-odds-and-ternarish, "goat" gobi-odd-and-ternarish, "moat" Mersenne-odds-and-ternarish, "oat" odd-and-ternarish, "shoat" shichibi-odds-and-ternarish and"stoat" sanbi-ternarish-odds-and-ternarish.

1074 Rin-Tin-Tin refers to the canine movie star of "The Man from Hell River" (1922). Dehyphenated "rintintin" describes rokubi-indexed-nibi-ternarish-indexed-nibi-ternarish-indexed-nibi. Some members preferred his nickname, "Rinty", which would simply be rokubi-indexed nibi times ten.

rinty	520	1,520	2,520	3,520	4,520	5,520	6,520	7,520	8,520	9,520	10,520
holiness	1	1	1	1	1	1	2	1	3	2	2

Similar decimatable sequences, divisible by ten, are: "banality", "basicity", "batty", "bearability", "beatability", "bitty", "busty", "butty", "calamity", "canty", "capability", "capacity", "capillarity", "catty", "causality", "chatty", "charity", "charitability", "circuity", "city", "clarity", "classifiability", "criminality", "cruelty", "crusty", "culpability", "curability", "curiosity", "cutty", "eatability", "elasticity", "erasability", "facility", "faculty", "fallibility", "fatality", "fatty", "fatuity", "fealty", "feasibility", "felicity", "fifty", "finality", "flinty", "footy", "fragility", "friability", "frugality", "fugacity", "fusty", "fruity", "fusibility", "futility", "futurity", "guaranty", "gullibility", "gusty", "gutty", "habitability", "hasty", "hearty", "heritiability", "hilarity", "hospitality", "humanity", "humility", "illimitability", "kissability", "kitty", "laity", "liability", "liftability", "likability", "litigability", "locality", "lusty", "malleability", "manipulability", "masulinity", "maturity", "minty", "miscibility", "mistakability", "misty", "mitigatability", "moiety", "molality", "multiplicity", "municipality", "muscularity", "musty", "nasty", "natality", "natty", "palatability", "nicety", "nifty", "nitty", "notability", "nullity", "nutty", "omissibility", "palpability", "parity", "party", "passability", "pasty", "patty", "payability", "peaceability", "perishability", "piety", "pitiability", "placability", "plasticity", "platy", "plausibility", "playability", "pliability", "plurality", "possibility", "potability", "publicity", "puerility", "punishability", "purity", "pusillanimity", "putty", "rapacity", "rascality", "rateability", "ratty", "reachability", "reality", "realty", "risibility", "rusty", "rutty", "sacristy", "sagacity", "salinity", "sanity", "saturability", "scanty", "scatty", "shifty", "shinty", "shirty", "shitty", "similarity", "simplicity", "singularity", "sinuosity", "slaty", "smutty", "society", "sooty", "stability", "sublimity", "suitability", "tasty", "tatty", "teachability", "tearability", "touristy", "tunability", "tutty", "usability" and "utility".

1075 sextilove

sextilove	111,110	111,120	111,130	111,140	111,150	111,160	111,170	111,180
holiness	1	1	1	1	1	2	1	3

1076 dextrilove

dextrilove	100,000	200,000	300,000	400,000	500,000	600,000	700,000	800,000
holiness	5	5	5	5	5	6	5	7

1077 dodrilove

dodrilove	1,000	2,000	3,000	4,000	5,000	6,000	7,000	8,000	9,000	10,000,010
holiness	3	3	3	3	3	4	3	5	4	6

1078 "I am my Beloved's and His desire is for me." (Sgs 7:10)

1079 sextichibi

sextichibi	200,001	200,021	200,031	200,041	200,051	200,061	200,071	200,081
holiness	4	3	3	3	3	4	3	5

dextrichibi	2,000,000,011	2,000,000,101	2,000,000,121	2,000,000,131
holiness	7	7	6	6

1080 bichibi

bichibi	101	121	131	141	151	161	171	181	191	211	311	411	511	611	711	811	911
holiness	1	0	0	0	0	1	0	2	1	0	0	0	0	1	0	2	1

1081 dodrichibi

dodrichibi	1,011	1,101	1,121	1,131	1,141	1,151	1,161	1,171	1,181	1,191	1,211
holiness	1	1	0	0	0	0	1	0	2	1	0

1082 sextinibi

sextinibi	100,002	100,012	100,032	100,042	100,052	100,062	100,072	100,082
holiness	4	3	3	3	3	4	3	5

1083 benibi

benibi	122	202	212	232	242	252	262	272	282	292	322	422	522	622	722	822
holiness	0	1	0	0	0	0	1	0	2	1	0	0	0	1	0	2

1084 dodrinibi

dodrinibi	100,002	100,012	100,032	100,042	100,052	100,062	100,072	100,082
holiness	4	3	3	3	3	4	3	5

1085 sextisanbi

sextisanbi	100,003	100,013	100,023	100,043	100,053	100,063	100,073	100,083
holiness	4	3	3	3	3	4	3	5

dextrisanbi	133,333	233,333	303,333	313,333	323,333	330,333	331,333	332,333
holiness	0	0	1	0	0	1	0	0

1086 besanbi

besanbi	133	233	303	313	323	343	353	363	373	383	393	433	533	633	733	833	933
holiness	0	0	1	0	0	0	0	1	0	2	1	0	0	1	0	2	1

1087 dodrisanbi

dodrisanbi	1,333	2,333	3,033	3,133	3,233	3,303	3,313	3,323	3,343	3,353	3,363	3,373
holiness	0	0	1	0	0	1	0	0	0	0	1	0

1088 sextiyongbi

sextiyongbi	100,006	100,016	100,026	100,036	100,046	100,056	100,076	100,086
holiness	5	4	4	4	4	4	4	6

dextriyongbi	166,666	266,666	366,666	466,666	566,666	606,666	616,666	626,666
holiness	5	5	5	5	5	6	5	5

1089 beyongbi

beyongi	144	244	344	404	414	424	434	454	464	474	484	494	544	644	744	844	944
holiness	0	0	0	1	0	0	0	0	1	0	2	1	0	1	0	2	1

1090 dodriyongbi

dodriyongi	1,444	2,444	3,444	4,044	4,144	4,244	4,344	4,404	4,414	4,424	4,434
holiness	0	0	0	1	0	0	0	1	0	0	0

1091 sextigobi

sextigobi	10,005	10,015	10,025	10,035	10,045	10,065	10,075	10,085	10,095	10,105
holiness	3	2	2	2	2	3	2	4	3	2

dextrigobi	15,555	25,555	35,555	45,555	50,555	515,555	525,555	535,555	545,555
holiness	0	0	0	0	1	0	0	0	0

1092 begobi

begobi	155	255	355	455	505	515	525	535	545	655	755	855	955	105,555	115,555
holiness	0	0	0	0	1	0	0	0	0	1	0	2	1	1	0

1093 dodrigobi

dodrigobi	1,555	2,555	3,555	4,555	5,055	5,155	5,255	5,355	5,455	5,505	5,515
holiness	0	0	0	0	1	0	0	0	0	1	0

1094 sextirokubi

sextirokubi	100,006	100,016	100,026	100,036	100,046	100,056	100,076	100,086
holiness	5	4	4	4	4	4	4	6

1095 berokubi

berokubi	166	266	366	466	566	606	616	626	636	646	656	766	866	966	100,066
holiness	2	2	2	2	2	3	2	2	2	2	2	2	4	3	5

1096 dodrirokubi

dodrirokubi	1,666	2,666	3,666	4,666	5,666	6,066	6,166	6,266	6,366	6,466	6,566
holiness	3	3	3	3	3	4	3	3	3	3	3

1097 sextishichibi

sextishichi	100,007	100,017	100,027	100,037	100,047	100,057	100,067	100,087
holiness	4	3	3	3	3	3	4	5

dextrishichibi	177,777	277,777	377,777	477,777	577,777	677,777	707,777	717,777

holiness	0	0	0	0	0	1	1	0

1098 beshichibi

beshichibi	177	277	377	477	577	677	707	717	727	737	747	757	767	787	797	877	977
holiness	0	0	0	0	0	0	1	0	0	0	0	0	1	2	1	2	1

1099 dodrishichibi

dodrishichibi	1,777	2,777	3,777	4,777	5,777	6,777	7,077	7,177	7,277	7,377	7,477
holiness	0	0	0	0	0	1	1	0	0	0	0

1100 sextihachibi

sextihachibi	100,008	100,018	100,028	100,038	100,048	100,058	100,068	100,078
holiness	6	5	5	5	5	5	6	5

dodrihachibi	188,888	288,888	388,888	488,888	588,888	688,888	788,888	808,888
holiness	10	10	10	10	10	11	10	11

1101 behachibi

behachibi	188	288	388	488	588	688	788	808	818	828	838	848	858	868	878	898
holiness	4	4	4	4	4	5	4	5	4	4	4	4	4	5	4	5

1102 dodrihachibi

dodrihachibi	1,888	2,888	3,888	4,888	5,888	6,888	7,888	8,088	8,188	8,288	8,388
holiness	6	6	6	6	6	7	6	7	6	6	6

1103 sextikyuubi

sextikyuubi	100,009	100,019	100,029	100,039	100,049	100,059	100,069	100,079
holiness	5	4	4	4	4	4	5	4

1104 bekyuubi

bekyuubi	199	299	399	499	599	699	799	899	100,099	100,199	100,299	100,399
holiness	2	2	2	2	2	3	2	4	5	4	4	4

1105 dodrikyuubi

dodrikyuubi	1,999	2,999	3,999	4,999	5,999	6,999	7,999	8,999	9,099	9,199	9,299
holiness	3	3	3	3	3	4	3	5	4	3	3

1106 "Fractal sequences and interspersions" by Clark Kimberling

1107 odious

odd cull	1	1	2	2	4	2	7	4	8	2	11	7	13	4	14	8	16	2	19	11	21	7	25	13	26
crescendo	1	1	2	1	2	4	1	2	4	7	1	2	4	7	8	1	2	4	7	8	11	1	2	4	7
decrescendo	1	2	1	4	2	1	7	4	2	1	8	7	4	2	1	11	8	7	4	2	1	13	11	8	7

panodd

odd cull	1	1	3	3	5	3	7	5	9	3	11	7	13	5	15	9	17	3	19	11	31	7	33	13
crescendo	1	1	3	1	3	5	1	3	5	7	1	3	5	7	9	1	3	5	7	9	11	1	3	5
decrescendo	1	3	1	5	3	1	7	5	3	1	9	7	5	3	1	11	9	7	5	3	1	13	11	9

odd numberdromes

podd	1	3	5	7	9	11	33	55	77	99	101	111	121	131	141	151	161	171
odd cull	1	1	3	3	5	3	7	5	9	3	11	7	33	5	55	9	77	3
crescendo	1	1	3	1	3	5	1	3	5	7	1	3	5	7	9	1	3	7
decrescendo	1	3	1	5	3	1	7	5	3	1	9	7	5	3	1	11	9	7

"Cull" without modifiers describes cube-unindexed luckier luckies.

cull	7	13	21	31	37	63	75	87	99	111	159	169	189	195	219	237	259

1108 crescendo

odd cull	1	1	1	1	2	1	1	2	2	1	3	1	1	2	2	2	3	1	4
crescendo	1	1	1	1	1	2	1	1	2	1	1	1	2	1	2	1	1	2	1
decrescendo	1	1	1	2	1	1	1	2	1	1	2	1	2	1	1	3	2	1	1

1109 decrescendo

odd cull	1	1	2	2	1	2	3	1	2	2	1	3	4	1	3	2	2	2	1
crescendo	1	1	2	1	2	1	1	2	1	3	1	2	1	3	2	1	2	1	3
decrescendo	1	2	1	1	2	1	3	1	2	1	2	3	1	2	1	1	2	3	1

1110

crescendo	1	3	1	3	5	1	3	5	7	1	3	5	7	9	1	3	5	7	9	11	1	3	5	7	9	11	13	1	3
decresendo	1	3	5	3	1	7	5	3	1	7	5	3	1	9	11	9	7	5	3	1	13	11	9	7	5	3	1	15	13

1111 ichibi

crescendo	1	21	1	21	41	1	21	41	61	1	21	41	61	81	1	21	41	61	81	101	1	21	41
decresendo	1	21	41	21	1	61	41	21	1	81	61	41	21	1	101	81	61	41	21	1	121	101	81

1112 linears

crescendo	1	4	1	4	7	1	4	7	11	1	4	7	11	14	1	4	7	11	14	17	1	4	7	11	14	17
decresendo	1	4	1	7	4	1	11	7	4	1	14	11	7	4	1	17	14	11	7	4	1	41	17	14	11	7

1113

crescendo sum	1	2	4	5	7	10	11	13	16	20	21	23	25	30	35	36	38	41	45	50	56	57	59	62	66

decrescendo sum	1	3	4	7	9	10	14	17	19	20	25	29	32	34	35	41	46	50	53	55	56	63	69

1114 cresendo

hypersum	1	12	124	1,245	12,457	1,245,710	124,571,011	12,457,101,113

decrescendo

hypersum	1	12	121	1,213	12,132	121,321	1,213,214	12,132,143

1115 GRUs

base 5	1	6	31	156	781	3,906	19,531	97,656	488,281
base 6	1	7	43	259	1,555	9,331	55,987	335,923	20,193,235
base 7	1	8	57	400	2,801	19,608	137,257	960,800	6,725,601
base 8	1	9	73	585	4,681	37,449	299,593	2,396,745	19,173,961
base 9	1	10	91	820	7,381	66,430	597,871	5,380,840	48,427,561
base 10	1	11	111	1,111	11,111	111,111	1,111,111	11,111,111	111,111,111
base 11	1	12	135	1,464	16,105	177,165	1,948,717	21,435,888	235,794,769
base 12	1	13	157	1,885	22,641	271,653	3,257,437	39,089,245	469,070,941

1116 more diagonals

2nd	1	3	13	85	781	9,331	137,257
3rd	1	4	21	156	1,555	19,608	299,593
4th	1	5	31	259	2,801	37,449	597,871
5th	1	6	43	400	4,681	66,430	1,111,111
6th	1	7	57	585	7,381	111,111	1,948,717
7th	1	8	73	820	11,111	177,165	3,257,437

1117 Even though the first four are prime, the last known double Mersenne prime was not discovered by Lucas until 1876.

M(M(p(4)))	170,141,183,460,469,231,731,687,303,715,884,105,727

1118

2nd diagonal	1	2	13	156	2,801	66,430	1,948,717

3rd	1	3	21	259	4,681	111,111	3,257,437
4th	1	4	31	400	7,381	177,165	960,800
5th	1	5	43	585	11,111	55,987	...

1119

base 6	1	55,987	> forty-three hundredplex
base 7	1	960,080	> eight-hundred-thousandplex
base 8	1	19,173,961	> seventeen-millionplex
base 9	1	435,848,050	> four-hundred-millionplex

1120 Catalan-Mersenne numbers

trimmed	1	2	6	16	6,030,072,353,581,206,205,762,260,477,304,616

1121 Fermat numbers

8th Fermat	340,282,366,920,938,463,463,374,607,431,768,211,457
holiness	17

1122 Four hyperdivides into two-two, twenty-two in base ten. It hyperdivides into one-two-eleven, since two-eleven (two hundred eleven) is a prime. One thousand, two hundred eleven hyperdivides into seven thousand, one hundred seventy-three. Adding a one as a head is called monocephalization. Doubling the head, whatever it may be, is called dicephalizatin and tripling it tricephalization.

4	2·2	1·2·11	7·173	1·3·3·797	3·103·433	337·9209
	22	1,211	7,173	133,797	3,103,433	3,379,209
5	1·5	3·5	1·1·5·7	13·89	1·1·3·463	3·3·71·801
	15	35	1,157	1,389	113,463	3,371,801
6	3·2	2·2·2·2·2	41·271·2	2·2·51589	29·77641	3·3·3·3·36761
	32	22,222	412,712	2,251,589	2,977,641	333,336,761
7	1·1·7	1·1·3·3·13	3·107·353	59·52667	7·7·41·2963	3·23·1121927
	117	113,313	3,107,353	5,952,667	77,412,963	3,231,121,927
8	2·2·2	2·3·37	3·19·41	1·3·3·7·13	1·3·3·83·179	41·191·1709
	222	2,337	31,941	133,713	13,383,179	411,911,709
9	3·3	3·11	1·3·11	3·19·23	3·3·3547	281·1187
	33	311	1,311	31,923	333,547	2,811,187

cephalation

mono-	110	111	112	113	114	115	116	117	118	119	120	121	122

bi-	110	111	112	113	114	115	116	117	118	119	220	221	222
tri-	1,110	1,111	1,112	1,113	1,114	1,115	1,116	1,117	1,118	1,119	2,220	2,221	2,222

1123

hyperdivided trimmed	1	4	11	12	14	16	22	24	0	2	6	18	111	112	120	122	124
ex-primes trimmed	1	4	0	2	6	8	12	18	30	32	36	42	48	30	40	72	78

Hyperadding a one as both a head and a tail is called augmentation as with the augmentable primes 9,091 and 90,901 and 5,882,353. Adding just one as a tail Joyce called "ugmentation".

augmented	111	121	131	141	151	161	171	181	191	1,101	1,111	1,121	1,131	1,141
ugmented	11	21	31	41	51	61	71	81	91	101	111	121	131	141

1124 factorial primes

tressub	1	2	3	4	6	7
broken-up	23	852	19,619	726,755	82,142,934	11,254,308,713
tressub	1	42	977	36,183	4,089,656	560,319,037
broken-down	1	24	553	20,485	2,315,358	317,224,531
broken	1	23	24	553	852	19,619
hypersum	23	2,337	233,753	23,375,373	23,375,373,113	23,375,373,113,137
tressub	1	116	11,638	1,163,791	1,163,791,299	1,163,791,299,308

1125 The first factorial prime factorial has a sub depth of fifty-two and the second of seventy-two and a holiness of twenty-three.

2nd factorial prime factorial	13,763,753,091,226,345,046,315,979,581,580,902,400,000,000

1126 superfactorial primes

tressub	1	33	1,229	65,121
broken-up	23	15,342	352,889	235,392,305
tressub	1	764	17,569	11,719,493
broken-down	1	24	553	368,875
broken	1	23	24	553
hypersum	23	23,667	2,366,724,679	23,667,246,791,307,987
tressub	1	1,178	117,832,283	1,178,322,834,077,960

1127 The phrase "bag and baggage" comes from the military meaning the soldiers' own individual possessions and that of the army as a whole. "Fag" describes Fibonacci-and-gobi, "flag" Fibonacci-luckies-and-gobi, "fleabag" Fibonacci-lucky-evens-and-binarish-and-gobi, "frag" Fibonacci-rokubi-and-gobi, "gasbag" gobi-and-sanbi-binarish-and-gobi, "hag" hachibi-and-gobi, "lag" luckies-and-gobi, "mag" Mersenne-and-gobi, "mailbag" Mersenne-and-ichibi-lucky-binarish-and-gobi, "nag" nibi-and-gobi, "rag" rokubi-and-gobi, "slag" sanbi-luckies-and-gobi, "snag" sorted-nibi-and-gobi, "stag" sanbi-ternarish-and-gobi and "stalag" sanbi-ternarish-and-luckies-and-gobi.

1128 Although the acronum "bag" describes exactly the same sequence as "gab", "cab" describes cubes-and-binarish and "cabbage" its complimentary sequence, "blab" binarish-luckies-and-binarish, "crab" cubed-rokubi-and-binarish, "fab" Fibonacci-and-binarish, "flab" Fibonacci-luckies-and-binarish, "lab" luckies-and-binarish, "nab" nibi-and-binarish, "prefab" Fibonacci-and-binarish downtown neighbors, "scab" sanbi- cubes-and-binarish, "slab" sanbi-luckies-and-binaeish, "stab" sanbi-ternarish-and-binarish and "tab" ternarish-and-binarish

1129 "Bar and grill" is said to have referred to the place where liquor was served and where it was stored under lock and key, but has come to mean where liquor and food, not only grilled food, are served. "Bear" describes binarish-evens-and-rokubi, '"blare" binarish-luckies-and-rokubi-evens, 'car" cubes-and-rokubi, "carfare" cubes-and-rokubi-Fibonacci-and-rokubi-evens, "fanfare" Fibonacci-and-nibi-Fibonacci-evens, "far" Fibonacci-and-rokubi, "flare" Fibonacci-luckies-and-rokubi-evens, "glare" gobi-luckies-and-rokubi-evens, "hare" hachibi-and-rokubi-even, "mar" Mersenne-and-rokubi, "pare" primes-and-rokubi-evens, "pear" prime-even-and-rokubi, "prepare" the downtown neighbors of "pare", "scar" sanbi-cubes-and-rokubi, "share" shichibi-and-rokubi-evens, "snare" sorted-nibi-and-rokubi-evens, "spare" sanbi-primes-and-rokubi-evens, "stare" sanbi-ternarish-and-rokubi-ebvens, "tare" ternarish-and-rokubi-evens and "tear" ternarish-evens-and-rokubi.

1130 "Rill" describes rokubi-indexed luckier luckies, "bill" binarish-indexed luckier luckies, "bluebill" binarish lucky-unindexed even binarish-indexed luckier luckies, "downhill" hachibi-indexed luckier luckies downtown neighbors, "frill" Fibonacci rokubi-indexed luckier luckies, "hill" hachibi-indexed luckier luckies, "ill" ichibi luckier luckies, "kill' kyuubi-indexed luckier luckies, "overkill" kyuubi-indexed luckier luckies uptown neighbors, "overspill" sanbi prime-indexed luckier luckies uptown neighbors, "sill' sanbi-indexed luckier luckies, "still" sanbi ternarish-indexed luckier luckies, "till" ternarish-indexed luckier luckies and "uphill" hachibi-indexed luckier luckies uptown neighbors.

1131 "Rash" describes rokubi-and-shichibi, "bash" binarish-and-shichibi, "calabash" cubes-and-luckies-and-binarsh-and-shichibi, "calash" cubes-and-luckies-and-shichibi, "callipash" cubes-and-luckier-luckies-indexed-primes-and-shichibi, "cash" cubes-and-shichibi, "clash" cubed-luckies-and-sjichibi, "crash" cubed-rokubi-and-shichibi, "flash" Fibonacci-luckies-and-shichibi, "gash" gobi-and-shichi, "goulash" gobi-odd-unindexed-luckies-and-shichibi, "hash" hachibi-and-shichibi, "lash" luckies-and-shichibi, "potash" prime-odd-ternarish-and-shichibi, "rash" rokubi-and-shichibi, "sash" sanbi-and-shichibi, "slash" sanbi-luckies-and-shichibi, "smash" sanbi-Mersenne-luckies-and-shichibi, "splash" sanbi-prime-luckies-and-shichibi, "stash" sanbi-ternarish-and-shichibi and "succotash" sanbi-unindexed-cubed-cubed-odd-ternarish-and-shichibi.

1132 "Boot camp" came from the Spanish-American war where sailors' leggings were called "boots" and the camps where recruits were trained became known as boot camps. "Clamp" describes cubed-luckies-and-Mersenne-primes, "cramp" cubed-rokubi-and-Mersenne-primes, "lamp" luckies-and-Mersenne-primes, "ramp" rokubi-and-Mersenne-primes, "scamp" sanbi-cubes-and-Mersenne-primes, "stamp" sanbi-ternarish-and-Mersenne-primes and "tamp" ternarish-and-Mersenne-primes.

1133 The complimentary sequence for "boo" is "hoo" from "boohoo" meaning in 1520 either laughter or noisy weeping, more recently a reference to the Hawaiian detective Harry Hoo ("The Amazing Harry Hoo" and "Hoo Done It" by Gerald Gardner and Dee Caruso).

"Baboo" describes "binarish-and-binarish-oddly-odds, "bugaboo" binarish-unindexed gobi-and-binarish-oddly-odds, "cashoo" cubes-and-shichibi-oddly-odds, "cockatoo" cubed-odd-cubed-kyuubi-and-ternarish-oddly-odds, "coo" cubed oddly odds, "cuckoo" cube-unindexed cubed kyuubi oddly odds, "goo" gobi oddly odds, "halloo" hachibi-and-luckier-lucky-oddly-odds, "hullabaloo" hachibi-unindexed luckier-lucky-and-binarish-and-lucky-oddly-odds, "loo" lucky oddly odds, "moo" Mersenne oddly odds, "shoo" shichibi oddly odds, "shampoo" shichibi-and-Mersenne-prime-oddly-odds, "taboo" ternarish-and-binarish-oddly-odds, "tatoo" ternarish-and-ternarish-oddly-odds, "too" ternarish oddly odds and "yahoo" yongbi-and-hoo.

1134 The complimentary sequence of "bo" is "peep" from "Little Po Peep". It was a hide-and-seek children's game refered to to in "King Lear" by William Shakespeare from the earlier meaning of "play bo peep" being sent to the pillary for public humiliation.

peep	8	88	888	898	8,888	8,898	8,988	8,998	89,988	89,989
holiness	2	4	6	5	8	7	7	6	8	7

1135 "Cash and carry" comes from FDR's policy for trade with belligerent nations before WWII restricting it to cash payment with delivery not included.

1136 "Cloaks" re-spelled as "cloax" can be superpluralized to "cloaxes".

cloaxes	19	91	191	199	271	279	291	299	391	399	491	499	591	599	691	699	791
holiness	1	1	1	2	0	1	1	2	1	2	1	2	1	2	2	3	1

"Daggers" were mated to "cloaks" by Charles Dickens in Barnaby Rudge in 1841, but set in 1780. In 1840 Henry Wadsworth Longfellow translated the French "cape et épée" and Spanish "capa y espada" into English. The superhero team Tyrone Johnson and Tandy Bowen took them as their noms de guerre. Triple agent Carlos Capa y Espada had them both naturally.

daggers	8	10	20	30	40	50	60	70	72	90	100	110	120	130	140	150	160	170	172
holiness	2	1	1	1	1	1	2	1	0	1	2	1	1	1	1	1	2	1	0

1137 "Corn" and "corny" describes the complimentary sequence of "cob". Those who find "corny" offensive like to re-spell "cob" as "Cobb" and so re-name the complimentary sequence "Ty" for baseball hall-of-famer Tyrone "Ty" Cobb.

Ty	8	878	8,632,368	8,969,698	8,628,669,368	8,665,366,698	8,996,996,998
holiness	2	4	6	9	11	10	12

1138 "Cocks"re-spelled "cox" can be superpluralized as "coxes".

coxes	47,045,881,631,441	63,144,147,045,881	631,441,594,823,321
holiness	6	6	4

1139 "En" describes even nibi, "beaten" binarish-even-and-ternarish-even-nibi, "befallen" binarish-even-Fibonacci-and-luckier-lucky-even-nibi, "chapfallen" cubed-hachibi-and-prime-Fibonacci-and-luckier-lucky-even-nibi, "chasten" cubed-hachibi-and-ternarish-even-nibi, "downfallen" Fibonacci-and-luckier-lucky-even-nibi downtown neighbors, "eaten" evens-and-ternarish-even-nibi, "fallen" Fibonacci-and-luckier-lucky-even-nibi, "fasten" Fibonacci-and-soprted-ternarish-even-nibi, "fen" Fibonacci even nibi, "cheapen" cubed-hachibi-evens-and-prime-even-nibi, "fifteen" Fibonacci-indexed Fibonacci ternarish even nibi, "glisten" gobi lucky-indexed sorted ternarish even nibi, "gluten" gobi lucky-unindexed ternarish even nibi, "laten" lucky-and-ternarish-even-nibi, "lenten" lucky even nibi ternarish even nibi, "lien" lucky-indexed even nibi, "linen" lucky-indexed nibi even nibi, "mien" Mersenne-indexed even nibi, "moisten" Mersenne odd-indexed sorted ternarish nibi, "neaten" nibi-even-and-ternarish-even-nibi, "oaten" odd-and-ternarish-even-nibi, "paten" prime-and-ternarish-even-nibi, "pen" prime even nibi (just one, two), "refasten" rokubi-even-Fibonacci-and-sorted-ternarish-even-nibi, "sateen" sanbi-and-ternarish-evenly-even-nibi, "teen" ternarish evenly even nibi, "ten" ternarish even nibi and "uneven" odd.

1140 Cogs are toothed wheels and sprockets toothed cylinders, particularly the futuristic Cogswell Cogs and Spacely Sprockets popularized on "The Jetsons" by Joseph Barbera and William Hanna. "Catalog" describes cubes-and-ternarish-and-lucky-odd-gobi, "clog" cubed lucky odd gobi, "flog" Fibonacci lucky odd gobi, "fog" Fibonacci odd gobi, "slog" sorted lucky odd gobi, "snog" sorted nibi odd gobi.

1141 "Billet" describes binarish-unindexed luckier lucky even ternarish, "bluet" binarish lucku-unindexed even ternarish, "cabinet" cubes-and-binarish-indexed nibi even ternarish, "cruet" cubed rokubi-unndexed even ternarish, "parapet" primes-and-rokubi-and-prime-even-ternarish, "planet" prime-lucky-and-nibi-even-ternarish, "spinet" sanbi-indexed nibi even ternarish and "suet" sanbi-unindexed even ternarish.

1142 "Cooks" can be re-spelled "coox" and superpluralized to "cooxes".

cooxes	117,649,729	729,117,649	704,969,729	729,704,969	2,146,689,729

holiness	3	3	5	5	6

1143 "Cookies" describe cubed oddly odd kyuubi-indexed evens and "cream" cubed-rokubi-evens-and-Mersennes. Malcolm Stoco Associates in "Who is Malcolm" credits Malcolm Stoco with mixing cookie crumbs into ice cream in 1976. Blue Bell Creameries popularized the ice cream flavor by 1980 and trademarked it as "Cookies 'n' Cream".

Sidney aka "Cookie Monster", a resident of Seseme Street since 1969, provides a better and irreplaceable complimentary sequence name.

monster	8,541	190,061	764,701	10,346,381		

"Babies" describes binarish-and-binarish-indexed-evens, "belies" binarish even luckiy-indexed evens, "buries" binarish-unindexed rokubi-indexed evens, "caries" cubes-and-rokubi-indexed-evens, "copies" cubed odd prime-indexed evens, "cosies" cubed odd sanbi-indexed evens, "cries" cubed rokubi-indexed evens, "curies" cube-unindexed-rokubi-indexed evens, "cuties" cube-unindexed ternarish-indexed evens, "easies" even-and-sanbi-indexed-evens, "facies" Fibonacci-and-cube-indexed evens, "flies" Fibonacci lucky-indexed evens, "fries" Fibonacci rokubi-indexed evens, "fogies" Fibonacci odd gobi-indexed evens, "furies" Fibonacci-unindexed rokubi-indexed evens, "lilies" lucky-indexed lucky-indexed evens, "pies" prime-indexed evens, "lunies" lucky-unindexed nibi-indexed evens, "mamies" Mersennes-and-Mersenne-indexed-evens, "molies" Mersenne odd lucky-indexed evens, "pities" prime-indexed ternarish-indexed evens, "plies" prime lucky-indexed evens, "polies" prime odd kyuubi-indexed evens, "rabies" rokubi-and-binarish-indexed evens, "ramies" rokubi-and-Mersenne-indexed evens, "shies" shichibi-indexed evens, "skies" sorted kyuubi-indexed evens, "spies" sanbi prime-indexed evens, and "ties" are ternarish-indexed evens."babies" describes binarish-and-binarish-indexed-evens, "rubies" rokubi-unindexed-binarish-indexed-evens, "sassabies" sanbi-and-sorted-sanbi-and-binarish-indexed-evens

1144 "Footstool" describes Fibonacci oddly odd ternarish sanbi ternarish oddly odd luckies, "overcool" cubed oddly odd lucky uptown neighbors, "stool" sanbi ternarish oddly odd luckies, "tool" describes ternarish oddly odd luckies and "uncool" non-cube oddly odd luckies.

1145 "Lukewarm" describes numbers higher than "cooler" but not as high as "warmer".

lukewarm	2,198	...	7,801	29,782	...	70,207	328,510	...	341,494	658,504	...
holiness	3	...	3	3	...	2	3	...	1	4	...

1146 "Flatfoot" describes Fibonacci-lucky-and-ternarish-Fibonacci-oddly-odd-ternarish, "foot" Fibonacci oddly odd ternarish, "galoot" gobi-and-lucky-oddly-odd-ternarish, "galloot" gobi-and-luckier-lucky-oddly-odd-ternarish, "loot" lucky oddly odd ternarish, "scoot" sorted cubed oddly odd ternarish, '"soot" sorted oddly odd ternarish and "underfoot" Fibonacci oddly odd ternarish downtown neighbors.

1147 Since all of the "old" numbers end in eight they are also "hold" numbers. "Bifold" describes old binarish-indexed Fibonacci, "billfold" old binarish-indexed luckier lucky Fibonacci, "cold" old cubes, "fold" old Fibonacci, "holdover" old uptown neighbors, "infold" old

Fibonacci downtown neighbors, "manifold" old Mersenne-and-nibi-indexed Fibonacci, "oversold" old sorted uptown neighbors, "scold" old sanbi cubes, "sold" old sorted, "undersold" old sorted downtown neighbors, "unfold" old non-Fibonacci and "uphold" old uptown neighbors.

1148 "Op" describes odd primes, "archbishop" binarish-indexed shichibi odd primes uptown neighbors, "bishop" binarish-indexed shichi odd primes, "helistop" hachibi-even-lucky-indexed sanbi ternarish odd primes, "hilltop" hachibi-indexed luckier lucky ternarish odd primes, "laptop" lucky-and-prime-ternarish-odd-primes, "lollipop" lucky odd luckier lucky-indexed prime odd primes, "stop" sanbi ternarish odd primes, "tiptop" ternarish-indexed prime ternarish odd primes, "unstop" non-sanbi ternarish odd primes,

1149 "Cops and Robbers" was an Italian film, originally titled "Guardie e Ladri" starring Prince Antonio Griffo Focas Flavio Angelo Ducas Comneno Porfirogenito Gagliardi De Curtis di Bisanzio better known as "Totò". His stage name translates pseudonumerologically to eleven, while his full name translates to a number higher than sixteenplex.

1150 "Blot" describes binarish lucky odd ternarish, bot" binarish odd ternarish, "cabot" cubes-and-binarish-odd-ternarish, "cachalot" cubed-and-cubed-hachibi-and-lucky-odd-ternarish, "challot" cubed-hachibi-and-luckier-lucky-odd-ternarish, "clot" cube lucky odd ternarish, "fusspot" Fibonacci-unindexed sorted sanbi prime odd ternarish, "galipot" gobi-and-lucky-indexed prime odd ternarish, "Iscariot" ichibi-sorted-cubes-and-rokubi-indexed odd ternarish, "lot" lucky odd ternarish, "maillot" Mersenne-and-ichibi-luckier-lucky-odd-ternarish, "mascot" Mersenne-and-sanbi-cubed-odd-ternarish, "picot" prime-indexed cubed odd ternarish, "pilot" prime-indexed lucky odd ternarish, "plot" prime lucky odd ternarish, "pot" prime odd ternarish, "riot" rokubi-indexed odd ternarish, "sabot" sanbi-and-binarish-odd-ternarish, "Scot" sanbi cubed odd ternarish, "shallot" shichibi-and-luckier-lucky-odd-ternarish, "sot" sanbi odd ternarish, "spot" sanbi prime odd ternarish, "subplot" prime lucky odd ternarish downtown neighbor and "teapot" ternarish-even-and-prime-odd-ternarish.

1151 "Ballute" describes binarish-and-luckier-lucky-unindexed ternarish evens, "chute" cubed hachibi-unindexed ternarish evens, "compute" cubed odd Mersenne prime-unindexed ternarish evens, "flute" Fibonacci lucky-unindexed ternarish evens, "hirsute" hachibi-indexed rokubi sorted-unindexed ternarish evens, "impute" ichibi Mersenne prime-unindexed ternarish evens, "lute" lucky-unindexed ternarish evens, "malamute" Mersenne-and-lucky-and-Mersenne-unindexed ternarish evens, "mute" Mersenne-unindexed ternarish evens, "parachute" prime-and-rokubi-and-cubed-hachibi-unindexed ternarish evens, "pollute" prime odd luckier lucky-unindexed ternarish evens, "refute" rokubi even Fibonacci-unindexed ternarish evens, "salute" sanbi-and-lucky-unindexed ternarish evens, "scute" sanbi cubed-unindexed ternarish evens, "shute" shichibi-unindexed ternarish evens, "solute" sanbi odd lucky-unindexed ternarish evens, "statute" sanbi-ternarish-and-ternarish-unindexed ternarish evens and "substitute" sanbi ternarish-indexed ternarish-unindexed ternarish even downtown neighbor.

1152 double-digit inflated

trimmed	1	3	5	7	0	1	3	5	7	1	11	13	15	17	10	101	103	105

1153 "Double-header" and "triple-header" are from baseball for two or three games played in one day. These can be exprapolated to quadruple-headed cranes, quintuple-headers, etc. Phoenix numbers, on the other hand, have a specific digit for a head.

quadruple-header	11,110	11,112	11,113	11,114	11,115	11,116	11,117
holiness	1	0	0	0	0	1	0
quintuple-header	111,110	111,112	111,113	111,114	111,115	111,116	111,117
holiness	1	0	0	0	0	1	0

1154 The "-head" and "-headed" suffixes describe a crane number[1235] with a certain frag as a head, just as "-morphic" describes a certain tail, "barehead" binarish-and-rokubi-even-headed cranes, "beachhead" binarish-even-and-cubed-hachibi-headed cranes, "beheaded" binarish-even-headed cranes, "bighead" binarish-indexed gobi-headed cranes, "billhead" binarish-indexed-luclier-lucky-headed cranes, "blackhead" binarish-lucky-and-cubed-kyuubi-headed cranes, "bulkhead" binarish-unindexed-lucky-kyuubi-headers, "bullhead" binarish-unindexed-luckier-lucky-headed cranes, "even-headed" even-headed cranes, "fathead" Fibonacci-and-ternarish headed cranes, "figurehead" Fibonacci-indexed-gobi-unindexed-rokubi-even-headed cranes, "flathead" Fibonacci-lucky-and-ternarish-headed cranes, "hashhead" hachibi-and-shichibi-headed cranes, "masthead" Mersenne-and-sanbi-ternarish-headed cranes, "nailhead" nibi-and-ichibi-lucky-headed cranes, "pinhead" prime-indexed nibi headed cranes, "pothead" prime odd ternarish-headed cranes, "railhead" rokubi-and-ichibi-lucky-headed cranes, "saphead" sanbi-and-prime-headed cranes, "spearhead" sorted-prime-even-and-rokubi-headed cranes, "subhead" sanbi-unindexed-binarish-headed cranes and "toolhead" ternarish-oddly-odd-lucky-headed cranes.

1155 The complimentary sequence to the Eckovers could be called Eckunders, starting with eight-one, eight hundred eight and one thousand ninety-four.

1156 "Huge factors of enormous numbers" by Walter Nissen. "Enormous" comes 1531 from the Latin for "out of the norm". Numbers getting higher even faster with to-the-third-z plus to-the-third-z-plus-one might be called "ginormous", a portmanteau of giant and enormous (1942). Those getting higher with to-the-fourth-z plus to-the-fourth-z-plus-one we could call "humongous" (1967) or in British "humungous", possibly from "huge" and "monstrous".

ginormous	17	7,625,597,485,003	> hundred-fifty-fourplex
humongous	65,537	> three-billionplex	> to-the-fourth-four

1157 "Carapace" describes cubes-and-rokubi-and-primes-and-cubed-evens, "enface" even-nibi-Fibonacci-and-cubed-evens, "fullface" Fibonacci-unindexed luckier lucky Fibonacci-and-cubed-evens, "glace" gobi-luckies-and-cubed-evens, "interface" Fibonacci-and-cubed-evens averages, "interlace" lucky-and-cubed-evens averages; "interrace" rokubi-and-cubed-evens averages, "lace" lucky-and-cubed-evens, "mace" Mersennes-and-cubed-evens, "millrace" Mersenne-indexed luckier lucky rokubi-and-cubed-evens, "misplace" Merseene-indexed sanbi-prime-lucky-and-cubed-evens, "outface" odd-unindexed ternarish-Fibonacci-and-cubed-evens, "outpace" odd-unindexed ternarish-primes-and-cube-evens, "pace" prime-and-cubed-evens, "palace" primes-and-lucky-and-cubed-evens, "peace" prime-even-and-cubed-evens, "place" prime-luckies-and-cubed-evens, "populace" prime-odd-prime-unindexed lucky-and-cubed-evens, "preface" Fibonacci-and-cubed-evens downtown neighbors, "race" rokubi-and-cubed-evens, "resurface" rokubi even sorted-unindexed rokubi Fibonacci-and-cubed-evens, "solace" sanbi odd lucky-and-cubed-evens, "space" sanbi-primes-and-cubed-evens,, "subsurface" sanbi-unindexed binarish sorted-unindexed rokubi Fibonacci-and-cubed-evens, "surface" sanbi-unindexed rokubi Fibonacci-and-cubed-evens and "unlace" not lucky-and-cubed-evens.

1158 "Antepast" describes prime-and-sanbi-ternarish downtown neighbors, "bast" binarish-and-sanbi-ternarish, "beast" binarish-evens-and-sanbi-ternarish, "blast" binarish-lucky-and-sanbi-ternarish, "boast" binarish-odds-and-sanbi-ternarish, "cast" cubes-and-sanbi-ternarish, "cast-off" cubes-and-sanbi-ternarish neighbors, "coast" cube-odds-and-sanbi-ternarish, "downcast" cubes-and-sanbi-ternarish downtown neighbors, "east" even-and-sanbi-ternarish, "forecast" cubes-and-sanbi-ternarish uptown neighbors, "hast" hachibi-and-sanbi-ternarish, "last" lucky-and-sanbi-ternarish, "least" lucky-evens-and-sanbi-ternarish, "mast" Mersanne-and-sanbi-ternarish, "outboast" binarish-odds-and-sanbi-ternarish uptown neighbors, "outlast" lucky-and-sanbi-ternarish uptown neighbors, "past" prime-and-sanbi-ternarish, "precast" cubes-and-sanbi-ternarish downtown neighbors, "recast" rokubi-even-cubes-and-sanbi-ternarish, "repast" rokubi-even-prime-and-sanbi-ternarish, "seacoast" sorted-even-and-cube-odds-and-sanbi-ternarish, "simulcast" sanbi-indexed Mersenne-unindexed lucky cubes-and-sanbi-ternarish, "toast" ternarish-odds-and-sanbi-ternarish, "topmast" ternarish odd prime Mersenne-and-sanbi-ternarish and "yeast" yongbi-evens-and-sanbi-ternarish.

1159 "The Foggy Dew" as the name of an Irish traditional song first appeared in *The Ancient Music of Ireland* by Edward Bunting in 1840.

dew	4	44	3,234	24,974	90,772,534	897,665,844	87,413,730,974	860,416,137,554

1160 "Fop" meant fool in 1440, but by the seventeenth Century became an overdressed fool, synonymous with coxcomb, fribble, popinjay and ninny and "foppish" appeared in 1599. The odd primes, of course, include all primes save two.

1161 The suffix -ish is a companion to -head and -morphic which refers to a frag on the front end of a number and -morphic which refers to one the end and applies to any place in a number.

Alternatively "foppish" could also describe the much rarer Fibonacci-odd-palindromic-prime-indexed shichibi, "banish" binarish-and-nibi-indexed shichibi, "batfish" binarish-and-ternarish-Fibonacci-indexed shichibi, "beamish" binarish-evens-and-Mersenne-indexed shichibi, "bearish" binarish-even-and-rokubi-indexed shichibi, "beanish" binarish-even-and-nibi-indexed shichibi, "bleakish" binarish-lucky-evens-and-kyuubi-indexed shichibi, "blimpish" binarish lucky-indexed Mersenne prime-indexed shichibi, "bluefish" binarish lucky-unindexed even Fibonacci-indexed shichibi, "blueish" binarish lucky-unindexed even-ibdexed shichibi, "boarish" binarish- odd-and-rokubi-indexed shichibi, "bookish" binarish oddly odd kyuubi-indexed shichibi, "boorish" binarish oddly odd rokubi-indexed shichibi, "buckish" binarish-unindexed cubed kyuubi-indexed shichibi, "bullish" binarish-unindexed luckier lucky-indexed shichibi, "bumpkinish" binarish-unindexed Mersenne prime kyuubi-indexed nibi-indexed shichibi, "catfish" cube-and-ternarish-Fibonacci-indexed shichibi, "clayish" cubed-lucky-and-yongbi-indexed shichibi, "cheapish" cubed-hachibi-evens-and-prime-indexed shichibi, "churlish" cubed hachibi-unindexed rokubi lucky-indexed shichibi, "clumpish" cubed lucky-unindexed Mersenne prime-indexed shichibi, "coalfish" cubed-odd-and-lucky-Fibonacci-indexed shichibi, "cockish" cubed odd cubed Fibonacci-indexed shichibi, "coltish" cubed odd lucky ternarish-indexed shichibi, "coolish" cubed oddly odd lucly-indexed shichibi, "elfish" even lucky Fibonacci-indexed shichibi, "famish" Fibonacci-and-Mersenne-indexed shichibi, "Fibonacci evenly even binarish lucky-indexed shichibi, "fetish" Fibonacci even ternarish-indexed shichibi, "finish" Fibonacci-indexed nibi-indexed shichibi, "Fibonacci-lucky-and-ternarish-Fibonacci-indexed shichibi, "Fibonacci lucky odd-unindexed rokubi-indexed shichibi, "folkish" Fibonacci odd luck kyuubi-indexed shichibi, "foolfish" Fibonacci oddly odd lucky Fibonacci-indexed shichibi, "foolish" Fibonacci oddly odd lucky-indexed shichibi, "garfish" gobi-and-rokubi-Fibonacci-indexed shichibi, "garish" gobi-and-rokubi-indexed shichibi, "girlish" gobi-indexed rokubi lucky-indexed shichibi, "goatfish" gobi-odd-and-ternarish-Fibonacci-indexed shichibi, "goatish" gobi-odd-and-ternarish-indexed shichibi, "hagfish" hachibi-and-gobi-Fibonacci-indexed shichibi, "hashish" hachibi-and-shichibi-indexed shichibi, "hellish" hachibi even luckier lucky-indexed shichibi, "hippish" hachibi-indexed palindromic prime-indexed shichibi, "impish" ichibi Mersenne prime-indexed shichibi, "languish" lucky-and-ng-unindexed/indexed shichibi, "latish" lucky-and-ternarish-indexed shichibi, "loutish" lucky odd-unindexed ternarish-indexed shichibi, "lumpish" lucky-unindexed Mersenne prime-indexed shichibi, "lumpfish" lucky-unindexed Mersenne prime Fibonacci-indexed shichibi, "lumpish" lucky-unindexed Mersenne prime-indexed shichibi, "manish" Mersenne-and-nibi-indexed shichibi, "marish" Mersenne-and-rokubi-indexed shichibi, "mopish" Mersenne odd prime-indexed shichibi, "mourish" Mersenne odd-unindexed rokubi-indexed shichibi, "mulish" Mersenne-unindexed lucky-indexed shichibi, "oafish" odd-and-Fibonacci-indexed shichibi, "palish" prime-and-lucky-indexed shichibi, "parish" prime-and-rokubi-indexed shichibi, "peakish" prime even-and-kyuubi-indexed shichibi, "planish" prime-lucky-and-nibi-indexed shichibi, "plumpish" prime lucky-unindexed Mersenne prime-indexed shichibi, "polish" prime odd lucky-indexed shichibi, "popish" prime odd prime-indexed shichibi, "publish" prime-unindexed binarish lucky-indexed shichibi, "puckish" prime-unindexed cubed kyuubi-indexed shichibi, "punish" prime-unindexed nibi-indexed shichibi, "pupfish" prime-unindexed prime Fibonacci-indexed shichibi, "rakish" rokubi-and-kyuubi-indexed shichibi, "raspish" rokubi-and-sanbi-prime-indexed shichibi, "ratfish" rokubi-and-ternarish-Fibonacci-indexed shichibi, "refinish" rokubi even Fibonacci-indexed nibi-indexed shichibi, "relish" rokubi even lucky-indexed shichibi, "runtish" rokubi-unindexed nibi ternarish-indexed shichibi, "sailfish" sanbi-and-ichibi-lucky-Fibonacci-indexed shichibi, "saltish" sanbi-and-lucky-ternarish-indexed shichibi, "scampish" sanbi-cubed-and-Mersenne-prime-indexed shichibi, "selfish" sorted even lucky Fibonacci-indexed shichibi, "shish" shichibi-indexed shichibi, "sickish" sanbi-indexed cubed kyuubi-indexed shichibi, "sissyish" sanbi-indexed sorted sanbi yongbi-indexed shichibi, "smallish" sanbi-Mersenne-and-luckier-lucky-indexed shichibi, "snappish" sorted-nibi-and-palindromic-prime-indexed shichibi, "sourish" sorted odd-unindexed rokubi-indexed shichibi, "starfish" sanbi-ternarish-and-rokubi-indexed shichibi, "stonish" sorted ternarish odd nibi-indexed shichibi, "sunfish" sanbi-unindexed nibi Fibonacci-indexed shichibi, "tallish" ternarish-and-luckier-lucky-indexed shichibi, "ticklish" ternarish-indexed cubed kyuubi lucky-indexed shichibi, "wolfish" lucky Fibonacci-indexed shichibi complimentary sequence,

1162 Unindexed luckies include: "babul" binarish-and-binarish-unindexed luckies, "bagful" binarish-and-gobi-Fibonacci-luckies, "bagsful" binarish-and-gobi-sorted-Fibonacci-unindexed luckies, "baleful" binarish-and-lucky-even-Fibonacci-unindexed luckies, "bashful" binarish-and-shichibi-Fibonacci-unindexed luckies, "blissful" binarish lucky-indexed sorted sanbi Fibonacci-unindexed luckies, "boastful" binarish odd-and-sanbi-ternarish-Fibonacci-unindexed luckies, "bountiful" binarish odd-unindexed nibi ternarish-indexed Fibonacci-unindexed luckies, "bulbul" binarish-unindexed lucky binarish-unindexed luckies, "canful" cubes-and-nibi-Fibonacci-unindexed luckies, "cansful" cubes-and-nibi-sorted-Fibonacci-unindexed luckies, "capful" cubes-and-prime-Fibonacci-unindexed luckies, "caracul" cubes-and-rokubi-and-cube-unindexed luckies, "careful" cubes-and-rokubi-even-Fibonacci-unindexed luckies, "carful" cubes-and-rokubi-Fibonacci-unindexed luckies, "cupful" cubes-unindexed prime Fibonacci-unindexed luckies, "cupsful" cubes-unindexed prime sanbi Fibonacci-unindexed luckies, "caul" cubes-and-unindexed-luckies, "earful" evens-and-rokubi-Fibonacci-unindexed luckies, "easeful" even-and-sorted-even-Fibonacci-unindexed luckies, "factful" Fibonacci-and-cubed-ternarish-Fibonacci-unindexed luckies, "fanciful" Fibonacci-and-nibi-cube-indexed Fibonacci-unindexed luckies, "fateful" Fibonacci-and-ternarish-even-Fibonacci-unindexed luckies, "fearful" Fibonacci-evens-and-rokubi-Fibonacci-unindexed luckies, "feastful" Fibonacci evens-and-sanbi-ternarish-Fibonacci-unindexed luckies, "fistful" Fibonacci-indexed sanbi ternarish Fibonacci-unindexed luckies, "fitful" Fibonacci-indexed ternarish Fibonacci-unindexed luckies, "fruitful" Fibonacci rokubi-unindexed ichibi ternarish Fibonacci-unindexed luckies, "glassful" gobi luckies-and-sorted-sanbi-Fibonacci-unindexed luckies, "guileful" gobi-unindexed/indexed lucky even Fibonacci-unindexed luckies, "hasteful" hachibi-and-sanbi-ternarish-Fibonacci-unindexed luckies, "hateful" hachibi-and-ternarish-even-Fibonacci-unindexed luckies, "hatful" hachibi-and-ternarish-Fibonacci-unindexed luckies, "hatsful" hachibi-and-ternarish-sorted-Fibonacci-unindexed luckies, "haul" hachibi-and-unindexed-luckies, "hushful" hachibi-unindexed shichibi Fibonacci-unindexed luckies, "inhaul" hachibi-and-unindexed luckies downtown neighbors, "insoul" sanbi odd-unindexed luckies downtown neighbors, "Istanbul" ichibi-sorted-ternarish-and-nibi-binarish-unindexed luckies, "karakul" kyuubi-and-rokubi-and-kyuubi-unindexed luckies, "lapful" luckies-and-prime-Fibonacci-unindexed luckies, "lifeful" lucky-indexed Fibonacci evens Fibonacci-unindexed luckies, "lustful" lucky-unindexed sanbi ternarish Fibonacci-unindexed luckies, "manful" Mersenne-and-nibi-Fibonacci-unindexed luckies, "maul" Mersennes-and-unindexed-luckies, "moanful" Mersenne-odd-and-nibi-Fibonacci-unindexed luckies, "moistful" Mersenne odd-indexed sanbi ternarish Fibonacci-unindexed luckies, "museful" Mersenne-unindexed sorted even Fibonacci-unindexed luckies, "overhaul" hachibi-and-unindexed-luckies uptown neighbors, "pailful" prime-and-ichibi-lucky-Fibonacci-unindexed luckies, "pailsful" prime-and-ichibi-luckies-sorted-Fibonacci-unindexed luckies, "panful" prime-and-nibi-Fibonacci-unindexed luckies, "Paul" primes-and-unindexed-luckies, "peaceful" prime-even-and-cubed-even-Fibonacci-unindexed luckies, "pitiful" prime-indexed ternarish-indexed Fibonacci-unindexed luckies, "plateful' prime-luckies-and-ternarish-even-Fibonacci-unindexed luckies, "platesful" prime lucky-and-ternarish-evens-sorted-Fibonacci-unindexed luckies, "playful" prime-luckies-and-yongbi-Fibonacci-unindexed luckies, "potful" prime odd ternarish Fibonacci-unindexed luckies, "rueful" rokubi-unindexed even Fibonacci-unindexed luckies, "Seoul" sorted even/odd-unindexed luckies, "shul" shichibi-unindexed luckies, "sinful" sanbi-indexed nibi Fibonacci-unindexed luckies, "skillful" sorted kyuubi-indexed luckier lucky Fibonacci-unindexed luckies, "skinful" sorted kyuubi-indexed luckier lucky Fibonacci-unindexed luckies, "sobful" sanbi odd binarish Fibonacci-unindexed luckies, "soulful" sanbi odd unindexed lucky Fibonacci-unindexed luckies, "spiteful" sanbi prime-indexed ternarish even Fibonacci-unindexed luckies, "tableful" ternarish-and-binarish-lucky-even-Fibonacci-unindexed luckies, "tablesful" ternarish-and-binarish-lucky-evens-sorted-Fibonacci-unindexed luckies, "tactful" ternarish-and-cubed-ternarish-Fibonacci-unindexed luckies, "tasteful" ternarish-and-sorted-ternarish-even-Fibonacci-unindexed luckies, "teacupful" ternarish-evens-and-cube-unundexed prime Fibonacci-unindexed luckies, "teacupsful" ternarish-evens-and-cubes-unindexed prime sorted Fibonacci-unindexed luckies, "tearful" ternarish-evens-and-rokubi-Fibonacci-unindexed luckies, "toilful" ternarish-odd-indexed lucky Fibonacci-unindexed luckies, "topful" ternarish odd prime Fibonacci-unindexed luckies, "tubful" ternarish-unindexed binarish Fibonacci-unindexed luckies, "tubeful" ternarish-unindexed binarish even Fibonacci-unindexed luckies, "tuneful" ternarish-unindexed nibi even Fibonacci-unindexed luckies, "unfruitful" non-Fibonacci rokubi-ubindexed ichibi ternarish Fibonaci-unindexed luckies, "unmanful" non-Mersenne-and-nibi-Fibonacci-unindexed luckies, "unsinful" non-sanbi-indexed nibi Fibonacci-unindexed luckies, "unskillful" non-sorted kyuubi-indexed luckier lucky"untactful" non-ternarish-and-cubed-ternarish-Fibonacci-unindexed luckies Fibonacci-unindexed luckies and "untasteful" non-ternarish-and-sorted-ternarish-even-Fibonacci-unindexed luckies.

"Foil", on the other hand, describes Fibonacci odd-indexed luckies, "boil" binarish odd-indexed luckies, "coil" cubed odd-indexed luclies, "milfoil" Mersenne-indexed lucky Fibonacci odd-indexed luckies, "moil" Mersenne odd-indexed luckies, "parafoil" prime-and-rokubi-and-Fibonacci-odd-indexed luckies, "preboil" binarish odd-undexed luckies downtown neighbors, "soil" sanbi odd-indexed luckies, "spoil" sanbi prime odd-indexed luckies, "subsoil" sanbi-unindexed binarish sanbi-indexed luckies, "tinfoil" ternarish-indexed nibi Fibonacci-indexed luckies, "toil" ternarish odd-indexed luckies, "topsoil" ternarish odd prime sanbi odd-indexed luckies and "uncoil" non-cubed odd-indexed luckies.

1163 "Bate" describes binarish-and-ternarish-evens, "bifurcate" binarish-indexed Fibonacci-unindexed rokubi-cubes-and-ternarish-evens, "bistate" binarish-indexed sanbi-ternarish-and-ternarish-evens, "bisulfate" binarish-indexed sanbi-unindexed lucky-Fibonacci-and-ternarish-evens, "califate" cubes-and-lucky-indexed-Fibonacci-and-ternarish-evens, "calculate" cubes-and-lucky-cube-unindexed luckies-and-ternarish-evens, "capacitate" cubes-and-primes-and-cube-indexed ternarish-and-ternarish-evens, "capsulate" cubes-and-prime-sanbi-unindexed luckies-and-ternarish-evens, "carate" cubes-and-rokubi-and-ternarish-evens, "castigate" cubes-and-sanbi-ternarish-indexed gobi-and-ternarish-evens, "celibate" cubed-even-lucky-indexed binarish-and-ternarish-evens, "chelate" cubed-hachibi-even-luckies-and-ternarish-evens, "circulate" cub-indexed rokubi cube-unindexed luckies-and-ternarish-evens, "climate" cubed lucky-indexed Mersenne-and-ternarish-evens, "coagulate" cubed-odds-and-gobi-unindexed lucky-and-ternarish-evens, "cogitate" cubed odd gobi-indexed ternarish-and-ternarish-evens, "collate" cubed-odd-luckier-luckies-and-ternarish-evens, "collimate" cubed odd luckier lucky-indexed Mersenne-and-ternarish-evens, "crate" cubed-rokubi-and-ternarish-evens, "create" cubed-rokubi-even-and-ternarish-evens, "culminate" cube-unindexed lucky Mersenne-indexed nibi-and-ternarish-evens, "cureate" cube-unindexed rokubi-evens-and-ternarish-evens, "curate" cube-unindexed rokubi-and-ternarish-evens, "elate" even-luckies-and-ternarish-evens, "eliminate" even lucky-indexed Mersenne-indexed nibi-and-ternarish-evens, "enate" even-nibi-and-ternarish-evens, "eructate" even rokubi-unindexed cubed ternarish-and-ternarish-evens, "escalate" even-sorted-cubes-and-luckies-and-ternarish-evens, "estate" even-sorted-ternarish-and-ternarish-evens, "estimate" even-sorted-ternarish-indexed Mersenne-and-ternarish-evens, "etiolate" even ternarish-indexed odd luckies-and-ternarish-evens, "facilitate" Fibonacci-and-cubes-indexed lucky-indexed ternarish-and-ternarish-evens, "fascinate" Fibonacci-and-sanbi-cube-indexed nibi-and-ternarish-evens, "fate" Fibonacci-and-ternarish-evens, "felicitate" Fibonacci even lucky-indexed cube-indexed ternarish-and-ternarish-evens, "figurate" Fibonacci-indexed gobi-unindexed rokubi-and-ternarish-evens, "fluctuate" Fibonacci lucky-indexed cubed ternarish-unindexed-and-ternarish-evens, "fossate" Fibonacci-odd-sorted-sanbi-and-ternarish-evens, "frigate" Fibonacci rokubi-indexed gobi-and-ternarish-evens, "fulgurate" Fibonacci-unindexed lucky gobi-unindexed rokubi-and-ternarish-evens, "fumigate" Fibonacci-unindexed Mersenne-indexed gobi-and-ternarish-evens, "glaciate" gobi-luckies-and-cubed-ichibi-and-ternarish-evens, "glutamate" gobi lucky-unindexed ternarish-and-Mersenne-and-ternarish-evens, "habilitate" hachibi-and-binarish-indexed lucky-indexed ternarish-and-ternarish-evens, "hate" hachibi-and-ternarish-evens, "hesitate" hachibi even sorted-indexed ternarish-and-ternarish-evens, "humiliate" hachibi-unindexed Mersenne-indexed lucky-ichibi-and-ternarish-evens, "illuminate" ichibi luckier lucky-unindexed Mersenne-indexed nibi-and-ternaris-evens, "incapacitate" cubes-and-primes-and-cube-indexed ternarish-and-ternarish-evens downtown neighbors,"inculcate" cube-unindexed lucky-cubes-and-ternarish-evens downtown neighbors, "incubate" cube-unindexed binarish-and-ternarish-evens downtown neighbors, "inculpate" cube-unindexed lucky-primes-and-ternarish-evens downward neighbors, "inflate" Fibonacci-luckies-and-ternarish-evens downtown neighbors, "initiate" ichibi-ternarish-and-ternarish-evens downtown neighbor, "insatiate" sanbi-and-ternarish-ichibi-and-ternarish-evens downtown neighbors, "insinuate" sanbi-indexed nibi-unindexed-and-ternarish-evens downtown neighbor, "instate" sanbi-ternarish-and-ternarish-evens downtown neighbors, "insulate" sanbi-unindexed luckies-and-ternarish-evens downtown neighbors, "intimate" ternarish-indexed Mersenne-and-ternarish-evens downtown neighbors, "kalifate" kyuubi-and-lucky-indexed Fibonacci-and-ternarish-evens, "karate" kyuubi-and-rokubi-and-ternarish-evens, "labiate" luckies-and-binarish-ichibi-and-ternarish-evens, "lacerate" lucky-and-cubed-even-rokubi-and-ternarish-evens, "lacinate" lucky-and-cube-indexed nibi-and-ternarish-evens, "lactate" luckies-and-cubed-ternarish-and-ternarish-evens, "laminate" luckies-and-Mersenne-indexed nibi-and-ternarish-evens, "late" luckies-and-ternarish-evens, "laureate" luckies-and-unindexed-rokubi-evens-and-ternarish-evens, "ligate" lucky-indexed gobi-and-ternarish-evens, "lineate" lucky-indexed nibi-evens-and-ternarish-evens, "litigate" lucky-indexed ternarish-indexed gobi-and-ternarish-evens, "locate" lucky odd cubes-and-ternarish-evens, "lunate" lucky-unindexed nibi-and-ternarish-evens, "macerate" Mersenne-and-cubed-even-rokubi-and-ternarish-evens, "machinate" Mersenne-and-cubed-hachibi-indexed nibi-and-ternarish-evens, "maculate" Mersenne-and-cube-unindexed luckies-and-ternarish-evens, "manipulate" Mersenne-and-nibi-indexed prime-unindexed luckies-and-ternarish-evens, "marinate" Mersenne-and-rokubi-indexed nibi-and-ternarish-evens, "masticate" Mersenne-

and-sanbi-ternarish-indexed cubes-and-ternarish-evens, "mate" Mersenne-and-ternarish-evens, "maturate" Mersenne-and-ternarish-unindexed rokubi-and-ternarish-evens, "micturate" Mersenne-indexed cubed ternarish-unindexed rokubi-and-ternarish-evens, "militate" Mersenne-indexed lucky-indexed ternarish-and-ternarish-evens, "ministate" Mersenne-indexed nibi-indexed sanbi-ternarish-and-ternarish-evens, "mitigate" Mersenne-indexed ternarish-indexed gobi-and-ternarish-evens, "mutate" Mersenne-unindexed ternarish-and-ternarish-evens, "mutilate" Mersenne-unindexed ternarish-indexed luckies-and-ternarish-evens, "nauseate" nibi-and-unindexed-sorted-evens-and-ternarish-evens, "nictate" nibi-indexed cubed-ternarish-and-ternarish-evens, "nictitate" nibi-indexed cubed-indexed ternarish-and-ternarish-evens, "nucleate" nibi-unindexed cubed-lucky-evens-and-ternarish-evens, "osculate" odd sanbi cube-unindexed luckies-and-ternarish-evens, "overrate" rokubi-and-ternarish-evens uptown neighbors, "overstate" sanbi-ternarish-and-ternarish-evens uptown neighbors, "paginate" primes-and-gobi-indexed nibi-and-ternarish-evens, "palate" primes-and-luckies-and-ternarish-evens, "palantinate" primes-and-luckies-and-nibi-ternarish-indexed nibi-and-ternarish-evens, "palliate" prims-and-luckier-lucky-ichibi-and-ternarish-evens, "palmitate" prime-and-lucky-Mersenne-indexed ternarish-and-ternarish-evens, "palmate" primes-and-lucky-Mersenne-and-ternarish-evens, "palpate" primes-and-lucky-primes-and-ternarish-evens, "palpitate" primes-and-lucky-primes-indexed ternarish-and-ternarish-evens, "papillate" primes-and-prime-indexed-luckier-luckies-and-ternarish-evens, "pate" primes-and-ternarish-evens, "peculate" prime-even-cubed-unindexed luckies-and-ternarish-evens, "pileate" prime-indexed lucky-evens-and-ternarish-evens, "pirate" prime-indexed rokubi-and-ternarish-evens, "pistillate" prime-indexed sanbi ternarish-indexed luckier-lucky-and-ternarish-evens, "placate" prime-luckies-and-cubes-and-ternarish-evens, "prelate" luckies-and-ternarish-evens downtown neighbors, "pulsate" prime-unindexed "pupate" prime-unindexed prime-and-ternarish-evens, "rate" rokubi-and-ternarish-evens, "restate" rokubi-even-sorted-ternarish-and-ternarish-evens, "ruinate" rokubi-unindexed/indexed-nibi-and-ternarish-evens, "ruminate" rokubi-unindexed Mersenne-indexed nibi-and-ternarish-evens, "rusticate" rokubi-unindexed sanbi ternarish-indexed cubes-and-ternarish-evens, "sate" sanbi-and-ternarish-evens, "satiate" sanbi-and-ternarish-ichibi-and-ternarish-evens, "saturate" sanbi-and-ternarish-unindexed rokubi-and-ternarish-evens, "senate" sorted-even-nibi-and-ternarish-evens, "separate" sorted even primes-and -rokubi-and-ternarish-evens, "shipmate" shichibi-indexed prime Mersenne-and-ternarish-evens, "sibilate" sanbi-indexed binarish-indexed luckies-and-ternarish-evens, "silicate" sanbi-indexed lucky-indexed cubes-and-ternarish-evens, "simulate" sanbi-indexed Mersenne-unindexed luckies-and-ternarish-evens, "skate" sorted-kyuubi-and-ternarish-evens, "slate" sanbi-luckies-and-ternarish-evens, "spate" sanbi-primes-and-ternarish-evens, "spatulate" sanbi-primes-and-ternarish-unindexed luckies-and-ternarish-evens, "spiculate" sanbi prime-indexed cube-unindexed luckies-and-ternarish-evens, "spinate" sanbi prime-indexed nibi-and-ternarish-evens, "staminate" sanbi-ternarish-and-Mersenne-indexed nibi-and-ternarish-evens, "stimulate" sanbi ternarish-indexed Mersenne-unindexed luckies-and-ternarish-evens, "stipulate" sanbi ternarish-indexed prime-unindexed luckies-and-ternarish-evens, "sublimate" sanbi-unindexed binarish lucky-indexed Mersenne-and-ternarish-evens, "sulfate" sanbi-unindexed lucky-Fibonacci-and-ternarish-evens, "sultanate" sanbi-unindexed lucky-ternarish-and-ternaris-evens, "supinate" sanbi-unindexed prime-indexed nibi-and-ternarish-evens, "suppurate" sanbi-unindexed palindromic primes-unindexed rokubi-and-ternarish-evens, "tabulate" ternarish-and-binarish-unindexed luckies-and-ternarish-evens, "tate" ternarish-and-ternarish-evens, "titillate" ternarish-indexed ternarish-indexed luckier-luckies-and-ternarish-evens, "tubulate" ternarish-unindexed bunarish-unindexed luckies-and-ternarish-evens, "uncrate" non-cubed-rokubi-and-ternarish-evens, "understate" sanbi-ternarish-and-ternarish-evens downtown neighbors, "unsaturate" non-sanbi-and-ternarish-unindexed rokubi-and-ternarish-evens and "upstate" sanbi-ternarish-and-ternarish-evens uptown neighbors,

1164 The pearly gates of New Jerusalem are mentioned by St. John in Revelation 21:21. "Early" describes the complimentary sequence to "late", and so is not a lucky yongbi like "bally"' binarish-and-luckier-lucky-yongbi, "billy" binarish-indexed luckier lucky yongbi, "blearily" binarish-lucky-evens-and-rokubi-indexed lucky yongbi, "blousily" binarish lucky odd-unindexed sanbi-indexed lucky yongbi, "bouncily" binarish odd-unimdexed nibi cubed-indexed lucky yongbi, "busily" binarish-unindexed sanbi-indexed lucky yongbi, "bully" binarish-unindexed luckier lucky yongbi, "cagily" cubes-and-gobi-indexed lucky yongbi, "casually" cubes-and-unindexed-sanbi-and-luckier-lucky-yongbi, "causally" cubes-and-unindexed-sanbi-and-luckier-lucky-yongbi, "chillily" cubed hachibi-indexed luckier lucky-indexed lucky yongbi, "chilly" cubed hachibi-indexed luckier lucky yongbi, "classily" cubed-lucky-and-sorted-sanbi-indexed lucky yongbi, "clumsily" cube lucky-unindexed Mersenne sanbi-indexed lucky yongbi, "coitally" cubed odd-indexed ternarish-and-luckier-lucky-yongbi, "craftily" cubed-rokubi-and-Fibonacci-ternarish-indexed lucky yongbi, "creakily" cubed-rokubi-even-and-kyuubi-indexed lucky yongbi, "creamily" cubed-rokubi-evens-and-Mersenne-indexed lucky yongbi,"crispily" cubed-rokubi-indexed sanbi prime-indexed lucky yongbi,"crustily" cubed rokubi-unindexed sanbi ternarish-indexed lucky yongbi, "cully" cube-unindexed luckier lucky yongbi, "eely" evenly even lucky yongbi, "epically" even prime-indexed cube-and-luckier-lucky-yongbi, "famously" Fibonacci-and-Mersenne-odd-unindexed sanbi lucky yongbi, "faultily" Fibonacci-and-unindexed-ternarish-indexed lucky yongbi, "faunally" Fibonacci-and-unindexed-nibi-and-luckier-lucky-yongbi, "filially" Fibonacci-indexed lucky-ichibi-and-luckier-lucky-yongbi, "fiscally" Fibonacci-indexed sanbi-cubes-and-luckier-lucky-yongbi, "fitfully" Fibonacci-indexed ternarish Fibonacci-unindexed luckier lucky yongbi, "flashily" Fibonacci-lucky-and-shichibi-lucky-yongbi, "flimsily" Fibonacci lucky-indexed Mersenne sanbi-indexed lucky yongbi, "fly" Fibonacci lucky yongbi, "freakily" Fibonacci-rokubi-evens-and-kyuubi-indexed lucky yongbi, "friskily" Fibonacci rokubi-indexed sorted kyuubi-indexed lucky yongbi, "frugally" Fibonacci rokubi-unindexed gobi-and-luckier-lucky-yongbi, "frumpily" Fibonacci rokubi-unindexed Mersenne prime-indexed lucky yongbi, "fully" Fibonacci-unindexed luckier lucky yongbi, "gilly" gobi-indexed luckier lucky yongbi, "glassily" gobi-lucky-and-sorted-sanbi-indexed lucky yongbi, "guiltily" gobi-unindexed/indexed lucky ternarish"gully" gobi-unindexed luckier lucky yongbi, "hilly" hachibi-indexed luckier lucky yongbi, "icily" ichibi cube-indexed lucky yongbi, "Italy" ichibi-ternarish-and-lucky-yongbi, "lilly" lucky-indexed luckier lucky yongbi, "lily" lucky-indexed lucky yongbi, "lumpily" lucky-unindexed Mersenne prime-indexed lucky yongbi, "lustily" lucky-unindexed sanbi ternarish-indexed lucky yongbi, "manfully" Mersenne-and-nibi-Fibonacci-unindexed lucky yongbi, "mangily" Mersenne-and-ng-indexed lucky yongbi, "manually" Mersenne-and-nibi-unindexed-and-luckier-lucky-yongbi, "milkily" Mersenne-indexed lucky kyuubi-indexed lucky yongbi, "mistily" Mersenne-indexed sanbi ternarish-indexed lucky yongbi, "mousily" Mersenne odd-unindexed sanbi-indexed lucky yongbi, "muckily" Mersenne-unindexed cubed kyuubi-indexed lucky yongbi, "mushily" Mersenne-unindexed shichibi-indexed lucky yongbi, "muskily" Mersenne-unindexed sorted kyuubi-indexed lucky yongbi, "mussily" Mersenne-unindexed sorted sanbi-indexed lucky yongbi, "mustily" Mersenne-unindexed sanbi ternarish-indexed lucky yongbi, "mutually" Mersenne-unindexed ternarish-unindexed-and-luckier-lucky-yongbi, "nasally" nibi-and-sanbi-and-luckier-lucky-yongbi, "nastily" nibi-and-sanbi-ternarish-indexed lucky yongbi, "natally" nibi-and-ternarish-and-luckier-lucky-yongbi, "neurally" nibi even-unindexed rokubi-and-luckier-lucky-yongbi, "nippily" nibi-indexed palindromic prime-indexed lucky yongbi, "oily" odd-indexed lucky yongbi, "overfly" Fibonacci lucky yongbi uptown neighbors, "pally" prime-and-luckier-lucky-yongbi, "papally" prime-and-prime-and-luckier-lucky-yongbi, "piously" prime-indexed odd-unindexed sorted lucky yongbi, "plaguily" prime-lucky-and-gobi-unindexed/indexed lucky yongbi, "pluckily" prime lucky-unindexed cubed kyuubi-indexed lucky yongbi, "plurally" "prime lucky-unindexed rokubi-and-luckier-lucky-yongbi, "plushily" prime lucky-unindexed shchibi-indexed lucky yongbi, "postally" prime-odd-sanbi-ternarish-and-luckier-lucky-yongbi, "publically" prime-unindexed binarish lucky-indexed cubes-and-luckier-lucky-yongbi, "pulpily" prime-unindexed lucky prime-indexed lucky yongbi, "pursily" prime-unindexed rokubi sorted-indexed lucky yongbi, "pushily" prime-unindexed shichibi-indexed lucky yongbi, "racially" rokubi-and-cube-ichibi-and-luckier-lucky-yongbi, "rally" rokubi-and-luckier-lucky-yongbi, "rascally" rokubi-and-sanbi-cubes-and-luckier-lucky-yongbi, "riskily" rokubi-indexed sorted kyuubi-indexed lucky yongbi, "ritually" rokubi-indexed ternarish-unindexed-and-luckier-lucky-yongbi, "ruefully" rokubi-unindexed even Fibonacci-unindexed luckier lucky yongbi, "rustically" rokubi-unindexed sanbi ternarish-indexed cubes-and-luckier-lucky-yongbi, "rurally" rokubi-unindexed rokubi-and-lucky-yongbi, "rustily" rokubi-unindexed sanbi ternarish-indexed lucky yongbi, "sally" sanbi-and-luckier-lucky-yongbi, "saltily" sanbi-and-lucky-ternarish-indexed lucky yongbi, "sappily" sanbi-and-palindromic-prime-indexed lucky yongbi, "sassily" sanbi-and-sorted-sanbi-indexed lucky yongbi, "scaly" sanbi-cubed-and-lucky-yongbi, "scantily" sanbi-cubes-and-nibi-ternarish-indexed lucky yongbi, "shakily" shichibi-and-kyuubi-indexed lucky yongbi, "shaly" shichibi-and-lucky-yongbi, "shiftily" shichibi-indexed Fibonacci ternarish-indexed lucky yongbi, "shilly" shichibi-indexed luckier lucky yongbi, "shinily" shichibi-indexed nibi-indexed lucky yongbi, "silkily" sanbi-indexed lucky kyuubi-indexed lucky yongbi, "sillily" sanbi-indexed luckier lucky-indexed lucky yongbi, "silly" sanbi-indexed luckier lucky yongbi, "sinfully" sanbi-indexed nibi Fibonacci-unindexed luckier lucky yongbi,"skimpily" sorted kyuubi-indexed Mersenne prime-indexed lucky yongbi, "slimily" sanbi lucky-indexed Mersenne-indexed lucky yongbi, "sloppily" sanbi lucky odd palindromic prime-indexed lucky yongbi, "slushily" sanbi lucky-unindexed shichibi-indexed lucky yongbi, "snappily" sorted-nibi-and-palindromic-primes-indexed lucky yongbi, "sneakily" sorted-nibi-even-and-kyuubi-indexed lucky yongbi, "snippily" sorted nibi-indexed palindromic prime-indexed lucky yongbi, "soapily" sanbi odds-and-prime-indexed lucky yongbi, "snootily" sorted-nibi-oddly-odd-ternarish-indexed lucky yongbi, "socially" sanbi odd cube-indexed-and-luckier lucky yongbi, "sootily" sorted oddly odd ternarish-indexed lucky yongbi, "spicily" sanbi prime-indexed cube-indexed lucky yongbi, "stagily" sanbi-ternarish-and-gobi-indexed lucky yongbi, "sulkily" sanbi-unindexed lucky kyuubi-indexed lucky yongbi, "sully" sanbi-unindexed luckier lucky, "sly" sorted lucky yongbi, "tackily" ternarish-and-cubed-kyuubi-indexed lucky yongbi, "tally" ternarish-and-luckier-lucky-yongbi, "tastily" ternarish-and-sanbi-ternarish-indexed lucky yongbi, "tipsily" ternarish-indexed prime sanbi-indexed lucky yongbi, "totally" ternarish odd-ternarish-and-luckier-lucky-yongbi, "tuftily" ternarish-unindexed Fibonacci ternarish-indexed lucky yongbi,

1165 Topo Gigio was the name of the talking Italian mouse made internationally famous by appearances on the "Ed Sullivan Show". "Boccaccio" describes binarish-odd-cubed-cubes-and-cubed-cuces-odds, "curio" cube-unibdexed rokubi-indexed odds, "fellatio" Fibonacci-even-luckier-luckies-and-ternarish-indexed odds, "helio" hachibi even lucky-indexed odds, "maleficio" Mersenne-and-lucky-even-Fibonacci-indexed-cube-indexed odds, "mustachio" Mersenne-unindexed-sanbi-ternarish-and-cube-hachibi-indexed odds, "patio" primes-and-ternish-indexed odds, "pistachio" prime-indexed-sanbi-ternarish-and-cubed-hachibi-indexed odds, "polio" prime odd lucky-indexed odds and "ratio" rokubi-and-ternarish-indexed odds.

1166 "Po", besides describing a pair of digits, described prime odds, as does "op", "campo" cubes-and-Mersenne-prime-odds and "hippo" hichibi-indexed palindromic prime odds.

1167 "Gigo", like "nino" (nothing in, nothing out) and "fish" (first in, still here) came after the original accounting acronyms, "lifo" (last in, first out) and "fifo" (first in, first out). The last three, like gigo, are all also "acronyums", acronyms that are also acronums. "Fish" describes Fibonacci-indexed shichibi, "lifo" flucky-indexed Fibonacci odds and "fifo" Fibonacci-indexed Fibonacci odds. "Bak" describes both back at keyboard and binarish-and-kyuubi, "basic" brothers and sister in Christ and binarish-and-sanbi-indexed cubes, "bf" best/boy friend and binarish Fibonacci, "bfn" bye for now and binarish Fibonacci nibi, "bi" business intelligence and binarish ichibi, "Bible" basic instruction before leaving Earth and bibarish-indexed binarish lucky evens, "bibo" bheer/beverage in bheer/beverage out and binarish-indexed binarish odd, "bif" basis in fact and binarish-indexed Fibonacci, "bil" brother-in-law and binarish-indexed luckies, "bio" bring it on and binarish-indexed odds, "bl" belly laugh and binarish luckies, "bnf" big name fan and binarish nibi Fibonacci, "bo" bug off and binarish odds, "bsf" but seriously folks and binarish sorted Fibonacci, "cas" crack a smile and cubes-and-sanbi, "cico" coffee in coffee out and cube-indexed cubed odds, "cil" check in later and cube-indexed luckies, "cio" check it out and cube-indexed odds, "cm" call me and cubed Mersennes, "cmf" count my fingers and cubed Mersenne Fibonacci, "cob" close of business and cubed odd binarish, "col" chuckling out loud and cubed odd luckies, "cos" change of subject and cubed odd sanbi, "cot" circle of trust and cubed odd ternarish, "crap" cheap redundant assorted products and cubed-rokubi-and-primes, "crat" can't remember a thing and cubed-rokubi-and-ternarish, "csl" can't stop laughing and cubed sanbi luckies, "csg" chuckle, snicker, grin and cubed sorted gobi, "cto" check this out and cubed ternarish odds, "cy" calm yourself and cubed yongbi, "eak" eating at keyboard and evens-and-kyuubi, "eft" electronic funds transfer and even Fibonacci ternarish, "fcol" for crying out loud and Fibonacci cubed odd luckies, "fe" fatal error and Fibonacci evens, "figs" French/Italian/German/Spanish and Fibonacci-indexed gobi sorted, "fil" father-in-law and Fibonacci-indexed luckies, "fine" fouled-up, insecure, neurotic, emotional and Fibonacci-indexed nibi evens, "fitb" fill in the blank(s) and Fibonacci-indexed ternarish/binarish, "flotus" first lady of the United States and Fibonacci lucky odd ternarish-unindexed sanbi, "foaf" friend of a friend and Fibonacci odds and Fibonacci, "foc" free of charge and Fibonacci odd cubes, "fomo" fear of missing out and Fibonacci odd Mersenne odds, "fouo" for official use only and Fibonacci odd-unindexed odds, "fsr" for some reason and Fibonacci sorted rokubi, "fubar" fouled up beyond all recognition and Fibonacci-unindexed binarish-and-rokubi, "fum" fouled up mess and Fibonacci-unindexed Mersennes, "fyc" for your consideration and Fibonacci yongbi cubes, "fye" for your edification and Fibonacci yongbi evens, "gab" getting a bheer/beverage and gobi-and-binarish, "gabi" grin and bear it and gobi-and-binarish-ichibi, "gal" get a life and gobi-and-luckies, "galgal" give a little/lot, get a little/lot and gobi-and-lucky-gobi-and-luckies, "gas" got a second and gobi-and-sanbi, "gf" girl friend anf gobi Fibonacci, "gist" great idea(s) for starting things and gobi-indexed sanbi ternarish, "gl" good luck/get lost and gobi luckies, "gol" giggling out loud and gobi odd luckies, "gsc" gimme some credit and gobi sorted cubes, "hago" have a good one and hachibi-and-gobi-odds, "hak" hugs and kisses and hachibi-and-kyuubi, "har" hit-and-run and hachibi-and-rokubi, "hay" how are you and hachibi-and-yongbi, "hf" hello friend/have fun/have faith and hachibi Fibonacci, "hig" how's it going and hachibi-indexed gobi, "hih" hope it helps and hachibi-indexed hachibi, "hippo" higher income people's personal opinion and hachibi-indexed palindromic prime odds, "hitaks" hang in there and keep smiling and hachibi-indexed ternarish-and-kyuubi-sorted, "hsik" how should I know and hachibi sorted-indexed kyuubi, "iac" in any case/I am confused/if anyone cares and ichibi-and-cubes, "iae" in any event and ichibi-and-evens, "iaits" it's all in the subject and ichibi-and-ichibi-ternarish-sorted, "ianac" I am not a crook and ichibi-and-nibi-and-cubes, "ianae" I am not an expert and ichibi-and-nibi-and-evens, "ianal" I am not a lawyer and ichibi-and-nibi-and-luckies, "iat" I am tired and ichibi-and-ternarish, "ic" independant contractor/in character and ichibi cubes, "iccl" I couldn't care less and ichibi cubed cubed luckies, "icihiccl" I couldn't imagine how I could care less and ichibi cube-indexed hachibi-indexed cubed cubed luckies, "ifab" found a bug and ichibi Fibonacci-and-binarish, "imao" in my arrogant opinion and ichibi-Mersennes-and-odds, "imco" in my considered opinion and ichibi Mersenne cubed odds, "imo" in my opinion and ichibi Mersenne odds, "iot" in order to and ichibi odd ternarish, "kc" keep cool and kyuubi cubed, "kc(a)co" keep cool (and) carry on and kyuubi cubed (and) cubed odd , "kir" keep it real and kyuubi-indexed rokubi, "kiss" keep it simple stupid and kyuubi-indexed sorted sanbi, "kit" keep in touch and kyuubi-indexed ternarish, "ko" knock out and kyuubi odds, "lg" life goal/long gone and lucky gobi, "lh" laughing hysterically and lucky hachibi, "lib" lying in bed and lucky-indexed binarish, "lig" let it go and lucky-indexed gobi, "lir" let it rest and lucky-indexed rokubi, "lis" laughing in silence and "lucky-indexed sanbi, "ll" living large and luckier luckies, "llap" live long and prosper and "luckier-luckies-and-primes, "llt" looks like trouble and luckier lucky ternarish, "lob" lying on bed and lucky odd binarish, "lpc" lead pipe cinch and lucky prime cubes, "ltic" laughing 'til I cry and lucky ternarish-indexed cubes, "mf" my friend and Mersenne Fibonacci, "mih" make it happen and Mersenne-indexed hachibi, "mil" mother-in-law and Mersenne-indexed luckies, "mirl" meet in real life and Mersenne-indexed rokubi luckies, "mis" make it so and Mersenne-indexed sanbi, "mitin" more into than I needed and Mersenne-indexed ternarish-indexed nibi, "mlas" my lips are sealed and Mersenne-luckies-and-sanbi, "mo" move on and Mersenne odds, "mubar" messed up beyond all recognition and Mersenne-unindexed binarish-and-rokubi, "nab" not a blonde and nibi-and-binarish, "nak" nursing at keyboard, "nascar" non-athletic sport centered around red-necks and nibi-and-sanbi-cubes-and-rokubi, "nato" no action, talk only and nibi-and-ternarish-odds, "nb" nota bene and nibi binarish, "nbfab" not bad for a beginner and nibi binarish Fibonacci-and-binarish, "nbif" no basis in fact and nibi binarish-indexed Fibonacci, "nfc" not favorably consided and nibi Fibonacci cubed, "nll" nice little lady and nibi luckier luckies, "ntb" not too bright and nibi ternarish/binarish, "oatus" on a totally unrelated subject and odds-and-ternarish-unindexed sanbi, "oc" original character and odd cubes, "of" old fart and odd Fibonacci, "ofap" old fart at play and odd-Fibonacci-and-primes, "oll" on-line love and odd luckier luckies, "oof" out of facility and oddly odd Fibonacci, "ooi" out of interest and oddly odd ichibi, "oos" out of stock/sight and oddly odd sanbi, "ost" on second thought and odd sorted ternarish, "ot" off topic and odd ternarish, "otfl" on the floor laughing and odd ternarish

Fibonacci luckies, "ots" on the scene/spot/off the shelf and odd ternarish sorted, "pal" peace and love and primes-and-luckies, "pans" pretty awesome new stuff and primes-and-nibi-sorted, "pc" personal computer/politically correct and primes cubed, "pi" politically incorrect, "picnic" problem in chair, not in computer and prime-indexed cubed nibi-indexed cubes, "pif" paid in full and prime-indexed Fibonacci, "pin" person in need and prime-indexed nibi, "plo" peace, love, out and prime lucky odds, "plur" peace, love, unity, respect and prime lucky-unindexed rokubi, "poak" passed out at keyboard and prime-odd-and-kyuubi, "potato" person over thirty acting twenty-one and prime-odd-ternarish-and-ternarish-odds, "potus" president of the United States and prime odd ternarish-unindexed sanbi, "pp" personal problem and palindromic primes, "ptl" praise the Lord and prime ternarish luckies, "rafo" read and find out and rokubi-and-Fibonacci-odds, "rc" remote control and rokubi cubed, "riyl" recommended if you like and rokubi-indexed yongbi luckies, "rl" real life and rokubi luckies, "rlf" real life friend and rokubi lucky Fibonacci, "sb" stand by and sanbi binarish, "sc" stay cool and sanbi cubed, "sf" science fiction and sanbi Fibonacci, "shit" sugar, honey in tea and shichibi-indexed ternarish, "sic" spelling is correct and sanbi-indexed cubes, "sil" sister-in-law and sanbi-indexed luckies, "sin" stop it now and sanbi-indexed nibi, "snag" sensitive, new age guy/gal and sorted-nibi-and-gobi, "snif" simple, nice index file and sorted nibi-indexed Fibonacci, "so" significant other and sanbi odds, "sol" sooner or later and sanbi odd luckies, "sop" standard operating procedure and sanbi odd primes, "spoc" single point of contact and sanbi-primes-odd-cubes, "tabom" ternarish-and-binarish-odd-Mersennes, "taf" that's all folks and ternarish-and-Fibonacci, "tah" take a hike and ternarish-and-hachibi, "taks" that's a knee slapper and ternarish-and-kyuubi-sorted, "tam" thanks a million and ternarish-and-Mersennes, "tansit" there's a new sheriff in town and ternarish-and-nibi-indexed ternarish, "tc" take care and ternarish cubes, "tf" too funny and ternarish Fibonacci, "tfs" thanks for sharing and ternarish Fibonacci sorted, "tic" tongue in cheek and ternarish-indexed cubes, "tilii" tell it like it is and ternarish-indexed lucky-indexed ichibi, "tisc" that is so cool and ternarish-indexed sanbi cubes, "tisnt" this is so not true and ternarish-indexed sorted nibi ternarish, "tisnf" this is so not fair and ternarish-indexed sorted nibi Fibonacci, "tobal" there oughta be a law and ternarish-odd-binarish-and-luckies, "tp" thanks pal and ternarish primes, "yaf" young angry female and yongbi-and-Fibonacci, "yam" young angry male and yongbi-and-Mersenne, "yahoo" you always have other options and yongbi-and-hachibi-oddly-odds and "yic" yours in Christ and yongbi-indexed cubes.

1168 Gödel also represented left parenthesis with a six prime exponent, right parenthesis with a seven prime exponent, addition with an eleven exponent and multiplication with a twelve exponent, but these do not at first produce more economical representations than successive successors. He also represented other mathematical and logical operators with exponents. For For NOT, ¬, Gödel used a 1, for FOR ALL, ɜɜ, a 2, for THEREFORE, ⊃, a 3, for OR, ∧, a 4, for AND, ∨, a 5, for EQUALS, =, a ten. Joyce preferred the so-called "Polish" notation of Jan Łukasiewicz for logical operators that avoids parentheses, A for OR, B for BECAUSE OF, C for CONDITIONALLY, D for NAND (not and), E for IFF (if and only if), J for XOR (or but not and), K for AND, L for NECESSARILY, M for POSSIBLY, N for NOT, O for FALSE, V for TRUE, X for NOR (not or), Π for FOR ALL and Σ for THERE EXISTS. (Seeger's Law: "Anything in parentheses can be ignored.")

The Gödel number for the inconsistency theorem would be higher than two-hundredduplex, NAPzz'NKPzz'Qz"z' = ¬(Pxy∧¬(Pxy∧Qzy)) =

$2^{1}3^{6}5^{12}7^{13}11^{16}13^{5}17^{1}19^{6}23^{12}29^{13}31^{16}37^{5}41^{15}43^{19}47^{16}53^{7}59^{7}$.

1169 "The Grays and the Blues" refers to the colors of the uniforms of the soldiers on the Confederate and the Union sides respectively in the War Between the States. Those that are on both sides are called either "spies" or "gays", gobi and yongbi combined.

spies	4	5	14	15	24	25	34	35	44	45	54	55	64	65	74	75	84	85	95	95	104	105

1170 "Gutter" describes gobi-unindexed ternarish with two of three trits.

gutters	10	12	21	100	101	110	112	122	200	211	212	220	1,000	1,001	1,010

1171 Hash is a mixture of chopped meat, potatoes and spices from the French for "chopped". When fresh meat was rationed during WWII corned beef, beef preserved with large "corned" salt grains, became popular.

corned beef	1	2	11	12	21	22	31	32	41	42	51	52	61	62	71	72	81	82	101	102	111

1172 Hue and cry are near synonyms, since hue means a shout, but "cry" is the complimentary sequence to hachibi-unindexed evens.

cry	1	3	5	7	11	13	15	17	19	21	25	27	29	31	33	35	37	39	41	45	47	49	51	53

1173 "Pyongbi" describes yongbi numberdromes.

pyongbi	44	404	414	424	434	444	454	464	475	484	494	4,004	4,114	4,224	4,334

1174 The amphisæna is a snake with a false head on its tail from *Natural History* by Pliny the Elder, so a number with the same digit as a head and a tail, even if not a numberdrome, is still an ambisæna number. They do not differ however until as high as thousands.

amphisæna	1,001	1,011	1,021	1,031	1,041	1,051	1,061	1,071	1,081	1,091	1,101
holiness	2	1	1	1	1	1	1	1	3	2	1

1175 "Phachibi" describes hachibi numberdromes.

phachibi	88	808	818	828	838	848	858	868	878	888	898	8,008	8,118	8,228
holiness	4	5	4	4	4	4	4	5	4	6	5	6	4	4
semiphachibi	44	404	409	414	419	424	429	434	439	444	449	4,004	4,059	4,114

1176 "Pshichibi" describes shichibi numberdromes.

pshichibi	77	707	717	727	737	747	757	767	777	787	797	7,007	7,117	7,227

1177 "Pnibi" describes nibi numberdromes.

pnibi	22	202	212	222	232	242	252	262	272	282	292	2,002	2,112	2,222	2,332

semipnibi	11	101	106	111	116	121	126	131	136	141	146	1,001	1,056	1,111	1,166

1178 "Psanbi" describes sanbi numberdromes.

psanbi	33	303	313	323	333	343	353	363	373	383	393	3,003	3,113

1189 "Gnumberdromes" describes gobi numberdromes.

gnumberdromes	55	505	515	525	535	545	555	565	575	585	595	5,005	5,115	5,225

1180 "Prokubi" describes rokubi numberdromes.

prokubi	66	606	616	626	636	646	656	666	676	686	696	6,006	6,116	6,226
holiness	2	3	2	2	2	2	2	3	2	4	3	4	2	2
semiprokubi	33	303	308	313	318	323	328	333	338	343	348	3,003	3,058	3,113

1181 "Knumberdromes" describes kyuubi numberdromes.

knumberdromes	99	909	919	929	939	949	959	969	979	989	999	9,009	9,119	9,229
holiness	2	3	2	2	2	2	2	3	2	4	3	4	2	2

1182 inflationary "inthree".

inthree	130	131	132	134	135	136	137	138	139	230	231	232	234	235	236	237	238
holiness	1	0	0	0	0	1	0	2	1	1	0	0	0	0	1	0	2

1183 "Honeydew list" a pun on "Honey do list", the list of things one's honey is expected to do.

1184 "Doe", on the other hand, describes the complimentary sequence of "buck", binarish-unindexed cubed kyuubi.

buck	6,859	24,389	59,319	117,649	205,379	328,509	493,039	704,969	2,146,689
holiness	4	3	2	2	2	4	3	4	5
doe	3,140	40,680	75,610	295,030	506,960	671,490	794,620	882,350	2,119,400
holiness	1	5	2	3	5	3	3	5	3

1185 "Less" describes the sequence with lucky even sorted sanbi, but "backless" without binarish or cubed kyuubi, "barless" without binarish or rokubi, "beakless" without binarish evens or kyuubi, "beamless" without binarish evens or Mersennes, "bibless" without binarish-indexed binarish, "bless" describes a sequence without binarish, "capless" without cubes or primes, "careless" without cubes or rokubi evens, "cashless" without cubes or shichibi, "casteless" without cubes or sorted ternarish evens, "chinless" without cubed hachibi-indexed nibi, "classless" without cubed luckies or sorted sanbi, "coalless" without cubed odds or luckies, "coatless" without cubes odds or ternarish, "costless" without cubed odd sangbi ternarish, "countless" without cubed odd-unindexed nibi ternarish, "cureless" without cube-unindexed robubi evens, "earless" without evens or rokubi, "eyeless" without even yongbi evens, "faceless" without Fibonacci or cubed evens, "fangless" without Fibonacci or ng, "fatless" without Fibonacci or ternarish, "fearless" without Fibonacci evens or rokubi, "featureless" without Fibonacci evens or ternarish-unindexed rokubi evens, "feeless" withgout Fibonacci evenly evens, "finless" without Fibonacci-indexed nibi, "fireless" without Fibonacci-indexed rokubi evens, "fishless" without Fibonacci-indexed shichibi, 'flagless" without Fibonacci luckies or gobi, "flapless" without Fibonacci luckies or primes, "foamless" without Fibonacci odds or Mersennes, "fogless" without Fibonacci odd gobi, "fruitless" without Fibonacci rokubi-unindexed ichibi ternarish, "funless" without Fibonacci-unindexed nibi, "furless" without Fibonacci-unindexed rokubi, "futureless" without Fibonacci-unindexed ternarish-unindexed rokubi evens, "gasless" without gobi or sanbi, "gateless" without gobi or ternarish evens, "giftless" without gobi-indexed Fibonacci ternarish, "ginless" without gobi-indexed nibi, "goalless" without gobi odds or luckies, "goatless" without gobi odds or ternarish, "guiltless" without gobi-unindexed ichibi lucky ternarish, "gunless" without gobi-unindexed nibi, "gutless" without gobi-unindexed ternarish, "hapless" without hachibi or prime luckies, "h(e)atless" without hachibi (evens) or ternarish luckies, "heirless" without hachibe even-indexed rokubi, "hiltless" without hachibi-indexed lucky ternarish, "hipless" without hachibi-indexed primes, "kingless" without kyuubi-indexed ng, "kinless" without kyuubi-indexed nibi, "leafless" without lucky evens or Fibonacci, "lifeless" without lucky-indexed Fibonacci evens, "limitless" without lucky-indexed Mersenne-indexed ternarish, "lineless" without lucky-indexed nibi evens, "lipless" without lucky-indexed primes, "luckless" without lucky-unindexed cubed kyuubi, "manless" without Mersenne or nibi, "mapless" without Mersenne or primes, "massless" without Mersenne or sorted sanbi, "mastless" without Mersenne or sanbi ternarish, "matless" without Mersenne or ternarish, "mateless" without Mersenne or ternarish evens, "napless" without nibi or prime luckies, "oarless" without odds or rokubi, "pilotless" without prime-indexed lucky odd ternarish, "pitless" without prime-indexed ternarish, "pitiless" without prime-indexed ternarish ichibi, "plan(e)less" without prime luckies or nibi (evens), "rayless" without rokubi or yongbi luckies, "ribless" without rokubi-indexed binarish, "riftless" without rokubi-indexed Fibonacci ternarish, "rimless" without rokubi-indexed Merennes, "ringless" without rokubi-indexed ng, "ruleless" without rokubi-undexed lucky evens, "rumless" without rokubi-unindexed Mersennes, "rumpless" without rokubi-unindexed Mersenne primes, "runless" without rikubi-unindexed nibi, "rungless" without rokubi-unindexed ng, "rustless" without rokubi-unindexed sanbi ternarish, "saltless" without sanbi or lucky ternarish, "sapless" without sanbi or primes, "scaleless" without sanbi cubes or lucky evens, "scarless" without sanbi cubes or rokubi, "shapeless" without shichibi or prime even, "shiftless" without shichibi-indexed Fibonacci ternarish, "sinless" without sanbi-indexed nibi, "sireless" without sanbi-indexed rokubi evens, "skinless" without sorted kyuubi-indexed nibi, "slitless" without sanbi lucky-indexed ternarish, "snapless" without sorted nibi or primes, "spaceless" without sanbi primes or cubed evens, "spanless" without sanbi primes or nibi, "spin(e)less" without sanbi prime-indexed nibi (evens), "spotless" without sanbi prime odd ternarish, "stalkless" without sanbi ternarish or lucky kyuubi, "stateless" without sanbi ternarish or ternarish evens, "sugarless" without sanbi-unindexed gobi or rokubi, "sunless" without sanbi-unindexed nibi, "tackless" without ternarish or cubed kyuubi, "tactless" without ternarish or cubed ternarish, "tasteless" without ternarish or sorted ternarish evens, "tintless" without ternarish-indexed nibi ternarish, "tipless" without ternarish-indexed primes, "tireless" without ternarish-indexed rokubi evens, "topless" without ternarish odd primes, "tubeless" without ternarish-unindexed binarish evens, "tuneless" without ternarish-unindexed nibi evens, "tuskless" without ternarish-unindexed sorted kyuubi, "womanless" without complimentary numbers of Mersennes or nibi.

1186
"Log-off" describes the neighbors of gobi luckies and "log-on" their complimentary sequence.

log-off	14	16	24	26	74	76	104	106	114	116	134	136	194	196	204	206	234
log-on	23	25	73	75	83	85	113	115	143	145	193	195	253	255	263	265	283

1187 "Balsam" describes binarish-and-luckies-and-Mersennes, "bantam" binarish-and-nibi-ternarish-and-Mersennes, "beam" binarish-evens-and-Mersennes, "cam" cubes-and-Mersennes, "cham" cubed-hachibi-and-Mersennes, "clam" cubed-luckies-and-Mersennes, "cr(e)am" cubed-rokubi-(evens-)and-Mersennes, "downstream" sorted-ternarish-rokubi-evens-and-Mersennes downtown neighbors, "flimflam" Fibonacci lucky-indexed Mersenne-Fibonacci-lucky-and-Mersennes, "foam" Fibonacci-odds-and-Mersennes, "gam" gobi-and-Mersennes, "gloam" gobi-lucky-odds-and-Mersennes, "ham" hachibi-and-Mersennes, "imam" ichibi-Mersennes-and-Mersennes, "inseam" sorted-evens-and-Mersennes downtown neighbors, "lam" luckies-and-Mersennes, "lingam" lucky-indexed ng-and-Mersennes, "loam" lucky-odds-and-Mersennes, "Malayalam" Mersennes-and-luckies-and-yongbi-and-luckies-and-Mersennes, "Nam" nibi-and-Mersennes, "r(e)am" rokubi-(evens-)and-Mersennes, "Sam" sanbi-and-Mersennes, "scam" sanbi-cubes-and-Mersennes, "scr(e)am" sorted-cubed-rokubi-(evens-)and-Mersennes, "sham" shichibi-and-Mersennes, "slam" sanbi-luckies-and-Mersennes, "steam" sorted-ternarish-evens-and-Mersennes, "sunbeam" sanbi-unindexed nibi-binarish-evens-and-Mesennes, "Surinam" sanbi-unindexed rokubi-indexed nibi-and-Mersennes, "tam" ternarish and Mersennes, "team" ternarish-evens-and-Mersennes, "upstream" sorted-ternarish-rokubi-and-Mersennes uptown neighbors, "wolfram" lucky-Fibonacci-rokubi-and-Mersennes complimentary numbers and "yam" yongbi-and-Mersennes.

1188 "Clop" describes cubed lucky odd primes, "gallop" describes gobi-and-luckier-lucky-odd-primes, "flop" Fibonacci lucky odd primes, "galop" gobi-and-lucky-odd-primes, "plop" palindromic lucky odd primes, "shallop" shichibi-and-luckier-lucky-odd-primes, "slop" sanbi lucky odd primes,

1189 The suffix -off describes neighbors, "blastoff" binarish-luckies-and-sanbi-ternarish neighbors, "boff" binarish neighbors, "castoff" cubes-and-sanbi-ternarish neighbors, "cutoff" cube-unindexed ternarish neighbpors, "falloff" Fibonacci-and-luckier-luckies neighbors, "faroff" Fibonacci-and-rokubi neighbors, "feoff" Fibonacci even neighbors, "kickoff" kyuubi-indexed cubed kyuubi neighbors, "layoff" luckies-and-yongbi neighbors, "liftoff" lucky-indexed Fibonacci ternarish neighbors, "payoff" primes-and-yongbi neighbors, "playoff" prime-luckies-and-yongbi neighbors, "putoff" prime-unindexed ternarish neighbors, "ripoff" rokubi-indexed prime neighbors, "runoff" rokubi-unindexed nibi neighbors, "scoff" sanbi cube neighbors, "setoff" sorted even ternarish neighbors, "shutoff" shichibi-unindexed ternarish neighbors, "spinoff'" sanbi prime-indexed nibi neighbors, "tipoff" ternarish-indexed prime neighbors and "toff" ternarish neighbors.

1190 Three and a hundred thirty-one are the only Ulams in the first twenty-eight billion to be the sum of consecutive Ulams. "Pulam" describes Ulam numberdromes.

Pulam	1	2	3	4	6	8	11	99	131	282	363	434	949	2,112	2,332	2,552	2,662	5,335	5,665

1191 The Generalized Markoff numbers extrapolate to four or five or more terms.

four	1	3	11	41	131	153	571	1,561	1,803	2,131	5,761	7,953	17,291	18,601
five	1	4	19	91	379	436	2,039	7,561	8,644	10,009	36,001	47,956	144,019	150,841

1192 Markons, of course, can also be generalized.

four	6	8	58	88	428	846	2,046	4,238	7,868	8,196	8,438	20,708	70,318	74,908
holiness	1	2	2	4	2	3	2	2	5	4	4	4	3	4
five	5	8	80	563	620	908	1,355	2,438	7,910	52,043	63,998	89,990	178,910	770,228

1193 It is not usually understood as the acronum nibi-rokubi-cube-indexed-sorted-sanbi-indexed-sanbi-ternarish-indexed-cubes. Other such acronums can be formed by adding the suffix -ic to the -ists.[997]

1194 "Et" describes even ternarish, "ballet" binarish-and-luckier-lucky-even-ternarish, "basinet" binarish-and-nibi-and-rokubi-indexed even ternarish, "bet" binarish even ternarish", "billet" binarish-indexed luckier lucky even ternarish, "bluet" binarish lucky-unindexed even ternarish, "bullet" binarish-unindexed luckier lucky even ternarish, "castanet" cubed-and-sanbi-ternarish-and-nibi-even-ternarish, "chalet" cubed-hachibi-and-lucky-even-ternarish, "cruet" cubed rokubi-unindexed even ternarish, "cullet" cubed-unindexed luckier lucky even ternarish, "facet" Fibonacci-and-cubed-even-ternarish, "feet" Fibonacci evenly even ternarish, "filet" Fibonacci-indexed lucky even ternarish, "fillet" Fibonacci-indexed luckier lucky even ternarish, "giblet" gobi-indexed binarish lucky even ternarish, "gullet" gobi-unindexed luckier lucky even ternarish, "inlet" lucky even ternarish downtown neighbors, "let" lucky even ternarish, "lunet" lucky-unindexed nibi even ternarish, "mallet" Mersenne-and-luckier-lucky-even-ternarish, "millet" Mersenne-indexed luckier lucky even ternarish, "mullet" Mersenne-unindexed luckier lucky even ternarish, "offset" sorted even ternarish neighbors, "pallet" primes-and-luckier lucky even ternarish, "pet" prime even ternarish, "planet" prime-luckies-and-nibi-even-ternarish, "preset" sorted even ternarish downtown neighbors, "pullet" prime-nindexed luckier lucky ternarish, "scilicet" sanbi cubed-indexed lucky-indexed cube even ternarish, "set" sorted even ternarish, "subset" sanbi-unindexed binarish sorted even ternarish, "sublet" sanbi-unindexed binarish sorted even ternarish, "suet" sanbi-unindexed even ternarish, "Tibet" ternarish-indexed binarish even ternarish, "toilet" ternarish odd-indexed lucky even ternarish.

1195 "Dragnet" comes from the Old English drægnet, a net to drag the bottom of a body of water in fishing, was used figurative in the 1640s and by the police by 1894. Beginning in

1949 "Dragnet" chronicled the cases of Los Angeles policeman Joe Friday on radio and continued on television until 1970. "Dragnet", the 1987 movie, featured Joe's nephew.

drag	7	77	87	877	887	897	7,777	7,787	7,797	7,877	7,887	7,897	7,977	7,987

1196 "Alley Oop" by Vincent Trout Hamlin chronicled the adventures of a time-traveling Moovian strongman, from the French "Allez oop" meaning "Up you go." "Coop" describes cubed oddly odd primes, "loop" lucky oddly odd primes, "poop" palindromic ooddly odd primes and "sloop" sorted lucky oddly odd primes.

1197 "Ey" describes even yongbi, "bluey" binarish lucky-unindexed even yongbi, "fey" Fibonacci even yongbi, "galley" gobi-and-luckier-lucky-even-yongbi, "gluey" gobi lucky-unindexed even yongbi, "gulley" gobi-unindexed luckier lucky even yongbi, "lacey" lucky-and-cubed-even-yongbi, "ley" lucky even yongbi, "maguey" Mersenne-and-gobi-unindexed even yongbi, "mousey" Mersenne odd-unindexed sorted even yongbi, "muley" Mersenne-unindexed lucky even yongbi, "paisley" prime-and-ichibi-sorted-lucky-even-yongbi, "plaguey" prime-lucky-and-gobu-unindexed even yongbi, "skiey" sorted kyuubi-indexed even yongbi, "spicey" sanbi prime-indexed even yongbi and "suey" sanbi-unindexed even yongbi, "tissuey" ternarish-indexed sorted sanbi-unindexed even yongbi, "unlet" non-lucky even ternarish, "unset" non-sorted even ternarish, "upset" sorted even ternarish uptown neighbors.

1198 *War and Peace* by Count Lev "Leo" Nikolayevich Tolstoy (1869). Since peace this side of Heaven just seems to be a temporary ceasefire in preparation for more war, half-peace is perhaps more appropriate.

half-peace	1	4	32	108	256	500	864	1,372	2,048	2,916	4,000	5,324	6,912

1199 "Bar" describes binarish-and-rokubi, "basilar" binarish-and-sanbi-indexed luckies-and-rokubi, "bear" binarish-evens-and-rokubi, "bilinear" binarish-indexed lucky-indexed nibi-evens-and-rokubi, "bipolar" binarish-indexed prime-odd-luckies-and-rokubi, "blear" binarish-licky-evens-and-rokubi, "boar" binarish-odds-and-rokubi, "bulbar" binarish-unindexed lucky-binarish-and-rokubi, "bursar" binarish-unindexed rokubi-sorted-and-rokubi, "caesar" cubes-and-evens-sorted-and-rokubi, "calamar" cubes-and-luckies-and-Mersennes-and-rokubi, "capsular" cubes-and-prime-sanbi-unindexed luckies-and-rokubi, "car" cubes-and-rokubi, "cellar" cubed-even-luckier-luckies-and-rokubi, "char" cubed-hachibi-and-rokubi, "cigar" cubeindexed gobi-and-rokubi, "cislunar" cube-indexed sanbi lucky-unindexed nibi-and-rokubi, "clear" cubed-lucky-evens-and-rokubi, "collinear" cubed odd luckier lucky-indexed nibi-evens-and-rokubi, "collar" cubed-odd-luckier-luckies-and-rokubi, "coplanar" cubed-odd-prime-luckies-and-nibi-and-rokubi, "copolar" cubed-odd-primes-odd-luckies-and-rokubi, "costar" cubed-odd-sanbi-ternarish-and-rokubi, "cougar" cubed-odd-unindexed gobi-and-rokubi, "cuticular" cube-unindexed ternarish-indexed cube-unindexed luckies-and-rokubi, "ear" evens-and-rokubi, "familiar" Fibonacci-and-Mersenne-indexed lucky-ichibi-and-rokubi, "far" Fibonacci-and-rokubi, "fear" Fibonacci-evens-and-rokubi, "fibular" Fibonacci-indexed binarish-unindexed luckies-and-rokubi, "filar" Fibonacci-indexed lucky-and-rokubi, "fistular" Fibonacci-indexed sanbi ternarish-unindexed luckies-and-rokubi, "flatcar" Fibonacci-luckies-

and-ternarish-cubes-and-rokubi, "foliar" Fibonacci-odd-lucky-ichibi-and-rokubi, "follicular" Fibonacci odd luckier lucky-indexed cube-unindexed luckies-and-rokubi, "funicular" Fibonacci-unindexed nibi-indexed cube-unindexed luckies-and-rokubi, "furcular" Fibonacci-unindexed rokubi cube-unindexed luckies-and-rokubi, "gar" gobi-and-rokubi, "guar" gobi-unindexed-and-rokubi, "guitar" gobi-unindexed ichibi-ternarish-and-rokubi, "hangar" hachibi-and-ng-and-rokubi, "hear" hachibi-evens-and-rokubi, "hussar" hachibi-unindexed sorted-sanbi-and-rokubi, "insofar" sanbi-odd-Fibonacci-and-rokubi downtown neighbors, "insular" sanbi-unindexed luckies-and-rokubi downtown neighbors, "interlunar" interlucky-unindexed nibi-and-rokubi, "interpolar" interprime odd luckies-and-rokubi, "kilobar" kyuubi-indexed lucky-odd-binarish-and-rokubi, "lacular" luckies-and-cube-unindexed luckies-and-rokubi, "laminar" luckies-and-Mersennes-indexed nibi-and-rokubi, "lascar" luckies-and-sanbi-cubes-and-rokubi, "lear" lucky-evens-and-rokubi, "liar" lucky-ichibi-and-rokubi, "linear" lucky-indexed nibi-evens-and-rokubi, "lobular" lucky odd binarish-unindexed luckies-and-rokubi, "lunar" lucky-unindexed nibi-and-rokubi, "macular" Mersenne-and-cube-unindexed luckies-and-rokubi, "mar" Mersenne-and-rokubi, "millibar" Mersenne-indexed luckier lucky-indexed binarish-and-rokubi, "minicar" Mersenne-indexed nibi-indexed cubes-and-rokubi, "mishear" Mersenne-indexed sorted-hachibi-evens-and-rokubi, "multipolar" Mersenne-unindexed lucky ternarish-indexed prime odd luckies-and-rokubi, "muscular" Mersenne-unindexed sanbi cube-unindexed luckies-and-rokubi, "near" nibi-evens-and-rokubi, "nebular" nibi even binarish-unindexed luckies-and-rokubi, "nonlinear" non-lucky-indexed nibi-evens-and-rokubi, "nonsecular" non-sorted even cube-unindexed luckies-and-rokubi, "oar" odds-and-rokubi, "ocular" odd cube-unindexed luckies-and-rokubi, "Oscar" odd-sanbi-cubes-and-rokubi, "oscular" odd sanbi cube-unindexed luckies-and-rokubi, "overbear" binarish-evens-and-rokubi uptown neighbors, "overhear" hachibi-evens-and-rokubi uptown neighbors, "papular" prime-and-prime-unindexed luckies-and-rokubi, "par" primes-and-rokubi, "patellar" primes-and-ternarish-even-luckier-luckies-and-rokubi, "pear" prime-even-and-rokubi, "peculiar" prime even cube-unindexed lucky-ichibi-and-rokubi, "peninsular" prime even nibi-indexed nibi sorted-unindexed luckies-and-rokubi, "pilar" prime-indexed luckies-and-rokubi, "pillar" prime-indexed luckier-luckies-and-rokubi, "planar" prime-luckies-and-nibi-and-rokubi, "polar" prime-odd-luckies-and-rokubi, "poplar" prime-odd-prime-luckies-and-rokubi, "popular" prime odd prime-unindexed luckies-and-rokubi, "postwar" war uptown neighbors, "premolar" Mersenne-odd-luckies-and-rokubi downtown neighbors, "prewar" war downtown neighbors, "pulsar" prime-unindexed lucky-sanbi-and-rokubi, "pupilar" prime-unindexed prime-indexed luckies-and-rokubi, "pustular" prime-unindexed sanbi ternarish-unindexed luckies-and-rokubi, "reappear" rokubi-evens-and-palindromic-prime-even-and-rokubi, "rear" rokubi-evens-and-rokubi, "scalar" sanbi-cubes-and-luckies-and-rokubi, "scapular" sanbi-cubes-and-prime-unindexed luckies-and-rokubi, "scar" sanbi-cubes-and-rokubi, "scimitar" sanbi cube-indexed Mersenne-indexed ternarish-and-rokubi, "sear" sorted-evens-and-rokubi, "secular" sorted even cube-unindexed luckies-and-rokubi, "semilunar" half lucky-unindexed nibi-and-rokubi, "similar" sanbi-indexed Mersenne-indexed luckies-and-rokubi, "simitar" sanbi-indexed Mersenne-indexed ternarish-and-rokubi, "singular" sanbi-indexed ng-unindexed luckies-and-rokubi, "sitar" sanbi-indexed ternarish-and-rokubi, "soar" sanbi-odds-and-rokubi, "sofar" sorted-odd-Fibonacci-and-rokubi, "solar" sanbi-odd-luckies-and-rokubi, "spar" sanbi-primes-and-rokubi, "spatular" sanbi-primes-and-ternarish-unindexed luckies-and-rokubi, "specular" sorted prime even cube-unindexed luckies-and-rocubi, "spectacular" sorted-prime-even-cubed-ternarish-and-cube-unindexed luckies-and-rokubi, "spicular" sanbi prime-indexed cube-unindexed luckies-and-rokubi, "star" sanbi-ternarish-and-rokubi, "sublunar" sanbi-unindexed binarish lucky-unindexed nibi-and-rokubi, "sugar" sanbi-unindexed gobi-and-rokubi, "tabular" ternarish-and-binarish-unindexed luckies-and-rokubi, "tar" ternarish-and-rokubi, "tear" ternarish-evens-and-rokubi, "tentacular" ternarish-even-nibi-ternarish-and-cube-unindexed luckies-and-rokubi, "titular" ternarish-indexed ternarish-unindexed luckies-and-rokubi, "tubular" ternarish-unindexed binarish-unindexed luckies-and-rokubi, "tutelar" ternarish-unindexed ternarish-even-luckies-and-rokubi, "unbar" non-binarish-and-rokubi, "unbear" non-binarish-evens-and-rokubi, "unclear" non-cubed lucky-evens-and-rokubi, "unfamiliar" non-Fibonacci-and-Mersenne-indexed lucky-ichibi-and-rokubi, "unpopular" non-prime odd prime-unindexed luckies-and-rokubi, "unspectacular" non-sorted-prime-even-cubed-ternarish-and-cube-unindexed luckies-and-rokubi,"uprear" rokubi-evens-and-rokubi uptown neighbors, "uproar" rokubi-odds-and-rokubi uptown neighbors and "year" yongbi-evens-and-rokubi.

1200 The complimentary sequence of the Phickovers could be called Phickunders.

Phickunders	34	79	8,536	82,874

1201 *The Mathematics of Oz* by Clifford Pickover, the complimentary sequence could be called "Pickunder".

Pickunders	3	71	758	66,210	71,975	88,293	8,473,199	26,845,172

1202 "Raenil" describes non-numberdrome reversible linears, "raenilivruc" non-numberdrome reversible curvilinears and "suoecavruc" non-palindromic reversible curvaceous.

raenil	14	17	41	47	71	74	114	117	141	144	147	171	174	177
raenilivruc	25	52	225	255	522	552	2,225	2,255	2,522	2,525	2,555	5,222	5,225	5,252
suoecavruc	30	36	38	39	60	63	68	69	80	83	86	89	90	93
holiness	1	2	2	1	2	1	3	2	3	2	3	3	2	1

1203 Extrapolating "quintered" describes a fifth of a hundred or with a frag of twenty,

quintered	20	120	220	320	420	520	620	720	820	920	1,020	1,120	1,220	1,320	1,420

1204 "Overdrawn" describes the uptown neighbors of drawn.

overdrawn	75	175	275	375	475	575	675	775	875	975	1,075	1,175	1,275	1,375	1,475

1205 The rags-to-riches story is that of David, Cinderella, Aladdin, Oliver Twist, Little Orphan Annie, the Foley sisters ("Rags to Riches" by Bernie Kukoff), etc. The Lord "raises up the lowly from the dust; from the dunghill He lifts up the poor." (Ps 113:7) The numbers going from rags to riches include tails seven through two, shichibi, hachibi, kyuubi, love, ichibi and nibi.

rags-to-riches	0	1	2	7	8	9	10	11	12	17	18	19	20	21	22	27	28	29	30	31	32	37	38	39

1206 There are also many riches-to-rags stories. (http://tvtropes.org/pmwiki/pmwiki.php/Main/RichesToRags). The riches-to-rags numbers include the tails three through six, sanbi, yongbi, gobi and rokubi.

riches-to-rags	3	4	5	6	13	14	15	16	23	24	25	26	33	34	35	36	43	44	45	46	53	54	55

1207 "Rat race" comes from jazz fans in the '30s meaning a fast dance and then any difficult, tiring, often meaningless, activity. "Race" describes rokubi-and-cubed-evens and so is not a good complimentary sequence. Better is "baby" from the equivalent "tar" from "Brer Rabbit and the Tar Baby" by Uncle Remus or the earlier *Cherokee Advocate* (1845). "Racial" describes rokubi-and-cubed-ichibi-and-luckies, "racy" rokubi-and-cubed-yongbi.

race	3	8	13	23	33	43	53	63	73	77	78	79	83	87	88	89	103	113	123	133	143

"Biracial" describes binarish-indexed rokubi-and-cubed-ichibi-and-luckies, "facial" Fibonacci-and-cubed-ichibi-and-luckies, "glacial" gobi-luckies-and-cubed-ichibi-and-luckies, "nonracial" non-rokubi-and-cubed-ichibi-and-luckies, "preglacial" gobi-luckies-and-cubed-ichibi-and-luckies downtown neighbors, "postglacial" gobi-luckies-and-cubed-ichibi-and-luckies uptown neighbors and "special" sorted-prime-even-cube-ichibi-and-luckies.

"Celibacy" describes cubed even lucky-indexed binarish-and-cubed-yongbi, "curacy" cube-unindexed rokubi-and-cubed-yongbi, "episcopacy" even prime-indexed sanbi-cubed-odd-primes-and-cubed-yongbi, "fallacy" Fibonacci-and-luckier-luckies-and-cubed-yongbi, "intimacy" ternarish-indexed Mersenne-and-cubed-yongbi downward neighbors, "intestacy" ternarish-even-sorted-ternarish-and-cubed-yongbi downtown neighbors, "lacy" luckies-and-cubed-yongbi, "lunacy" lucky-unindexed nibi-and-cubed-yongbi, "papacy" primes-and-prime-and-cubed-yongbi, "piracy" prime-indexed rokubi-and-cubed-yongbi, "prelacy" luckies-and-cubed-yongbi downtown neighbors and "testacy" ternarish-even-sorted-ternarish-and-cubed-yonbi.

1208 Rain and shine were mated by George Wither in *Philarette* (1622).

rainy	37	137	237	337	437	537	637	737	837	1,137	1,237	1,337	1,437	1,537	1,637

1209 "Sitter" describes sanbi-indexed ternarish with two of three trits.

sitter	10	212	1,121	1,222	10,110	12,221	21,112

1210 Skin and bones were mated in *Hymns to the Virgin and Christ* (1430).

boney	17	117	217	317	417	517	617	717	817	1,017	1,117	1,217	1,317	1,417	1,517	1,617

1211 "Body" describes the complimentary sequence to "soul" and "spirit" everything else. "Now may the God of peace himself sanctify you completely, and may your whole spirit and soul and body be kept blameless at the coming of our Lord Jesus Christ."(1 Th 5:23)

body	6	20	24	26	30	32	48	56	62	66	74	84	86	90	104	108	110	114	126	128
holiness	1	1	0	1	1	0	2	1	1	2	0	2	3	2	1	3	1	0	1	2

spirit	1	2	4	5	7	8	10	11	12	14	16	17	18	19	21	22	23	27	28	29

1212 "Inner space" and "outer space" describe the downtown and uptown neighbors of "space". The Kármán line named for Theodore von Kármánat 100 kilometers above sea level marks the height at which an aircraft must travel at escape velocity to remain airborne.

inner	2	7	12	22	42	52	63	72	82	102	112	162	172	192	215	222	262	282	292
outer	4	8	14	24	44	54	65	74	84	104	114	164	174	194	217	224	264	284	294

1213 "Timeless" describes numbers without sanbi-prime-and-cubed-even mates, "timeout" and "overtime" and "uptime" sanbi-prime-and-cubed-even mates uptown neighbors, "downtime" sanbi-prime-and-cubed-even mates downtown neighbors, "maritime" Mersenne-and-rokubi-indexed sanbi-prime-and-cubed-even mates and "teatime" ternarish-evens-and-sanbi-prime-and-cubed-even-mates.

1214 Spic and span were mated as early as 1579 as "spick and span", meaning "new as a recently made spike and chip of wood". In 1926 "Spic and Span" was trademarked by the Whistle Bottling Company, but when Proctor and Gamble advertised the all-purpose cleaner invented by Elizabeth "Bet" MacDonald and Naomi Stenglein on daytime television dramas they became known as "soap operas".

1215 "Nasp" also describes nibi-and-sanbi-primes, "clasp" cubed-luckies-and-sanbi-primes, "crisp" cubed rokubi-indexed sanbi primes, "cusp" cube-unindexed sanbi primes, "gasp" gobi-and-sanbi-primes, "hasp" hachibi-and-sanbi-primes, "lisp" lucky-indexed sanbi primes, "rasp" rokubi-and-sanbi-primes, "tieclasp" ternarish-indexed even-cubed-lucky-and-sanbi-primes and "unclasp" non-cubed-luckies-and-sanbi-primes.

1216 "Topper" describes ternarish odd primes with two of three trits and alludes to Cosmo Topper haunted by Marion and George Kirby in *Topper* by Thorne Smith.

topper	101	211	2,111	2,221	12,211	21,121	101,111	111,121	111,211	112,111
holiness	1	0	0	0	0	0	1	0	0	0
kirbies	788	898	7,778	7,888	78,878	87,788	887,878	887,888	888,788	888,878
holiness	4	5	2	6	6	6	8	10	10	10

1217 Since they are all even, we also have "half-bottoms".

half-bottom	44	394	449	3,889	3,944	3,894	4,399	4,489	38,944	39,389	39,394

1218 This can be extrapolated to quadruple-digit or higher inflation.

quadruple	4	8	12	16	20	24	28	32	36	40	44	48	412	416	420	424	428
quintuple	5	10	15	20	25	30	35	40	45	50	55	510	515	520	525	530	535
sextuple	6	12	18	24	30	36	42	48	54	60	66	612	618	624	630	636	642

holiness	1	0	2	0	1	1	0	2	0	2	2	1	3	1	2	2	1
septuple	7	14	21	28	35	42	49	56	63	70	77	714	721	728	735	742	749
octuple	8	16	24	32	40	48	56	64	72	80	88	816	824	832	840	848	856
holiness	2	1	0	0	1	2	1	1	0	3	4	3	2	2	3	4	3

1219 Numbers that are not ugly, i. e., with prime factors higher than five, cannot be called beautiful, though many are prime, but they can truthfully be called "unugly".

unugly	7	11	13	14	17	19	21	22	23	26	28	29	31	33	34	35	37	38	39	41	42	43	44

1220 Subtraction and addition can also be combined by adding the infix -ddi- to get addictive "subtraddiction". Subtracting the second digit from the third and adding that to the first.

substraddictive	103	107	113	137	139	157	167	211	223	227	311	313	337	359	421	431
holiness	**1**	**1**	**0**	**0**	**1**	**0**	**1**	**0**	**0**	**0**	**0**	**0**	**0**	**1**	**0**	**0**
substradded	3	7	3	5	7	3	2	2	3	7	3	5	7	7	3	2

1221 Four-digit

subtraddictive	1,021	1,031	1,051	1,087	1,093	1,097	1,153	1,163	1,181	1,291	1,361
holiness	1	1	1	3	2	2	0	1	2	1	1
substradded	2	3	5	2	5	3	2	3	7	7	3

1222 Five-digit

subtraddictive	10,037	10,067	10,079	10,103	10,111	10,133	10,169	10,177	10,211
holiness	2	3	3	2	1	1	3	1	1
primes	5	2	3	5	2	2	5	2	3

1223 Four- and five-digits

exponentiadditive	1,013	1,019	1,117	1,201	1,213	1,217	1,319	1,321	1,361	1,409
holiness	1	2	0	1	0	0	1	0	1	2
exponentiadded	2	2	2	2	2	2	2	3	7	2

exponentiadditive	10,009	10,039	10,061	10,069	10,079	10,091	10,093	10,099
holiness	4	3	3	4	3	3	3	4
exponentiadded	11	11	3	11	11	3	5	11

1224 Four- and five-digits

primes	1,151	1,171	1,231	1,321	1,451	1,471	1,531	1,571	1,621	1,721	1,831
empowered	5	7	3	2	5	7	3	7	2	2	3

primes	10,007	10,103	10,133	10,303	10,607	10,613	10,711	10,903	11,003
empowered	7	3	3	3	7	3	7	3	3

1225 Ben Jahveri communicates with Sandy Banatori in "Short Circuit 2" via dial tone tunes.

California Here I Come	8	8	8	8	9	1	3	3	3	3	6	3	4
holiness	2	2	2	2	1	0	0	0	0	0	1	0	0
sum		16	24	32	41	42	45	48	51	54	60	63	67

Hey Hey We're the Monkees	6	6	5	4	6	6	4	4	4	4	5	5	5	5	6	4	4
holiness	1	1	0	0	1	1	0	0	0	0	0	0	0	0	1	0	0
sum		12	17	21	27	33	37	41	45	49	53	58	63	68	74	78	82

99 Bottles of Beer	6	6	6	1	1	1	6	6	6	6	9	9	9	2	2	2	9	3	3	3
holiness	1	1	1	0	0	0	1	1	1	1	1	1	1	0	0	0	1	0	0	0
sum		12	18	19	20	21	27	33	39	45	54	63	72	74	76	78	87	90	93	96

Rain, Rain Go Away	6	4	6	6	4	6	6	4	9	6	6	4	6	6	4	4	6	6
holiness	1	0	1	1	0	1	1	0	1	1	1	0	1	1	0	0	1	1
sum		10	16	22	26	32	38	42	51	57	63	67	76	83	87	91	97	103

Three blind mice	6	8	4	6	8	4	9	5	5	4	9	5	5	4
holiness	1	2	0	1	2	0	1	0	0	0	1	0	0	0
sum		14	18	24	32	36	45	50	55	59	68	73	78	82

1226 Some ennea phoenixes can be beheaded into a eight-headed hoenix.

okta hoenix	98	980	981	982	983	984	985	986	987	988	989	9,800	9,801	9,802	9,803
holiness	3	4	3	3	3	3	3	4	3	5	4	5	4	4	4

1227 Some okta phoenixes can be beheaded into a seven-headed hoenix.

hepta hoenix	87	870	871	872	873	874	875	876	877	878	879	8,700	8,701	8,702	8,703
holiness	2	3	2	2	2	2	2	3	2	4	3	4	3	3	3

1228 Some hepta phoenixes can be beheaded into a six-headed hoenix.

hexa hoenix	76	760	761	762	763	764	765	766	767	768	769	7,600	7,601	7,602
holiness	1	2	1	1	1	1	1	2	1	3	2	3	2	2

1229 Some hexa phoenixes can be beheaded into a five-headed hoenix.

penta hoenix	65	650	651	652	653	654	655	656	657	658	659	6,500	6,501	6,502
holiness	1	2	1	1	1	1	1	2	1	3	2	3	2	2

1230 Some penta phoenixes can be beheaded into a four-headed hoenix.

tetra hoenix	54	540	541	542	543	544	545	546	547	548	549	5,400	5,401	5,402

1231 Some tetra phoenixes can be beheaded into a three-headed hoenix.

tri hoenix	43	430	431	432	433	434	435	436	437	438	439	4,300	4,301	4,302	4,303

1232 Some tri phoenixes can be beheaded into a two-headed hoenix.

duo hoenix	32	320	321	322	323	324	325	326	327	328	329	3,200	3,201	3,202	3,203

1233 Some duo phoenixes can be beheaded into an one-headed hoenix.

mono hoenix	21	210	211	212	213	214	215	216	217	218	219	2,100	2,101	2,102	2,103

1234 Mono phoenixes either lays an l'oef (egg), are transformed phoenix-like into a new phoenix or into a headless tail.

beheadable	10	100	101	102	103	104	105	106	107	108	109	1,000	1,001	1,002	1,003
holiness	1	2	1	1	1	1	1	2	1	3	2	3	2	2	2
beheaded	0	0	1	2	3	4	5	6	7	8	9	0	1	2	3
holiness	1	1	0	0	0	0	0	1	0	2	1	1	0	0	0

1235 The Headless Horseman threatened Ichabod Crane in "The Legend of Sleepy Hollow" by Washington Irving, so the opposite number to a headless number would be one with at least one head analogous to the -morphic numbers,[1154] like ten-headed primes.

10-headed primes	101	103	107	109	1,009	1,013	1,019	1,021	1,031	1,033	1,039
holiness	1	1	1	2	3	1	2	1	1	1	2

Eleven cannot be included in the eleven-headed primes since it's heartless, "all head and no heart", the "heart" being between head and tail.

11-headed primes	113	1,103	1,109	1,117	1,123	1,129	1,151	1,153	1,163	1,171	1,181

12-headed primes	127	1,201	1,213	1,217	1,223	1,229	1,231	1,237	1,249	1,259	1,277

hundred-headed primes	1,009	10,037	10,039	10,061	10,067	10,069	10,079	10,091

thousand-headed primes	10,007	10,009	100,019	100,043	100,049	100,057	100,069

million-headed primes	100,000,037	100,000,039	100,000,049	100,000,073

1236 The prefix up- usually describes uptown neighbors, "upbear" binarish-evens-and-rokubi uptown neighbors, "upchuck" cubed hachibi-unindexed cubed kyuubi uptown neighbors, "upleap" lucky-evens-and-primes uptown neighbors, "uplift" lucky-indexed Fibonacci ternarish uptown neighbors, "upmost" Mersenne odd sanbi ternarish uptown neighbors, "upping" prime-indexed ng uptown neighbors, "uppish" prime-indexed sorted hachibi uptown neighbors, "uprise" rokubi-indexed sorted evens uptown neighbors, "uprisen" rokubi-indexed sorted even nibi, "uprising" rokubi-indexed sanbi-indexed ng uptown neighbors, "upscale" sanbi-cubed-and-lucky-evens uptown neighbors, "upshift" shichibi-indexed Fibonacci ternarish uptown neighbors and "uptilt" ternarish-indexed lucky ternarish uptown neighbors.

1237 The prefix "down-" and suffix "-down" both usually describe downtown neighbors, "crackdown" cubed-rokubi-and-cubed-kyuubi downtown neighbors, "cutdown" cube-unindexed ternarish downtown neighbors, "downbeat" binarish-evens-and-ternarish downtown neighbors, "downcast" cubes-and-sanbi-ternarish downtown neighbors, "downer" even rokubi downtown neighbors, "downfall" Fibonacci-and-luckier-luckies downtown neighbors, "downfallen" Fibonacci-and-luckier-lucky-even-nibi downtown neighbors, "downhill" hachibi-indexed luckier luckies downtown neighbors, "downplay" prime-luckies-and-yongbi downtown neighbors, "downs" sanbi downtown neighbors (nibi), "downshift" shichibi-indexed Fibonacci ternarish downtown neighbors, "downstate" sanbi-ternarish-and-ternarish-even downtown neighbors, "downy" yongbi downtown neighbors (sanbi), "facedown" Fibonacci-and-cubed-evens downtown neighbors, "hoedown" hachibi-odd/even downtown neighbors, "letdown" lucky even ternarish downtown neighbors, "pulldown" prime-unindexed luckier luckies downtown neighbors, "putdown" prime-unindexed ternarish downtown neighbors, "rubdown" rokubi-unindexed nibi downtown neighbors, "rundown" rokubi-unindexed nibi downtown neighbors, "showdown" show downtown neighbors, "shutdown" shichibi-unindexed ternarish, "splashdown" sanbi-prime-luckies-and-shichibi downtown neighbor, "sundown" sanbi-unindexed nibi downtown neighbors, "teardown" ternarish-even-and-rokubi downtown neighbors and "touchdown" ternarish odd-unindexed cubed hachibi downtown neighbors.

1238 "Peak" alternatively could describe prime-even-and-kyuubi, "beak" binarish-evens-and-kyuubi, "bleak" binarish-luckies-evens-and-kyuubi, "creak" cubed-rokubi-evens-and-kyuubi, "freak" Fibonacci-rokubi-evens-and-kyuubi, '"leak" lucky-evens-and-kyuubi, "outback" odd-unindexed ternarish downtown neighbors, "sneak" sorted-nibi-evens-and-kyuubi, "steak" sorted-ternarish-evens-and-kyuubi and "teak" ternarish-evens-and-kyuubi.

"TriPeaks" is a solitary card game, but could describe a number with three peaks. "BiPeaks" would be a numner with two peaks. Two peaks of equal size would be "twin peaks".

TriPeaks	1,201,020	1,201,021	1,202,020	1,202,021	1,202,120	1,202,121	1,212,020
holiness	3	2	1	2	1	1	0
BiPeaks	12,020	12,021	12,120	12,121	12,131	12,132	13,140
holiness	2	1	1	0	0	0	1
twin peaks	12,021	12,121	13,031	13,131	13,232	14,141	14,142
holiness	1	0	1	0	0	0	0

1239 A valley can also be multiple.

valleys	10,101	10,201	10,202	10,212	10,301	10,302	10,303	10,312	10,313	10,323
holiness	2	2	2	1	2	2	2	1	1	2

1240 A pass can also be multiple.

passes	21,212	21,213	21,214	21,215	21,216	21,217	21,218	21,219	21,312	21,313

1241 A plateau can be nested. The French plural plateaux can also be superpluralized to "plateauxes".

plateaux	122,110	133,110	133,220	133,221	144,110	144,220	144,221	144,330
holiness	1	1	1	0	1	1	0	1

plateauxes	122,110,133,110	122,110,133,220	122,110,133,221	122,110,144,110
holiness	2	2	1	2

1242 Like the plateau the basin can be multiple.

basins	100,112	100,113	100,114	100,115	100,116	100,117	100,118	100,119
holiness	2	2	2	2	3	2	4	3

1243 Everything not undulating is by backformation "dulating".

1244 The zebra is actually black with white stripes, so zebra stripes describe alternating white stripes, i. e., zeroes.

zebra-striped	10	20	30	40	50	60	70	80	90	101	102	103	104	105	106	107	108	109	201
holiness	1	1	1	1	1	2	1	3	2	1	1	1	1	1	2	1	3	2	1

Theodore "Dr. Seuss" Geisel wrote about the gargelorum in *On Beyond Zebra.* The singular form, "gargeli" is eleven-to-the-fifty-first squared or higher than a hundred-sixplex, 16,674,540,931,184,946,922,783,187,978,228,460,499,279,840,752,046,544,384,107,237,-838,107,532,103,558,292,289,097,540,299,192,481,699,966,121

1245 The tiger, unlike the zebra, has black stripes on an orange coat. Tiger stripes therefore describes alternating black stripes, i. e., nines.

tiger-striped	19	29	39	49	59	69	79	89	90	91	92	93	94	95	96	97	98	190	191
holiness	1	1	1	1	1	2	1	3	2	1	1	1	1	1	2	1	3	2	1

1246 The lady and tiger were mated by Frank R . Stanton in "The Lady, or the Tiger?" in 1882. The question is would the barbarian princess rather have her lover killed or married to another? In "The Discourager of Hesitancy" (1885) a prince chooses his unknown bride between a smiling and a frowning lady. The answer to the second choice would reveal the answer to the first, but hesitancy would be fatal.

Tigger refers to that bouncy anthropomorphic tiger from the Winnie-the-Pooh books (1926-8) by Alan Alexander Milne. Other friends of Christopher Robin were timid Piglet, mother Kanga and her baby Roo, melancholic Eeyore, verbose Owl and Pooh aka Edward Bear.

1247 Joyce's exclamation "To infinity and beyond!" was popularized by Buzz Lightyear of "Buzz Lightyear of Star Command".

1248 In John von Neumann's set notation zero was represented by the empty set, { }, and one the by the set including the empty set and the set that includes only the empty set, {{ }, {{ }}}. Infinity would therefore be the set higher than all other possible finite sets, {{ }, {{ }}, {{}, {{ }}}, ...}. Conway's notation refines the notation by defining the set between the leftwardly empty set and the rightwardly empty set, { | }. One would be the number to the right of that, {{ | }| } and minus one the number to the left { |{ | }}. The number to the right of an infinite number of numbers to the right would be {{{...{ | }| }...}, our 1\10, and higher than that would be {{{.{..{ | }| }...}| }, our 1\5.

Conway has also suggested that the names for the days of the week be renamed, counting from zero: Noneday or Sansday, Oneday, Twosday, Trebleday, Foursday, Fiveday and Six-a-day. Alternatively the digits zero, one and three through six could be renamed sun, mon, wedne, thur, fri and satur.

Tolkien translated the Hobbitish names of the week as Sunday, Monday, Trewesday, Hevensday, Mersday, Highday and Sterday. The AJAP naturally had its meetings on Highday,

[NOTE: Seven days without God makes one weak: Sinday, Mournday, Tearsday, Wasteday, Thirstday, Frightday and Shatterday.]

Steven John Robinson has defined "infiniti" as infinity squared plus one over infinity, what Joyceans call "garoopliperoo".

1249 reptile numbers

retropentile	5\1 = 02 = 1\0 + 2
retrosemile	2\1 = 05 = 1\0 + 5
retrovigitile	20\1 = 050 = 1\0 + 50
retroquadrile	4\1 = 052 = 1\0 + 52
retroctile	8\1 = 0521 = 1\0 + 521
retrosedecile	16\1 = 0526 = 1\0 + 526
17\1	07,467,114,925,328,855
retrododecile	12\1 = 380 = [(1\0 - 1)/3] + 47
19\1	012,486,374,987,513,625
retroquadrodecile	14\1 = 5,824,170 = 18\1 + 259
retroctodecile	18\1 = 50 = [((1\0 - 1)5)/9] - 5
47\1	0,716,398,419,135,524,043,278,736,015,808,644,759,567,212
retroquintadecile	15\1 = 60 = (2(1\0 - 1)/3 - 6
retrosextile	6\1 = 61 = 15\1 + 1
retrotridecile	13\1 = 703,296
29\1	0,139,731,427,155,698,602,685,728,443
retroseptile	7\1 = 758,241

1250 Angelic numbers are those with an infinite number of terminating zeroes, in other words, multiples of Aleph-nullplex, named by Matthew Fox in *The Physics of Angels: Exploring the Realm Where Science and Spirit Meet* by Rupert Sheldrake and Matthew Fox. Joyce named them more specifically: ijooplex = 2\0, vooplex = 5\0, xooplex = 10\0, looplex = 50\0, cooplex = 100\0, clooplex = 150\0, dooplex = 500\0, mooplex = 1,000\0, etc.

1251 *Mpossibilities* 68:1 published by the Fortean Mysteries SIG of American Mensa. The operation that transforms a finite or rational about the decimal point (or radix point) into an enfinite (verbalized nfinite, between finite and infinite) or arrational (verbalized rrational, between rational and irrational) one is termed enversion (verbalized beheaded inversion). Infinitely many even numberdromic infinitely left and right.

1\9 + 1/9	...,111,111,111,111,111.11'111'111'111'111'111,...
1\7 + 1/7	...758,241,758,241.14'285'714'285'714'285'714...
1\7+ 1/7 - 1\9 - 1/9	...647,130,647,130,647130.03'174'603'174'603'174'...

π\1 + π	...2,397,985,356,295,144.44'159'265'358'979'323'846...

1252 retrotriviginile

23\5	5,659,680,628,743,403,193,712
23\15	5,968,062,874,340,319,371,256
23\2	6,287,434,031,937,125,659,680
23\12	6,596,806,287,434,031,937,125
23\21	6,806,287,434,031,937,125,659
23\19	7,434,031,937,125,659,680,628
23\16	8,062,874,340,319,371,256,596
23\6	8,743,403,193,712,565,968,062
23\3	9,371,256,596,806,287,434,031
23\9	9,371,256,596,806,287,434,031
23\13	9,680,628,743,403,193,712,565
23\18	4,340,319,371,256,596,806,287

1253 Retroroots

$2^{1\backslash3}$	...9,940,129,952.1
$2^{1\backslash4}$	...511,702,981.1
$2^{1\backslash5}$	...553,896,841.1
$2^{1\backslash6}$	...3,840,264,221.1
$2^{1\backslash7}$	...7,315,980,401.1
$2^{1\backslash8}$	...7,237,795,090.1
$2^{1\backslash9}$	...9,837,950,080.1

1255 The cumlative weights can be helpful with retroroots as well.

[$2^{1\backslash2}$] wt.	4	5	9	11	12	15	20	26	28	32
[$7^{1\backslash2}$] wt.	6	10	15	22	27	28	31	32	33	34
[$3^{1\backslash2}$] wt.	7	10	12	12	17	17	25	25	32	37
[$8^{1\backslash2}$] wt.	8	10	18	22	26	33	34	36	40	47
[$5^{1\backslash2}$] wt.	2	5	11	11	17	23	32	39	46	51

1256 Almost-pointless sequences can be thought of as generated by surreal numbers.

half-phi	1	3	0	8	1	4	0	9	...90,418,031
collapsed	1	11	1	4	13	1	4	2	...2,411,341,111
root two	1	4	1	2	1	3	5	6	...65,312,141

half-pi	1	5	7	0	7	9	6	3	...36,970,751
collapsed	1	5	14	9	6	3	2	6	...623,691,451

1257 "A List of Saints Names" ed. by Rev. John P. O'Connell

1258 pseudonumerological names
20 Anysia, Eunice (even-unindexed nibi-indexed cubed evens), Inez, Neysa; **21** Andy; **22** Ann, Anna, Anne, Hannah, Nan, Nina, Nona; **24** Honore, Honoria, Nora; **26** Ang; **27** Enoch, Hank, Inigo; **30** Maisie; **32** Mona; **34** Mary, Maura, Moira, Myra; **35** Mihel (Mersenne-indexed hachibi even luckies); **37** Mike; **40** Iris, Rosa, Rose; **41** Art, Hart, Howard, Rod, Ward, Wright; **42** Aireen, Irene, Renne; **43** Remy; **45** Israel, Raoul (rokubi-and-odd-unindexed-luckies); **46** Alicia; **47** Eric, Roch; **49** Herb, Rob, Rube; **50** Aloysia, Alice, Alys, Elisa, Eloise, Elsa, Heloise, Ilse, Lisa, Lois (lucky odd-indexed sanbi), Louisa, Lucy (lucky-unindexed cubed yongbi); **51** Ewald, Walt; **52** Alena, Aline, Eileen (even-indexed lucky evenly even nibi), Elaine, Helen, Ilona, Lena, Leona; **53** Elmo; **54** Aulaire, Hilaire, Hilary, Laura, Lora; **55** Howell, Loyola, Ollie (odd luckier lucky-indexed evens), Will, Willie; **57** Alec, Luke; **58** Olive; **59** Ailbe; **60** Joyce; **61** Chad, Jude; **62** Jean, Joan, Joanie, Juana, June; **63** Jim; **65** Joel; **67** Jake; **71** Guido; **72** Eugenia, Gwen; **75** Gale, Gil (gobi-indexed lucky); **79** Gabe (gobi-and-binarish-evens); **81** Vito; **84** Vera; **85** Philo (prime hachibi-indexed lucky odds), Val; **87** Vic; **88** Viv; **91** Bede, Pete; **92** Abina; **95** Abel, Paul; **98** Bev, Davia; **99** Bob (binarish odd binarish); **100** Eudosis, Tess; **104** Desiree; **110** Theodosia; **114** Deidra, Theodora; **120** Athansia, Dionysia; **122** Donna, Etienne (even ternarish-indexed even nibi nibi even); **125** Daniel; **131** Timothy; **132** Adamina, Damiana, Dominia; **133** Timmy, Tommy; **140** Doris, Teresa, Theresa, Therese; **141** Edward; **142** Trina; **147** Derek; **149** Darby; **152** Adelina; **170** Eudoxia; **181** David; **182** Daphne; **191** Thibaud; **194** Debora; **212** Antonia, Nadine; **214** Andrea; **215** Anatole; Wendel; **220** Nancy; **222** Nino, Nonna; **238** Nympha; **242** Norine; **275** Angelo; **277** Nick (nibi-indexed cubed kyuubi); **311** Matt, Matthew; **322** Minnie; **325** Manuel; **327** Mungo (Mersenne-unindexed ng odds); **340** Marius (Mersenne-and-rokubi-indexed/unindexed-sanbi), Maurice; **341** Murtaugh; **342** Amarna, Maran (Mersenne-and-rokubi-and-nibi), Maureen, **343** Miriam; **347** Emeric, Marco, Mark; **352** Melanie; **357** Malachy; **372** Imogene; **375** Michael (Mersenne-indexed-cubed-hachibi-and-even-luckies); **377** Mickey; **394** Imperia; **400** Raissa; **404** Rosaria; **411** Roddy; **412** Ariadne; **421** Randy; **432** Hermione, Ramona, Romaine, Romana; **447** Rurik (rokubi-unindexed rokubi-indexed kyuubi); **451** Harold; **455** Rollo; **472** Regina; **477** Reggie; Ricky; **485** Raphael (rokubi-and-palindromic-hachibi-and-even-luckies); **491** Herbert; **492** Robin; **499** Robbie; **502** Alison, Luciana, Lucina; **512** Lidwina; **517** Ludwig; **520** Leonce; **524** Eleanor, Lenore; **525** Lionel; **542** Loraine; **547** Alaric, Ulric; **551** Elliot; **552** Ellen, Lilian; **553** Wilhelm, William; **570** Alexia, Alix; **582** Alvina; **584** Alvira, Elvira; **608** Josepha; **620** Janice, **622** Jeanne, Jenny, Joanna, Johanna; **632** Irmina; **633** Jimmy; **634** Chris; **641** Jared, Sherwood; **642** Sharon; **643** Jeremy, Jerome; **646** Georgia; **652** Juliana; **673** Joachim; **677** Chuck (cubed hachibi-unindexed cubed kyuubi), Jack; **679** Jacob; **700** Cassie (cubes-and-sorted-sanbi-indexed-evens); **703** Cosmo; **718** Octavia; **720** Agnes; **721** Canute (cubes-and-nibi-unindexed-ternarish-evens), Knute; **726** Ignacia; **740** Grace; **741** Curt, Kurt; **742** Carina, Karen; **745** Carl (cubes-and-rokubi-luckies), Carlo, Karl, Karol; **747** Greg; **751** Claude, Kent; **752** Aquilina; **753** Clem, Colum; **754** Clare (cubed-luckies-and-rokubi-evens), Gloria; **759** Caleb; **822** Fanny, Yvonne; **841** Fred; **842** Verona; **843** Ephrem; **851** Vlad; **852** Evelyn; **854** Flora, Valery; **856** Felicia; **858** Flavia; **859** Philip (palindromic hachibi-indexed lucky-indexed

primes); **872** Iphigenia; **882** Vivian; **900** Bessie; **905** Basil (binarish-and-sanbi-indexed-luckies); **921** Hubert; **941** Bart, Bert, Hobart; **943** Abraham, Bram; **951** Plato (prime-luckies-and-ternarish-odds); **952** Pauline; **955** Bill (binarish-indexed-luckier-luckies); **990** Babs (binarish-and-binarish-sanbi); **992** Bibiana; **999** Bobbie; **1072** Toscana; **1147** Dietrich, Theodoric; **1251** Donald; **1302** Thomasine; **1321** Edmund; **1327** Dominic; **1382** Dymphna; **1412** Edwardine; **1414** Deirdra; **1422** Adrianne; **1470** Dorcas; **1477** Dirck; **1540** Dolores; **1582** Adelphina, Delphine; **1731** Dermot; **1851** Theobald; **2010** Anastasia, Anstice; **2053** Anselm; **2125** Nathanael; **2155** Wendell; **2752** Angelina; **3152** Madelina; **3241** Maynard, Meinrad; **3325** Emmanuel, Immanuel; **3352** Emmiline; **3405** Marcel; **3411** Meredith; **3412** Martina; **3417** Murdoch; **3422** Marianne; **3452** Marlene; **3465** Marshal, Martial; **3500** Melissa, Milissa; **3702** Maxine; **3742** Macrina; **3941** Humbert; **4022** Rosanne, Rozanne; **4034** Rosemary; **4052** Roslyn; **4127** Harding; **4147** Roderigo; **4201** Ernest; **4215** Randal; **4251** Arnold, Ronald; **4321** Armand, Raymond; **4521** Orlando, Roland; **4641** Richard; **4702** Roxana; **4741** Ricardo; **4827** Irving; **4941** Robert, Rupert; **5214** Leonard, Leonardo; **5251** Reynard; **5410** Lourdes; **5420** Lorenza; **5426** Laurentia, **5532** Wilhelmina; **5574** Allegra; **5740** Lecrece; Lecreza; **5746** Lecretia; **5820** Alfonsa, Alphonsa; **5841** Alfred, Alfredo, Wilfred; **5842** Alvernia; **5941** Albert, Alberto, Leopard, Wilbert; **5947** Alberic; **5947** Leopold; **6012** Justina; **6052** Jocelyn; **6082** Josephine; **6099** Giuseppe; **6288** Genevieve; **6423** Geronimo; **6432** Charmaine, Germaine; **6441** Gerard, Gerhard, Girard; **6451** Gerald, Geraldo, Jarlath; **6452** Georgiana; **7142** Catherine, Katherine, Kathryn, Katrina; **7152** Catalina, Kathleen; **7158** Guadalupe; **7207** Gonzaga; **7217** Caradoc; **7221** Kenneth; **7241** Conrad, Konrad; **7253** Kenelm; **7355** Gamaliel; **7410** Caritas (cubes-and-rokubi-indexed-ternarish-and-sanbi); **7432** Carmen; **7437** Cormac; **7452** Careen, Caroline; **7462** Gretchen; **7512** Claudine; **7540** Clarice (cubed-lucky-androkubi-indexed-cubed-evens), Gloriosa; **7941** Egbert; **7945** Gabriel; **8005** Vassily; **8012** Faustina; **8101** Vedast; **8200** Vanessa; **8427** Frank; **8441** Everard; **8465** Virgil; **8472** Virginia; **8532** Philomena; **8559** Phillip (palindromic hachibi-indexed luckier lucky-indexed primes); **8714** Victoria; **8822** Vivienne; **9075** Pascal (prime-and-sanbi cubes-and luckies); **9120** Patience) primes-and-ternarish-indexed-even-nibi-cubed-evens); **9140** Beatrice; **9146** Patricia; **9147** Padraic; **9415** Bartel; **9420** Berenice, Bernice; **9421** Briant; **9429** Barnaby; **9443** Borromeo; **9494** Barbara; **9501** Placid; **9526** Blanche; **9552** Apollina, Apolline; **9771** Becket; **10021** Toussaint; **15941** Adalbert, Adelbert; **24000** Narcissa; **24941** Norbert; **32841** Manfred; **34655** Marshall; **35753** Malcolm; **37152** Magdalene; **39402** Ambrosine; **40055** Russell; **41321** Redmond; **41477** Roderick; **42012** Ernestine; **46951** Archibald; **47251** Reginald; **52051** Lancelot; **53941** Lambert; **57900** Alexis; **58202** Alphonsine; **58727** Wolfgang; **59471** Albrecht; **62284** Jennifer; **64012** Christina; **64512** Geraldine; **67752** Jacqueline; **70941** Gaspard; **71941** Cuthbert; **72152** Gwendolyn; **74012** Kerstin; **74455** Carroll; **74881** Griffith; **75914** Cleopatra; **75941** Gilbert; **82021** Vincent; **84200** Frances; **85022** Felicenne; **85212** Valentina; **85420** Florence; **85941** Philbert; **87142** Victorine; **91470** Beatrix; **91477** Patrick; **92171** Benedict; **92211** Bennett; **94094** Prosperia; **94120** Prudence; **94143** Bertram; **94151** Berthold; **94153** Bartholomew; **94212** Bernadine; **94241** Barnard, Bernard; **95212** Blandina; **95271** Plunket; **95749** Polycarp; **99101** Baptiste; **155941** Ethelbert; **500214** Alessandra; **539471** Lambrecht; **570214** Alexandria; **700214** Cassandra; **720120** Constance; **720912** Conception; **740013** Chrysostom; **753212** Clemintine; **841221** Ferdinand; **841477** Frederick; **854212** Florentina; **941177** Braddock; **941421** Bertrand; **5149421** Hilderbrand; **5702142** Alexandrina

1259 Like "Lola" by Ray Davies in "Lola vs. Powerman and the Moneygoround, Part 1" by the

Kinks.
11 Dot, Edith, Etta, Hedda, Huetta; **12** Aidan, Dan, Dion, Don, Edwin, Tony; **14** Thiery; **15** Adela, Del, Delia, Ethel, Odile; **17** Hedwig; **18** Dave; **19** Deb; **20** Aeneas, Hans; **21** Anita, Nita, Wanda, Wendy; **24** Henry; **26** Ang; **30** Amos; **31** Amata, Maitha, Maude; **32** Eamon; **33** Emma, Mame, Mimi; **34** Amory, Emery, Mario (Mersenne-and-rokubi-indexed-odds), Maur, More; **35** Amelia, Emilia, Emily; **41** Rita; **42** Aaron, Arnie, Ernie, Rene, Ron; **43** Erma, Irma; **44** Harry, Rory; **45** Aurelia; **46** Richy, Rog; **47** Erica; **48** Harvey, Rafe (rokubi-and-Fibonacci-evens); **50** Elias, Louis (lucky odd-unindexed/indexed sanbi); **51** Alodia, Hilda, Holda, Hulda, Lydia; **52** Alaine, Alan, Leon; **53** Alma; **54** Hilaire; **55** Ella, Eulalia, Illa, Lelia, Lily (lucky-indexed-lucky-yongbi), Lola; **57** Helga, Olga; **58** Alf, Olaf; **59** Alba, Lupe; **60** Josue; **62** Jean; John, Juan; Shane (shichibi-and-nibi-evens), Shaun, Shawn; **65** Julia, Julie, Sheila; **72** Egan, Eugene, Ken, Quin; **74** Igor; **75** Cleo (cubed luckies even/odd), Gale (gobi-and-luckie-evens); **80** Ives, Yves; **81** Faith; **82** Evan, Ivan; **83** Euphemia; **84** Ivor; **85** Viola; **89** Phoebe; **90** Pius; **91** Beata, Beth; **92** Ben; **95** Paula; **97** Peg; **102** Declan, Hudson; **104** Desire; **107** Tosca; **110** Titus (ternaris-indexed-ternarish-unindexed-sanbi); **111** Dotty, Odette; **114** Theodore, Tudor; **115** Ottilia; **120** Denis; **121** Donata; **122** Denny, Donnie; **125** Daniela; **130** Thomas; **131** Timothea; **132** Damian, **134** Othmar; **141** Dorothy; **142** Adrian, Hadrian; **151** Adelaide, Tilda; **155** Della, Dolly, Tillie (ternarish-indexed luckier lucky-indexed evens); **158** Adolf, Adolph; **170** Hodges; **174** Edgar; **175** Thecla; **181** Devota; **190** Tobias; **191** Tabitha; **199** Debbie; **211** Netta; **212** Anthony, Anton, Antony, Nathan; **214** Andre, Andrew; **215** Natalie (nibi-and-ternarish-and-lucky-indexed-evens); **220** Johannes; **241** Honorata; **247** Enrica; **250** Niles; **255** Nell (nibi even luckier luckies); **270** Aengus, Angus; **275** Angela; **300** Moses; **310** Amadeus; **321** Amanda; **325** Manuela; **327** Monica; **340** Marius, Maurice; **341** Martha; **342** Marion, Myron; **343** Miriam; **347** Margo; **350** Miles, Myles; **351** Humility; **352** Emilian; **355** Aemillia, Milly (Mersenne-indexed luckier lucky yongbi), Molly; **370** Max, Micas (Mersenne-indexed-cubes-and-sanbi); **374** Macaire; **375** Michaela; **377** Maggie; **384** Humphrey; **385** Amabel, Mabel; **400** Russ (rokubi-unindexed sorted sanbi); **402** Orson; **405** Rosalie, Ursula; **413** Artemia; **414** Arthur; **422** Ronan, Ronny; **424** Werner; **430** Hermes; **432** Armon, Herman, Ramon, Romain; **441** Harriet; **442** Warren; **451** Imelda; **452** Aurelian; **458** Ralph, **474** Roger, **465** Rachel; **470** Rex; **480** Rufus (rokubi-unindexed Fibonacci-unindexed sanbi); **482** Irvin; **485** Raphaela; **492** Robin, Reuben; **500** Aloysius; **502** Wilson; **505** Lucile (lucky-unindexed cube-indexed lucky evens); **511** Letty, Lottie, Yolette; **514** Lothaire, Lothar, Lothario, Walter; **520** Alonso, Alonzo, Lance (lucky-and-nibi-cubes-even), Leonce; **521** Leonita, Yolanda; **534** Elmer; **538** Olympia; **540** Lars; **544** Larry; **547** Ulrica; **550** Ellis (even luckier lucky-indexed sanbi); **551** Lolita; **552** Allan, Allen; **555** Louella, Luella; **570** Alex, Eligius, Lucus (lucky-unindexed cube-unindexed sanbi); **571** Electa; **572** Alcuin; **574** Alger, Elgar, Liguori; **577** Alacoque; **582** Alvin; **584** Oliver; **592** Alban, Albin; **594** Wilber; **599** Libby; **600** Jesse, Josias; **601** Justa; **602** Jason; **508** Josef, Joseph; **511** Judith; **520** Jonas; **521** Janet, Juanita; **622** Johnny; **630** James, Shamus (shichibi-and-Mersenne-unindexed-sanbi); **633** Gemma; **640** Chris, Joris; **641** Charity; **642** Sherwin; **644** Jerry; **646** George; **650** Giles, Julius; **651** Juliet; **652** Julian; **655** Jill; **672** Joaquin; **688** Jeff; **701** Augusta; **703** Cosima; **711** Kitty; **712** Gedeon, Gideon, Godwin; **714** Gauthier; **718** Octave; **720** Aquinas, Ignace, **722** Conan, Kenny; **724** Conor; **740** Garcia; **741** Gerty, Greta; **742** Cornie, Kieran (kyuubi-indexed-even-rokubi-and-nibi; **745** Carol; **750** Claus; **751** Claudia, Gilda; **752** Colin; **754** Clarey; **782** Gavin, Kevin; **810** Vitus; **811** Yvette; **813** Fatima; **822** Finian; **840** Valerius; **841** Freda; **845** Averil; **846** Virg; **851** Violet; **852** Evelyn; **858** Flavio; **859** Felipa; **874** Fiacre; **877** Vicky; **892** Fabian; **895** Fabiola;

910 Boethius; **911** Betty, Patty; **914** Pedro, Peter, Pietro; **921** Benita; **922** Benny; **927** Bianca; **940** Boris, Brice, Bruce, Percy, Piers; **941** Bertha, Bride, Eberta; **942** Barney, Bernie, Brian, Bruno, Byron; **944** Barr, Barry; **945** Beryl, Pearl; **947** Burga; **949** Barb; **950** Blaise; **952** Blaine; **955** Bella, Polly; **977** Becky, Peggy; **1030** Dismas; **1110** Thaddeus; **1200** Athanasius, Dionysius; **1211** Danette; **1220** Dennis; **1255** Danielle; **1314** Dmitri; **1321** Edmunda; **1327** Dominique; **1420** Terence; **1422** Tiernan; **1521** Ethelreda; **1555** Tallulah; **1947** Edburga; **2050** Weneslaus; **2074** Ansgar; **2174** Notker; **2201** Nunciata; **2211** Annette, Nanette, Ninette; **2411** Henrietta; **2741** Ingrid; **2750** Nicholas; **2757** Angelica, Angelique; **2841** Winifred; **2952** Napoleon; **3101** Modesa; **3110** Matthias; **3151** Mathilda; **3211** Minette; **3412** Martin; **3462** Marcian; **3470** Americus, Marcus (Mersenne-and-rokubi-cubed-unindexed-sanbi); **3471** Margot; **3472** Morgan; **3574** Maelchior; **3655** Michelle; **3720** Magnus; **3940** Ambrose; **3955** Amabella; Maybelle; **4010** Orestes; **4030** Erasmus (even-rokubi-and-sanbi-Mersenne-unindexed-sanbi); **4059** Rosalba; **4120** Hortenius; **4130** Artemas, Artemus; **4147** Roderica; **4158** Rudolf; **4251** Ronalda; **4411** Harrietta; **4521** Orlanda; **4641** Richarda; **4941** Roberta; **4977** Rebecca; **5011** Lisette; **5014** Alastair; **5021** Lucinda; **5040** Lazarus; **5055** Lucilla, Lucille (lucky-unindexed cube-indexed luckier lucky evens); **5091** Elizabeth, Elsbeth, Elpeth; **5187** Ludovica; **5214** Leander; **5241** Leonarda; **5410** Lauritz; **5411** Laurette, Loretta; **5420** Laurence, Lorenz; **5700** Alexis; **5701** Lucasta; **5841** Alfreda, Elfreda, Wilfreda; **5852** Loughlan; **5941** Alberta; **5947** Walburga; **5951** Leopolda; **6007** Jessica; **6010** Justus; **6012** Justin; **6014** Chester; **6021** Jacinta; **6094** Jasper; **6211** Jeanette; **6412** Jordan; **6430** Jeremias; **6432** Germain; **6450** Charles; **6451** Geralda; **6472** Jurgen; **6480** Gervais, Gervase, Jarvis; **6511** Juliette; **6822** Giovanni; **6884** Geoffrey, Jeffrey; **7010** Augustus; **7012** Augustine, Gaston; **7018** Gustave; **7030** Cosmas; **7034** Casimir (cube-and-sanbi-indexed-Mersenne-indexed-robubi); **7094** Caspar (cubes-and-sanbi-primes-and-rokubi), Gaspar (gobi-and-sanbi-primes-and-rokubi); **7180** Octavius; **7182** Octavian; **7184** Godfrey; **7205** Consuelo; **7210** Cantius, Quintus; **7211** Candida; **7212** Quentin; **7214** Gunther; **7260** Ignatius; **7355** Camilla; **7392** Campion; **7410** Quartus; **7415** Cordelia; **7422** Cronan; **7435** Carmela; **7450** Carlos; **7460** Gratius; **7474** Gregory; **7488** Griff; **7500** Gelasius; **7510** Gildas; **7511** Colette; **7512** Coleman; **7539** Columba; **7541** Clareta; **7550** Achilles; **7552** Killian; **7580** Clovis; **7612** Cajetan; **8014** Foster; **8151** Fidelity; **8211** Venetta; **8294** Finbar; **8427** Veronica; **8470** Fergus; **8501** Felicity; **8502** Felician; **8511** Violette; **8520** Valens; **8532** Philemon; **8542** Florian, Valerian; **8570** Felix; **8582** Flavian; **8599** Philippa; **8714** Victor; **9012** Bastien; **9014** Webster; **9280** Boniface; **9412** Bertin, Burton; **9417** Brudgey; **9421** Brenda; **9440** Porres; **9471** Brigid; **9491** Perpetua; **9511** Paulette; **9512** Baldwin; **9845** Beverly; **9911** Babette; **11471** Attracta; **12012** Dunstan; **13140** Demetrius; **14012** Thurstan; **14055** Drusilla; **14070** Tarsicius; **14420** Terrence; **17014** Dexter; **20100** Anastasius; **21211** Antoinette; **22020** Innocence; **22201** Annunciata; **24000** Narcissus; **27511** Nicolette; **31010** Modestius; **32000** Manasses (Mersenne-and-sorted-sanbi/even-sanbi); **33751** Immaculata; **34741** Marguerite; **35141** Mildred; **40321** Rosamund; **40521** Rosalind; **42158** Randolph; **43521** Ermelinda; **51050** Ladislas; **51355** Ludmilla; **51741** Hildegard; **53941** Lamberta; **55552** Llewellyn; **58200** Alphonsus; **64141** Gertrude; **64511** Charlotte; **64611** Georgette; **73550** Camillus (cubes-and-Mersenne-indexed-luckier-lucky-unindexed-sanbi); **74003** Carissima; **74012** Christian; **74012** Kursten; **74250** Cornelius; **74255** Cornelia; **74351** Carmelita; **74511** Carlotta; **75111** Claudette; **75420** Clarence; **75501** Callista; **75941** Gilberta; **79455** Gabrielle; **84221** Fernande; **85134** Vladimir; **85212** Valentine; **92171** Benedicta; **92632** Benjamin; **94094** Prosper; **94151** Bertild; **94155** Bertilla; **94171** Bridget; **94212** Brandan, Brendan; **94222**

Brennan; **94290** Barnabus; **95050** Boleslaus; **95104** Balthasar; **99101** Baptista; **340550** Marcellus; **570214** Alexander; **640195** Christabel; **721742** Kertigern; **740184** Christopher, **740761** Crescentia; **753920** Columbanus; **842607** Francesca; **851050** Vladislaus; **914255** Petronilla; **927410** Pancratius; **940055** Priscilla; **942111** Bernadette; **946320** Berchmans; **3703552** Maximillian; **7201212** Constantine; **9218214** Bonadventure

1260
0.10 Eustace (even-unindexed-sanbi-ternarish-and-cubed-even), Stasia; **0.12** Austin, Sidney, Stan (sanbi-ternarish-and-nibi); **0.12050** Stanislaus; **0.125** Stanley; **0.14** Asteria, Esther, Hester, Isadore; **0.140** Esdras; **0.141** Astrid; **0.155** Estelle, Eustella, Stella; **0.182** Stephanie, Stephen; **0.21** Cynthia, Hyacinth, Sandy; **0.214** Sandor, Sandra; **0.22** Susanna, Suzanne; **0.26** Sancho; **0.271** Sancta; **0.2754** Sinclair; **0.302** Samson; **0.32** Simeon, Simon, Simona; **0.321** Osmond; **0.35** Samuel, Samuela; **0.40** Azarius, Cyrus; **0.45** Cyril, Israel; **0.455** Cyrilla; **0.47** Serge; **0.470** Sergius; **0.482** Seraphina; **0.501** Celeste; **0.5012** Celestina; **0.51** Oswald; **0.52** Celine, Selena; **0.53** Salome, Selma; **0.532** Solomon; **0.55** Sally; **0I58** Silvio; **0.58014** Silvester; **0.5814** Salvador; **0.59** Silvia; **0.7010** Sixtus; **0.70321** Sigmund; **0.74** Oscar, Zachary; **0.740** Zacharias; **0.75** Ezechiel; **0.75017** Scholastica; **0.77** Zack; **0.7841** Siegfried; **0.82** Savina; **0.84** Xavier; **0.9012** Sebastian; **0.9130** Septimus; **0.92** Sabina; **0.942** Cyprian; **0.95** Isabel, Sibyl; **0.955** Isabella

1261 see http://www-history.mcs.st-and.ac.uk/Chronology/30000BC_500BC.html

150,000,000 BC ants count paces (*The Math Book*)
30,000,000 BC primates count
1,000,000 BC non-prime number cicadas eaten by composite-number-eater
35,000 BC Lebombo tally bone
18,000 BC Ishango bone with primes to 19
8500 BC 6th prime discovered, Lebembo Mts., Swaziland
5000 BC Caral, Peru, quipu and Burnt City, Iran, dice and backgammon, decimal system in Egypt
3000 BC Babylonians use base 60, geometry, abacus
2200 BC lo-shu magic square
1950 BC Babylonians solve quadratic equations, $5^2/2^3 < \pi < (4/3)^4$, $\pi \approx 4{,}073/1{,}296$
1850 BC Ahmes $\pi \approx 2^8/3^4$
1800 BC Plimpton 322 tablet with Pythagorean triplets
1747 BC Jacob offers Esau 220 sheep and goats (amicable no. Gen 32:14)
1650 BC Rhind papyrus with prob. 79: 7-houses-cats-mice-ears-measures-grain by Ahmes
1300 BC tic-tac-toe
530 BC Pythagoreans telepathically communicate with numbers 1 to 10
517 BC Pythagoras' students, Pholoclese and Dionesa discover irrational numbers
500 BC Pingala, Hindu, describe Fibonacci numbers, "Pythagorean triplets" in *Sulvasutras*
465 BC Hippasus describes dodecahedron
425 BC Theodorus of Cyrene discovers irrational root to $\sqrt{17}$
335 BC Dinostratos squares circle with quadratrix
300 BC Euclid *Elements*, Babylonian counting board
275 BC Euclid, Greek mathematician, discovers 496 and 8,128 perfect

250 BC Archimedes discovers pi 223/71 > π > 22/7 (π ≈ 3,123/994), *Sand Reckoner* $10^{202,544}$, Dzu Tse, Han dynasty Taoist, developed Jootsy calculus; Sun-tsi Chinese Remainder Theorem, Diophantus of Alexandria writes *Arithmetica*
230 BC Eratosthenes of Cyrene invents prime sieve
211 BC Archimedes of Syracuse, b. 287 BC, dies trying to prove his Last Theorem with last words, "Don't disturb my circles!", son of astronomer Phidias, etc., invented high number system based on myriad, posed bovinum problema unsolved for 2,165 years famous for "Eureka!" ("I have discovers it!")
190 BC Apollonius of Perga "the Great Geometer" wrote *Conics*, invented superscription notation for higher numbers -- in Roman numerals
150 BC Hypsicles writes *On the Ascension of Stars,* divides Zodiac into 360°
100 BC Nichomachus of Gerasa publishes *Introduction Arithmetica* $\Sigma(2z + 1) = z^3$,
1 AD Liu Hsin uses decimal fractions.
125 AD Chiag Hong *The Mathematical Art* $\pi \approx \sqrt{10}$, zero
150 AD Claudius Ptolemy *Syntaxis Mathematica, Almagest,* law of sines, π ≈ 377/120,
250 AD Mayan base-20
264 AD Liu Hsin calculates π ≈ 3.14159 via 3,072-gon
320 AD Iamblichus of Chalcis amicable numbers
350 AD Bakhshali manuscript
380 AD Hindus π ≈ 3,927/1,250
415 AD Hypatia, Greek female mathematician killed by mob, scholars flee Alexandria
460 AD Zu Changzhi π ≈ 355/113
499 AD Aryabhata π ≈ 62832/20000
540 AD figurate numbers
594 AD zero used in India
628 AD Brahmagupta writes *The Opening of the Universe*
700 AD Mayans use zero
800 AD Muhummad ibn Musa al-Khwarizmi uses zero (sifr), Flaccus Albinus Alcuinus of York
830 AD Al-Khwarizmi's *Algebra*
850 AD Thabit ibn Qurra amicable number formula, magic squares, algebra; Mahavira's pearl problem $1161/(1/2)^7 \approx 10^{24}$
900 AD Pathodius teachs pataphysics based on Methodius' metaphysics
950 AD Gerbert of Aurillac (future Pope Sylvester II) reintroduces abacus and Arabic numerals into Europe
953 AD Abu Hasan Ahmid ibn Ibrahim al-Uqlidisi used decimal point
990 AD Abu Bakr ibn Muhammed ibn al-Hussayn Al-Karaji's [Pascal] triangle
1000 AD nepohualtzitzin (Aztec treaded corn kernel abacus)
1070 BC Omar Khyyyam's binomial theorem, non-Euclidean geometry
1150 AD Bhaskara *Lilivaki,* Ibn Yahya al-Maghribi al-Samawal's *The Dazzling in Algebra,* $x^0 = 1$, $n(n+1)(2n+1)/6$
1200 AD suan-pan (Japanese abacus)
1202 AD Fibonnacci writes *Liber abaci,* π ≈ 3.141818
1206 AD Yang Hui's "Pascal triangle"
1256 AD chessboard doubling tale retold $2^{64} - 1 > 10^{19}$
1258 AD Bagdad House of Wisdom destroyed by Mongols, Tigris runs black with bookblood
1303 AD Zhu Shijie *The Precious Mirror of the Four Elements* (Pascal's triangle)

1310 AD Ibu al-Banna discovers 8th amicable pair
1350 AD Nicole Osresme discovers harmonic series divergence, $\sum 1/n = \infty$
1360 AD Nicole Oresme *Algorismus Proportionum* fractional exponents
1427 AD Ghivath al-Din Jamshid Massud al-Kashi discovers law of cosines
1429 AD Ghivath al-Din Jamshid Massud al-Kashi discovers $\pi 10^{16}$
1478 AD *Treviso Arithmetic*, 1st printed recreational math
1488 AD Nicholas Chuquet invents é*chelle courte* (short count) number system
1489 AD Johannes Widman uses plus, +, and minus, - signs
1491 AD Filippo Calandri long division
1499 AD *Steganographia* by Johannes Trithemius, cryptology mistaken for black magic
1500 AD Nilakantha Somayaji publishes *Tantrasangraha*, $\pi/4 = \sum 1 \pm 1/(2n+1)$
1509 AD Fra Luca Bartolomeo de Pacioli's *Divina Porportion* on ϕ
1525 AD Christoff Rudolff *Die Cross* uses squareroot sign, √, Scipione del Ferro solves $z^3 + mz = n$
1526 AD Gerolamo Cardano's *Liber de ludo aleae* probability
1535 AD Nicholo di "Tartaglia" Brescia solved $z^3 + mz^2 = n$
1536 AD Hudalrichus Regius discovers $2^{12}(2^{13} - 1)$ perfect
1540 AD Lodovivo Ferrari solves quartic equation, Rafael Bombelli disc. imaginaries
1542 AD Robert Recorde *The Grounde of Arts, Teachyng the Works and Practice of Arithmetike*
1545 AD imaginary numbers in *Artis magnae* by Gerolamo Cardano (named by Descarte, i named by Euler)
1549 AD Jacques Pelletier invents *échelle longue* (long scale)
1555 AD J. Scheybl discovers $2^{16}(2^{17} - 1)$ prime
1556 AD Nicholo di "Tartaglia" Brescia uses parentheses
1557 AD Robert Recorde popularizes equality sign, =
1559 AD Johannes Buteo *De quadratura circuli* history of π
1572 AD Rafael Bombelli's *Algebra*, imaginary numbers
1573 AD Valentinus Otho discovers $\pi 10^{7}$
1576 AD Gerolamo Cardano commits suicide
1579 AD François Viete $\pi = 1/\sqrt{½}(\sqrt{½} + ½(\sqrt{½}(\sqrt{½} + ½(\sqrt{½} ...)))$ and via 393,216-gon
1585 AD Simon Stevin uses decimal fractions
1593 AD Adriaan van Roomen discovers $\pi 10^{15}$ via 2^{30}-ogon, Simon Duchesne $\pi \approx (39/22)^2$, Christopher Clavius uses dot for multiplication
1596 AD Ludolph von Ceulen $\pi 10^{20}$
1588 AD Cataldi discovers 10- and 12-digit perfects
1600 AD Yazdi discovers 104th amicable pair
1605 AD John Napier invents first calculator
1610 AD Ludolph van Ceulen discovers $\pi 10^{35}$ via 2^{62}-ogon
1611 AD Johann Kepler re-discovers "Fibonacci numbers", conjectures max. packing = $\pi/\sqrt{18} \approx 74\%$ (proved 1998)
1614 AD John Napier publishes logarithm table
1617 AD John Napier *Rabdologiae*, popularized decimal point, 1st mechanical calculataor
1619 AD John Napier's *Mirifici logarithmorum canonis constructio* published posthumously
1621 AD slide rule rules (until 1948)
1624 AD Wilhelm Schickardt's calculating machine, Henry Briggs coins mantissa and

characteristic
1630 AD Johannes Kepler died, German astronomer uses base 3, Grienberger discovers $\pi 10^{39}$
1631 AD Marin Mersenne poses discovering multiple perfects, William Oughtred uses x for multiplication, Thomas Harriot uses > and <
1636 AD Pierre de Fermat rediscovers amicables 17,296 and 18,416 (previously discovers by al-Banna bef. 1321)
1637 AD Fermat's Last Theorem (proved 1994)
1638 AD Rene Descartes discovers amicables 5020 and 5564
1640 AD Pierre de Fermat $4z + 1 = n^2 + m^2 = p$
1641 AD Evangelista Torricelli's (or Gabriel's) finite-volume/infinite-area trumpet
1642 AD Blaise Pascal adding machine
1644 AD Fr. Marin Mersenne discovers primes, 8,191 and 131,071 and 524,287 and predicted other Mersenne primes
1650 AD John Wallis $\pi = 2(\prod(2z)^2/\prod(2z + 1)^2)$
1654 AD Pierre de Fermat discovers laws of probability, Blaise Pascal's triangle
1655 AD John Wallis uses ∞ and negative exponents
1657 AD William Neile discovers arc length of semicubic $x^{3/2}$
1659 AD Johann H. Rahn uses
1665 AD Isaac Newton discovers $\pi 10^{16}$
1666 AD Thomas Morland invents multiplying machine
1669 AD iterative approximation $x_{n+1} = x_n - f(x_n)/f(-x_n)$
1670 AD Fermat's son publishes his marginal notes including Fermat's Last Theorem
1674 AD Gottfried Leibnitz $\pi = 4\Sigma(1 \pm (2z + 1))$, astroid $x^{2/3} + y^{2/3} = r^{2/3}$ ($L = 6r$, $A = 3\pi r^2/8$)
1679 AD Gauss uses binary
1684 AD Gottfried Leibnitz invents differential calculus
1685 AD John Wallace logarithm = exponent
1686 AD Adamas Kochansky magic cubes, Gottfried Leibnitz invents integral calculus
1693 AD Bernard Frenicie de Bessy discovers all 880 4x4 magic squares
1699 AD Abraham Sharp discovers $\pi 10^{71}$
1706 AD John Machen discovers $\pi 10^{100}$, William Jones uses π
1713 AD Jacob Bernouli's Law of Large Numbers $\sum x_n/n = E(x)$
1719 AD Roger Cotes discovers $e10^7$, DeLagny discovers $\pi 10^{112}$
1722 AD Takebe $\pi 10^{41}$ via 1024-gon
1727 AD Leonhard Euler names e, John Hill discovers smallest pandigital square $11{,}826^2$
1730 AD James Sterling's *Methodus Differentialis*, approximation $n! \approx n^n e^{-n}\sqrt{(2\pi n)}$
1732 AD Leonard Euler $F_5 = 232 + 1 = 641(6{,}700{,}417) \neq p$
1733 AD Abraham de Moivre bell curve $(\exp(\int x-\mu)^2/x_0^{\,2}))/\sigma\sqrt{2\pi}$
1738 AD Daniel Bernouli's St. Petersburg paradox, $V(x) = \sum xP(x) = \sum 2^n/2n = \infty$
1739 AD Matsanuga discovers $\pi 10^{50}$
1742 AD Christian Goldbach's conjecture $p = p_1 + p_2 + p_3$
1746 AD Kruger's 100,000 primes list
1747 AD Euler discovers 76th amicable pair
1750 AD Leonhard Euler discovers 31st Mersenne prime, $M_{31} = p$
1751 AD Jedediah Buxton (2,3145,789(5,642,732)54,965 yd^3 in in^3 =

37,040,672,287,483,822,689,500,528,640; Euler's dissection no. $E_n = \Pi(4n-10)/(n-1)!$, V - E + F = 2
1753 AD Leonhard Euler $a^3 + b^3 = c^3$
1754 AD Abraham De Moivre dies when lim(awaketime/sleeptime) = 0
1761 AD Thomas Bayes's probability theorem P(A|B) = P(B|A)P(A)/P(B)
1766 AD Euler goes blind
1769 AD Benjamin Franklin's bent magic squares
1770 AD Leonhard Euler *Complete Introduction to Algebra*, Edward Waring *Meditationes algebraicae*
1772 AD Leonhard Euler develops integral calculus, discovers 19-digit perfect, convenient numbers, $p = n^2 - n + 41$
1773 AD Leonhard Euler proves existence of God with $(a + b)^n/n = x \rightarrow$ Gott
1776 AD Johann Heinrich Lambert 408,000 factor table, Antonio Felkel prime factors to 2,000,000
1761 AD Johann Heinrich Lambert proves $\pi \neq n/n'$
1777 AD Comte de Buffon discovers π probabilistically
1794 AD Georg Vega discovers $\pi 10^{140}$
1795 AD Carl Frederick Gauss, 18, discovers least squares
1796 AD Carl Frederick Gauss, 19, proves $n = \Delta + \Delta + \Delta$
1797 AD Georg Vega primes to 400,031
1801 AD Carl Frederick Gauss, 24, *Disquisitiones arithmeticae,* modular arithmetic
1807 AD Jean Baptiste Joseph Fourier, $f(x) = \sum\sin(y) + \cos(z)$
1808 AD Christian Kramp uses ! for factorials
1811 AD Chernac prime factors to 1,020,000
1816 AD Burkhardt least factors table to 3,036,000
1819 AD pseudo-primes discovered
1821 AD P. L. Tchebychev proves $n > p > 2n$
1822 AD Charles Babbage invents Difference Engine
1824 AD William Rutherford discovers $\pi 10^{152}$, Johann Dase human calculator
1825 AD Sophie Germain prime $p' = 2p + 1$
1829 AD Nicholai Lobachevsky's non-Euclideean geometry
1831 AD Marie-Sophie Germain died, aka "M. le Blanc"
1832 AD Tuman Safford coins "complex number", Évariste Galois, 20, shot in duel
1841 AD Karl Weierstrass uses absolute value signs
1844 AD von Strassnitsky and Zacharias Dase discovers $\pi 10^{200}$
1847 AD Thomas Clausen discovers $\pi 10^{248}$
1943 AD Sir William Rowan Hamilton's quaterions
1851 AD Joseph Liouville number 0.110001000000000000000000010...
1853 AD William Rutherford discovers $\pi 10^{440}$
1854 AD George Boole promotes Boolean algebra in *Investigation of the Laws of Thought*
1855 AD Karl Friedrich Gauss died, defined "measurable infinity", to-the-fourth-nine, $^4 9$, Gaussian Mersenne numbers, Richter discovers $\pi 10^{500}$
1856 AD Crelle 6 million primes
1858 AD August Ferdinand Möbius strip at 68
1859 AD George Friedrich Bernard Riemann's $\zeta(x) = \Sigma(1/n)^x$, finite for $x > 1$
1867 AD Arthur Cayley matrix algebra

1861 AD Zacharias Dase 9 million primes
1863 AD Yakov Kulik died, compiled 4,212-page list of smallest factor of integers to 100,330,200
1866 AD Nicolo Paganini, 16, discovers 65th amicable number pair 1,184 and 1,210 missed by Euler
1870 AD Ernst Meisel $\#(p < 10^8) = 5{,}761{,}455$
1872 AD Karl Theodor Wilhelm Weirstrass undifferentatable function $\Sigma a^k\cos(b^k\pi x)$, $ab > 1 + 3\pi/2$ (a fractal), Fifteen Puzzle by Noyes Palmer Chapman
1873 AD William Shanks discovers $\pi 10^{527}$
1874 AD William Shanks reunit primes, Georg Cantor institutionalized over transfinites
1875 AD Franz Reauleaux's constant-width triangle, $A = (\pi - \sqrt{3})r^2/2$ (used in square-hole drill)
1876 AD François Edouard Anatole Lucas, discovered 77-digit perfect, Edouard Lucas defined Lucas primes; Lord Kelvin's harmonic analyzer, $f(x) = A_n\sin(nx) + B_n\cos(nx)$
1879 AD E. B. Escott $n^2 - 79n + 1{,}601$
1880 AD F. Landry factors F_6, Venn diagrams
1881 AD Benford's law disc. by Simon Newcomb, $P(1^{st}\text{ digit}) = \log(1 + 1/n)$
1882 AD Ferdinand Lindemann proves π transcendental; Felix Klein's bottle
1883 AD I. M. Pervouchine discovered M_9 missed by Euler; François Édouard Anatole Lucas's tower of Hanoi, $2^n - 1$ moves
1884 AD Edward Abbott's *Flatland*
1888 AD Charles Howard Hinton's tesseract
1889 AD Giuseppe Peano's *A New Era of Thought* axioms: $0 = n_0$, $S(n) \neq 0$, $S(n_i) = n_i+1$, $S(n_i) = S(n_j) \equiv n_i = n_j$
1890 AD discovers transdimensional space-filling curve
1891 AD Evgraf Stephanovich Fedorov studies 17 repeating tilings (see 13th century Alhambra)
1893 AD Ernst Eduard Kummer died, German mathematician gave name to Kummer numbers, D. H. Lehmer 50,847,534 primes
1894 AD Eugéne Charles Catalan died, Belgian mathematician, gave name to Catalan numbers
1896 AD Johann Carl Friedrich Gauss, 15, discovers $\pi(n) \approx n/\ln(n)$
1899 AD R. Perrin popularized Perrin pseudoprimes
1900 AD Reichenbacher $\pi \approx (2143/22)^{1/4}$, Karl Pearson's observation/expection comparison, $\chi^2 = \Sigma(O_i - E_i)^2/E_i$
1901 AD Bertrand Russell's Barber paradox
1903 AD Frank Cole factored M_{67}
1904 AD R. Chartres relative prime probability $P(n/n' \neq n'') = 6/\pi^2$,
1905 AD Fr. James Cullen, S. J., discovered Cullen numbers
1906 AD Axel Thue and Marston Morse's aperiodic, self-referential sequence (1 → 10, 0 → 01)
1909 AD C. A. Laisant discovers $^39 > 10^{369,693,100}$, Félix Edourd Justin Borel's normal numbers
1911 AD R. E. Powers discovers M_{10}
1912 AD R. D. Carmichael's Carmichael number product of three odd primes, Norman

Anning's Cocoanut solution
1913 AD J. N. Muncey prime magic square, E. Fauquembergue M_{11}, Alfred North Whitehead's
1914 AD D. N. Lehmer, author of *List of prime numbers from 1 to 10,006,721*, R. E. Powers discovered 65-digit perfect
1918 AD Paul Poulet 5-link sociable chain, Felix Hausdorff explores interdimensionality
1919 AD Viggo Brun's constant, $\sum(1/p + 1/p+2) \approx 1.368827$
1925 AD Zbigmiew Moroń found 9-tile 33x32 dissection; David Hilbert's infinite hotel
1926 AD D. H. Lehmer discovers $e10^{709}$, Karl Menger sponge, $D \approx 2.73$, Ben Ames Williams new Cocoanut problem, 1,024n = 1,562n + 11,529 (see 1912)
1932 AD Stanisław Marcin Ulam and Lothar Collatz's 3 +1 (Hailstone) problem
1933 AD Skewes no. $\approx {}^{31}10^{34}$; David Gawen Champernowne's transcendental nos. 0.1234567891011... and 0.11011100101...
1934 AD Alexander Gelfond discovers e^{π} transcendental
1931 AD Kurt Gödel theorem
1935 AD Nicolas Bouraki's secret under-50 mathematician society
1936 AD Heinz Voderberg's nonagonal tile
1937 John Robert Reuel Tolkien popularized eleventy in *The Hobbit,* I. Vinogradoff $n > 10^{350,000} = \Sigma 3$ odd primes
1939 AD B. H. Brown discovers smallest odd amicables 12,285 and 14,595
1942 AD E. S. Selmer and G. Nesheim list twin primes to 200,000, Hex invented by Piet Hein
1944 AD Paul Poulet factors M_{135}
1945 AD D. F. Ferguson discovers $\pi 10^{527}$
1946 AD D. F. Ferguson discovers $\pi 10^{620}$, E. B. Escott lists 390 amicables; PRNG (pseudorandom no. generator) $x_{n+1} = 10^5 x_n{}^2 - [10^5 x_n{}^2] - [10^{15} x_n{}^2]/10^{10}$
1947 AD D. F. Ferguson discovers $\pi 10^{808}$, H. Tietze lists twin primes to 300,000, W. H. Mills A^3^n = p, 14X14 Hex invented by John Forbes Nash, Jr. (see 1942), reflected Frank Gray code 0, 1, 11, 10, 11, 110, 111, 101, 100, 1100, 1101,
1949 AD John Wrench, Jr., and Levi Smith discovers $\pi 10^{1120}$ on mechanical calculator, Reitwieser on ENIAC discovers $\pi 10^{2037}$, D. R. Kaprekar discovers constant 6174 from ananum differences, Capt. Benson discovers 32-order tri-magic square; Szolem Mandelbrojt leaves Bourbaki Society
1950 AD D. H. Lehmer smallest even pseudo-prime 161,038, Benot Mandelbrot $D \approx \log L/\log l$
1951 AD Miller used EDIAC computer to discovers prime
1952 AD Robinson discovered 13th through 17th perfects, the last 1,372 digits, including 1st hectoperfect and kiloperfect, 313 and 1,326 digits respectively, D. H. Wheeler on ILLIAC discovers $e10^{60,000}$
1953 AD Goldberg discovers Wilson prime 563
1954 AD A. L. Brown adds 200 new multiple perfects; Henri Cartan leaves Bourbaki Society
1955 AD Nicholson and Jeenel on NORC discovers $\pi 10^{3089}$
1956 AD Stanisław Marcin Ulam of Los Alamos discovers lucky numbers; André Weil leaves Bourbaki Society
1957 AD Hans Riesel, Swede, on BESK (Binary Electronic Sequence Calculator) discovered 1,936-digit perfect, Felton on PEGASUS discovers $\pi 10^{7480}$
1958 AD Cullen number $C_{141} = p$
1959 AD F. Genuys on IBM 704 discovers $\pi 10^{10,000}$, PEGASUS discovers $\pi 10^{10,021}$, IBM 704

discovers $\pi 10^{16,167}$, 455,052,511 primes < 10^{10}, Claude Chevalley leaves Bourbaki Society
1960 AD George Gamow authored *One, Two, Three ... Infinity*, Sierpinski's no. $78577(2^n+1) \neq p$
1961 AD Alexander Hurwitz discovered 19th and 2,662-digit 20th perfect numbers, Daniel Shanks and John Wrench on IBM 7090 discovers $\pi 10^{100,265}$
1962 AD Wilhelm Ackermann died, contributed to epsilon calculus, set theory and logic, noted generalized exponential is non-primitive recursive function, defined tower numbers; S. Knapowski defined Knapowski's number, Harry Langman order 7 perfect magic cube, Solomon W. Golomb's rep-tiles, including sphinx pentagon
1963 AD Donald B. Gillies discovered 21st to 23rd perfects, last 6,750-digit; Stanisław Marcin Ulam's spiral
1964 AD J. A. H. Hunter 17-digit automorphics, Rolf discovers 13th amicable pair
1965 AD R. A. German solves bovinum problema with H. C. Williams and C. R. Zarnke, Piet Hiet's superegg, $|x|^{2/a} + | y|^{2/a} + |z|^{2/b} = 1$
1966 Jean Gilloud and J. Fillatoire on IBM 7030 discovers $\pi 10^{250,000}$, Dr. Matrix predicts $\pi 10^{1,000,000} = 5 \pmod{10}$, Chen Jing-Runi proves $p = p_1 + p_2 p_3$
1967 AD Gilloud and Dichampt on CDC 6600 discovers $\pi 10^{500,000}$, Lander and Parkin 6-term prime chain 121,174,811 + 30n
1969 AD Henri Cohen discovers 7 new sociable chains
1970 AD Morrison and Brillhart factor F_7, Gödel proves existence of God
1971 AD Bryant Tuckerman discovers 12,003-digit perfect, David Singmaster discovers 3,003 most common Pascal triangle number
1972 AD scientific pocket calculator makes sliderule antiquated
1973 A D N. J. A. Sloane's *A Handbook of Integral Sequences*, Roger Penrose's quasicrystalline tiles
1974 AD Ernő Rubik's cube 43,252,003,274,489,856,000 possibilities, John Horton Conway's surreal nos.
1975 A D Rivest, Shamir and Adleman large number factoring, fractals named, Mitchel Jay Feigenbaum's constant
1976 AD Jean Gilloud and Martine Bouyer on CDC 7600 discovers $\pi 10^{1,000,000}$, published by French AEC
1977 A D Hugh C. Williams and Eric Seah discovers first new repunit prime $[10^{318}/9]$ in 50 years, Weintraub prime chain 999,900,067,719,989 + 30n, Ivan Niven's Niven numbers, RSA public key biprime pairs
1978 AD Laura Ariel Glenn nee Nickel with Landon Cole Noll, highschoolers co-discovers M_{25} and 13,066-digit perfect, M. Yorinaga discovers 8 13-factor Carmichaels; A. W. J. Duivestijn found 21-tile square dissection; Gödel dies by starvation
1979 AD Harry L. Nelson, Livermore, Cal., puzzle developer, discovered 26,790-digit perfect; Landon Cole Noll discovers M_{26} and 13,973-digit perfect, Cormack and Williams discovers titanic prime $25(2^{3,314}) - 1$, Aiken and Rickert discover twin primes $1,159,142,985(2^{2,304}) \pm 1$, M. Penk factors $2^{257} - 1$
1980 AD Keller factors $F_{9,448}$
1981 AD Joe P. Buhler discovered factorial prime, Richard E. Crandall, founder of Perfectly Scientific, Inc., discovered 1,115 -digit primordial prime, R. A. Knoebel, first used left superscription for tetration, Michael A Peck, discovered 1,115-digit primorial prime and 1051-digit factorial prime, Kazunori Miyoshi and Nakayama on FACOM M-200 discovers $\pi 10^{2,000,038}$,

R. P. Brent and John Pollard factor F_8, A Symposium of Two-dimensional Science and Technology
1982 AD David Slowinski on Cray 1 discovers M_{28} prime, Tamura and Yasumasa Kanada on HITAC M-280H discover $\pi 10^{4,194,293}$, Albert Wilansky's Smith numbers, D. Woods and J. Huenemann discover 432-digit Carmichael
1983 AD Douglas Hofstader promoted Luring lottery in *Scientific American*; Hugh C. Williams, U. of Manitoba prof , discovered 3,021-digit palindromic prime, Tamura and Yasumasa Kanada on HITAC M-280H discovers $\pi 10^{16,777,216}$, David Slowinski on Cray 1 discovers M_{30} prime
1984 AD Yates' 110 titanic primes list, Keller discovers prime $17^2(2^{18,502}) + 1 = (17(2^{9,251})^2 + 1$ and $C_{4,713}$, $C_{5,795}$, $C_{6,611}$ and $C_{18,496}$, Harvey Dubner discovers 1281-digit binarish prime
1985 AD Bill Gosper discovers $\pi 10^{17,000,000}$, Yates 581 titanic prime list, David Slowinski on Cray X-MP 24 discovers M_{31}, Williams discovers $2(10^{3,020}) - 1$ prime, Keller factors $F_{23,471}$, Harvey Dubner discovers 1057-digit Carmichael, Dorin Andrica's prime gap conjecture $g_n < 2\sqrt{p_n} + 1$, David Masser and Joseph Osterl�square-free nos.
1986 AD Shri Dattathreya Ramachandra Kaprekar died, Indian mathematician, gave name to Kaprekar numbers, Bailey, Borwein and Borwein on Cray 2 discovers $\pi 10^{29,360,000}$, Yasumasa Kanada discovers $\pi 10^{33,000,000}$, 2,592,699-digit Smith, H. Williams and Harvey Dubner discover repunit prime $[10^{1,032}/9] -1$, and $150,093(10^{8,000}) + 1$, Keller discovers Sophie Germain prime $3,9051(2^{6'001}) - 1$, Harry Nelson and David Slowinski discover 1,000,000!, John Horton Conway's audioactive sequence
1987 AD Yasumasa Kanada on NEC SX-2 discovers $\pi 10^{134,217,728}$, Jeff Young and Duncan Buell on Cray 2 factor F_{22}, Young and Fry 20-term prime chain 214,861,583,621 + 1,884,649,770z, Smith brothers 7289 and 729, and Psmith 12,345,554,321, Yasumasa Kanada discovers $\pi 10^{134,217,728}$, Michael Keith's repfigit 7,913,837
1988 AD Walter N. Colquitt, and Luther Luke Welsh, Jr., co-discover M_{29} and 66,530-digit perfect, Manfred Toplic, Klagenfurt, Austria, system administrator, discovered 10-prime linear sequence, Yasumasa Kanada on Hitachi discovers $\pi 10^{201,326,000}$, Yates 876 titanic prime list, Young and Buell factor F_{20}, Harvey Dubner discovers primes $6(10^{4,333}) -1$ and $1,358(10^{3,821}) - 1$ and 1104-digit primish prime, Brent and Moran factor F_{11} and F_{20}
1989 AD John Brown, discovered 65,087-digit prime, Bodo Parady, physicist/computer engineer, discovered 65,087-digit prime, Mikl� Laczkovich proves squaring of circle possible with 10^{50} pieces, Gene and Joel Smith, brothers, Berkeley number theorist and software engineer, co-discovered 65,087-digit prime, Landon Cole Noll discovers 65,087-digit prime; Jeffrey Young, Cray systems analyst, discovered 91,241-digit prime; Sergio Zarantonello, U. of Wisconsin math prof ,co-discovered 65,087-digit prime, David and Gregory Chudnovsky on Cray 2 discovers $\pi 10^{1,011,196,691}$, Loh discovers 12-term Sophie Germain chain and 13-term Cunningham chain beginning 758,083,947,856,951, John Brown, etal; discover prime $391,581(2^{216,193}) -1$ and twins $1,706,595(2^{11,235}) \pm 1$, Clifford Pickover finds repdigits higher eightplex, Harvey Dubner discovers 3,710-digit Carmichael
1990 AD Yates' 1,426 titanic prime list, H. J. te Riele on NEC-SX-2 factors 101-digit number, Jaesche lists 10^{12} Carmichaels, B. K. Parody, etal., discover twins $571,305(2^{7,701}) \pm 1$, Robert Dubner 10^{10} primes in *The Book of Primes*, Harvey Dubner palindromic prime $10^{11,310} + (4,661,664(10^{56,752})) + 1$, Pollard factors F_9, Colin Percival discovers $5(10^{12})$ bit of π^2
1992 AD Yvonne Riggs inspires John Conway and Martin Kruskal to define surreal numbers, Isaac Asimov died, prolific author of fiction and non-fiction, invented T-T- (or teratera-) number

system; Chris Caldwell, U. of Tenn. professor, discovered factorial primes (and again in 1993, 1998); primorial primes (in 1992, 1993), 196,486-digit prime (2005), David Slowinski and Paul Gage on Cray 2 discover M_{32} prime and 455,663-digit perfect, Harvey Dubner discovers twin primes $1{,}691{,}232(1{,}001)10^{4{,}020} - 1$ and $4{,}650{,}828(1{,}001)10^{3{,}429} - 1$
1993 AD Ken Davis, computer scientist from Canberra College, discovered primorial prime; Theoni Pappas authored *Fractals, Googols and Other Mathematical Tales*, Fred Helenius discovers many more multiperfects

ADDENDUM

"Let Us Remember André Joyce"
by Gabe Hoya
(reprinted from *Clarion* 9/2/1985)

There is an opening now for BMOC, at least in the Math Department, with the retirement of Prof. André Joyce. Although the good professor celebrated his 75th birthday on the Fourth of July this past summer, the administration had considered making an exception to the mandatory retirement age in his case since the date of his birth is technically unknown. He was visiting the United States when Adolf Hitler came to power in 1933 and did not go back to Germany, where he had been working on a post-doctorate at the Berlin Academy of Sciences.

Joyce was born to an English father, Ferdinand, and a French mother, Vivianne from Toulouse, but the records of exactly where and when he was born were lost during the war. When asked when asked to give his birth date he patriotically gave his adopted country's birthday.

We do know that he attended Catholic school at from the age of five. In his teens Joyce clashed with school authorities and resented the school's regimen and teaching method. He later wrote that the spirit of learning and creative thought were lost in strict rote learning.

Joyce's wife, Bea Cuisinier, was the daughter of Fanny Cuisinier, cousin of Joyce's mother, Vivianne, whose maternal grandfather Rudolf Joyce was the son of Raphael

Joyce, and a brother of Joyce's paternal grandfather, thus they were second cousins. Their relationship developed into a friendship and the friendship into romance, as they read books together on extra-curricular pataphysics in which Joyce was taking an increasing interest. In late 1939, Bea failed her own oral examination. To cheer her Joyce gave her an engagement ring for an early Christmas present. There have been claims that Bea collaborated with Joyce on his doctorial thesis, but historians of pataphysics who have studied the issue have not been able to confirm it.

During the war Joyce worked at the Government Code and Cypher School, Bletchley Park, Milton Keynes, Buchinghamshire. Sir Harry Hinsley official historian of British Intelligence in World War II, said of Joyce and his co-workers, that they shortened the war "by not less than two years and probably by four years."

After the war the Joyces settled far away from Europe, far from Washington, in East Dakota. Over about forty years Prof. Joyce has grown to become the big man in the math department, although he has multiple times refused the chair of the department. Most of his publications have been in rather obscure French journals, and so far untranslated, so he is little known outside of our own campus.

When asked if he would miss teaching, said, "It has been tremendously interesting to be a teacher, to watch my students grow up and help them along; to see their characters develop and what they become when they leave school and the world gets hold of them. I don't see how you could ever get old in a world that's always young."

When asked if this meant that the he would spend his final years in even greater obscurity, since he and Bea had no children, he said, "But you're wrong. I have. Thousands of them. Thousands of them."

Joyce final words were to us, his "children", "Remember me sometimes. I shall always remember you."

He had seen hundreds of his "children" go off and die much too young during the Korean and Vietnam Wars. Let those of us who survived, remember the lessons our dear prof taught us, and pass on those memories to our children and grandchildren.

www.ingramcontent.com/pod-product-compliance
Lightning Source LLC
Chambersburg PA
CBHW052307220526
45472CB00001B/5
* 9 7 8 1 7 1 9 3 3 9 9 4 0 *